Cyanobacterial Harmful Algal Blooms

ADVANCES IN EXPERIMENTAL MEDICINE AND BIOLOGY

Recent Volumes in this Series

Volume 610
TARGET THERAPIES IN CANCER
Edited by Francesco Colotta and Alberto Mantovani

Volume 611
PETIDES FOR YOUTH
Edited by Susan Del Valle, Emanuel Escher and William D. Lubell

Volume 612
RELAXIN AND RELATED PETIDES
Edited by Alexander I. Agoulnik

Volume 613
RECENT ADVANCES IN RETINAL DEGENERATION
Edited by Joe G. Hollyfield, Matthew M. LaVail and Robert E. Anderson

Volume 614
OXYGEN TRANSPORT TO TISSUE XXIX
Edited by Kyung A. Kang, David K. Harrison and Duane F. Bruley

Volume 615
PROGRAMMED CELL DEATH IN CANCER PROGRESSION AND THERAPY
Edited by Roya Khosravi-Far and Eileen White

Volume 616
TRANSGENIC MICROALGAE AS GREEN CELL FACTORIES
Edited by Rosa León, Aurora Gaván and Emilio Fernández

Volume 617
HORMONAL CARCINOGENESIS V
Edited by Jonathan J. Li

Volume 618
HYPOXIA AND THE CIRCULATION
Edited by Robert H. Roach, Peter Hackett and Peter D. Wagner

Volume 619
CYANOBACTERIAL HARMFUL ALGAL BLOOMS: STATE OF THE SCIENCE AND RESEARCH NEEDS
Edited by H. Kenneth Hudnell

A Continuation Order Plan is available for this series. A continuation order will bring delivery of each new volume immediately upon publication. Volumes are billed only upon actual shipment. For further information please contact the publisher.

H. Kenneth Hudnell
Editor

Cyanobacterial Harmful Algal Blooms: State of the Science and Research Needs

Springer

Editor
H. Kenneth Hudnell
United States Environmental Protection Agency
Triangle Park, NC
USA

ISBN: 978-0-387-75864-0 e-ISBN: 978-0-387-75865-7
DOI: 10.1007/ 978-0-387-75865-7

Library of Congress Control Number: 2007939828

Printed on acid-free paper

9 8 7 6 5 4 3 2 1

springer.com

Contents

Causes, Prevention, and Mitigation Workgroup

Toxins Workgroup

Analytical Methods Workgroup

Ecosystem Effects Workgroup

Risk Assessment Workgroup

Preface

Interagency ISOC-HAB Symposium Introduction

This symposium was held to assess the state-of-the-science and identify research needed to address the increasing risks posed by freshwater harmful algal blooms to human health and ecosystem sustainability. Information obtained through the symposium will help form the scientific basis for developing and implementing strategies to reduce these risks.

All chapters in this book are based on platform sessions or draft workgroup reports that were presented at ISOC-HAB. All chapters were completed after the conclusion of ISOC-HAB. Each chapter was critically reviewed by at least two peers with expertise in the subject matter, revised based on those reviews, and reviewed by the editor before being accepted for publication.

Grateful acknowledgment is given to the National Science and Technology Council's Committee on the Environment and Natural Resources in the Executive Office of the President for providing guidance, to the sponsoring agencies, to the agency representatives named below who organized the symposium, to the international scientific community members who participated in the symposium, and to EC/R of Durham, NC, the contracting organization that provided logistical support for the symposium and this monograph.

Interagency ISOC-HAB Organizing Committee

H. Kenneth Hudnell, Lead Organizer & Symposium Director
U.S. Environmental Protection Agency
Office of Research and Development
National Health and Environmental Effects Research Laboratory
Research Triangle Park, NC 27711
Neurotoxicologist, 1984–2007

Currently: Vice President & Director of Science SolarBee, Inc.
(http://www.SolarBee.com)
105 Serrano Way
Chapel Hill, NC 27517
(877) 288-9933; (919) 932-7229
kenhud@SolarBee.com
Main Office & Service Center
3225 Hwy. 22, PO Box 1930, Dickinson, ND 58602
(866) 437-8076 . (701) 225-4495 . Fax (701) 225-0002

Lorrie Backer, CDC
Brenda Boutin, EPA
Armah de la Cruz, EPA
Ed Dettmann, EPA
Joyce Donohue, EPA
Quay Dortch, NOAA
Al DuFour, EPA
TJ Evens, USDA
Gary Fahnenstiel, NOAA
John Fournie, EPA
Jim Goodrich, EPA
Sherwood Hall, FDA
Elizabeth Hilborn, EPA
Michelle Hooth, NIH/NIEHS
Karl Jensen, EPA
John R Kelly, EPA
Beth LeaMond, EPA
Alan Lindquist, EPA
Brian Melzian, EPA
Michael Meyer, USGS
Bruce Mintz, EPA
Tonya Nichols, EPA
Nena Nwachuku, EPA
Jennifer Orme-Zavaleta, EPA
Rex Pegram, EPA
Gina Perovich, EPA
Joel Scheraga, EPA
Jim Sinclair, EPA
Cynthia Sonich-Mullin, EPA
Jeffery Steevens, USACE
Bruce Vogt, NOAA

ISOC-HAB Executive Advisory Committee

Paul Berger, EPA
Bob MacPhail, EPA
Gerard Stelma, EPA
Barbara Walton, EPA
Harold Zenick, EPA

Invited Participants

Occurrence Workgroup

Workgroup Members

Sherwood Hall, Co-chair
US Food & Drug Administration
shall@cfsan.fda.gov
301-210-2160

Jim Sinclair, Co-chair
USEPA
Sinclair.james@epa.gov

Gregory L Boyer
College of Environmental Science and Forestry
State University of New York
glboyer@esf.edu
315-470-6825

Julie Berkman
US Geological Survey
jberkman@usgs.gov
614-430-7730

JoAnn M Burkholder
Center for Applied Aquatic Ecology, NCSU
joann_burkholder@ncsu.edu
919-515-3421

John Burns
PBS&J
jwburns@pbsj.com
904-477-7723

Wayne W. Carmichael
Wright State University
wayne.carmichael@wright.edu
937-775-3173

Al DuFour
USEPA
dufour.alfred@epa.gov

William Frazier
City of High Point
bill.frazier@highpointnc.gov
336-883-3410

Tony Fristachi
NCEA CIN
fristachi.anthony@epa.gov
513-569-7144

Steve Morton
NOAA
steve.morton@noaa.gov

Eric O'Brien
University of Iowa
eobrien@igsb.uiowa.edu
319-560-6128

Steven Walker
Nebraska Department of Environmental Quality
steve.walker@ndeq.state.ne.us
402-471-4227

Invited Speakers on Occurrence

Gregory L Boyer
(see above)
John Burns
(see above)

Dr. Wayne W. Carmichael
(see above)
Steven Walker
(see above)

Causes, Prevention, and Mitigation

Workgroup Members

Jim Goodrich, Co-chair
USEPA
goodrich.james@epa.gov
513-569-7605

Gina Perovich, Co-chair
National Center for Environmental Research, USEPA
Perovich.gina@epa.gov
202-343-9843

Quay Dortch, Co-chair
NOAA Coastal Ocean Program
quay.dortch@noaa.gov
301-713-3338 ext 157

Paul S. Berger
USEPA (retired)
rainchoir@aol.com
703-751-6742

Justin Brooks
SA Water Centre for Water Science and Systems
CRC for Water Quality and Treatment
Justin.Brookes@sawater.com.au
+61 8 8259 0222

Terence J. Evens
USDA-ARS
TEvens@ushrl.ars.usda.gov
772-462-5921

Chris Gobler
Marine Science Research Center
Long Island University
Christopher.Gobler@liu.edu

Jennifer Graham
U.S. Geological Survey
jlgraham@usgs.gov
785-832-3511

James Hyde
New York Department of Health
jbh01@health.state.ny.us

Dawn Karner
Wisconsin State Laboratory of Hygiene
608-224-6230
dkarner@mail.slh.wisc.edu

Kevin O'Shea
Florida International University
osheak@fiu.edu

Valerie Paul
Smithsonian Marine Station
paul@sms.si.edu
772-465-6630x140

Causes, Prevention, and Mitigation

Workgroup Members

Hans W. Paerl
UNC - Chapel Hill
Institute of Marine Sciences
hans_paerl@unc.edu
(252) 222-6346

Barry Rosen
U. S. Fish & Wildlife Service
barry_rosen@fws.gov
772-562-3909

Pat Tester
NOAA Center for Coastal Fisheries and Habitat Research
Pat.Tester@noaa.gov
252-728-8792

Michael Piehler
UNC - Chapel Hill
Institute of Marine Sciences
mpiehler@email.unc.edu
(252) 726-6841 ext. 160

Mary Santelmann
Oregon State University
santelmm@onid.orst.edu
541-737-1215

Dr. Judy Westrick
Chemistry, CRW 318
Lake Superior State University
jwestrick@lssu.edu
(906) 635-2165

Invited Speakers on Causes, Prevention, and Mitigation

Dr. Hans W. Paerl
(see above)

Dr. Valerie J. Paul
(see above)

Michael Piehler
(see above)

Dr. Judy Westrick
(see above)

Cyanotoxins Workgroup

Workgroup Members

Tonya Nichols, Co-chair
USEPA
Nichols.tonya@epa.gov

Stacey Etheridge
FDA HFS-426 BRF
stacey.etheridge@cfsan.fda.gov
301-210-2162

Rex Pegram, Co-chair
USEPA
pegram.rex@epa.gov

Andrew Humpage
CRC for Water Quality and Treatment
andrew.humpage@sawater.com.au
+61 8 8259 0222

Cyanotoxins Workgroup

Workgroup Members

Susan LeBlanc
Department of Biology, University of Ottawa
30 Marie Curie, P.O. Box 450, Station A
Ottowa, Canada K1N 6N5

Brett Neilan
Microbiology
University of New South Wales
b.neilan@unsw.edu.au
612 9385 3235

Maria Runnegar
University of Southern California
323 442 3231
runnegar@usc.edu

Adam Love
Lawrence Livermore National Laboratory
love5@llnl.gov
925-422-4999

Stephan Pflugmacher
Leibniz Institute of Freshwater Ecology and Inland
pflug@IGB-Berlin.de
0049-30-64181639

Robert Thacker
University of Alabama at Birmingham
thacker@uab.edu
205-956-0188

Invited Speakers on Cyanotoxins

Andrew Humpage
(See above)

Adam Love
(See above)

Brett Neilan
(See above)

Analytical Methods Workgroup

Workgroup Members

Dr. Armah A. de la Cruz, Co-chair
USEPA
delacruz.armah@epa.gov
513-569-7224

Kathy Echols
USGS
Columbia Env. Research Center
573-876-1838
kechols@usgs.gov

Dr. Michael Meyer, Co-chair
US Geological Survey
mmeyer@usgs.gov
785 832-3544

Ambrose Furey
Cork Institute of Technology
afurey@cit.ie
00353-21-4326701

Analytical Methods Workgroup

Workgroup Members

Dr. James Hungerford
FDA Seafood Products Research Center
James.Hungerford@fda.gov
425-483-4894

Rosemonde Mandeville
Biophage Pharma Inc.
rosemonde.mandeville
@biophagepharma.com
514-496-1488

Dr. Parke Rublee
Professor of Biology
UNC Greensboro
parublee@uncg.edu
Phone: 336 256-0067

Dr. Gerard Stelma
USEPA
National Exposure Research Laboratory
stelma.gerard@epa.gov
513-569-7384

Paul Zimba
USDA
pzimba@msa-stoneville.ars.usda.gov

Linda Lawton
The Robert Gordon University
+44 1224 262823
l.lawton@rgu.ac.uk

Jussi Meriluoto
Abo Akademi University
jussi.meriluoto@abo.fi
+358-2-2154873

Kaarina Sivonen
Helsinki University
Department of Applied Chemistry
and Microbiology
kaarina.sivonen@helsinki.fi
+358-9-19159270

Dr. Steven W. Wilhelm
Department of Microbiology
The University of Tennessee
wilhelm@utk.edu
865-974-0665

Invited Speakers on Analytical Methods

Dr. Jussi Meriluoto
(see above)
Dr. Linda Lawton
(see above
Dr. Kaarina Sivonen
(see above)
Dr. Steven W. Wilhelm
(see above)

Human Health Effects Workgroup

*Workgroup Members**

John W Fournie, Co-chair
USEPA
fournie.john@epa.gov

Elizabeth D Hilborn, Co-chair
USEPA
hilborn.e@epa.gov

Sandra Azevedo
Federal University of Rio de Janeiro
sazevedo@biof.ufrj.br

Neil Chernoff
USEPA
chernoff.neil@epa.gov

Ian Falconer
Pharmacology Department
University of Adelaide
+62 2 6251 1345

Michelle Hooth
National Institute of Environmental Health Science
hooth@niehs.nih.gov

Karl Jensen
USEPA
Jensen.karl@epa.gov

Robert MacPhail
Neurotoxicology
USEPA
macphail.robert.epa.gov

Ian Stewart
Research Fellow
School of Public Health
Griffith University
and
Research Officer
Organic Chemistry
Queensland Health Scientific Services
39 Kessels Road
COOPERS PLAINS
QLD 4108, Australia

*Michael Gage, Ellen Rogers, and Glen Shaw also contributed to the Workgroup Report.

Invited Speakers on Human Health Effects

Ian Falconer
(see above)

Ian Stewart
(see above)

Louis Pilotto
Faculty of Medicine
University of New South Wales
+61 (2) 69335111
l.pilotto@unsw.edu.au

Ecosystem Effects Workgroup

Workgroup Members

John W Fournie, Co-chair
(see above)

Elizabeth D Hilborn, Co-chair
(see above)

Geoff Codd
University Of Dundee
G.A.Codd@Dundee.Ac.Uk

Dr. Michael Coveney
St. Johns River Water Management District
mcoveney@sjrwmd.com
386-329-4366

Julie Dyble
NOAA
juli.dyble@noaa.gov

Karl Havens
Department of Fisheries and Aquatic Sciences
University of Florida / IFAS
352-392-9617 ext. 232
khavens@ifas.ufl.edu

Bas Ibelings
Netherlands Institute of Ecology
b.ibelings@nioo.knaw.nl
+ 31 294239349

Jan Landsberg
Florida Fish and Wildlife Conservation Commission
jan.landsberg@fwc.state.fl.us
727-896-8626

Wayne Litaker
National Ocean Service
NOAA
wayne.litaker@noaa.gov
252-728-8774

Invited Speakers on Ecosystem Effects

Dr. Bas Ibelings
(see above)
Karl Havens
(see above)

Risk Assessment Workgroup

Workgroup Members

Joyce Donohue, Co-chair
USEPA
donohue.joyce@epa.gov
202-566-1098

Jennifer Orme-Zavaleta, Co-chair
USEPA
orme-zavaleta.jennifer@epa.gov
919-541-5680

Michael Burch
Cooperative Research Centre for Water Quality and Treatment
SA Water
mike.burch@sawater.com.au
61 8 82590352

Dr. Daniel Dietrich
SSPT, GSPT, EUROTOX, FATS
University of Pittsburgh
Daniel.Dietrich@uni-konstanz.de
0049-7531-883518

Belinda Hawkins
USEPA
hawkins.belinda@epa.gov
513 569-7523

Tony Lloyd
Drinking Water Inspectorate (Retired)
AL Consultants
julony@btinternet.com
00441424754013

Wayne Munns
USEPA
munns.wayne@epa.gov
401-782-3017

Jeffery Steevens
US Army Corps of Engineers
steevej@wes.army.mil
601-634-4199

Dennis Steffenson
Cooperative Research Centre for Water Quality and Treatment
SA Water
dennis.steffensen@sawater.com.au
61 8 82590326

Dave Stone
Oregon Health Services
Dave.Stone@state.or.us
971-673-0444

Peter Tango
Maryland Department of Natural Resources
ptango@dnr.state.md.us
410-260-8651

Invited Speakers on Risk Assessment

Michael D Burch
(see above)

Daniel Dietrich
(see above)

Wayne Munns
(see above)

Jennifer Orme-Zavaleta
(see above)

Dennis Steffensen
(see above)

Agency Disclaimers

EPA

This symposium was funded in part by the U.S. Environmental Protection Agency. This document has been reviewed in accordance with the U.S. Environmental Protection Agency policy and approved for publication. Approval does not signify that the contents reflect the views of the Agency, nor does mention of trade names or commercial products constitute endorsement or recommendation for use.

USACE

This work was supported, in part, by the Aquatic Nuisance Species Research Program at the U.S. Army Engineer Research and Development Center (Dr. Al Cofrancesco, Technical Director). Permission has been granted by the Chief of Engineers to publish this material.

CDC

The findings and conclusions in this report are those of the author(s) and do not necessarily represent the views of the Centers for Disease Control and Prevention/the Agency for Toxic Substances and Disease Registry.

NIH/NIEHS

This research was supported in part by the Intramural Research Program of the NIH, National Institute of Environmental Health Sciences. The content of this paper reflects the opinions and views of the authors and does not represent the official views or policies of NIEHS, NIH, or NTP.

Chapter 1: An Overview of the Interagency, International Symposium on Cyanobacterial Harmful Algal Blooms (ISOC-HAB): Advancing the Scientific Understanding of Freshwater Harmful Algal Blooms

H Kenneth Hudnell, Quay Dortch, Harold Zenick

Abstract

There is growing evidence that the spatial and temporal incidence of harmful algal blooms is increasing, posing potential risks to human health and ecosystem sustainability. Currently there are no US Federal guidelines, Water Quality Criteria and Standards, or regulations concerning the management of harmful algal blooms. Algal blooms in freshwater are predominantly cyanobacteria, some of which produce highly potent cyanotoxins. The US Congress mandated a Scientific Assessment of Freshwater Harmful Algal Blooms in the 2004 reauthorization of the Harmful Algal Blooms and Hypoxia Research and Control Act. To further the scientific understanding of freshwater harmful algal blooms, the US Environmental Protection Agency (EPA) established an interagency committee to organize the Interagency, International Symposium on Cyanobacterial Harmful Algal Blooms (ISOC-HAB). A theoretical framework to define scientific issues and a systems approach to implement the assessment and management of cyanobacterial harmful algal blooms were developed as organizing themes for the symposium. Seven major topic areas and 23 subtopics were addressed in Workgroups and platform sessions during the symposium. The primary charge given to platform presenters was to describe the state of the science in the subtopic areas, whereas the Workgroups were charged with identifying research that could be accomplished in the short- and long-term to reduce scientific uncertainties. The proceedings of the symposium, published in this monograph, are intended to inform policy determinations and the mandated Scientific Assessment by describing the scientific knowledge and areas of uncertainty concerning freshwater harmful algal blooms.

Background

There is growing concurrence among scientists, risk assessors, and risk managers that the incidence of harmful algal blooms (HABs) is increasing in spatial and temporal extent in the US and worldwide. HABs occur in marine, estuarine, and freshwater ecosystems. A National Plan that primarily targets HABs and their toxins in marine and estuarine waters has been developed, Harmful Algal Research and Response: A National Environmental Science Strategy 2005-2015, (HARNESS 2005), but an analogous plan for freshwater HABs has not been developed. Although many algal groups form HABs within a range of salinity levels, dinoflagellates comprise the majority of marine and estuarine HABs, whereas cyanobacteria are the predominant source of freshwater HABs. The Interagency, International Symposium on Cyanobacterial Harmful Algal Blooms (ISOC-HAB) focused on cyanobacterial HABs (CHABs) because characterization of the state of the science and identification of research needs is essential for the development of a freshwater research and response plan. CHABs and their highly potent toxins, collectively known as cyanotoxins, pose a potential risk to human health. Ecosystem sustainability is compromised by CHABs due to toxicity, pressures from extreme biomass levels, and the hypoxic conditions that develop during CHAB die offs and decay. Some of these risks are described in the World Health Organization's guidelines for CHABS (WHO 1999). However, current data in the US are insufficient to unequivocally confirm an increased incidence or to fully assess the risks of CHABs, thereby complicating Federal regulatory determinations and the development of guidelines, Water Quality Criteria and Standards, and regulations. As a result, state, local, and tribal authorities are placed in the quandary of responding to CHAB events by developing and implementing risk management procedures without comprehensive information or Federal guidance. This dilemma was recognized by the US Congress and expressed in the 2004 reauthorization and expansion of the 1998 Harmful Algal Blooms and Hypoxia Research and Control Act (HABHRCA). Whereas HABHRCA originally targeted harmful algal blooms in the oceans, estuaries and the Great Lakes, the reauthorized Act mandated a Scientific Assessment of Freshwater Harmful Algal Blooms, which will: 1) examine the causes, consequences, and economic costs of freshwater HABs throughout the US; 2) establish priorities and guidelines for a research program on freshwater HABs; and 3) improve coordination among Federal agencies with respect to research on HABs in freshwater environments.

The US Environmental Protection Agency (EPA) is authorized to protect human health and the environment from contaminants in drinking and recreational waters through the mandates of the Safe Drinking Water Act, last amended in 1996 (SDWA 1996), and the Clean Water Act, last amended in 2002 (CWA 2002). The National Oceanographic and Atmospheric Administration (NOAA), EPA and other Federal agencies recognize that cyanotoxins in freshwaters may present a risk to human health through the potential for exposure from recreational waters, drinking water, fish and shellfish consumption, and other vectors. The Federal agencies also recognize that cyanobacteria and cyanotoxins threaten the viability of aquatic ecosystems through alteration of the habitats that sustain plants, invertebrates and vertebrates. EPA's Office of Water listed cyanobacteria and cyanotoxins on the first drinking water Contaminant Candidate List (CCL) of 1998 and the second, CCL2, of 2005 (CCL 2006). Risk assessments, regulatory determinations, and risk management procedures can be informed by research that further clarifies: 1) the spatial extent and temporal frequency of freshwater CHABs, both toxic and non-toxic; 2) dose-response relationships describing the effects of individual cyanotoxins and commonly occurring cyanotoxin mixtures in humans and other species at risk; and 3) cost effective means to prevent, control, and mitigate CHABs in surface waters.

EPA's National Health and Environmental Effects Research Laboratory, a component of the Office of Research and Development, invited other Federal and state entities to co-sponsor a CHAB symposium, ISOC-HAB. The purpose of the Symposium was to characterize the state of the science and to identify research needs, thereby informing EPA's Office of Water and the HABHRCA-mandated Scientific Assessment of Freshwater Harmful Algal Blooms. NOAA and seven other Federal entities, the Food and Drug Administration, Department of Agriculture, Centers for Disease Control and Prevention, Army Corps of Engineers, US Geological Survey, National Institutes of Health, and National Institute of Environmental Health Sciences, as well as the University of North Carolina Institute of Marine Sciences joined EPA in co-sponsoring ISOC-HAB. An interagency organizing committee of 32 members and a five member executive advisory committee (see Organizing Committee page) were assembled to develop an operational structure for ISOC-HAB.

Theoretical Framework for Cyanobacterial Harmful Algal Blooms

The ISOC-HAB Organizing Committee developed a theoretical framework of interrelationships between factors that may influence the development of CHABs and be impacted by CHABs to help identify the major topic areas and subtopics of the symposium (Fig. 1). Both natural forces and human activities may be promoting CHABs through habitat alteration (Causes, Prevention and Mitigation Workgroup Report this volume). The natural forces may include an upswing in temperature cycles that allow tropical genera of cyanobacteria to flourish in subtropical regions, the evolution of new strains of cyanobacteria that can better compete for survival and dominance, a decline in predatory populations that limit cyanobacteria growth, and age-related eutrophication of surface waters. Anthropogenic pressures may be major sources of ecological change that promote CHABs. There is evidence that greenhouse gasses are increasing global temperatures, thereby allowing temperature limited genera and species to expand spatially and temporally (Paul this volume). Excessive levels of nitrogen and phosphorus in surface waters from point and non-point sources promote the development of CHABs, and their ratios may determine which species dominate blooms (Paerl this volume). Waters that are high in phosphorus and relatively low in nitrogen are typically dominated by species that contain heterocysts, specialized cells to collect and fix nitrogen into useable forms. Non-heterocyst containing species often dominate blooms in waters that are high in nitrogen. The incidence of CHABs may be increased by pollutants, such as pesticides and metals in storm-water runoff and other sources that disrupt the balance between cyanobacteria and their predators, or lead to the rise of more resilient strains of cyanobacteria through natural selection. The introduction of non-native organisms into surface waters also may promote CHABs. The recent resurgence of CHABs in the Great Lakes is associated with the invasion of Asiatic Zebra muscles, *Dreissena polymorpha,* that may selectively filter-feed non-toxic phytoplankton (Occurrence Workgroup Report this volume). The combined pressures from natural forces and human activities on surface waters may provide a competitive advantage to cyanobacteria over their predators, leading to an increase in the spatial and temporal extent of CHABs.

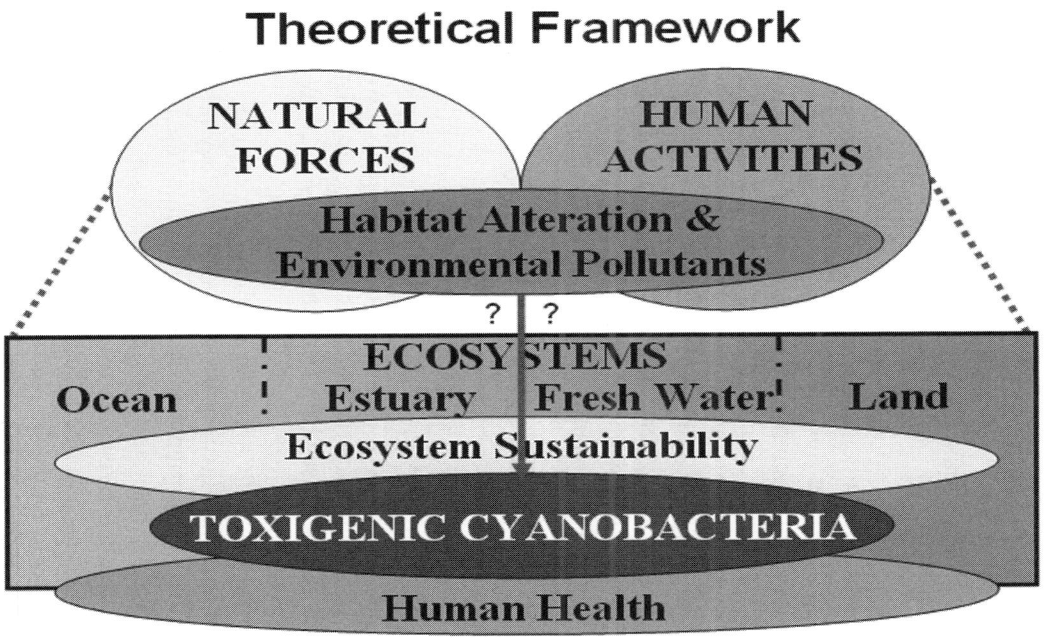

Fig. 1. Both natural forces and human activities may alter habitats in ways that promote the occurrence of cyanobacterial harmful algal blooms, increasing the potential for adverse effects on ecosystem sustainability and human health.

Although CHABs primarily occur in fresh and estuarine waters, there is increasing recognition that cyanobacteria blooms in oceans are threatening the sustainability of some marine ecosystems (Ecosystem Effects Workgroup Report this volume). The recent and unprecedented decline in viable coral reefs worldwide is due in part to marine CHABs (Paul this volume). Species of toxigenic *Lyngbya* adapted to high salinity environments can form benthic mats that expand over an area equivalent to a football field within an hour, causing ecological damage and endangering human health (Australian Environmental Protection Agency 2003).

Cyanotoxins also are found in terrestrial environments where they may pose a risk to human and animal health. Surface waters are increasingly used for field irrigation in agricultural production. Water drawn from sources experiencing toxigenic CHABs is sprayed on crops, producing cyanotoxin-containing aerosols that may be inhaled by humans and other animals, and absorbed by crops. Cyanobacteria can form a symbiotic relationship with terrestrial plants which may biomagnify cyanotoxins. Cyanobacteria of the genus *Nostoc* form colonies on the roots of cycad plants in Guam where for more than 30 years scientists have tried to unravel the genesis of the mysterious neurodegenerative disease that afflicts the native Chamorro population. An amino acid cyanotoxin produced by *Nostoc*, beta methylamino-alanine (BMAA), accumulates in cycad seeds. The seeds are eaten by a species of bat that accumulates high levels of BMAA in its tissues. The bat is a traditional food source for the Chamorro. Analyses detected BMAA in brain tissues of Chamorro victims, leading to the hypothesis that BMAA causes neurodegeneration that may manifest with features of amyotrophic lateral sclerosis, Parkinson's disease, and Alzheimer's dementia. Recent evidence indicates that BMAA is produced by most types of cyanobacteria, and that it may be associated with neurodegenerative diseases elsewhere (Human Health Effects Workgroup Report this volume).

Cyanobacteria and cyanotoxins are clearly hazardous to human health and ecosystem sustainability, but the degree of risk they present is unclear (Risk Assessment Workgroup Report, this volume). Research is needed to accurately assess the risks and provide risk managers with cost effective options for reducing the risks as warranted. A Scientific Assessment of Freshwater HABs can describe a comprehensive approach toward understanding the interconnections between the causes of blooms and toxin production, the characteristics and magnitude of the risks they pose, and the means for reducing the risks through prevention and mitigation strategies. Meeting these objectives requires that relationships between CHABs, humans, and the environment be viewed as a system of interconnected components.

A Systems Approach to Cyanobacterial Harmful Algal Blooms

The concept of a systems approach can be traced back to ancient Greece when Aristotle proclaimed that "The whole is more than the sum of its parts." A system is generally defined today as a dynamic process that provides the functionality required by users of the system. In engineering, a systems approach integrates multidisciplinary groups into a unified team that develops and implements a process from concept to operation. The application of a systems approach to risk assessment and management issues requires several fundamental components.

- Integration of discovery (i.e., descriptive) science with hypothesis-driven science
- A cross-disciplinary team to develop and implement the system
- Development of new approaches and technologies coupled with tools for data acquisition, storage, integration, and analysis

Whereas a systems approach to CHABs is appropriate, a broad perspective is required to accommodate the stochastic nature of biological and ecological processes. That is, the causes, occurrences, production of hazardous materials, routes of exposure, dosage of hazardous materials, and effects of a CHAB can be viewed as an ordered collection of random variables whose values change over space and time. These components and their interconnections, the processes by which one component at least partially determines the qualities of the next component, form the CHAB pathway. The combination of the CHAB pathway, risk assessment, policy determination, and risk management forms a systems approach to CHABs. A systems approach to CHABs provides the perspective that ecosystems partially determine human well-being, and that humans partially determine ecosystem well-being. To produce the tools required to manage the risks that CHABs impose on humans and ecosystems, it is necessary to characterize the components and their interconnections. Successful risk management tools may target the components and interconnections of the CAHB pathway for disruption to reduce risk.

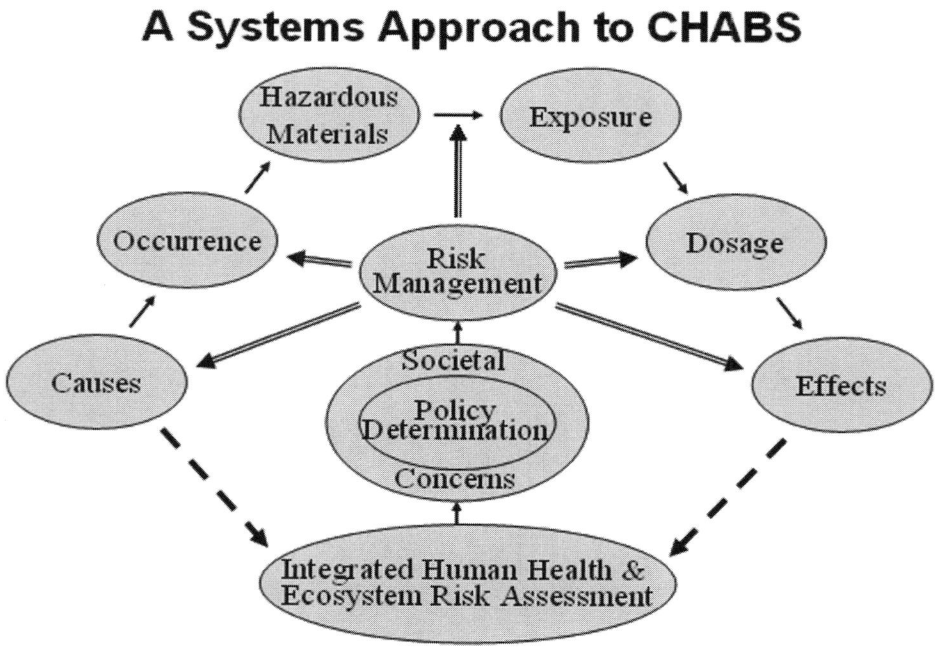

Fig. 2. A systems approach to Cyanobacterial Harmful Algal Blooms. This diagram illustrates the system's components (ovals), their interconnections (thin arrows), the consideration of the entire CHAB pathway during integrated risk assessment (dashed arrows), and some of the intervention targets that, if disrupted, could reduce risk (thick arrows).

The concept of a systems approach for managing the risks of CHABs is illustrated in Figure 2. The nine ovals identify components of a system for characterizing and managing CHABs. The thin arrows between the ovals represent the interconnections between components. The dashed arrows signify the incorporation of all characteristics of the CHAB pathway, from causes to effects, into an integrated approach to the assessment of risks that CHABs pose to human health and ecosystem sustainability. The risk assessment in conjunction with societal concerns such as laws, legal decisions, public values, available technologies, and economic, social, and political factors inform the policy and regulatory process, potentially resulting in the development and implementation of a risk management plan. The thick arrows radiating from the risk management component indicate some of the potential targets for risk management interventions. A system is formed by combining the components and interconnections along the CHAB pathway with the components and interconnections of the risk assessment, policy determination, and risk management processes into a functional unit. Implementation of such a system will provide a dynamic process that helps to prevent, predict, and respond to CHABs to protect human health and the ecosystem.

A starting point for the development of a system to manage CHABs is the identification of areas of uncertainty within the CHAB pathway as shown in Figure 2. Many environmental factors that contribute to the development of CHABs are known (Paerl this volume, Paul this volume). However, the threshold levels of individual factors, the dependence of thresholds on the magnitude of other stressors, and the processes whereby the integration of stressors triggers CHABs are not well characterized. Although actions can be taken to minimize the contribution of known stressors (Piehler this volume), research that better characterizes the interdependence of stressors will enable the development of more targeted and effective risk management tools. Actions also can be taken to terminate CHABs and reduce levels of free toxin in water, but research is needed to more fully characterize the unexpected and untoward environmental impacts of these actions, and to develop interventions that have minimal adverse effects. For example, the use of copper sulfate to terminate CHABs causes high levels of cyanotoxins and potentially toxic levels of copper in water, and the use of flocculants to bind toxins and transport them to the bottom stresses benthic dwellers. Most CHABs produce an extreme biomass associated with hypoxia (Ibelings this volume, Havens this volume), but a CHAB does not necessarily indicate the presence of toxins (Cyanotoxins Workgroup Report this volume, Carmichael this volume). Research that characterizes the processes that trigger toxin production may lead to the development of methods to minimize their production. Field-ready tests that rapidly and inexpensively identify and quantify a broad array of cyanobacteria and cyanotoxins (Analytical Methods Workgroup Report this volume, Meriluoto this volume, Wilhelm this volume, Lawton this volume, Sivonen this volume) are needed to identify the hazardous materials in CHABs to help assess risks so that risk managers can prevent exposure through actions such as public notification. Cyanotoxins occasionally are present in finished drinking water (Burns this volume), indicating the need to develop effective water treatment processes (Westrick this volume). Mathematical models that integrate physical, chemical, and biological variations over space and time are needed to predict the occurrence of CHABs and toxin production to expand the window of time for risk management actions. Medical interventions may reduce the dosage of toxins that reach target sites, and the duration that toxins circulate in exposed humans and animals (Human Health Effects Workgroup Report this volume, Hudnell 2005). Validation of medical interventions to eliminate cyanotoxins and the development of other treatments for affected individuals are needed to supplement the current standard of care in medical practice, supportive therapy. The assessment of risks from toxic exposures requires extensive information on dose-response relationships. CHABs often contain a mixture of cyanotoxins (Humpage this volume), presenting a formidable challenge to cyanotoxin risk assessment (Burch this volume). Equally challenging is the need to quantify CHAB risks to humans and ecosystems holistically, so that risk management actions can be identified that are effective, efficient, and without unintended consequences (Risk Assessment Workgroup Report this volume, Orme-Zavaleta and Munns this volume).

Characterization of the CHAB pathway as the interconnections between CHABs, humans, and the environment provides a basis for integrated risk assessment. The characterization of integrated risks and the development of cost effective interventions will support policy determinations and risk management processes. The primary goals of ISOC-HAB, discussed below, were to describe the state of the science of CHAB system components and interconnections, and to identify research needed to reduce scientific uncertainties and improve risk management processes.

ISOC-HAB Organization, Charges to Speakers and Workgroups, and Products

The ISOC-HAB Organizing Committee identified seven major topic areas and 23 subtopics to be addressed during the symposium (Table 1). Each subtopic was addressed by an invited participant during a platform session.

The primary charge given to each speaker was to describe the state of the science in the assigned area. The major topic areas were addressed by Workgroups for which specific charges also were developed. The primary charges for each Workgroup were to identify research needed to reduce scientific uncertainties and to develop processes that ultimately will provide risk managers with cost-effective tools to prevent and mitigate the effects of CHABs.

Table 1. The seven major topic areas and the 23 subtopics addressed at ISOC-HAB.

Occurrence of CHABS • A US & World Overview • The Florida Experience • The Nebraska Experience • The New York & Great Lakes Experience	Analytical Methods • Sample Preparation • Laboratory Methods • Field Methods • Emerging High Throughput Analyses
Causes, Prevention, & Mitigation • Nutrients and Other Causes • Global Climate Change • Watershed Mangement • Drinking Water Treatment	Human Health Effects • Laboratory Exposures • Environmental Exposures • Epidemiology
Cyanotoxin Characteristics • Types, Toxicokinetics & Toxicodynamics • Genomics & Proteomics • Bioterrorism Potential	Ecosystem Effects • Aquatic Vertebrates • Trophic Status & Ecological Conditions
	Risk Assessment • Economic Impact • Toxic Microbes & Mixtures • Human & Ecological Integration

In addition to the primary charge for each topic area, additional charges were given to each Workgroup. The Organizing Committee realized that there was overlap between some of the topic areas, largely due to the interconnections between components on the CHAB pathway. Similarities among some of the charges given to different Workgroups were intended to promote characterization of the interconnections from a broader diversity of perspectives. All Workgroups were asked to identify factors needed in models to predict the occurrence of events along the CHAB pathway, not for the immediate construction of predictive models, but to help identify research needed to reduce scientific uncertainties. Highlights of the Workgroups' charges are described below.

Occurrence of CHABs

The Occurrence Workgroup was charged with identifying trends in: 1) the spatial and temporal incidence of CHABs in the US and worldwide; 2) the prevalence of specific genera and species in fresh, estuarine, and marine water CHABs; 3) the percentage of CHABs that produce cyanotoxins; 4) the types and mixtures of cyanotoxins that most commonly occur in CHABs; (5) the health and ecological risk potentials of CHABs in recreational and drinking water reservoirs; and (6) the development of guidelines and standards by state and local governments. The Workgroup's primary charge was to identify research needed to remove impediments to the collection of CHAB occurrence data in the US, including implementation of the EPA Office of Water's Unregulated Contaminant Monitoring Rule for drinking water (UCMR 1999).

Causes, Prevention, & Mitigation

The Causes, Prevention & Mitigation Workgroup was charged with identifying research needed to better characterize or develop: 1) the natural and anthropogenic causes of CHAB occurrence and toxin production; 2) watershed management and other tools that reduce the probability of CHAB occurrence; 3) methods for terminating CHABs and removing cyanotoxins from source and drinking waters; and 4) methods for potential inclusion in recreational and drinking water risk management guidelines. The Workgroup also considered factors needed in models to predict cost and benefit relationships for methods that prevent CHABs and remove cyanotoxins from water.

Cyanotoxin Characteristics

Charges for the Cyanotoxin Characteristics Workgroup included identifying research needed to better characterize: 1) methods for rapid and cost effective identification and quantification of known and novel cyanotoxins; 2) the pharmacokinetic properties of cyanotoxin absorption, distribution, and metabolism in animals; 3) the toxicodynamics of cyanotoxin modes of action in the production of adverse health effects; 4) factors that increase and decrease susceptibility to adverse health effects; 5) the genomics and proteomics of cyanobacteria and cyanotoxin production; and 6) methods to reduce the potential risk of cyanotoxin use in bioterrorism. The identification of factors needed in models to predict the production of cyanotoxins during CHABs also was included in the Workgroup's charges.

Analytical Methods

The Analytical Methods Workgroup was charged with identifying and evaluating current methods for detecting and quantifying cyanobacteria, single cyanotoxins, and mixtures of cyanotoxins. The Workgroup's charges also included identifying research needed to develop rapid and cost-effective: 1) field screening kits to identify and quantify cyanobacteria and cyanotoxins; 2) field screening kits to detect genes responsible for cyanotoxin production; 3) laboratory methods to identify and quantify cyanobacteria and cyanotoxins; and 4) laboratory methods to identify the genes responsible for cyanotoxin production. The Workgroup's charges also included the identification of methods needed to produce bulk cyanobacteria and cyanotoxins of known quality, as well as certified toxin standards for use by a broad scientific community.

Human Health Effects

The Human Health Workgroup's charges included the identification of research needed to further characterize: 1) human health effects associated with exposure to particular cyanobacteria genera and species; and 2) human health effects associated with exposure to particular cyanotoxins, individually and in mixtures. The Workgroup's charges also included the identification of existing and needed infrastructure to better assess exposure and effect relationships, including exposure monitoring and health surveillance programs, and internet-based data management and distribution systems. The Workgroup also considered factors needed in models to predict exposure and effect relationships in human populations.

Ecosystem Effects

The Ecosystem Effects Workgroup was charged with identifying research needed to better characterize: 1) effects on biota in land and water ecosystems associated with exposure to particular cyanobacteria genera and species; and 2) effects on biota in land and water ecosystems associated with exposure to particular cyanotoxins, individually and in mixtures. The Workgroup's charges also included the identification of existing and needed infrastructure to better assess exposure-and-effect relationships, including exposure monitoring and surveillance indicators such as sentinel species. The Workgroup also considered factors needed in models to predict exposure-and-effect relationships in aquatic populations, land animals, and ecosystem indicators.

Risk Assessment

The Risk Assessment Workgroup was charged with identifying research needed to: 1) support guideline, criteria and standards, and regulation development; 2) develop tiered monitoring and response systems for fresh, estuarine, and marine waters; 3) develop an integrated human health and ecosystem sustainability risk assessment process; and 4) develop a framework for making policy determinations that encompasses CHAB type, overall risk, and cost/benefit optimization. Also included in the Workgroup's charges were the identification of factors needed in models to predict the cost-and-benefit relationships of risk management tools, and the need to revise or produce new risk management guidelines and regulations.

ISOC-HAB Product & Goals

This monograph contains the proceedings of ISOC-HAB, a series of chapters that describe:

- An overview of ISOC-HAB (this chapter);
- A synthesis of research needed to improve risk assessments and management;
- Seven Workgroup Reports on short- and long-term research needs in the topic areas;
- Twenty-three Speaker Reports on the state of the science in the subtopic areas;
- Fourty-two poster abstracts that describe emerging research in the topic areas.

The monograph is divided into sections corresponding to the major topic areas. Each section contains the Workgroup Report, Speaker Reports, and poster abstracts that address the topic area.

Publication of the ISOC-HAB proceedings in this monograph, and the ongoing publication of materials on the EPA website (http://www.epa.gov/cyano_habs_symposium/) are intended to further the scientific understanding of freshwater harmful algal blooms and provide a resource for:

- Developing the products mandated by Congress through HABHRCA;
- Developing an interagency National Research Plan for CHABs;
- Integrating academic, industrial, local, state, and Federal CHAB research;
- Informing EPA's Office of Water and other Federal institutions;
- Informing states, Indian tribes, and local governments;
- Informing industries, academic institutions, and non-governmental institutions;
- Informing other countries confronting the risks posed by CHABs

References

Australian Environmental Protection Agency (2003) http://www.epa.qld.gov.au/environmental_management/coast_and_oceans/marine_habitats/lyngbya_management_strategy/

Clean Water Act (2002) US Congress reauthorization http://www.epa.gov/region5/water/pdf/ecwa.pdf

Contaminant Candidate List (2006) US Environmental Protection Agency, Office of Water, http://www.epa.gov/safewater/ccl/index.html

Harmful Algal Blooms and Hypoxia Research and Control Act (2004) US Congress reauthorization http://www.cop.noaa.gov/pubs/habhrca/2004_publ456.108.pdf

HARRNESS (2005) Harmful Algal Research and Response: A National Environmental Science Strategy 2005-2015. JS Ramsdell, DM Anderson and PM Glibert (eds), Ecological Society of America, Washington DC, 82 http://www.esa.org/HARRNESS/harrnessReport10032005.pdf

Hudnell HK (2005) Chronic biotoxin-associated illness: Multiple-system symptoms, a vision deficit, and effective treatment, Neurotoxicol Teratol 27:733 – 743. http://www.sciencedirect.com/

Safe Drinking Water Act, 1996. US Congress reauthorization, http://epa.gov/safewater/sdwa/text.html

Unregulated Contaminant Monitoring Rule (1999) http://www.epa.gov/safewater/methods/unregtbl.html

World Health Organization (1999) Toxic cyanobacteria in water: A guide to their public health consequences, monitoring and management. I Chorus and J Bartram (eds.), E & FN Spon, New York, 416 http://www.who.int/water_sanitation_health/resourcesquality/toxicyanbact/en/

Chapter 2: A Synopsis of Research Needs Identified at the Interagency, International Symposium on Cyanobacterial Harmful Algal Blooms (ISOC-HAB)

H Kenneth Hudnell and Quay Dortch

Abstract

Evidence indicates that the incidence of cyanobacterial harmful algal blooms (CHABs) is increasing in spatial extent and temporal frequency worldwide. Cyanobacterial blooms produce highly potent toxins and huge, noxious biomasses in surface waters used for recreation, commerce, and as drinking water sources. The Interagency, International Symposium on Cyanobacterial Harmful Algal Blooms (ISOC-HAB) characterized the state of the science and identified research needed to address the risks posed by CHABs to human health and ecosystem sustainability. This chapter provides a synopsis of CHAB research needs that were identified by workgroups that addressed charges in major topic areas. The research and infrastructure needed are listed under nine categories: 1) Analytical Methods; 2) CHAB Occurrence; 3) CHAB Causes; 4) Human Health; 5) Ecosystem Sustainability; 6) CHAB Prevention; 7) CHAB Control and Mitigation; 8) Risk Assessment and; 9) Infrastructure. A number of important issues must be addressed to successfully confront the health, ecologic, and economic challenges presented by CHABs. Near-term research goals include the development of field-ready tests to identify and quantify cells and toxins, the production of certified reference standards and bulk toxins, formal assessments of CHAB incidence, improved understanding of toxin effects, therapeutic interventions, ecologically benign means to prevent and control CHABs, supplemental drinking water treatment techniques, and the development of risk assessment and management strategies. Long-term goals include the assimilation of CHAB databases into emerging U.S. and international observing systems, the development of quantitative models to predict CHAB occurrence, effects, and management outcomes, and economic analyses of CAHB costs and management benefits. Accomplishing further infrastructure development and freshwater HAB research is discussed in relationship to the Harmful Algal Blooms and Hypoxia Research and Control Act and existing HAB research programs. A sound scientific basis, the integration of CHAB infrastructure with that of the marine HAB community, and a systems approach to risk assessment and management will minimize the impact of this growing challenge to society.

Introduction

The Interagency, International Symposium on Cyanobacterial Harmful Algal Blooms (ISOC-HAB) characterized the state of the science and identified research needed to address the risks posed by cyanobacterial harmful algal blooms (CHABs) to human health and ecosystem sustainability. The state of the science was described by invited experts who addressed specific charges for CHAB subtopics in platform sessions and authored 23 chapters of this monograph. The research needed to develop a systems approach toward the assessment and management of CHAB risks (Hudnell *et al.* this volume) were identified in seven workgroups whose members addressed specific charges and summarized their findings in additional chapters of this monograph. The workgroups were organized to address the major topic areas of: 1) Analytical Methods; 2) CHAB occurrence; 3) CHAB causes, prevention, and mitigation; 4) Cyanotoxin characteristics; 5) Human health effects; 6) Ecosystem effects and; 7) Risk assessment. The Organizing Committee realized that there was overlap between some of the topic areas, largely due to the interconnections between components on the CHAB pathway (Hudnell *et al.* this volume). Similarities among some of the charges given to different workgroups were intended to promote characterization of the interconnections from a broader diversity of perspectives. Table 1 presents the research and infrastructure needs identified by the workgroups. Research in each of the Priority Areas is briefly discussed below. More detailed discussions of the state of the science and research needs are presented in the speaker and workgroup report chapters.

Research in the nine Priority Areas identified in Table 1 was considered to be high priority over the long term. The workgroup reports designate each research need as a near-term or long-term goal. The near-term goals are those that do not require other research to be accomplished prior to addressing those goals, whereas the long-term goals are dependent upon the completion of near-term goals or require an extended time period to complete. The need for certified analytical methods and readily available reference standards was generally acknowledged by the workgroups to be of highest priority because many other goals are dependent of the availability of methods and materials. Methods and materials were similarly given the highest priority in the HABs report, Harmful Algal Research and Response: A National Environmental Science Strategy 2005-2015, (HARNESS 2005).

The U.S. Congress reauthorized and expanded the Harmful Algal Blooms and Hypoxia Research and Control Act (HABHRCA 2004). Whereas HABHRCA originally targeted harmful algal blooms in the oceans, estuaries and the Great Lakes, the reauthorized Act mandated a Scientific Assessment of Freshwater Harmful Algal Blooms, which will: 1) examine the causes, consequences, and economic costs of freshwater HABs throughout the U.S.; 2) establish priorities and guidelines for a research program on freshwater HABs; and 3) improve coordination among Federal agencies with respect to research on HABs in freshwater environments. The research topics discussed below are intended to help identify issues that should be addressed in order to fully meet the mandates of HABHRCA.

Analytical Methods

Standardized and certified methods for collecting field samples are needed to ensure that the samples consistently represent the existing environmental conditions, and that results can be compared across time and between collectors. The samples generally consist of water, plankton, invertebrates, vertebrates, or sediments. Standardized methods also are needed for sample processing, including filtration, stabilization, transportation, and storage, as well as for the extraction of cyanotoxins from complex matrices such as biological tissues and sediments.

A tiered approach toward screening environmental samples for cyanobacteria and cyanotoxins is needed to accommodate a variety of settings and purposes, and to make efficient use of resources. Initial screening methods should be designed for field settings such as water utilities or recreational water management facilities. The identification and quantification of organisms traditionally has been accomplished through microscopy, a time consuming method that requires a high level of training. Genetically-based methods should be further developed for the identification of cyanobacteria to the species level, and for the detection of genetic sequences involved in toxin production. Automated cell counting methods are needed for quantification. Although standard methods exist for analyzing some cyanotoxins (Meriluoto and Codd 2005), improved methods are needed for rapid, inexpensive, and reliable analyses in field settings. Enzyme linked immunosorbent assays (ELISA) or other emerging methods are needed to measure cyanotoxin levels. The methods should be sensitive to a wide variety of cyanotoxin analogues or congeners. A combination of toxin level measurements and bioassays for toxicity will indicate the total potential for toxicity from environmental exposures. A long-term goal for field analyses is to produce real time, *in situ* monitors coupled with data transmission systems. Remote sensing systems will provide early indicators of environmental conditions that favor the emergence of CHABs, as well as information on the initiation, development, and senescence of CHABs. Finally, specialized laboratories are needed to verify field results, validate results from developing techniques, and identify novel toxins. These laboratories, which may require sophisticated and expensive equipment (for example liquid chromatographs/mass spectrometers) and high levels of technical expertise, should be shared-use facilities due to budget constraints. These facilities should be capable of operating on an emergency basis to provide a rapid response to situations endangering public health.

Table 1. Synopsis of Research Needs for Cyanobacterial Harmful Algal Blooms

Priority Area	Needs	Results - Improved Understanding, Methods, Products & Prediction
Analytical Methods	• Standardized Methods	• Sample Collection, Filtration, Stabilization, Transport, Storage Toxin Extraction from Complex Matrices
	• Tiered Screening	• Strategies Adaptable to Location & Purpose
	• Field Methods	• Probes for Organism Identification & Toxin Production Multiple Analogue Sensitive Toxin Identification & Quantification
	• Laboratory Methods	• Improved & New Techniques for Known & Novel Toxins
CHAB Occurrence	• Consensus Taxonomy	• Consistent Taxonomic Identification to Species Level
	• Nationwide Survey	• CHAB & Toxin Occurrence in Source Water using UCMR CHAB & Toxin Occurrence in Recreational Water
	• Long-term Monitoring	• CHAB Occurrence Trends in Source & Recreational Waters Remote Sensing Methods & Coupling with Global Observing Systems
	• Toxin Transport & Fate	• Environmental Transport, Accumulation & Degradation
	• Predictive Models	• Local CHAB Occurrences Toxin Production, Environmental Transport, Accumulation & Fate

CHAB Causes	• Retrospective Data Analyses	• Physical, Chemical & Biological Variations Over Space & Time
	• Controlled Studies - Lab, Field, Microcosom & Mesocosom	• CHAB Responses to Controlled Environmental Variables Identification of Toxin Production Triggers
	• Ecosystem Monitoring	• CHAB Dynamics & Environmental Interactions CHAB Expansion with Climate Change & Other Stressors Driver Thresholds that Destabilize Ecosystems & Induce CHABs
	• Predictive Models	• Factors Controlling CHAB Initiation, Dynamics & Toxin Production
Human Health	• Human Health Effects	• Bioindicators of Human Exposure & Effect Toxicokinetics, Toxicodynamics, Dose-Response Relationships Epidemiology, Repeated Recreation & Drinking Water Exposures
	• Predictive Models	• Routes & Quantities of Human Exposure to Toxins Quantitative Structure-Activity Relationships
Ecosystem Sustainability	• Ecosystem Effects	• Toxin & Concurrent Stressor Effects on Key Biota & Communities Bioaccumulation, Bioconcentration & Biomagnification in Food Web Eutrophication, CHAB & Turbidity Relationships
	• Predictive Models	• Eutrophication & CHAB Driven Ecosystem Alterations & Fate
CHAB Prevention	• Nutrient Source Identification	• External Inputs Versus Internal Nutrient Recycling
	• Watershed Management	• Methods to Reduce External Nutrient Input
	• Water Management	• Methods to Increase Flow, Destratisfy & Increase Competitive Forces
	• Predictive Models	• Relative Effectiveness of Prevention Strategies

CHAB Control & Mitigation	• Bloom & Toxin Destruction	• Environmentally Benign Methods to End CHABs & Degrade Toxins
	• Drinking Water Treatment	• Detect Presence of Cell Fragments & Toxins During Water Processing Methods to Remove Fragments, Toxins, Taste & Odor Compounds
	• Predictive Models	• Relative Effectiveness of Control & Mitigation Strategies
Risk Assessment	• Integrated Human Health & Ecosystem Risk Assessment	• Interdependence of Human Health & Ecosystem Sustainability
	• Reduce Data Uncertainties	• Assessing Risks from CHAB Biomass Risk from Single Toxins Cyanotoxin, Analogue Toxicity Equivalence Factors & QSAR Total Risk from Common Toxin Mixtures & Crude Cell Extracts Susceptibility Factors
	• Accidental & Intentional Toxin Release	• Safe Production, Storage & Transportation Airborne Dispersal & Route of Exposure Relative Toxicity
	• Predictive Models	• Local & National Total Cost of CHABs to Society CHAB Cost Versus Prevention, Control & Mitigation Cost/Benefit
	• Effectiveness Measures	• Pre- Versus -Post Prevention, Control & Mitigation Cost/Benefit

Infrastructure	• Shared Centralized Facility & Service	• High Complexity Equipment, Analyses, Training & Certification Produce & Provide Certified Toxin Standards & Bulk Toxins U.S. Surveillance, Databases, & Coupling with International Systems
	• Coordination	• Improved Federal, Stakeholder & International Coordination Standing Fresh to Marine Advisory Committee & International Link
	• Education	• Train Volunteer, Industry, Utility, Government, Academic, Post Doc Public & Stakeholder Education Services

CHAB Occurrence

Although CHABs have been reported worldwide and in most or all states in the U.S., there is no international or national database that contains records of all CHAB events. The degree to which states or local governments record CHABs is highly variable. Therefore, no definitive information is available on the incidence of CHABs over time and space in the U.S., the genera and species involved, toxin production, transportation and fate, environmental conditions, or effects on humans and ecosystems. There is widespread concurrence among scientists, risk assessors, and risk managers, however, that the incidence of CHABs is increasing in spatial and temporal extent in the U.S. and worldwide. The Occurrence Workgroup Report and the chapters describing CHABs in Florida, Nebraska, and New York and the Great Lakes contained in this monograph support the hypothesis of increasing CHAB occurrence. A major impediment to the development of a national database is the lack of a consensus on taxonomy for cyanobacteria. Most field taxonomists rely on the traditional morphology–based botanical approach for phylogenetic classification because molecular data are not available for many species. Research is needed to develop a consensus on taxonomy for cyanobacteria based on genetic fingerprints or an array of characteristics potentially including morphological, molecular, physiological, bioinformatic, and biogeochemical information to classify algal communities with depth and precision.

Nationwide surveys to describe CHAB occurrence will become practical as improved analytical methods to identify species and quantify cyanotoxins become available. Surveys of CHABs in drinking water sources and recreational waters are needed because both types of surface waters present human health risks during CHABs. The EPA has the regulatory authority to implement the Unregulated Contaminant Monitoring Rule (UCMR) that requires a subset of large municipal water utilities to conduct surveys for substances potentially hazardous to human health. Implementation of the UCMR for cyanobacteria and their toxins is not being considered by the EPA at this time because of the need for less expensive, more reliable, accurate, and field-ready analytical methods for quantifying single toxins and multiple analogues. The EPA also could undertake, require, or encourage CHAB monitoring in recreational waters. The BEACH Act, which amends the Clean Water Act, requires EPA to ensure state adoption of recreational water quality standards, revise water quality criteria, publish beach monitoring criteria, and maintain a beach database (EPA 2006). Occurrence data are a primary requirement for the Agency to make regulatory determinations concerning the development of CHAB regulations or guidelines.

Long-term monitoring is the only method by which trends in CHAB occurrence over time and space can be identified. In addition to identifying changes in CHAB incidence, long-term monitoring programs can provide data needed to assess a variety of issues including CHAB dynamics, environmental interactions, relationships to global climate change, and effects. An understanding of the interactions between cyanotoxins and environmental factors is needed to assess the potential for exposure of human and other biota. Of particular interest is the transportation of cyanotoxins through aerosols, biota, and water, the accumulation and magnification of the toxins in biota and inorganic matrices, and environmental processes through which cyanotoxins are degraded. Only long-term monitoring of CHABs and their toxins in combination with ecological survey data can reveal the cumulative effects of CHABs on ecosystem diversity and population dynamics.

A long-term goal is integration of CHAB monitoring with emerging earth observation systems - the U.S. integrated earth and ocean observing systems (IEOS, IOOS), the Global Oceans Observing System (GOOS) - which culminate in the Global Earth Observing System of Systems (GEOSS; Oceanus 2007). The goal of these observation systems is the routine and continuous delivery of quality controlled data and information on current and future environmental conditions in forms and at rates required by decision makers to address societal goals such as human health protection and ecosystem sustainability. The systems combine remote and *in situ* monitoring data, data management and communication subsystems, and data analysis and modeling components to deliver near real-time and forecasted information to primary users. The combination of *in situ* and remotely sensed data (e.g., aircraft and satellite detection of photopigment type and quantity), and incorporation into U.S. observing systems, will provide a sustainable system for monitoring CHABs and delivering useable information to risk managers.

Forecasts of imminent CHABs will require the development of predictive models that incorporate near real-time data on physical, chemical, and biological conditions at specific locations. As our understanding of CHAB dynamics and environmental interactions increases, it may become possible to not only predict occurrence, but also to predict toxin production, environmental transport, accumulation, and fate. The validation and iterative development of predictive models can be based both on hindcasts derived from datasets not used in model development, and on empirical evidence collected at predicted times and locations. Models that forecast CHABs will provide a window of time for local officials to take risk management actions such as public notification to prevent exposure or installation of equipment to vertically mix the water column to disrupt bloom formation.

CHAB Causes

CHABs occur in a wide variety of aquatic environments, and the general conditions associated with the initiation of CHABs are known. CHABs require nutrients, particularly nitrogen and phosphorus, and sunlight, and tend to occur in warm, slow moving waters that lack vertical mixing. However, the dynamics of cyanobacterial interactions with environmental factors involved in bloom formation, and the factors that trigger toxin production, are poorly understood. Retrospective analyses of long-term datasets can identify associations between physical, chemical, and biological variations over space and time and the occurrence of CHABs and toxins. Improved understanding of the complex interactions that promote blooms and toxin production will enable the development of hypotheses that can be tested under controlled conditions in laboratory, microcosom, mesocosom, and perhaps field studies. Issues such as the role of trace metals in bloom and toxin production can be addressed most directly through controlled studies. The responses of cyanobacteria to experimentally controlled variables will provide insights into bloom initiation and toxin production that may lead to the development of improved and environmentally benign strategies for controlling CHABs.

Ecosystem monitoring can be used to test hypotheses derived through retrospective data analyses and controlled studies, and to address location-specific issues. The dynamics of CHAB initiation, sustainment, and termination may vary through interactions with location-specific factors. For example, cyanobacteria predators and infectious agents may be or become abundant in some areas, causing relative rapid termination of CHABs. Monitoring may detect CHABs in previously unaffected water bodies as land use practices shift and global climate change raises temperatures and alters hydrologic conditions. A particularly important issue to address through monitoring is threshold levels of environmental factors at which ecosystems undergo long-term phase shifts that promote CHABs and are difficult or impossible to reverse. Only long-term monitoring can reveal trends in the spatial and temporal incidence of CHABs.

The development of mathematical models of CHAB dynamics and interactions with environmental factors will provide a basic framework for relating causative factors to bloom occurrence, toxin production, bloom maintenance and termination. Models of CHAB dynamics and environmental interactions will form integral components of models that ultimately will be developed to predict local CHAB occurrences and the relative efficiency of local control, mitigation, and prevention strategy options.

Human Health

Information on the human health effects of cyanotoxins is largely limited to characterizations of effects from single, high-level exposures. Many animal studies describe the dose (usually intraperitoneal or oral gavage dosing) of single cyanotoxins that causes lethality in 50% of the animals in a study (LD_{50}). Such studies are useful in that they demonstrate that cyanotoxins are among the most potent toxins known, and in identifying the organ system in which failure is the primary cause of death. However, these studies leave many important questions unanswered. They do not address many issues likely to be of importance in human environmental exposures, such as: 1) the relative potency of cyanotoxins through different routes of exposure (i.e., inhalation, dermal absorption, ingestion; 2) the effects of repeated, low-level exposures; 3) the combined effects from exposure to commonly occurring cyanotoxin mixtures (including additive and synergistic effects); and 4) factors that increase or decrease the susceptibility of individuals (and other animal species) to adverse effects from exposure. A combination of well controlled animal studies and both retrospective and prospective epidemiological studies is needed to provide the scientific basis for developing human health risk assessments for exposure to cyanotoxins.

A significant impediment to both animal and human studies is the lack of rapid, reliable, and inexpensive biomarkers of exposure and effect. Analytical methods are needed to quantify multiple cyanotoxin analogues and metabolites in blood and other biological tissues. Protein and DNA adduct measurements may also be useful in characterizing exposure. Accurate characterizations of cyanotoxin exposure present one of the most difficult challenges to human health research. The primary reason is that CHABs often produce several types of cyanotoxins and numerous analogues that vary in toxicologic properties and potencies. Additionally, animal and *in vitro* studies invariably indicate that crude cell extracts are more potent than dose-equivalent quantities of cyanotoxins observed in the cells. The cause of this superpotency may be potentiation of the cyanotoxin modes of actions by other cellular components, or an outcome of exposure to the cyanotoxins and other components not recognized to be toxic.

The characterization of effects from cyanotoxin exposures presents another difficult challenge to human health research. The biological mechanisms through which cyanotoxins cause acute effects may differ from those that cause delayed or chronic effects because different biological systems vary in their ability to repair tissue damage. Cumulative damage may result from repeated exposures in systems with less efficacious repair or compensatory processes. Also, acute effects are more likely to involve direct effects of toxins, whereas chronic effects may results from secondary toxin actions such as triggering an inflammatory response in the immune system. Research is needed to identify biomarkers of cyanotoxin effects in multiple organ systems to characterize the array of effects that may arise from exposure, particularly repeated, low-level exposures. Biomarkers of effect should include biochemical, behavioral, and other indicators of function in all biological systems that may be affected by the direct or indirect actions of cyanotoxins.

Central to the assessment of health risks from cyanotoxin exposure is characterization of toxicokinetic, toxicodynamic, and dose-response relationships in animal models. Toxicokinetic research is needed to characterize toxin uptake through various routes of exposure, the metabolism of the parent compounds and degradation products, distribution of those compounds in tissues, the time-course of toxin retention in tissues, and the pathways of toxin elimination. Toxicodynamic research is needed to describe the modes of action by which the cyanotoxins and metabolites interact with biological tissues to alter physiology and function within affected organ systems. Dose-response research is needed to describe relationships between toxicokinetic parameters of exposure and adverse health outcomes. It is critical to assess relationships in a variety of animal species, to model inter- and intra-species differences, and to validate the ability of the animal models to predict comparable relationships in humans. The inclusion of potentially susceptible subpopulations in the studies, such as fetuses (through *in utero* exposures), the young, and the aged, is needed to reduce scientific uncertainties and improve the accuracy of risk assessments.

Human exposures to cyanotoxins are most likely to occur through contact with recreational waters and drinking water, although the risk for exposure through food consumption is not well characterized. The probability of high-level exposures through water ingestion is less than that for repeated, low-level exposures through recreational or drinking water contact (e.g., ingestion, dermal absorption, inhalation). However, much less is known about the health risks posed by repeated, low-level exposures. Lower level exposures may cause acute illness characterized by nonspecific symptoms such as gastro-intestinal distress, skin rashes, respiratory difficulty, and flu-like symptoms. Lower level exposures also may cause chronic illness in some individuals, such as that reported following the acute-phase of ciguatera seafood poisoning (Palafox and Buenconsejo-Lum 2005). Whereas acute-phase illness is characterized by gastrointestinal and respiratory distress, the chronic-phase is characterized by sustained fatigue, muscle and joint pains, and severe neurologic symptoms that persist indefinitely. Clinical research is needed to describe modes of action in human illness, and to develop therapeutic interventions beyond the current standard-of-care, supportive therapy. Methods are needed to greatly enhance toxin elimination rates, as are toxin antidotes. Other evidence indicates that chronic conditions such as neurodegenerative diseases and delayed illnesses such as cancer may be associated with repeated exposures to cyanotoxins. Both animal and epidemiologic research is needed to characterize the health risks associated with repeated, low-level exposure to cyanotoxins. Retrospective epidemiologic studies may be able to use existing datasets to explore potential linkages between repeated exposures to cyanotoxins and health outcomes. However, prospective epidemiologic studies are needed for more definitive evidence on causal relationships between repeated cyanotoxin exposures and health outcome. The validation of animal models of cyanotoxin exposure-effect relationships also is largely dependent on the availability of epidemiologic data.

Quantitative models are needed to predict dosages of cyanotoxins to which people may be exposed through contact with contaminated water. Both recreational and drinking water contact provides the opportunity for inhalation, ingestion, and dermal exposures to cyanotoxins. The dosage of cyanotoxins depends on the activities in which people are involved, the durations of those activities, and the concentration of cyanotoxins in the water among other factors. Quantitative models that predict the dosages to which people may be exposed based on these factors will assist risk managers in making decisions to ensure that humans are not exposed to dosages that present a health risk.

Quantitative structure-activity relationship (QSAR) models will assist risk assessors by predicting the dosages of cyanotoxins that pose a health risk when no data are available on the cyanotoxin in question, but data on similarly structured molecules are available. For example, few data are available on the effects of repeated dosing with anatoxin-a(s), an organophosphate cyanotoxin that inhibits acetylcholinesterase. However, anatoxin-a(s) is structurally and functionally similar to organophosphate pesticides such as parathion and malathion. It may be possible to develop a QSAR model that predicts the toxicity of anatoxin-a(s) based on the literature describing the toxicity of organophosphate pesticides. A similar approach may be useful in predicting the toxicity of the multiple analogues of cyanotoxins. There are now over 80 known analogues of microcystins.

QSAR models may be able to predict the toxicity of the analogues, obviating the need for extensive toxicity testing for each analogue.

Ecosystem Sustainability

Although most attention and research has focused on the threats to human and domestic animal health, CHABs also disrupt ecosystems by a variety of mechanisms, most of which are poorly characterized. Toxins produced by CHABs have the potential to affect organisms at a variety of trophic levels. Little is known about cyanotoxin impacts on most aquatic biota or the extent to which CHAB toxins transfer between trophic levels. It is frequently noted that CHABs are not grazed by either planktonic or benthic filter feeders. For example, the return of CHAB blooms to some areas of the Great Lakes is hypothesized to have resulted from the introduction of zebra mussels, which graze other algae, but not colonial CHABs, such as *Microcystis*. Grazing inhibition may be due to toxins, the occurrence of CHABs in large mucoid colonies, or the overall unpalatability of many CHABs. A potentially related problem in shallow water bodies undergoing eutrophication and an increasing frequency of CHABs is a persistent shift from a clear to turbid state. Turbid waters are associated with increased populations of disease causing microorganisms, and declines in primary producer populations, including phytoplankton, benthic algae and vascular plants. Research is needed to clarify relationships between CHAB toxins and other stressors, and the development of adverse ecological conditions such as grazing inhibition, turbidization, and disruption of both benthic and plankton food webs.

Freshwater cyanobacteria are often the predominant phylum of plankton in extremely eutrophic waters. Problems caused during CHAB development and maturation include shading and overgrowth of other algae and aquatic vegetation. The huge amounts of organic material produced by CHABs harm ecosystems, even in the absence of toxin production, following bloom senescence. Bloom die-offs result in large amounts of decaying biomass on the benthos that produce hypoxic and anoxic conditions. The lack of dissolved oxygen stresses and kills many benthic dwellers, resulting directly in the loss of benthic biological diversity and weakening the primary producer end of the food web. This impact extends throughout the food web as biota at intermediate and upper levels increasingly lack sufficient nutritional sources. The impacts culminate in the loss of biological diversity at all levels, including the depletion of populations on which humans depend for food sources, recreational activities, and food stock for the production of other nutritional sources. The loss of biological diversity also may impact ecosystem sustainability by allowing the expansion of less desirable populations, including toxigenic cyanobacteria, due to the lack of competitive forces. This situation is exacerbated when CHABs also produce toxins.

Finally, CHABs generally occur in areas of poor water quality, environments impaired by multiple stressors. CHABs are both a response to the stressors and an additional stressor. The combined effects of the stressors are additive or synergistic, resulting in such extensive ecological changes that the ecosystem may loose its resilience and reach an ecological threshold beyond which it is difficult to return to a more pristine state. An array of studies are needed that focus first on the impacts of CHABs and their toxins on individual components of ecosystems, and then on interactions between the components. This research can be accomplished through a series of culture, microcosms, mesocosms, and ecosystems field studies. An ultimate goal is to develop models that not only provide comprehensive descriptions of current conditions in local areas, but also predict the impacts of future environmental changes, and assess the efficacy of risk management actions.

CHAB Prevention

Changes in land use practices that increase nutrient input into surface waters have been associated with an increased incidence of CHABs in many locations. Nutrient sources include storm water runoff from fertilized and other nutrient rich lands, discharges from sewage treatment plants and large confined animal feeding operations, as well as airborne depositions of nitrogen from organic sources. The development of effective strategies for preventing CHABs requires an understanding of local conditions. For example, CHAB initiation and duration may be dependent on nutrient input in some surface waters, where as benthic/pelagic coupling or the cycling of nutrients between sediment, biota and the water column may sustain CHABs in eutrophic environments. Research that characterizes nutrient input rates and benthic/pelagic coupling in local areas is needed to develop land and water management plans that will reduce CHAB incidence and promote aquatic ecosystem sustainability.

An increasing incidence of CHABs is an indication of decreasing water quality, usually due to increasing nutrient loads and/or decreasing flow rates. The ultimate driver of decreasing water quality is land use practices. Watershed management plans are increasingly developed and implemented to improve land use practices such that nutrient input into surface waters are reduced and controlled. Research is needed to develop more efficient and cost-effective methods for reducing non-point source inputs of nutrients and processing point source nutrients in an ecologically sustainable manner.

Watershed management techniques provide a long-term strategy for reducing CHAB incidence and improving water quality. The implementation of water management plans may provide near-term improvements that prevent or terminate CHABs. Increased flow rates and decreased water temperatures improve water quality and reduce the probability of a CHAB occurrence. Where as flow rate and water temperature control are impractical for many surface waters, evidence indicates that artificial destratification of the water column may be a highly efficient means of preventing and terminating CHABs. The use of bubble systems to vertically mix the water column may be effective on a small scale if the density of bubblers is sufficient, but upscaling to large areas is usually impractical. Alternatively, floating solar powered platforms can host pumps that draw in water from above the benthos (to avoid nutrient resuspension from sediment) but below levels at which CHABs occur. The water is discharged at the surface, creating a vertical mixing loop over areas as large as 35 acres. Research is needed to further assess the effectiveness of bubble and pump vertical mixing systems at controlling CHABs, and to identify the mode(s) of action by which vertical mixing inhibits CHABs. Vertical mixing may inhibit CHABs by disrupting cyanobacteria's ability to regulate its position in the water column, or by inducing the dispersion and amplification of microbial colonies that effectively prey upon or infect cyanobacteria. Additional research may reveal other methods by which the competitive forces against cyanobacteria can be increased.

A long-term goal is the development of models that predict the abilities of land and water management techniques to improve water quality and prevent CHABs at specific locations. These models would incorporate area-specific data on CHAB occurrence and dynamics, the physical, chemical, and biological conditions associated with CHABs, and the relative abilities of land and water management techniques to improve those conditions. These models will provide risk managers with powerful tools for developing cost effective strategies to prevent CHABs.

CHAB Control & Mitigation

Environmentally benign methods are needed to terminate CHABs and neutralize cyanotoxins. Algaecides such as copper sulfate have been used for many years to terminate CHABs, and chelated copper compounds have been used in more recent times to extend the period during which algaecidal activity in the water column is retained. However, algaecides, and to a lesser extent algaestats, have several drawbacks. First, algaecides cause cell lysis, resulting in rapid release of cyanotoxins and high concentrations in the water column. High concentrations of cyanotoxins in the water column increase the health risk for humans in recreational waters. High concentrations also may increase the probability of cyanotoxin accumulation in the food web, as well as the probability that drinking water treatment processes will be insufficient to reduce cyanotoxin concentrations to safe levels. Second, algaecides quickly precipitate out of the water column, presenting a health risk to aquatic biota such as benthic invertebrates. The loss of benthic invertebrates disrupts the food chain, eliminates a pathway for phosphorus uptake, and greatly impedes the decomposition of detritus organic matter, each of which threatens ecosystem sustainability. Third, the use of copper algaecides selects for cells tolerant of copper, leading to the development of copper resistant populations. Fourth, algaecides induce rapid bloom collapse, resulting in large biomass deposition on the benthos. Cellular decomposition often depletes dissolved oxygen, thereby producing anoxic or hypoxic conditions that exacerbate the stress already confronting aquatic biota due to high cyanotoxin concentrations. Oxygen depletion also causes the uncoupling of phosphorus from iron oxides in sediment, resulting in resuspension of phosphorus in the water column and increased probability of new CHABs. Research is needed to determine if algaecides and algaestats, as well as other bloom termination techniques such as ultrasound, can be developed that do not have untoward effects on human health and ecosystem sustainability. Research should evaluate the efficacy and ecosystem impacts of vertical mixing techniques, discussed above under prevention, relative to those of algaecidal and other bloom termination techniques.

Research on processes to neutralize cyanotoxins in surface water has met with limited success and may have drawbacks similar to those described above for algaecides. These processes have included the use of compounds such as alum to coagulate or flocculate cyanotoxins, potassium permanganate or titanium dioxide to oxidize cyanotoxins, and electrocoagulation techniques. In addition to problems involving deposition on the benthos, cyanotoxin neutralization techniques are costly, do not inhibit new CHABs, and are impractical for large scale applications. Research on cyanotoxin neutralization techniques may be more applicable to drinking water than surface water treatment.

The need for research on methods to detect and remove cyanobacteria and cyanotoxins during drinking water treatment is particularly great. Evidence indicates that cyanotoxin levels in finished water can be higher than that in raw water, presumably because water pressure on cells during filtration causes lysis and the release of toxins. Additionally, cyanobacterial taste and odor compounds, particularly geosmin and 2-methylisoborneol (MIB), frequently necessitate that water treatment utilities expend considerable resources to remove these compounds to reduce the frequency of customer complaints. Water utilities commonly rely on the observation of surface scums to detect CHABs, but because CHABs may occur without visible evidence on the surface, they may go undetected at the utilities. Research and development efforts are needed that result in automated, reliable, and inexpensive techniques to detect and quantify cyanobacteria and cyanotoxins in source waters and water treatment systems. Early detection will enable utility managers to implement specialized treatment processes to remove the cells and toxins before they enter the finished water stream.

Many methods have been used in attempts to remove cyanobacteria and cyanotoxins during drinking water treatment, including various oxidation, absorption, coagulation, sedimentation, traditional filtration, and membrane filtration techniques. To date, none of these techniques has proven to be generally effective for intact cell and cell fragment or toxin removal under the variety of conditions that occur during water treatment. The efficacy of cyanotoxin removal techniques depends not only on the concentration and molecular configuration of toxins and analogues, but also on environmental factors such as pH, temperature, and dissolved organic matter concentration, and on processing parameters such as concentrations and contact durations of oxidants and absorbents with the toxins. Research is needed to assess and further develop techniques for cell removal and toxin degradation during drinking water processing that are applicable under a variety of conditions and that maximize cost and benefit ratios. Assessments should also include analyses of treatment process impacts on the production of toxic disinfection byproducts.

Long-term goals include the development of models that predict the effectiveness, cost, and adverse impacts of control and mitigation techniques. For the control of CHABs in surface waters, the models would include factors describing CHAB species and toxin production, physical characteristics of the water body, ecosystem structure and services (including economic benefit), and functional, cost, and outcome characteristics of alternative control options. For the mitigation of cells and toxins entering source water intakes for drinking water processing, the models would predict the degree to which the contaminants would enter the finished water stream without altering treatment processes, and the outcomes, costs, and benefits resulting from the implementation of supplemental treatment options.

Risk Assessment & Management

Assessments of the risk posed to human health and ecosystem sustainability by toxic substances in the environment have traditionally been done separately. The rationale has been that the analyses will indicate whether the risk is greatest for human health or ecosystem sustainability at the lowest concentration which adversely impacts either humans or the environment, and that regulatory determination will be based on the one that is most susceptible to those impacts. The conceptual flaw to this approach is that it fails to capture the inherent relationships between humans and the environment; ecosystems partially determine human wellbeing, and humans partially determine ecosystem wellbeing. This intimate interconnection is perhaps nowhere clearer than with HABs. The eutrophication of surface waters due to human activities is the single greatest cause of the worldwide increase in the incidence of CHABs, and CHABs have the potential to irrevocably damage ecosystems and to have severe consequences for human health. Research is needed to develop an integrated approach to the assessment of risks posed by CHABs to human health and ecosystem sustainability. An integrated approach to risk assessment and supporting risk analyses has the potential to characterize the total impact of CHABs on society, thereby better informing risk mangers as they develop strategies to reduce risks.

The risks posed by CHABs are complex due to the large variety of CHAB species and toxins, the unpredictability of toxin production, the adverse impacts of CHAB biomass independent of toxins, the impacts of CHAB toxins on aquatic biota, and many other issues that have not been sufficiently clarified through research. Research must fully address several issues in order to reduce scientific uncertainties and improve CHAB risk assessments. First, reduced ecosystem sustainability due to lost biological diversity from excess CHAB biomass and toxins must be assessed. Research is needed to better characterize the total impact on CHABs on ecosystem sustainability and the ecological services that are provided to humans.

Second, risk assessments usually identify concentrations at which single toxins do not pose a risk for adverse health effects over a lifetime of exposure. Accordingly, the only World Health Organization (WHO 1999) guideline on cyanotoxins in drinking water is for Microcystin LR, 1 μg/l. That assessment assumed that Microcystin LR is noncarcinogenic, implying that there is a threshold level of exposure below which there is no adverse biological effect. However, there is recent evidence indicating that Microcystin LR may be a carcinogen. The regulatory determination process typically incorporates data on the occurrence, dose-response functions, and practical means for reducing risk. Existing evidence has been deemed insufficient to support risk assessments for Microcystin LR and every other cyanotoxin in the U.S. In addition to microcystins, cylindrospermopsins, anatoxins, and to a lesser extent saxitoxins, are considered to be priority cyanotoxins in the U.S. Research is needed to support risk assessments and regulatory determinations for individual cyanotoxins.

Third, the actual risks posed by toxigenic CHABs are not characterized by risk assessments for single cyanotoxins such as Microcystin LR because of several complicating factors. Toxigenic CHABs rarely if ever produce only a single analogue of a single cyanotoxin type. CHABs often produce multiple analogues of a cyanotoxin type, and the analogues may differ in toxic potency. In recognition of this fact, several countries such as Australia have produced guidelines for the total sum of Microcystins in a bloom, 1.3 μg/l. That guideline level is based on the equivalent toxicity of a number of analogues to that of Microcystin LR. In order to produce guidelines for mixtures of cyanotoxin analogues, data are needed that express the toxicity of each analogue to that of a "reference" analogue, as was produced for some of the analogues of Microcystin LR. Toxicity equivalence factors can be derived in two ways: 1) through direct comparisons of the toxicity of each analogue to that of the "reference" analogue as indicated by a bioassay test or; 2) by the development of QSAR models that estimate the relative toxicity of the analogues through an understanding of structure-activity relationships. Therefore, research is needed that supports the development of risk assessments for priority cyanotoxin types based on toxicity equivalence factors.

Fourth, another complicating factor that impedes the characterization of the actual risk posed by toxigenic CHABS is that many blooms contain more than one type of cyanotoxin. Many cyanobacteria species produce several types of cyanotoxins, and some blooms are composed of more than one cyanobacteria genera, each of which may produce different types of cyanotoxins. Yet risk assessment processes have not been developed that can characterize the risk of mixtures of different types of toxins, particularly when the mixtures can vary significantly over time and between locations, as with CHABs. Further complicating the situation is the observation that crude cell extracts are invariably more toxic than the toxins isolated from those cells. Innovative approaches are needed to assess the actual risks posed by mixtures of toxins produced by CHABs.

Fifth, risk assessments are generally designed to be protective of the most vulnerable members of a population. Factors that may render humans more susceptible to toxic exposures include developmental stage (i.e., *in utero*), life stage (e.g., the young and elderly), preexisting illnesses, past exposures, and concurrent stressors. These and other factors may also pertain to the vulnerability of aquatic biota. Research is needed to better describe factors that increase the susceptibility of humans and other life forms to injury from CHABs.

In addition to the risks posed by CHABs in surface waters, there is need to address the potential risks from cyanotoxin release into other media through accidental or intentional causes. Research on cyanobacteria and cyanotoxins is largely dependent on the production, storage, and transportation of these substances. There is also an increasingly frequent need for transportation of cyanotoxin standard reference materials between suppliers (e.g., commercial and academic) and users (e.g. laboratories, utilities, environmental and health agencies). The need for cyanotoxin-related materials must be balanced with the inherent risk of accidental release of these highly toxic substances. Therefore, cyanobacteria and cyanotoxins are increasingly coming under the regulatory control of international agencies such as the International Air Transport Association (IATA 2006). It is critical that the suppliers and users know of and comply with regulations concerning cyanobacteria and cyanotoxins. On the other hand, it is essential that the regulations are not so cumbersome that they overly inhibit research and risk management processes, thereby increasing the risks from the natural occurrence of these substances. Research is needed to better characterize the risks from accidental release of cyanotoxins during production, storage, and transportation in order to produce regulatory controls that optimize risk reduction.

The potential for weaponization of cyanotoxins has been recognized (USA Patriot Act 2001, ATCSA 2001). Cyanotoxins are high potency and relatively low molecular weight compounds that can be extracted and purified from cultures and in some cases synthesized in laboratories. These features provide the potential for stockpiling large quantities of cyanotoxins for illegitimate purposes. Large populations could be exposed to cyanotoxins through several routes, including municipal water supplies, foodstuffs, and airborne dispersion. Acts as simple as deliberately dumping truckloads of fertilizer into municipal water supplies to trigger CHABs could lead to major societal disruptions, health risks, and economic losses. Yet the risk for cyanotoxins, and other biotoxins, to be used as chemical/biological weapons has not been well characterized. For example, few studies have compared the relative potency of cyanotoxins when delivered through different routes of exposure. Little is known about the potential for preparing cyanotoxins for airborne dispersion, either through the generation of aerosols or production of ultrafine particles from crystallized cyanotoxins. Inhalation exposure to the alkaloid neurotoxins is likely to be much more dangerous than exposure through ingestion due to increased absorption and rapid access to brain. Research is needed to better characterize the risks and develop risk management strategies to protect against the use of cyanotoxins as weapons.

Quantitative models are needed to estimate the total cost of CHABS on local and national scales to inform regulators and improve risk management strategies. The costs of CHABs can generally be separated into direct costs of CHABs and the costs of CHAB prevention, control, and mitigation. Some direct costs of CHABs may be relatively easy to quantify, such as costs borne by fisheries due to lost productivity, and decreased revenues to merchants from declines in tourism and recreational activities of local populations. Many other direct costs may be difficult to quantify or even go unrecognized. For example, although it might be generally agreed that the costs of CHABs are incurred in categories such as human health and quality of life, ecosystem services, ecological decline, property values, and animal death, it may be difficult to recognize all factors within these categories and to quantify their costs. Innovative approaches are needed to identify all areas in which CHABs have direct impacts and to quantify the costs of those impacts. Assessments of the direct costs of CHABs are needed for comparison with the costs of CHAB prevention, control, and mitigation strategies.

The costs of CHAB prevention, control, and mitigation options can be determined more readily. For example, the costs of implementing watershed management practices or installing vertical mixing apparatus to prevent or control CHABs can be assessed using standard economic approaches. Likewise, costs of mitigation practices such as implementing supplemental water treatment processes or obtaining alternative sources of drinking water can be calculated with reasonable precision. Quantitative models of the direct costs of CHABs and the costs of prevention, control, and mitigation strategies will improve risk management decisions by minimizing costs to the public and maximizing public benefits.

It is critical to measure the efficiency and effectiveness of risk management decisions to provide accountability to risk management systems. As with all governmental initiatives, public support of risk management practices is dependent on the ability to clearly demonstrate that the benefits exceed the costs. Economic models should be developed to quantify the costs of CHABs and CHAB risk management strategies, and to estimate the economic benefits derived from risk management. Application of the models before and after implementation of risk management strategies provides a method for assessing the efficiency and effectiveness of risk management decisions. Efficiency and effectiveness can be maximized over time through an iterative process as improved models and risk management practices are developed.

Infrastructure

As stated in HARRNESS (2005), "Development of infrastructure will be key to the success of the National Plan, and will ensure that the new strategy is responsive to the needs of scientists, managers, public health coordinators and educators." The need for infrastructure is also emphasized in the Congressionally mandated, Scientific Assessment of Freshwater Harmful Algal Blooms (FASHAB 2007). These documents and the ISOC-HAB workgroups identify a number of roles for a distributed network of HAB research centers that are crucial in developing a systems approach to assessing and managing the risks posed by CHABs. Central to the concept of CHAB infrastructure is shared facilities that provide high complexity equipment and certified analytical procedures, certified toxin standards and other reference materials, bulk cyanotoxins, database development and management, assistance in the coordination of CHAB research, and educational services. A stable infrastructure will increase the efficiency of CHAB risk assessment and management efforts by reducing redundancies in CHAB research and providing state of the art services to stakeholders.

HAB infrastructure centers ideally would be shared between marine and freshwater research and management communities. The centers would provide the expertise and equipment for culturing and storing cells, extracting and purifying toxins, identifying novel toxins, certifying analytical methods, and validating tests to be used in the field to identify and quantify cells and toxins. The certification of analytical methods might best be done in association with organizations such as the Association of Analytical Communities, International (AOAC 2007). An essential function of the centers would be to provide cell, certified toxin standards, and other reference materials of a known type or degree of purity for use in developing methods and calibrating equipment. Bulk cyanotoxins are needed to characterize the effects of exposure in animal models. The centers should serve as conduits for the integration of many existing databases on cyanobacteria properties, environmental characteristics, and CHAB events for both research purposes and the development of visualization and prediction tools. The databases should be standardized using the IOOS Data Management and Communications (DMAC) structure for reporting and distributing data so that they can be integrated into emerging observation system networks. The integration and standardization of databases for research and model development will yield products that can be utilized by a wide user community. HABHRCA (2004) specifically calls for improved coordination of Federal CHAB research. The FASHAB (2007) report addresses Federal research coordination and recognizes the value of improved coordination across all levels of U.S. and International CHAB research. The centers could take a lead role in integrating perspectives of the U.S. freshwater HAB research community into the National HAB Committee, a group previously dedicated to the coordination of marine HAB research (HARRNESS 2005). A similar role in international coordination could be taken by the centers through interaction with groups such as CYANONET (2007). The centers also could take the lead role in providing educational services and products to stakeholders and the general public. Educational services could range from the training of volunteers and employees of industry, utilities, state and local governments in the use of equipment to monitor for CHAB cells and toxins to the formal education of students pursuing careers in related fields. Educational products could range from pamphlets and website materials to inform the general public of CHAB risks and management strategies to standardized health advisories for state and local public health officials to issue when CHABs are predicted or in progress. A strong and stable infrastructure provides an efficient foundation to support development of a systems approach toward the assessment and management of CHAB risks.

Aligning the Infrastructure and Research Needs with HABHRCA

The 2004 reauthorization of HABHRCA calls for a "competitive, peer-reviewed, merit-based interagency research program as part of the Ecology and Oceanography of Harmful Algal Blooms (ECOHAB 2007) project, to better understand the causes, characteristics, and impacts of harmful algal blooms in freshwater locations…" The interagency (NOAA, EPA, NSF, NASA, ONR) ECOHAB program and the NOAA Monitoring and Event Response for Harmful Algal Blooms (MERHAB 2007) programs currently meet this requirement. However, the freshwater focus has been on Great Lakes and upper reaches of estuaries because those areas are within the purview of NOAA. No equivalent program exists for inland waters other than the Great Lakes, whereas HABHRCA calls for an examination of the causes, consequences, and economic costs of freshwater HABs throughout the U.S. The causes, characteristics, and impacts of HABs in smaller freshwater bodies are likely to differ significantly from those of marine and Great Lakes HABs. In addition, research related to HABs in drinking source waters and treatment options, an important public health issue, falls out of the scope of the programs mentioned above. Finally, although there are programs that provide assistance to state and local management agencies in responding to marine HABs to protect human health and coastal economies, there are no such programs for inland freshwater HABs. Thus, state, local, and tribal governments are left to develop management strategies without an adequate scientific basis or Federal guidance. This problem could be solved either by broadening the scope of existing programs or establishing programs specifically for freshwater/inland HABs within agencies having the appropriate mandate.

References

AOAC (2007) The Association of Analytical Communities, International. http://www.aoac.org/

ATCSA (2001) Anti-Terrorism, Crime and Security Act, The Stationery Office Ltd, London, UK 118 pp. (Chapter 24).

Australian Environmental Protection Agency (2003)
http://www.epa.qld.gov.au/environmental_management/coast_and_oceans/marine_habitats/lyngbya_management_strategy/

CYANONET (2007) A Global Network for the Hazard Management of Cyanobacterial Blooms and Toxins in Water Resources. http://www.cyanonet.org/

ECOHAB (2007) Ecology and Oceanography of Harmful Algal Blooms. http://www.cop.noaa.gov/stressors/extremeevents/hab/current/fact-ecohab.html

FASHAB (2007) Scientific Assessment of Freshwater Harmful Algal Blooms. Joint Subcommittee on Ocean Science and Technology, in press.

Harmful Algal Blooms and Hypoxia Research and Control Act (2004) U.S. Congress reauthorization. http://www.cop.noaa.gov/pubs/habhrca/2004_publ456.108.pdf

HARRNESS (2005) Harmful Algal Research and Response: A National Environmental Science Strategy 2005-2015. JS Ramsdell, DM Anderson and PM Glibert (eds.), Ecological Society of America, Washington DC, 82. http://www.esa.org/HARRNESS/harrnessReport10032005.pdf

IATA (2006) Dangerous Goods Regulations (47th edition), International Air Transport Association, Montreal, Canada, 815.

MERHAB (2007) Monitoring and Event Response for Harmful Algal Blooms. http://www.cop.noaa.gov/stressors/extremeevents/hab/current/fact-merhab.html

Meriluoto J, Codd GA (eds) (2005) Toxic: Cyanobacterial monitoring and cyanotoxin analysis. Abo Academie University Press.

Ocean.us (2007) What is IOOS? http://www.ocean.us/what_is_ioos

Palafox and Buenconsejo-Lum (2005) Ciguatera fish poisoning: review of clinical manifestations, Toxin Reviews 20/2:141-160.

US Environmental Protection Agency (2006) Beach Monitoring and Notification: Report to Congress. http://www.epa.gov/waterscience/beaches/report/

USA Patriot Act (2001) Uniting and Strengthening America by Providing Appropriate Tools Required to Intercept and Obstruct Terrorism (USA Patriot Act) Act of 2001. H.R. 3162 in the senate of the United States, 342

World Health Organization (1999) Toxic cyanobacteria in water: A guide to their public health consequences, monitoring and management. I Chorus and J Bartram (eds.), E & FN Spon, New York, 416 http://www.who.int/water_sanitation_health/resourcesquality/toxicyanbact/en/

Chapter 3: Occurrence of Cyanobacterial Harmful Algal Blooms Workgroup Report

Edited by Anthony Fristachi and James L Sinclair

Workgroup Co-Chairs:

James L Sinclair, Sherwood Hall

Workgroup Members[1]:
Julie A Hambrook Berkman; Greg Boyer; JoAnn Burkholder; John Burns; Wayne Carmichael; Al DuFour; William Frazier; Steve L Morton; Eric O'Brien; Steven Walker

Acknowledgements

Appreciation is given to all those whose efforts made the production of this book possible. Thanks are due to the editors, Anthony Fristachi, US Environmental Protection Agency, Office of Research and Development, Cincinnati, Ohio and Jim Sinclair, PhD, US Environmental Protection Agency. Thanks are also due to Dr. Jeffrey Johansen of John Carroll University for his review and revisions to the Section 4 discussion of the current status of taxonomy of cyanobacteria.

Special thanks to the workshop co–chairs: Jim Sinclair, PhD, US Environmental Protection Agency, Office Water, Cincinnati, Ohio and Sherwood Hall, PhD, Food and Drug Administration, CFSAN, Laurel, Maryland, who co–managed the process of preparing the manuscript.

---

[1] See Workgroup members and affiliations in Invited Participants section.

Introduction

Freshwater cyanobacteria periodically accumulate, or bloom, in water bodies across the United States (US). These blooms, also known as cyanobacterial harmful algal blooms (CHAB), can lead to a reduction in the number of individuals who engage in recreational activities in lakes and reservoirs, degrade aquatic habitats and potentially impact human health. In 1998, Congress passed the 1998 Harmful Algal Bloom and Hypoxia Research and Control Act (HABHRCA) to address CHABs that impacted living marine resources, fish and shellfish harvests and recreational and service industries along US coastal waters. In 2004, as part of its reauthorization, HABHRCA requires federal agencies to assess CHABs to include freshwater and estuarine environments and develop plans to reduce the likelihood of CHAB formation and to mitigate their damage (NOAA 2004). Many federal agencies recognize the potential impacts of CHABs and share risk management responsibilities; an interagency task force was established and charged to prepare a scientific assessment of the causes, occurrence, effects and economic costs of freshwater. The United States Environmental Protection Agency (EPA) has included "cyanobacteria (blue–green algae), other freshwater algae, and their toxins" in its Contaminant Candidate List (CCL) as one of the microbial drinking water contaminants targeted for additional study, but it does not specify which toxins should be targeted for study (EPA 2005b). Based on toxicological, epidemiology and occurrence studies, the EPA Office of Ground Water and Drinking Water has restricted its efforts to 3 of the over 80 variants of cyanotoxins reported, recommending that Microcystin (MC) congeners LR, YR, RR and LA, Anatoxin–a (AA) and Cylindrospermopsin (CY) be placed on the Unregulated Contaminant Monitoring Rule (UCMR) (EPA 2001). The EPA uses the UCMR program to collect data for contaminants suspected to be present in drinking water that do not have health–based standards set. This monitoring supplies information on the nature and size of populations exposed to cyanotoxins through tap water use.

Various federal agencies are mandated to address CHABs and their impacts, which commonly have been managed on a case–by–case, somewhat fragmented basis. A new national US plan, the Harmful Algal Research and Response National Environmental Science Strategy (HARNESS), is "designed to facilitate coordination by highlighting and justifying the needs and priorities of the research and management communities and by suggesting strategies or approaches to address them" (HARNESS 2005). CHABs cross all four critical areas identified for harmful algal research and response: bloom ecology and dynamics; toxins and their effects; food webs and fisheries; and public health and socioeconomic impacts. HARNESS (2005) also noted that "research on freshwater harmful algal blooms [mostly formed by cyanobacteria] has lagged behind efforts to address marine harmful algal blooms [in general], and there is no comprehensive source of information on the occurrence and effects of freshwater harmful algal blooms in the US". This is ironic, considering that CHABs have been documented to be highly responsive to nutrient pollution (Glibert and Burkholder 2006), and toxic to livestock, wildlife and humans in the US and worldwide (Chorus and Bartram 1999).

This chapter will describe the occurrence of bloom–forming cyanobacteria and their toxins. Cyanobacteria water blooms are defined here as the visible coloration of a water body due to the presence of suspended cells, filaments and/or colonies and, in some cases, subsequent surface scums (surface accumulations of cells resembling clotted mats or paint–like slicks). They are a common occurrence in the US and throughout the world. Over 2000 species of cyanobacteria, many associated with nuisance blooms, are currently known (e.g. Komárek and Anagnostidis 1999, Komárek 2003). Freshwater HABs can assume many forms and while most are composed of prokaryotic cyanobacteria (blue–green algae), eukaryotic species may be present in blooms. While surface scums are commonly associated with blooms of Microcystis, blooms of other species such as Cylindrospermopsis in Arizona, Indiana and Iowa can be characterized by lower cell numbers events. Although not all blooms are characterized by high biomass events, high biomass blooms of toxic species/strains can have significant impacts on ecosystems by adding potentially large amounts of toxins (Glibert and Burkholder 2006). In addition, high–biomass blooms, whether of toxic or nontoxic species, also increase in oxygen demand during blooms and bloom decline, leading to localized hypoxia/anoxia and fish kills. These blooms can also cause habitat loss and food web changes. For example, surface blooms can block the light from reaching the benthos, leading to changes in attached plant communities and increased re–suspension of nutrients from the sediments. High–biomass blooms also commonly lead to economic loss through their negative impacts on recreation activities, and tourism and their production of substances that cause taste–and–odor problems and increased treatment costs in potable water supply plants. Blooms of all types can lead to a decreased quality of life by causing a negative perception of the state of health of the water body.

Multiple interacting physical, chemical and biological factors lead to the formation of CHABs. Planktonic cyanobacteria are a natural component of the phytoplankton in most surface waters of the world. Toxigenic species (capable of producing toxins, although not all populations or strains do so) are all naturally occurring members of freshwater and sometimes brackish water phytoplankton. The toxins that they produce can be compared with other naturally produced toxins (Carmichael 1992). Since the first report of toxic cyanobacteria in the late 19th century, all continents except Antarctica have reported toxic blooms (Carmichael 1992). Toxic blooms can be especially problematic and are characterized by their production of hepatotoxins, neurotoxins and acute dermatotoxic compounds (Chorus and Bartram 1999). In the US, blooms of cyanobacteria have been associated with the death of both wildlife and domestic animals (Carmichael 1998). To date, no human fatalities in the US can unambiguously be attributed to cyanobacterial toxins, though in at least one instance, this connection has been made in the popular press (Behm 2003; Campbell and Sargent 2004). Human health effects within the US have primarily included gastrointestinal illness in Pennsylvania (Lippy and Erb 1976), swimmers itch in Florida and Hawaii (Williams et al. 2001) and skin rashes, nausea, and other gastrointestinal disorders in Nebraska (Walker 2005). In Brazil, a gastroenteritis epidemic was associated with the construction of the Itaparica Dam in Brazil (Teixeira et al. 1993), and the hospitalization of 140 children when supplied with drinking water from a reservoir containing a bloom of Cylindrospermopsis raciborskii in Australia (Chorus and Bartram 1999). The use of cyanotoxin–contaminated water in dialysis equipment in Brazil accounted for the only reported human fatalities (Carmichael et al. 2001a; Azevedo et al. 2002).

Distribution of CHABs across the US

Blooms in freshwater environments

Freshwater CHABs are ubiquitous throughout the US and Canada (Fig. 1). Cyanobacteria perform many roles that are vital for the health of ecosystems, especially as photosynthetic organisms, but they may also cause harm through either excessively dense growth or release of toxins or other harmful metabolites. CHABs span habitats ranging from smaller eutrophic prairie ponds to larger more oligotrophic regions of the Great Lakes. This broad distribution is expected given the considerable diversity in habitats occupied by cyanobacteria and the widespread distribution of potential toxicity in the more common genera.

Blooms in estuarine and marine environments

While toxigenic cyanobacteria are generally associated with freshwater blooms, there is increasing recognition that, in certain regions of the world, blooms in estuarine and marine environments may also have an important impact on human health. The Baltic Sea and the Gulf of Finland often sustain massive blooms of toxic *Nodularia spumigena* that are easily visible in satellite images of the region (Sivonen et al. 1989). These blooms affect the natural biota and can spoil recreational opportunities on the coastline. Similarly, large blooms of benthic cyanobacterium *Lyngbya majuscule* in Moreton Bay, Australia (Albert et al. 2005), and off the Hawaiian Islands (Moikeha and Chu 1971; Moikeha et al. 1971) have caused skin, eye and respiratory effects on recreational users of these waters. Not all detrimental effects are directly associated with toxins; for example, increased cyanobacterial biomass in Florida Bay and other estuaries has been associated with sea grass die–offs and the decline of the adjacent coral reef systems (Williams et al. 2001).

Both *Microcystis aeruginosa* and *Cylindrospermopsis raciborskii* have only limited tolerance to salinity (< 2–7 ppt) and this may limit the growth of these common toxic species in brackish waters environments (Barron et al. 2002). Increasing salinity in oligohaline and mesohaline estuaries results in increased osmotic stress and aggregation of cyanobacterial cells (Sellner et al. 1988). Many cyanobacterial species rapidly drop out from phytoplankton assemblages as salinity concentrations exceed their thresholds and thus are limited to freshwater tidal portions of these major rivers (Sellner et al. 1988). Nevertheless, increased freshwater inputs into rivers from increased runoff dilute the estuarine environment and can promote blooms of these toxic cyanobacterial (Sellner et al. 1988). For example, Lehman et al. (2005) recently documented a large bloom of *M. aeruginosa* in the upper San Francisco Bay Estuary. The bloom was widespread throughout 180 km of waterways; microcystins were detected at all stations sampled, and were also found in zooplankton and clam tissues. Other species of toxigenic cyanobacteria readily bloom in full salinity conditions and can dominate these environments. Non–toxic *Aphanizomenon flos–aquae* and toxic *Nodularia spumigena* form high biomass blooms in the Baltic Sea, and have increased in abundance (Sivonen et al. 1989; Finn et al. 2001). Anabaena (*Aphanizomenon*) *aphanizomenoides* occasionally has been reported to be abundant in the Neuse Estuary in North Carolina (Lung and Paerl 1988).

Changes in the Distribution of Toxic Cyanobacteria

In considering long–term trends of both the frequency of occurrence and distribution of cyanobacterial toxins in the US and Canada, it is important to recognize the remarkable advances that increased interest and awareness that has brought to the discipline over the last few decades. Both the number of known cyanobacterial toxins and toxigenic species has dramatically increased with the increased focus. While few, if any, states had monitoring programs for cyanobacterial toxins in the mid 1970's in most cases, toxic cyanobacteria were identified as being responsible for livestock or wildlife fatalities. By 2005, a number of states had monitoring programs in place, resulting in an increase in the number of CHABs reported. It is difficult to determine if this increasing number of reports represents an increased incidence of CHABS, or is simply a reflection of increasing awareness and monitoring efforts.

Several geographical regions have recently introduced "regional" surveys for the occurrence of cyanobacterial toxins. These regional studies have documented that the occurrence of cyanobacterial toxins has expanded outside of the midwestern prairie states (Fig. 3, see Color Plate 1). Studies in New York, Nebraska, New Hampshire and Florida have shown an increase in the abundance of cyanobacterial blooms. Sasner et al. studied 50 New Hampshire lakes and reported that all had detectable levels of microcystins (Sasner Jr. et al. 1981). Particulate concentrations of microcystins in the bloom spanned over four orders of magnitude and ranged from 0.8 to 31,470 ng/g^{-1} wet weight; concentrations were generally higher in the summer than in the spring. In a similar study of 180 New York lakes and rivers, over 50% of the samples contained easily detectable levels of particulate microcystins in the water column, ranging from 0 to over 1000 μg/L (Boyer et al. 2004). Results of a short–term occurrence study by Carmichael et al. (unpublished 2006) in the Pacific Northwest states of Washington, Oregon and Northern California (Fig. 4, see Color Plate 1), for the years 2002–2005, show that producers of microcystin and anatoxin–a occur on regular basis.

In warmer climates, both the intensity and duration of these bloom events increase. For example, cyanobacterial blooms are common in Florida lakes, rivers, streams and ponds; data has been collected on a limited basis since 1999. Approximately 20 bloom–forming cyanobacteria, including *Microcystis*, *Cylindrospermopsis*, *Anabaena*, *Aphanizomenon*, *Lyngbya* and *Planktothrix*, are found distributed throughout the state with cyanotoxins detected in surface waters and post–treated drinking water (Williams et al. 2001).

Microcystin concentration was most often detected within the 0.1 to 10 $\mu g\ L^{-1}$ range with a maximum concentration of 107 $\mu g\ L^{-1}$; detected in post–treated drinking water at a maximum concentration of ~10 $\mu g\ L^{-1}$ (Burns 2005). Anatoxin–A was not detected in most samples collected, but a maximum concentration of 156 $\mu g\ L^{-1}$ was detected in one sample; detected in post–treated drinking water at ~10 $\mu g\ L^{-1}$. Cylindrospermopsin was detected between 10 and 100 $\mu g\ L^{-1}$ (Burns 2005). The maximum cylindrospermopsin concentration detected in surface waters was 202 $\mu g\ L^{-1}$. The maximum concentration of cylindrospermopsin detected in post–treated drinking water was ~100 $\mu g\ L^{-1}$ (Burns 2005). The first report of *cylindrospermopsis* in North America is mapped in Fig. 5.

While there is some evidence for seasonal changes in CHABS occurrence (Graham et al. 2004), overall there is a paucity of data related to the temporal distribution of CHABs.

Do invasive species change the occurrence of toxic species?

Zebra mussels and Lake Erie (Vanderploeg et al. 2001)

The Great Lakes and especially Lake Erie have experienced many problems with invasive species over the past 100 years. The dreissenid zebra mussel, *Dreissena polymorpha* established itself in Lake Erie in the late 1980's and by 1995 blooms of the toxic cyanobacterium, *Microcystis aeruginosa* had become a problem. Studies done since the late 90's have led to the conclusion that zebra mussel selectively filtrate toxic *Microcystis aeruginosa,* promoting toxic *Microcystis* blooms. This has resulted in new efforts to understand and control CHABs in the Great Lakes.

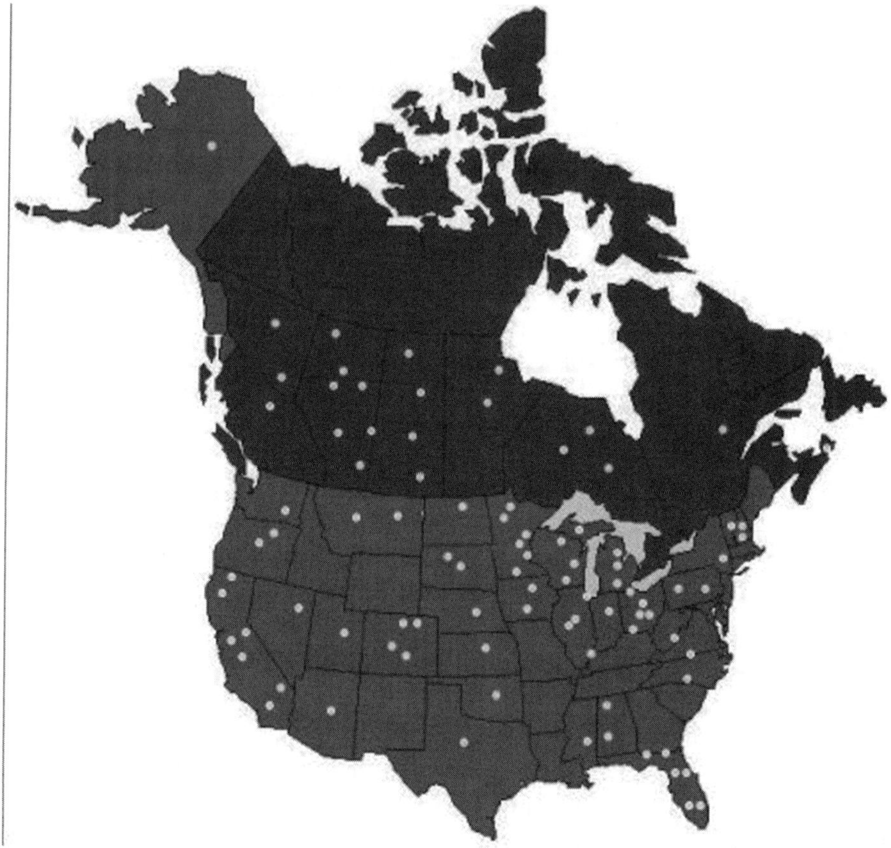

Fig. 1. Distribution of Documented CHAB outbreaks in North America. Unpublished data (Carmichael 2006).

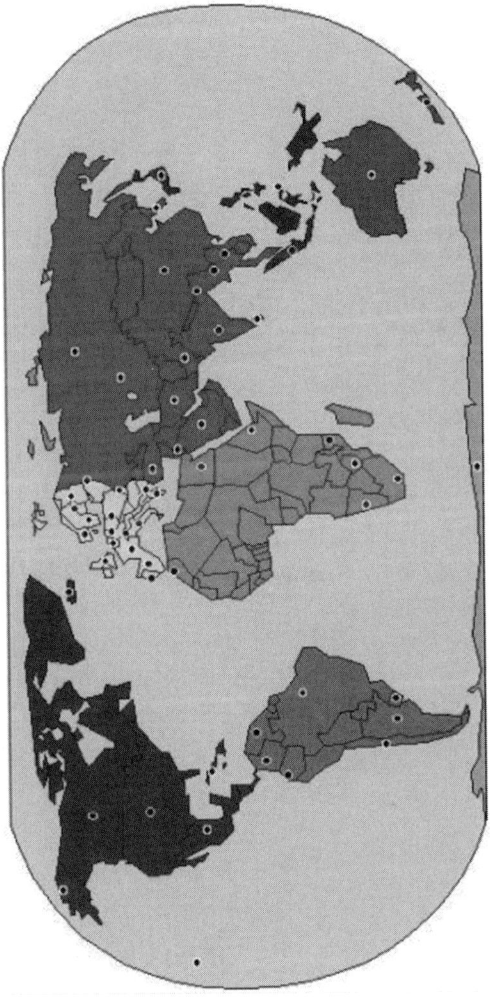

Fig. 2. Countries Reporting One or More CHAB Outbreaks. Unpublished data (Carmichael 2006).

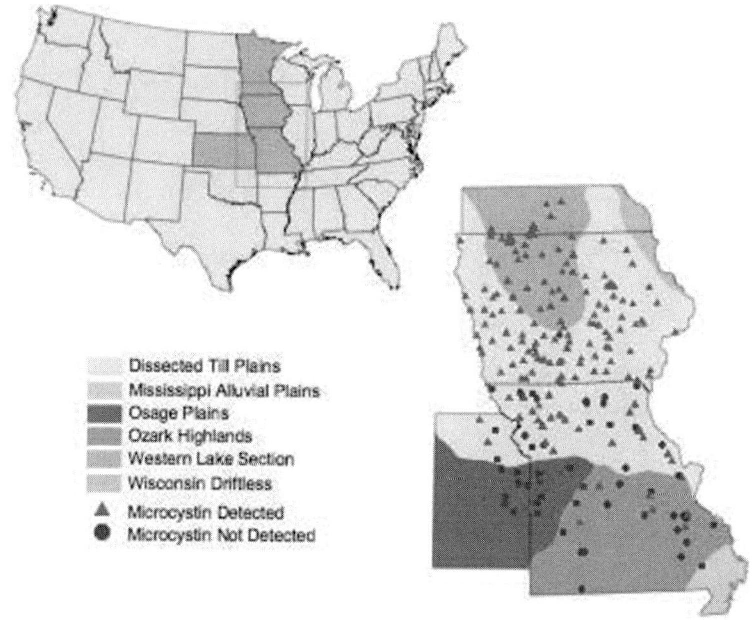

Fig. 3. Trends in Microcystin Occurrence in Midwestern Lakes. Taken from Graham et al. 2004. (See Color Plate 1).

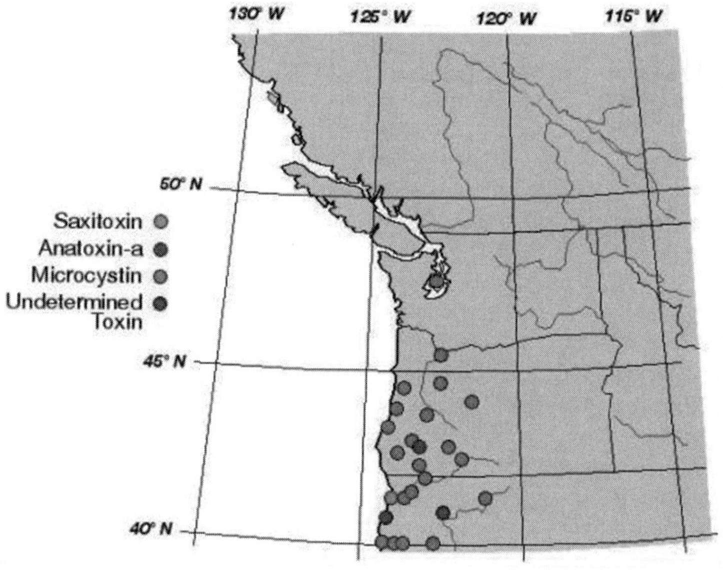

Fig. 4. Cyanotoxin Events in the Pacific Northwest (2001–2005). Taken from Carmichael 2006. (See Color Plate 1).

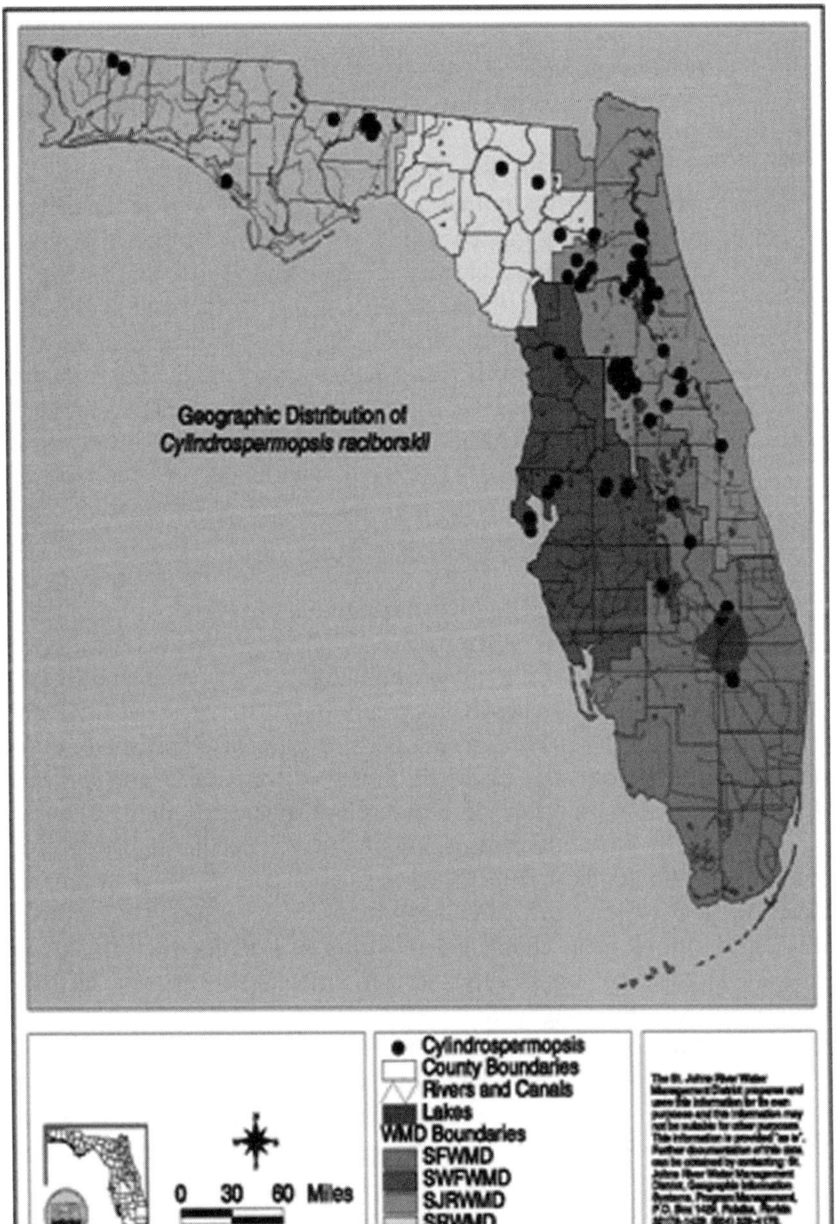

Fig. 5. Florida Distribution of *Cylindrospermopsin.* (Williams et al. 2001).

Cyanotoxin Production

The taxonomy of cyanobacteria, often imprecise, is currently under a state of revision. In general, cyanobacteria cells may be unicellular or filaments, existing singularly or in colonies. Morphological features can be distinct or variable, which has lead to considerable uncertainty when using the botanical taxonomic system that is based on morphological characteristics. For example, an early guide by Geitler (1932) and others listed more than 2000 species of cyanobacteria. Drouet and Daily, in a series of monograms published in the 1950s and 1960s (e.g. Drouet and Daily 1956; 1957; Drouet 1968), revised this flora using only the simplest of markers to lump everything together in only 62 species. Other taxonomists such as Stanier (1971), Castenholz (1992) and others proposed a taxonomic scheme for cyanobacteria based not on the botanical system, but using the provisions of the International Code of Nomenclature for Bacteria (Lapage et al. 1992). This system uses axenic clonal cultures and morphology as well as modern molecular and biochemical techniques to define the species. Despite the age of these proposals, only one taxon of cyanobacteria is currently described under the International Code of Nomenclature of Prokaryotes or ICNP (Euzeby 2004). Bergey's Manual of Systematic Bacteriology contains only a few of the many described cyanobacterial genera, and no species. Recently, Castenholz (2001) reversed his earlier position and stated "I believe that the current rules of Nomenclature (botanical and bacteriological) should be mostly disregarded" and indicated that he preferred a moratorium on naming any species or undertaking any taxonomic revisions for 50 years or more so that molecular data sets could accrue (Hoffmann 2005).

The botanical system for cyanobacteria has recently been revised using traditional morphological characters resulting in a major revision of species, especially for the non–heterocystous filamentous forms such as *Oscillatoria* (Komárek and Anagnostidis 1999, 2005). The revised system resurrects many of the older botanical names for genera. While cell structure and ultrastructure (thylakoids, cell wall formation, cellular dimensions) and molecular data have been congruent, sheath characteristics, which are critical in definition of both species and genera, are not always congruent with the molecular data. Unfortunately, ultrastructure, biochemical, and molecular studies, such as RAPD fingerprinting and 16S rRNA gene sequences are only available for a small portion of cyanobacterial taxa.

The ecological diversity observed in cyanobacteria is grossly underestimated if one looks only at those taxa in culture collections (the taxa which are the source of the molecular and biochemical data). In some cases, molecular sequencing produces similar phylogenies. However, it is also quite easy to generate quite distinct phylogenetic lineages, depending on the particular part of the genome or trait that was used to generate the phylogeny. Most field taxonomists still rely on the traditional morphology–based botanical approach because molecular data have not yet been available for many species (Burkholder 2002), although the use of molecular based techniques is rapidly increasing, especially in those species that lack or contain confusing morphological features. Fortunately, toxic cyanobacteria have received concentrated attention, and so the taxonomy of these taxa may be among the first to be revised based on polyphasic character sets.

Toxic cyanobacteria blooms are associated with a wide diversity of species and environments. Chorus and Bartram (1999) (1999), in their classic monogram for the World Health Organization, *Toxic Cyanobacteria in Water: a Guide to Their Public Health Consequences, Monitoring and Management,* reviewed which species of cyanobacteria have been associated with the production of toxins based on observations and reports since the mid 1980's. Since that time, there have numerous new observations, including new toxigenic species such as the unknown *Stigonematales* species responsible for avian vacuolar myelinopathy (AVM) and new toxins such as β–N–methylamino alanine (BMAA) produced by a wide number of species. *Phormidium* sp., which produces microcystin, has been found in shallows of reservoirs. Microcystin LA is reported to account for the protein phosphatase inhibitory activity of the terrestrial cyanobacterium, *Hapalosiphon hibernicus*. Other terrestrial isolates produce the toxin β–methylamino alanine (BMAA). With the recent report of microcystins from lichen–associated *Nostoc*, it becomes apparent that non–aquatic environments must also be considered when discussing cyanobacterial toxicity. Table 1 lists 22 genera and 42 species associated with toxin formation. Most of the data was obtained from planktonic environments, although there are a number of toxic benthic species. Where appropriate, we have updated the nomenclature to the taxonomic scheme of Komárek and Anagnostidis (Komárek and Anagnostidis 1999, 2005).

Table 1. Major Toxigenic Cyanobacteria Species, their Toxins and Distribution.

Organism	Toxin	Location	Reference
Anabaena spp.	microcystins	Denmark, Egypt, Finland,	
	anatoxin–a	Germany, Ireland, Japan, US	
A. circinalis	microcystins	Finland, France, Australia	
	saxitoxins		
A. flos–aquae	microcystins	Canada, Finland, Norway,	
	anatoxin–a	Canada	
A. lemmermannii	microcystins	Finland	
	saxitoxins		
A. planktonica	anatoxin–a	Italy	
A. variabilis	BMAA	US	
Aphanizomenon spp.	anatoxin–a	Finland, Germany	
A. (flos–aquae) issatschenkoi	saxitoxins	US	
A. flos–aquae	anatoxin–a(s)	Canada	
A. gracile	LPS endotoxins	US	
A. lemmermannii	saxitoxins	Portugal, Denmark	
	anatoxin–a(s)		
A. ovalisporum	cylindrospermopsin	Israel	
Arthrospira fusiformis	microcystins	Africa, Spain	
Anabaenopsis millerii	microcystins	Greece	

Organism	Toxin	Location	Reference
Cylindrospermum spp.	anatoxin–a	Finland	
Cylindrospermospis raciborskii, C. phillipinensis	saxitoxins cylindrospermopsin BMAA	Brazil, Australia, Hungary, US, Australia	
Haphalosiphon hibernicus	microcystins	US	
Lyngbya spp. *L. majuscula*	pahayokolide–a BMAA	US Africa, Australia, US, Pacific Is.	
L. wollei (Plectonema wollei) Farlowe ex Gomont	dermatotoxins saxitoxins	US	
Microcystis spp.	microcystins anatoxin–a BMAA	Worldwide	
M. aeruginosa	microcystins	Worldwide	
M. viridis	microcystins	Japan	
M. botrys	microcystins	Denmark	
Nodularia spumigena *N. spumigena*	nodularins nodularins	Canada Australia, New Zealand, Baltic Sea	
Nostoc spp.	microcystins	Finland, England and US	

Organism	Toxin	Location	Reference
	BMAA		
Oscillatoria spp.	anatoxin–a	Finland, Scotland, Ireland and US	
O. brevis (Phormidium breve) (Kützing ex Gomont) Agnostidis et Komárek	LPS endotoxins	Switzerland	
O. limosa Agardh ex Gomont	microcystins	US	
O. nigroviridis (O. nigro–viridis) Twaites ex. Gomont sensu Parakutty	dermatotoxins	US	
O. tenuis	LPS endotoxins	US	
Phormidium spp.	microcystins BMAA saxitoxins	US and Italy	
Planktothrix	microcystins, anatoxin–a	US, Finland	
P. agardhii	Microcystins BMAA	Denmark, Finland, Norway, Ireland	
P. flavosum	anatoxin–a	France	
*P. formosa**	homoanatoxin–a	Norway	
P. mougeotii (P. isothrix) (Skuja) Komárek and Komarkova*	microcystins	Denmark	
Plectonema sp.	BMAA	Unknown origin	
Prochlorococcus marinus	BMAA	Sargasso Sea	

Organism	Toxin	Location	Reference
Raphidiopsis spp.			
R. curvata	cylindrospermopsin	Japan	
R. mediterranea	anatoxin–a, homoanatoxin–a		
Schizothrix calcicola	aplysiatoxins LPS (nontoxic)	Pacific Islands and US	
Stigonematales sp.	avm	US	
Synechococcus sp.	BMAA	US	
Trichodesmium thiebautii	BMAA saxitoxin	Sargasso Sea	
Umezakia natans	cylindrospermopsin	Japan	

*denotes a species designation that is considered invalid using the current classification scheme

Cyanobacteria toxicity is very common in natural blooms with estimates of toxicity approaching as high as 50% (WHO 2003). In the late 1990s, in an attempt to systematically estimate the abundance of toxic cyanobacteria, approximately 200 strains of the Pasteur Collection of Cyanobacteria (PCC) were screened for microcystins using ELISA and HPLC–PDA techniques. G. Codd and R Rippka (unpublished) confirmed that more than 20 species produce cyanobacterial toxins, of which 11 strains of *Microcystis* and 1 strain of *Nostoc* produced concentrations of more than 100 μg microcystins per gram dry weight (Boyer 2006). Many of the strains in the PCC have also been screened for the genetic potential for microcystin production using PCR techniques (Dittmann and Borner 2005). Similar results were also obtained from the NIVA culture collection in Norway, where a large percentage of the nearly 500 strains of cyanobacteria in the collection appeared to produce toxins or other bioactive substances (Skulberg 1984). Microcystin and harmful hepatotoxic peptides were produced by many, but not all strains of the cyanobacteria belonging to the genera *Anabaena, Microcystis, Planktothrix* and *Nostoc* (Hisbergues et al. 2003).

Toxigenic cyanobacteria are generally defined as species that have toxic strains (populations), which are capable of producing neurotoxic, hepatotoxic or dermatotoxic compounds (Burkholder and Glibert 2006). This term is important to note, since within a given species of toxigenic cyanobacteria, there commonly occur both toxic and nontoxic strains (Chorus and Bartram 1999; Burkholder and Glibert 2006). If one expands the definition of toxins to include bioactive peptides with protease inhibition activity such as the micropeptins, cyanopeptolins, microviridins, oscillapeptins, oscillamides, nostopeptins, aeruginosins, aeuginopeptins anabaenopeptilides, anabaenopepins and the cytotoxic compounds from marine origin, then the list of toxin–producing cyanobacteria species becomes quite large. Recently, the PCC were screened for the presence of non–ribosomal peptide synthetase (NRPS) genes, those genes responsible for the synthesis of many bioactive peptides such as the microcystins, microveridins and anabaenapeptins (Christiansen et al. 2001). This biosynthetic activity is widespread throughout the cyanobacteria taxa, with NRPS activity found in over 75% of the 146 strains tested. While the presence of this gene activity does not directly correlate to toxicity, it indicates the potential for production of bioactive peptides. The non–ribosomal peptide synthetases used for microcystin toxin biosynthesis in cyanobacteria allow for considerable structural diversity (Christiansen et al. 2001). More than 400 individual peptides have been identified, mostly from *Anabaena, Lyngbya, Nostoc* and *Microcystis* species (Krishnamurthy et al. 1986;

Luesch et al. 2000; Pluotno and Carmeli 2005). Many of these peptides have protease inhibition activity and can lead to detrimental effects if consumed as part of the food web.

In addition to the production of toxins, cyanobacteria have often been associated with the production of taste and odor compounds in with drinking water; the two most are geosmin and 2–methylisoborneol (MIB), although a number of different hydroxyketones and β–ionone derivatives can also add to the off–flavors associated with freshwater cyanobacteria (Izaguirre and Taylor 2004). While these taste and odor compounds are not actually toxic, they are of concern to the consuming public when their drinking water has an "off" flavor, resulting in considerable monitoring and treatment costs to water supply providers. Taste and odor issues are complex and need not be associated solely with the presence of cyanobacteria. Many actinomycetes bacteria such as *Actinomyces* and *Streptomyces* species, aquatic fungi such as *Basidiobolus ranarum*, and myxobacteria such as *Nannocystis exedens* can also produce these compounds (Niemi et al. 1982). Table 2 illustrates the large number of species associated with the production of compounds that affect taste and odor. These species may or may not be the same species reported to produce toxins. The biosynthetic pathways for taste and odor compounds are separate and unrelated to the toxin biosynthetic pathways (Carmichael 2001).

Table 2. Major Cyanobacterial Taste and Odor Producing Species.

Organism	Compound
Anabaena spp.	Geosmin
A. circinalis	
A. laxa	
A. macrospora	
A. scheremetievi	
A. spiroides	
A. viguieri	
Aphanizomenon flos–aquae	Geosmin
Calothrix sp. *(C. parietina)*	Ketones, Ionones
Fischerella muscicola	Geosmin
Jaaginema geminatum	MIB
(Oscillatoria geminata) (Menehgnin ex Gomont Komárek and Anagnostidis	
Lyngbya spp.	
L. aestuarii (Lyngbya aestuarii) Liebman ex Gomont	MIB
L. cryptovaginata (Planktothrix cryptovaginalt) (Škorbatov) Anagnostidis et Komárek	Geosmin
Oscillatoria spp.	
O. chalybea (Phormidium chalybeum) (Mertens ex Gomont) Anagnostidis et Komárek	MIB
O. raciborskii (Planktothrix raciborskii) (Woloszyńska) Anagnostidis et Komárek	
O. agardii (Planktothrix agardhii) (Gomont) Anagnostidis et Komárek	Geosmin
O. amoena (Phormidium amoenum) (Kützing 1843) Anagnostidis et Komárek	
O. amphibia (Geitlerinema amphibium) (Agardh ex Gomont) Anagnostidis	
O. bornetii (Tychonema bornetii) (Zuka) Anagnostidis et Komárek	

Organism	Compound
O. brevis (Phormidium breve) (Kützing ex Gomont) Anagnostidis et Komárek	
O. variabilis (Oscillatoria variabilis) (Rao)	
O. prolifica (Planktothrix prolifica ([Grevelle] Gomont) Anagnostidis et Komárek	
O. simplicissima (Phormidium simplicissimum) (Gomont) Anagnostidis et Komárek	
O. splendida (Geitlerinema splendidum) (Greville ex Gomont) Anagnostidis	
O. cortiana (Phormidium cortiamum) (Meneghini ex Gomont) Anagnostidis et Komárek	Geosmin, MIB
O. curviceps Agardh ex Gomont	
O. limosa Agardh ex Gomont	
O. tenuis Agardh ex Gomont	
Phormidium sp.	
P. autumnale (Phormidium autumnale) [Agardh] Trevisan ex Gomont	Geosmin
P. calcicola (Phormidium calcicola) Gardner	Geosmin, MIB
P. tenue (Phormidium tergestinum) (Kützing) Anagnostidis et Komárek	MIB
Plectonema sp.	Ketones, Ionones
P. notatum (Leptolyngbya notata) (Schmidle) Anagnostidis et Komárek	
Pseudanabaena catenata	
P. catenata	MIB
P. limnetica (Oscillatoria limnetica) (Lemmermann) Komárek and Anagnostidis	
Rivularia spp.	Ketones, Ionones
Schizothrix muelleri	Geosmin
(Symplocastrum muelliri) (Nägeli ex Gomont) Anagnostidis	
Symploca muscorum Gomont ex Gomont	Geosmin
Tolypothrix distorta	Ketones, Ionones

Cyanotoxin Occurrence and Distribution

Cyanobacterial toxins fall into several diverse categories and are usually divided based on their biological activity (hepatotoxins versus neurotoxins) or chemical structure (peptide toxins versus alkaloids) (Table 3). This division is becoming more complicated as new toxins are being identified. For example, in 1988, there were 10 reported microcystins; as of 1998, there are over 60 different chemically identified microcystins (Chorus and Bartram 1999). In the sections below, we summarize the basic classes of cyanobacterial toxins, their source organisms and their biological effects.

Microcystins

As shown in Table 1, microcystins have been reported in *Microcystis*, *Anabaena*, *Oscillatoria*, *Planktothrix*, *Nostoc*, *Hapalosiphon* and *Anabaenopsis* with new sources appearing yearly (Huisman and Hulot 2005). Microcystins (Fig. 6) contain 7 amino acids in their cyclic peptide ring system, are closely related to nodularin (Fig. 7) or nodulapeptins, which have 5 amino acids in the ring system, as well as anabaenopeptins, aeruginopeptin and other bioactive peptides (Sivonen and Rapala 1998; Harada 2004). Most of these peptides are hepatotoxins, though their LD–50 depends on the specific amino acids present.

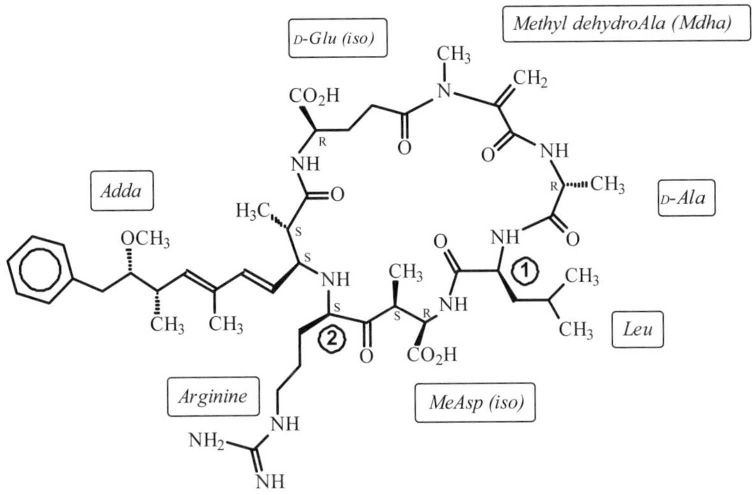

Fig. 6. The generic structure of a microcystin.

Variations occur primarily at positions 1 and 2. For example, microcystin–LR contains the amino acids leucine (L) and arginine (R) at positions 1 and 2 respectively; microcystin–RR has arginine at both positions. Nodularins are similar with the five amino acids Adda–γGlu–Mdhb–βMeAsp–Arg making up the core ring system (Harada et al. 1996).

Table 3. Commonly observed cyanotoxins in US fresh, estuarine and marine waters. (adapted from Codd et al. 2005).

Toxin	Type and (# congeners)	Mode of Action
Hepatotoxins		
Microcystins	Cyclic heptapaptides (>70)	Protein phosphatase inhibitor, tumor promoter
Nodularins	Cyclic penta–peptide (9)	Protein phosphatase inhibitor,
Cylindrospermopsin	Guanidine alkaloid (3)	Protein synthesis inhibitor, genotoxic and necrotic injury to liver and other organelles.
Neurotoxins		
Anatoxin–a	azobicyclic alkaloid (5)	Postsynaptic neuromuscular blocking agent and Acetylcholinesterase agonist
Anatoxin–a(S)	Guanidine organophosphate (1)	Acetylcholinesterase inhibitor
Saxitoxin	Carbamate alkaloid (>20)	Sodium channel blocker
B–methylamino alanine	Modified amino acid	Neurodegenerative agent
Dermatotoxins		
Lyngbyatoxin–a	Indole Alkaloid (1)	Inflammatory agent and Protein Kinase C activator
Aplysiatoxin	Polyacetate Alkaloid (2)	Inflammatory agent and Protein Kinase C activator
Endotoxins		
Lipopolysaccharides	Lipopolysaccharides	Gastrointestinal irritants

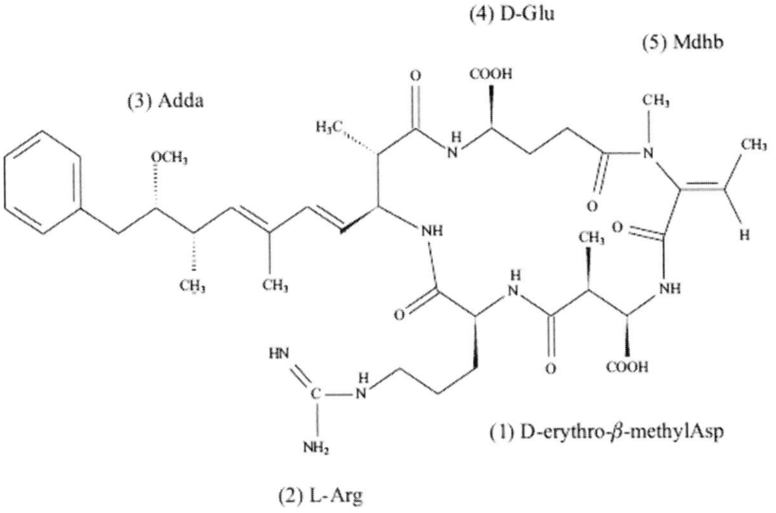

Fig. 7. The generic structure of a Nodularin.

Cylindrospermopsin

Separate from the hepatotoxic cyclic peptides is the hepatotoxic alkaloid cylindrospermopsin (Fig. 8a). This toxin is usually produced by *Cylindrospermopsis raciborskii* and an outbreak of this organism in the drinking water supply on Palm Island, Queensland Australia led to a severe outbreak of hepatoenteritis among the inhabitants of the island (WHO 2002). Cylindrospermopsin has also been reported from *Umezakia natans* in Japan, *Aphanizomenon ovalisporum* in Israel (Sivonen and Jones 1999), and *Raphidiopsis Curuata* in China (Li et al., 2001). It was generally assumed that the toxigenic cylindrospermopsin–forming species only bloomed in tropical and arid environments and *Cylindrospermopsis raciborskii* blooms are commonplace in the warmer Florida drinking water reservoirs. Increasing reports of *C. raciborskii* occurring in temperate zones of Europe and the U.S have been documented (Padisak 1997). Cylindrospermopsin–producing blooms can be both high and low–biomass events, leading to difficulties in identifying the source of this toxin.

Anatoxin–a

The neurotoxic cyanobacterial toxins consist of anatoxin, anatoxin–a(S), saxitoxin and related analogs. The most important of these compounds from an environmental health aspect is probably anatoxin–a (Fig. 8b) (Carmichael et al. 1975). Anatoxin–a was originally isolated from *Anabaena flos–aquae* (Devlin et al. 1977), but is also reported in *A. planktonica*, *Oscillatoria* spp., *Planktothrix* spp., and *Cylindrosperum*. Homoanatoxin–a, a toxic homologue with a propyl group replacing the acetyl group, was isolated from *Phormidium* (*Oscillatoria*) *formosa*. Both are potent nicotinic agonists and act as neuromuscular blocking agents.

Anatoxin–a(S)

Anatoxin–a(S) (Fig. 8c) is a naturally occurring organophosphate, with no reported structural variants, produced by *Anabaena flos–aquae* strain NRC 525–17 and more recently by *Anabaena lemmermannii* (Henriksen et al. 1997; Onodera et al. 1997). It is distinct from the neurotoxic anatoxin–a in both its chemical structure and biological mode of action (Mahmood and Carmichael 1986b). Anatoxin–a(S) acts as an acetyl–cholinesterase inhibitor and its name was derived from the fact that the intoxicated animals often suffered from extreme salivation (Matsunaga et al. 1989). Anatoxin–a(S) is commonly reported in the prairie states of the US (Kenefick et al. 1992); this may lead to an underestimate of its occurrence since, lacking a clear cyanobacterial source, death of livestock or pets from anatoxin–a(S) would likely be attributed to pesticide intoxication.

Saxitoxins and related analogs

Saxitoxins (STX) (Fig. 8d) are representative of a large toxin family referred to as the Paralytic Shellfish Poisoning (PSP) toxins. These include the N–1 hydroxysaxitoxin or neosaxitoxin (NeoSTX), 11–sulfate analogs of STX and NeoSTX called gonyautoxins (GTX) and 6 N21 sulfo derivatives of STX, neoSTX and GTX toxins B1, B2, C1 to C4 (Mahmood and Carmichael 1986a; Chorus and Bartram 1999). Not all of these 18 PSP analogs have been identified in marine dinoflagellates and shellfish. These toxins are identical to those produced by some toxigenic marine dinoflagellates that accumulate in shellfish that feed on those algae (Anderson 1994). STX, neoSTX, C1, C2 and GTX have all been reported in freshwater cyanobacteria including *Aphanizomenon* spp., (Pereira et al. 2004, Li and Carmichael, 2003, Li et al. 2000) *Anabaena circinalis, Lyngbya wollei* and a Brazilian isolate of *Cylindrospermopsis raciborskii* (Lagos et al. 1999).

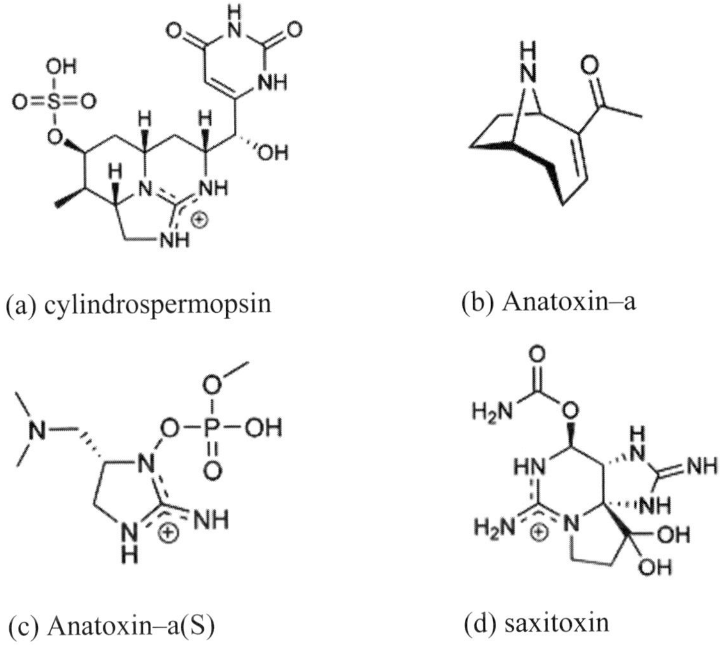

Fig. 8a–d. Structures of the alkaloid cyanobacterial toxins.

β–methylamino alanine (BMAA)

A recently discovered cyanobacterial neurotoxin is non–protein amino acid β–N–methylamino–L–alanine (BMAA). BMAA, originally isolated from a Guam cycad, *Cycas circinalis*, has been implicated as the causative agent of amyotrophic lateral sclerosis or Parkinsonism dementia (Murch et al. 2004b). Symbiotic *Nostoc* species associated with the roots of cycads produce BMAA and it accumulates in the seeds of the plant. The indigenous Chamorro people of Guam feed on the flying foxes, which consume these seeds, and are subject to a classic biomagnification of the toxin up through the higher trophic levels (Cox et al. 2005). BMAA may also be incorporated into peptides or proteins, both increasing its biomagnification and acting as a reservoir to slowly releasing toxins upon peptide hydrolysis (Murch et al. 2004a).

Cox et al. (2005) surveyed 30 different strains of cyanobacteria obtained from culture collections that spanned the entire spectrum of cyanobacterial morphology and taxonomy for the presence of BMAA. Although BMAA is most common in *Nostoc* species, with 7 out of 10 isolates containing high levels of either the free or protein–bound amino acid, it is ubiquitous in its distribution, being found in freshwater, marine, brackish and terrestrial environments. Only 1 of the 3 species tested, a freshwater *Synechocystis* species, lacked BMAA. Algal strains in culture often produce much lower levels of toxins than found *in situ* suggesting that there is potential for widespread human exposure to this toxin (Cox et al. 2005). BMAA was recently detected in the brains of nine Canadian Alzheimer patients, but was not found in the brains of 14 other Canadians that died from causes unrelated to neuro–degeneration (Murch et al. 2004a, 2004b). Since cycads are not part of the normal Canadian flora, cyanobacteria may be the source of BMAA in these patients. Drinking water contaminated by cyanobacterial blooms has been suggested as a potential pathway for human exposure to BMAA and a topic of future data needs.

Lyngbyatoxins, Aplysiatoxin and Debromaplysiatoxin.

Lyngbyatoxins, contact irritants, are mainly marine water toxins but their occurrence in freshwaters has not been fully assessed. *Lyngbya*, an aggressive genus capable of widespread growth, has been found in freshwater supplies in the US. *Lyngbya* blooms are common in Florida, forming mats in freshwater springs and overgrowing some coastal reefs, where low levels of *Lyngbya*–toxin–A, debromoaplysiatoxin and saxitoxin have been detected (Williams et al. 2001). More data are needed to determine its health effects and occurrence in freshwater supplies. The inflammatory activity of *Lyngbya* is caused by aplysiatoxin and debromoaplysiatoxin, which are tumor promoters and protein kinase–C activators (Berry et al. 2004). Debromoaplysiatoxin along with other toxic compounds has also been isolated from other members of the family Oscillatoriaceae such as *Schizothrix calcicola* and *Oscillatoria nigroviridis* (Chorus and Bartram 1999).

Other Bioactive Compounds

Many cyanobacteria including *Microcystis*, *Planktothrix* and *Anabaena* species have the ability to produce cyclic and linear peptide protease inhibitors that are potentially toxic (Harada 2004). Examples of these compounds include the micropeptins, cyanopeptolins, microviridin, oscillapeptin, oscillamide, nostopeptin, aeruginosins, aeruginopeptins, anabaenaopeptilides, anabaenopeptins and circinamide (papain inhibitor). Two such compounds isolated from *Microcystis aeruginosa*, aeruginopeptins and cyanopeptolins, are shown in Fig. 9. The effects of these compounds on the toxicity of cyanotoxins is not well known.

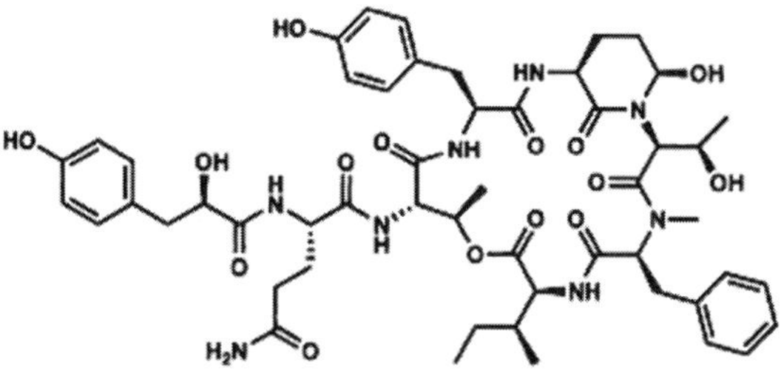

aeruginopeptin 228–a

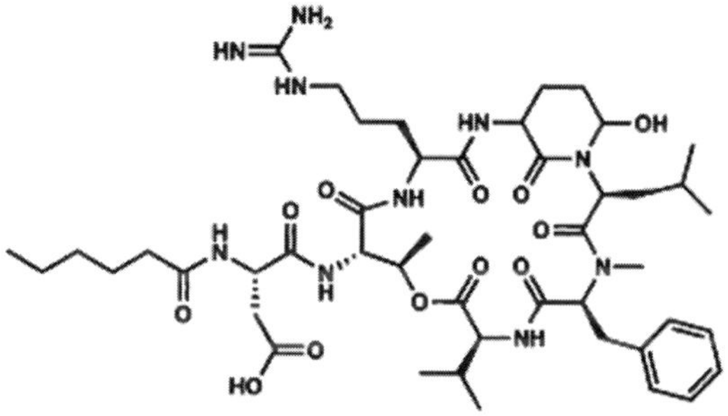

cyanopeptolin–a

Fig. 9. Structures of two bioactive peptides isolated from Microcystis aeruginosa.

Impacts on Waterbody Health and Ecosystem Viability

The factors that lead to CHAB formation and induce toxin production are not well known (Chorus and Bartram 1999). Although multiple studies have examined the empirical relationship between environmental variables and toxin concentrations, few have measured the many different variables necessary to compile a comprehensive data set for analysis, (Kneale and Howard 1997). Even fewer have analyzed the data based on sound statistical methods recommended by EPA and other agencies (EPA 2003). Temporal and spatial patterns of CHAB formation are often complex, and the mechanisms of bloom progression are largely unexplained. Additionally, CHABs can include both toxic and non–toxic strains, and many strains can be present during CHABs in specific lakes (Vezie et al. 1998; Jacoby et al. 2000; Pawlik–Skowronska et al. 2004). A major source of variation in toxin concentration appears to be the strain–dependant influence of environmental factors on toxin release (Rapala and Sivonen 1998; Vezie et al. 2002) and successive replacement of toxic and non–toxic species (Kardinal and Visser 2005). No single modeling study has attempted to predict the dynamics of toxin releases and/or toxin levels, reflecting a limited understanding of factors influencing toxin release.

Several recent reviews (Paerl 1996; Soranno 1997; Carmichael et al. 2001b; Saker and Griffiths 2001; Landsberg 2002; Codd et al. 2005; Huisman and Hulot 2005). provide a comprehensive discussion of individual environmental factors that may be responsible for CHABs and their correlated effects on ecosystem health. Physical factors include temperature, light availability and meteorological conditions. Hydrologic factors include alteration of water flow, turbidity and vertical mixing. Chemical factors include nutrient loading (principally nitrogen [N] and phosphorous [P]), pH changes and trace metals, such as copper, iron and zinc.

Freshwater Habitats

Most of the world's population relies on surface freshwaters as their primary source for drinking water. The drinking water industry is constantly challenged with surface water contaminants that must be removed to protect public health. Recent reports suggest that CHABs are an emerging issue in the US because of increased source–water nutrient pollution causing eutrophication. Although CHABs historically have been strongly correlated with nutrient over–enrichment in many freshwater lakes with low abiotic turbidity (e.g Schindler 1987, Whitton and Potts 2000), flowing waters can also develop planktonic CHABs in slowly flowing segments (e.g Pinckney et al. 1997). Intermediate systems – run–of–river impoundments or reservoirs – generally have faster flushing and higher abiotic turbidity from suspended sediment loads than natural lakes (Wetzel 2001). In these systems light, rather than nutrients, often is the most important resource limiting cyanobacteria growth (Cuker et al. 1990). In addition, because of the surface characteristics of their cells, cyanobacteria can be especially susceptible to being attracted to sediment particles (Avnimelech et al. 1982, Søballe and Threlkeld 1988, Burkholder 1992). When that occurs, they become too heavy to remain in the water column and settle out of the water column (Burkholder and Cuker 1991). Nevertheless, during intervals between episodic sediment loading events from land erosion and runoff, dense cyanobacteria blooms can develop in reservoir systems, and nutrients – both nitrogen and phosphorus – can become important in controlling their production (Touchette et al. (submitted), Cuker et al. 1990, Burkholder and Cuker 1991); Fig. 10.

Many eutrophic lakes and impoundments (reservoirs) have documented CHAB problems, for example: The Metropolitan Water District of Southern California (MWDSC) has a long history of algal problems, in the form of planktonic blooms and benthic proliferations (Izaguirre 2005). The main concern is taste and odor, specifically the compounds geosmin and MIB. All of the reservoirs listed below have experienced algal blooms of one kind or another, including in some cases known toxigenic species. In addition, Lakes Mathews, Skinner, Perris and Diamond Valley have developed benthic mats that have resulted in severe off–flavor problems (Izaguirre 2005). MWDSC, is composed of 26 member agencies and supplies drinking water to about 18 million people in six counties in the coastal plain of southern California; its two sources of water are the Colorado River and water from northern California delivered through the California aqueduct feeding into three reservoirs in Riverside County: Lake Mathews, Lake Skinner and Diamond Valley Lake (Izaguirre 2005). In North Carolina, a large, previously undetected, sewer spill that occurred over several months was positively correlated with *Lyngbya* blooms (Frazier 2006).

As research progresses, new toxins from cyanobacteria in freshwater systems are being discovered (Codd et al. 2001; Huisman and Hulot 2005), as well as new potential roles in their impacts on food webs. For example, avian vacuolar myelopathy (AVM), a neurological disease in waterfowl and their predators, was first described in the mid–1990s, with unknown cause (Thomas et al. 1998). AVM causes characteristic lesions in the myelin of the brain and spinal cord. Affected birds have abnormal flying and swimming because their muscle coordination is compromised. The causative agent of AVM is associated with submersed aquatic vegetation such as the aquatic weed hydrilla (Birrenkott et al. 2004). At present, the prime "suspect" is a toxin producing, yet unnamed cyanobacterium within the order Stigonematales that colonizes the leaves of hydrilla and other submersed plants (Wilde et al. 2005). Thus, the working hypothesis being examined is that waterfowl such as coots consume submersed vegetation covered with the toxic cyanobacteria, bioaccumulate the toxin(s), and are impaired in their ability to escape from predators. The affected waterfowl are caught and consumed by higher predators such as bald eagles, which then also develop AVM.

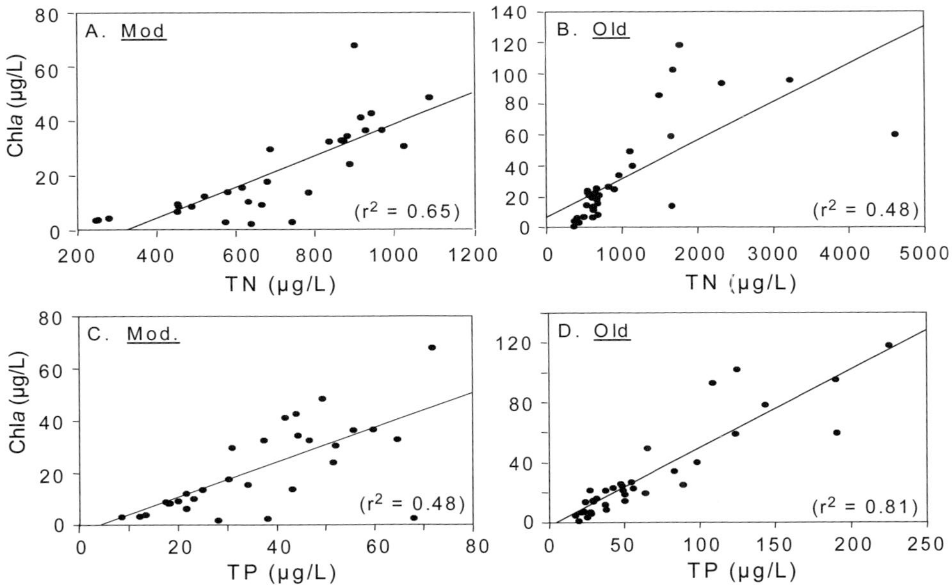

Fig. 10. Relationship between chlorophyll a (chla) and total nitrogen (TN, upper panels) or total phosphorus (TP, lower panels). *(A, C) moderate–age reservoirs (mod., Group 1, 20–30 years post–fill, $n = 5$) and (B, D) old reservoirs (Group 2, 60–85 years post–fill, $n = 6$) during summer months when cyanobacteria were dominant* (Touchette et al. (submitted)).

Estuarine Habitats

Estuaries are comprised of high population densities of microbes, plankton, benthic flora and fauna. These organisms tend to be highly vulnerable to human activities and a wide array of human impacts that can compromise their ecological integrity. Many of the factors responsible for CHABS as well as the CHABs themselves affect estuarine habitat health. Anthropogenic stress in the form of increased nutrient input significantly impacts estuaries as well as other water bodies. Modeling studies suggest the importance of nutrient loading to water quality and CHAB probability and severity. Nitrogen (N) and phosphorus (P) enrichment are considered cyanobacteria bloom precursors. Many cyanobacteria are nitrogen–fixers, making N a limiting factor in competition over other phytoplankton species (Burkholder 2002). Although nutrient influences are widely studied factor linked to stimulation of CHABs, the nutrient(s) that are most important in limiting cyanobacteria growth are still in debate, and likely depend upon the habitat and history of nutrition (e.g. Lenes et al. 2001; Glibert et al. 2005, 2006).

Cyanobacteria blooms were noted in the Potomac River and the upper Chesapeake Bay during the 1950s and 1960s coincident with the invasion of water milfoil. Since 1985, cyanobacteria blooms have been documented in the tidal tributaries of Chesapeake Bay almost annually during summer months by the Maryland Department of Natural Resources (MD–DNR) long–term comprehensive water–quality monitoring program. During September 2000, an extensive late summer bloom of *Microcystis* on the Sassafras River, however, was among the first blooms tested for cyanotoxins in Maryland and results were positive for elevated concentrations of microcystin. The microcystin levels (591.4–1041 $\mu g\ g^{-1}$ dry wt) led to the Kent County Health Department closing a public beach in the bloom area for the remainder of the year, the first beach closure in the history of the state due to detected levels of cyanotoxins (Tango et al. 2005). Dr. Harold Marshall, a renowned phytoplankton specialist in the Chesapeake Bay area, has also documented significant increases in cyanobacteria biomass, mostly as *Microcystis aeruginosa*, from long–term datasets in tributaries of lower Chesapeake Bay (Fig. 11).

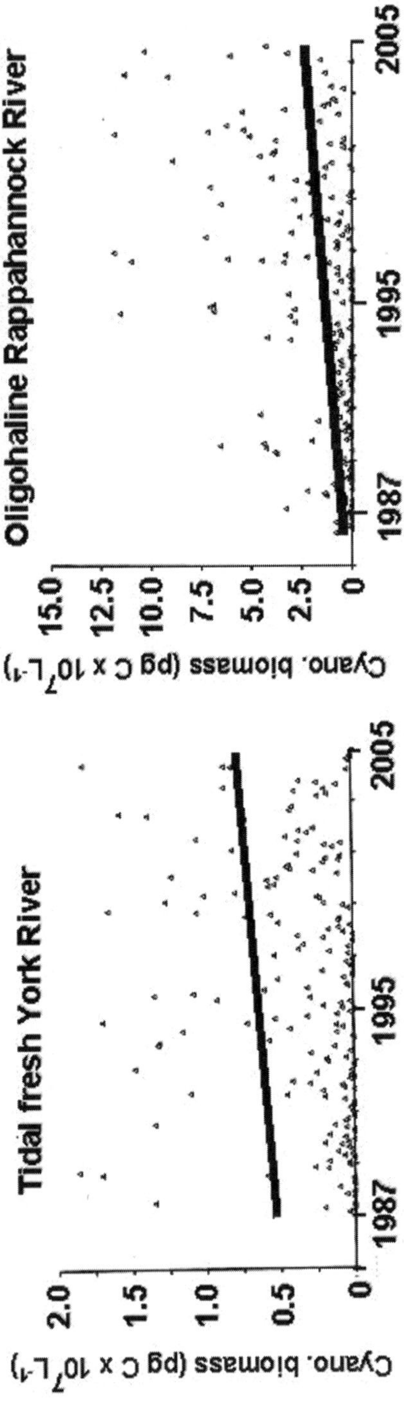

Fig. 11. Long–term datasets (1987–2005) show increasing cyanobacteria biomass in tributaries of lower Chesapeake Bay.

Marine Habitats

In comparison to fresh and estuarine waters, relatively little is known about harmful cyanobacteria in marine ecosystems, except for their negative effects on coral reefs. For example, the toxigenic, benthic cyanobacterium *Lyngbya majuscula* proliferates in warm marine waters enriched with nutrients (especially phosphorus or iron) (e.g., Watkinson et al. 2005). Nicknamed "stinging seaweed" (Sims and Zandee Van Rilland 1981), *L. majuscula* produces an array of toxins that adversely affect human health (Osborne et al. 2001) and have been implicated as causes of tumors and other diseases in marine fauna (e.g. Arthur et al. 2006). It also has caused serious ecological impacts in smothering coral reefs and seagrass meadows. For example, a recently described bloom expanded to 8 km^2 of coverage within 55 days, and its dense accumulation (average biomass 210 grams dry weight $\cdot$ m^{-2}) led to declines in beneficial seagrasses (Watkinson et al. 2005). As another example, the cyanobacterium *Phormidium corallyticum* is a member of a microbial consortium that causes black band disease and death of corals (Richardson 2005).

Although cyanobacteria in many saltwater ecosystems have not been rigorously examined for toxicity, limited available evidence suggests that cryptic toxigenic species can be abundant in these habitats. For example, a survey of the landlocked, brackish to hypersaline, Salton Sea revealed that 85% of all water samples tested from areas with high incidence of grebe deaths contained low but detectable levels of microcystins produced by picoplanktonic *Synechococcus* and benthic *Oscillatoria* (Carmichael and Li 2006). Moreover, the livers of affected grebes contained microcystins at potentially lethal levels.

Possible interactions between cyanobacteria and marine harmful algal blooms: *Trichodesmum* spp. are marine cyanobacteria that fix nitrogen gas into inorganic nitrogen forms that are used to make proteins not only by the cyanobacteria, but by other organisms as nitrogen leaks from the cyanobacteria cells. The impacts are currently being explored and are not necessarily bad, but the added available nitrogen has been hypothesized to help support blooms of harmful algae such as *Karenia brevis*, the toxic 'red tide' organism of the Gulf of Mexico (Lenes et al. 2001, Glibert et al. 2006). This hypothesis is presently under debate (Hu et al. 2006).

Potential Chronic, Insidious Impacts on Human Health

Present understanding about the effects of toxigenic cyanobacteria on human health are mostly limited to obvious, acute impacts, and thus do not include chronic, subtle or insidious impacts that are happening but not yet detected, and potential impacts where hazards exist in remote areas where health impacts have not yet been sustained. It is expected that the risk from cyanobacterial CHABs to drinking and recreational waters will increase with increased stress and demand on those systems (Glibert and Burkholder 2006) (e.g. Fig. 12, see Color Plate 2 and Fig. 13, see Color Plate 2). The magnitude of the threat will depend on many variables including the increase in population and its exposure to toxins, the requirement for new drinking water and recreational sources and the speed and extent with which effective management strategies are implemented.

It is expected that the demand for drinking water in the US will increase dramatically in the future. Although many utilities in the US use underground water supplies, many also use surface water and that proportion is expected to increase in many areas (e.g. Martin 2001). Increasing population has placed increased demand on surface water supplies, either through their anthropogenic inputs or through their increased demand for drinking water. Conventional water treatment appears to be ineffective in removing cyanotoxins from drinking water (Hoeger et al. 2004, 2005); cyanobacterial cells lyse during filtration and chlorine does not appear to readily oxidize the resulting MC (Wannemacher Jr. et al. 1993; Hitzfeld et al. 2000). To control HABS, algaecides, especially copper sulfate, are commonly added to surface drinking water sources (Chorus and Bartram 1999), but this leads to cell lysis and a substantial release of cyanotoxins into the water, as well as potential copper toxicity (Kenefick et al. 1993). Powdered and granular activated carbon adsorption, with reported removal efficiencies of greater than 90%, removes toxins as well as intact cells and appears to be a more effective and reliable methods for removing cyanotoxins than Cl (Hamann et al. 1990; Donati et al. 1994). A similar increased in risk is also expected for recreational waters where, for humans, the primary route of exposure is oral from accidental or deliberate ingestion of recreational water, inhalation of aerosolized cells or from dermal contact.

OCCURRENCE OF CYANOBACTERIAL BLOOMS IN THE LAKE PONTCHARTRAIN ESTUARY, LOUISIANA

Toxic cyanobacterial blooms have been reported in Lake Pontchartrain, an oligohaline lake bordering New Orleans, Louisiana. Lake Pontchartrain is a tidally influenced estuary that connects to the Mississippi River via the Bonnet Carré Spillway and to the Gulf of Mexico by the Rigolets strait, Chef Menteur Pass, and Lake Borgne. It represents the second largest salt–water lake in the US (1630 km^2), after the Great Salt Lake in Utah, and the largest lake in Louisiana.

Toxic cyanobacterial blooms (e.g., *Microcystis* & *Anabaena*) have occurred on Lake Pontchartrain following the discharge of water from the Mississippi River to the estuary via the Bonnet Carré Bonnet Spillway. The spillway was constructed to divert floodwater through the lake before it reaches the city of New Orleans. The floodgate between the spillway and the river is only opened during severe flood events, but water can seep into the spillway through the closed gate (Lane et al. 2001). The seeping river water leaks brown clouds of sediment and nutrients into Lake Pontchartrain. When the floodgates have been opened in the past, the influx of fresh, nutrient–rich water has been followed by fish kills and toxic cyanobacterial blooms.

The 1997 opening of the Bonnet Carré Spillway is one of the better–documented events where toxic cyanobacterial booms occurred on the lake following the release of water from the river to the lake (Dortch et al. 1998). The algal bloom became apparent over large parts of the lake during May and produced high levels of hepatotoxins, prompting health advisories by the Louisiana Department of Health and Hospitals.

Fig. 12. Anabaena bloom on Lake Pontchartrain.

Fig. 12. (cont.) Anabaena bloom on Lake Pontchartrain (Photo courtesy of John Burns. See Color Plate 2).

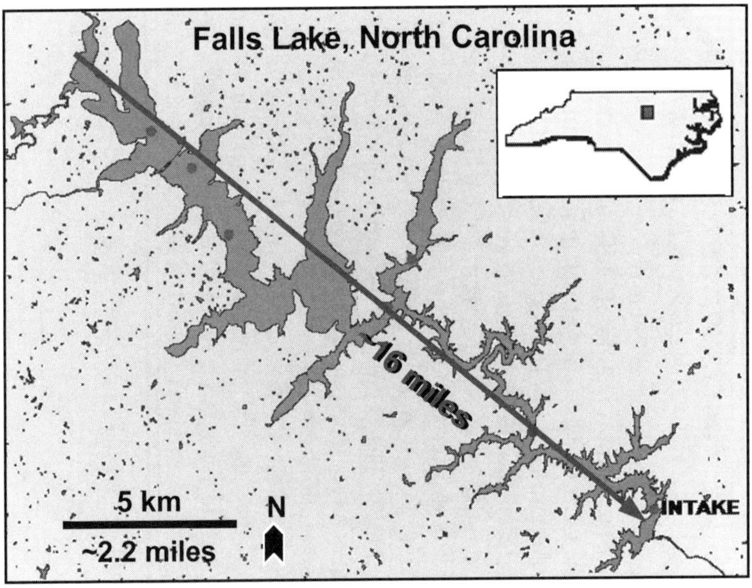

Fig. 13. Map of Falls Lake, NC, indicating six sampling stations and the location of the intake for the water treatment plant of the capital city, Raleigh. *The watershed is sustaining rapid human population growth and associated increased nutrient runoff into receiving waters. From Burkholder (2006b).* (See Color Plate 2).

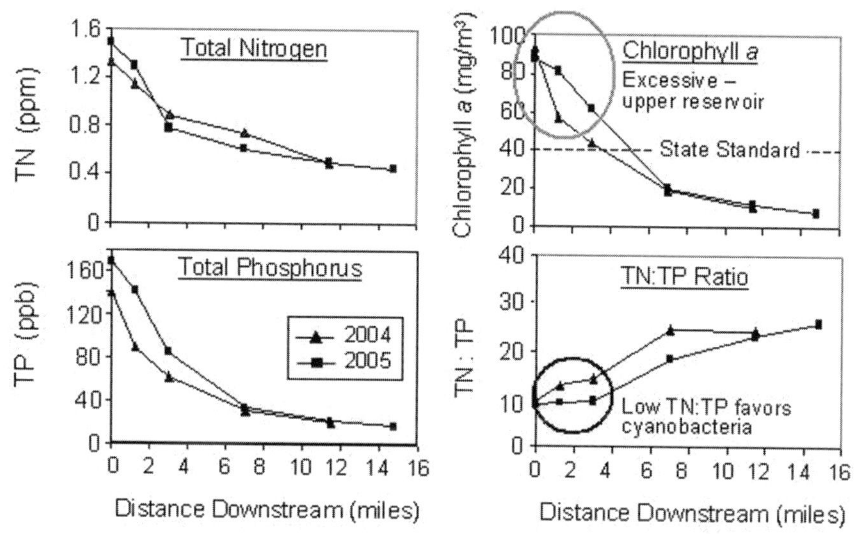

Fig. 14. Total N, total P, chlorophyll a (indicator of algal/cyanobacterial biomass), and total N : total P ratios in Falls Lake plotted against distance downstream from the headwaters to the dam.

The last station (intake) is the location of the intake for the City of Raleigh water treatment plant. About 75% of the phytoplankton cells in this reservoir during summer months are cyanobacteria, including various toxigenic species based on data from 2002–2005 (Burkholder 2006). The lower end of the reservoir where the water treatment plant is located presently has good water quality, but there is concern that as nutrient loading to the reservoir continues to increase, water quality will become more degraded throughout the reservoir and cyanobacteria will be further stimulated.

Frequency and Severity Ranking.

Both our exposure and our knowledge follow the hierarchy of frequency and severity ***freshwater > estuarine (brackish) > marine***. While adverse impacts of cyanobacteria in eutrophic freshwaters and (less so) estuaries are known worldwide (Whitton and Potts 2000), impacts on marine ecosystems have only recently begun to be recognized. Thus, the indicated rankings may change as additional information becomes available.

Data Needs

To improve our ability to measure the incidence and predict the occurrence of CHABS, we need to make full use of existing information that is currently available through a variety of sources. State and regional investigators have conducted a number of surveys of cyanotoxins that needs to be compiled into a readily accessible database. Information that has been collected on the factors related to bloom occurrence should be compiled so that it can be related to toxin occurrence. This includes land use data, nutrient concentration in water, and weather conditions. Information on robust analytical methods for cyanobacteria and toxin quantification in environmental waters needs also needs to be compiled.

Short—term Objectives

Research Needs

There is currently a lack of agreement on species level taxonomy for cyanobacteria. Because of this lack of agreement, cyanobacteria can only be identified easily to the genus level, which means that information may be lost in surveys when only genera are identified. Additionally, changing taxonomy can be a confounding factor for older surveys. Modern molecular techniques offer one solution to this problem but even these techniques require a consensus on the definition of toxigenic cyanobacterial species.

Considerable information on cyanobacterial species and their toxins is also available in non–traditional literature sources. A comprehensive literature survey, including the grey literature, is needed. Both planktonic and benthic cyanobacteria should be included in this survey. These and newer references need to be reviewed and summarized for occurrence of toxigenic species. A working bibliography on cyanobacteria compiled by the EPA can be found at http://www.epa.gov/safewater/ucmr/ucmr1/docs/meetings_umcr1_cyanobacteria_references.doc.

An improved understanding is needed of the environmental factors that are responsible for cyanobacterial growth and bloom formation. Additionally, a better understanding is needed of when cyanobacteria do and do not produce toxins, and factors responsible for production of toxins. This information can be obtained from both field studies and supporting laboratory experiments. Perhaps the best information currently available is for *Microcystis aeruginosa*; the genes involved in toxin production have been identified, and various environmental factors including light, nitrogen, phosphorus, and iron nutrient supplies have been shown to influence toxin production (Zurawell et al. 2005).

Research is needed to determine the importance of very small (picoplanktonic) cyanobacterial species in the production of cyanobacterial toxins. Similar research is needed for many benthic species. Among the many questions that remain unanswered are: how significant are levels of toxins produced by planktonic and benthic species? How does their toxin production vary over time as compared to historically recognized bloom species? What kinds of toxins are produced by these cyanobacteria?

BMAA originating from cyanobacteria on the roots of cycads has been reported to be associated with Alzheimer's disease in patients who have been exposed to BMAA from the food chain. Additional studies have detected BMAA in cyanobacteria from a variety of geographical locations (Cox et al. 2005). Because of the seriousness of Alzheimer's disease, plus the apparent widespread distribution of BMAA, research is needed to determine the occurrence of BMAA in water in the US, including the fate of this toxin during water treatment for drinking water production.

The conventional wisdom for cyanobacteria is that cylindrospermopsin is released extracellularly from healthy cells, whereas other commonly occurring cyanotoxins are contained in healthy cells, and only released if cells are damaged or dying. Recent evidence suggests that this generalization may not hold in some situations; thus, research is needed to verify that how toxins typically are distributed in commonly occurring species. These studies should use more than one strain of each species, because as mentioned, within a given species of cyanobacteria are many nontoxic as well as toxic strains (Burkholder and Glibert 2006).

Research is also needed to determine whether cyanotoxins are accumulated in the tissues of fish and shellfish. These studies should be done for the most important cyanotoxins which fish and shellfish are likely to be exposed to in both natural bodies of water as well as in aquaculture (e.g. nodularin in some estuarine/marine environments). The exposure of people to cyanotoxins originating from spray or aerosols should be investigated.

Analytical Needs

There is an immediate need for rapid techniques for both screening and confirmatory methods for the cyanotoxin analyses, including simple, inexpensive methods for rapid screening of samples during sampling of blooms. Different types of methods are needed for different applications and each must be targeted towards its end users. Several ELISA screening methods are available for microcystins, but none are currently manufactured for cylindrospermopsin and other toxins anatoxin–a. On–line detectors would be useful for water utilities for continuous monitoring; instrumental methods are needed for use where quantitation and specificity are important. Methods including LC/MS and LC/MS/MS have been developed, but none have been accepted as validated US EPA methods or consensus organization methods. There is a need for standardized instrumental methods that can be run by commercial labs. Many of the instrumental methods for cyanotoxins are dependent on sample cleanup methods that remove interfering substances. To be successful, these methods must perform in a wide variety of water types, including those with high levels of total organic carbon.

In addition to identifying the toxins, there is a need for additional tools targeted towards identifying a potentially toxigenic organism. Species–specific and toxin–specific molecular probes, microarrays and other technology are needed for studies to determine the occurrence and distribution of potentially toxigenic cyanobacteria. To be most useful, these detection methods should be cost–effective, rapid and reliable.

Standardized methods for quantifying cells and estimating biomass are also needed. These methods should be developed in conjunction with taxonomic identification methods, since identification of species may depend on characteristics that may be altered in some cell quantification schemes. When developed, these methods should be validated. Since cell count methods using microscopy generally are tedious and labor–intensive, alternate, accurate methods to determine cell density or biomass are needed, such as improved techniques to quantify photosynthetic pigments, total suspended solids or particulate organic carbon. Unfortunately, these methods at present all fall short of accurately estimating cyanobacteria biomass, and cannot provide information about toxicity or the dominant taxa involved. Improvements in technology are needed to rapidly track toxigenic species and formation of blooms. These technologies include remote sensing, remote–operated vehicles (ROVs) and automated sampling stations located on vulnerable bodies of water (Glasgow et al. 2004).

Monitoring Needs

There is an immediate need for a nation–wide survey(s) on the occurrence and distribution of toxigenic cyanobacteria and their toxins throughout the US. Some surveys of cyanobacterial toxins have been conducted; however, information on cyanotoxin occurrence is still fragmented and limited. Any information of cyanotoxin occurrence is useful, but in particular, national surveys, consistently conducted, are needed to assess cyanotoxin occurrence on a larger scale.

The US EPA is planning to conduct a national survey of drinking water for cyanotoxins, under its Unregulated Contaminant Monitoring Rule (UCMR) when the suite of analytical methods needed becomes available. Monitoring will occur for one year at approximately 2,800 large public water systems (PWS) and a representative sample of 800 (out of 66,000) small PWS. Transient water systems are not required to monitor nor are systems that purchase 100% of their water (EPA 2005a). The monitoring results from these systems will be used to form an initial nationally representative occurrence assessment of the most commonly occurring cyanotoxins in the US. A similar survey is needed for ambient water sources, including recreational and agricultural water. Currently, no program exists for conducting this kind of survey, which should include both randomly selected locations as well as targeted locations where blooms historically have been reported, or are reported to be in progress. This design would assure that both typical occurrences and unusual bloom events would be considered and could be used to determine risk to exposed people or animals.

In many locations in the US, cyanobacterial blooms are common events. Nevertheless, they can be sporadic in occurrence and difficult to predict. Even targeting locations that recently have experienced blooms may not mean that blooms will occur in these areas in the future; on the other hand, in developing watersheds blooms may begin to develop in previously "bloom–free" locations. Undetected or missed blooms would result in an underestimate of the risk of exposure. A central reporting website needs to be established to assist in documenting blooms. This website needs to be publicized to utility operators, state department of environmental quality officials and others who would be aware of blooms as they occur. These reports would allow researchers who have the ability to analyze water samples for cyanotoxins to sample these events while they are occurring.

Long—term Objectives

Research Needs

As a longer–term project, an easily accessed database (i.e. website) on the historic occurrence of toxigenic strains is critically needed, including user–friendly summaries for the general citizenry. This could be incorporated into a national database of toxin occurrence data that can be routinely updated. Currently, there is no central location for cyanotoxin data produced by states or other organizations. Long–term monitoring needs to be done to determine how cyanotoxins occur over time. This monitoring should also be done over a wide geographical area to determine how cyanotoxins are distributed spatially and to establish baseline information for assessing how blooms and cyanotoxin distribution are changing over time. This research would clarify how commonly blooms resulting in high concentrations of cyanotoxins occur, and how this is related to geographical location. It is well known that not all cyanotoxins have been described, so the distribution of known cyanotoxins may be greater presently understood. Thus, studies are also needed to identify new cyanotoxins.

Analytical Needs

For continual monitoring and development of research tools, there is a long–term need for a stable supply of toxic organisms and toxins. Toxic strains of many species commonly have lost their ability to make toxins over time in the highly artificial conditions represented by laboratory cultures. To avoid the loss of these sources of toxins, culture collections are needed which can supply a continuing source of toxin–producing strains of cyanobacteria recently isolated from natural habitats. Similarly, the development of analytical methods and other studies have been hampered by lack of availability of analytical chemistry standards for cyanotoxins. Information on concentration or purity of available standards is unavailable or ambiguous. Manufacturers need to produce larger supplies of at least the more common toxins, and to certify their purity. These toxins should be made commercially available.

Monitoring Needs

For long–term priorities, the same types of surveys are needed that were described for short–term needs. These surveys would include sampling at locations with typical levels of cyanobacteria as well as targeted surveys that sample blooms in progress. To respond to the appearance of cyanobacterial toxins in a manner that will minimize public health risk, rapid analyses are needed which will permit a response before drinking water consumers or recreational water users are exposed to the toxins. Rapid assays or screening methods are needed for this purpose. Data on long–term occurrence of cyanotoxins and cyanobacteria that produce the toxins are needed to assess how commonly cyanotoxins occur, at what levels, and whether there are any long–term changes in their occurrence.

Summary

Distribution of CHABs across the US

- Freshwater blooms are ubiquitous across the US and Canada.
- Most cyanobacterial blooms are in freshwater, but some are found in estuarine or marine waters.
- More detections of cyanotoxins have been reported recently, however, increased awareness had led to more monitoring programs.
- Blooms of cyanobacteria tend to be greater and last longer in warmer climates.
- Detections in ambient waters have been observed to vary from no detection to 31,470 μg/L for microcystin, no detection to 156 μg/L for anatoxin–a, and no detection to 202 μg/L for cylindrospermopsin. Detections occurred in treated water but concentrations were lower.
- Few data are available on temporal changes in cyanotoxin occurrence.

Cyanotoxin production

- Determination of taxonomic groups which produce toxins has been hampered by uncertainty in the taxonomic definitions for cyanobacteria.
- Table 1 lists 22 genera and 42 species of cyanobacteria associated with toxin formation.
- Table 2 lists taste and odor–producing species of cyanobacteria.
- Table 3 lists the commonly produced types of cyanotoxins in freshwater.
- Microcystin has been reported to be produced by *Microcystis, Anabaena, Oscillatoria (Planktothrix), Nostoc, Hapalosiphon* and *Anabaenopsis*.
- Cylindrospermopsin is produced by *Cylindrospermopsis raciborskii, Aphanizomenon ovalisporum,* and *Uzmekia natans*.
- Anatoxin–a is produced by *Anabaena flos–aquae, A. planktonica, Oscillatoria* species, and *Cylindrosperum*. Homoanatoxin–a, was isolated from *Phormidium Formosa*.

- Anatoxin–a(s) was reported from A. flos–aquae strain NRC 525–17 and more recently by *A. lemmermannii.*
- Saxitoxin and related toxins are made by *Aphanizomenon flos–aquae, Anabaena circinalis, Lyngbya wollei* and a Brazilian isolate of *Cylindrospermopsis raciborskii.*
- BMAA is produced by *Nostoc* species and many others.
- Lyngbyatoxins aplysiatoxin and debromoaplysiatoxin are made by *Lyngbya*, Oscillatoriaceae such as *Schizothrix calcicola* and *Oscillatoria nigroviridis*.
- Cyanobacteria also produce other bioactive compounds.

Impacts on waterbody health and ecosystem viability

- The factors that lead to bloom formation and toxin production are not well understood. Light, temperature, hydrological and chemical factors may be involved.
- In freshwater, blooms in Metropolitan Water District's reservoirs have led to taste and odor problems. A sewage spill in North Carolina led to a *Lyngbya* bloom.
- In estuarine habitats including the upper Chesapeake Bay, cyanobacterial blooms occurred with an invasion of water milfoil. Cyanobacterial blooms with fish kills have occurred in brackish Lake Ponchartrain, Louisana from freshwater or nutrient influxes.
- Ecological effects have occurred as a result of cyanobacterial blooms that are caused by effects other than toxin production.

Data Needs

- A consensus is needed for species level taxonomy for cyanobacteria.
- Existing information needs to be compiled into an accessible form.
- New surveys should be conducted (including UCMR monitoring of drinking water), especially for newly emerging toxins in raw water and drinking water.
- A national monitoring program is needed for recreational water.

- Analytical methods are needed including screening methods, instrumental methods, and online and remote detection. Sample preparation methods and analytical chemistry standards are needed to support analytical methods.
- A better understanding is needed of what factors lead to bloom formation and cyanotoxin production.
- The relation between toxin retention and release for different toxins needs to be clarified.
- Accumulation of cyanotoxins in the tissues of fish and shellfish needs study.
- A website is needed for reporting of cyanotoxin occurrence information.

Glossary

Akinete – thick–walled, often enlarged cell that can survive adverse conditions and form a new filament; often refractive under light microscopy because it is filled with food reserve (cyanophycean starch); also often orangish in color from abundant, protective carotenoid pigments. This specialized cell is found in some members of the Order Nostocales.

Anaerobic – Without oxygen.

Anoxic – Referring to habitats with little or no oxygen, such as the bottom waters of many eutrophic lakes in summer

Avian vacuolar myelinopathy [AVM] – a recently discovered neurological disease affecting water birds and their predators, primarily American coots and bald eagles, in the southern US. AVM causes characteristic lesions in the myelin of the brain and spinal cord.

Benthic – Living in or on the bottom of a body of water, or growing attached to various substrata including other organisms (e.g. the noxious aquatic weed, hydrilla), but not free–floating in the water column. Note: Many benthic cyanobacteria can be moved into the plankton by wind/waves or other disturbance, and can thrive in the water column.

Brackish – Somewhat salty, as in brackish (estuarine) water.

Buoyant – Floats easily, and is less dense than water.

Chlorophyll *a* – The "universal plant pigment", found in all organisms that undergo plant–like photosynthesis with production of oxygen. The concentration of this pigment is often used as an indicator or surrogate measure for algal biomass or production.

Cyanobacteria – Formerly (and sometimes still) called blue–green algae, cyanophytes or mixophytes (means "slime plants"). In present–day taxonomy, considered a type of bacteria ("blue bacteria") because of their primitive cell structure and other features, although they have the "universal plant pigment" that is directly involved in "higher plant" photosynthesis, chlorophyll *a*.

Estuary – A semi–enclosed body of water where saltwater mixes with freshwater.

Eutrophic – Referring to aquatic freshwater and/or estuarine/coastal ecosystems that are nutrient–enriched and highly productive.

Filamentous – Group of cells aligned in a row; very thin, threadlike.

Freshwater – Aquatic ecosystem with salinity lower than 0.1 ppt (see below).

Genus (plural: genera) – A taxonomic group below family and above species

Habitats – The places where plants and animals live.

Harmful algal bloom – A proliferation of microscopic or macroscopic algae that are capable of causing disease or death of humans or beneficial aquatic life through production of toxins, or that adversely affect aquatic ecosystems through other mechanisms (e.g. by depleting oxygen needed to sustain fish, or by reducing light that is essential for growth of beneficial submersed plants).

Heterocyte (formerly heterocyst) – thick–walled cell, usually pale in color compared to the other cells, used for nitrogen fixation in some cyanobacteria. This specialized cell is found in some members of the Order Nostocales.

Heterotroph – An organism that depends on consumption of other organisms and organic material for energy.

Hypersaline – referring to water with salinity higher than average seawater salinity.

Indicator species – A species whose status provides information on the overall health of the ecosystem and of other species in that ecosystem. Indicator species reflect the quality and changes in environmental conditions as well as aspects of community composition.

Macrophyte – Usually referring to a rooted submersed aquatic plant.

Mesohaline – Moderately salty water (salinity range 5–18 ppt).

Mixotroph – An organism that uses both photosynthesis and particulate or dissolved organic substances made by other organisms for its energy. Many cyanobacteria are mixotrophs.

Oligohaline – Low salinity (0.5–5 ppt).

Phytoplankton– Microscopic algae that live and reproduce suspended in the water column of freshwater, estuarine, and marine ecosystems. Phytoplankton form the base of the food web in many aquatic ecosystems.

Picoplankton – Very small phytoplankton in aquatic ecosystems, including cyanobacteria (diameter 0.2 – 2 μm).

Polyhaline – Referring to waters that are variable in salinity, often highly brackish water with a salinity of 18–30 ppt.

Ppt – Parts per thousand, a measure of salinity. For example, the salinity of the open ocean is generally about 35 ppt; freshwaters are < 0.1 ppt.

Reservoir (impoundment) – an artificial (man–made) lake, formed by damming a river.

Specialized cell – akinete, heterocyte (see above).

Strain – A population within a given species of cyanobacteria or microscopic algae.

Terrestrial – Living or growing on land or in the soil.

Tidal freshwater – Freshwater (0–0.5 ppt) that is tidally influenced.

Toxigenic – potentially toxic; referring to a species of cyanobacteria that is known to have some toxic strains. (Note: toxigenic algae have both nontoxic and toxic strains.)

References

Albert S, O'Neil JM, Udy JW, Ahern KS, O'Sullivan CM, Dennison WC (2005) Blooms of the Cyanobacterium *Lyngbya Majuscula* in Coastal Queensland, Australia: Disparate Sites, Common Factors. *Mar Poll Bull.* 51 428–37.

Anderson DM (1994) Red Tides. *Scientific American.* August 52–58.

Arthur KE, Limpus CJ, Roelfsema CM, Udy JW, Shaw GR (2006) A Bloom of *Lyngbya Majuscula* in Shoalwater Bay, Queensland, Australia: An Important Feeding Ground for the Green Turtle (*Chelonia Mydas*). *Harmful Algae.* 5 251–65.

Azevedo SM, Carmichael WW, Jochimsen EM, Rinehart KL, Lau S, Shaw GR, Eaglesham GK (2002) Human Intoxication By Microcystins During Renal Dialysis Treatment in Caruaru—Brazil. *Toxicology.* 181–182 (661) 441–46.

Barron S, Weber C, Marino R, Davidson E, Tomasky G, Howarth R (2002) Effects of Varying Salinity on Phytoplankton Growth in a Low–Salinity Coastal Pond Under Two Nutrient Conditions. *Biol Bull.* 260–61.

Behm D (2003) Coroner Cites Algae in Teen's Death – Experts Are Uncertain About Toxin's Role. *Milwaukee Journal Sentinel*, 2/5/04 2003.

Berry JP, Gantar M, Gawley RE, Wang ML, Rein KS (2004) Pharmacology and Toxicology of Pahayokolide a, a Bioactive Metabolite From a Freshwater Species of *Lyngbya* Isolated From the Florida Everglades. *Comp Biochem Physiol–C: Phamacol Toxicol.* 139 231–38.

Birrenkott AH, Wilde SB, Hains JJ, Fischer JR, Murphy TM, Hope CP, Parnell PG, Bowerman WW (2004) Establishing a Food–Chain Link Between Aquatic Plant Material and Avian Vacuolar Myelinopathy in Mallards. *J Wildlife Dis.* 40 485–92.

Boyer G (2006) Personal Communication.

Boyer GL, Watzin MC, Shambaugh AD, Satchwell MF, Rosen BH, Mihuc T (2004) The Occurrence of Cyanobacterial Toxins in Lake Champlain. In *Lake Champlain: Partnerships and Research in the New Millennium*, eds. Boyer, GL, Watzin, MC, Shambaugh, AD, Satchwell, MF, Rosen, BH, and Mihuc, T, New York: Kluwer Academic/Plenum Publ. 241–57.

Burkholder JM (2006) Falls Lake, North Carolina – a Major Potable Water Supply Reservoir Poised for Increased Cyanobacteria Blooms. *Lakeline.*

Burkholder JM, Cuker BE (1991) Response of Periphyton Communities to Clay and Phosphate Loading in a Shallow Reservoir. *Journal of Phycology.* 27 373–84.

Burkholder JM, Glibert, PM (2006) Intraspecific Variability: An Important Consideration in Forming Generalizations About Toxigenic Algal Species. *South Africa Journal of Marine Science.* (in press).

Burkholder JM (2002) Cyanobacteria. In *Encyclopedia of Environmental Microbiology*, ed. Bitton, G, New York: Wiley Publishers. 952–81.

Burns J (2005) Cyanohabs – the Florida Experience. *International Symposium on Cyanobacterial Harmful Algal Blooms* Durham, NC. 09/6–09/10.

Campbell R, Sargent R (2004) Wisconsin Teen's Death a Wake–Up Call About Toxic Algae. *The Orlando Sentinel, 5 March 2004*, 2004.

Carmichael WW (2006) Personal Communication. 07/26.

Carmichael WW (1992) *A Status Report on Planktonic Cyanobacteria (Blue Green Algae) and Their Toxins*. Cincinnati, OH: Environmental Monitoring Systems Laboratory, Office of Research and Development, US Environmental Protection Agency.

Carmichael WW (1998) Algal Poisoning. In *The Merck Veterinary Manual*, ed. Aiello, S, Whitehouse Station, New Jersey: Merck & Co., Inc. 2022–23.

Carmichael WW (2001) *Assessment of Blue–Green Algal Toxins in Raw and Finished Drinking Water*. Denver, Colorado: AWWA Research Foundation and American Water Works Assoc. ed.

Carmichael WW, Azevedo SMFO, An JS, Molica RJR, Jochimsen EM, Lau S, Rinehart KL, Shaw GR, Eaglesham GK (2001a) Human Fatalities From Cyanobacteria: Chemical and Biological Evidence for Cyanotoxins. *Environmental Health Perspectives.* 109 (6448) 663–68.

Carmichael WW, Biggs DF, Gorham PR (1975) Toxicology and Pharmacological Action of *Anabaena Flos–Aquae* Toxin. *Science.* 187 (118) 542–44.

Carmichael WW, Li RH (2006) Cyanobacteria Toxins in the Salton Sea. *Saline Systems.* doi:10.1186/1746–1448–2–5

Carmichael WW, Sirenko LA, Klochenko PD, Shevchenko TF (2001b) A Comparative Assessment of the Toxicity of Algae and Cyanobacteria in Water Bodies of Ukraine. *Phycologia.* 40 (10338) 15.

Castenholz RW (1992) Species Usage, Concept, and Evolution in the Cyanobacteria (Blue–Green Algae). *Phycology.* 28 737–45.

Castenholz RW (2001) Phylum Bx. Cyanobacteria. Oxygenic Photosynthetic Bacteria. In *Bergey's Manual of Systematic Bacteriology*, eds. D.R., B, and R.W., C, New York: Springer. 473–599.

Chorus I, Bartram J (1999) *Toxic Cyanobacteria in Water: A Guide to Their Public Health Consequences, Monitoring and Management*. London: E & FN Spon. (ed.)

Christiansen G, Dittmann E, Ordorika LV, Rippka R, Herdman M, Börner T (2001) Nonribosomal Peptide Synthetase Genes Occur in Most Cyanobacterial Genera as Evidenced By Their Distribution in Axenic Strains of the PCC. *Archives of Microbiology.* 176 (388) 452–58.

Codd GA, Morrison LF, Metcalf JS (2005) Cyanobacterial Toxins: Risk Management for Health Protection. *Toxicology and Applied Pharmacology.* 203 264–72.

Codd GA, Metcalf JS, Ward CJ, Beattie KA, Bell SG, Kaya K, Poon GK (2001) Analysis of Cyanobacterial Toxins By Physicochemical and Biochemical Methods. *J Assoc Off Analyt Chem Int.* 84 (307) 1626–35.

Cox PA, Banack SA, Murch SJ, Rasmussen U, Tien G, Bidigare RR, Metcalf JS, Morrison LF, Codd GA, Bergman B (2005) Diverse Taxa of Cyanobacteria Produce Beta–N–Methylamino–L–Alanine, a Neurotoxic Amino Acid. *Proc Natl Acad Sci.* 102 5074–78.

Cuker BE, Gama P, Burkholder JM (1990) Type of Suspended Clay Influences Lake Productivity and Phytoplankton Response to P Loading. *Limnology and Oceanography*. 35 830–39.

Devlin JP, Edwards OE, Gorham PR, Hunter NR, Pike RK, Stavric B (1977) Anatoxin–a, a Toxic Alkaloid From *Anabaena Flos–Aquae* Nrc–44H. *Can J Chem*. 55 (177) 1367–71.

Dittmann E, Borner T (2005) Genetic Contributions to the Risk Assessment of Microcystin in the Environment. *Toxicology and Applied Pharmacology*. 203 192–200.

Donati C, Drikas M, Hayes R, Newcombe G (1994) Microcystin–LR Adsorption By Powdered Activated Carbon. *Water Research*. 28 (5328) 1735–42.

Dortch Q, Peterson J, Turner R (1998) Algal Bloom Resulting From the Opening of the Bonnet Carre Spillway in 1997. *Fourth Bi–annual Basics of the Basin Symposium* 28–29.

Drouet F, Daily WA (1956) Revision of the Coccoid Myxophyceae. *Butler University Botanical Studies*. 12 1–218.

Drouet F, Daily WA (1957) Revision of the Coccoid Myxophyceae: Additions and Corrections. *Trans Am Microsc Soc*. 76 219–22.

Drouet F (1968) *Revision of the Classification of Oscillatoriaceae. Monograph of the Academy of Natural Sciences*. Philadelphia 15.

EPA (2001) *Creating a Cyanotoxin Target List for the Unregulated Contaminant Monitoring Rule. Summary of Meeting Held in Cincinnati, Oh on May 17–18, 2001*. Washington, DC: Office of Water, US Environmental Protection Agency. URL: http://www.epa.gov/OGWDW/ucmr/pdfs/meeting_ucmr1_may2001.pdf. Accessed: 06/15/05.

EPA (2003) *Statistical Analysis Methods for Biological Assessment*. Washington, DC: US Environmental Protection Agency. URL: http://www.epa.gov/owow/monitoring/tech/appdixe.html. Accessed: 12/16/05.

EPA (2005b) *Drinking Water Contaminant Candidate List 2*. Washington, DC: Office of Water, US Environmental Protection Agency. URL: http://www.epa.gov/safewater/ccl/ccl2_list.html. Accessed: 3/15/05.

EPA (2005a) *Unregulated Contaminant Monitoring Regulation (UCMR) for Public Water Systems Revisions*. 40 CFR Part 141. Washington, DC: US Environmental Protection Agency. 49093–138.

Euzeby J (2004) Validation of Publication of New Names and New Combinations Previously Effectively Published Outside of the Ijsem. *International Journal of Systematic and Evolutionary Microbiology*. 54 1–2.

Finn T, Kononen K, Olsonen R, Wallström K (2001) The History of Cyanobacteria Blooms in the Baltic Sea. *Ambio*. 30 172–78.

Frazier W (2006) Personal Communication. 07/11.

Geitler L (1932) Cyanophyceae. In *Rabenhorst's Kryptogamenflora*, Leipzig: Akad. Verlagsges.

Glasgow HB, Burkholder JM, Reed RE, Lewitus AJ, Kleinman JE (2004) Real–Time Remote Monitoring of Water Quality: A Review of Current Applications, and Advancements in Sensor, Telemetry, and Computing Technologies. *Journal of Experimental Marine Biology and Ecology*. 300 409–48.

Glibert PM, Burkholder JM (2006) The Complex Relationships Between Increasing Fertilization of the Earth, Coastal Eutrophication, and Hab Proliferation, Chapter 24. In *The Ecology of Harmful Algae*, eds. Granéli, E, and Turner, J, Springer–Verlag: New York. (in press).

Graham JL, Jones JR, Jones SB, Downing JA, Clevenger, TE (2004) Environmental Factors Influencing Microcystin Distribution and Concentration in the Midwestern United States. *Water Research*. 38 4395–404.

Hamann CL, McEwen JB, AG M (1990) Guide to Selection of Water Treatment Processes. In *Water Quality and Treatment – a Handbook of Community Water Supplies*, ed. Pontius, FW, New York: American Water Works Association. 157–87.

Harada K (2004) Production of Secondary Metabolites By Freshwater Cyanobacteria. *Chemical & Pharmaceutical Bulletin*. 52 889–99.

Harada KI, Fujii K, Hyashi K, Suzuki M, Ikai Y, Oka H (1996) Application of D,L–Fdla Derivatization to Determination of Absolute Configuration of Constituent Amino Acids in Peptide By Advanced Marfey's Method. *Tetrahed Lett*. 37 (1386) 3001–04.

HARNESS (2005) *Harmful Algal Research and Response: A National Environmental Science Strategy 2005–2015*. Edited by Ramsdell, JS, Anderson, DM, and Glibert, PM. Washington, DC: Ecological Society of America. ed.

Henriksen P, Carmichael WW, An JS, Moestrup Ø (1997) Detection of an Anatoxin–a(S)–Like Anticholinesterase in Natural Blooms and Cultures of Cyanobacteria/Blue–Green Algae From Danish Lakes and in the Stomach Contents of Poisoned Birds. *Toxicon*. 35 (1111) 901–13.

Hisbergues M, Christiansen G, Rouhiainen L, Sivonen K, Börner T (2003) PCR–Based Identification of Microcystin–Producing Genotypes of Different Cyanobacterial Genera. *Archives of Microbiology*. 180 (10438) 402–10.

Hitzfeld BC, Hoeger SJ, Dietrich DR (2000) Cyanobacterial Toxins: Removal During Drinking Water Treatment, and Human Risk Assessment. *Environmental Health Perspectives*. 108 (6) 113–22.

Hoeger SJ, Hitzfeld BC, Dietrich D (2005) Occurrence and Elimination of Cyanobacterial Toxins in Drinking Water Treatment Plants. *Toxicology and Applied Pharmacology*. 203 231–42.

Hoeger SJ, Shaw G, Hitzfeld BC, Dietrich DR (2004) Occurrence and Elimination of Cyanobacterial Toxins in Two Australian Drinking Water Treatment Plants. *Toxicon*. 43 (14248) 639–49.

Hoffmann L (2005) Nomenclature of Cyanophyta/Cyanobacteria: Roundtable on the Unification of the Nomenclature Under the Botanical and Bacteriological Codes. *Algological Studies*. 117 13–29.

Hu C, Muller–Karger FE, Swarzenski PW (2006) Hurricanes, Submarine Groundwater Discharge, and Florida's Red Tides. *Geophysical Research Letters*. 33 doi:10.1029/2005GL025449.

Huisman J, Hulot FD (2005) Population Dynamics of Harmful Cyanobacteria. In *Harmful Cyanobacteria,* Aquatic Ecology Series. eds. Huisman, J, Matthijs, HCP, and Visser, PM, Dordrecht, The Netherlands: Springer. 143–76.

Izaguirre G (2005) Harmful Algal Blooms and Cyanotoxins in Metropolitan Water District's Reservoirs. *International Symposium on Cyanobacterial Harmful Algal Blooms* Durham, NC. 09/6–09/10.

Izaguirre G, Taylor WD (2004) A Guide to Geosmin–and Mib–Producing Cyanobacteria in the United States. *Water Science and Technology.* 49 (14448) 19–24.

Jacoby JM, Collier DC, Welch EB, Hardy FJ, Crayton M (2000) Environmental Factors Associated With a Toxic Bloom of Microcystis Aeruginosa. *Can J Fish Aquat Sci.* 57 231–40.

Kardinal WEA, Visser PM (2005) Dynamics of Cyanobacterial Toxins. In *Harmful Cyanobacteria,* Aquatic Ecology Series. eds. Huisman, J, Matthijs, HCP, and Visser, PM, Dordrecht, The Netherlands: Springer.

Kenefick SL, Hrudey SE, Peterson HG, Prepas EE (1993) Toxin Release From *Microcystis Aeruginosa* After Chemical Treatment. *Water Science and Technology.* 27 (1316) 433–40.

Kenefick SL, Hrudey SE, Prepas EE, Motkosky N, Peterson HG (1992) Odorous Substances and Cyanobacterial Toxins in Prairie Drinking Water Sources. *Water Science and Technology.* 25 (1317) 147–54.

Kneale PE, Howard A (1997) Statistical Analysis of Algal and Water Quality Data. *Hydrobiologial.* 349 59–63.

Komárek J (2003) Coccoid and Colonial Cyanobacteria, Chapter 4. In *Freshwater Algae of North America – Ecology and Classification,* eds. Wehr, JD, and Sheath, RG, Academic Press: Boston.

Komárek, J, Anagnostidis K (1999) Cyanoprokaryota. 1. Chroococcales. In *Suesswasserflora Von Mittleeuropa 19/1,* Stuttgart: Fischer Verlag. 548.

Komárek J, Anagnostidis K (2005) Cyanoprokaryota. 2. Oscillatoriales. In *Suesswasserflora Von Mittleeuropa 19/2,* Stuttgart: Fischer Verlag. 759.

Krishnamurthy T, Carmichael WW, Sarver EW (1986) Toxic Peptides From Freshwater Cyanobacteria (Blue–Green Algae) I. Isolation, Purification and Characterization of Peptides From *Microcystis Aeruginosa* and *Anabaena Flos–Aquae. Toxicon.* 24 (414) 865–73.

Lagos N, Onodera H, Zagatto PA, Andrinolo D, Azevedo SMFO, Oshima, Y (1999) The First Evidence of Paralytic Shellfish Toxins in the Freshwater Cyanobacterium *Cylindrospermposis Raciborskii,* Isolated From Brazil. *Toxicon.* 37 (11858) 1359–73.

Landsberg JH (2002) The Effects of Harmful Algal Blooms on Aquatic Organisms. *Rev Fish Sci.* 10 (505) 113–390.

Lane RR, Day JW, Kemp PG, Demcheck DK (2001) The 1994 Experimental Opening of the Bonnet Carre Spillway to Divert Mississippi River Water Into Lake Pontchartrain, Louisiana. *Ecological Engineering.* 17 (4) 411–22.

Lapage SP, Sneath PHA, Lessel EF (1992) *International Code of Nomenclature of Bacteria and Statutes of the International Committee on Systematic Bacteriol-*

ogy and Statutes of the Bacteriolog: Bacteriological Code. Washington, DC: ASM Press. ed.

Lehman PW, Boyer G, Hall C, Waller S, Gehrts K (2005) Distribution and Toxicity of a New Colonial *Microcystis Aeruginosa* Bloom in the San Francisco Bay Estuary, California. *Hydrobiologia*. 541 87–99.

Li RH, Carmichael WW (2003) Morphological and 16S rRNA Gene Evidence for Reclassification of the Paralytic Shellfish Toxin Producing *Aphanizomenon flos-aquae* LMECYA 31 as *Aphanizomenon issatschenkoli J Phycol*. 39 814–818.

Li RH, Carmichael WW, Brittain S, Eaglesham GK, Shaw GR, Liu Y, Watanabe MM (2001) First Report of the Cyanotoxins Cylindrospermopsin and deoxycylindrospermopsin from *Raphidiopsis curuata* (Cyanobacteria). *J Phycol*. 37(6) 1121–1126.

Li R, Carmichael WW, Liu Y, Watanabe MM (2000) Taxonomic Re-evaluation of *Aphanizomenon flos-aquae* NH-5 Based Upon Morphology and 16S rRNA Gene Sequences. *Hydrobiologia*. 438(1) 99–105.

Lippy EC, Erb J (1976) Gastrointestinal Illness At Sewickley, Pa. *J Am Water Works Assoc*. 68 (462) 606–10.

Luesch H, Yoshida WY, Moore RE, Paul VJ (2000) Apramides a–G, Novel Lipopeptides From the Marine Cyanobacterium *Lyngbya Majuscula*. *J Natural Products*. 63 (544) 1106–12.

Lung WS, Paerl, HW (1988) Modeling Blue Green Algal Blooms in the Lower Neuse River. *Water Research*. 22 (9648) 895–905.

Mahmood NA, Carmichael WW (1986a) Paralytic Shellfish Poisons Produced By the Freshwater Cyanobacterium *Aphanizomenon Flos–Aquae* Nh–5. *Toxicon*. 24 (480) 175–86.

Mahmood NA, Carmichael WW (1986b) The Pharmacology of Anatoxin–a(S), a Neurotoxin Produced By the Freshwater Cyanobacterium *Anabaena Flos–Aquae* Nrc 525–17. *Toxicon*. 24 (481) 425–34.

Martin E (2001) Drying Up – North Carolina's Unquenchable Thirst Could Lead to Something More Than a Drinking Problem. *Business North Carolina*. Oct 46–59.

Matsunaga S, Moore RE, Niemczura WP, Carmichael WW (1989) Anatoxin–a(S), a Potent Anticholinesterase From *Anabaena Flos–Aquae*. *Journal of the American Chemical Society*. 111 (494) 8021–23.

Moikeha SN, Chu GW (1971) Dermatitis–Producing Alga *Lyngbya Majuscula* Gomont in Hawaii. II. Biological Properties of the Toxic Factor. *J Phycol*. 7 (542) 8–13.

Moikeha SN, Chu GW, Berger LR (1971) Dermatitis–Producing Alga *Lyngbya Majuscula* Gomont in Hawaii. I. Isolation and Chemical Characterization of the Toxic Factor. *J Phycol*. 7 (543) 4–8.

Murch SJ, Cox PA, Banack SA (2004a) A Mechanism for Slow Release of Biomagnified Cyanobacterial Neurotoxins and Neurodegenerative Disease in Guam. *Proc Natl Acad Sci*. 101 12228–31.

Murch SJ, Cox PA, Banack SA, Steele JC, Sacks OW (2004b) Occurrence of Beta–Methylamino–L–Alanine (BMAA) in Als/Pdc Patients From Guam. *Acta Neurol Scand*. 110 267–69.

Niemi RM, Knuth S, Lundstrom K (1982) Actinomycetes and Fungi in Surface Waters and in Potable Water. *Applied and Environmental Microbiology.* 43 (599) 378–88.

NOAA (2004) *Harmful Algal Bloom and Hypoxia Research and Control Reauthorization.* Public Law 108–456, 01/06. US Department of Commerce, National Oceanic and Atmospheric Administration.

Onodera H, Oshima Y, Henriksen P, Yasumoto T (1997) Confirmation of Anatoxin–a(S), in the Cyanobacterium *Anabaena Lemmermannii,* as the Cause of Bird Kills in Danish Lakes. *Toxicon.* 35 (4098) 1645–48.

Osborne NJT, Webb PM, Shaw GR (2001) The Toxins of *Lyngbya Majuscula* and Their Human and Ecological Health Effects. *Environ Int.* 27 (328) 381–92.

Padisak J (1997) *Cylindrospermopsis Raciborskii* (Woloszynska) Seenayya Et Subba Raju, an Expanding, Highly Adaptive Cyanobacterium: Worldwide Distribution and Review of Its Ecology. *Arch Hydrobiol/Suppl.* 107 (4) 563–93.

Paerl HW (1996) Microscale Physiological and Ecological Studies of Aquatic Cyanobacteria: Macroscale Implications. *Microscopy Res Technique.* 33 (3658) 47–72.

Pawlik–Skowronska B, Skowronski T, Pirszel J, Adamczyk A (2004) Relationship Between Cyanobacterial Bloom Composition and Anatoxin–a and Microcystin Occurrence in the Eutrophic Dam Reservoir (Se Poland). *Polish J Ecology.* 52 479–90.

Pereira P, Li R, Carmichael WW, Dias E, Franca S (2004) Taxonomy and Production of Paralytic Shellfish Toxins by the Freshwater Cyanobacterium *Aphanizomenon gracile* LMECYA40. *European Journal of Phycology.* 39(4) 361–369.

Pinckney JL, Millie DF, Vinyard BT, Paerl HW (1997) Environmental Controls of Phytoplankton Bloom Dynamics in the Neuse River Estuary, North Carolina, USA. *Canadian Journal of Fisheries and Aquatic Science.* 54 2491–501.

Pluotno A, Carmeli S (2005) Banyasin a and Banyasides a and B, Three Novel Modified Peptides From a Water Bloom of the Cyanobacterium *Nostoc Sp. Tetrahed.* 61 575–83.

Rapala J, Sivonen K (1998) Assessment of Environmental Conditions That Favor Hepatotoxic and Neurotoxic *Anabaena* Spp. Strains Cultured Under Light Limitation At Different Temperatures. *Microbial Ecol.* 36 (8038) 181–92.

Richardson LL (2005) Coral Diseases: What is Really Known? *Trends in Ecology and Evolution.* 13 438–43.

Saker ML, Griffiths DJ (2001) Occurrence of Blooms of the Cyanobacterium *Cylindrospermopsis Raciborskii* (Woloszynska) Seenaya Et Subba Raju in a North Queensland Domestic Water Supply. *Mar Freshwater Res.* 52 (320) 907–15.

Sasner Jr. JJ Ikawa M, Foxall TL, Watson WH (1981) Studies on Aphantoxin From *Aphanizomenon Flos–Aquae* in New Hampshire. In *The Water Environment: Algal Toxins and Health*, ed. Carmichael, WW, New York: Plenum Press. 389–403.

Schindler DW (1987) Determining Ecosystem Responses to Anthropogenic Stress. *Canadian Journal of Fisheries & Aquatic Sciences.* 44 (Suppl. 1) 6–25.

Sellner KG, Lacouture RV, Parrish CR (1988) Effects of Increasing Salinity on Cyanobacteria Bloom in the Potomac River Estuary. *Journal of Plankton Research.* 10 (1) 49–61.

Sims JK, Zandee Van Rilland RD (1981) Escharotic Stomatitis Caused By the "Stinging Seaweed" *Microcoleus Lyngbyaceus* (Formerly *Lyngbya Majuscula):* Case Report and Literature Review. *Hawaii Med J.* 40 (839) 243–48.

Sivonen K, Jones G (1999) Cyanobacterial Toxins. In *Toxic Cyanobacteria in Water: A Guide to Their Public Health Consequences, Monitoring and Management*, eds. Chorus, I, and Bartram, J, London: E & FN Spon. 41–111.

Sivonen K, Kononen K, Carmichael WW, Dahlem AM, Rinehart KL, Kiviranta J, Niemelä SI (1989) Occurrence of the Hepatotoxic Cyanobacterium *Nodularia Spumigena* in the Baltic Sea and the Structure of the Toxin. *Applied and Environmental Microbiology.* 55 (855) 1990–95.

Sivonen K, Rapala J (1998) Assessment of Environmental Conditions That Favor Hepatotoxic and Neurotoxic *Anabaena* Spp. Strains Cultured Under Light Limitations At Different Temperatures. *Microbial Ecol.* 36 (3538) 181–92.

Skulberg OM (1984) *Directory to Toxic Cyanophyte Literature From Norden.* Oslo: Norsk institutt for vannforskning (NIVA). ed.

Soranno PA (1997) Factors Affecting the Timing of Surface Scums and Epilimnetic Blooms of Blue–Green Algae in a Eutrophic Lake. *Can J Fish Aquat Sci.* 54 1965–75.

Stanier RY, Kunisawa R, Mandel M, Cohen–Bazire G (1971) Purification and Properties of Unicellular Blue–Green Algae (Order Chroococcales). *Bacteriol Rev.* 35 (906) 171–205.

Tango P, Butler W, Michael B (2005) Cyanotoxins in the Tidewaters of Maryland's Chesapeake Bay: The Maryland Experience. *International Symposium on Cyanobacterial Harmful Algal Blooms* Durham, NC. 09/6–09/10.

Teixeira MDGLC, Costa MDCN, Carvalho VLPD, Pereira MDS, Hage E (1993) Gastroenteritis Epidemic in the Area of the Itaparica Dam, Bahia, Brazil. *Bull Pan Am Health Org.* 27 (11918) 244–53.

Thomas NJ, Meteyer CU, Sileo L (1998) Epizootic Vacuolar Myelinopathy of the Central Nervous System of Bald Eagles *(Haliaeetus Leucocephalus)* and American Coots *(Fulica Americana). Vet Pathol.* 35 (2838) 479–87.

Touchette BW, Burkholder JM, Alexander JL, Allen EH, James J, Britton, CH (submitted) Eutrophication and Cyanobacteria Blooms in Run–of–River Impoundments in North Carolina, USA. *Lake and Reservoir Management.*

Vanderploeg HA, Liebig JR, Carmichael WW, Agy MA, Johengen TH, Fahnenstiel GL, Nalepa TF (2001) Zebra Mussel *(Driessena Polymorpha)* Selective Filtration Promoted Toxic *Microcystis* Blooms in Saginaw Bay (Lake Huron) and Lake Erie. *Can J Fish Aquat Sci.* 58 (6598) 1208–21.

Vezie C, Brient L, Sivonen K, Bertru G, Lefeuvre J–C, Salkinoja–Salonen M (1998) Variation of Microcystin Content of Cyanobacterial Blooms and Isolated Strain in Lake Grand–Lieu (France). *Microbial Ecol.* 35 (2978) 126–35.

Vezie C, Rapala J, Vaitomaa J, Seitsonen J, Sivonen K (2002) Effect of Nitrogen and Phosphorus on Growth of Toxic and Nontoxic *Microcystis* Strains and on Intracellular Microcystin Concentrations. *Microbial Ecol.* 43 (492) 443–54.

Walker SR (2005) Nebraska Experience. *International Symposium on Cyanobacterial Harmful Algal Blooms* Durham, NC. 09/6–09/10.

Wannemacher Jr. RW, Dinterman RE, Thompson WL, Schimdt MO, Burrows WD (1993) *Treatment for Removal of Biotoxins From Drinking Water*. Technical Report 9120. Frederick, MD: US Army Biomedical Research and Development Laboratory at Fort Detrick.

Watkinson AJ, O'Neill JM, Dennison WC (2005) Ecophysiology of the Marine Cyanobacterium, *Lyngbya Majuscula* (Oscillatoriaceae) in Moreton Bay, Australia. *Harmful Algae.* 4 697–715.

Wetzel RG (2001) *Limnology*. New York: Academic Press. 3 ed.

Whitton BA, Potts M (2000) *The Ecology of Cyanobacteria: Their Diversity in Time and Space*. Dordrecht; Boston: Kluwer Academic Publishers. ed.

WHO (2002) *Guidelines for Drinking Water Quality. Addendum: Microbiological Agents in Drinking Water*. Geneva, Switzerland: World Health Organization. 142.

WHO (2003) *Algae and Cyanobacteria in Coastal and Estuarine Waters: Guidelines for Safe Recreational Water Environments– Vol. 1 Coastal and Fresh Waters*. Geneva, Switzerland: World Health Organization. 128–35.

Wilde SB, Murphy TM, Hope CP, Habrun SK, Kempton J, Birrenkott A, Wiley F, Bowerman WW, Lewitus AJ (2005) Avian Vacuolar Myelinopathy Linked to Exotic Aquatic Plants and a Novel Cyanobacterial Species. *Environmental Toxicology*. 20 348–53.

Williams CD, Burns J, Chapman A, Flewelling L, Pawlowicz M, Carmichael WW (2001) *Assessment of Cyanotoxins in Florida's Lakes, Reservoirs, and Rivers*. Palatka, FL: St. Johns River Water Management District.

Zurawell RW, Chen HR, Burke JM, Prepas EE (2005) Hepatotoxic Cyanobacteria: A Review of the Biological Importance of Microcystins in Freshwater Environments. *J Toxicol Environ Health–Part B*. 8 1–37.

Chapter 4: A world overview — One-hundred-twenty-seven years of research on toxic cyanobacteria — Where do we go from here?

Wayne Carmichael

Department of Biological Sciences, Wright State University, Dayton, Ohio 45435 U.S.A.

Introduction

Both marine and freshwater Harmful Algal Blooms (HABs) have been observed throughout history. The first literature reference for toxic cyanobacteria (CyanoHABs) was in 1878. George Francis issued a report on sheep and cattle deaths from the brackish water cyanobacteria *Nodularia spumigena* in *Nature* called "Poisonous Australian Lake". For the marine HABs, a 1928 report in the Journal of Preventive Medicine described human intoxication from mussel poisoning cases in the San Francisco area during July of 1927. This led to work that described the first phycotoxin group, Saxitoxins, by Edward Schantz in the 1950's. In response the occurrence of red tides in New England during 1972, the First International Conference on Toxic Dinoflagellate Blooms was held in 1974. Currently there are twelve international marine HAB conferences. The First International Conference on Toxic Cyanobacteria proceedings, "The Water Environment Algal Toxins and Health", was published in 1981. The 7th such conference is scheduled to be held in Brazil in 2007. United States (U.S.) HAB response resulted in the development of a U.S. National Plan for Marine Biotoxins and Harmful Algae (Anderson et al. 1993), which led to "The Harmful Algal Bloom and Hypoxia Research and Control Act of 1998 written by the U.S. Senate-Subcommittee on Oceans and Fisheries. In the 2003 revision of this act the U.S. house of representatives included freshwater algae especially the harmful cyanobacteria. The current national plan is called Harmful Algal Research and Response; a National Environmental Science Strategy 2005-2015 (HARRNESS) (Ramsdell et al. (eds) 2005). This CyanoHAB Overview will focus on occurrences of cyanobacteria and cyanotoxins that have been observed in freshwater, drinking water, recreational water, estuaries and marine water, as well as the impacts on health and/or ecosystem viability in the U.S. and worldwide.

Charge 1

What occurrences have been observed, and what were the conditions at the time that might provide insight into conditions that promoted these CyanoHABs? Highlight any differences between the U.S. and elsewhere in the world.

The terms Blue-greens, blue-green algae, *Myxophyceae*, *Cyanophyceae*, *Cyanophyta* and Cyanobacteria all refer to a group of freshwater cyanobacteria, found in a diverse range of water bodies in the United States (U.S.); most are photosynthetic autotrophs that share some properties with algae (i.e. they possess *chlorophyll-a* and liberate oxygen during photosynthesis). Cyanobacteria are widespread and found in a diverse range of environments, including soils, seawater and, most notably, freshwater environments. Some environmental conditions that can promote growth include sunlight, warm weather, low turbulence and high nutrient levels. Some taxa of cyanobacteria can release significant quantities of toxins (cyanotoxins) into water bodies as part of their normal lifecycle during or immediately following a planktonic (surface) bloom or benthic bloom. As a water basin ages, a condition called eutrophication occurs, primarily as a result of an increase in nutrients, in biological activity (productivity), and in sediments and organic matter from the watershed that fill the water basin. It is now accepted that cultural eutrophication--eutrophication from human activity--is a significant factor in the aging process of the world's water bodies. Some lakes are naturally eutrophic, but in many others cultural eutrophication has accelerated the aging process to such an extent that the term hypereutrophy (extreme eutrophy) has become synonymous with the affect of human activities on the aging process of water bodies. Human activities, include the discharge of wastewater (agricultural, municipal and domestic), impounding of rivers, off-river storage impoundments and other activities that alter the normal course of natural waters.

The occurrence and control of cyanobacteria and their toxins in water resources can result in animal and human exposure and cause adverse health effects. It is widely believed that the incidence of bloom formation and the threat posed by toxins has increased world-wide during the past 30 years as a result of cultural eutrophication. Health problems attributed to the presence of such toxins in drinking water have been reported world-wide. Cyanobacterial blooms and cyanotoxins been linked to the deaths of wild and domestic animals all over the world. They constitute a health-risk for human beings worldwide via recreational or drinking water through the production of a range of hepatotoxins and neurotoxins.

The relationship between the presence of cyanotoxins and human illness is unclear, and for studies to better identify and assess human health impacts have been identified in some reports are needed. The level of cyanotoxin exposure through consumption of drinking water remains largely unknown due to a lack of studies that report cyanotoxin levels in drinking water. The types of published surveys for cyanobacteria and cyanotoxins are primarily ecological and biogeographical. Early surveys in a number of countries involved toxicity testing of bloom or scum samples by mouse bioassay. More recent surveys have employed more sensitive and definitive characterization of the toxins by chromatographic or immunological methods.

Two main factors appear to affect the potential for exposure to undesirable levels of cyanobacteria: (Anderson et al. 1993) the heterogeneity of cyanobacterial population distributions and (Ramsdell et al. (eds.) 2005) the human activities being undertaken. This heterogeneous distribution is a consequence of a number of factors:

- colonial growth patterns where "bloom" type growth can result in large colonies or filaments being formed
- buoyancy regulation that promotes aggregation of cells and colonies in the upper layers of the water column
- scum formation by the trapping of buoyant colonies on the water surface, initially, at least, by surface tension
- wind-driven accumulations and distribution which can result in the beaching of scums but can also be responsible for the mixing of cyanobacteria and for the establishment of more homogenous distributions

Since human activity is the primary cause of cultural eutrophication, the patterns of CyanoHAB occurrence and risks are basically the same throughout the world. Climate, topography and degree of cultural eutrophication have led to some geographical and cultural differences in type and degree of toxic occurrences. In North America, the major effect from CyanoHABs has been:

1. The loss of wild and domestic animals in man-made and natural water bodies
2. Domestic animal and human toxicity in recreational waters
3. Effects on aquacultured species
4. Increased expense for water treatment facilities to remove toxigenic cells and monitor for cyanotoxins.

This pattern has been seen worldwide, especially those warmer climes that have a higher degree of eutrophication, a longer bloom growing season; and tend to experience more problems with direct human exposure to CyanoHAbs (Chorus and Bartram (eds.) 1999, Codd et al. 2006, Yoo et al. 1995).

Charge 2

If possible, describe occurrences in freshwater, drinking water, recreational water, estuaries and marine water, and impacts on health and/or ecosystem viability in the U.S. and elsewhere in the world.

An individual description of each CyanoHAB event is beyond the scope of this expanded summary. A chronological listing of North American CyanoHAB events is given in Appendix A located in the Conclusion and Summary section of this paper. In addition, Fig. 1 shows those countries where cyanobacteria waterblooms producing cyanotoxins have occurred. Fig. 2 shows occurrence by U.S. EPA region, and Fig. 3 shows the North American occurrence of CyanoHAB events. Occurrences are based upon published reports of CyanoHAB events in "CyanoHAB Search," a reference data of all known CyanoHAB publications (Carmichael 2004).

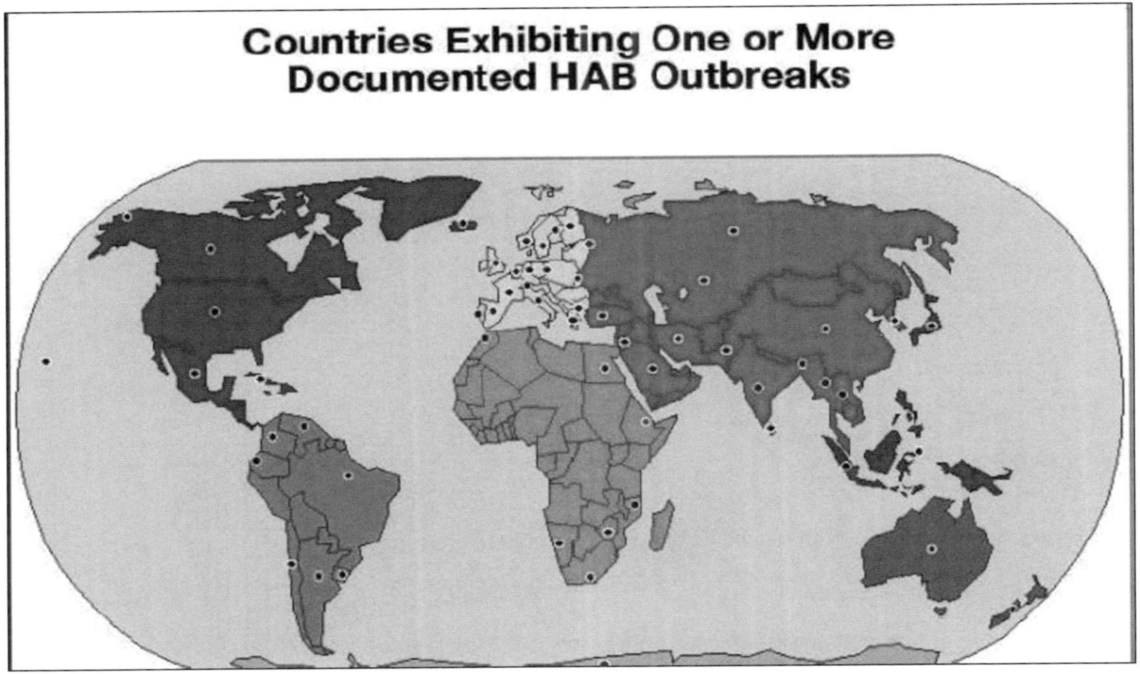

Fig. 1. Countries where cyanobacteria waterblooms producing cyanotoxins have been documented. Dots do not indicate number of cyanoHAB occurrences. In most cases there have been multiple occurrences but not every occurrence has had a poisoning event documented.

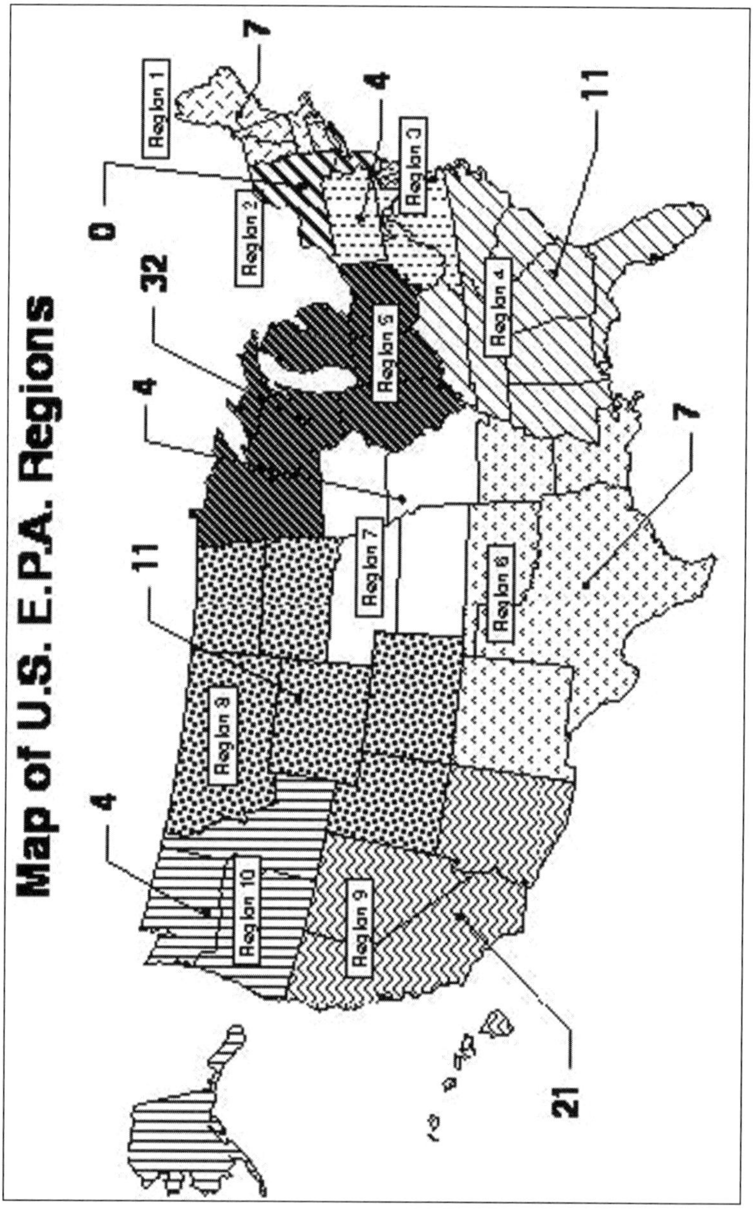

Fig. 2. Number of reported CyanoHAB events for each U.S. EPA region. Numbers in bold represent the number of articles published in the CyanoHAB database for each EPA Region.

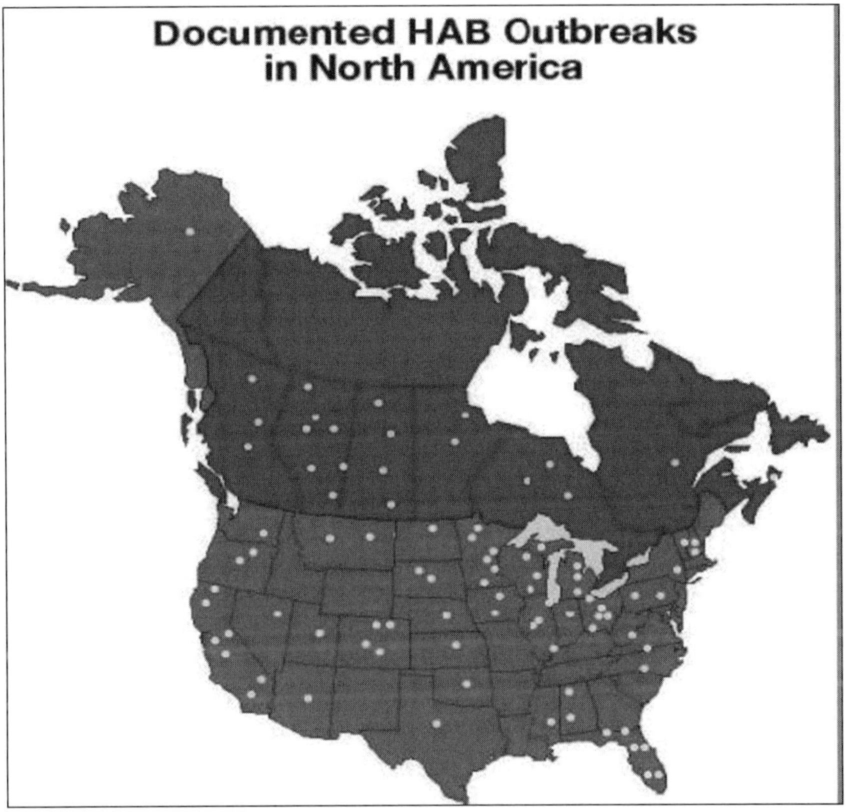

Fig. 3. North American frequency and approximate location of States and Provences reporting a CyanoHAB event.

Charge 3

If possible, address how common these occurrences are in the U.S. and elsewherein the world. Are they frequent, or rare, do they repeat at a given location, or occur sporadically at different locations?

An accurate statement of how common CyanoHAB events are is beyond the degree of understanding we currently have. A safe statement would be that CyanoHABs are an infrequent but repeated occurrence in all areas of the U.S. and around the world. They are seasonal occuring most often during the warmer, dryer times of the year. Frequency and intensity of the CyanoHAB waterbloom depends a lot on the conditions which selected for the waterbloom-i.e. nutrient status for the water body, use patterns, climate and degree of animal and human contact with the toxic waterbloom. (Carmichael and Stukenberg 2006, Carmichael 2001, Chorus and Bartram (eds.) 1999).

A summary of CyanoHABs for the U.S. is excerpted below from Carmichael and Stukenberg (2006). The earliest reference to toxic cyanobacteria was in the late 1800's from Minnesota (Arthur 1883). The earliest documented investigation in the U.S., into the poisonous potential of blue-green algae recorded a 1925 outbreak in Big Stone Lake in South Dakota when a farmer lost 127 hogs and 4 cows after they drank from lake water. The livestock deaths were attributed to algae poisoning (Wilmot Enterprise 1925).

The first described instance of *human* illness due to algal toxins occurred in Charleston, West Virginia, in 1931 (Tisdale 1931). A massive *Microcystis* bloom in the Ohio and Potomac Rivers caused intestinal illness in an estimated 5,000 to 8,000 people. According to the article, though the drinking water taken from these rivers was treated by precipitation, filtration and chlorination, these treatments were not sufficient to remove the toxins. Very few actual algal outbreaks were reported during the next two decades

The 1950s showed a marked increase in the number of reported blue-green algae studies in the United States, publishing a total of nine articles including one outbreak case concerning some domestic animal deaths from algal poisoning in Illinois (Scott 1955). Again, there was a decrease in reported cases published from the 1960s through the 1980s. In the 1990's, published accounts of algal outbreaks jumped to 19 in the 1990s. It should be acknowledged that there is a potential reporting artifact in these statistics, as reporting such events in the United States has always been voluntary.

Since 1971, however, in a USEPA and CDC cooperative effort, surveillance data from waterborne outbreaks have been more consistently compiled and more conclusively upheld as cyanobacteria-related. On August 25th, 1975, for example, a gastrointestinal outbreak occurred in Sewickley, Pennsylvania. Though the CDC's epidemiological study ruled out blue-green algae as the contaminant at the time, a subsequent analysis posited *Schizothrix* and *Calcicola* as causative agents (Lippy and Erb 1976).

In the survey *Outbreaks of Waterborne Disease in the United States: 1989-90* one "cyanobacteria-like body" outbreak was reported out of 26 total outbreaks due to ingestion of water intended for drinking. Protozoologists later confirmed this case was actually caused by the protozoan *Cyc-*

lospora and not a cyanobacterium. In the 1990s, 19 individual studies were published on U.S. cyanobacterial cases, in Alabama, Arizona, Arkansas, Colorado, Florida, Illinois, Kansas, Michigan, Mississippi, Nevada, Ohio, Oklahoma, Vermont, Washington and Wisconsin.

Clearly, through not only increased occurrence, but also through a combination of increased awareness, vigilance and reporting, more and more incidents of cyanobacterial outbreaks have come to be published. More than fifteen times as many articles on individual United States cyanobacterial outbreaks were published in the 1900s as in the 1800s; as many in the 1990s as in the 2 decades that preceded it combined; and already fifteen articles on individual outbreaks of cyanobacteria in the United States have been published since between 2000 and 2004. Since the number of articles in the entire CyanoHAB Search© database (Carmichael 2004) has increased more than 3 fold in the last ten years, it is likely that the subset of individual outbreak incidents reported will increase proportionally over the next ten years, not only in the United States, but worldwide.

Charge 4

Over the last 10-20 years, has the incidence of cyanobacterial HABs increased,remained constant, or decreased on spatial and temporal scales in the U.S. and elsewhere in the world?

Based upon the number of published papers and CyanoHAB event reports their incidence has increased dramatically since the 1960's. Fig. 4 shows this dramatic trend for all countries based upon published reports. The U.S. pattern is very much the same as the rest of the world. In all areas the increase in reported CyanoHAB events coincides with the increased realization that eutrophication of natural waters was a significant factor in decreased water quality around the world.

Charge 5

Are HABs of cyanotoxin-producing cyanobacteria a regional or worldwide problem? Identify areas in the U.S. and the rest of the world that appear to have the highest incidence of toxigenic HABs.

CyanoHABs occur worldwide and therefore should be considered a worldwide problem. Based upon a 2004 literature search (Carmichael 2004) it is possible to make some general conclusions about the occurrence of CyanoHABs. The reference is called "CyanoHAB Search" and it evolved from over 30 years of studies of toxic blue-green algae, and contains 3,063 references citing 705 journals written by 4,687 authors and editors. Figs. 4 and 5 detail some the findings from a search of references in this database.

Highest occurrences of CyanoHABs in the U.S. seem to coincide with two main factors:

1. areas of highest eutrophication coupled with a climate most favorable to extended waterblooms i.e. Florida
2. areas of driest summer conditions that favor longer retention times within the water body coupled with eutrophication i.e. southwest and west.

Other factors that favor a CyanoHAB event in a given area include mainly local weather and water use patterns along with eutrophication.

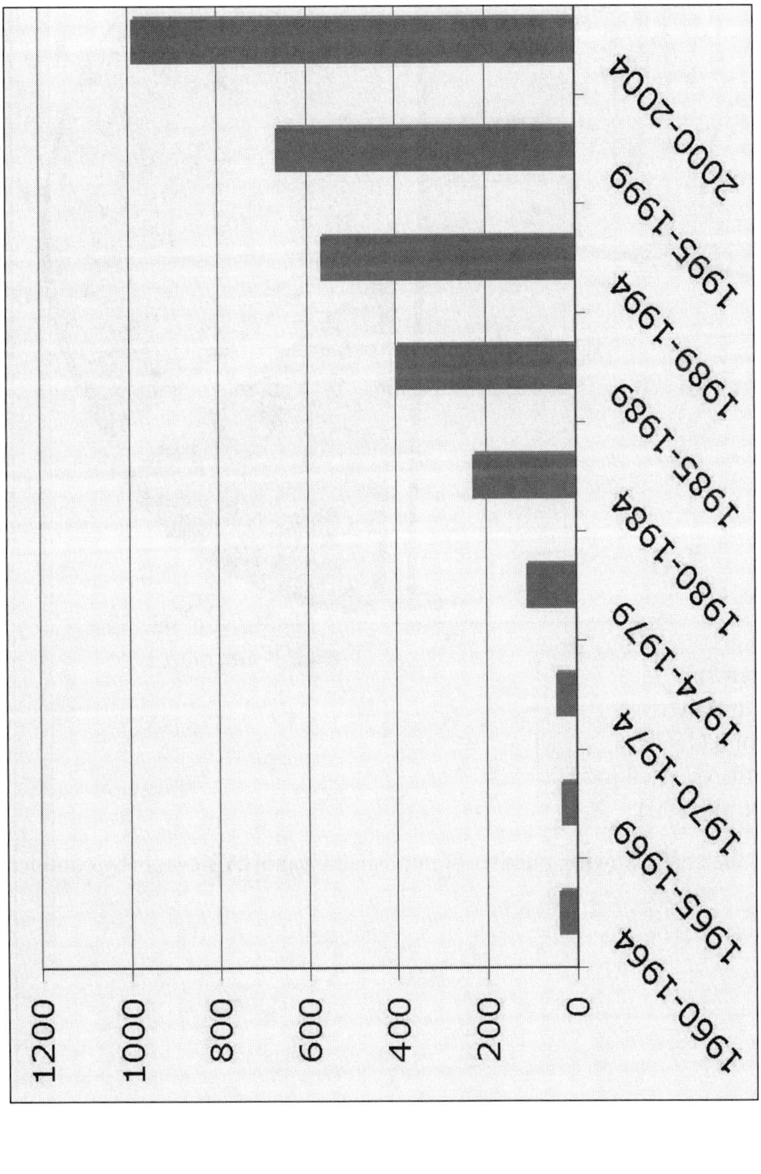

Fig. 4. Histogram of published CyanoHAB events by date.

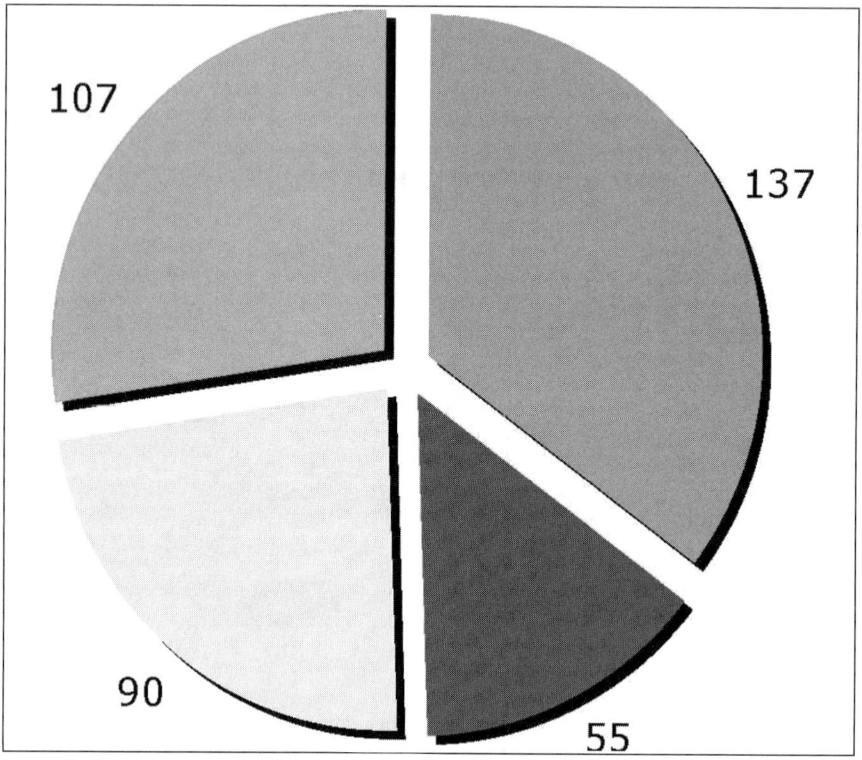

Legend:
North America (107)
Europe (137)
South America (55)
Australia (90)

Fig. 5. Pie chart showing number of published CyanoHAB events by continent.

Charge 6

Do cyanobacterial HABs that present a health risk occur often enough to warrant regulation, or are general guidelines or advisories on an as-needed basis adequate?

Animal health risks from CyanoHABs are much easier to define than risks for humans as animals are more likely to come in contact with cyanotoxin-contaminated water. Action levels for CyanoHABs have been defined in Chorus and Bartram (1999), but these levels certainly need reassessment as more information becomes available. Nonetheless they are a good starting point for the U.S. to begin assessing risk to humans in the U.S. Human exposures range from accidental ingestion of water from recreational activities, ingestion of cyanotoxin-contaminated drinking water, low level consumption of cyanotoxins in algal nutritional products and from water used in medical treatments (i.e. medical dialysis). At this time insufficient information is available on exposure risk for the various cyanotoxins to warrant regulation. The number and type of documented exposures however do warrant guidelines and advisories for all the major groups of cyanotoxins identified in U.S. waters, (*i.e.* microcystins, cylindrospermopsins, anatoxins and saxitoxins). Table 1 is a summary of known human exposures from CyanoHABs in U.S. waters (Carmichael and Stukenberg 2006, Codd et al. 2006, Yoo et al. 1995), while Appendix A is a chronological list of published articles on CyanoHAB outbreaks in the U.S.

Table 1. Acute intoxications of humans from cyanobacteria U.S. and Canada

	U.S.
1931	A massive Microcystis bloom in the Ohio and Potomac Rivers caused illness in 5,000 to 8,000 persons whose drinking water was taken from these rivers. Drinking water treatment by precipitation, filtration and chlorination was not sufficient to remove the toxins.
1968	Numerous cased of gastrointestinal illness after exposure to mass developments of cyanobacteria in drinking water were compiled by Schwimmer and Schwimmer.
1975	Endotoxic shock of 23 dialysis patients in Washington, D.C. is attributed to a cyanobacterial bloom in a drinking water reservoir.
1975	"Gastrointestinal illness at Sewickley, Pa, Cyanobacterial waterborne disease outbreak which struck 62 percent of the population served by the Sewickley water utility.
1980	Water-associated human illness in northeast Pennsylvania and its suspected association with blue-green algae blooms.
1992	Carmichael unpublished - case studies on nausea, vomiting, diarrhea, fever and eye, ear and throat infections after exposure to mass developments of cyanobacteria in drinking water.
	Canada
1959	In spite of livestock deaths and warnings against recreational use, people did swim in a lake infested with cyanobacteria. Thirteen persons became ill (headaches, nausea, muscular pains, painful diarrhea). In the excreta of one patient—a medical doctor who had accidentally ingested 300mL of water—numerous cells of Microcystis spp. And some tricomes of Anabaena circinalis could be clearly identified.

Conclusion and Summary

CyanoHABs are an intermittent but repeated occurrence throughout the world. They have been documented to impact water quality through production of compounds affecting taste and odor. Selected species within about 40 genera can produce potent toxins that can cause chronic, acute and acute lethal poisonings to wild and domestic animals and humans. Direct and indirect evidence indicates that the incidence, duration and intensity of CyanoHABs is increasing throughout the world, due largely to a decline in the quality of water supplies. While significant effort has gone into defining the chemistry and toxicology of cyanotoxins, a worldwide effort from scientists and environmental agencies to map their occurrence and environmental health risk is needed. An unprioritized listing for those efforts is suggested below (Ramsdell et al. (eds.) 2005):

1. Improved Ability to Detect HAB Species and Analyze CyanoHAB Toxins.
2. Improved Capability for Monitoring and Forecasting CyanoHABs in a Cost Effective and Timely Manner.
3. Improved Protection of Human Health.
4. Improved Protection of Endangered Species and Improved Ecological Health.
5. Improved Prevention and Mitigation Strategies.
6. Improved Economic Cost Estimates of CyanoHAB Events.
7. Improved Economics for Aquaculture and Shellfish Safety
8. An Educated and Informed Public.

References

Anderson DM, Galloway SB, Joseph JD (1993) Marine Biotoxins and Harmful Algae: A National Plan. Woods Hole Oceanographic Institution-Technical Report. WHOI-93-02.

Arthur JC (1883) Some algae of Minnesota supposed to be poisonous. Bull. Minn. Acad. Sci. 2: 1-12.

Carmichael WW (2004) CyanoHAB Search: A List of Toxic Cyanobacteria References, USEPAWebsite: http://www.epa.gov/safewater/standard/ucmr/main.html#meet.

Carmichael WW (2001) Assessment of Blue-Green Algal Toxins in Raw and Finished Drinking Water. Denver, Colorado, AWWA Research Foundation and American Water Works Assoc.

Carmichael WW, Stukenberg M (2006) North American CyanoHABs. In Codd GA, Azevedo SMFO, Bagchi SN, Burch MD, Carmichael WW, Harding WR, Kaya K, Utkilen HC (2005) CYANONET, a Global Network for Cyanobacterial Bloom and Toxin Risk Management: Initial Situation Assessment and Recommendations. IHP-VI Technical Documents in Hydrobiology No. 76. UNESCO, Paris.

Chorus I, Bartram J (eds) (1999) Toxic Cyanobacteria in Water: A Guide to Their Public Health Consequences, Monitoring and Management. World Health Organization, E&FN Spon, Routledge, London.

Codd GA, Azevedo SMFO, Bagchi SN, Burch MD, Carmichael WW, Harding WR, Kaya K, Utkilen HC (2006) CYANONET, a Global Network for Cyanobacterial Bloom and Toxin Risk Management: Initial Situation Assessment and Recommendations. IHP-VI Technical Documents in Hydrobiology No. 76. UNESCO, Paris.

Lippy EC, Erb J (1976) "Gastrointestinal illness at Sewickley, Pa." J Am Water Works Assoc 68: 606-610.

Ramsdell JS, Anderson DM, Glibert PM (Eds) HARRNESS (2005) Harmful Algal Research and Response: A National Environmental Science Strategy 2005-2015, Ecological Society of America, Washington DC, 82 pp.

Scott RM (1955) Domestic animal deaths attributed to algal toxins. *Case History Report from Illinois State Department of Public Health*, Illinois State Department of Public Health.

Tisdale ES (1931) "Epidemic of intestinal disorders in Charleston, W. Va., occurring simultaneously with unprecedented water supply conditions." Am. J. Pub. Health 21: 198-200.

Wilmot Enterprise (1925) One hundred twenty seven hogs, 4 cows die after drinking water from (Big Stone) lake, stock stricken, last Saturday, all die in short time, lake water sent in for analysis. *Wilmot Enterprise*. Wilmot, South Dakota.

Yoo RS, Carmichael WW, et al. (1995) Cyanobacterial (Blue-Green Algal) Toxins: A Resource Guide. Denver, Colorado. AWWA Research Foundation and AWWA Association.

Chapter 4 Appendix A

Chronology of Published Articles on Cyanobacterial Outbreaks in the U.S. 1883-2003 (Carmichael and Stukenberg 2006)

Date	Article Title
1883	Some algae of Minnesota supposed to be poisonous.
1887	Some algae of Minnesota supposed to be poisonous.
1887	Second report on some algae of Minnesota supposed to be poisonous
1887	Investigation of supposed poisonous vegetation in the waters of some of the lakes of Minnesota.
1903	Observations upon some algae which cause water bloom(Fergus Falls, Minnesota)
1925	Farmer tells some news (on stock poisoning in Big Stone Lake) (South Dakota)
1925	One hundred twenty seven hogs, 4 cows die after drinking water from (Big Stone) lake, stock stricken, last Saturday, all die in short time, lake water sent in for analysis. (South Dakota)
1927	Plants of Michigan poisonous to livestock.
1931	Epidemic of intestinal disorders in Charleston, W. Va., occurring simultaneously with unprecedented water supply conditions.
1934	Waterbloom as a cause of poisoning in domestic animals (Minnesota)
1939	Toxic algae in Colorado.
1940	Toxic algae in Colorado.
1943	Sheep poisoning by algae(Montana)
1947	Waterbloom as a cause of poisoning in livestock in North Dakota.
1948	A heavy mortality of fishes resulting from the decomposition of algae in the Yahara River, Wisconsin
1952	Illustrations of fresh water algae toxic to animals. Cincinnati, Ohio.
1953	Toxic algae in Iowa lakes.
1953	Rice fields study report: blue-green algae—a possible anti-mosquito measure for rice fields (California)
1954	Blue-green algae control at Storm Lake. (Iowa)
1955	Further studies during 1954 on blue-green algae—a possible anti-mosquito measure for rice fields (California)
1955	Domestic animal deaths attributed to algal toxins. (Illinois)
1956	Present knowledge concerning the relationship of blue-green algae and mosquitoes in California rice fields.
1959	A dermatitis-producing alga in Hawaii.
1959	Dermatitis escharotica caused by a marine alga.(Hawaii)
1960	Water poisoning — a study of poisonous algae blooms in Minnesota.
1960	Algae and other interference organisms in the waters of the South-Central United States
1961	Klamath Lake, an instance of natural enrichment. (Oregon)
1964	Blue-greens. Waterfowl Tomorrow. (Minnesota)
1966	Toxicity of a Microcystis waterbloom from an Ohio pond.
1971	Dermatitis-producing alga Lyngbya majuscula Gomont in Hawaii. I. Isolation and chemical characterization of the toxic factor.
1971	Dermatitis-producing alga Lyngbya majuscula Gomont in Hawaii. II. Biological properties of the toxic factor.

Date	Article Title
1975	Pyrogenic reactions during hemodialysis caused by extramural endotoxin. (Washington, D.C.)
1976	Gastrointestinal illness at Sewickley, Pa.
1977	Toxic blooms of blue-green algae. (New Hampshire)
1977	Are algae toxic to honey bees? (Arizona)
1978	Dermatitis from purified sea algae toxin (debromoaplysiatoxin) (Hawaii)
1979	Lytic organisms and photooxodative effects: influence on blue-green algae (cyanobacteria) in Lake Mendota, Wisconsin.
1980	Blue-green algae and selection in rotifer populations (Florida)
1981	A toxic bloom of Anabaena flos-aquae in Hebgen Reservoir Montana in 1977.
1981	Studies on aphantoxin from Aphanizomenon flos-aquae in New Hampshire.
1981	Water-associated human illness in northeast Pennsylvania and its suspected association with blue-green algae blooms.
1982	Seaweed itch on Windward Oahu.
1984	Toxic algae. Montana Water Quality
1984	Antineoplastic evaluation of marine algal extracts (Hawaii)
1986	Toxicity of a clonal isolate of the cyanobacterium (blue-green alga) Microcystis aeruginosa from Lake Erie. (Ohio)
1987	Blue green algae (Microcystis aeruginosa) hepatotoxicosis in cattle (Illinois)
1988	Modeling blue green algal blooms in the lower Neuse River.(North Carolina)
1988	Anticholinesterase poisonings in dogs from a cyanobacterial (blue-green algae) bloom dominated by Anabaena flos-aquae (South Dakota)
1989	Consistent inhibition of peripheral cholinesterases by neurotoxins from the freshwater cyanobacterium Anabaena flos-aquae: studies of ducks, swine, mice and a steer (EPA Region 5)
1990	Isolation, characterization and detection of cyanobacteria (blue-green algae) toxins from freshwater supplies.(Ohio)
1990	Blue-green algae toxicosis in Oklahoma.
1992	Identification of 12 hepatotoxins from a Homer Lake bloom of the cyanobacteria Microcystis aeruginosa, Microcystis viridis, and Microcystis wesenbergii: nine new microcystins.(Illinois)
1992	Neurotoxic Lyngbya wollei in Guntersville Reservoir, Alabama.
1992	Outbreaks of waterborne disease in the United States: 1989-90
1993	Chemical study of the hepatotoxins from Microcystis aeruginosa collected in California.
1993	Toxicosis due to microcystin hepatotoxins in three Holstein heifers. (Michigan)
1994	Algal toxins in drinking water? Research in Wisconsin.
1995	Cascading disturbances in Florida Bay, USA: cyanobacteria blooms, sponge mortality and implications for juvenile spiny lobsters, Panulirus argus.

Date	Article Title
1995	Seven more microcystins from Homer Lake cells: application of the general method for structure assignment of peptides containing a-b-dehydroamino acid unit(s). (Illinois)
1996	Assessment of blue-green algal toxins in Kansas
1996	Aplysiatoxin and debromoaplysiatoxin as the causative agents of a red alga: Gracilaria coronopifolia poisoning in Hawaii.
1997	Mechanisms of ecosystem change: the role of zebra mussels in Saginaw Bay.(Michigan)
1997	Evidence for paralytic shellfish poisons in the freshwater cyanobacterium Lyngbya wollei (Farlow ex Gomont) comb. nov.(Alabama)
1997	Recent appearance of Cylindrospermopsis (cyanobacteria) in five hypereutrophic Florida lakes.
1997	Occurrence of the black band disease cyanobacterium on healthy coral of the Florida Keys.
1998	Blue-green algae toxicosis in cattle. (Colorado)
1999	Effect of surface water on desert Bighorn sheep in the Cabeza-Prieta National Wildlife Refuge, Southwestern Arizona.
1999	Spread of toxic algae linked to zebra mussels. (Ohio)
2000	New malyngamides from the Hawaiian cyanobacterium Lyngbya majuscula.
2000	Harvesting of Aphanizomenon flos-aquae Ralfs ex. Born. and Flah. var. flosaquae (cyanobacteria) from Klamath Lake for human dietary use. (Oregon)
2000	Desert Bighorn sheep mortality due to presumptive type-C botulism in California.
2001	Microcystin algal toxins in source and finished drinking water. (Wisconsin)
2001	Confirmation of catfish, Ictalurus punctatus (Rafinesque) mortality from Microcystis toxins (South Central U.S.)
2001	Assessment of Blue-Green Algal Toxins in Raw and Finished Drinking Water.Denver, Colorado
2001	Zebra mussel (Driessena polymorpha) selective filtration promoted toxic Microcystis blooms in Saginaw Bay (Lake Huron) and Lake Erie
2002	Possible importance of algal toxins in the Salton Sea, California.
2002	Clinical and necropsy findings associated with increased mortality among American alligators of Lake Griffin, Florida.
2002	Removal of pathogens, surrogates, indicators and toxins using riverbank filtration (California)
2002	Dreissenid mussels increase exposure of benthic and pelagic organisms to toxic microcystins. (Michigan)
2003	Natural algacides for the control of cyanobacterial-related off-flavor in catfish aquaquaculture (South Central U.S)
2003	A synoptic survey of musty/muddy odor metabolites and microcystin toxin occurrence and concentration in southeastern USA channel catfish (Ictalurus punctatus Rafinesque) production ponds

Date	Article Title
2003	Variants of microcystin in south-eastern USA channel catfish (Ictalurus punctatus Rafinesque) production ponds
2003	Cyanobacterial toxicity and migration in a mesotrophic lake in western Washington,

Chapter 5: Toxic Cyanobacteria in Florida Waters

John Burns

Florida Lake Management Society, 506 Emmett Street, Palatka, Florida 32177

Cyanobacteria represent a common component of algal assemblages in Florida's lakes, springs, estuaries, and other marine environments. The frequency of cyanobacterial bloom reports has recently become more common and algal toxin data are just beginning to be collected with the advent of monitoring efforts and the availability of laboratories in the state capable of conducting algal toxin analyses.

Given Florida's growing population, subtropical environment, and the eutrophic nature of its lakes, rivers, and estuaries, it is not surprising that cyanobacteria are a common feature of aquatic ecosystems. However, cyanobacterial blooms may have always been common in some regions that include water bodies influenced by underlying phosphate deposits. Phosphate deposits cover much of peninsular Florida and are mined throughout the west–central part of the state (FIPR 2006; FDEP 2006).

Although there is little doubt that the phenomenon of cyanobacterial blooms predates human development in Florida, the recent acceleration in population growth and associated changes to surrounding landscapes has contributed to the increased frequency, duration, and intensity of cyanobacterial blooms and precipitated public concern over their possible harmful effects to aquatic ecosystems and human health. Toxic cyanobacterial blooms in Florida waters represent a major threat to water quality, ecosystem stability, surface drinking water supplies, and public health. Many of Florida's largest and most important lakes, rivers, and estuaries are increasingly impacted by toxic cyanobacterial blooms, including the St. Johns River (SJRWMD 2001), Lake Okeechobee (Havens et al. 1996), St. Lucie River (FFWCC 2005), Caloosahatchee River (Gilbert et al. 2006), and the Harris Chain of Lakes (LCWA, personal communication). Cyanobacteria, such as *Lyngbya*, are also common in many of Florida's springs and along coastal reef tracts where blooms form extensive mats and blanket corals and submersed aquatic vegetation. Shifts in phytoplankton composition to potentially toxic cyanobacteria taxa have also occurred in some eutrophic Florida lakes (Chapman and Schelske 1997). Historically, atopic sensitivity to cyanobacteria has been reported following exposure to algae in lakes. For example, Heise (1949) found blue–green algae responsible for seasonal rhinitis following exposure to algae while swimming in lakes. Human sensitivity to cyanobacteria may be related to a hereditary predisposition toward developing certain hypersensitivity reactions when exposed to specific antigens.

There is a growing need in Florida and the other areas of the United States to define the specific relationships among freshwater cyanobacterial blooms, the production of secondary blooms in estuaries and marine coastal systems, and potential ecological and human health consequences associated with prolonged toxic bloom events. Moreover, because Florida lacks sufficient research and biomonitoring programs for toxic cyanobacteria, and little coordination exists between surface water managers and public health officials, relationships between toxic cyanobacteria and their environmental consequences remain largely in the realm of incidental observation and speculation.

Historically, reports of cyanobacterial blooms in Florida are often associated with fish kill events in freshwater and estuarine systems. These reports often lacked specific algal identification information and toxin data was not collected. The earliest known record of algal toxins in Florida was reported following the death of cattle near Lake Okeechobee (Carmichael 1992). Toxic cyanobacterial blooms in Florida were first recorded by Wayne Carmichael in Lake Okeechobee (1987, 1989) and Lake Istokpoga (1988). Dead cattle, signs of poisoning in laboratory mice, and contact irritation were found associated with Anabaena and Microcystis blooms. The toxin microcystin and an unidentified neurotoxin were attributed to the toxic effects found in Florida lake samples. Although toxic cyanobacterial blooms had become a major concern throughout the world, and the World Health Organization had set provisional guidelines for the consumption of microcystin– LR (WHO 1998), little information on toxic cyanobacteria in Florida waters had been published since those first toxic events identified in 1992.

No Florida survey of cyanobacterial blooms or cyanobacterial toxins had been conducted until the advent of the creation of the Florida Harmful Bloom Task Force in 1999. The Florida Legislature created the Harmful Algal Bloom Task Force in Chapter 370 F. S. "The Harmful–Algal–Bloom Task Force shall: (a) Review the status and adequacy of information for monitoring physical, chemical, biological, economic, and public health factors affecting harmful algal blooms in Florida; (b) Develop research and monitoring priorities for harmful algal blooms in Florida, including detection, prediction, mitigation, and control; (c) Develop recommendations that can be implemented by state and local governments to develop a response plan and to predict mitigate, and control the effects of harmful algal blooms; and (d) Make recommendations to the Florida Marine Research Institute by October 1, 1999, for research detection monitoring, prediction, mitigation, and control of harmful algal blooms in Florida." In March 1999, a technical advisory group to the Harmful Algal Bloom Task Force produced a report that identified known existing harmful algal species in the state and made recommendations for research and monitoring priorities (Steidinger et al. 1999). The report provided the following recommendations to the Task Force regarding toxic blue–green algae: 1) Determine distribution of toxic and non–toxic strains in Florida waters; 2) Develop epidemiological studies to determine what public health threats are involved; 3)Develop economic impact studies to properly evaluate losses by locale or industry; and 4) Determine the roles of nutrient enrichment and managed freshwater flow in bloom development; 5) Investigate the applicability and efficacy of control and mitigation methods.

Following the above recommendations, the Task Force funded an investigation of the occurrence and distribution of cyanobacteria and their toxins in 1999. The goals of the state–wide survey for the cyanobacteria survey included: 1) identification of toxic cyanobacteria throughout Florida; 2) identify and characterize level of cyanotoxins; 3) assist with the development of analytical capability for algal toxin analysis within the Florida Department of Health laboratory in Jacksonville, Florida; and evaluate the presence of cyanobacterial toxins at water treatment plants that utilize surface water resources.

The FHABTF, Florida Marine Research Institute, and St. Johns River Water Management District initiated a collaborative study with the Florida Department of Health and Wright State University to identify potential cyanobacterial toxins in Florida's lakes, rivers, reservoirs, and estuaries. Samples were collected and analyzed during 1999 and then extended to better understand the potential impacts of cyanobacterial toxins detected in waters currently utilized for drinking water or identified as a potential future drinking water source.

Methods employed to identify alga taxa included microscopic examination and epifluorescence. Algal toxins were characterized and quantified by enzyme linked immunosorbant assay (ELISA), protein phosphatase inhibition assay (PPIA), HPLC–Fl, HPLC–UV, and LC/MS/MS. Mouse bioassays were used to characterize toxicity by intraperitoneal injection of freeze–dried sample extracts into ICR–Swiss male mice.

With the assistance of numerous state and local agencies in 1999, a total of 167 samples were collected throughout Florida, eighty–eight of these samples, representing 75 individual water bodies, were found to contain cyanotoxins. Approximately 80% of the samples containing cyanotoxins were found to be lethal to mice following intraperitoneal injection. Most bloom forming cyanobacteria genera were distributed throughout the state, but water bodies such as Lake Okeechobee, Harris Chain of Lakes, Lower St. Johns River, Calooshatchee River, Lake Seminole, Lake George, Crescent Lake, Doctors Lake, and the St. Lucie River were water bodies that supported extensive cyanobacterial biomass. Seven genera of cyanobacteria were identified from water samples collected. *Microcystis* (43.1%), *Cylindrospermopsis* (39.5%), and *Anabaena* (28.7%) were observed most frequently and in greatest concentration. *Planktothrix* (13.8%), *Aphanizomenon* (7.2%), *Coelosphaerium* (3.6%) and *Lyngbya* (1.2%) were found less frequently, but at times accounted for a significant proportion of the planktonic and macroalgal species composition. *Aphanizomenon* and *Anabaenopsis* were also found consistently during the 2000 survey. Cyanobacterial blooms were common throughout the state, some of which form long lasting or continuous blooms in eutrophic and hypereutrophic systems. Many of the water bodies affected by cyanobacterial blooms were identified by water management agencies as areas of current concern or were being addressed by ongoing or proposed restoration efforts.

Algal toxins identified from bloom material during the study included hepatotoxic microcystins, neurotoxic anatoxin–A, and the cytotoxic alkaloid cylindrospermopsin. Subsequent identification of lyngbyatoxin–A and debromoaplysiatoxin were found associated with *Lyngbya* blooms collected from Florida springs and coastal embayments. Microcystins were the most commonly found toxins in Florida waters, occurring in all 87 samples analyzed during 1999. During the 2000 survey, microcystins were detected in pre– and post–treated drinking water. Finished water microcystin concentration ranged from below detection levels to 12.5 $\mu g\ L^{-1}$. Microcystins are considered the most frequently found cyanobacterial toxins around the world. Over 60 structural variants of this cyclic peptide have been reported, causing considerable concern due to their high chemical stability, high water solubility, environmental persistence and exposure to humans in surface water bodies. The World Health Organization has set a provisional consumption limit of 1 $\mu g\ L^{-1}$ for microcystin–LR (WHO 1998). The mammalian toxicity of microcystin occurs by active transport across membrane boundaries and is mediated through binding to protein phosphatases (Runnegar et al. 1991; Falconer et al. 1992). Analysis of protein phosphatase inhibition activity, an index of microcystin bioactivity, was found to be positive in 44 (69%) of 64 Florida samples tested. There is limited evidence of tumor promotion (Sato et al. 1984; Falconer 1991, Nishiwaki–Matsushima et al. 1992, Wang and Zhu 1996) and clastogenic dose–related increases in chromosomal breakage by microcystin (Repavich et al. 1990), but no mutagenic evidence has been reported (Runnegar and Falconer 1982, Repavich et al. 1990).

Anatoxin–a is a potent neurotoxic alkaloid that has been frequently implicated in animal and wildfowl poisonings (Ressom et al. 1994). It is considered a nicotinic agonist that binds to neuronal nicotinic acetylcholine receptors which leads to depolarization and a block of electrical transmission in the body (Soliakov et al. 1995). At sufficiently high doses (oral LD50 = >5,000 $\mu g\ kg^{-1}$ body weight), it can lead to paralysis, asphyxiation, and death (Fitzgeorge et al. 1994, Carmichael 1997). Although alternative hypotheses exist to help explain alligator mortalities in Florida lakes, unexplained bird and alligator mortality events during cyanobacterial blooms may be due to exposure to neurotoxic compounds produced by species of *Anabaena*, *Aphanizomenon*, and *Cylindrospermopsis*. Anatoxin–a was found in three finished water samples and in tissues from Blue Tilapia and one White Pelican during surveys in 2000. Anatoxin–a was found in the gut and liver of a White Pelican and in Blue Tilapia (0.51 to 43.3 $\mu g\ g^{-1}$) and in finished drinking water (below detectable limits to 8.46 $\mu g\ L^{-1}$).

All 1999 samples containing the organism *Cylindrospermopsis* were positive for the toxin cylindrospermopsin. Nine (9) finished drinking water samples collected during the 2000 survey were positive for cylindrospermopsin and ranged in concentration from 8.07 to 97.12 $\mu g\ L^{-1}$. Identification of the algal toxin cylindrospermopsin during this study represented the first record of this hepatotoxic alkaloid in North America. This toxin primarily affects the liver, but extracts given orally or injected in mice also induce pathological damage to kidneys, spleen, thymus, and heart (Hawkins et al. 1985, 1997). Cylindrospermopsin is a potential important contaminant of drinking waters in Australia, Central Europe, South America, and the United States. The toxin was identified after 138 children and 10 adults were poisoned following a *Cylindrospermopsis* bloom and copper sulfate applications in a water supply reservoir on Palm Island, Australia (Hawkins et al. 1985). Over 69% of the affected individuals required intravenous therapy for electrolyte imbalance, and the more severe cases for hypovolemic and acidotic shock (Byth 1980). The oral toxicity or lethal dose of cylindrospermopsin has been reported between 4.4 and 6.9 $mg\ kg^{-1}$ mouse body weight with death occurring 2–3 days after treatment (Humpage and Falconer 2002). In experiments where cell free extract of *Cylindrospermopsis* was administered to mice in drinking water over 90 days, no pathological symptoms were recorded up to a maximum dose of 150 mg kg^{-1} day^{-1} (Shaw et al. 2001). Humpage and Falconer (2002) suggest a Tolerable Daily Intake and Guideline Value for cylindrospermopsin in drinking water of 1 µg L^{-1}based on an oral No Observed Adverse Effect Level of 30 µg kg^{-1} day^{-1} and a Lowest Observed Adverse Effect Level of 60 µg kg^{-1} day^{-1}.

Lyngbyatoxin–a and debromoaplysiatoxin have been identified from *Lyngbya* samples collected from Florida springs and marine embayments, respectively (J. Burns, N. Osborne, and G. Shaw, unpublished data). The aplysiatoxins and lyngbyatoxins are considered dermatotoxic alkaloids, causing severe dermatitis among swimmers and other recreational users of water bodies that come into direct contact with the organism (Mynderse et al. 1977, Fujiki et al. 1982). Aplysiatoxins are lethal to mice at a minimum dose of 0.3 mg kg^{-1} (Moore 1977). Aplysiatoxins and lyngbyatoxins are also considered potent tumor promoters and protein kinase C activators (Fujiki et al. 1990). Osborne et al. (2001) reviewed the human and ecological effects of *Lyngbya majuscula* blooms and reported acute contact dermatitis in Hawaii, Japan, and Australia. One potential death due to exposure via ingestion of turtle meat containing lyngbyatoxin– a was reported by Yasumoto (2000). Severe dermatitis has also been reported in Florida following recreational activities in waters supporting *Lyngbya* blooms in Florida's springs (John Burns, Ian Stewart and G. Shaw, unpublished data). *Lyngbya* mats have been detected along coral reef tracts adjacent to the southeast coast of Florida near Dade and Broward counties. Although no toxins have been detected from the limited number of samples analyzed, thick mats of *Lyngbya* have smothered corals, causing severe damage to the reef.

To date, toxic cyanobacterial blooms continue to occur throughout Florida and no state–wide monitoring program for cyanobacteria or cyanobacterial toxins exists. No Florida guidelines for recreational exposure to toxic cyanobacteria or cyanobacterial toxins in drinking water are available. However, several independent monitoring efforts for cyanobacteria and their toxins have been initiated and the Florida Harmful Algal Bloom Task Force has helped fund the following efforts, "Cyanobacteria Automated Detection Workshop", "*Cylindrospermopsis* Culture for Production of Cylindrospermopsin", and "Cyanobacteria Public Health Issues: Education and Epidemiologic Study Monitoring and response efforts have been initiated by the St. Johns River Water Management District (Lower St. Johns River Basin), South Florida Water Management District (Lake Okeechobee), Lake County Water Authority (Harris Chain of Lakes), Florida Lake Watch, and several water utilities in the state that utilize surface waters for drinking water supply. One of the most severe blooms ever recorded in Florida occurred in the Lower St. Johns River during the summer of 2005 (Fig. 1,2,3 see Color Plate 3), extending from Lake George to the mouth of the river at Mayport, Florida. Large rafts of toxic algal scum were slowly transported north through the city of Jacksonville to the Atlantic Ocean by tide. *Microcystis* and *Cylindrospermopsis* were the dominant bloom forming species, with microcystin detected as high as ~1,400 $\mu g\ L^{-1}$ (SJRWMD 2005). One human death was reported during the bloom event following recreational contact (i.e., jet skiing) with surface algal scums. It was reported that a young female with an open leg wound contracted a lethal *Vibrio* infection following recreational contact with waters of the St. Johns River near Jacksonville, Florida. Although the *Vibrio* infection was not related to algal toxin poisoning, it is important to recognize that the presence of cyanobacterial blooms and the concentration of algal scums along shorelines may increase the likelihood of human exposure to other bacteria and pathogens that thrive in such conditions. The Florida Department of Health released a public health advisory that warned the public to refrain from use of the river during the bloom event. During the summer of 1995, toxic *Microcystis* blooms also occurred in Lake Okeechobee, Calooshatchee River, St. Lucie River, and the West Palm Beach Canal (C–51). Canal gates near the entrance to an existing water supply for a water treatment plant in south Florida were temporarily closed during the bloom event to protect the quality of existing surface water supplies.

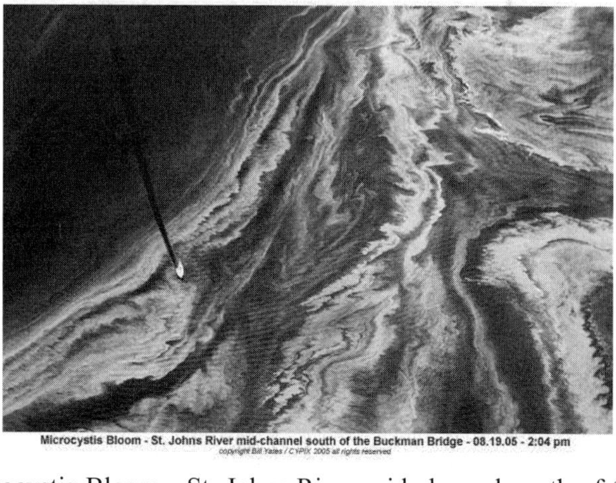

Fig. 1. Mycrocystis Bloom—St. Johns River mid-channel south of the Buckman Bridge. (See Color Plate 3).

Fig. 2. Microcystis Bloom—I295 (Buckman Bridge) over the St. Johns River. (See Color Plate 3).

Fig. 3. Microcystis Bloom – East bank of the St. Johns River – Mandarin. (See Color Plate 3).

Summary

The occurrence of toxic cyanobacterial blooms in Florida waters have become more prominent following increased growth, declining groundwater supplies, and identification of impaired surface waters as future drinking water sources. Cyanobacterial toxins have been identified in source waters used for drinking water supply and in post–treated drinking water during algal bloom events. Algal toxin concentrations in post–treated drinking water have exceeded existing and proposed World Health Organization guidelines for the oral consumption of microcystin and cylindrospermopsin. Severe dermatitis has also been reported by swimmers in Florida springs where *Lyngbya* mats have expanded.

The prevalence and toxicity of cyanobacteria should be considered when developing appropriate Total Maximum Daily Loads for impaired Florida waters that do not currently meet their designated use. It could also support further efforts to characterize potential ecological and human health risks due to toxic cyanobacterial blooms. Identification of algal toxins in finished drinking water and reports of severe skin irritation following contact with toxic cyanobacteria should be utilized for justification and implementation of increased monitoring of potentially toxic cyanobacterial blooms by surface water managers and water utilities. Epidemiological studies may also be required in Florida to assess potential human health risks due to algal toxin consumption at the tap and for those exposed to cyanotoxic blooms during recreational use of lakes, springs and rivers.

Without adequate water treatment and coordinated state–wide monitoring efforts, it is anticipated that the likelihood for human exposure to cyanobacteria and their toxins will increase as Florida becomes more dependent upon surface waters to supply a growing population and an expanding urban environment. Coordination and communication between surface water managers and public health officials at the local level will be critical to the overall protection of the environment and public health during toxic cyanobacterial bloom events.

References

Byth S (1980) Palm island mystery disease. Med. J. Aust. 2:40–42.

Carmichael WW (1997) The cyanotoxins. Adv. Botan. Res. 27:211–256.

Carmichael W (1992) A Status Report on Planktonic Cyanobacteria (blue–green algae) and their Toxins. EPA/600/R–92/079.

Chapman AD, Schelske CL (1997) Recent appearance of *Cylindrospermopsis* (Cyanobacteria) in five hypereutrophic Florida lakes. Journal of Phycology 33:191–195.

FFWCC (2005) http://www.floridaconservation.org/fishing/health.html.

Falconer IR (1991) Tumor promotion and liver injury caused by oral consumption of cyanobacteria. Environ. Toxicol. Water Qual., 6(2):177:184.

Falconer IR, Choice A, Hosja W (1992) Toxicity of edible mussels (*Mytilus edulis*) growing naturally in an estuary during a water bloom of the blue–green alga Nodularia spumigena. J. Environ. Toxicol. Water Qual. 7, 119–123.

FDEP (2006) http://www.dep.state.fl.us/water/wastewater/iw/phosphate.htm

FIPR (2006) http://www.fipr.state.fl.us/index.html

Fitzgeorge RB, Clark SA, Keevil CW (1994) Routes of intoxication. In: G.A. Codd, T. M. Jeffries, C.W. Keevil and E. Potter [Eds] *1st International Symposium on Detection Methods for Cyanobacterial (Blue–Green Algal) Toxins,* Royal Society of Chemistry, Cambridge, UK, 69–74.

Fujiki H, Suganuma M, Nakayasu M, Hoshino H, Moore RE, Sugimura T (1982) The third class of new tumor promoters, polyacetates (debromoaplysiatoxin and aplysiatoxin), can differentiate biological actions relevant to tumor promoters. Gann 73(3):495–497.

Fujiki H, Suganuma M, Suguri H, Yoshizawa S, Takagi K, Nakayasu M, Ojika M, Yamada K, Yasumoto T, Moore RE, Sugimura T (1990) New tumor promoters from marine natural products. In: S. Hall and G. Strichartz (eds.) Marine Toxins: Origin, Structure and Molecular Pharmacology, American Chemical Society, Washington D.C., pp. 232–240.

Havens KE, Aumen NG, James RT, Smith VH (1996) Rapid ecological changes in a large subtropical lake undergoing cultural eutrophication. Ambio. Stockholm, 25(3):150–155.

Hawkins PR, Runnegar MTC, A.R.B. Jackson, and I.R. Falconer. 1985. Severe hepatotoxicity caused by the tropical cyanobacterium (blue–green alga) *Cylindrospermopsis raciborskii* (Woloszynska) Seenaya and Subba Raju isolated from a domestic water supply reservoir. Appl. Environ. Microbiol. 50(5):1292–1295.

Hawkins PR, Chandrasena NR, Jones GJ, Humpage AR, Falconer IR (1997) Isolation and toxicity of *Cylindrospermopsis raciborskii* from an ornamental lake. Toxicon 35(3):341–346.

Humpage AR, Falconer IR (2002) Oral Toxicity of Cylindrospermopsin: No Observed Adverse Effect Level Determination in Male Swiss Albino Mice. Cooperative Research Center for Water Quality and Treatment, Salisbury, South Australia, Australia.

Gilbert PM, Harrison J, Heil C, Seitzinger S (2006) Escalating worldwide use of urea – a global change contributing to coastal eutrophication. Biogeochemistry, 77(3):441–463.

Moore RE (1977) Toxins from blue–green algae. Bioscien. 27:797–802.

Mynderse, J.S., R.E. Moore, M. Kashiwagi, T.R. Norton. 1977. Antileukemia activity in the Oscillatoriaceae: isolation of debromoaplysiatoxin from *Lyngbya.* Science 196:538–539.

Nishiwaki–Matsushima R, Ohta T, Nishiwaki S, Suganuma M, Kohyama K, Ishikawa T, Carmichael WW, Fujiki H (1992) Liver tumor promotion by the cyanobacterial cyclic peptide toxin microcystin LR. J. Cancer Res. Clin. Oncol. 118(6):420–424.

Osborne NT, Webb P, Shaw GR (2001) The toxins of *Lyngbya majuscula* and their human and ecological health effects. Environment International 27:381–392.

Repavich WM, Sonzogni WC, Standridge JH, Wedepohl RE, Meisner LF (1990) Cyanobacteria (blue–green algae) in Wisconsin waters: acute and chronic toxicity. Water Res. 24(2):225–231.

Ressom R, Soong FS, Fitzgerald J, Turczynowicz L, El Saadi O, Roder D, Maynard T, Falconer I (1994) Health Effects of Toxic Cyanobacteria (Blue–Green Algae). Australian National Health and Medical Research Council, Looking Glass Press.

Runnegar MTC, Falconer IR (1982) The *in vivo* and *in vitro* biological effects of the peptide hepatotoxin from the blue–green alga *Microcystis aeruginosa*. South African J. Sci. 78:363–366.

Runnegar MTC, Gerdes RG, Falconer IR (1991) The uptake of the cyanobacterial hepatotoxin microcystin by isolated rat hepatocytes. Toxicon 29(1):43–51.

SJRWMD (2000) District Water Supply Plan, Executive Summary. Special Publication SJ2000–SP1. St. Johns River Water Management District, Palatka, Florida.

SJRWMD (2001) Assessment of Cyanotoxins in Florida's Lakes, Rivers and Reservoirs. Prepared for the Florida Harmful Algal Bloom Task Force, Florida Marine Research Institute, St. Petersburg, Florida.

Sato K, Kitahara A, Satoh K, Ichikawa T, Tatematsu M, Ito N (1984) The placental form of glutathione S–transferase as a new marker protein for preneoplasia in rat chemical carcinogenesis. Gann, 75:199–202.

Shaw GR, Seawright AA, Moore MR (2001) Toxicology and human health implications of the cyanobacterial toxin cylindrospermopsin; in: Mycotoxins and Phycotoxins in Perspective at the Turn of the Millennium. Eds, W. J. Dekoe, R.A. Samson, H.P. van Egmond, J. Gilbert, M. Sabino. IUPAC & AOAC International Brazil pp. 435–443.

Soliakov L, Gallagher T, Wonnacott S (1995) Anatoxin–a–evoked [3H} dopamine release from rat striatal synaptosomes. Neuropharmacology 34(11):1535–1541.

Steidinger KA, Landsberg JH, Tomas CR, Burns JW (1999) Harmful Algal Blooms in Florida. Submitted to Florida's Harmful Algal Bloom Task Force by the Harmful Algal Bloom Task Force Technical Advisory Group. Florida Marine Research Institute, St. Petersburg, Florida.

WHO (1998) Guidelines for Drinking–water Quality. Second Edition, Addendum to Volume 2, Health Criteria and Other Supporting Information. World Health Organization, Geneva.

Wang HB, Zhu HG (1996) Promoting activities of microcystins extracted from water blooms in SHE cell transformation assay. Biomed. Environ. Sci., 9:46–51.

Yasumoto T (2000) The chemistry and biological function of natural marine toxins. Chem Rec 1(3):228–242

Chapter 6: Nebraska Experience

Walker SR[1], Lund JC[1], Schumacher DG[1], Brakhage PA[1], McManus BC[1], Miller JD[2], Augustine MM[2], Carney JJ[3], Holland RS[3], Hoagland KD[4], Holz JC[4], Barrow TM[4], Rundquist DC[4], Gitelson AA[4]

[1]Nebraska Department of Environmental Quality (NDEQ), [2]Nebraska Department of Health and Human Services System (NHHSS), [3]Nebraska Game and Parks Commission (NGPC), [4]University of Nebraska-Lincoln (UNL)

Abstract

Nebraska agencies and public health organizations collaboratively addressed cyanobacterial issues for the first time after two dogs died within hours of drinking water from a small private lake south of Omaha on May 4, 2004. A necropsy on one of the dogs revealed that the cause of death was due to ingestion of Microcystin toxins. Within two weeks after the dog deaths, state and local officials jointly developed strategies for monitoring cyanobacterial blooms and issuing public health alerts and advisories. Weekly sampling of public lakes for microcystin toxins and cyanobacteria was initiated during the week of May 17, 2004. ELISA laboratory equipment and supplies were purchased to achieve a quick turnaround time for measuring weekly lake samples for total microcystins so that public health advisories and alerts could be issued prior to each weekend's recreational activities. A conservative approach was selected to protect human health, pets, and livestock, which included collecting worst-case samples from cyanobacterial blooms; freezing and thawing of samples to lyse algal cells and release toxins prior to laboratory analysis; and using action levels of 15 ppb and 2 ppb of total microcystins, respectively, for issuing health alerts and health advisories. During 2004, five dog deaths, numerous wildlife and livestock deaths, and more than 50 accounts of human skin rashes, lesions, or gastrointestinal illnesses were reported at Nebraska lakes. Health alerts were issued for 26 lakes and health advisories for 69 lakes. Four lakes were on health alert for 12 or more weeks. The primary cyanobacterial bloom-forming genera identified in Nebraska lakes were Anabaena, Aphanizomenon, and Microcystis. Preliminary assessments of lake water quality data indicated that lower lake levels from the recent drought and low nitrogen to phosphorus ratios may have contributed, in part, to the increased numbers of cyanobacterial complaints and problems that occurred in 2004.

Background

Over the past two decades, occasional pet, livestock, and wildlife deaths, and human skin rashes and gastrointestinal illnesses have been associated with lakes and ponds in Nebraska, but rarely were cyanobacterial blooms suspected as the cause. In October 2003, a workshop taught by Dr. Russell Rhodes of Southwest Missouri State University for several Nebraska agencies, raised awareness of the frequency and magnitude of cyanobacterial problems in the United States and throughout the world. On May 4, 2004, after two dogs died within a few hours of drinking water from Buccaneer Bay Lake near Plattsmouth, Nebraska agencies began to actively address cyanobacterial issues for the first time. A water sample and a necropsy on one of the dogs revealed that the dog deaths were likely due to high levels of the cyanobacteria toxin Microcystin LR. The Microcystin LR concentration of the water was 69.4 ppb. These dog deaths were reported in the national news and investigated by the Centers for Disease Control and Prevention. During this same timeframe, three more dog deaths were reported at two other lakes with cyanobacterial blooms. Meetings were held between the Nebraska Department of Environmental Quality (NDEQ), Nebraska Health and Human Services System (NHHSS), Nebraska Game and Parks Commission (NGPC), and the University of Nebraska-Lincoln (UNL). Excellent cooperation and quick action were demonstrated by these agencies in developing joint strategies for cyanobacterial monitoring and public notification within two weeks after the dog deaths occurred at Buccaneer Bay Lake.

Methods

NDEQ purchased Enzyme-Linked Immunosorbent Assay (ELISA) laboratory test kits for analyzing the levels of total microcystins in Nebraska lakes. ELISA kits provided a low cost, semi-quantitative analytical method for measuring concentrations of total microcystins, which are the most common toxins released by cyanobacteria. In 2004, analysis of 748 samples using ELISA test kits instead of High Performance Liquid Chromatography (HPLC) or Liquid Chromatography/Mass Spectrometry (LC/MS) analyses resulted in an estimated savings of $77,000. Another advantage of analyzing water samples with ELISA kits was the quick turnaround time, which allowed weekly updates of lake microcystin conditions and public health alerts and advisories prior to each weekend's recreational activities. NDEQ initiated a weekly microcystin monitoring program on May 17, 2004, which targeted public and private lakes with known or suspected cyanobacterial problems. Citizen complaints were also important in providing information on lakes where blooms were occurring. UNL coordinated a volunteer monitoring program for private lakes, and upon request, supplied lake homeowners with a sample kit. Samples returned to UNL were analyzed under a microscope for cyanobacterial genera and relative biomass estimates. UNL referred samples with a high biomass of cyanobacterial genera to NDEQ for analysis of total microcystins. Likewise, NDEQ referred samples with high levels of total microcystins to UNL for microscopic identifications and biomass estimates. A weekly routine was established in which water samples were collected and delivered to the laboratory on Monday and Tuesday, processed using freeze-thaw methods on Wednesday, and analyzed on Thursday. Sample results were reported on Thursday, and, if necessary, warning signs were posted at lakes on Friday.

Because of its initial unfamiliarity with cyanobacterial issues, Nebraska chose to err on the side of safety by selecting conservative approaches for protecting public health, which included measuring worst-case cyanobacterial bloom conditions and human exposure risks, and using ELISA kits to analyze samples. The ELISA kits measured all microcystin congeners, not just the LR congener. Therefore, sample results measuring all microcystin congeners were compared to the World Health Organization (WHO) action level for only one of the congeners, Microcystin LR. Also, lake samples were frozen and thawed three times prior to analysis with the ELISA kits to lyse the cyanobacterial cells and release the toxins. Thus, the sample results were likely higher than the free microcystin levels in the lakes. This was done to simulate the exposure risk that might exist from ingestion of lake water and the release of toxins from cyanobacterial cells in the stomach. These conservative procedures provided a safety margin in case grab samples failed to measure the highest concentrations of total microcystins in a lake, or other cyanobacterial toxins such as anatoxins, saxitoxins, or cylindrospermopsins, which are not measured by the ELISA kit, were present. In 1998, WHO recommended that Microcystin LR concentrations of 20 ppb or higher should trigger further action for recreational uses. Nebraska chose an initial, more conservative action level of 15 ppb of total microcystins for issuing health alerts in 2004, and a level of 2 ppb of total microcystins for issuing health advisories. The action level of 15 ppb of total microcystins was selected because it was more protective than the 20 ppb of Microcystin LR recommended by WHO. Methods used to notify the public about potential health concerns from exposure to cyanobacteria included the development of a fact sheet about cyanobacteria (Fig. 1); weekly updates to microcystin sampling results and health alerts on the NDEQ web site (Fig. 2); emails to interested agencies and organizations;

news releases and interviews with newspapers, radio, and TV stations; and posting of warning signs at lake beaches and boat ramps (Fig. 3 see Color Plate 4).

Results and Discussion

On Monday, July 12, 2004, the east swimming beach at Pawnee Lake near Emerald, Nebraska was sampled for total microcystins after a dense algal bloom was observed during routine E. coli monitoring. E. coli concentrations were slightly elevated (276/100ml compared to the single sample criterion of 235/100 ml), and the most recent five-sample geometric mean of 179/100 ml exceeded the geometric mean criterion of 126/100 ml; however, E. coli concentrations of this magnitude are relatively common in Nebraska lakes. No documented complaints about gastrointestinal illnesses after swimming in Nebraska lakes had ever been received by state officials prior to the weekend of July 17-18, 2004. Levels of total microcystins at the east swimming beach exceeded 15 ppb on July 12, 2004, and a health alert was issued for Pawnee Lake on Thursday, July 15. Local authorities were asked to post signs at the boat ramp and at both swimming beaches on Pawnee Lake prior to the weekend of July 17-18. Unfortunately, due to the short notice and logistics of mass-producing warning signs, only one sign was posted at the east swimming beach, and no signs were posted at the boat ramp or west swimming beach. Heavy public use of the lake occurred that weekend, and more than 50 calls were received from the public, complaining about symptoms such as skin rashes, lesions, blisters, vomiting, headaches, and diarrhea after swimming or skiing in Pawnee Lake. Although unfortunate, this incident provided evidence that the initial health alert action level adopted in Nebraska was indicative of a serious health threat, and that the state needed to do a better job of informing the public. During 2004, in addition to five dog deaths and the Pawnee Lake human health problems, several livestock and wildlife deaths and additional complaints of skin rashes and gastrointestinal illnesses were reported at other lakes. No cyanobacterial toxin data were collected in Nebraska prior to 2004; therefore, trends regarding the incidence of cyanobacterial harmful algal blooms could not be determined. However, the numbers of cyanobacterial problems reported in 2004 were unprecedented. A total of 671 microcystin samples (748 including QC samples) were collected from 111 different lakes in 2004, resulting in health alerts for 26 lakes and health advisories for 69 lakes. A total of 22 of the 26 health alert lakes (84.6%) were located in the eastern one-third of Nebraska (Fig. 4). Most of the state's population lives in eastern Nebraska, which is intensively farmed with many areas of high to very high erosion potential. Most lakes in this area typically receive high nutrient loadings. Also, housing developments are common at sandpit lakes along the Platte River in eastern Nebraska, and many of these are suspected of having failing septic systems, which may contribute to the nutrient levels of these lakes.

High concentrations of total microcystins (>15 ppb) were measured in Nebraska lakes from May through December, although health alerts and advisories occurred most frequently during the months of July, August, and September. Preliminary remote sensing data indicated that cyanobacterial succession in lakes varied significantly throughout the year, even among lakes located in close proximity to one another (e.g. Fremont State Lakes). Four lakes (Carter Lake near Omaha, Swan Creek Lake (5A) near Tobias, Pawnee Lake near Emerald, Iron Horse Trail Lake near Humboldt) were on health alert status for 12 or more weeks during 2004. The most common cyanobacterial genera identified in Nebraska lake samples were Anabaena (32.0%) and Microcystis (30.0%), followed by Oscillatoria (14.3%) and Aphanizomenon (4.9%). A total of 24.8% of the lake samples analyzed in response to citizen complaints had a high or very high biomass of cyanobacteria. Preliminary assessments of concentrations of total microcystins and ancillary data indicated that lower water levels from the recent drought conditions and lower nitrogen to phosphorus ratios may have contributed, in part, to the increased numbers of cyanobacterial complaints and problems that occurred in 2004. To date, no problems with levels of total microcystins have been documented in drinking water sources in Nebraska. However, 85% of the state's residents use groundwater as their source of drinking water and the remaining 15% primarily obtain their drinking water from the Missouri River.

New Developments

Several changes to the microcystin sampling and public notification protocols were made in 2005. The health alert action level was raised from 15 to 20 ppb to correspond more closely with the WHO recommended action level for recreation. However, it should be noted that the Nebraska action level of 20 ppb of total microcystins is still more protective than the WHO action level of 20 ppb, which is based solely on Microcystin LR concentrations. Lakes were placed on health alert status in 2005 if the weekly total microcystins concentration equaled or exceeded the 20 ppb action level. However, once a health alert was issued for a lake, the weekly total microcystins concentration had to fall below 20 ppb for two consecutive weeks before the lake was removed from health alert status. In contrast, lakes were dropped from health alert status in 2004 based on weekly changes in levels of total microcystins. In addition, the health advisory action level of 2 ppb was dropped in 2005 because using both an advisory level and an alert level was confusing to the public. In 2005, swimming beach sampling protocols were changed to better represent the lake conditions that most adults and children are exposed to. Grab samples were collected at the mid-point of designated swimming areas in knee-deep water. In 2004, worst-case samples were often collected from surface scums along the shorelines.

Special studies were initiated in 2005 to better identify causes and ecological consequences of cyanobacterial blooms. NDEQ contracted with UNL to conduct a cyanobacterial remote sensing project. This project will evaluate the potential of remote sensing as a tool for early detection of cyanobacterial harmful algal blooms based upon analysis of images acquired by aircraft overflights and in-situ monitoring. Spectrally based algorithms will be developed for detecting and quantifying in real time the presence of pigments associated with cyanobacteria, including phycocyanin (a blue pigment found in cyanobacteria) and chlorophyll a. A preliminary map of phycocyanin mask images for the Fremont State Lakes showing the presence of cyanobacteria is displayed in Fig. 5. Images or maps of cyanobacterial blooms may also be useful in guiding field sampling efforts. Another objective of this project is to develop a model for predicting cyanobacterial blooms based on parameters such as zooplankton, phytoplankton, total microcystins, nutrients, turbidity, pH, water temperature, depth, stratification, air temperature, wind, cloud cover, and precipitation. A study of the concentrations of total microcystins in fish tissue will be conducted in 2005 to determine if fish caught from health alert lakes are safe to eat. Fish fillets and entrails of various game fish species from three lakes, Pawnee Lake, Fremont Lake #20, and Carter Lake, which have most frequently been on health alert status in 2004 and 2005, will be analyzed for levels of total microcystins. Nebraska currently advises the public not to eat whole fish from health alert lakes and to consider practicing catch and release as a safety precaution. The results of this study will be used to modify existing fish consumption advisories. Future studies will analyze lake samples for individual cyanotoxins such as Anatoxin-a; Cylindrospermopsin; Microcystin LR, LA, LF, LW, and RR; and chemical and physical water quality parameters and plankton population dynamics, which may help explain why some lakes have persistent cyanobacterial problems.

In 2005, an interagency workgroup was formed to discuss and recommend potential methods for preventing, controlling, and mitigating cyanobacterial blooms in different types of lakes including large flow-through impoundments ≥ 25 acres (i.e. reservoirs built for flood control and irrigation), small flow-through impoundments < 25 acres (i.e. farm and urban ponds), and closed systems (i.e. oxbow lakes, sand and barrow pits). Preliminary recommendations include reducing nutrient inputs; installing watershed treatments to reduce nutrient runoff; dredging to remove in-lake sediments, nutrients, and increase depth; controlling rough fish populations; applying alum treatments to inactivate phosphorus; and applying algacides. The effectiveness and cost of these options are limited by factors such as lake size, watershed size, outlet structure, sources of water, and remediation costs.

NEWS RELEASE

Issued jointly from the
Nebraska Department of Health and Human Services
Regulation and Licensure
Nebraska Department of Environmental Quality
Nebraska Game and Parks Commission
UN-L Water Quality Extension Program

Contact

Brian McManus, Department of Environmental Quality, (402) 471-4223
Kathie Osterman, Communications and Legislative Services, (402) 471-9313
Jim Carney, Game and Parks Commission, (402) 471-5547
Tadd Barrow, UN-L Water Quality Extension, (402) 472-7783

FACT SHEET:

Precautions and facts regarding toxic algae

What is toxic blue-green algae?

Although it technically is not a true algae, what is commonly referred to as toxic blue-green algae refers to certain strains of cyanobacteria that produce toxins. These toxins were found in a number of Nebraska lakes in 2004.

Toxic blue-green algae can dominate the algal populations of a lake under the right combinations of water temperature, low water depths, and nutrients (such as high nitrogen and phosphorus concentrations from wastewater discharges and/or runoff from agricultural land and communities).

What should I look for to avoid toxic algae?

The toxic strains of blue-green algae usually have heavy surface growths of pea-green colored clumps, scum or streaks, with a disagreeable odor and taste. It can have a thickness similar to motor oil and often looks like thick paint in the water. Algae blooms usually accumulate near the shoreline where pets and toddlers have easy access and the water is shallow and more stagnant. It is important to keep a watchful eye on children and pets so that they do not enter the water. Aspects to watch out for include:

Water that has a neon green, pea green, blue-green or reddish-brown color.
Water that has a bad odor.
Foam, scum or a thick paint-like appearance on the water surface.
Green or blue-green streaks on the surface, or accumulations in bays and along shorelines.

Fig. 1. Fact sheet about cyanobacteria (toxic algae)

What are the risks and symptoms?

Pets and farm animals have died from drinking water containing toxic blue-green algae (or licking their wet hair/fur/paws after they have been in the water). Blue-green algae toxins have been known to persist in water for several weeks after the bloom has disappeared.

The risks to humans come from external exposure (prolonged contact with skin) and from swallowing the water. Symptoms from external exposure are skin rashes, lesions and blisters. More severe cases can include mouth ulcers, ulcers inside the nose, eye and/or ear irritation and blistering of the lips. Symptoms from ingestion can include headaches, nausea, muscular pains, central abdominal pain, diarrhea and vomiting. Severe cases could include seizures, liver failure, respiratory arrest – even death, although this is rare. The severity of the illness is related to the amount of water ingested, and the concentrations of the toxins.

Are some people more at risk?

Yes. Some people will be at greater risk from toxic blue-green algae than the general population. Those at greater risk include:
Children. Toddlers tend to explore the shoreline of a lake, causing greater opportunity for exposure. Based on body weight, children tend to swallow a higher volume of water than adults, and therefore could be at greater risk.
People with liver disease or kidney damage and those with weakened immune systems.

Here are some tips on what you can do, and things to avoid:
Be aware of areas with thick clumps of algae and keep animals and children away from the water.
Don't wade or swim in water containing visible algae. Avoid direct contact with algae.
Make sure children are supervised at all times when they are near water. Drowning, not exposure to algae, remains the greatest hazard of water recreation.
If you do come in contact with the algae, rinse off with fresh water as soon as possible.
Don't boat or water ski through algae blooms.
Don't drink the water, and avoid any situation that could lead to swallowing the water.

Fig. 1. (cont.) Fact sheet about cyanobacteria (toxic algae)

Is it safe to eat fish from lakes that are under a Health Alert?

The toxins have been found in the liver, intestines and pancreas of fish. Most information to date indicates that toxins do not accumulate significantly in fish tissue, which is the meat that most people eat. It is likely that the portions of fish that are normally consumed would not contain these toxins. However, it is ultimately up to the public to decide whether they want to take the risk, even if it is slight. Fishing is permitted at lakes that are under a Health Alert, but anglers may want to consider practicing catch and release at these lakes.

Where can I find out more information about lake sampling for toxic algae?

The Nebraska Department of Environmental Quality is conducting weekly and monthly sampling at select public lakes that are either popular recreational lakes, or have historically had toxic algae problems. This information is updated weekly on the agency web site, www.deq.state.ne.us.

What should I do if I have concerns regarding a private lake?

As part of the University of Nebraska Water Quality Extension Program, UN-L has developed a "Volunteer Monitoring Program" and lake test kits that will be sent to interested lake associations, owners, etc. so they can collect a sample and send it to UN-L for analysis. To obtain more information and a test kit please contact the program at (402) 472-7783, or (402) 472-8190.

If I think a public lake has a toxic algae bloom, who do I call?

Please contact the Department of Environmental Quality's Surface Water Section at (402) 471-0096, or (402) 471-2186.

If I am experiencing health symptoms, who do I call?

If you experience health symptoms, notify your physician, and also report it to the Nebraska Health and Human Services System at (402) 471-2937. You can also contact the Nebraska Regional Poison Center at 800-222-1222 for more information.

For more information, contact
MoreInfo@NDEQ.state.NE.US
Nebraska Department of Environmental Quality
1200 "N" Street, Suite 400
PO Pox 98922
Lincoln, Nebraska 68509
(402) 471-2186 FAX (402) 471-2909

Fig. 1. (cont.) Fact sheet about cyanobacteria (toxic algae)

TOXIC BLUE-GREEN ALGAE SAMPLING RESULTS

DEQ is conducting weekly and monthly sampling for toxins at a number of public recreational lakes across Nebraska, and the results will be updated weekly, usually on Fridays. Depending on circumstances, there may be additional lakes added to the weekly sampling schedule.

Samples taken: September 19 and 20, 2005, unless noted otherwise
Analysis completed: September 23, 2005

The analysis of recent sampling shows that the following lakes are currently considered in Health Alert status:

Conestoga Reservoir
Kirkman's Cove
Pawnee Reservoir

An explanation of Health Alert: This designation means that the state believes that the level of toxins in the water make it unsafe for full-body recreational activities, such as swimming.

During a Health Alert at a public lake, signs will be posted advising the public to use caution. Affected swimming beaches will be closed. Boating and other recreational activities will be allowed, but the public will be advised to use caution and avoid prolonged exposure to the water, particularly avoiding any activity that could lead to drinking the water.

The level established in 2005 for a Health Alert is 20 parts per billion of total microcystins. Lakes under Health Alert will be sampled weekly, and the Health Alert will stay in effect until the level stays below 20 parts per billion for two consecutive weeks.

Fig. 2. Weekly updates to total microcystins sampling results and health alerts on the NDEQ web site (**www.deq.state.ne.us**).

The chart below shows the lakes that were sampled, and the level of toxin found at the lake.

Lake Name	County	Level of Total Microcystins (ppb)	Is it on alert?
Bluestem Res. at swimming beach	Lancaster	0.11	No
Carter Lake at north boat ramp*	Douglas	Area 10 beach – 3.45 Lieber's Point beach – 4.68	No
Conestoga Res. at boat ramp	Lancaster	>30	YES
Fremont Lake #9 at Fremont Beach	Dodge	0.44	No
Fremont Lake #10 at Brick's Bay Beach	Dodge	0.09	No
Fremont Lake #20	Dodge	East beach – 3.80 West Beach – 3.59	No
Iron Horse Trail at swimming beach	Pawnee	0.45	No
Kirkman's Cove	Richardson	15.18	**YES**—must remain below 20 ppb for 2 cons. Wks
Louisville Lake #2 at swimming beach	Cass	3.61	No
Maskenthine Lake at steel pier	Stanton	0.97	No
Pawnee Reservoir	Lancaster	East beach – >30 West beach – not sampled	**YES**
Summit Lake at swimming beach	Burt	2.18	No
Wagon Train Lake at swimming beach	Lancaster	0.00	No

*The north boat ramp is located on the Nebraska side of Carter Lake.

Fig. 2. (cont.) Weekly updates to total microcystins sampling results and health alerts on the NDEQ web site (**www.deq.state.ne.us**).

If a private lake is under a Health Alert, the Nebraska Department of Environmental Quality will inform the head of the lake association or other lake representative, and ask them to take the appropriate measures to ensure users of the lake are informed of the potential hazards.

For more information about toxic blue-green algae, please see the related Fact Sheet.
For more information, contact
MoreInfo@NDEQ.state.NE.US

Nebraska Department of Environmental Quality
1200 "N" Street, Suite 400
PO Box 98922
Lincoln, Nebraska 68509
(402) 471-2186 FAX (402) 471-2909
Nebraska.gov | Security, Privacy & Accessibility Policy

Fig. 2. (cont.) Weekly updates to microcystins sampling results and health alerts on the NDEQ website (**www.deq.state.ne.us**).

Fig. 3. Posting of warning signs at lake beaches and boat ramps. (See Color Plate 4).

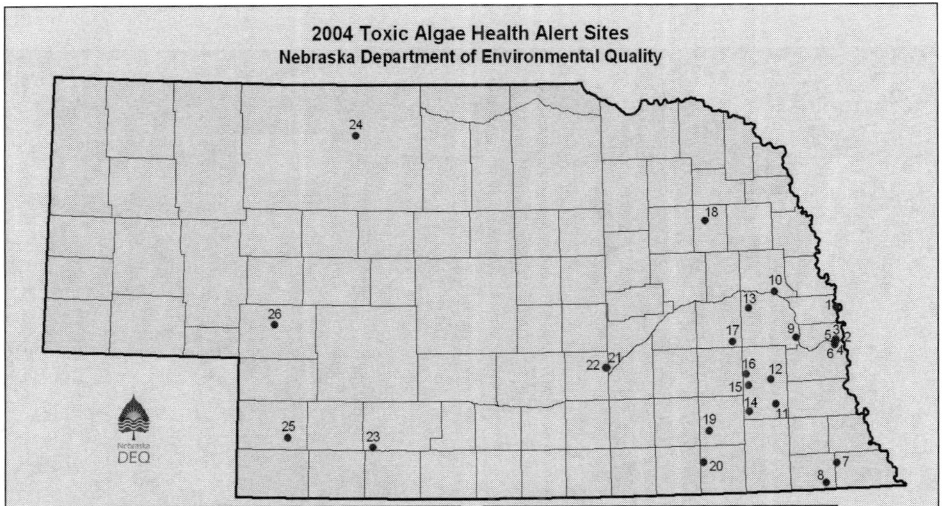

Index	Water Body	Index	Water Body
1	Carter Lake north of Omaha	14	Olive Creek Reservoir near Kramer
2	Bellevue Rod & Gun Club south lake near Bellevue	15	Conestoga Reservoir near Denton
3	Hanson Lake #4 south of Bellevue	16	Pawnee Reservoir near Emerald
4	Hanson Lake #1 south of Bellevue	17	Redtail WMA near Dwight
5	Hanson Lake #2 south of Bellevue	18	Maskentine Reservoir near Stanton
6	Buccaneer Bay Lake near Plattsmouth	19	Swan Lake #5A near Tobias
7	Kirkman's Cove near Humboldt	20	Alexandria SRA West Lake
8	Iron Horse Trail Reservoir near DuBois	21	Hidden Lake near Grand Island
9	Thomas Lake near Ashland	22	Arends Lake near Grand Island
10	Fremont Lake #20 near Fremont	23	Red Willow Reservoir near McCook
11	Wagon Train Reservoir near Hickman	24	Merritt Reservoir near Valentine
12	Wedgewood Lake in Lincoln	25	Enders Reservoir near Imperial
13	Czechland Reservoir near Praque	26	Lake McConaughy near Ogallala

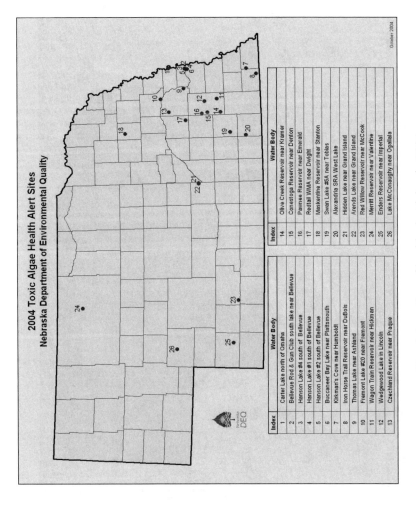

Fig. 4. Location of health alert lakes in 2004.

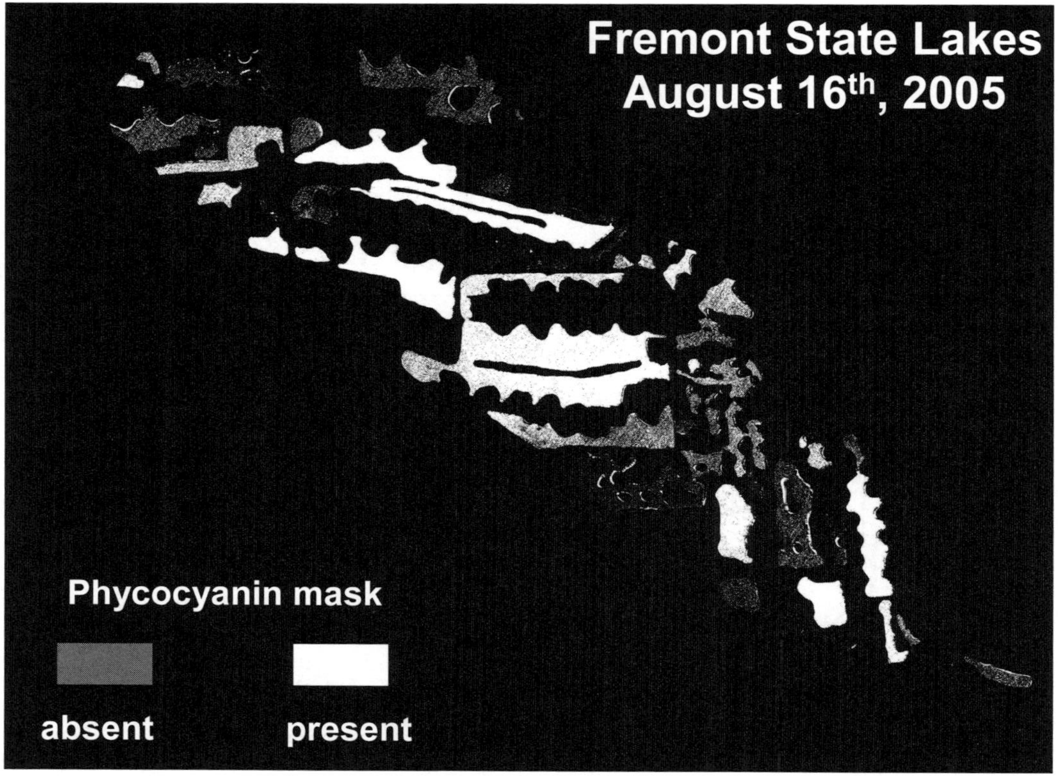

Fig. 5. Map of phycocyanin mask images for the Fremont State Lakes showing the presence of cyanobacteria.

Chapter 7: Cyanobacterial Toxins in New York and the Lower Great Lakes Ecosystems

Gregory L Boyer

Department of Chemistry, State University of New York, College of Environmental Science and Forestry, Syracuse NY 13210

Abstract

Toxic cyanobacterial blooms are an increasing problem in the lower Laurentian Great Lakes. To better understand their occurrence and distribution, samples for particulate toxin analysis were collected from more than 140 New York Lakes including Lakes Erie, Champlain and Ontario. Microcystins were of most importance and were detected in nearly 50% of the samples. Anatoxin-a, cylindrospermopsin and the paralytic shellfish toxins occurred much less frequently (0-4%). The implications for the management of cyanobacterial harmful algal blooms are discussed.

Introduction

The North American Great Lakes located between the United States and Canada collectively provide drinking water for more than 22 million people. In addition, numerous smaller lakes provide recreational opportunities for inhabitants and visitors to the northeastern US. Historically, very little is know historically about the occurrence of cyanobacterial toxins in New York State waters, despite the early isolation of a paralytic shellfish toxin (PST)–producing strain (*Aphanizomenon schizoide* aka *Aph. flos–aquae*) from nearby Vermont in the 1960s (Sawyer et al. 1968), and extensive work by Paul Gorham and coworkers at the National Research Council of Canada in the 1950s and 1960s documenting toxic cyanobacterial blooms in the nearby Canadian lakes and ponds (Gorham 1962, Gorham et al.

1964). This situation achieved a much higher awareness following the large outbreak of toxic *Microcystis* bloom in the western basin of Lake Erie in the mid 1990's (Brittain et al. 2000). Additional outbreaks have occurred in subsequent years where concentrations of microcystins have reached levels as high as 20 µg microcystin–LR equivalents (MC–LR_{eq}) L^{-1} (Oullette et al. 2005, Rinta Kanto et al. 2005). In 1998 and 1999, several dogs died at public campgrounds located along the shores of Lake Champlain after coming in contact with algal scums washed up along the shoreline (Boyer et al. 2004). Preliminary investigations indicated that the cause of death in these animals was likely due to ingestion of the neurotoxin anatoxin–a. In response to these highly publicized events, the National Oceanographic and Atmospheric Administration (NOAA), through their Monitoring and Event Response for Harmful Algal Blooms (MER-HAB) program, the Centers for Disease Control (CDC) and the NOAA's Sea Grant programs initiated extensive field studies to better understand the occurrence and distribution of these toxins in New York waters and the lower Great Lakes. The results from five years of these field studies spanning from 2000 – 2004 are summarized here.

Methods

Sampling methodology has steadily evolved over the 5 years of this study. More than 2500 samples were collected from over 1000 sites at 81 different New York lakes. This includes large lakes such as Lake Erie, Lake Ontario and Lake Champlain, intermediate–sized bodies of water such as the New York Finger Lakes and Oneida Lake, and smaller impoundments and lakes such as Lake Neatahwanta and Labrador Pond. Samples were collected at a variety of times throughout the growing season (June – October) with the bulk of the samples collected in late July and early September, times of peak cyanobacterial abundance. Early in the study (year 2000), 1 liter grab samples were collected from a depth near the surface of the water body and vacuum filtered through a 47mm 945AH glass fiber filter. These filters were immediately frozen on dry ice in the field and returned to the laboratory for extraction and analysis. The aim of these studies was to achieve large geographical coverage and samples were collected from 58 different lakes and over 150 sites, but generally only once during the season in late July and early August. In subsequent years, sampling focused more on single bodies of water (Lake Erie, Lake Ontario, Oneida Lake, Lake Champlain, etc.) with multiple samples taken throughout the growing season. To achieve a greater sensitivity in oligotrophic waters such as the Great Lakes, the volume filtered in later years was also increased. Samples were collected from 0.5–1 m in depth and rapidly filtered through a 90 mm 934AH glass fiber filter using a peristaltic pump until either 20 L passed through the filter or the filter plugged due to particulate material in the water column. Filters were again frozen immediately in the field and returned frozen to the lab for later toxin analysis. Additional samples were collected for chlorophyll determination, DNA extraction, nutrients and, in selected cases, dissolved toxins. Only the results of the particulate toxin analysis are reported here.

Upon return to the lab, the filters were extracted in 5–10 ml of 50% methanol containing 1% glacial acetic acid using ultrasonic disruption. Control experiments showed that this extraction protocol solubilized more than 95% of the microcystin–LR, –RR, –LF, anatoxin–a (ATX) and the PST saxitoxin from the cells. Microcystins were measured using a combination of assays including inhibition of the protein phosphatase 1A (PPIA, Carmichael and An, 1999), Enzyme linked immunoassays (ELISA) and by high performance liquid chromatography (HPLC) coupled with either photodiode array (PDA) or mass selective (MS) detection (Harada, 1996). Anatoxin–a was determined by HPLC after derivatization with 7–fluoro–4–nitro–2,1,3–benzoxadiazole (NBD–F) (James et al., 1998) and confirmed by HPLC–MS of the free or NBD–derivatized toxin. The PST toxins (saxitoxin, neosaxitoxin, and gonyautoxins 1–4) were measured by HPLC with fluorescent detection after either chemical (PCRS: Oshima 1995) or electrochemical (ECOS: Boyer and Goddard 1999) post–column derivatization. Cylindrospermopsin was measured by HPLC using PDA detection and confirmed by HPLC–MS (Li et al. 2001).

Results

New York Overview

Between 2000 and 2004, more than 1000 samples were collected from 81 different lakes scattered across New York State. The distribution of those samples between lakes and the toxins that they were analyzed for is shown in Table 1. The lower Great Lakes (Lake Erie and Lake Ontario) plus Lake Champlain accounted for a large number of those samples (65%), especially in the later sample years due to the Great Lakes focus of the MERHAB sampling program (Boyer et al. 2004a). Samples from these large lakes consisted of a mixture of open water and "coastal" sites and were obtained from both sampling cruises as part of the MERHAB–Lower Great Lakes and Microbial Ecology of the Lake Erie Ecosystem (MELEE) scientific programs (i.e. see section on Lake Erie below) on the CCGS Limnos and RV Lake Guardian, as well as from smaller boats and shore samples. Samples collected from the smaller lakes (Oneida Lake, Finger Lakes and other NY Lakes) were either from targeted studies by our own lab (Oneida Lake, Onondaga Lake), from the NY–Department of Environmental Conservation (Finger Lakes) or as part of the broad shotgun survey across NY State conducted in 2000. Only in a few cases were these samples collected in direct response to a reported toxic algal bloom.

Table 1. Sample Distribution between New York Lakes and the combined number of samples tested for the different cyanobacterial toxins during the 2000–2004 sample seasons.

Task	Total #	Lake Ontario	Lake Erie	Lake Cham–plain	Oneida Lake	Finger Lakes	Other NY Lakes
# Samples collected:	2513	736	308	590	314	138	427
Analyzed for MC's*:	2286	561	293	579	302	137	414
Analyzed for ATX:	2307	589	286	572	315	138	407
Analyzed for PSTs:	1078	258	174	314	163	29	240
Analyzed for CYL:	366	104	79	32	34	0	117

*Abbreviations: MC's = microcystins, ATX = anatoxin–a, PST = saxitoxin + neosaxitoxin, CYL = cylindrospermopsin.

Not all samples were analyzed for all toxins, though more than 90% of them were analyzed for microcystins by at least the PPIA and for anatoxin–a by HPLC–FD. Fewer samples were analyzed for the less common toxins such as the PST toxins and cylindrospermopsin (43% and 15% respectively).

The determination if the sample would be considered toxic was dependant on how "toxic" was defined. Table 2 shows the results for microcystin analysis using three different thresholds for toxicity: the World Health Organization's (WHO) advisory limit of 1 μg MC–LR equivalents L^{-1}, an arbitrary guideline value of 0.1 μg L^{-1}, and roughly the detection limit of the PPIA assay of 0.01 μg MC–LR equiv. L^{-1}. This detection limit is based on collection of a 10 L sample via filtration and its extraction/concentration into 10 ml of solvent. Approximately 14% of the samples collected statewide exceeded the WHO advisory limit of 1 µg MC–LR$_{eq}$ L^{-1}. As expected, the bulk of these samples (239 or 73%) came from smaller, more eutrophic, water bodies. Samples from the more oligotrophic Great Lakes, Lake Erie and Lake Ontario, exceeded the WHO advisory limit of 1 µg L^{-1} only 15 times during the five year time period. In contrast, samples from Lake Champlain exceeded the 1 µg L^{-1} advisory limit more than 71 times during this same time period. Most of these toxic Lake Champlain samples were collected from the eutrophic Missisquoi Bay region which is characterized by high phosphorus inputs and expansive *Microcystis* blooms during the summer growing season. Similarly, more than 25% of the samples collected from Oneida Lake with its well established cyanobacterial blooms exceeded the WHO advisory limit of 1 µg L^{-1}. Equally informative was the abundance of detectable levels of microcystin–LR$_{eq}$ in the water. Over 50% of the total samples from New York Lakes, including 40% of the samples collected from Lake Erie and 28% of the samples collected from Lake Ontario had easily detectable levels of microcystins that exceeded the 0.01 µg L^{-1}. This dropped to 11% and 29% respectively for Lake Ontario and Lake Erie when that threshold was raised to the more reasonable 0.1 µg L^{-1}.

Table 2. Percentage of samples collected during the 2000 – 2004 sample season that exceeded a 1.0, 0.1 and 0.01 µg L^{-1} threshold for microcystins as measured using the protein phosphatase inhibition assay.

Threshold concentration MC–LR$_{eq}$ L^{-1}	Total # all lakes	Lake Ontario	Lake Erie	Lake Cham–plain	Oneida Lake	Finger Lakes	Other NY Lakes
# Samples Analyzed:	2513	736	308	590	314	138	427
>1.0 µg:	326	4	11	71	71	1	168
(%)	(14%)	(1%)	(4%)	(12%)	(24%)	(1%)	(41%)
>0.1 µg:	829	61	84	190	207	23	264
(%)	(36%)	(11%)	(29%)	(33%)	(69%)	(17%)	(64%)
>0.01 µg:	1223	155	117	296	249	113	293
(%)	(53%)	(28%)	(40%)	(51%)	(82%)	(82%)	(71%)

* The protein phosphatase inhibition assay is an activity–based technique that calculates toxicity from a microcystin–LR –derived standard curve. Results are given in MC–LR equivalents, e.g. that concentration of microcystin–LR which exhibits the equivalent toxicity to the sample. This is similar, but not identical, to the microcystin–LR equivalents derived from structural–based assays such as ELISA.

The neurotoxic cyanobacteria toxins such as anatoxin–a (ATX) and paralytic shellfish toxins (PSTs) saxitoxin and neosaxitoxin, and hepatotoxic cylindrospermopsin (CYL) were found much less frequently. Significant numbers of these toxins were detected only when the threshold of analysis was set at or near the detection limit (Table 3). Anatoxin–a was the second most common cyanobacterial toxin observed in New York State lakes with 11 samples (<1%) exceeding the 1 µg L^{-1} threshold, 29 samples (1%) exceeded the 0.1 µg ATX L^{-1} threshold and 74 samples (3%) exceeding the most stringent 0.01 µg L^{-1} threshold. A large proportion of those toxic samples were from either the northern regions of Lake Champlain where animal fatalities from anatoxin–a have been reported in the past (Boyer et al., 2004b), highly eutrophic Lake Agawam on Long Island (Gobler, unpublished), or the eutrophic embayments along the New York coast of Lake Ontario (Yang et al., 2005). High concentrations of ATX were not observed in either Lake Erie or the offshore waters of Lake Ontario. Interestingly, the peak periods of anatoxin–a in Lake Champlain did not coincide with the large microcystin–producing algal blooms. The initial dog intoxication event at Au Sable State Park occurred in June of 1999, a time when the lake was generally cooler and the cyanobacterial biomass was low. In this event, there was very little in the way of surface scums in the water itself, but sizable accumulations did occur along the shore.

Table 3. Occurrence of anatoxin–a, paralytic shellfish toxins and cylindrospermopsin in samples collected during the 2000 – 2004 sample season that exceeded a predefined threshold. Percentage represents the percent of those samples from that particular lake that exceeded the threshold.

Task	Total # all lakes	Lake Ontario	Lake Erie	Lake Cham– plain	Oneida Lake	Finger Lakes	Other NY Lakes
>1.0 µg ATX L^{-1}:	11	0	0	7	0	0	4
%	(<1%)	(0%)	(0%)	(1%)	(0%)	(0%)	(1%)
>0.1 µg ATX L^{-1}:	29	2	2	12	2	2	9
%	(1%)	(<1%)	(1%)	(2%)	(1%)	(1%)	(2%)
>0.01 µg ATX L^{-1}:	74	14	14	24	4	2	16
%	(3%)	(2%)	(5%)	(4%)	(1%)	(1%)	(4%)
>0.01 µg PST L^{-1}:	2	0	1	0	0	0	1
%	(0%)	(0%)	(1%)	(0%)	(0%)	(0%)	(1%)
>0.01 µg CYL L^{-1}:	8	1	2	0	0	–	5
%	(2%)	(1%)	(3%)	(0%)	(0%)	–	(4%)

The other two classes of cyanobacterial toxins, namely the PSTs and cylindrospermopsin, occurred very rarely if at all. PST toxicity was detected in only two samples of nearly 1100 tested during the 5–year time period with the maximum concentration of only 0.09 $\mu g\ L^{-1}$. At these low concentrations, its identification was tentative and remains to be confirmed. This low occurrence was despite the common occurrence of high biomass blooms of *Aphanizomenon flos–aquae*, and the fact that the original PST–producing strain was isolated from nearby Vermont. Cylindrospermopsin was detected in only 8 samples of 366 tested. Most of those samples (5) occurred during an August bloom on Lake Agawam, Long Island, which coincided with a period of relative nitrogen limitation as evidenced by nutrient addition experiments (Gobler et al. 2006). In all cases, the maximum concentration of these cylindrospermopsin was low ($< 0.25\ \mu g\ L^{-1}$). Confirmation of its identification using more advanced (HPLC–MS–MS and the polymerase chain reaction) techniques is in progress and until that time, its identification in New York water should also be considered tentative.

For the toxins other than microcystins, the cyanobacterial species responsible for their production is unknown. However, many of the embayments in Lake Champlain and Lake Ontario with easily measurable anatoxin–a concentrations also had significant co–occurring blooms of *Anabaena* species. The predominate organism responsible for microcystin formation in New York waters is likely to be *Microcystis aeruginosa*. However *Microcystis* is not the only species capable of producing microcystin in these waters. Significant blooms of *Planktothrix* and *Anabaena* species, both known microcystin producers routinely occur in New York State waters. In the absence of physically isolating and culturing the responsible organism from a toxic bloom, molecular techniques are now routinely being used to determine what organism(s) are likely to be responsible for the observed toxicity. For example, the western basin of Lake Erie routinely experiences toxic blooms of *Microcystis aeruginosa*, however *Planktothrix* appears to be the species responsible for microcystin production in the adjacent Sandusky Bay of that lake (Rinta Kanto et al.. 2005, 2007). Blooms of toxic cyanobacteria in Oneida Lake also show very different PCR banding patterns in their microcystin biosynthetic genes, suggesting that genetic differences in microcystin–producing organisms can occur within the same water body (Hotto et al. 2004 unpublished).

Lake Erie and Lake Ontario

With the isolation of toxic *Microcystis* from Lake Erie by Brittain and co-workers, there has been considerable effort to determine the occurrence and distribution of cyanobacterial toxins in Lake Erie and nearby Lake Ontario. Table 4 summarizes the results of five different research cruises on Lake Erie during 2000–2004 that specifically targeted cyanobacterial toxins. Most of the sampling was focused in the highly eutrophic western basin of Lake Erie. Toxicity was highly variable in this basin with some years showing high levels of toxicity (2003, >20 μg L^{-1}) and other years showing little to no toxicity (2002). Part of the explanation for this variability is the time of sampling. The availability of large ships for sampling on Lake Erie is limited and cruise dates were often determined by ship availability rather than optimum time for cyanobacterial blooms. In addition, the timing of the blooms was variable in themselves and occurred as early as July and as late as October. Nor was the western basin of Lake

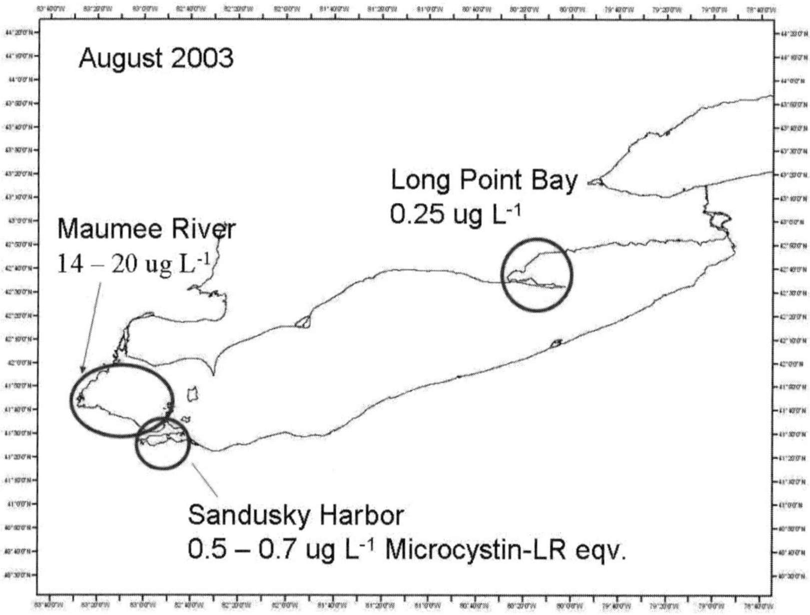

Fig. 1. Distribution of microcystin toxicity during the 2003 research cruises on Lake Erie. Highest concentrations were obtained near the Maumee River in August, but two distinct blooms also occurred near Long Point Bay and in Sandusky Harbor. Toxicity was measured using the PPIA and is expressed in terms of μ $MC–LR_{eq}$ L^{-1}.

Erie the only site of cyanobacterial toxicity. Significant levels of microcystin toxicity were also observed in Sandusky Harbor and near Long Point Bay in the eastern basin (Fig. 1). The source of toxicity in the different regions was quite different. In both the western basin and at Long Point, well defined blooms of *Microcystis aeruginosa* most likely accounted for the observed toxicity. In contrast, Sandusky Harbor often contained a dense cyanobacterial flora that was usually dominated by *Aphanizomenon flos–aquae* and *Anabaena* species. *Microcystis* is present but often a low abundance. Recent molecular analysis has indicated that the source of microcystin toxicity in this system was more likely due to *Planktothrix* species present in the understory of the bloom (Rinta–Kanto and Wilhelm, submitted) rather than *Microcystis* itself. This has important implications for monitoring protocols based on cell abundance. Neither is Lake Erie the only lower Great Lake with demonstrated levels cyanobacterial toxins. In 2003, a large bloom of *Microcystis aeruginosa* formed in the eastern basin of Lake Ontario near Oswego, New York. Microcystin concentration in this bloom near the Onondaga County drinking water intakes approached the WHO advisory limit of 1 μg MC–LR$_{eq}$ L^{-1}.

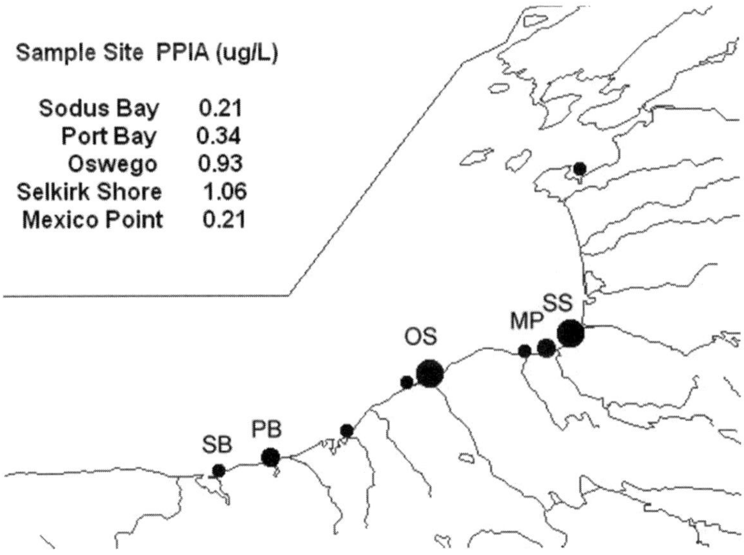

Fig. 2. The amount and distribution of toxicity off Oswego NY during the August 2003 *Microcystis* bloom in eastern Lake Ontario. Toxicity was measured using the PPIA and is expressed in terms of μ MC–LR$_{eq}$ L^{-1}.

Table 4. Recent Occurrence and Distribution of Cyanobacterial Toxins in Lake Erie

Cruise and Date	# samples	% containing toxin	Highest measured value	Comments
Brittain et al. Sept 1996	44	MC's ~10%*	3.4 µg L^{-1} MC	Western basin only
MELEE–VII July 2002	119	MC's 7 % ATX 14% PST's 0 %	0.7 µg L^{-1} MC 0.04 µg L^{-1} ATX	Whole lake survey with highest values at Sandusky, Long Point and Rondeau Bays
MELEE–VIII July 2003	59	MC's 41% ATX 5%	0.65 µg L^{-1} MC 0.11 µg L^{-1} ATX	Whole–lake survey with highest values in the western basin and Sandusky Bay
Lake Guardian & OSU, August 2003	48	MC's 60% ATX 4%	21 µg L^{-1} MC 0.2 µg L^{-1} ATX	Western basin only, Highest values were obtained near the Maumee River
MELEE–IX July 2004	40	MC's 38% ATX 33% CYL 0%	>1 µg L^{-1} MC 0.6 µg L^{-1} ATX	Highest values near the Maumee River and in Sandusky Bay
RV Limnos August 2004	13	MC's 85% ATX 31% CYL 15%	2.4 µg L^{-1} MC 0.07 µg L^{-1} ATX 0.18 µg L^{-1} CYL	Western basin only

*These results are extrapolated from Fig. 2 in Brittain et al. (2000). Abbreviations: MC's = microcystins, ATX = anatoxin–a, PST = saxitoxin + neosaxitoxin, CYL = cylindrospermopsin.

Conclusions and Discussion

Cyanobacteria blooms are common throughout New York State waters however historically, cyanobacteria toxins were not measured. In recent years, several widely publicized animal fatalities have occurred in New York waters due to cyanobacterial toxins. These include dog deaths in Lake Champlain in 1999 due to anatoxin–a and in 2000 due to microcystin toxicity, as well as a dog and water fowl deaths in Lake Neatahwanta in 2004. The presumptive toxin was identified based on the occurrence of toxins in the water, but the causative organisms was not identified and cultured. Several recreation closures due to cyanobacterial toxins have occurred throughout the region and toxic cyanobacterial blooms routinely occur near the water intakes for major drinking water facilities located along Lake Erie, Lake Ontario and Lake Champlain. To date, these toxins have not been observed in the water distribution system. These events have raised the awareness of cyanobacterial toxins in NY waters and as a result, several water treatment facilities and local health departments now sporadically monitor for cyanobacterial toxins.

Detailed analysis of more than 2500 samples indicates that microcystin toxins were the most common toxin encountered in the state with approximately 15% of the samples collected state–wide exceeding the WHO advisory limit of 1 $\mu g\ L^{-1}$ and nearly 60% of the samples containing detectable levels of microcystins. These numbers represents a fairly unbiased estimate of the occurrence of cyanobacterial toxins in New York waters as most of the samples were not collected in direct response to a cyanobacterial bloom but rather as part of the ongoing MERHAB or MELEE sampling program whose sample locations were chosen to guarantee broad spatial coverage regardless of the cyanobacterial density. They are also in good agreement with other large regional surveys from the midwestern United States and Europe (see these proceedings). Some bias in sample collection did exist as samples were generally taken in late summer when cyanobacterial species were more likely to predominate and some long–term sampling sample sites such Oneida Lake were chosen for study because of their well established cyanobacterial blooms. In those highly eutrophic systems where cyanobacteria form dense accumulations, the likelihood of finding toxicity was much greater than in the more oligotrophic waters.

Neither were the larger lakes immune from cyanobacterial toxicity. Both Lake Erie and Lake Ontario have well established *Microcystis* populations that routinely produce toxic blooms that exceed the WHO threshold of 1 μg L^{-1}. Lake Champlain has also experience consist toxic blooms with animal fatalities from both anatoxin–a intoxication and microcystin intoxication occurring in the region. Despite the fact that many of these larger lakes are current serving as drinking water supplies for large metropolitan areas, most of the reported health impacts for cyanobacterial toxins has been due to contact with surface contact with highly toxic scums. It is not uncommon for the wind–borne accumulation of toxic cyanobacteria to yield surface scums that have particulate microcystin concentrations in excess of 500 μg L^{-1} and higher values exceeding 1000 μg L^{-1} have been reported for highly eutrophic embayments such as Missisquoi Bay on Lake Champlain (Watzin and Boyer, unpublished). The health impacts due to recreational contact with the highly toxic surface scums remains to be determined.

Acknowledgements

This work was supported by New York Sea Grant under #NA16RG1645 and by the NOAA's Coastal Ocean Program under MERHAB grant # NA16OP2788. I gratefully acknowledge the many students and technicians that played an integral role in the collection and analysis of this data.

References

Brittain SM, Wang J, Babcock–Jackson L, Carmichael WW, Rinehart KL, Culver DA (2000) Isolation and characterization of microcystins, cyclic heptapeptide hepatotoxins from a Lake Erie Strain of *Microcystis aeruginosa*. J. Great Lakes Res. 26:241–249.

Boyer GL, Goddard GD (1999) High Performance Liquid Chromatography (HPLC) coupled with Post–column electrochemical oxidation (ECOS) for the detection of PSP toxins. Natural Toxins. 7:353–359.

Boyer GL, Makarewicz JC, Watzin M, Mihuc T (2004a) Monitoring strategies for harmful algal blooms in the lower great lakes; Lakes Erie, Ontario and Champlain, USA. Abstracts, 11th Internat. Conference on Harmful Algae. Capetown, South Africa, November 15th, 2004.

Boyer G, Watzin MC, Shambaugh AD, Satchwell MF, Rosen BR, Mihuc T (2004b) The occurrence of cyanobacterial toxins in Lake Champlain. In: "Lake Champlain: Partnerships and Research in the New Millennium. T. Manley, P. Manley, T. Mihuc, Eds., Kluwer Acad, p 241–257.

Carmichael WW, An J (1999) Using an enzyme linked immunosorbent assay (ELISA) and a protein phosphatase inhibition assay (PPIA) for the detection of microcystins and nodularins. Natural Toxins. 7:377–385.

Gobler, C. J., T. W. Davis, K. J. Coyne, and G. L. Boyer (2006) Interactive influences of nutrient loading, zooplankton grazing, and microcystin synthetase gene expression on cyanobacterial bloom dynamics in a eutrophic New York lake. Harmful Algae. *in press.*

Gorham PR (1962) The toxin produced by waterblooms of the blue–green algae. Am. J. Public Health, 52: 2100–2105.

Gorham PR, McLachlan JL, Hammer UT, Kim UK (1964) Isolation and culture of toxic strains of *Anabaena flos–aquae* (Lyngb.). Verh. Int. Verein. theor. Angew. Limnol. 15:796–804.

Harada K (1996) Chemistry and detection of microcystins. In: "Toxic *Microcystis*" M. F. Watanabe, K. Harida, W. W. Carmichael, and H. Fujiki, Eds., CRC Press, Boca Raton, FL, pp. 103–148.

Hotto A, Satchwell M, Boyer G (2004) Seasonal Production and Molecular Characterization of Microcystins in Oneida Lake, New York, USA. Environ. Toxicol. 20:243–248.

James KJ, Furey A, Sherlock IR, Stack MA, Twohig M, Caudwell FB, Skulberg OM (1998) Sensitive determination of anatoxin–a, homoanatoxin–a and their degradation products by liquid chromatography with fluorimetric detection. J. Chromatogr. A. 798:147–157.

Li R, Carmichael WW, Brittain S, Eaglesham GK, Shaw GR, Mahakhant A, Noparatnaraporn N, Yongmanitchai W, Kaya K, Watanabe MM (2001) Isolation and identification of the cyanotoxin cylindrospermopsin and deoxy–cylindrospermopsin from a Thailand strain of *Cylindrospermopsis raciborskii* (Cyanobacteria). Toxicon. 39:973–980.

Oshima Y (1995) Postcolumn derivatization liquid chromatographic method for paralytic shellfish toxins. J. AOAC Int. 78:528–532.

Ouellette AJA, Handy SM, Wilhelm SW (2005) Toxic *Microcystis* is widespread in Lake Erie: PCR detection of toxin genes and molecular characterization of associated cyanobacterial communities. Microbial Ecology. submitted.

Rinta–Kanto JM, Ouellette AJA, Twiss MR, Boyer GL, Bridgeman T, Wilhelm SW (2005) Quantification of toxic *Microcystis* spp. during the 2003 and 2004 blooms in western Lake Erie using quantitative real–time PCR. Environ. Sci. Technol. 39:4198–4205.

Rinta-Kanto, J. M., and S. W. Wilhelm (2006) Diversity of microcystin-producing cyanobacteria in spatially isolated regions of Lake Erie. Appl Environ. Microbiol. 72:5083-5085.

Sawyer PJ, Gentile JH, Sasner Jr. JJ (1968) Demonstration of a toxin from *Aphanizomenon flos–aquae*. Can. J. Microbiol. 14:1199–1204.

Yang X, Boyer GL (2005) Occurrence of the cyanobacterial neurotoxin, anatoxin–a, in lower Great Lakes. Abstracts, International Assoc. Great Lake Research. Annual Meeting, Ann Arbor MI, May 2005.

Chapter 8: Occurrence Workgroup Poster Abstracts

Delaware's Experience with Cyanobacteria in Freshwater Ponds

Humphries EM,[1] Savidge K,[2] Tyler RM[1]

[1] Delaware Department of Natural Resources and Environmental Control (DNREC), Div. of Water Resources, Environmental Laboratory, Dover, DE 19901, [2]University of Delaware College of Marine Studies, Lewes DE 19958.

Introduction

With the increased emphasis on HABs in estuarine environments and their potential to impact natural resources and human health, the State DNREC in 2001 initiated a program to evaluate commercially available Microcystin kits and to measure Microcystin concentrations in select freshwater ponds. Although no environmental or human health impacts associated with blue-green algal blooms have been reported, Delaware does have a large number of private and public freshwater ponds which are accessible to humans, domestic animals, and wildlife.

A Standard Operating Procedure was developed for the EnviroLogix analysis tools in order to insure accurate repeatable results. The EnviroLogix kits are based on the presence of Microcystin LR in combination with Microcystin LA, Microcystin RR, Microcystin YR and Nodularin.

Hypothesis

Since records, compiled by the State of Delaware Division of Fish and Wildlife, indicate that 14 out of 38 ponds have historically exhibited blue-green algae blooms, some of which included extensive masses of scum; it appeared possible that Microcystin concentrations could be elevated during bloom events. Since hepato-toxins produced by blue-green algae *Microcystis* and *Anabaena* had not been tested previously in Delaware freshwater ponds, it was necessary to collect data in order to demonstrate a need for routine monitoring of these surface waters.

Methods

Surface water samples were collected from 6 freshwater ponds within the State which historically exhibited blue-green algal blooms. Samples were collected adjacent to the shoreline and in surface scum when present. Presence of scum anywhere in the pond on the day of collection was recorded. Ambient water samples were analyzed for the predominance of *Microcystis* and *Anabeana* via light microscopy (100X using a 0.1 Palmer Cell Counter) and for Microcystin concentrations using commercially available Microcystin analysis kits: EnviroLogix Microcystin Tube Kit ET022 (years 2002 & 2003), Strategic Diagnostic EnviroGard Microcystin Plate Kit (2002), and EnviroLogix Microcystin Plate Kit EP022 in conjunction with the Bio Tek μQuant Spectrophotometer Plate Reader Flx800 using KC4 software (2003). Intracellular Microcystin was released by freezing the water samples for a minimum of 24 hours but less than 1 month.

Results

- High concentrations of the organism *Microcystis* is NOT necessarily a good indicator of the level of Total Microcystin in water sample.
- Two out of 88 samples from 9 ponds showed Dissolved Microcystin > 3 ppb and 3 out of 88 showed concentrations > .5 ppb but < 3 ppb using the EnviroLogix Tube Kit. This occurrence was noted in 5 separate ponds all of which were sampled on September 24, 2002.
- Five out of 18 samples analyzed with the EnviroGard Microcystin Plate Kit showed Total Microcystin > 1.00 ppb, this is the provisional upper limit of Microcystin LR established by the World Health Organization for finished drinking water. Samples showing this exceedence (1.28 ppb to 3.28 ppb)

were collected in late September (Sept. 26 and 30, 2002) at 4 separate ponds.

- One out of 33 samples from 10 ponds showed Total Microcystin concentrations > 1 ppb with the EnviroLogix Plate Kit., and 1 out of 33 samples showed Dissolved Microcystin >0.5 but < 3.0 ppb. These results are from late August and mid-September respectively and from 2 separate ponds in 2003.

Conclusions

Surface scum of freshwater ponds showed high concentrations of *Microcystis* 88% of the time, but had measurable levels of Microcystin (>0.5 ppb) only 17% of the time.

Total Microcystin measures both the Microcystin free in the water-column and that sequestered in the blue-green algae cells and as such its determination is recommended if Dissolved Microcystin concentrations approach 1 ppb.

Investigation of microcystin concentrations and possible microcystin–producing organisms in some Florida lakes and fish ponds

Yilmaz M, Phlips EJ

Department of Fisheries and Aquatic Sciences, University of Florida, Gainesville, Florida 32653 USA

Introduction

Occurrence of potentially toxic cyanobacteria blooms is very common in Florida lakes. These include blooms of *Microcystis, Anabaena, Oscillatoria* and *Cylindrospermopsis*. Among these the first three are potential microcystin producers. However, the toxicity data for many lakes are not available. We investigated 10 lakes, ranging from 4 to 424 acres in Hillsborough County, Tampa, FL over a two month period. In addition we tested samples from Lake George, St. Johns River system and two fish ponds. Here we present the data obtained from microcystin ELISA and PCR with specific primers targeted to the condensation domain of *mcyA* gene, which were designed to detect microcystin–producing *Microcystis, Anabaena* and *Oscillatoria* strains.

Hypothesis

We propose that many of the lakes in Florida that have high cyanobacteria populations will have microcystins.

Methods

Water samples were collected from each lake with a vertical integrating sampling tube. Sub–samples were taken for phytoplankton counts, chlorophyll measurements, ELISA and PCR. ELISA was performed with the Envirologix Microcystin Plate Kit. Primer pairs, mcyA–CD 1F and mcyA–Cd 1R were used in the PCR reaction.

Results

In the case of Hillsborough County lake samples, microcystins were detected in 3 out of 10 lakes. These were Cedar Lake, Lake Brant and a scum sample on the shore of Lake Magdelene. Microcystin concentrations were 0.11, 1.84 and 18.58 $\mu g.l^{-1}$, respectively. ELISA measurements are compared with the results obtained from the PCR analyses. The possible organisms that might be producing microcystins in these samples are discussed.

Conclusion

In addition to taxonomic identification and assessment of population dynamics of cyanobacteria present in water bodies, it is also vital to determine toxin concentrations and toxin producing organisms. ELISA and PCR are two fast and inexpensive methods that can perform these functions.

Potentially toxic cyanobacteria in Chesapeake Bay estuaries and a Virginia lake

Marshall, H.G.[1], Burchardt, L.[2], Egerton, T.A.[1], Stefaniak, K.[2], and Lane, M.[1]

[1]Department of Biological Sciences, Old Dominion University, Norfolk, Virginia 23529-0266, U.S.A., [2]Department of Hydrobiology, Adam Mickiewicz University, 61-614 Poznan, Poland

Introduction

Since 1985, phytoplankton populations have been monitored monthly in Chesapeake Bay (U.S.A) and its tidal estuaries. This surveillance has identified 29 potential toxin producing phytoplankters in these waters which includes representative taxa of dinoflagellates, diatoms, cyanobacteria, and raphidophytes (Marshall 1995, Marshall et al. 2005). Although toxic events attributed to these taxa have been rare, their wide spread presence is noted, with evidence for increased seasonal bloom events indicated for several species in our records. Local studies of HAB's in Virginia's regional lakes have been minimal. However, annual blooms of *Microcystis aeruginosa* have occurred annually in the Potomac River, a major tributary of Chesapeake Bay. This species produced a bloom (>10^6 cells ml^{-1}) in 2004 that lasted from June through August and included an extensive area of the river, with microcystin levels consistently recorded at >3 ppb. Decomposition products and wind blown algal masses accumulated along the river shore; these conditions and the high microcystin levels temporarily closed recreational usage in some regions of the river.

Results

M. aeruginosa is a common algal component of the tidal rivers in this region, with significant concentrations annually present in the tidal fresh and oligohaline river sections. Its development farther downstream and into Chesapeake Bay increases during the summer months and is enhanced by periods of extended rain and increased river flow. Other potentially toxic cyanobacteria identified in these rivers have included: *Anabaena affinis*, *A. recta, A. solitaria, Aphanizomenon flos-aquae, A. issatschenkoi, Microcystis firma, Planktothrix agardhii, P. limnetica,* and *P. limnetica* f. *acicularis*.. Long term trend analysis from 1985-2004 indicates there are significant increases in biomass and abundance of cyanobacteria within the James, Rappahannock, and York Rivers in Virginia, and as well in Chesapeake Bay. In addition, Lake Burnt Mills, a shallow reservoir of 288 ha located in southeastern Virginia experienced an extensive cyanobacteria bloom in July 2005. This bloom persisted for several days and was produced by *Microcystis aeruginosa* and *M. wesenbergii*, with concentrations of 3.5×10^5 and 22.9×10^5 cell ml^{-1} respectively. During this period microcystin levels near shore exceeded 3 ppb.

Summary

Long term trend analysis of phytoplankton populations in three Virginia rivers and Chesapeake Bay indicate an increase in the abundance of biomass of cyanobacteria has occurred in these waters since 1985. Among these cyanobacteria are populations of potentially toxic species. The most common bloom producing taxon within this group is *Microcystis aeruginosa*. Microcystin concentrations have been associated with these blooms in the Potomac River. Continual monitoring of these rivers and Chesapeake Bay will continue for the presence harmful algal species. In addition, greater surveillance of these bloom producers in regional lakes has taken place and will continue in the future.

Expanding existing harmful algal blooms surveillance systems: canine sentinel

Chelminski AN, Williams CJ, Hunter JL, Shehee MW

North Carolina Division of Public Health

Introduction

There have been several episodes in the United States in the past several years in which dogs have died after having been exposed to cyanobacterial (blue green algae) blooms. Reports of their deaths preceded any reports of human illness related to exposure to these blooms. Investigations of these as well as other animal deaths have shown them to be related to exposure to cyanobacterial toxins in these blooms. Documented deaths have occurred in both domestic animals (pets and livestock) and in wildlife. Some of the factors that may account for the susceptibility of dogs to illness and death from cyanobacterial toxins may include increased exposure to bloom waters during the summer months and the size of the animals relative to the dose of toxin that they received. While the poisoning and death of these dogs is disheartening, the reporting of canine deaths associated with exposure to bloom waters serves as an important tool for preventing human exposure and for reducing further animal exposure to cyanobacterial toxins.

Hypothesis

The reporting of dead dogs will not detect potentially harmful algal blooms.

Methods

During the past year, the North Carolina Harmful Algal Blooms (HAB) and the Veterinary Public Health Programs coordinated a sentinel surveillance system to detect acute lethal poisonings of dogs. This program encourages practicing veterinarians to voluntarily report any deaths of dogs that they think might be related to exposure to blue green algae blooms. Veterinarians were notified of this program by use of an internet–based " Listserv" communication tool.

Results

Over 800 veterinarians were informed about the rationale for reporting dog deaths and they were provided with an educational flyer to further document and reinforce this message. No dogs deaths have been reported to date. One dead waterfowl incident was reported by a concerned citizen, but the presence of a harmful algal bloom or toxins was not determined.

Conclusions

At this time the HAB and Veterinary Public Health programs are planning to expand the Canine Sentinel Surveillance Program to include additional animals (livestock and wildlife), other state and local public health agencies, and to encourage other states to take part in this program. The North Carolina Cooperative Extension Program is one example of another governmental agency that is being encouraged to participate in this surveillance program. The Canine Sentinel Surveillance Program will be used to help detect occurrences of deaths in animals. This program should also reduce the number of animal poisonings due to algal toxins by increasing public awareness of this potential problem and through preventive measures to reduce the exposure of animals to algal toxins in public recreational waters of North Carolina. The increased public awareness may also reduce the potential for human exposure to algal toxins.

Use of embedded networked sensors for the study of cyanobacterial bloom dynamics

Stauffer, BA,[1,3] Sukhatme GS,[2,3] Oberg C,[2,3] Zhang, B[2,3]; Dhariwal, A[2,3]; Requicha, A[2,3]; Caron, DA[1,3]

[1]Department of Biological Sciences, University of Southern California;
[2]Department of Computer Science, University of Southern California;
[3]Center for Embedded Networked Sensors (CENS)

Introduction

Traditional monitoring techniques are limited in the development of predictive models for aquatic microbial populations that requires very fine spatial and temporal resolution of data. The need for continuous (or real-time remote) monitoring of the environment combined with the desire for directed (intelligent or autonomous) sampling has prompted the development of a sensor network. The network incorporates low-energy demand, and highly adaptable sensors which exploit recent advances in computer networking and robotics to process sensor data and ensure high data fidelity. The coordination of stationary sensor nodes and mobile sensing using a sampling robot allow for efficient collection of samples from features of interest, as exemplified in a recent study of a cyanobacterial bloom in Lake Fulmor, California.

Hypothesis

The application of Embedded Networked Sensing (ENS) technology to monitor cyanobacterial bloom in a lake environment will provide new observational capabilities with unique information on the distribution and/or behaviors of planktonic assemblages.

Methods

The sensor network (NAMOS: Networked Aquatic Microbial Observing System) consists of 10 stationary buoys and one mobile robotic boat. Each buoy is equipped with a computer, sensor suite, and wireless communication. They are networked and communicate with each other and a shore-based station via wireless ethernet. Onboard sensors include a thermistor array for measuring water temperature to 3 m depth and a fluorometer capable of detecting chlorophyll (chl) *a* concentrations from 0.5-500 μg/L. The robotic boat is equipped with similar sensors and processing capabilities in addition to a water sampler capable of taking six 4-ml samples. The robotic boat is autonomously controlled using information obtained from the network.

Results

Over the course of a 4-day NAMOS deployment, the chl *a* concentration showed high temporal variability. Cyclic daily variations in subsurface chl *a* fluorescence were observed with a peak between the hours of midnight and 5 am. During this time period, chl *a* concentrations increased from a day-time average of ≈2.5 μg/L to >6 μg/L. *Spirulina* sp. strongly dominated the phytoplankton community. The sensor network also detailed the spatial distribution of photosynthetic organisms along the length of Lake Fulmor, indicating increased concentrations of chl *a* towards the southwest end of the lake.

Conclusion

The presence of daily variations in chl *a* concentration at all static node stations implies a strong vertical migratory behavior of phytoplankton in the lake, most likely *Spirulina*. Accumulations of chl *a* in the southwest corner of the lake suggest reduced mixing or increased nutrients in this deeper, more protected area. And finally, the ability to resolve this trend at several points along Lake Fulmor and over the course of several days, and the combination of these data with autonomously collected water samples, demonstrates a marked improvement over traditional point sampling techniques.

Bloom and toxin occurrence

Suseela MR

Algology laboratory, National Botanical Research Institute, Lucknow, India

Fresh water is a renewable but a finite and vulnerable resource and essential to sustain life, development and the environment. In fact 97.3 percent of earth's water is saline and 2.7 per cent is fresh water. 70 per cent of this fresh water is frozen in ice caps of Antarctica and green land and balance is present either as soil moisture or underground moisture. As a result less than 1 per cent of all the water on earth is accessible for direct human use. This water is found in lakes, rivers, reservoirs etc. Only this tiny portion of the planet's water is recycled by nature's cycle.

Water is getting scarce due to rising population, rapid urbanization and growing industrial demands. People dump wastes, untreated sewage and chemical discharges, which pollute the sources of water like rivers, lakes, ponds and even underground resources. The most significant issues of water ecosystem is the eutrophication and deteriorating water quality including the development of numerous harmful algal blooms. Significant impacts of these blooms are high biomass, visible surface scums, loss of submerged aquatic vegetation and benthic habitat. Harmful cyanobacterial blooms produce toxins and affect commercial species like fish etc. Non–toxic blooms affect the benthic flora and fauna due to decreased light penetration. These blooms also affect the recreational activities of humans.

Microcystis aeruginosa is the most common cyanobacterial HAB not only in US but also all over the world in fresh, eusturine and marine waters. Other toxic blooms formed by *Lyngbya majuscula, Schozothrix calcicola, Oscillatoria nigroviridis* cause swimmer's itch and these are commonly found in tropical and sub–tropical sea waters. *Anabaena flosaquae, Aphanozomenon flosaquae* are the common cyanobacterial toxic blooms. *Gleotrichea intermedia, Aphanothece gelatinosa, Anabaena iyengarii, Cylindrospermum stagnale, Scytonema javanicum, Scytonema simplex, Oscillatoria princes, Nodularia, Lyngbya martinsiana, Phormidium anomala, Nostoc commune* are the other common toxin producing cyanobacterial blooms.

Cyanobacterial blooms which are toxic in fresh water may not be necessarily toxic marine environment or vice versa. In general toxic cyanobacterial blooms are same or similar all over the world. According to the literature 25 per cent of cyanobacterial blooms produce toxins. Cellular target phytoplankton toxins are ichyotoxins, neurotoxins, hepatotoxins, hemolysins and cytotoxins. Most commonly observed toxins all over the world are:

- Paralytic shellfish poisoning toxins (PSP) – water soluble neurotoxins.
- Amnestic shellfish poisoniong toxins (DSP) –water soluble neurotoxins
- Neurotoxic shellfish poisoning toxins (NSP) _lipid soluble brevetoxins
- Ciguatera fish poisoning toxins (CFP) –lipid soluble heat stable.

These are the most commonly observed toxins in US and all over world in marine and eustarine waters. If the trend of eutrophication continues in the same manner cyanobacterial HABs will increase proportionately and pose a greater threat not only to natural ecosystems but also to the human health. Therefore, there should be legislative actions to ensure that efforts to achieve nutrient reduction and establish a water quality standard.

Public education is one of the major tools other than scientific research in efforts to minimize the impacts of Cyanobacterial HABs and their toxins in marine, eustarine and fresh waters.

Cyanotoxins in the tidewaters of Maryland's Chesapeake Bay: The Maryland Experience

Tango P,[1] Butler W,[2] Michael B[1]

[1]Maryland Department of Natural Resources, 580 Taylor Avenue, D-2, Annapolis, MD 21401. [2]Maryland Department of Natural Resources, 1919 Lincoln Drive, Annapolis, MD 21401.

Introduction

Cyanobacteria blooms were noted in the Potomac River and the upper Chesapeake Bay during the 1950s and 1960s coincident with the invasion of water milfoil. Since 1985, cyanobacteria blooms have been documented in the tidal tributaries of Chesapeake Bay almost annually during summer months by the Maryland Department of Natural Resources (MDNR) long-term comprehensive water quality monitoring program. During September 2000, an extensive late summer bloom of *Microcystis* on the Sassafras River, however, was among the first blooms tested for cyanotoxins in Maryland and results were positive for elevated concentrations of microcystin. The microcystin levels (591.4-1041 ug*g^{-1} dry wt) led to the Kent County Health Department closing a public beach in the bloom area for the remainder of the year, the first beach closure in the history of the state due to detected levels of cyanotoxins.

Hypotheses

We hypothesized that elevated cyanobacterial toxin levels were common features of the annual blooms in the tidal tributaries and multiple toxins would be present. Such findings of an increased diversity of toxic HABs and CyanoHABs would represent expanded management concerns regarding human water-related activities and living resources effects for the Bay.

Methods

From 2002 to 2004, MDNR conducted cyanotoxin surveys working with Dr. Wayne Carmichael (Wright State University, Dayton, OH) and Dr. Greg Boyer (State University of New York College of Environmental Science and Forestry, Syracuse, NY). We examined water samples from tidal regions of the Chesapeake Bay with elevated concentrations (>10,000 cells/ml) of cyanobacteria.

Results

Microcystin, anatoxin-a and saxitoxin were detected from tributaries throughout Maryland tidewaters at wet weight concentrations of 0.34-657.9 ug*L^{-1} (n=40), 0.009-3 ug*L^{-1} (n=6) and 0.003 ug*L^{-1} (n=1), respectively. Mean and median concentration were 35.24 ug*L^{-1} and 5.04 ug*L^{-1} for microcystin and 0.54 ug*L^{-1} and 0.05 ug*L^{-1} for Anatoxin-a. In 100% of *Microcystis* bloom samples tested (concentrations > 10,000 cells*ml^{-1}) there were detections of microcystin. Anatoxin-a and saxitoxin testing has been uncommon.

Conclusions

1. Microcystin concentrations exceeded the WHO drinking water standard of 1 ug*L^{-1} with 85% of test samples. Anatoxin-a and saxitoxin have also been detected in the open waters of the tidal tributaries of the Chesapeake Bay system.
2. The findings increase the range of habitats where potential human health and living resource threats due to aquatic born toxins must be considered by management agencies in Maryland.
3. County health departments again closed beaches in 2003 and 2004 in response to recommendations from Maryland's Interagency Harmful Algae Task Force regarding the detected levels of cyanotoxins. State resource agency efforts to alert the public regarding timing and location of bloom waters as well as potential risks to human health, pets and livestock included 1) HAB webnews articles on the State resource agency websites 2) State Press Releases linked with MDNRs "Eyes on the Bay" water quality monitoring website, and 3) print, radio and TV news coverage of the issues.

Harmful Algal Blooms and Cyanotoxins in Metropolitan Water District's Reservoirs

Izaguirre G

The Metropolitan Water District of Southern California (MWDSC) supplies drinking water to about 18 million people in six counties in the coastal plain of southern California. MWDSC is composed of 26 member agencies, which are cities or regional water agencies. Its two sources of water are the Colorado River and water from northern California, called State Project Water (SPW), delivered through the California Aqueduct. MWDSC operates three reservoirs in Riverside County: Lake Mathews, Lake Skinner and Diamond Valley Lake. The former is the terminal reservoir of the Colorado River Aqueduct; the other two reservoirs are supplied with a blend of the two waters. In addition, the state Department of Water Resources owns and operates Silverwood Lake, Lake Perris and Castaic Lake, three combined drinking water and recreational lakes that receive State Project water. Metropolitan regularly receives water from Castaic Lake (northwest of Los Angeles) and Silverwood (in the San Bernardino Mountains), and occasionally also uses water from Lake Perris.

Metropolitan has a long history of algal problems, in the form of planktonic blooms and benthic proliferations. The main concern is taste and odor, specifically the compounds geosmin and 2-methylisoborneol (MIB), which impart a disagreeable flavor to the water and cannot be easily removed by conventional treatment methods. All of the reservoirs listed above have experienced algal blooms of one kind or another, including in some cases known toxigenic species. In addition, Lakes Mathews, Skinner, Perris, and Diamond Valley have developed benthic mats that have resulted in severe off-flavor problems.

In 1996, an AWWARF project on the occurrence of algal toxins in raw and treated waters in the United States and Canada was initiated, perhaps the first serious effort by the U.S. drinking water industry to assess the extent of the problem. This study showed that at least one type of cyanotoxin, the microcystins, can be found in many water sources in the U.S., sometimes even in treated waters, albeit at low concentrations (Carmichael, 2001).

Metropolitan was a participating utility in this study, and the results indicated that microcystins could be found in cyanobacterial bloom material from various source-water reservoirs, and in the corresponding plant influents and in some cases even the effluents. However, the levels were generally low, and no further monitoring was done on our system until the summer of 2001, when there was a severe bloom of *Aphanizomenon* in Silverwood Lake and of *Microcystis* in Lake Skinner. Samples of bloom material were sent to the laboratory of Dr. Gregory Boyer at SUNY in Syracuse, NY. Two of the *Microcystis* samples had relatively high microcystin levels, while the *Aphanizomenon* samples had no significant levels of any of the tested toxins. These results, though not surprising (the bloom samples giving the highest results were fairly concentrated samples), prompted concern regarding the need for more regular monitoring of these compounds in the water. In view of the likelihood that cyanotoxin monitoring will be required under the UCMR, this concern would appear justified.

The results of the 2001 samples prompted the development of a cyanotoxin monitoring program at Metropolitan's Water Quality Laboratory. This monitoring utilized two ELISA test kits for microcystin (Envirologix Inc., Portland, ME), a plate kit and a tube kit. The former was used primarily for testing water samples, while the latter was used for screening bloom samples, benthic algal samples and cultures. In addition, many samples were sent to Dr. Boyer's lab under a contract with Metropolitan. This was for confirmation of microcystin and identification of the variant. Also, since we are unable to test for toxins other than microcystin, the contract lab was needed to test for these other toxins, e.g., cylindrospermopsin and anatoxin-a.

We now have two years' worth of data on cyanotoxins in our system. Microcystin has been found in varying concentrations in surface water from all six reservoirs that were sampled. The concentrations ranged from 0.116 μg/L to 55.27 μg/L, although most of the samples were closer to the lower end of this range. The highest values were all from samples of concentrated bloom material, usually dominated by *Microcystis*. However, the majority of the water samples tested had no detectable microcystin or were just barely over the detection limit of 0.147 μg/L. The WHO guideline for drinking water is 1.0 μg/L.

In the summer of 2003, we also began testing benthic algal material from the shallows of various reservoirs, and in the process of this testing found that a cyanobacterium that is very common in three of those reservoirs produces microcystin. Eighteen isolates of this organism (a *Phormidium* sp.) were sent to Dr. Boyer's lab for verification, quantitation and identification of the toxin, and twelve of them were confirmed as strong microcystin producers. The variant was microcystin-LR in most cases. In addition, seven sediment samples were analyzed, and some of them had relatively high concentrations of microcystin, the highest being 287.95 μg/g dry wt, and the lowest 1.23 μg/g dry wt. A paper on this work is in the process of being reviewed by two collaborators in preparation for submission to a journal.

The benthic microcystin producer is the only source of the toxin found to date in our system other than *Microcystis*, and may be significant in being a more permanent "inhabitant" than the more transient and seasonal planktonic sources. Moreover, this organism grows intermingled with cyanobacteria that produce odorous compounds like geosmin and 2-methylisoborneol. These benthic proliferations are periodically treated with copper sulfate to control taste-and-odor problems. Application of the algicide to these organisms can in theory release the microcystin into the water, potentially affecting water supplied to several treatment plants.

No other toxins have been found in significant levels in Metropolitan's waters. Blooms of *Anabaena flos-aquae* and *A. lemmermanii* have been tested for anatoxin (at Dr. Boyer's lab), but all were negative. However, an *Anabaena* isolated from Castaic Lake in 1999 was found to produce anatoxin-a.

Chapter 9: Causes, Prevention, and Mitigation Workgroup Report

Workgroup Co-chairs: Gina Perovich, Quay Dortch, James Goodrich

Workgroup Members and Authors[1]**:** Paul S Berger, Justin Brooks, Terence J Evens, Christopher J Gobler, Jennifer Graham, James Hyde, Dawn Karner, Dennis (Kevin) O'Shea, Valerie Paul, Hans Paerl, Michael Piehler, Barry H Rosen, Mary Santelmann, Pat Tester, Judy Westrick

Introduction

Cyanobacteria (blue-green algae) are estimated to have evolved 3.5 billion years ago, at which time they began to add oxygen to the existing anaerobic atmosphere, actually changing the chemistry of the planet and allowing new life forms to evolve. These ubiquitous microbes are capable of tolerating desiccation, hypersalinity, hyperthermal conditions, and high ultraviolet radiation, often for extensive periods of time. Recently, cyanobacteria have responded to human alterations of aquatic environments, most notably nutrient-enhanced primary production, or eutrophication. In fact, cyanobacterial blooms are now viewed as widespread indicators of freshwater, brackish and marine eutrophication.

Due to the complex interactions between physical and ecological processes, it is difficult to point to any single, definitive ***cause*** for the development and proliferation of these blooms. In reality, cyanobacterial harmful algal blooms (CHABs) likely result from a combination of factors, including hydrology, available nutrients, sunlight, temperature, and ecosystem disturbance; any number of which must interact in precisely the right combination to create optimal conditions for growth. Thus, it should come as no surprise that successful ***prevention*** (inhibiting bloom formation

---

[1]See workgroup member affiliations in Invited Participants section.

through the manipulation of causative factors) and ***mitigation*** (ameliorating the effects of and/or controlling the blooms themselves) strategies for dealing with CHABs, may require correspondingly complex approaches.

This document examines possible causative factors that have been implicated in the initiation and maintenance of cyanobacterial blooms, methods for bloom prevention, and techniques for managing the risks posed by cyanobacterial blooms and their toxins. The material presented reflects the views of the workgroup participants about the current state of CHAB knowledge and identifies the research needs and priorities necessary for addressing the problem. All identified research needs are considered to be "high-priority", and are classified as "near-term" and "long-term" because the completion of certain research tasks is dependent upon the completion of other. Therefore, the numerical ordering of research goals below does not indicate prioritization, but is for ease of reference only. This product is a direct response to the charges that were received by the workgroup, and it is organized according to four topic areas as shown below:

- **Causes** - Identify and prioritize research needed to better describe factors contributing to CHAB initiation, maintenance, and termination in fresh, estuarine and marine waters, recognizing the interconnectivity of systems.

- **Prevention (pre-bloom efforts)** - What are the best prevention options available right now for use by managers? Are there easily identifiable factors that will improve their efficacy? Identify and prioritize research to improve land and water management strategies for preventing or minimizing CHABS, focusing both on techniques that are currently available and emerging approaches in fresh, estuarine, and marine water environments with the intent of providing guidelines to managers.

- **Control/Mitigation (post-bloom efforts)** - Can draft guidelines for controlling and mitigating CHABs be developed now for use by managers? If not, identify and prioritize research to improve processes for removing cells and toxins from fresh, estuarine, and marine water bodies and drinking water, focusing both on current and emerging techniques with the intent of providing guidelines to managers.

- **Economic Analysis** - What ecological factors should be included in models used to predict the relative costs and benefits of processes used to prevent or reduce the occurrence of CHABS in water bodies and cyanotoxins in drinking water?

Overarching Themes

In this workgroup's initial discussions, it quickly became evident that there were a number of overarching issues - areas of concern that were necessary to take into account when addressing the CHAB problem, but that did not fit neatly into the four categories of charges above. Instead, these issues extend across the topics and charges, such that they are a vital part of the entire framework for examining the causes, prevention, and mitigation of CHABs.

The Freshwater – Marine Continuum

There is general agreement that CHABs are a world-wide, rapidly-expanding water quality, human health, and ecosystem problem (Paerl, 1988; Chorus and Bertram, 1999). Furthermore, planktonic and benthic CHAB proliferation is not necessarily confined to any particular type of aquatic habitat, with recent expansions noted across a spectrum of ecosystems ranging from previously pristine mountain streams and springs to large lakes, estuaries and coastal seas (Paerl and Fulton, in press). One commonality of systems impacted by CHAB expansion is that their water and air-sheds are experiencing accelerating human population growth accompanied by increased nutrient and other pollutant loadings, as well as hydrologic modifications. There is little doubt that there is strong linkage between human use and modification of water and air-sheds and CHAB dynamics. In addition, climatic changes, including changes in large storm (i.e., hurricane), flood and drought frequencies and intensities have impacted nutrient and hydrologic regimes of these systems. Identifying causes and formulating prevention and mitigation strategies must take human and climatic factors and drivers into consideration. Because the impacted aquatic ecosystems and their drainage basins are inextricably linked, research, monitoring and management approaches must address the CHAB issue across the freshwater-marine continuum. In most instances, CHAB dynamics are not confined to a single component system and often cross a salinity gradient. Thus, causes and effects must also be traced and addressed across this continuum. Accordingly, the operational scale at which CHAB issues should be assessed and ultimately managed is that of this continuum, which, depending on the specific problem may include habitat, ecosystem, and regional levels.

Relevant Scales for Research and Management

CHAB occurrence is influenced by interacting meteorological, hydrological, physiochemical, and biological factors on a variety of spatiotemporal scales. For example, discernable ecological patterns are often influenced by the specific spatiotemporal resolution associated with the study, and results may change with scale. Additionally, global, regional, and local factors all influence the timing, magnitude, and duration of CHABs. Thus, macro-, meso-, and micro-scale studies, as well as studies involving the incorporation of more than one of these scales, are necessary to adequately describe the regulatory factors driving CHAB formation and to identify the most effective scale at which to implement management efforts. Key spatiotemporal scales include: a) global studies to assess long-term (decadal) changes in climate, environmental patterns, and eutrophication with respect to the apparent increase in CHAB development over the last fifty years; b) ecoregion and air/watershed studies to assess the impacts of moderate to long-term (years) changes in emission of pollutants, land-use practices, and watershed/waterbody management on CHAB formation; and c) ecosystem, community, and population studies to assess the influence of near-term (hours to seasons) changes in the environmental and physiological factors that influence CHAB occurrence, magnitude, and duration.

Monitoring

The first step in understanding the prevalence and severity of CHABs is monitoring. Monitoring includes a variety of techniques that range from collecting discrete water samples that are analyzed for the presence of cyanobacteria, to continuous-flow instruments that measure pigments that are unique to this group of algae. The types and quantities of cyanobacteria are an indicator for the types and quantities of cyanotoxin that may be present. Monitoring also provides early warning so that human health can be protected before serious problems develop. The ultimate aims are to determine: 1) whether toxin-producing cyanobacteria are present; 2) the concentration of organisms of concern; 3) if the population and/or toxin level(s) are increasing or decreasing; 4) the current level of impact; and 5) the development, transport, and dissipation of the bloom and bloom impacts.

The design of a monitoring program should be tailored to the specific issues in a waterbody. For example, monitoring a reservoir used for drinking water should focus on the areas near the water intake structure(s) or the actual raw water brought into the facility. The timing of sampling in reference to the seasonal occurrence of the cyanobacterial bloom should also be considered, with little or no sampling needed in cold months, and more intense sampling during warmer seasons and bloom development. If feasible, information on the conditions that initiate or sustain bloom formation should be collected so that models can be developed to better understand and predict blooms. Understanding the causes of blooms will ultimately lead to better methods of prevention.

For recreational waters, wind-driven currents often cause buoyant cyanobacterial blooms to amass on shorelines. These accumulations contain orders of magnitude more cyanobacterial cells than blooms in open water areas, thus presenting more of a health risk to humans and animals. Monitoring the spatial location of these accumulations is important in determining the maximum potential exposure during a bloom; however, it is difficult to quantify the number of organisms present. Many cyanobacteria are capable of regulating their buoyancy on a diurnal basis as a function of their internal physiology. Thus, integrated water sampling throughout the entire water column may be needed in open waters. For example, *Cylindrospermopsis,* which does not form a surface scum, inhabits mid-water depths at lower light intensities, necessitating integrated water sampling.

At the core of many monitoring systems are the sensors and the platforms on which they are deployed. One of the greatest needs in this field is the development of quantitative, easy-to-use, rapid sensors of CHAB cells and toxins in natural planktonic and benthic assemblages for use across hydrologically and geographically variable ecosystems at multiple scales. Although microscopic identification and counts are needed for verification, they are time consuming, require expertise often not available, and in the case of filamentous and colonial taxa, (e.g. *Microcystis*, *Nodularia*, *Anabaena*, *Aphanizomenon*) lack quantitative rigor. New techniques for monitoring, such as the use of diagnostic photopigments and various molecular approaches, are under development and will provide additional tools for helping to understand the potential health risks associated with cyanobacterial blooms. Satellite images and spectra from airborne radiometers showing ocean color and sea surface temperature have been used for detecting and tracking cyanobacterial blooms. However, there are other potential remote sensing techniques that can be developed and implemented as currently available satellites are phased out. There is a need for sensors that provide additional information about other environmental factors, such as nutrients, in order to understand the role of human activities in causing blooms. There is also a particular need for sensor systems applicable to large lake, estuarine and coastal ecosystems. These environments require large numbers of samples and rapid throughput, necessitating the development and deployment of sensors on unattended monitoring platforms, including buoys, AUVs (automated underwater vehicles), aircraft, ferries, and other "ships of opportunity" as well as satellites.

Modeling

Mathematical models are needed to better understand CHABs and their relationships to the surrounding environment. Models will help to establish a basic framework for relating potential causative factors to the occurrence of CHABs, improve our understanding of the conditions that initiate toxin production, characterize the factors that maintain CHABs under various conditions, and to both develop and measure the efficacy of various management strategies.

In order to develop these models, whole watershed research projects are needed in which nutrient exports from controlled agricultural watersheds are monitored and compared to those that have methods in place to aggressively reduce nutrient export from non-point source pollution. These should be long-term research/monitoring sites at which hydrology as well as nutrient and sediment export can be studied over decadal time frames, since system response to land use change occurs over many years. This type of critical, watershed-level research would be analogous to the research done at Hubbard Brook which led to improved understanding of the consequences of acid rain and harvest practices on ecosystem biogeochemistry. Such studies could be undertaken in partnership with the National Science Foundation in coordination with targeted funding from other agencies, such as the Environmental Protection Agency, the US Geological Survey, the National Oceanic and Atmospheric Administration, and the National Aeronautics and Space Administration, for important ecosystem-level research such as that proposed for hydrologic observatories and/or national observation networks. Long-term data is crucial to the development, refinement, and validation of explanative and predictive models.

Modeling and mapping of regional risk levels could help target locations for monitoring programs and process-based research on causes and prevention of CHABs. Methodology could include the use of enhanced geodatabases of harmful algal bloom (HAB) occurrence and attributes of the systems in which they occur (developed by water body type) in conjunction with statistical exploratory data analysis methods to assist in initial model development and threshold determination (e.g., CHAID, Chi-squared Automatic Interaction Detector).

Infrastructure Needs

The ability to understand the causes of CHABs, so that they can be predicted, monitored, controlled, and prevented is dependent on the availability of critical infrastructure. These infrastructure needs were laid out in detail for all HABs in HARRNESS (2005). Reference materials and shared-use analytical facilities are needed so that scientists and managers can quickly and easily identify HAB species and toxins in cells, water, air, and other organisms. Certified toxin standards and HAB-specific probes must be readily available for routine use at reasonable cost. Taxonomic training for identifying HABs, using both morphological and new molecular methods, must be widely available at levels suitable for a range of expertise from local managers to expert researchers. Culture collections and tissue banks can be used to archive newly isolated species and samples from exposed animals for later study. Regional observing systems, now in the planning stages, should be developed with the capability of monitoring for CHABs. Platforms for remote sensing of HABs, such as satellites and in situ moorings, equipped with HAB cell or toxin-specific sensors, need to be developed. Finally, old HAB data must be rescued before it is lost; common data management plans must be developed and data repositories must be established so that raw data, as well as associated metadata, can be shared and large scale/long-term analyses can be conducted.

In the long-term, research programs cannot provide funding to support infrastructure. However, as HABs occur more often, and their impacts continue to intensify, the need for national and regional infrastructure support programs with a long-term funding base will also become more urgent (HARRNESS 2005).

Sample Guidelines (modified from the Wisconsin Division of Public Health's fact sheet on cyanobacteria, and their toxins, and health impacts)

Source: http://dhfs.wisconsin.gov/eh/Water/fs/Cyanobacteria.pdf

- Never drink untreated surface water, whether or not algal blooms are present. Boiling the water will not remove toxins. Owners should always provide alternative sources of drinking water for domestic animals and pets, regardless of the presence of algae blooms.
- If washing dishes in untreated surface water is unavoidable, rinsing with bottled water may reduce possible residues.
- People, pets and livestock should avoid contact with water where algae are visible (e.g., pea soup, floating mats, scum layers, etc.) or where the water is discolored. Do not swim, dive, or wade in this water. Do not use the water to fill a pool or for an outdoor shower.
- Always rinse off yourself and your pet after swimming in any ponds, lakes or streams, regardless of the presence of visible algae blooms. Pay close attention to the bathing suit area and pet's fur.
- Contact your local health department or department of natural resources office to report any large algae blooms on public or private lakes, streams or ponds.
- Never allow children or pets to play in or drink scummy water. Do not allow pets to eat dried scum or algae on the shoreline.
- Do not water-ski or jet-ski over algae mats.
- Do not use algaecides to kill the cyanobacteria. When the cyanobacteria cells die, the toxins within the cells are released.
- Obey posted signs for beach closings. Wait at least one to two weeks after the disappearance of cyanobacteria before returning to the water for wading, bathing or other activities.

Outreach/Education

In order to empower individuals and communities to act on environmental issues, it is necessary to increase environmental awareness through coordinated environmental education efforts. The public must be informed about known or potential cyanobacterial problems in their recreational waters or drinking water supplies, as well as the risks associated with them, so that they can make educated decisions based on that information. Some states have produced short informational documents that are available on the internet and/or disseminated via kiosks at or near impacted recreational areas. These informational documents usually explain what cyanobacteria or blue-green algae are, what algal blooms are, how to recognize a bloom, where blooms occur, and what conditions promote bloom development. They also describe the health effects that could occur with exposure to cyanotoxins, including typical symptoms. Other important information should be included about routes of exposure, so the public will understand how they or their animals are exposed to the toxin, such as the swallowing of a surface scum, contact with the skin, or inhalation of aerosols during swimming, bathing or showering in contaminated waters. This should be followed by suggested ways in which people can avoid or limit their exposure. Finally the public needs to be educated about the causes of HABs and informed about steps that concerned citizens or their representatives can take to prevent blooms or minimize their impacts.

Charge 1: Causes

Identify and prioritize research needed to better describe factors contributing to CHAB bloom initiation, maintenance, and termination in fresh, estuarine and marine waters, recognizing the interconnectivity of systems.

From research and management perspectives, identifying environmental factors which cause and sustain CHABs is key to developing an understanding of how to predict, prevent, and control these unwanted occurrences. While we know that nutrient and hydrologic conditions strongly influence harmful planktonic and benthic CHAB dynamics in aquatic ecosystems, and observations have shown that increased urbanization, agricultural and industrial development have led to increased nitrogen (N) and phosphorus (P) discharge, there are many other factors that remain unidentified, unquantified or unexplored. For example, additional factors such as N:P ratios, organic matter availability, light attenuation, temperatures, freshwater discharge, flushing rates (residence time) and water column stability likely play interactive roles in determining CHAB composition (i.e., N_2 fixing vs. non-N_2 fixing taxa) and biomass. Human activities may influence these factors either directly, by controlling hydrologic, nutrient, sediment and toxic discharges, or indirectly, by influencing climate.

Nutrients

Among the nutrient elements required for aquatic plant growth, N and P are often most stimulatory, because requirements are high relative to availability. It follows that N and P enrichment are often most effective in stimulating and supporting blooms in receiving waters (Fogg 1969, Reynolds and Walsby 1975). These elements have, and continue to be, the focus of efforts aimed at controlling blooms (Likens 1972, Schindler 1975, Shapiro 1990), although there is increasing interest in the role of trace metals in some systems. The most notable of these trace metals is iron (Fe) in its soluble form Fe^{++}. Iron is required for the synthesis and activity of photosynthetic, N_2 fixing and N assimilatory enzymes. Unlike N and P, Fe inputs are not strongly linked to human activities, such as agriculture, urbanization and most industrial activities. Rather, Fe availability is more often controlled by natural weathering or rocks, aeolian processes (dust transported by wind), and within-system oxygen (e.g., hypoxia) and biogeochemical (redox) cycling.

Excessive P (as orthophosphate) loading has been shown to promote potentially-toxic nitrogen (N_2) fixing genera (i.e., *Anabaena*, *Aphanizomenon*, *Cylindrospermopsis*, *Nodularia*), while excessive P and N (as dissolved inorganic N; nitrate and ammonium) loading can stimulate toxic blooms of non-N_2 fixing genera (*Microcystis*, *Lyngbya*, *Planktothrix*). From a supply standpoint, both the absolute amounts and relative proportions of these nutrients play important roles in determining the composition, magnitude, and duration of CHABs. There is also evidence that the production of toxic substances by CHABs is at least in part determined by the amounts and ratios of nutrients and trace metals supplied to affected water bodies (Sivonen, 1996; Skulberg *et al.,* 1994; Giani *et al.,* 2005).

Nutrient supply rates strongly interact with other environmental factors, including light, turbulence and flushing rates, temperature, pH (and inorganic C availability), salinity, and grazing pressure to determine; 1) *if* a specific water body is susceptible to CHAB formation, 2) the extent (magnitude, duration) to which CHABs may dominate planktonic and or benthic habitats, and, 3) whether an affected water body is amenable to management steps aimed at minimizing or eliminating CHABs.

Further research is needed to investigate the diversity of CHABs, including N_2 fixing and non-N_2 fixing groups, different cyanobacterial species, and toxic and non-toxic strains.

Near-term Research Priorities

1. Conduct retrospective analyses of long-term changes in eutrophication and CHABs especially in areas where eutrophication has been reversed by management actions (e.g., Lake Washington in the US, Lake Erken in Sweden, and other European water bodies). The purpose is to provide an independent approach to understanding the role of nutrients with regard to blooms and how to reverse the impacts.
2. Determine the response of a variety of N_2 fixing and non-N_2 fixing CHAB organisms to nutrients under controlled conditions in the laboratory and in the field with water enclosures (micro- and mesocosm experiments).
   - Examine and evaluate the selective impacts of various forms of nitrogenous (nitrate, ammonium, organic N) and phosphorus (orthophosphate, organic P), and nutrient ratios on CHAB growth, bloom dynamics and toxin production.
   - Determine the role(s) of iron and trace metals, alone or in combination with macronutrients, on CHAB growth, bloom dynamics and toxin production.
   - Determine the role of other environmental conditions, such as light intensity and quality, temperature, and water column stability on nutrient utilization and toxin production.
3. Conduct ecosystem-scale field studies, combined with use of monitoring data to determine how nutrient supplies and ratios interact with other anthropogenic stressors, the rest of the biota, hydrology, light, local weather patterns and climatic changes, vertical mixing, residence time, and benthic/pelagic coupling to control CHAB dominance and bloom occurrence and toxicity.

Long-term Research Priority

1. Develop models that can be used to predict CHAB bloom events and evaluate the effectiveness of preventive measures, such as setting Total Maximum Daily Loads based on current nutrient and turbidity conditions.

Climate Change

Considerable evidence indicates that the Earth and the oceans have warmed significantly over the past four decades, suggesting long-term climate change. Increasing temperatures and changing rainfall patterns have been documented. Cyanobacteria have a long evolutionary history, with their first occurrence dating back to at least 2.7 billion years ago. They evolved under anoxic conditions and are well adapted to environmental stressors including UV exposure, high solar radiation, high temperatures, and fluctuations in nutrient availability. These environmental conditions favor the dominance of cyanobacteria in many aquatic habitats, from freshwater to marine ecosystems. The responses of cyanobacteria to changing environmental patterns associated with global climate change are important subjects for future research. Results of this research will have ecological and biogeochemical significance as well as management implications.

Near-term Research needs

1. Retrospective analyses of existing literature, long-term observational programs and datasets of environmental patterns coupled with cyanobacterial abundance to examine the relationships between global change parameters and regional expansion of CHABs along temperature, precipitation, and nutrient gradients.
2. Studies of physiological conditions for cyanobacterial bloom dynamics and toxin production in different species and strains in relation to a variety of abiotic factors (e.g., temperature, light, UV, CO_2, pH).
3. Experimental (mesocosm, manipulative) studies to decouple various climatic and anthropogenic factors (e.g., temperature, light, UV, CO_2, nutrients).

Long-term Research Needs

1. Couple above-mentioned approaches, techniques and indicators to ongoing and developing observational programs (IOOS, IEOS, Coastal GOOS, etc.).
2. Develop models for predicting influence of climate change on occurrence and toxicity of CHABs.

Food Webs

Food web interactions may impact cyanobacteria bloom dynamics both positively and negatively. For example, outbreaks of cyanobacteria blooms in some US lakes appear to be partly stimulated by the arrival of recently established zebra mussel populations (*Dreissena* sp.; Vanderploeg et al. 2001, Raikow et al. 2004). However, the relationship between zebra mussels and toxic cyanobacteria has been temporally and spatially inconsistent. Some reports indicate that zebra mussel invasions have yielded increased cyanobacteria bloom occurrences (Vanderploeg et al. 2001, Raikow et al. 2004) while others state that zebra mussels have decreased densities of toxic cyanobacteria in NY waters (Caraco et al. 1997, Smith 1998). Moreover, many freshwater systems without zebra mussels experience very intense blooms of toxic cyanobacteria (Chorus and Bartram 1999). Intense grazing by herbivorous or planktivorous fish could also directly or indirectly promote blooms either through a trophic cascade or by consuming competing algae. However, these questions have yet to be investigated.

Laboratory experiments and field-work from ecosystems around the globe have indicated that grazing by some zooplankton can be disrupted by toxic cyanobacteria (Lampert 1987, de Bernardi and Giussani 1990, Sellner et al. 1993, Boon et al. 1994, Christoffersen 1996, Paerl et al. 2001) or negatively influenced by the secondary metabolites produced by cyanobacteria (Pennings et al. 1997, Nagle and Paul 1998 and 1999, Capper et al. 2006). Thus, the chemical defenses of cyanobacteria may play a critical role in bloom formation and persistence by limiting the grazing activity of potential consumers. In many systems which experience dense and/or toxic blooms, both cladocerans and copepods can be impacted, experiencing reduced feeding, reduced food assimilation or even mortality (Lampert 1987, de Bernardi and Giussani 1990, Paerl et al. 2001). However, the degree to which zooplankton and other consumers graze cyanobacteria blooms can be influenced by many factors including toxin concentrations, strains of cyanobacteria species, species of herbivore, various environmental conditions (Paerl 1988, Sellner et al. 1993, Boon et al. 1994, Christoffersen 1996, Nagle et al. 1998) and, perhaps, prior exposure to toxins (Walls et al. 1997, Hairston et al. 2001, Sarnelle and Wilson 2005).

Research is needed to clarify the role of food web interactions in the occurrence of CHABs. Research should strive to understand how these interactions impact the proliferation of different CHAB species and strains (i.e. toxic and non-toxic), as well as co-occurring, non-HAB species.

Near-term Research needs

1. Describe the ability of various clades of zooplankton (protozoa, cladocerans, copepods) and other consumers to graze individual CHABs species and strains relative to non-HAB species.
2. Determine how alteration of upper trophic level predator densities (e.g. fish) ultimately impacts the development of CHABs via alteration of aquatic food webs.
3. Determine the impact of benthic filter feeders on the development of CHABs and how this impact may vary with ecosystem trophic status.
4. Assess the impact of invasive species on the occurrence (development or prevention) of CHABs.
5. Assess the interactive effects of nutrients and climate change on food web interactions relative to the occurrences of CHABs.

Charge 2: Prevention (pre-bloom) through *Watershed Management*

What are the best prevention options available right now for use by managers? Are there easily identifiable factors that will improve their efficacy? Identify and prioritize research to improve land and water management strategies for preventing or minimizing CHABS, focusing both on techniques that are currently available and emerging approaches in fresh, estuarine, and marine water environments with the intent of providing guidelines to managers.

Eutrophication has long been known to be a major causal factor producing HABs (Chorus and Muur 1999, Paerl this volume). Current measures to reduce eutrophication and CHABs address the source, transport and fate of nutrients and include: 1) land management for reduction of nutrient export (USDA SCS 1996); 2) water management to minimize nutrient transport (e.g., hydrologic manipulation and water management practices such as removal or routing of water for irrigation, timing and extent of flow released from dams, etc.); and 3) water management to minimize impact of available nutrients in ambient water. These three approaches are discussed in detail in the following sections. All can be utilized both to prevent CHABs from occurring and to reverse the effects of excess nutrients in existing eutrophic systems.

In general, research in the area of prevention through watershed management must include research components across two major temporal scales of relatively equal importance; first, research into effective means for reducing and reversing eutrophication of aquatic systems over the long-term, and second, research into methods to reduce the impacts of existing eutrophication and decrease the likelihood of CHABs in the interim. Research at multiple spatial scales is needed to guide design and placement of strategies to reduce and prevent eutrophication, and to inform efforts to mitigate and reverse effects of excess nutrients in existing eutrophic systems. There is also a need to develop generalizable (across system) intrinsic indicators of land use and land use change that can be applied as predictive tools for CHAB potential (e.g., influence on stream flow and hydrodynamics).

Here, we outline and prioritize research needs in the area of CHAB prevention. In addition to the charge as described above, we also asked, what important data must be gathered to assure that prevention plans are as sound and successful as possible? Following are the priorities for research to improve watershed management strategies aimed at preventing or minimizing CHABs. The priorities focus both on currently available techniques and emerging approaches in fresh, estuarine, and marine water environments with the intent of providing guidelines to managers.

External vs. Internal Nutrient Control

Nutrient supplies often control the growth of CHABs in aquatic systems and managing nutrients is a common approach to preventing CHAB proliferation. Because nutrients are available through both internal and external sources, strategies must account for both sources and consider which type is more appropriate for management. Consideration of internal supplies is also crucial because many aquatic systems will not respond to reduction of external nutrients because of enormous internal stores. Triggers for internal loading are often distinct from those for external loading. External loading is often triggered by precipitation or river loading, whereas internal loading can be triggered by diverse mechanisms including changes in sediment redox potential and wind events.

Near-term Research Needs

1. Improve the understanding of the importance of internal and external supplies of nutrients in the full range of aquatic systems.

Long-term Research Needs

1. Of particular need are cost-benefit models that accurately predict the relative benefits and feasibility of internal and external nutrient supply management.

Land Management

The development of regionally appropriate land management practices for reduction in nutrient applications and exports draws upon two bodies of prior research. First, the US EPA has developed a classification scheme for dividing the US into regions on the basis of their geologic, physiographic, hydrologic, and water quality characteristics (Omernik 1987, Griffith et al. 1994). This framework can be used to guide the development of regional research plans and strategies for reducing the occurrence of CHABs. Second, the USDA (USDA SCS 1996) along with agronomists and agricultural specialists (Keeney 1990) have been working for decades to develop sustainable agricultural practices that will reduce the environmental impact of agriculture, and in particular, decrease erosion and sediment and nutrient export from cropland. The USGS programs for monitoring streamflow and water quality in multiple basins across the US are absolutely critical to understanding the interaction of management practices and water quantity and quality in producing CHABs. These USGS programs (such as the National Water-Quality Assessment Program) must be supported and continued if we are to have the long-term hydrologic data needed to understand existing conditions and trends, and to inform predictive models.

Arrays of methods already exist for decreasing nutrient exports through land management practices on agricultural lands (USDA NRCS 1997, SCS 1996). Mitsch et al. (2001) outlined the nature and extent of management practices needed to substantially reduce N export from the Mississippi River Basin, whereas authors of other studies (Vache et al. 2002, Santelmann et al. 2004, Boody et al. 2005) designed and evaluated watershed-specific alternative future landscapes to estimate the impact of the practices in the various alternative designs on flow, N (as nitrate), and erosion or sediment export. Most of the existing studies focus on N as the nutrient of interest. However, P export (often strongly influenced by sediment) and

N:P ratios have been found to be extremely important in determining the occurrence of CHABs. Research is needed to better understand the effects of various land management practices on P and sediment as well as N.

In order to guide policy and quantify the benefits of programs to reduce eutrophication, research is needed to decide what practices should be employed, where they should be implemented, and to what extent they should be used. Multi-scale, systems-level research is needed to understand and quantify watershed (areas on the order of 10,000 to 100,000 ha.) and basin (areas on the order of 100,000 to 1,000,000 ha) response to land use and management measures, as well as uncertainty and variability inherent in the system response. Because the appropriate methods to reduce nutrient exports will vary among regions, research is needed to design effective approaches and to implement methods appropriate to the region in which they are to be used (Santelmann et al. 2001). Finally, all future research should involve regional experts and stakeholders who can: 1) identify regional goals, benchmarks and timetables for achieving those goals; 2) develop regional strategies to achieve those goals; and 3) ultimately decide how to best implement strategies that include specific practices that either reduce nutrient applications, enhance nutrient uptake/ removal, or prevent movement of excess nutrients to aquatic systems.

Near-term Research Needs

1. Describe the effects of various land management practices on P, N, and sediment, export from land to waterbodies.
2. Assess the effectiveness of constructed wetlands as a management strategy for CHABs, to determine their ability to remove nutrients, and to determine the effects that resulting N:P rations will have on downstream aquatic systems.
3. Perform basin-scale research to help optimize site-selection for restored or constructed wetlands in watersheds.
4. Similarly, in lotic systems, describe the potential effectiveness of practices such as riparian plantings along streams and rivers to shade and cool streams.

Water Management

Water management activities to effectively reduce the occurrence of CHABs must focus on methods for mitigating effects of nutrients already present in such systems (e.g., dynamics of sediment-bound P), and efforts to manage the hydrodynamics of these systems to dilute nutrients at key points in the season during which low flows increase the probability of CHABs.

An entire mosaic of aquatic habitats is affected by water management, global climate change, and relative sea level rise. Freshwater supplies downstream will be affected by changes in consumptive uses, releases from reservoirs, and changes in precipitation. Increases in relative sea level rise will push salt water farther upstream. Successful research in this area will require a broad national campaign, conducted within each major eco-region, with priority given to regions in which the occurrence of CHABs is increasing, to examine the interactive effects of freshwater flow modification on nutrient supplies, hydrologic properties, and temperature. For example, increasing freshwater discharge has been shown to reduce the prevalence of CHABs in some water bodies by decreasing residence time and destratifying the water column. Among the methods for increasing freshwater discharge is removing impediments to flow, including dams and reservoirs. The impacts of removal of these structures on nutrient transport downstream, perhaps through enhanced discharge or changes in nutrient inputs, should be modeled prior to implementation and assessed post-implementation. This research will assist in the development of models to forecast the net implications for CHABs that may result from the removal of in-stream obstructions. System responses and practices considered appropriate for the region will vary from region to region, but general principles should be developed.

Additionally, trophic interactions and processes within aquatic systems can be managed to prevent movement of nutrients from compartments in which they are less likely to induce harmful algal blooms (e.g., sediments or deep in the hypolimnion) to compartments in which they are more likely to promote algal blooms (water column or metalimnion). Alternatively, ecosystem trophic structures may have been altered by human activities so that top-down controls on CHABs are no longer effective. In the near-term, although we may not be able to rapidly reverse the eutrophication of aquatic systems which has proceeded over the past half-century, we may be able to manipulate the system to favor the growth of aquatic macrophytes and algal species that are more amenable to removal or remediation, or less harmful while progress is made in efforts to reduce eutrophication. According to Chorus and Muur (1999), measures addressing light availability or targeting aquatic community trophic structure (e.g., biomanipulation) tend to be most successful in less eutrophic situations, but such measures may also accelerate restoration in highly eutrophic water bodies.

Near-term Research Needs

1. Develop tools to predict changes in aquatic habitat that will result from a variety of global climate change and water management scenarios, and to predict the net effects of such changes on the occurrence of CHABs.
2. Models that describe CHAB dynamics in relation to causative factors and ecosystem trophic structure are needed to evaluate prevention and mitigation measures.
3. Determine the community dynamics of CHABs at the wetland-water body interface, the potential interactions among CHABs and wetland vascular plants, and trophic interactions that might be manipulated to reduce the probability of CHAB occurrence.
4. Describe the conditions under which the aforementioned mitigation measures can be employed effectively.

Unintended consequences

It is often easier to focus on one problem at a time, but because CHABs likely result from a variety of causes, care must be taken in regard to unintended or secondary effects when manipulating some causative factors. For example, current regulations and monitoring schemes have focused largely on control of N. Phosphorus, however, has been shown to be a key nutrient in the determination of CHAB occurrence. If regulatory strategies focus exclusively on N, reductions in N concentrations could shift N:P ratios toward the conditions that strongly favor cyanobacteria rather than other, less harmful algal taxa (Piehler et al. 2002).

Near-term Research Need

1. Evaluate existing management practices and regulatory measures for their potential impacts not only on N, but on P and sediment (and perhaps other nutrients such as iron) as well, so that they can be optimized to prevent CHABs.

Establishing thresholds: uncertainty and phase shifts

Significant attention has been focused on identifying threshold levels of physical and chemical drivers that cause major destabilizing shifts in previously stable ecosystems (Scheffer et al. 2001). This response in ecosystem function to forcing mechanisms can have enormous ecological, economic, and management implications. For example, better constraints on ecosystem thresholds provide context in which to interpret long-term monitoring data such as nutrient levels and CHAB prevalence. It also has bearing on remediation efforts because phase shifts and hysteresis may make it impossible to return a system to its original 'pristine' state.

Near-term Research Need

1. Determine the relationship between drivers (e.g., nutrients, hydrologic modification) and CHABs responses (e.g., productivity, toxicity) in the freshwater-marine continuum to improve our understanding of threshold attainments to cause sudden changes in the response variables.

Sociological Impediments

The implementation of management strategies for environmental change often requires difficult societal decisions. Thus, an understanding of public perception regarding the benefits and values associated with controlling CHABs is critical in designing successful management plans. The choices required to control CHAB prevalence may carry significant costs, and without an understanding of the full range of benefits that result from CHAB control, a true cost-benefit analysis cannot be presented to inform policy determinations. Targeting funding to watershed-based programs where residents are actively working to improve water quality and ecosystem function may help develop "model watersheds" that will assist in quantifying the economic benefits of CHAB control. Programs should be selected to provide an array of watersheds that represent variety of different land use and management practices.

Long-term Research Needs

1. Describe societal obstacles to changes that would reduce nutrient inputs to water bodies and sociopolitical strategies to help overcome these obstacles in order to implement any of the practical solutions to the problems that have produced CHABs.
2. Develop comprehensive and accurate cost-benefit analyses of CHAB control that are based on both pre- and post-implementation evaluations of CHAB control practices.

Charge 3: Control/Mitigation (post-bloom)

Can draft guidelines for controlling and mitigating CHABs be developed now for use by managers? If not, identify and prioritize research to improve processes for removing cells and toxins from fresh, estuarine, and marine water bodies and drinking water, focusing both on current and emerging techniques with the intent of providing guidelines to managers.

Cyanobacterial toxins in recreational and drinking waters have become an increasingly visible public health and environmental issue, both nationally and internationally. Newspaper headlines, recreational water closings, and reported animal deaths have contributed to this greater visibility. To control the risk from cyanobacteria and their toxins, it is important to implement a multi-barrier approach. Control strategies should take into account the most important factors influencing cyanobacterial growth and toxin production in ambient waters, especially nutrient types and levels, and water temperature. At the drinking water treatment facility, a better understanding is needed on the effectiveness of various widely used treatment processes (e.g., coagulation, sedimentation, filtration, oxidation, granular activated carbon) for controlling various types of harmful cyanobacterial cells and their toxins. While much work has already been published on this topic, especially for the microcystins, important knowledge gaps remain. These are identified in the research needs listed below.

Bloom Control and Toxin Fate

In addition to understanding the causes and prevention of CHABs, it is also important to have mechanisms in place to control them once they occur. Numerous techniques already exist for managing blooms. However, techniques have often not been explicitly evaluated and optimized for use in the control of CHABs, particularly when toxins are present. The following research needs are designed to fill existing gaps and identify appropriate CHAB control options in ambient water.

Near-term Research Needs

1. Artificial destratification–The use of artificial destratification for cyanobacterial control has had mixed success. A potential exists to improve this process by modifying the configuration. Are there

different configurations of water mixers that would increase basin scale circulation and disrupt stratification in the surface layer (e.g., an additional bubble line at a shallower depth than the first bubble plume line). This would directly target mixing in the surface zone where the cyanobacteria grow and facilitate transport of cells deeper into the water column where they may become light-limited.

2. Increasing flushing rates by enhancing freshwater discharge -- Reducing water residence time by increasing flushing has bee shown to decrease the dominance by relatively slow-growing cyanobacteria (c.f. Paerl et al., 2001). If upstream freshwater supplies are available, this could be a "low tech" approach to minimizing the opportunities for CHABs to form and dominate in otherwise susceptible (i.e., nutrient-enriched) downstream waters.

3. Ultrasound–Determine the effectiveness of commercially available ultrasound units to control CHABs. Assess the mode of action of ultrasound on cells and the best methods for applying ultrasound in large reservoirs.

4. Electrocoagulation–Unlike most commonly used coagulation processes, electrocoagulation adds few metals into the water system. Determine the effectiveness of commercially available units to control CHABs.

5. Evaluate new and existing coagulants for recreational sources (metal and non-metal based). Evaluate the ultimate fate of the coagulated and/or sedimented cyanobacterial cells and their toxins, and assess impacts of the coagulants.

6. Evaluate new and innovative algicidal or algistatic compounds that show promise for the control of CHABs. It is also necessary to identify unintended impacts.

Drinking Water Treatment

Drinking water utilities ensure potable water by using a source-to-tap multi-barrier approach to water treatment. The multi-barrier approach recognizes the importance of individual procedures, unit processes, and tools, and their inter-relationships, to control drinking water contamination. A short review of the efficacies various treatment processes and the multi-barrier approach to remove and inactivate cyanotoxins is presented in Westrick (this volume).

Near-term Research Needs

1. Determine if the enhanced coagulation technology (currently required under recent revisions to the U.S. EPA's Surface Water Treatment Rule) for removal of dissolved organic carbon, is also effective in removing harmful cyanobacterial cells. Also determine how enhanced coagulation could be improved to better remove cyanobacterial cells and their toxins.
2. Determine the efficacy of widely used water filtration processes in controlling cyanotoxins, including the biodegradation of cyanotoxins on sand filters.
3. Develop methods for real-time monitoring of source waters and drinking water intakes for the presence of harmful cyanobacterial cells and their dissolved toxins or their easily measured surrogates.
4. Determine the CT values (disinfectant concentration x time) needed to inactivate important freshwater cyanotoxins (including emerging cyanotoxins) for all widely used water disinfectants.
5. Determine the extent to which the control of cyanotoxins by disinfection will increase the levels of those toxic disinfectant byproducts regulated under U.S. EPA's drinking water regulations.

Long-term Research Needs

1. Assess the applicability of new technologies for degrading cyanotoxins, such as advanced oxidation with titanium oxides catalysts.
2. Determine the toxicity of major byproducts resulting from the interaction between widely used water disinfectants and various cyanotoxins.
3. Develop and evaluate the utility of smart sensors for assessing system performance and coupling sensors to models for prediction.

Charge 5: Economic Analysis

What ecological factors should be included in models used to predict the relative costs and benefits of processes used to prevent or reduce the occurrence of CHABS in water bodies and cyanotoxins in drinking water?

Today's environmental resource managers are faced with increasingly difficult decisions that require a balance between the responsible use of public funds and the maximization of environmental benefits. In making such decisions, managers traditionally consider many objectives, including environmental quality, and threats to ecosystem integrity. However, benefits to the natural environment alone are not enough to encourage the changes in public behavior that may be needed to manage a threatened resource. A demonstration of the societal benefits associated with a given management strategy is also required, and this is usually done through economic analysis. Thus, environmental managers need comprehensive decision-making tools to help them evaluate the tradeoffs between different management scenarios – tools that take into account economic impacts and benefits in addition to environmental ones.

What follows is a discussion of *ecological factors* that the group felt must be considered in order to accurately predict the relative costs and benefits of various management strategies for CHABs. This is meant to provide input on the types of information that should be incorporated in economic models. It is not a strategy for doing so, but rather a starting point from which to engage experts in both disciplines (ecology and economics) in a productive dialogue, to begin addressing the difficulties in incorporating ecological complexity and multi-dimensional effects into economic valuations.

Designing Assessments

While there is widespread agreement that eutrophication is responsible for the initiation of CHABs (Chorus and Muur 1999, Paerl this volume), there is less agreement as to the most appropriate, mitigation and prevention activities. Once eutrophication has occurred, the system may respond slowly to efforts to reduce additional nutrient input. In the near-term, harmful algal blooms are still likely to occur. In contrast, the mitigation and prevention measures required to reduce nutrient input to aquatic systems may have other immediate and substantial economic impacts.

Managers must balance these immediate economic impacts (both positive and negative) with the future benefits of nutrient reduction, such as fewer algal blooms. No solution to the problem of CHABs will be effective without reductions in nutrient input, particularly P loading. In order to be ultimately effective in preventing CHABs, management activities must address both the long-term and critically important goal of nutrient reduction and removal from the system, and the implementation of effective near-term practices that mitigate existing eutrophication.

Assessment of the economic benefits and costs of management practices designed to reduce algal blooms should include all of the benefits and costs resulting from their implementation. Also, since reducing nutrients inputs into water bodies will impact other ecosystem services, any other ecological benefits that result from these practices should be included in the economic assessment, as well. For instance, a management practice might generally improve water quality which can lead to increased recreational opportunities, improved wildlife habitat and health, and reductions in drinking water treatment costs. Since costs of such management measures may occur in the near-term, while benefits might occur in the long-term, all benefits and costs should be discounted to the present so they can be compared equally. Although the near-term costs may be greater than the near-term environmental benefits, the inclusion of other indirect benefits may make the *total* benefits larger than the *total* costs in the long run. Additionally, benefits and costs that accrue to those beyond the immediate management area should be included as appropriate for the scale of the analysis. For instance, in a national or regional analysis, all benefits and costs occurring in the nation or region should be included if they result from management actions. It will also be critical to link any of the physical and biological model outputs to economic endpoints that will change as a result of the management measures. Thus, in order to design a good assessment, it is necessary to ensure that the correct biological endpoints are being monitored and matched with the appropriate economic endpoints. In order to achieve this, the connection between changes in the biological and physical components of the system and changes in the economic components must be well-understood.

In designing an assessment, managers may want to compare the benefits and costs of a number of practices that could be implemented and to choose the one that provides the highest net benefits (benefits-costs) or the highest benefit/cost ratio. Alternatively, if a given level of nutrient reduction is desired, the assessment could be designed to achieve the given level at the least cost. In this case, standard linear programming models could be used to determine the best solution.

Requirements of Models across Multiple Scales

Local Watershed Scale

At the small watershed scale (5-10,000 ha), the economic impacts and environmental benefits of practices and programs designed to decrease nutrient export could be modeled based on the usual factors included in farm enterprise budgets and models that calculate crop yields and water quality response to various practices (such as Erosion/Productivity Impact Calculator - EPIC or Soil and Water Assessment Tool - SWAT). At this scale, compensation to producers from USDA environmental improvement programs could offset some of the costs paid for by producers. Examples of economic and environmental modeling efforts at this scale are published in Coiner et al. (2000), Santelmann et al. (2004) and Boody et al. (2005). None of the studies employed at this scale to date have been able to explore sufficient numbers of alternatives to discern whether the response of the system is linear or whether there may be thresholds of response. For example, a nutrient reduction in surface water may be very gradual until a threshold level is reached at which the response rate increases significantly. Further efforts to explore the linearity or non-linearity of watershed response to measures implemented at this scale could be quite valuable in helping to inform policy concerning the minimal extent of changes that must be made in order to achieve significant, measurable results across multiple environmental objectives.

County and Basin Scale

At the spatial extent of counties and small river basins, the economic impacts of additional components should be considered. Some negative economic impacts of existing practices are already occurring, such as impacts of environmental degradation on regional infrastructure (costs of water treatment, costs of road and bridge replacement over eroding stream channels, episodic environmental disasters from failure of manure containment facilities, and a decline in property values surrounding large confinement feeding operations). Thus, an accurate modeling of economic impacts at this scale must consider the benefit of reducing the negative consequences of existing practices. Another component to consider at regional (and national) scales would be the potential for banking of credits for carbon sequestration, wetland mitigation and habitat restoration can improve surface or groundwater quality. These elements could all be factored into models developed for larger regions. Capturing the value of such ecosystem services over longer time frames (25-50 years) would be necessary to demonstrate the economic value of changes in land use and management over time (e.g., so that the different impacts of 100 year flood events or drought under different management regimes could be quantified). Again, exploration of response thresholds (in terms of both spatial and temporal extent) over which significant measurable change can be expected, would be important.

Among the costs that might be expected to emerge at this scale are losses in revenue to rural communities and counties for land enrolled in set-asides, such as decreased commodity production, the sales of agricultural supplies, and any decline in need for services or losses of jobs that might occur in response to changes in practices.

Regional Scale

Models developed to evaluate alternatives at the regional scale should include some key factors or "pressure points" whose alteration may change the behavior of the modeled system and thereby have implications for the selection of nutrient-control techniques. For example, it would be extremely valuable develop models for evaluating the influence of changes in energy costs on the relative costs of control processes, or to explore the costs and benefits of alternative control processes as the system response changes with key climatic shifts in precipitation or temperature patterns.

National Scale

At the national scale, models should incorporate the effects of implementing specific programs or guidelines which, when accumulated at a national scale, could influence the balance of trade by increasing commodity supply for export. Additionally, models at this scale must consider the effects of potential changes in commodity supply and demand on prices, unintended consequences of specific agricultural programs (e.g., "slippage" in response to set-aside programs), as well as the environmental benefits that accrue at a national scale outside the region in which the land use practices are implemented (decreased need for some forms of water treatment, improved flood control along major rivers, carbon or pollutant trading credits etc.).

Temporal Scale

Progress in the near-term will be required in order to produce demonstrable results from management activities that will satisfy public desire for measurable progress, a key requirement for sustained funding for remedial programs. Metrics for measuring progress should include both reduction of nutrients within the aquatic system and decrease in the extent, duration, and toxicity of algal blooms. However, it must be acknowledged up front that significant improvements in the response of nutrient export from these systems to changes in agricultural practices may take years. McIsaac et al. (2001) modeled the influence of nitrogen applications in the Mississippi River Basin (MRB) on nutrient export into the Gulf of Mexico, and found that nutrient export at the mouth of the Mississippi reflected loading in the Basin from the previous 9 years, with strongest influence being from loading in the previous 1-5 year interval. Given these and other studies indicating time lags in system response to reductions in nutrient applications to land, it is unlikely that we will see improvements in either nutrient loading or control of algal blooms from watershed management efforts alone until 5-10 years after these measures are implemented. Yet, we also know that unless we implement such measures now, we will see continued eutrophication and degradation of these systems, which will be more and more difficult to reverse over time.

Societal Considerations

The need for informing the general public (especially those stakeholders of whom the greatest sacrifices will be demanded), concerning the needs and goals of any programs designed to decrease CHABs will be critically important. In addition, efforts must be made to ensure that the changes in land use and management that are proposed to meet environmental goals are culturally acceptable as well as perceived as fair and equitable across groups (Santelmann et al. 2001). For example, both actual and perceived fairness in regulation of various entities and activities in meeting regional water quality goals must be considered in order to achieve acceptance and participation. If urban point sources are allowed to continue to release large quantities of phosphorus and total dissolved solids into water while agricultural enterprises are stringently regulated, or vice versa, this will jeopardize the atmosphere of compliance and “we’re all in this fight together” that will be needed to make an impact on reversing current trends of eutrophication in aquatic systems.

References

Boody G, Vondrack B, Andow D, Krinke M, Westra J, Zimmerman J, Welle P (2005) Multifunctional agriculture in the United States. Bioscience 55(1): 27-37.

Boon PI, Bunn SE, Green JD, Shiel RJ (1994) Consumption of cyanobacteria by fresh-water zooplankton – implications for the success of top-down control of cyanobacterial blooms in Australia. Australian J Mar Fresh Res 45: 875-887

Capper A, Cruz-Rivera E, Paul VJ, Tibbetts IR (2006) Chemical deterrence of a marine cyanobacterium against sympatric and non-sympatric consumers. Hydrobiologia 553: 319-326

Caraco NF, Cole JJ, Raymond PA, Strayer DL, Pace ML, Findlay SEG, Fischer DT (1997) Zebra mussel invasion in a large, turbid river: Phytoplankton response to increased grazing. Ecology 78: 588-602

Chorus I, Bartram J (1999) Toxic cyanobacteria in water: a guide to their public health consequences, monitoring and management. World Health Organization. E&FN Spon, Routledge, London.

Chorus I, Muur L (1999) Remedial measures. In: Toxic Cyanobacteria in Water: A guide to their public health consequences, monitoring and management. I. Chorus and J. Bartram, eds. Published by WHO, UNESCO, and UNEP. London, UK.

Christoffersen K (1996) Effect of microcystin on growth of single species and on mixed natural populations of heterotrophic nanoflagellates. Natural Toxins 4: 215-220

Coiner C, Wu J, Polasky S (2001) Economic and Environmental Implications of Alternative Landscape Designs in the Walnut Creek Watershed of Iowa, Ecological Economics 38(1):119-139

De Bernardi R, Giussani G (1990) Are blue green algae a suitable food for zooplankton? A review. Hydrobiologia 200/201: 29-41

Fogg GE (1969) The physiology of an algal nuisance. Proc R Soc London B. 173:175-189

Giani A, Bird DF, Prairie YT, Lawrence JF (2005) Empirical study of cyanobacterial toxicity along a trophic gradient of lakes. Can J Fish Aquat Sci 62: 2100-2109

Griffith GJ, Omernik T, Wilton, Pierson S (1994) Ecoregions and subregions of Iowa: A framework for water quality assessment and management. Jour. Iowa Acad. Sci. 101(1): 5-1

Hairston Jr. MG, Holtmeier CL, Lampert W, Weider LJ, Post DM, Fisher JM, Caceres CE, Fox JA, Gaedke U (2002) Natural selection for grazer resistance to toxic cyanobacteria: Evolution of phenotypic plasticity? Evolution 55:2203-2214.

HARRNESS (2005) Harmful Algal Research and Response: A National Environmental Science Strategy 2005–2015.

Keeney D (ed.) (1990) Farming systems for Iowa: Seeking alternatives. Leopold Center for Sustainable Agriculture Ames Iowa USA.

Lampert W (1987) Laboratory studies on zooplankton-cyanobacteria interactions derived from enclosure studies. N.Z. J. Mar Freshwater Res. 21: 483-490

Likens GE (ed) (1972) Nutrients and Eutrophication. American Soc Limnol Oceanogr Special Symp

McIsaac G, David M, Gertner G, Goolsby D (2001) Nitrate flux in the Mississippi River. Nature 414: 166-167

Mitsch W, Day J, Gilliam J, Groffman P, Hey D, Randal G, Wang N (2001) Reducing nitrogen loading to the Gulf of Mexico from the Mississippi River Basin: Strategies to counter a persistent ecological problem. Bioscience 51: 373-388.

Nagle DG, Camacho FT, Paul VJ (1998) Dietary preferences of the opisthobranch mollusk *Stylocheilus longicauda* for secondary metabolites produced by the tropical cyanobacterium *Lyngbya majuscula*. Mar Biol 132: 267-273

Nagle DG, Paul VJ (1998) Chemical defense of a marine cyanobacterial bloom. J. Exp Mar Biol Ecol 225: 29-38

Nagle DG, Paul VJ (1999) Production of secondary metabolites by filamentous tropical marine cyanobacteria: ecological functions of the compounds. J. Phycol 35: 1412-1421

Omernik JM (1987) Ecoregions of the conterminous United States. Annals of the Association of American Geographers 77(1): 118-125.

Paerl HW (1988) Nuisance phytoplankton blooms in coastal, estuarine, and inland waters. Limnol Oceanogr 33:823-847

Paerl HW, Fulton RS, Moisander PH, Dyble J (2001) Harmful freshwater algal blooms with an emphasis on cyanobacteria. The Scientific World 1: 76-113

Paerl HW, Fulton III RS (2006) Ecology of harmful cyanobacteria. Pp. 95-109, In, E. Graneli and J. Turner [Eds.]. Ecology of Harmful Marine Algae. Ecological Studies, Vol. 189Springer-Verlag, Berlin .

Pennings SC, Pablo SR, Paul VJ (1997) Chemical defenses of the tropical benthic, marine cyanobacterium *Hormothamnion enteromorphoides*: diverse consumers and synergisms. Limnol Oceanogr 42: 911-917.

Piehler MF, Dyble J, Moisander PH, Pinckney JL, Paerl HW (2002) Effects of modified nutrient concentrations and ratios on the structure and function of the native phytoplankton community in the Neuse River Estuary, North Carolina USA. Aquatic Ecology 36: 371-385.

Raikow DF, Sarnelle O, Wilson AE, Hamilton SK (2004) Dominance of the noxious cyanobacterium Microcystis aeruginosa in low-nutrient lakes is associated with exotic zebra mussels. Limnol. Oceanogr. 49:482-487.

Ramsdell, J.S., D.M. Anderson and P.M. Glibert (Eds.), Ecological Society of America, Washington DC, 96 pp. http://www.cop.noaa.gov/stressors/extremeevents/hab/current/harrness.html

Reynolds CS, Walsby AE (1975) Water blooms. Biol Rev 50:437-481

Sample Guidelines (modified from the Wisconsin Division of Public Health's fact sheet on cyanobacteria, and their toxins, and health impacts) http://dhfs.wisconsin.gov/eh/Water/fs/Cyanobacteria.pdf

Santelmann M, White D, Freemark K, Nassauer J, Clark M, Coiner C, Cruse R, Danielson B, Eilers J, Polasky S, Vache K, Sifneos J, Rustigian H, Debinski

D, Wu J (2004) Assessing alternative futures for agricultural watersheds. Landscape Ecology 19: 357-374

Santelmann M, Freemark K, White D, Nassauer J, Clark M, Danielson B, Eilers J, Cruse R, Polasky S, Vache K, Galatowitsch S, Wu J (2001) Applying Ecological Principles to Land-Use Decision Making in Agricultural Watersheds. *In*: V.H. Dale and R. Haueber (eds.) *Applying Ecological Principles to Land Management*. Springer-Verlag, NY.

Sarnelle O, Wilson AE (2005) Local adaptation of Daphnia pulicaria to toxic cyanobacteria. Limnol Oceanogr 50: 1565-1570

Scheffer M et al. (2001) Catastrophic shifts in ecosystems. Nature 413: 591-96.

Schindler DW (1975) Whole-lake eutrophication experiments with phosphorus, nitrogen and carbon. Verh Int Verein Theor Angew Limnol 19:3221-3231

Sellner KG, Brownlee DC, Buundy MH, Brownlkee SG, Braun KR (1993) Zooplankton grazing in a Potomac River cyanobacteria bloom. Estuaries 16: 859-872

Shapiro J (1990) Current beliefs regarding dominance of blue-greens: The case for the importance of CO_2 and pH. Int Verein Theor Angew Limnol Verh 24:38-54

Sivonen K (1996) Cyanobacterial toxins and toxin production. Phyclogia 35(6):12-24

Skulberg, OM, Underdal B and Utkilen H (1994) Toxic waterblooms with cyanophytes in Norway - Current knowledge. Alogol Studies 75:279-289

Smith TE, Stevenson RJ, Caraco NF, Cole JJ (1998) Change in phytoplankton community structure during zebra mussel invasion of the Hudson River. J. Plankton Res 20: 1567-1579

U. S. Department of Agriculture Natural Resources Conservation Service (1997) Profitable pastures. USDA-NRCS, Des Moines, Iowa. 19 pp.

U. S. Department of Agriculture Soil Conservation Service (1996) Conservation Choices: Your guide to 30 conservation and environmental farming practices. USDA SCS St. Paul, MN. 34 pp.

Vache K, Eilers JM, Santelmann M (2002) Water quality modeling of alternative agricultural scenarios the U.S. Corn Belt. JAWRA 38(3):773-787.

Vanderploeg HA, Liebig WW, Carmichael WW, Agy MA, Johengen TH, Fahnenstiel GL, Nalepa TF (2001) Zebra mussel (*Dreissena polymorpha*) selective filtration promoted toxic *Microcystis* bloom in Saginaw Bay (Lake Huron) and Lake Erie. *Canadian Journal of Fisheries and Aquatic Sciences* 58(6):1208-1221

Walls M, Laurenmaatta C, Ketola M, OhraAho P, Reinikainen M, Repka S (1997) Phenotypic plasticity of Daphnia life history traits: the roles of predation, food level and toxic cyanobacteria. Freshwater Biol. 38: 353-36

Chapter 10: Nutrient and other environmental controls of harmful cyanobacterial blooms along the freshwater–marine continuum

Hans Paerl

University of North Carolina at Chapel Hill, Institute of Marine Sciences, Morehead City, NC 28557 (hpaerl@email.unc.edu)

Abstract

Nutrient and hydrologic conditions strongly influence harmful planktonic and benthic cyanobacterial bloom (CHAB) dynamics in aquatic ecosystems ranging from streams and lakes to coastal ecosystems. Urbanization, agricultural and industrial development have led to increased nitrogen (N) and phosphorus (P) discharge, which affect CHAB potentials of receiving waters. The amounts, proportions and chemical composition of N and P sources can influence the composition, magnitude and duration of blooms. This, in turn, has ramifications for food web dynamics (toxic or inedible CHABs), nutrient and oxygen cycling and nutrient budgets. Some CHABs are capable of N_2 fixation, a process that can influence N availability and budgets. Certain invasive N_2 fixing taxa (e.g., *Cylindrospermopsis*, *Lyngbya*) also effectively compete for fixed N during spring, N–enriched runoff periods, while they use N_2 fixation to supplant their N needs during N–deplete summer months. Control of these taxa is strongly dependent on P supply. However, additional factors, such as molar N:P supply ratios, organic matter availability, light attenuation, freshwater discharge, flushing rates (residence time) and water column stability play interactive roles in determining CHAB composition (i.e. N_2 fixing vs. non–N_2 fixing taxa) and biomass. Bloom potentials of nutrient–impacted waters are sensitive to water residence (or flushing) time, temperatures (preference for >15 °C), vertical mixing and turbidity. These physical forcing features can control absolute growth rates of bloom taxa. Human activities may affect "bottom up" physical–chemical modulators either directly, by controlling hydrologic, nutrient, sediment and toxic discharges, or indirectly, by influencing climate. Control and management of cyanobacterial and other phytoplankton blooms invariably includes nutrient input constraints, most often focused on N and/or P. While single nutrient input constraints may be effective in some water bodies, dual N and P input reductions are usually required for effective long–term control and management of blooms. In some systems where hydrologic manipulations (i.e., plentiful water supplies) are possible, reducing the water residence time by flushing and artificial mixing (along with nutrient input constraints) can be effective alternatives. Blooms that are not readily consumed and transferred up the food web will form a relatively large proportion of sedimented organic matter. This, in turn, will exacerbate sediment oxygen demand, and enhance the potential for oxygen depletion and release of nutrients back to the water column. This scenario is particularly problematic in long–residence time (i.e., months) systems, where blooms may exert a strong positive feedback on future events. Implications of these scenarios and the confounding issues of climatic (hydrologic) variability, including droughts, tropical storms, hurricanes and floods, will be discussed in the context of developing effective CHAB control strategies along the freshwater–marine continuum.

Introduction

CHABs and Eutrophication

The accumulation of cyanobacterial biomass as bright green, yellow–brown and red blooms in fresh, brackish or saline waters is one of the most obvious and problematic symptom of anthropogenic nutrient enrichment, or eutrophication (Fogg 1969, Reynolds and Walsby 1975, Paerl 1988) (Fig. 1, see Color Plate 4). Cyanobacterial blooms, which often culminate in an unsightly, odoriferous mess, can also cause harm from ecological and health perspectives. Ecologically, blooms may be inedible or toxic to consumer species, causing food web alterations, with potentially detrimental effects on nutrient cycling, biodiversity and fisheries (Fogg 1969, Paerl et al. 2001). Because they may not be consumed, blooms can accumulate as thick scums and mats, which when decomposed cause excessive oxygen consumption (hypoxia), a major factor in the decline or elimination of fish, shellfish, invertebrate and plant habitats (Diaz and Solow 1999). In addition, N_2 fixing cyanobacterial blooms can constitute significant sources of "new" nitrogen (N), potentially impacting N–driven eutrophication and N cycling (Horne 1977, Paerl 1988). From animal and human health perspectives, blooms produce a variety of odor and taste compounds (geosmins, DMIB), rendering affected waters unsuitable for drinking, swimming and other recreational purposes. Lastly, numerous cyanobacterial bloom species (Fig. 2, see Color Plate 5) produce alkaloid, peptide and other compounds that can be toxic upon ingestion or contact with affected waters (Codd and Bell 1996, Carmichael 1997, Chorus and Bartram 1999).

From research and management perspectives, identifying environmental factors causing and sustaining harmful cyanobacterial blooms (CHABs) is key to developing an understanding of how to control these unwanted manifestations of man–made nutrient, sediment and hydrologic alterations. Cyanobacterial blooms have accompanied human modification of watersheds for agricultural, urban and industrial development for centuries. One line of evidence is the paintings of Holland's agricultural landscapes by the 17th century Dutch Masters, which show surface algal scums diagnostic of nutrient over–enrichment (Fig. 3, see Color Plate 5).

Fig. 1. Harmful cyanobacterial blooms in a range of nutrient–enriched aquatic ecosystems. Upper left. A bloom of the non–N_2 fixing genera *Microcystis aeruginosa* and *Oscillatoria* sp. in the Neuse River, NC (Photo, H. Paerl). Upper right. A mixed *Microcystis* sp. and *Anabaena* spp. (N_2 fixers) bloom in the St. Johns River, Florida (Photo, J. Burns). Lower left. A bloom of the benthic filamentous N_2 fixer *Lyngbya wollei* in Ichetucknee Springs, Florida (Photo H. Paerl). Lower right. A massive bloom of *Microcystis* sp. and *Anabaena* spp. in Lake Pontchartrain, Louisiana (Photo J. Burns). (See Color Plate 4).

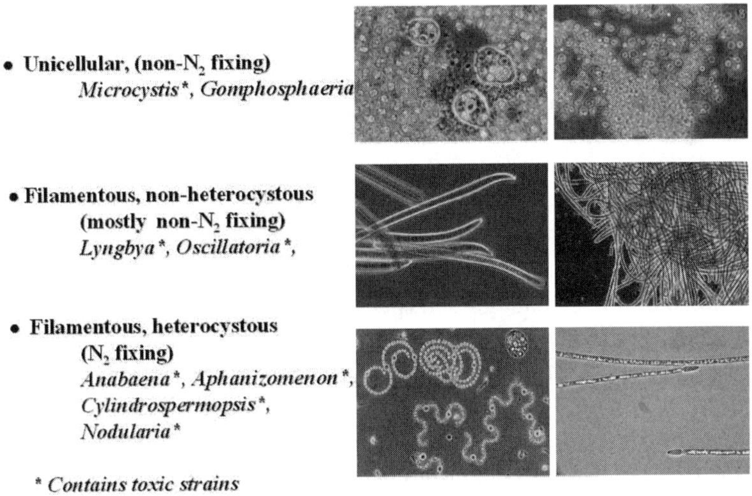

Fig. 2. Photomicrographs of genera representing the three major CHAB morphological groups, including coccoid, filamentous non–heterocystous and filamentous heterocystous types. (See Color Plate 5).

Fig. 3. Painting of Haarlemmermeer, a shallow, eutrophic lake in the Netherlands. Jan van Gooyen, ca. 1650. Note the surface scums characterizing the lake. (See Color Plate 5).

Nutrients and Hydrology: Key Controls of CHABs

Among the nutrient elements required for aquatic plant growth, N and phosphorus (P) are often most stimulatory, because requirements are high relative to availability. It follows that N and P enrichment are often most effective in stimulating and supporting blooms in receiving waters (Fogg 1969, Reynolds and Walsby 1975). These elements have, and continue to be, the focus of efforts aimed at controlling blooms (Likens 1972, Schindler 1975, Shapiro 1990). While N and P are generally considered the main "culprits" of freshwater and marine eutrophication, they are by no means the *only* environmental factors controlling bloom formation, duration and proliferation. Other natural and anthropogenically–influenced factors also play roles in controlling bloom dynamics. These include; 1) sedimentation which can alter both the nutrient and light environments, and 2) hydrology, specifically freshwater discharge, flushing and residence time, which affect both nutrient delivery to and cycling in affected waters. Nutrient inputs or loads may synergistically or antagonistically interact with sedimentation, freshwater discharge and water column stability (vertical mixing regime) to determine; 1) if a specific water body is susceptible to CHAB formation, 2) the extent (magnitude, duration) to which CHABs may dominate planktonic and or benthic habitats, and, 3) whether an affected water body is amenable to management steps aimed at minimizing or eliminating CHABs (Fig. 4, see Color Plate 6). Here, I will discuss the roles N and P play as nutrients controlling planktonic and benthic CHABs, and the interactive roles light, hydrologic and hydrodynamic conditions play in modulating CHABs. Finally, nutrient and other management options for controlling CHABs will be explored.

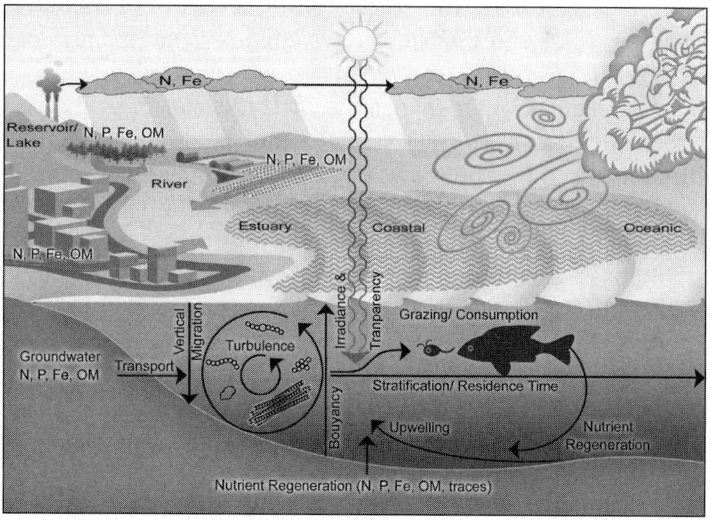

Fig. 4. Conceptual diagram, showing the interactive physical, chemical and biotic controls of cyanobacterial blooms along the freshwater–marine continuum. (See Color Plate 6).

Nitrogen and Phosphorus: Linking Their Inputs to CHAB Dynamics

In freshwater ecosystems, excessive P loading has been most frequently linked to eutrophication (Likens 1972, Paerl 1988). P–driven eutrophication can be a prerequisite to N_2–fixing or non–N_2 fixing CHABs. This is often exacerbated by low freshwater discharge (low flushing rates, long residence time), elevated water temperatures >20°C, and strong vertical stratification (Fogg 1969, Reynolds and Walsby 1975, Paerl 1988, Shapiro 1990). In some instances, organic matter–enriched conditions may also favor the CHAB dominance (Pearsall 1932, Fogg 1969). Whether or not N_2 fixers dominate depends on several co–occurring factors, the most important of which is the availability of biologically utilizable N relative to P (Paerl 1990). Freshwater systems having low molar ratios of both total and soluble (biologically–available) N to P (<15) are most likely to experience cyanobacterial dominance (Smith 1983, 1990). Conversely, waters having molar N:P ratios in excess of 20 are more likely to be dominated by eukaryotic algal taxa (Smith 1983). This rule has proven broadly applicable to periodically stratified, long residence (> 30 days) temperate and tropical freshwater systems (Downing et al. 2001).

There are exceptions to the N:P rule. These include; 1) systems in which both N and P loadings are very large (i.e., hypereutrophic systems in which N and P inputs exceed the assimilative capacity of the phytoplankton), and 2) highly–flushed, short residence time systems, in which the flushing rate exceeds growth or doubling rates of cyanobacteria (generally >1 d^{-1}). In N and P enriched systems, N:P ratios may readily exceed 20, but since both N and P are being supplied at close to non–limiting rates, factors other than nutrient limitation (e.g., light, vertical mixing, residence time, salinity, organic matter content) may control algal community activity, biomass and composition. Under these conditions, N_2 fixation confers little if any advantage, and non–N_2 fixing taxa predominate. Often, these conditions favor high rates of primary production and biomass accumulation. This may severely reduce clarity, restricting transmittance of photosynthetically–active radiation (PAR; 400–700 nm), providing a niche for buoyant, surface–dwelling, "nutriphilic" phytoplankton, most notably the bloom–forming non–N_2 fixing cyanobacterial genus *Microcystis*. *Microcystis* and other non–diazotrophic nuisance genera (*Oscillatoria,* some *Lyngbya* and *Planktothrix* species) often co–dominate under these conditions.

Moderately N and P–enriched waters tend to support mixed assemblages of diazotrophic and non–diazotrophic species. This condition often occurs in systems receiving sequential pulse loadings of either high N or P. One scenario is springtime elevated N–laden surface runoff, which favors the establishment of non–diazotrophic bloom species. During summer, when runoff subsides, externally–supplied P loads (from point sources such as wastewater treatment plants, municipal and industrial sources) or internally–generated P loads released from hypoxic sediments, tend to become more prominent components of nutrient loading. P enrichment (i.e., declining N:P ratios) frequently selects for the establishment of N_2 fixing species (Paerl 1982, 1988) (Fig. 5, see Color Plate 6). Once N_2 fixers are established, non–diazotrophic species can remain a significant fraction of the phytoplankton, because they are able to utilize fixed N produced and released by N_2 fixing species (Paerl 1990). Co–existing diazotrophic and N–requiring bloom species are capable of buoyancy regulation, and thus a near–surface existence, in highly productive, turbid waters. Typically, *Anabaena, Aphanizomenon* and *Microcystis* (the notorious trio, "Annie, Fannie and Mike") co–occur under these circumstances. In clearer waters where light reaches the bottom, benthic N_2 fixing and non–fixing assemblages (e.g., *Lyngbya*, some *Oscillatoria*, *Microcoleus*, *Scytonema*, *Phormidium*) can predominate. Mixed assemblages often persist as a bloom "consortium" during summer and fall (Paerl 1983, 1986, 1987), until unfavorable physical conditions such as cooling (<15°C) and water column turnover take place.

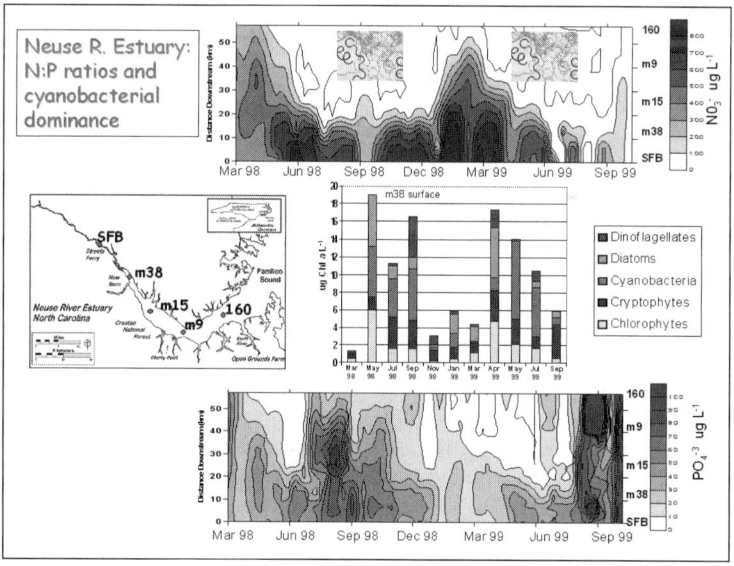

Fig. 5. Relationship, in space and time, of nitrate and phosphate concentrations, and relative dominance by cyanobacteria in the Neuse River Estuary, NC. Phytoplankton composition along a transect of 5 locations ranging from the upper oligohaline to lower mesohaline segments of the estuary was determined using high performance liquid chromatographic (HPLC) analysis of diagnostic (for major algal groups) photopigments (see Paerl et al. 2003). The period during which N_2 fixing cyanobacteria were present is indicated by the photomicrograph in the upper frame. (See Color Plate 6).

Some CHABs thrive under the relatively low light conditions caused by increased turbidity accompanying eutrophication. Members of the non–N_2 fixing filamentous genus *Oscillatoria* can form metalimnetic blooms in nutrient– (N and P) enriched lakes and reservoirs. Odor and taste producing *Oscillatoria* spp. and *Lyngbya* spp. can be particularly problematic in drinking water reservoirs, where their ability to adapt to low light conditions provides them a niche and allows them to coexist with surface–dwelling genera, including *Anabaena*, *Aphanizomenon*, and *Microcystis*. The toxic N_2 fixing, filamentous heterocystous species *Cylindrospermopsis raciborskii* appears to be taking similar advantage in nutrient enriched, eutrophying subtropical and tropical freshwater ecosystems. This low light adapted species, has, within a matter of a decade, invaded eutrophying inland waters of central Florida, USA by blooming as cloudlike masses throughout the water column (Chapman and Schelske 1997). Even though *C. raciborskii* is able to fix N_2 to meet its N requirements (under P sufficient conditions), this CHAB also effectively competes with eukayotes and other cyanobacteria for combined N when available (Padisak 1997). Furthermore, it is capable of intracellularly storing P as polyphosphate bodies. This cockroach–like CHAB appears to be able to effectively exploit altered nutrient cycling and optical conditions resulting from eutrophication in Florida, other US Southeastern and Midwest regions. Nutrient addition bioassays indicate that *both* N and P reductions are likely needed arrest the explosive growth and expansion of this CHAB (Fig. 6).

Because planktonic and benthic bloom assemblages may have complex nutritional requirements, efforts aimed at reducing CHAB dominance by manipulating N:P ratios have met with mixed results. P input constraints are often the most feasible and least costly approach in freshwater systems. In certain cases, P cutbacks can be highly effective on their own (without parallel N removal), because; 1) they may reduce total P availability enough to reduce growth of *all* bloom taxa, and 2) they may increase N:P ratios enough to provide eukaryotic algae a competitive advantage over cyanobacteria. There are noteworthy examples where exclusive P reductions have led to dramatic declines in cyanobacterial dominance and bloom control. These include Lake Washington, WA, USA, where reduction of sewage–based P inputs led to profound reversal of eutrophication (Edmondson and Lehman 1981), Lake Erie (Laurentian Great Lakes) (Likens 1972), and Himmerfjärden fjord, Sweden, where reduction of wastewater, agricultural and industrial P discharges caused a rapid decline in CHABs (Elmgren and Larsson 2002). P reduction efforts were helped by a phosphate detergent ban in the mid–1980's (Paerl et al. 2004). Similarly, reductions in wastewater and agricultural P inputs have led to decreased cyanoHAB bloom activities in large European and Asian lakes (e.g., Lakes Constance and Lucerne,

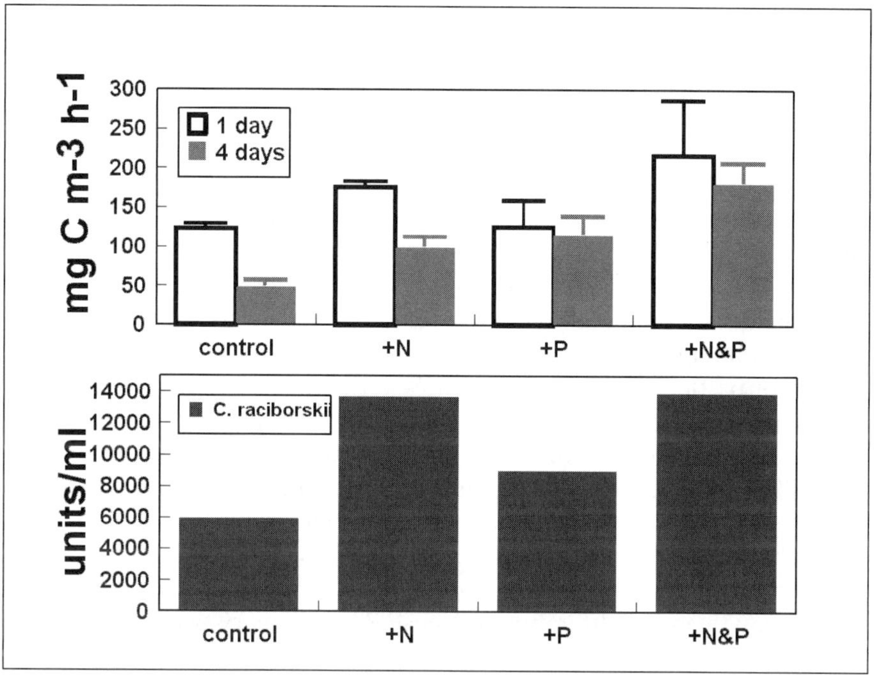

Fig. 6. Upper frame. Results from *in situ* nutrient addition bioassay, showing the effects of N (as 20 μM NO_3^-) and P (as 5 μM PO_4^{3-}), added singly or combined, on primary productivity of Lake George, a lake located in the upper St Johns River system, Florida. This lake has supported blooms of the invasive CHAB *Cylindrospermopsis raciborskii*. Lower frame shows the effects of these nutrient additions on *C raciborskii* biomass as numbers of filaments or "units" per ml. These bioassay results indicate that growth of the entire phytoplankton community and *C raciborskii* is stimulated by N and P additions individually and combined, support for a dual nutrient reduction strategy. (See Chapter 10: "Nutrient and other environmental controls of harmful cyanobacterial blooms along the freshwater–marine continuum" [Paerl this volume]).

Germany–Switzerland; Lake Trummen, Sweden; Lago Maggiorre, Italy; Lake Biwa, Japan). In other cases, parallel N and P reductions have been needed to reduce bloom potentials. In ecosystems where large amounts of previously–supplied and/or naturally–occurring P reside in the sediments, both N and P reductions are required to reduce the size and duration of blooms (Vollenweider and Kerekes 1982).

In contrast to P–limited inland waters, brackish estuarine and full salinity coastal waters tend to be N–limited (oligohaline regions of estuaries can be N and P co–limited) (Ryther and Dunstan 1971, Nixon 1986, 1995). N–enriched estuarine and coastal waters have experienced a recent upsurge in algal blooms (Paerl 1988, Hallegraeff 1993, Richardson 1997). Reducing N inputs has been recommended as a means of stemming coastal eutrophication (Vollenweider et al. 1992, Elmgren and

Larsson 2002, Boesch et al. 2002). In the Neuse River Estuary, North Carolina, USA, deteriorating water quality has prompted calls for an N input "cap" and an overall 30% reduction in N loading (Paerl et al. 1995, 2004). However, changes in N loading alone may result in shifts in ratios of dissolved N to P supply rates in this and other N–sensitive estuaries (Paerl et al 2004). Altered N:P inputs impact microalgal communities far beyond a simple reduction in productivity and biomass, including shifts in species composition and possible selection for low N:P adapted species (Smith 1983, Tilman and Kiesling 1984). In particular, the phytoplankton community could become dominated by N_2 fixing cyanobacteria that may circumvent N–limitation imposed by N reductions. This is of particular concern in shallow estuarine and coastal waters having rich (marine) repositories of P stored in the sediments that can be recycled to the water column during periods of hypoxia. Recently (summer 1997), we observed N_2–fixing *Anabaena* strains in mesohaline (5–15 psu) segments of the Neuse River Estuary (Piehler et al. 2002). In a parallel laboratory study (Moisander et al. 2000), two toxic Baltic Sea *Nodularia* strains and native *Anabaenopsis* and *Anabaena* species were capable of growth and proliferation in Neuse River Estuary water over a wide range of salinities, demonstrating the potential for CHAB expansion in estuaries (Fig. 7).

Both diazotrophic and non–diazotrophic CHABs can utilize diverse forms of combined N (inorganic and organic) (Paerl 1988). This nutritional flexibility may provide a key competitive advantage in response to anthropogenic N loading events. Large pulses of non–point source N loading have increased and are drivers of freshwater and marine eutrophication (Nixon 1995, Vitousek et al. 1997, Paerl 1997, 1998). Cyanobacterial growth and bloom responses in N–limited North Carolina estuaries closely track (in time and space) such events (Pinckney et al. 1997, 1998). In particular, organic N and ammonium–enriched conditions may favor cyanobacteria (Pinckney et al. 1997) and toxicity of bloom genera (Paerl and Millie 1996). Earlier observations of such correlations in nature (Pearsall 1932, Fogg 1969) have been largely overlooked, but might be relevant.

Non–N_2 fixing planktonic and benthic CHAB genera, including *Microcystis*, *Lyngbya*, and *Oscillatoria*, can also exploit these N loading scenarios. *Microcystis* blooms tend to be confined to oligohaline waters, while *Oscillatoria* and *Lyngbya* can thrive in seawater salinities. These genera thrive under relatively low N:P ratios, as long as adequate P supplies exist (Smith 1990, Paerl 1990). Estuaries with relatively abundant P supplies (natural or anthropogenic) and growing (non–point) N inputs are potential targets for these CHABs. In particular, systems susceptible to bottom water anoxia accompanied by sediment N and P release events may be vulnerable.

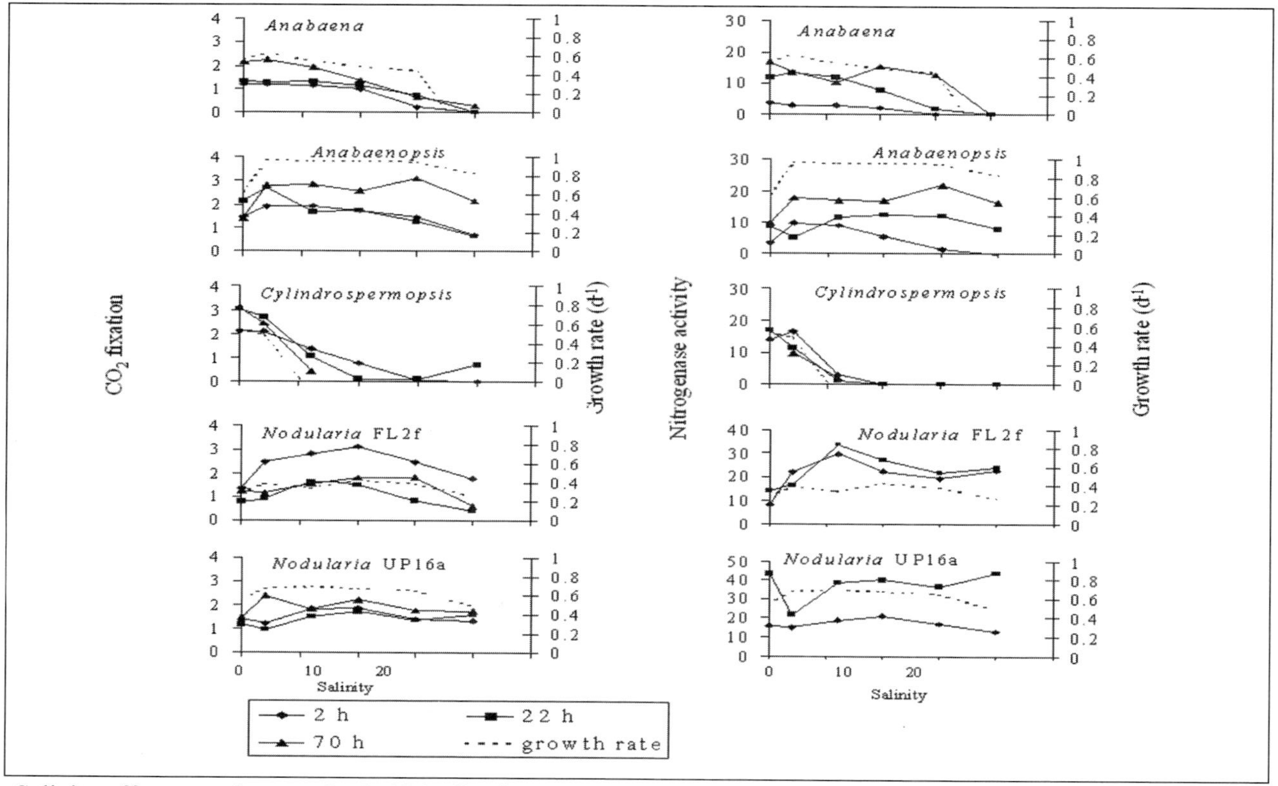

Fig. 7. Salinity effects on photosynthesis (CO_2 fixation), growth, and nitrogen fixation (nitrogenase activity) of some common N_2 fixing CHABS. Figure adapted from Moisander et al. (2002a).

Options for CHAB Control

As shown above, ecosystem level, physical, chemical and biotic regulatory variables often co–occur and may interact synergistically and antagonistically to control the activities (N_2 fixation, photosynthesis) and growth of CHABs (Paerl 1988, Paerl and Millie 1996). Thus, overriding, easily–executed controls are desirable.

Means of controlling blooms include; 1) applications of algacides, the most common of which is copper sulfate, 2) nutrient input reduction and manipulation (of N:P ratios), 3) disrupting vertical stratification, through either mechanically or hydrologically induced vertical mixing, 4) reducing retention time (increasing flushing) of bloom–impacted waters, and 5) biological manipulation. Option 1 has been used in small impoundments, such as ponds and small reservoirs. This approach is not advised in larger ecosystems, or any waters to be used for fishing, drinking water and other animal and human use purposes, unless the system is drained after algacide application, flushed several times and refilled with algacide–free water. If the bloom–affected water body is small and accessible enough for installing destratification equipment, option (3) may be feasible. If abundant water supplies (i.e., upstream reservoirs) are available for hydrodynamic manipulative purposes, option (4) may be possible. Biological manipulation (5) encompasses a number of approaches to change the aquatic food web to increase grazing pressure on cyanobacteria or to reduce recycling of nutrients. Biomanipulation approaches can include introducing fish and benthic filter feeders capable of consuming cyanobacteria, or introduction of lytic bacteria and viruses. However, the most common biomanipulation approaches are intended to increase the abundance of herbivorous zooplankton by removing zooplanktivorous fish or introducing piscivorous fish. Alternatively, removal of benthivorous fish can reduce resuspension of nutrients from the bottom sediments. Questions have been raised about the long–term efficacy of curtailing cyanobacterial blooms by increasing grazing pressure, because this may lead to dominance by ungrazable or toxic strains (McQueen 1990; Ghadouani et al. 2003). Presently, biomanipulation is viewed as one component of an integrated approach to water quality management in circumstances in which nutrient reductions alone are insufficient to restore water quality (Moss et al. 1996, Scheffer 1998, Elser 1999). Otherwise, option 2 is the most practical, economically feasible, environmentally–friendly, long–term option. Below we will consider P and N management options for mitigating CHABs.

Phosphorus

Phosphorus inputs to aquatic ecosystems are dominated by; 1) non point source surface runoff, and 2) point sources such as effluents from wastewater treatment plants, industrial and municipal discharges, and 3) subsurface drainage from septic systems and groundwater. Among these, point sources have been the focus of P reductions. In many watersheds, targeting point sources is attractive, as they can account for a highly significant share of P loading. Point sources are readily identifiable, accessible, and hence from a regulatory perspective, easiest to reduce and manage.

In agricultural and urban watersheds, non–point surface and subsurface P inputs are of increasing concern. Increased P fertilizer use, generation and discharge of animal waste, soil disturbance and erosion, conversion of forests and grasslands to row–crop and other intensive farming operations, and the proliferation of septic systems accompanying human population growth are rapidly increasing non–point P loading. In agricultural and urbanizing watersheds, non–point sources can account for at least 50% of annual P loading. Because of the diffuse nature of these loadings, they are more difficult to identify and address from a nutrient management perspective.

As with nitrogen, the manner in which P is discharged to P–sensitive waters pays a role in controlling CHABs. Considerations include; 1) total annual (i.e., chronic) P loading, 2) shorter–term seasonal and event–based pulse (i.e., acute) P loadings, 3) particulate vs. dissolved P loading, and 4) inorganic vs. organic P loading. With respect to ecosystem P budgets and long–term responses to P loadings (and reductions), annual P inputs are of fundamental importance. However, when considering CHAB dynamics, seasonal and shorter–term acute loading events are of critical, and at times, overriding importance. When and where P enrichment occurs can determine the difference between bloom–plagued vs. bloom–free conditions. For example, if a large spring P discharge event precedes a summer of dry, stagnant (stratified) conditions in a relatively long residence time water body, the spring P load will be available for summer bloom development and persistence. Effective exchange and cycling between the water column and bottom sediments can retard P transport and hence retain P. As a result, acute P inputs during high flow periods may be retained longer than estimated based on water flushing time alone. In effect, water bodies exhibit both rapid biological responses to and a "memory" for acute P loads.

Unlike N, P exists in relatively few dissolved and particulate forms in natural waters. No gaseous forms of P are common, although under anaerobic conditions, trace amounts of the unstable gas phosphine (PH_3) may be generated. Overall, the main concern is with dissolved vs. particulate forms of inorganic and organic P. Dissolved inorganic P (DIP) exists as orthophosphate (PO_4^{3-}), which is readily assimilated by all CHAB taxa. Many CHABs can accumulate assimilated P intracellularly as polyphosphates. Polyphosphates can serve as internal stores of P, for subsequent use in the event of ambient P depletion (Healy 1982). Dissolved organic P (DOP) can be a significant fraction of the total dissolved P pool. DOP can be assimilated by bacteria, microalgae and cyanobacteria, although not as rapidly as PO_4^{3-} (Lean 1973). A large fraction of the assimilated DOP is microbially recycled to DIP, enhancing P availability. The role of particulate P (as inorganic or organic forms) in aquatic production and nutrient cycling dynamics is less well understood. Particulate P (PP) may provide a source of DIP and DOP via desorption and leaching, and it may serve as a sorption/precipitation site for DIP. PP therefore exists in dynamic equilibrium with the dissolved phases of P. It is safe to assume that some fraction of the PP can serve as a source of biologically–available P and hence play a role in CHAB dynamics. On the ecosystem–scale, sedimented PP serves as an important source of stored P for subsequent release, especially during hypoxic/anoxic periods. It is prudent to include *both* dissolved and particulate P when formulating and managing P inputs and N:P ratios.

Nitrogen

Nitrogen exists as dissolved, particulate and gaseous forms. Many of these forms are biologically–available and readily exchanged within and between the water column and sediments. In addition, biological nitrogen (N_2) fixation and denitrification control the exchange between inert gaseous atmospheric N_2 and biologically–available combined N forms. Combined forms of N include dissolved inorganic N (DIN; including ammonium (NH_4^+), nitrate (NO_3^-) and nitrite (NO_3^-)), dissolved organic N (DON; e.g., amino acids and peptides, urea, organo–nitrates), and particulate organic N (PON; polypeptides, proteins, organic detritus). These sources can be supplied as non–point and point sources. Non–point sources include surface runoff, atmospheric deposition and groundwater, while point sources are dominated by municipal, agricultural and industrial wastewater. In rural and agricultural settings, non–point N inputs tend to dominate (>50% of total N loading), while in urban centers, point sources often dominate. All sources contain diverse organic and inorganic N species in dissolved and particulate forms; representing a mixture of biologically–available DIN, DON and PON that plays a critical role in the eutrophication process. Depending on sources, chemical makeup, delivery mechanisms and spatial distribution of N inputs, ecosystem response can vary dramatically.

N inputs are dynamic, reflecting land use, population and economic growth. The means and routes by which human N sources impact and mediate estuarine and coastal eutrophication are changing. Among the most rapidly–growing (amount and geographic scale) sources of human N loading are surface runoff, groundwater and atmospheric deposition. Atmospheric N loading is an often–overlooked, but expanding source of *new* N loading to N–sensitive waters. In eastern North Carolina, the combined emissions of fossil fuel combustion (NO_X) and volatilization of NH_3 from stored animal waste (lagoons and land–applied) are a major (>30% of *new* N loading), rapidly–growing source of biologically–available N. Surface and groundwater N releases from expanding animal operations and urbanization are of additional concern.

The contribution of groundwater and atmospheric N to coastal watershed and oceanic N budgets will increase substantially as we enter the next century, when nearly 70% of North American and European populations will reside within 50 km of the coast (Vitousek et al. 1997). Globally, it is estimated that AD–N accounts for ~40 Tg N y^{-1}, compared to ~30 Tg N y^{-1} from riverine discharge, ~10 Tg N y^{-1} for groundwater and ~20 Tg N y^{-1} for biological nitrogen fixation (Paerl and Whitall 1999). A significant fraction of atmospheric N is directly deposited to N–sensitive estuarine and coastal waters, bypassing the estuarine N "filter". In many locations, including the US Eastern Seaboard, Europe, and East Asia atmospheric N is among the dominant sources of anthropogenic N to the coastal zone. When and where anthropogenic N inputs are intercepted are critical determinants of ecosystem sensitivity, water quality responses and resourcefulness in response to N enrichment. The ramifications of this previously "out of sight out of mind" but growing new N source in algal bloom, including CHAB, dynamics should be investigated.

There is increasing emphasis on reducing N inputs to control estuarine and coastal eutrophication (c.f. Boesch et al. 2001, Elmgren and Larsson 2002). While this is undoubtedly a step in the right direction, the ramifications of reducing N relative to P in coastal waters with regard to CHAB bloom potential needs to be carefully assessed. North Carolina's Neuse River Estuary has been the site of periodic massive blooms of the toxic, surface scum–formers *Microcystis aeruginosa* and *Oscillatoria* spp. Dominance by these non–N_2 fixers attests to the current N "overload situation" (Paerl 1987, NC DENR 1988). Dilution bioassays and historic N loading trend data indicate that watershed loading of N would need to be reduced by at least 30% to obtain N–limited conditions during the critical spring bloom initiation period (Paerl et al. 1995). Accordingly, a 30% N reduction has been legislatively mandated (since 1997), and is the target of a Total Maximum Daily (N) Load (TMDL) imposed by the US EPA (NC DENR 2001).

A complicating aspect nitrogen reduction strategies in the Neuse River is the presence of N_2 fixing cyanobacteria, *Anabaena, Aphanizomenon*, and *Anabaenopsis*. This, combined with evidence that P loading also is excessive (Paerl 1987, Stow et al. 2001), suggests that if N loading is reduced by 30% without parallel P reductions, there may be potential for replacing non–N_2 fixing *Microcystis* with N_2–fixing *Anabaena* or *Aphanizomenon* blooms (Piehler et al., 2002). Similarly, there is concern that reducing N without maintaining strict reductions on P inputs to control eutrophication in Sweden's Himmerfjärden, may allow CHABs to regain dominance (Elmgren and Larsson 2001). Indeed, initial reductions of N in this fjord draining to the Baltic Sea have led to an increase in cyanobacterial biomass (Elmgren and Larsson 2001). Detailed and timely monitoring of phytoplankton community trends as N reductions proceed to stem cultural eutrophication in these and other brackish water systems will enable managers to formulate N and P loadings aimed at de–eutrophication without promoting CHABs.

Conclusions

CHABs are globally distributed and regulated by an interplay of geographically– and ecologically–diverse environmental variables. The long evolutionary history of bloom taxa has led to both tolerance and adaptability to short–term (i.e., diel, seasonal, decadal) and longer–term (geological) environmental change, making these photosynthetic prokaryotes a "group for all seasons".

In this contribution, I have explored the interactive physical, chemical and biotic factors implicated in the development, proliferation and expansion of CHABs. Despite their seemingly infinite adaptation to environmental change on both geological and biological time scales, cyanobacterial nuisance characteristics (e.g., large anoxia–generating and toxic blooms) are impacted by human alterations of aquatic environments. The most notable and controllable alterations include; 1) nutrient (especially N and P) enrichment, 2) hydrological changes, including freshwater diversions, the construction of impoundments such as reservoirs, water use for irrigation, drinking, flood control, all of which affect water residence time or flushing rates. In smaller waterbodies (<50 hectares) destratification of bloom–impacted waters by mechanically–induced vertical mixing or bubbling, and application of algacides such as copper salts and herbicides are options. Algacide applications are only feasible in small impoundments that are not to be used for consumption (either water, fish or shellfish) or recreational purposes (bathing). Furthermore, algacides need to be applied repeatedly, and thus unlike nutrient reductions, are not an effective long–term bloom reduction strategy. Biological controls, such as the introduction and manipulation of grazers (from zooplankton to fish), lytic bacteria, viral cyanophages, and antibiotics have been proposed as controls, but remain highly experimental and have not been broadly effective. Hence, they are not discussed in detail here.

Effective long term control and management of nuisance, particularly toxin–producing, CHABs should consider the interactive nature of abovementioned physical, chemical and biotic factors known to play regulatory roles. In addition, knowledge of the ecological and physiological adaptations that certain taxa possess to circumvent specific environmental controls is of central importance. These include; 1) the ability of N_2 fixing taxa to exploit N–limited conditions, 2) the ability of certain buoyant taxa to counteract mixing and other means of man–induced destratification aimed at minimizing cyanobacterial dominance, 3) specific mutualistic and symbiotic associations that cyanobacteria have with other microorganisms, plants and animals, which may affect CHAB community structure and function.

Lastly, we should strive to better understand the potential roles toxins might play in bloom dynamics. Progress in identifying and understanding the roles toxins and other metabolites play in the physiology and ecology of bloom–forming cyanobacteria may be achieved by integrating physiological, toxicological and ecological perspectives and expertise. This includes hypothesis testing and problem solving using interdisciplinary experimental, monitoring and assessment approaches. In addition, the synthesis of well defined laboratory experimental work with ecosystem–level studies utilizing similar techniques and measurements will prove invaluable in unraveling the complexity of environmental regulation of cyanobacterial blooms. We are at the threshold of more holistic approaches to environmental problem solving. In this regard, the advent and incorporation of novel analytical and molecular identification and characterization techniques in environmental biology and management will prove invaluable and indispensable.

Acknowledgements

I appreciate the technical assistance and input of A. Joyner, R. Fulton, P. Moisander, L. Valdes and B. Peierls. Research discussed in this chapter was partially supported by the National Science Foundation (DEB 9815495, 0452324, and OCE 0327056), the US Dept. of Agriculture NRI Project 00–35101–9981, U.S. EPA STAR Projects R82–5243–010 and R82867701, NOAA/North Carolina Sea Grant Program R/MER–43, the North Carolina Dept. of Natural Resources and Community Development/UNC Water Resources Research Institute (Neuse River Estuary Monitoring and Modeling Project, ModMon), and the St. Johns Water Management District, Florida. This is NOAA ECOHAB Contribution #177.

References

Boesch DF, Burreson E, Dennison W, Houde E, Kemp M, Kennedy V, Newell R, Paynter K, Orth R, Ulanowicz R (2001) Factors in the decline of coastal ecosystems. Science 293: 629–638

Carmichael WW (1997) The cyanotoxins. Advances in Botanical Research 27:211–256

Carmichael WW (1998) Microcystin concentrations in human livers, estimation of human lethal dose–lessons from Caruaru, Brazil. In: Paerl HW, Carmichael WW (eds) Proc of the 4th International Conference on Toxic Cyanobacteria, Beaufort, NC, Sept 1998

Chapman AD, Schelske CL (1997) Recent appearance of *Cylindrospermopsis* (Cyanobacteria) in five hypereutrophic Florida lakes. J Phycol 33:191–195.

Chorus I, Bartram J (eds) (1999) Toxic Cyanobacteria in Water. E&F Spon, London

Codd GA, Bell SG (1996) The occurrence and fate of blue–green algal toxins in freshwaters. National Rivers Authority, R & D Report 29, London

Diaz RJ, Solow A (1999) Ecological and Economic Consequences of Hypoxia. Topic 2 Report for the Integrated Assessment of Hypoxia in the Gulf of Mexico. NOAA Coastal Ocean Program Decision Analysis Series No. 16. NOAA COP, Silver Spring, MD, 45pp.

Downing JA, Watson SB, McCauley E (2001) Predicting Cyanobacteria dominance in lakes. Can J Fish Aquat Sci 58: 1905–1908

Edmondson WT, Lehman JT (1981) The effect of changes in the nutrient income and conditions of Lake Washington. Limnol Oceanogr 26:1–29

Elmgren R, Larsson U (2001) Nitrogen and the Baltic Sea: Managing nitrogen in relation to phosphorus. The Scientific World 1(S2): 371–377

Elser JJ (1999) The pathway to noxious cyanobacteria blooms in lakes: The food web as the final turn. Freshwater Biol 42: 537–543

Fogg GE (1969) The physiology of an algal nuisance. Proc R Soc London B. 173:175–189

Ghadouani A, Pinel–Alloul B, Prepas EE (2003) Effects of experimentally induced cyanobacterial blooms on crustacean zooplankton communities. Freshwater Biol 48: 363–38

Hallegraeff GM (1993) A review of harmful algal blooms and their apparent global increase. Phycologia 32:79–99

Healy FP (1982) Phosphate. In: Carr NG, Whitton BA (eds) The Biology of Cyanobacteria. Blackwell Scientific Public, Oxford, pp. 105–124.

Horne AJ (1977) Nitrogen fixation.– a review of this phenomenon as a polluting process. Progr Wat Technol 8:359–372

Kononen K, Kuparinen J, Mäkelä J, Laanemets J, Pavelson J, Nõmmann S (1996) Initiation of cyanobacterial blooms in a frontal region at the entrance to the Gulf of Finland, Baltic Sea. Limnol Oceanogr 41:98–112

Konopka A (1984) Effect of light–nutrient interactions on buoyancy regulation by planktonic cyanobacteria. In: Klug MJ, Reddy CA (eds) Current Perspectives in Microbial Ecology. Am Soc Microbiol, Washington, DC, pp 41–48

Konopka A, Brock TD, Walsby AE (1978) Buoyancy regulation by planktonic blue–green algae in Lake Mendota, Wisconsin. Arch Hydrobiol 83:524–537

Lean DRS (1973) Movement of phosphorus between its biologically–important forms in lakewater. J Fish Res Bd Canada 30:1525–1536

Likens GE (ed) (1972) Nutrients and Eutrophication. American Soc Limnol Oceanogr Special Symp 1

McQueen DJ (1990) Manipulating lake community structure: where do we go from here? Freshwater Biol 23: 613–620

Moisander PH, Paerl HW (2000) Growth, primary productivity, and nitrogen fixation potential of *Nodularia* spp. (Cyanophyceae) in water from a subtropical estuary in the United States. J Phycol 36:645–658

Moisander PH, McClinton III E, Paerl HW (2002a) Salinity effects on growth, photosynthetic parameters, and nitrogenase activity in estuarine planktonic cyanobacteria. Microb Ecol 43:432–442

Moisander PH, Hench JL, Kononen K, Paerl HW (2002b) Small–scale shear effects on heterocystous cyanobacteria. Limnol Oceanogr 47: 108–119

Moss B, Madgwick J, Phillips JG (1996) A guide to the restoration of nutrient–enriched shallow lakes. W.W. Hawes, United Kingdom

Niemi A (1979) Blue–green algal blooms and N:P ratio in the Baltic Sea. Acta Bot Fenn 110:57–61

Nixon SW (1986) Nutrient dynamics and the productivity of marine coastal waters. In: Halwagy R, Clayton D, Behbehani M (eds) The Alden Press, Oxford, pp 97–115

Nixon SW (1995) Coastal eutrophication: A definition, social causes, and future concerns. Ophelia:199–220

North Carolina Department of Environment and Natural Resources (1996) Report of the NC State Senate Select Committee on Fish Kills and Water Quality, Raleigh, NC

North Carolina Dept. of Environment and Natural Resources–Div. Water Quality (2001) Phase II of the Total Maximum Daily Load for Total Nitrogen to the Neuse River Estuary, NC. NCDENR–DWQ, Raleigh, NC

Padisak J (1997) *Cylindrospermopsis raciborskii* (Woloszynska) Seenayya et Subba Raju, an expanding, highly adaptive cyanobacterium: worldwide distribution and review of its ecology. *Archive fur Hydrobiologie Suppl.* 107:563–593.

Paerl HW (1983) Environmental factors promoting and regulating N_2 fixing blue–green algal blooms in the Chowan River, NC. Univ of North Carolina Water Resources Research Instit Report No. 176, 65 pp

Paerl HW (1986) Growth and reproductive strategies of freshwater blue–green algae (cyanobacteria), In: Sandgren CD (ed) Growth and Reproductive Strategies of Freshwater Phytoplankton. Cambridge Univ Press

Paerl HW (1987) Dynamics of blue–green algal blooms in the lower Neuse River, NC: Causative factors and potential controls. Univ of North Carolina Water Resources Research Institute Report No 229, 164 pp

Paerl HW (1988) Nuisance phytoplankton blooms in coastal, estuarine, and inland waters. Limnol Oceanogr 33:823–847

Paerl HW (1990) Physiological ecology and regulation of N_2 fixation in natural waters. Adv Microbiol Ecol 11:305–344

Paerl HW (1996) A comparison of cyanobacterial bloom dynamics is freshwater, estuarine and marine environments. Phycologia 35(6):25–35

Paerl HW (1997) Coastal eutrophication and harmful algal blooms: Importance of atmospheric deposition and groundwater as "new" nitrogen and other nutrient sources. Limnol Oceanogr 42:1154–1165

Paerl HW, Mallin MA, Donahue CA, Go M, Peierls BL (1995) Nitrogen loading sources and eutrophication of the Neuse River Estuary, NC: Direct and indirect roles of atmospheric deposition. UNC Water Resources Research Instit Report No 291. Raleigh, NC

Paerl HW, Millie DF (1996) Physiological ecology of toxic cyanobacteria. Phycologia 35(6):160–167.

Paerl HW, Whitall DR (1999) Anthropogenically–derived atmospheric nitrogen deposition, marine eutrophication and harmful algal bloom expansion: Is there a link? Ambio 28:307–311.

Paerl HW, Fulton RS, Moisander PH, Dyble J (2001) Harmful Freshwater Algal Blooms, With an Emphasis on Cyanobacteria. The Scientific World 1:76–113.

Paerl HW, Kuparinen J (2002) Microbial aggregates and consortia. In: Bitton G (ed) Encyclopedia of Environmental Microbiology, Vol. 1. John Wiley and Sons, Inc New York, NY, pp 160–181

Paerl HW, Valdes LM, Pinckney JL, Piehler MF, Dyble J, Moisander PH (2003) Phytoplankton photopigments as Indicators of Estuarine and Coastal Eutrophication. BioScience 53(10) 953–964.

Paerl HW, Valdes LM, Piehler MF, Lebo ME (2004) Solving problems resulting from solutions: The evolution of a dual nutrient management strategy for the eutrophying Neuse River Estuary, North Carolina, USA. Environmental Science & Technology 38:3068–3073

Pearsall W (1932) Phytoplankton in the English Lakes. 2. The composition of the phytoplankton in relation to dissolved substances. Journal of Ecology 20:241–262

Piehler MF, Dyble J, Moisander PH, Pinckney JL, Paerl HW (2002) Effects of modified nutrient concentrations and ratios on the structure and function of the native phytoplankton community in the Neuse River Estuary, North Carolina USA. Aquat Ecol 36: 371–385

Pinckney JL, Millie DF, Vinyard BT, Paerl HW (1997) Environmental controls of phytoplankton bloom dynamics in the Neuse River Estuary (North Carolina, USA). Can. J. Fish. Aquat. Sci. 54:2491–2501.

Reynolds CS (1987) Cyanobacterial water blooms. Adv Bot Res 13:67–143

Reynolds CS, Walsby AE (1975) Water blooms. Biol Rev 50:437–481

Richardson K (1997) Harmful or exceptional phytoplankton blooms in the marine ecosystem. Adv Mar Biol 31:302–385

Ryther JH, Dunstan WM (1971) Nitrogen, phosphorus and eutrophication in the coastal marine environment. Science 171:1008–1112

Schindler DW (1975) Whole–lake eutrophication experiments with phosphorus, nitrogen and carbon. Verh Int Verein Theor Angew Limnol 19:3221–3231

Shapiro J (1990) Current beliefs regarding dominance of blue–greens: The case for the importance of CO_2 and pH. Int Verein Theor Angew Limnol Verh 24:38–54

Scheffer M (1998) Ecology of shallow lakes. Chapman and Hall, London

Smith VH (1983) Low nitrogen to phosphorus ratios favor dominance by blue–green algae in lake phytoplankton. Science 221:669–671

Smith VH (1990) Nitrogen, phosphorus, and nitrogen fixation in lacustrine and estuarine ecosystems. Limnol Oceanogr 35:1852–1859.

Stow CA, Borsuk ME, Stanley DW (2001) Long–term changes in watershed nutrient inputs and riverine exports in the Neuse River, North Carolina. *Water Research* 35:1489–1499.

Tedder SW, Sauber J, Ausley L, Mitchell S (1980) Working Paper: Neuse River Investigation 1979. Division of Environmental Management, North Carolina Department of Natural Resources and Community Development, Raleigh, North Carolina.

Tilman D, Kiesling RL (1984) Freshwater algal ecology: Taxonomic tradeoffs in the temperature dependence of nutrient competitive abilities. In: Klug MJ, Reddy CA (eds) Current Perspectives in Microbial Ecology. American Soc Microbiol Washington, DC, pp 314–320

Vincent WF (ed) (1987) Dominance of bloom forming cyanobacteria (Blue–green algae). NZ Jour Mar and Freshwat Res 21(3):361–542

Vitousek PM, Mooney HA, Lubchenko J, Mellilo JM (1997) Human domination of Earth's ecosystems. Science 277:494–499

Vollenweider RA, Kerekes JJ (1982) Eutrophication of waters: Monitoring, assessment and control. OECD, Paris

Vollenweider, RA, Marchetti R, Viviani R (eds) (1992) Marine Coastal Eutrophication. Elsevier Science, New York.

Chapter 11: Global warming and cyanobacterial harmful algal blooms

Valerie J Paul

Smithsonian Marine Station at Fort Pierce, 701 Seaway Drive, Fort Pierce, FL 34949 paul@sms.si.edu

Abstract

The Earth and the oceans have warmed significantly over the past four decades, providing evidence that the Earth is undergoing long-term climate change. Increasing temperatures and changing rainfall patterns have been documented. Cyanobacteria have a long evolutionary history, with their first occurrence dating back at least 2.7 billion years ago. Cyanobacteria often dominated the oceans after past mass extinction events. They evolved under anoxic conditions and are well adapted to environmental stress including exposure to UV, high solar radiation and temperatures, scarce and abundant nutrients. These environmental conditions favor the dominance of cyanobacteria in many aquatic habitats, from freshwater to marine ecosystems. A few studies have examined the ecological consequences of global warming on cyanobacteria and other phytoplankton over the past decades in freshwater, estuarine, and marine environments, with varying results. The responses of cyanobacteria to changing environmental patterns associated with global climate change are important subjects for future research. Results of this research will have ecological and biogeochemical significance as well as management implications.

Introduction

Global warming

The Intergovernmental Panel on Climate Change published its third assessment report in 2001(IPCC 2001). The report concluded that the Earth is warming, which is causing regional climate changes and influencing many physical, biological and chemical processes (Walther et al. 2002, Treydte et al. 2006). The global average surface temperature increased by $0.6 \pm 0.2^{o}C$ during the 20th century, with most of the increase observed over the past four decades (Fig. 1). Instrumental measurements and proxy data show that the rate and duration of warming in the 20th century has been greater than in any of the previous nine centuries, and the 1990s were the warmest decade of the millennium (IPCC 2001). The patterns of warming, which influence trends in temperature and precipitation, are regionally highly variable. This can have striking effects on ecological responses to climate change (Walther et al. 2002). Some of the implications of global change for freshwater ecosystems have been reviewed (Carpenter et al. 1992). Evidence shows a lengthened freeze-free season at many mid- and high-latitudes, the retreat of mountain glaciers in non-polar regions, a decrease in snow cover by about 10%, and a decrease of about two weeks in the annual duration of lake and river ice cover in the Northern Hemisphere (IPCC 2001).

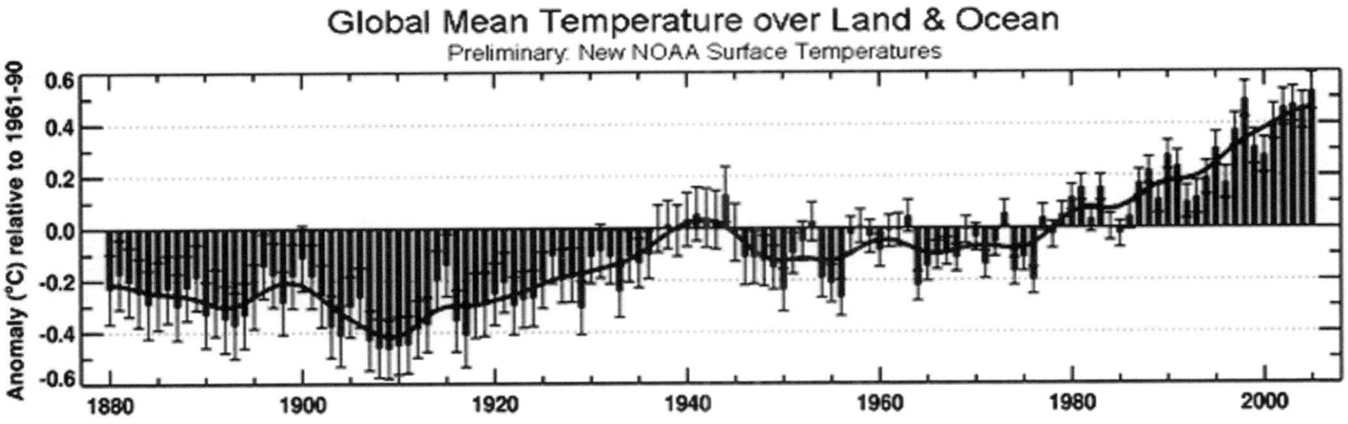

Fig. 1. Time series approximates the temperature of the Earth and how it has been changing through time. Taken from NOAA website. Source: http://www.ncdc.noaa.gov/img/climate/research/2005/ann/global-blended-temp-pg.gif

Most of the heating of the earth since the 1950s has occurred in the oceans. During 1955-1998 ocean heat content in the upper 3000 m of the world's oceans increased by 14.5 x 10^{22} J (a mean temperature increase of 0.037°C) (Levitus et al. 2005). This has led to thermal expansion of the uppermost 700 m layer of the oceans (Antonov et al. 2005), and may lead to greater stratification of the ocean (Hegerl and Bindoff 2005). There is good evidence from modeling studies compared with actual temperature records that atmospheric heat trapped by rising greenhouse gases is responsible for warming the oceans (Barnett et al. 2005). Other impacts of global warming related to sea surface temperature increases may be observed. For example, evidence suggests the destructiveness and number of intense (categories 4 and 5) tropical cyclones has increased over the past 35 years (Emanuel 2005, Webster et al. 2005).

Global climate change also involves stratospheric-ozone depletion, which leads to increased levels of biologically damaging UV-radiation reaching the Earth's surface, and an increasing concentration of atmospheric CO_2. Increasing CO_2 levels contribute to sea-level rise and alteration of ocean chemistry (Hallock 2005). Changes in ocean chemistry due to increasing atmospheric CO_2 lead to acidification of the ocean (Sabine et al. 2004), which can have profound impacts on calcifying organisms such as corals, calcareous algae, formaniferans, and mollusks and their rates of calcification (Feely et al. 2004, Hallock 2005).

Evolutionary history of cyanobacteria

Cyanobacteria have been on Earth a very long time-- billions of years. Fossil evidence for the presence of oscillatoreacean cyanobacteria over 3 billion years ago can be found in the 3.5 billion year-old Apex chert deposits in Western Australia (Schopf 2000). Well preserved fossil cyanobacteria are nearly indistinguishable in morphology from their extant relatives and can be found in intertidal and shallow marine environmental settings like those inhabited by cyanobacteria today. Modern cyanobacteria produce 2-methylbacteriohopanepolyols. The geologically stable derivatives, 2α-methylhopanes, serve as biomarkers in sediments for cyanobacteria, which provide further evidence that cyanobacteria were abundant at least 2.5 billion years ago (Summons et al. 1999).

Cyanobacteria can be distinguished from all other prokaryotes by their ability to carry out oxygen-producing photosynthesis. Evidence of both the reactants of oxygenic photosynthesis (H_2O and CO_2) and the products (reduced organic carbon and O_2) can be found in the early rock record of the Apex Chert (Schopf 2000). Although oxygen was being released by cyanobacteria at least 2.7 billion years ago, a low-O_2 environment is thought to have persisted for at least another 300 million years based on several lines of evidence (Canfield 1999, Kerr 2005). Banded iron formations, iron-oxide rich sediments, are widespread in geologic formations at least 2.5 billion years ago, indicating that molecular oxygen was removed from the system by its reaction with iron during the early Precambrian (Schopf 2000). Sulfur isotope methods applied to rocks of different ages indicate a major change in the sulfur cycle between 2.4 and 2.0 billion years ago (Farquhar et al. 2000), suggesting the appearance of atmospheric oxygen at levels of at least 1 ppm during that time (Kerr 2005). However, it was not until the end of the Proterozoic (600 million years ago) that oxygen rose to near modern levels and multicellular animals appeared (Kerr 2005). The history of the rise of oxygen on the early Earth is the subject of debate among geochemists, paleobiologists and astrobiologists; however, it is clear that cyanobacteria played a significant role in the early evolution of Earth's atmosphere (probably to their own detriment, once multicellular life evolved) (Canfield 1999, Kasting and Siefert 2002, Kerr 2005).

Environmental conditions on the early Earth under which cyanobacteria evolved and thrived were very different from today (Kasting and Siefert 2002). Anoxia, higher UV exposure, higher temperatures, and high levels of iron, sulfide and methane were all factors that influenced early life on Earth. Additionally, cyanobacteria experienced little or no grazing pressure or competition from higher organisms. This may explain why cyanobacteria can thrive under conditions of environmental stress and in extreme environments where they are able to out-compete other organisms. For example, in coral reef habitats cyanobacteria are becoming increasingly dominant on degraded reefs because of their ability to tolerate the environmental conditions associated with anthropogenic impacts and global climate change (Hallock 2005).

Cyanobacteria greatly increased in abundance following mass extinction events (Copper 1994, Hallock 2005). Sheehan and Harris (2004) documented a resurgence of microbialites following the late Ordovician extinction event in western North America approximately 440 million years ago. Microbialites, including stromatolites and other microbial mats that bind sediments, precipitate minerals, and form crusts, increased in size, abundance, and morphological diversity following the late Ordovician extinction event. The period of microbialite resurgence lasted about 5 million years and corresponded with low faunal diversity during the post-extinction recovery period. Similar microbial expansion has been documented after the Permian-Triassic faunal mass extinction (Xie et al. 2005), approximately 250 million years ago, which eliminated over 90% of all marine species (Jin et al. 2000). A survey of microbial biomarkers in sedimentary rocks in South China spanning the Permian-Triassic (P/TR) boundary was conducted and compared with faunal extinction patterns (Xie et al. 2005). Both cyanobacterial biomarker and invertebrate fossil records showed evidence for two periods of extinction and ecosystem change across the P/Tr boundary. Dramatic changes in the cyanobacterial community followed the two episodes of faunal mass extinction. Two maxima in the 2α-methylhopane (2-MHP) indices following faunal extinctions indicated either a change in the cyanobacterial community (because not all cyanobacteria produce 2-MHPs) or an increase in the abundance of cyanobacteria. Xie et al. (2005) suggested that cyanobacteria dominated following the mass extinction because of low faunal diversity and lack of grazing pressure.

In a review of expansion and collapse of reef ecosystems during the Paleozoic era and examination of reefs existing during periods of less than optimal environmental conditions, Copper (1994) concluded that periodic collapse of reefs and extinction of reef fauna serve as a good indicator of global change. The disappearance of reefs in the geologic record often preceded global mass extinction events by 0.5-1 million years. Copper (1994) termed "disaster species" as groups of organisms that are extinction resistant or ecological opportunists that can survive under marginal conditions. "Disaster species" include cyanobacteria, calcareous algae, encrusting foraminiferans and bryozoans, and these organisms are major components of reef communities during times of reef collapse (Copper 1994). The final remnants of reefs undergoing extinction were covered by skeletal, carbonate-secreting cyanobacteria, and reefs that developed after extinction events featured these and other "disaster species". Environmentally tolerant cyanobacteria, which are resistant to strong solar radiation, higher temperatures, abundant nutrients, and many generalist grazers, may be useful indicators of stress conditions in reef ecosystems today (Hallock 2005).

In this introduction I have set the stage for further discussion of the influence of climate change on cyanobacteria in aquatic and marine habitats. Global climate change is occurring, which is already causing changes in terrestrial and marine ecosystems (Walther et al. 2002). Based on geological records, paleobiological evidence, and physiological and ecological studies, cyanobacteria seem likely to benefit from environmental changes associated with global warming. I review specific physiological and ecological examples below as well as the few studies that have directly examined the effects of climate change on cyanobacterial abundance.

Environmental influences on cyanobacterial growth and toxicity

Chemical defenses

Toxicity in cyanobacteria has been reported since the late 19th century, mostly from poisonings in freshwater environments (Carmichael et al. 1990). These reports describe sickness and death of livestock, pets and wildlife following ingestion of water containing blooms of toxic algae. Cyanobacterial toxins and other bioactive compounds have been the subject of many investigations by toxicologists and natural products chemists (Chorus 2001, Gerwick et al. 2001). These substances are of considerable importance because of their public health implications (Codd et al. 2005) but have also been examined for possible biomedical uses as antitumor compounds (Burja et al. 2001). The majority of natural products from cyanobacteria are lipopeptides, cyclic peptides and depsipeptides, and alkaloids (Moore 1996, Gerwick et al. 2001, Chorus 2001).

Cyanobacterial natural products can also serve as chemical defenses against grazers and competitors (Kirk and Gilbert 1992, Christoffersen 1996, Nagle and Paul 1999, Paul et al. 2001, Landsberg 2002). Freshwater cyanotoxins such as microcystins can have negative physiological effects (Fulton and Pearl 1987, DeMott et al. 1991, Rohrlack et al. 2001, 2005) and negative fitness consequences (Gustafsson et al. 2005) on zooplankton grazers such as *Daphnia* spp. Cyanobacteria can inhibit feeding by copepods (Fulton and Paerl 1987, Sellner et al. 1994). Although rapid increases in the abundance of cyanobacteria may be caused by factors similar to those that influence other algal blooms, such as nutrient enrichment, the interactive effects of eutrophication and herbivory on cyanobacterial populations are not well-understood (Thacker et al. 2001, Raikow et al. 2004).

Tsuda and Kami (1973) suggested that selective browsing by herbivorous fishes on macroalgae removed potential competitors and favored the establishment of unpalatable benthic cyanobacteria in tropical marine environments. Crude extracts and isolated secondary metabolites of several benthic marine cyanobacteria such as *Lyngbya* spp. have been tested in assays for feeding deterrence and are usually deterrent to generalist herbivores (Pennings et al. 1997, Nagle and Paul 1998, 1999, Capper et al. 2006). Thus, the chemical defenses of cyanobacteria may play a critical role in bloom formation and persistence by limiting the grazing activity of potential consumers.

UV tolerance

Many cyanobacteria are UV-tolerant and have evolved various mechanisms to counter UV radiation (Castenholz and Garcia-Pichel 2000, He and Häder 2002, He et al. 2002, Xue et al. 2005). For example, the UV-absorbing mycosporine-like amino acids (MAAs) are abundant in marine, freshwater, and terrestrial cyanobacteria (Garcia-Pichel and Castenholz 1993, Castenholz and Garcia-Pichel 2000, Liu et al. 2004). Shinorine, mycosporine-glycine, porphyra-334, and asterina 330 are among the MAAs identified in cyanobacteria (Figure 2). These colorless, low molecular weight, water soluble, UV-absorbing compounds are present under all growth conditions. They were found to increase in *Gleocapsa* when cells were exposed to UV radiation at 310-320 nm (Garcia-Pichel et al. 1993). Cyanobacteria also produce sheath pigments such as scytonemin, a yellow to brown-colored UV-absorbing pigment found in many different cyanobacteria (Figure 2). Exposure to UV radiation induces the production of scytonemin, which absorbs maximally at the longest wavelengths of the UV (approx. 380 nm) (Castenholz and Garcia-Pichel 2000). The UV-absorption of scytonemin complements the UV-absorption of the MAAs, which have absorption maxima ranging from 310 to 360 nm (Garcia-Pichel and Castenholz 1993). Other mechanisms of dealing with UV stress include antioxidant enzymes, such as superoxide dismutase, catalase, and glutathione peroxidase, and antioxidant molecules, including ascorbate, carotenoids, and tocopherols (Paerl et al. 1985, He and Häder 2002, Xue et al. 2005).

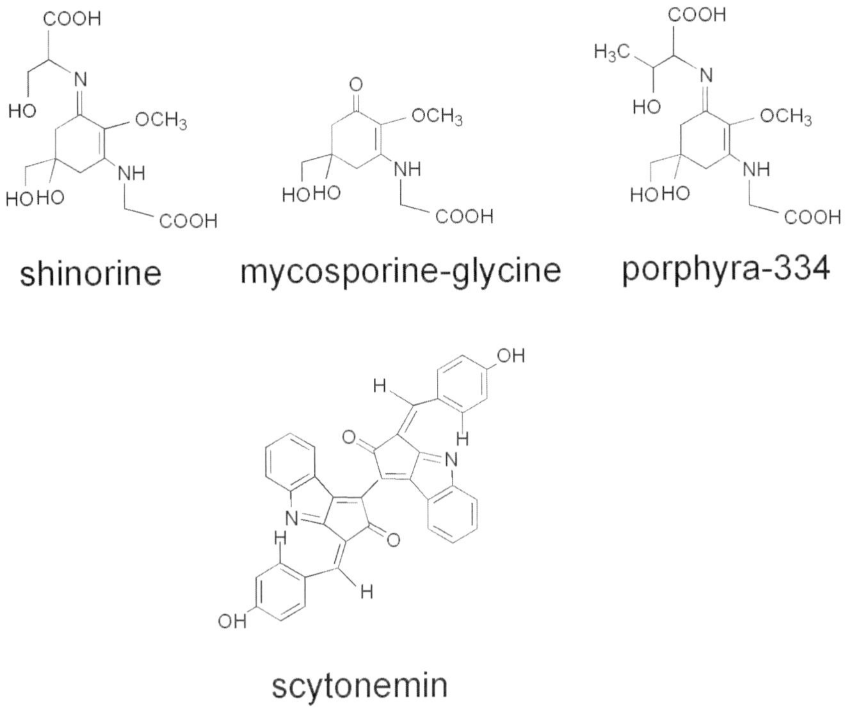

Fig. 2. MAAs (e.g. shinorine, mycosporine-glycine, porphyra-334) and the sheath pigment scytonemin are UV-absorbing compounds commonly found in cyanobacteria.

Temperature and light

Cyanobacteria often favor warm water temperatures and high light environments (Paerl et al. 1985, Robarts and Zohary 1987). In a study designed to understand the invasive behavior of the toxic, bloom forming cyanobacterium *Cylindrospermopsis raciborskii,* Briand et al. (2004) examined the growth of 10 strains of this cyanobacterium under different light intensities and temperatures. All 10 strains grew under a broad range of temperatures and light intensities, suggesting that the invasion of *C. raciborskii* into freshwater ecosystems at mid-latitudes may result from its ability to tolerate a variety of environmental conditions. Regional and global warming could provide this species with better environmental conditions for optimal growth, which occurs at temperatures approximately 30°C (Briand et al. 2004).

Trichodesmium species are globally significant, nitrogen-fixing marine cyanobacteria found throughout the tropical and subtropical oceans. *Trichodesmium* species inhabit low-nutrient, clear surface waters and are adapted to a high light regime. Active growth occurs largely at temperatures above 20°C. Pigment composition and photosynthetic capabilities indicate maximum photosynthetic efficiency and effective photoprotection in high light environments (Capone et al. 1997). These cyanobacteria prefer warm water temperatures. For example, a bloom of *T. erythraeum* occurred in the Canary Islands Archipelago during August 2004, the warmest period recorded since 1912 (27.5°C) (Ramos et al. 2005). Blooms of *T. erythraeum* had not been previously recorded anywhere along the coast bordering the NW African upwelling zone, and this bloom event was thought to be associated with exceptionally warm weather and dust storms observed in the area (Ramos et al. 2005). Enhanced vertical stratification arising from increased warming of oceanic surface waters could also favor blooms of this bouyant genus.

Cyanobacteria have also been shown to be dominant in the warmer months in subtropical estuaries (Pensacola Bay, FL) (Murrell and Lores 2004). Cyanobacterial abundance (*Synechococcus*) peaked in the upper estuary during summer months. Their abundance at the freshwater end of the Bay was very low, suggesting they were not delivered to the estuary by freshwater. Changes in phytoplankton seemed to be related to changes in zooplankton composition, suggesting important trophic implications of a shift to cyanobacterial dominance.

Benthic marine cyanobacteria, such as *Lyngbya* species, have been documented to bloom in coastal environments with increasing frequency during the past decade (Paul et al. 2001, O'Neil and Dennison 2005, Paul et al. 2005). Large blooms of *Lyngbya majuscula* have been reoccurring seasonally in Moreton Bay, Australia, for several years. In an attempt to understand factors influencing bloom dynamics, Watkinson et al. (2005) assessed various environmental parameters before and during a bloom. A pulse of rainfall activity preceded the *Lyngbya* bloom they were monitoring, and a period of high incident light, water temperatures above 24°C, and calm weather conditions coincided with bloom initiation. The high temperature requirements for this cyanobacterium are similar to those reported for other cyanobacteria (Robarts and Zohary 1987). Nutrients (especially iron) carried by creeks flowing into the study region may have also contributed to bloom formation (Watkinson et al. 2005). Other studies have also documented the potential roles of phosphorus and iron in stimulating *Lyngbya majuscula* productivity and growth (Kuffner and Paul 2001, Elmetri and Bell 2004, Albert et al. 2005).

The black band disease cyanobacterium *Phormidium corallyticum*, a major component of the microbial consortium that causes this common coral disease, photosynthesizes maximally under aerobic conditions at water temperatures of 30-37°C (Richardson and Kuta 2003). Field observations report that black band disease is most common during the summer months, and these data suggest that water temperature is one of the most important factors in the seasonality of black band disease (Richardson and Kuta 2003). Elevated water temperatures >30°C can cause bleaching and mortality among many coral species (Donner et al. 2005); impaired coral physiology might also contribute to black band disease occurrence.

Light and temperature influence growth rates as well as toxin production for many species of cyanobacteria (Sivonen 1990, Rapala et al. 1997). Several studies have shown that toxin production in *Microcystis aeruginosa* increases with irradiance under light-limited conditions (Watanabe and Oishi 1985, Utkilen and Gjolme 1992, Weidner et al. 2003). Microcystin composition of *Planktothrix agardhii* changed toward a higher proportion of a more toxic variant with increased light intensity (Tonk et al. 2005). Pangilinan (2000) found that light levels influenced growth and production of the compound pitipeptolide A by *Lyngbya majuscula*. The cyanobacterium grew twice as much and contained significantly more pitipeptolide A at higher light levels of 135-169 μmol photon m^{-2} s^{-1} than at light levels approximately half that amount after four weeks. Temperature is also an important factor affecting both growth and secondary metabolite accumulation in cyanobacteria (Watanabe and Oishi 1985, Sivonen 1990, Rapala et al. 1997, Lehtimaki et al. 1994). In recent years, understanding of the genetics of cyanotoxin biosynthesis has advanced considerably (Tillett et al. 2000, Dittmann et al. 2001, Kaebernick et al. 2002, Edwards and Gerwick 2004). This facilitates molecular approaches to studying how light and other environmental factors may influence cyanotoxin production (Kaebernick et al. 2000, Kaebernick and Neilan 2001).

Temperature, light, and nutrients may have interactive effects on growth and toxin production, with consequent effects on aquatic food webs. One laboratory study showed that increases in water temperature could increase the susceptibility of rotifers to toxic effects of anatoxin-a produced by *Anabaena flos-aquae* (Gilbert 1996). Moss et al. (2003) studied the effects of warming, nutrient addition, and fish on phytoplankton composition in mesocosms designed to mimic macrophyte-dominated shallow lake environments. Warming 3°C above ambient had smaller effects on the phytoplankton community than did the presence or absence of fish (sticklebacks, *Gasterosteus aculeatus*) or addition of nutrients. Contrary to expectation, warming did not increase the abundance of cyanobacteria in the experimental treatments.

Nutrients- rainfall patterns

The amounts and chemical composition of nutrients sources (especially N and P) to freshwater and nearshore coastal waters are well known to influence cyanobacterial bloom formation and dynamics. This topic has been reviewed elsewhere in these proceedings (Pearl 2006). Some climatic effects of global change, including variation in rainfall patterns (Walther et al. 2002, Treydte et al. 2006), floods, droughts, dust storms (Prospero and Lamb 2003), tropical storms, and intensity of hurricanes (Webster et al. 2005) can synergistically (along with nutrients) impact cyanobacterial and algal communities and bloom dynamics. The consequences of these changing patterns, their impact on nutrient entry and utilization in the freshwater-marine continuum (Carpenter et al. 1992), and the interactions between nutrients and other environmental factors on cyanobacterial bloom formation and cyanotoxin production have largely not been investigated.

Ecological and ecosystem consequences

Few specific examples address the ecological consequences of global warming on cyanobacterial blooms and their dominance in aquatic ecosystems. Variable results have been obtained from the few studies that address this topic (Moss et al. 2003, Briand et al. 2004, Ramos et al. 2005).

Cyanobacterial abundance can increase during warmer years, suggesting that global warming will have long term effects on phytoplankton communities. A 42-year record of primary productivity in a small, subalpine lake (Castle Lake, northern California) showed that *Daphnia* and cyanobacterial biomass were higher during warmer years (Park et al. 2004). Increasing water temperatures were accompanied by increasing summer cyanobacterial biovolume, whereas other phytoplankton groups did not show significant trends with water temperature. The authors suggested that changes in air temperature and precipitation as a result of global warming, the Pacific Decadal Oscillation, and El Niño Southern Oscillation could influence primary productivity and plankton communities in North American dimictic lakes (Park et al. 2004). Other studies have also shown correlations between climate variation, often related to the North Atlantic Oscillation (NAO), and spring plankton dynamics and clear water phases in shallow, polymictic European lakes (Gerten and Adrian 2000, Scheffer et al. 2001). Cyanobacteria can be abundant in polymictic shallow lakes, with rapid and intense blooms reported in spring and summer (Abrantes et al. 2006).

Lake Tanganyika, a large rift valley lake in east Africa, has been surveyed and records of its nutrients and temperatures have been published several times in the past century. Temperatures have increased in the past century by 0.9°C at 100 m, with 50% of the heat gained by the lake in the upper 330 m. The lake records show a century-long warming trend, and impacts on its planktonic ecosystem are reported (Verburg et al. 2003). A sharpened density gradient, a consequence of surface warming, has slowed vertical mixing and reduced primary production. The phytoplankton biomass in 2000 was lower than in 1975 by 70%, and the composition of the phytoplankton also changed. Cyanobacteria comprised a larger portion of the total biomass in March to April 2001 than they did during the same time in 1975 (Verburg et al. 2003).

Conclusions

Examination of the evolutionary history of cyanobacteria, studies of their ecophysiology, and recent investigations of phytoplankton dynamics and community structure in response to global climate change all suggest that cyanobacteria will probably thrive under environmental conditions associated with global warming. Most studies addressing cyanobacterial responses to changes in irradiance, temperature, and nutrients have been done in the laboratory; few studies have addressed these important topics under field conditions. Clearly, studies of changes in cyanobacterial community composition, bloom dynamics, and toxicity in response to increasing temperatures and other environmental factors associated with global warming are of great importance to our understanding of harmful cyanobacterial bloom dynamics. Interactions among anthropogenic effects, including eutrophication and food web alterations, and changing environmental conditions are likely to be complex. A much better understanding of these issues is urgently needed.

Acknowledgements

My research on cyanobacterial blooms has been largely supported by the ECOHAB program through funding from US EPA (R82-6220) and NOAA through the University of North Carolina at Chapel Hill (NA05NOS-4781194). I am grateful to two anonymous reviewers for comments that helped improve this manuscript. This is contribution #661 of the Smithsonian Marine Station at Fort Pierce.

References

Abrantes N, Antunes SC, Pereira MJ, Goncalves F (2006) Seasonal succession of cladocerans and phytoplankton and their interactions in a shallow eutrophic lake (Lake Vela, Portugal). Acta Oecologia 29: 54-64

Albert S, O'Neil JM, Udy JW, Ahern KS, O'Sullivan CM, Dennison WC (2005) Blooms of the cyanobacterium *Lyngbya majuscula* in coastal Queensland, Australia: disparate sites, common factors. Mar Poll Bull 51: 428-437

Antonov JI, Levitus S, Boyer TP (2005) Thermosteric sea level rise, 1955-2003. Geophys Res Lett 32:L12602

Barnett TP, Pierce DW, AchutaRao KM, Gleckler PJ, Santer BD, Gregory JM, Washington WM (2005) Penetration of human-induced warming into the world's oceans. Science 309: 284-287

Briand J-F, Leboulanger C, Humbert J-F (2004) *Cylindrospermopsis raciborskii* (Cyanobacteria) invasion at mid-latitudes: selection, wide physiological tolerance, or global warming. J Phycol 40: 231-238

Burja AM, Banaigs B, Abou-Mansour E, Burgess JG, Wright PC (2001) Marine cyanobacteria—a prolific source of natural products. Tetrahedron 57: 9347-9377

Canfield DE (1999) A breath of fresh air. Nature 400: 503-504

Capper A, Cruz-Rivera E, Paul VJ, Tibbetts IR (2006) Chemical deterrence of a marine cyanobacterium against sympatric and non-sympatric consumers. Hydrobiologia 553: 319-326

Capone DG, Zehr JP, Paerl HW, Bergman B, Carpenter EJ (1997) *Trichodesmium*, a globally significant marine cyanobacterium. Science 276: 1221-1229

Carmichael WW, Mahmood NA, Hyde EG (1990) Natural toxins from cyanobacteria (blue-green algae). In: Hall S, Strichartz G (eds.) Marine toxins: origin, structure, and molecular pharmacology. American Chemical Society, Washington, D.C. Pp. 87-106

Carpenter SR, Fisher SG, Grimm NB, Kitchell JF (1992) Global change and freshwater ecosystems. Annu Rev Ecol Syst 23: 119-139

Castenholz RW, Garcia-Pichel F (2000) Cyanobacterial responses to UV-radiation. In: Whitton BA, Potts M (eds) The ecology of cyanobacteria. Kluwer Academic Publishers, Dordrecht, pp. 591-611

Chorus I (ed) (2001) Cyanotoxins: occurrence, causes, consequences. Springer-Verlag, Berlin, 357 pp

Christoffersen, K (1996) Ecological implications of cyanobacterial toxins in aquatic food webs. Phycologia 35: 42-50

Codd GA, Morrison LF, Metcalf JS (2005) Cyanobacterial toxins: risk management for health protection. Toxicol App Pharmacol 203: 264-272

Copper P (1994) Ancient reef ecosystem expansion and collapse. Coral Reefs 13:3-11

DeMott WR, Zhang Q-X, Carmichael W W (1991) Effects of toxic cyanobacteria and purified toxins on the survival and feeding of a copepod and three species of *Daphnia*. Limnol Oceanogr 36: 1346-1357

Dittmann E, Neilan BA, Börner T (2001) Molecular biology of peptide and polykeide biosynthesis in cyanobacteria. Appl Microbiol Biotechnol 57: 467-473

Donner SD, Skirving WJ, Little CM, Oppenheimer M, Hoegh-Guldberg O (2005) Global assessment of coral bleaching and required rates of adaptation under climate change. Global Change Biol 11: 2251-2265

Edwards DJ, Gerwick WH (2004) Lyngbyatoxin biosynthesis: sequence of biosynthetic gene cluster and identification of a novel aromatic phenyltransferase. J Am Chem soc 126: 11432-11433

Elmetri I, Bell PRF (2004) Effects of phosphorus on the growth and nitrogen fixation rates of *Lyngbya majuscula*: implications for management in Moreton Bay, Queensland. Mar Ecol Prog Ser 281: 27-35

Emanuel K (2005) Increasing destructiveness of tropical cyclones over the past 30 years. Nature 436: 686-688

Farquhar J, Bao H, Thiemens M (2000) Atmospheric influence of Earth's earliest sulfur cycle. Science 289: 756-758

Feely RA, Sabine CL, Lee K, Berelson W, Kleypas J, Fabry VJ, Millero FJ (2004) Impact on anthropogenic CO_2 on the $CaCO_3$ system in the oceans. Science 305: 362-366

Fulton RS III, Paerl HW (1987) Toxic and inhibitory effects of the blue-green alga *Microcystis aeruginosa* on herbivorous zooplankton. J Plank Res 9: 837-855

Garcia-Pichel F, Castenholz RW (1993) Occurrence of UV-absorbing, mycosporine-like compounds among cyanobacterial isolates and an estimate of their screening capacity. Appl Env Microbiol 59: 163-169

Garcia-Pichel F, Wingard CE, Castenholz RW (1993) Evidence regarding the UV sunscreen role of a mycosporine-like compound in the cyanobacterium *Gloeocapsa* sp. Appl Env Microbiol 59: 170-176

Gerten D, Adrian R (2000) Climate-driven changes in spring plankton dynamics and the sensitivity of shallow polymictic lakes to the North Atlantic Oscillation. Limnol Oceanogr 45: 1058-1066

Gerwick WH, Tan LT, Sitachitta N (2001) Nitrogen-containing metabolites from marine cyanobacteria. In:Cordell GA (ed) lkaloids: Chemistry and Biology, Vol. 57 Academic Press, NY, pp 75-184

Gilbert JJ (1996) Effect of temperature on the response of planktonic rotifers to a toxic cyanobacterium. Ecology 77: 1174-1180

Gustafsson S, Rengefors K, Hansson L-A (2005) Increased consumer fitness following transfer of toxin tolerance to offspring via maternal effects. Ecology 86: 2561-2567

Hallock P (2005) Global change and modern coral reefs: new opportunities to understand shallow-water carbonate depositional processes. Sedimentary Geology 175: 19-33

He Y-Y, Häder D-P (2002) Involvement of reactive oxygen species in the UV-B damage to the cyanobacterium *Anabaena* sp. J Photochem Photobiol B: Biol 66: 73-80

He Y-Y, Klisch M, Häder D-P (2002) Adaptation of cyanobacteria to UV-B stress correlated with oxidative stress and oxidative damage. Photochem Photobiol 76: 188-196

Hegerl GC and Bindoff NL (2005) Warming the world's oceans. Science 309: 254-255

Jin YG, Wang Y, Wang W, Shang QH, Cao CQ, Erwin DH (2000) Pattern of marine mass extinction near the Permian-Triassic Boundary in South China. Science 289: 432-436.

Kaebernick M, Dittmann E, Börner T, Neilan BA (2002) Multiple alternate transcripts direct the biosynthesis of microcystin, a cyanobacterial nonribosomal peptide. Appl Environ Microbiol 68: 449-455

Kaebernick M, Neilan BA (2001) Ecological and molecular investigations of cyanotoxin production. FEMS Microbiol Ecol 35: 1-9

Kaebernick M, Neilan BA, Börner T, Dittmann E (2000) Light and the transcriptional response of the microcystin biosynthesis gene cluster. Appl Environ Microbiol 66: 3387-3392

Kasting JF, Siefert JL (2002) Life and the evolution of Earth's atmosphere. Science 296: 1066-1068.

Kerr RA (2005) The story of O_2. Science 308: 1730-1732

Kirk KL, Gilbert JJ (1992) Variation in herbivore response to chemical defenses: zooplankton foraging on toxic cyanobacteria. Ecology 73: 2208-2217

Kuffner IB, Paul VJ (2001) Effects of nitrate, phosphate and iron on the growth of macroalgae and benthic cyanobacteria from Cocos Lagoon, Guam. Mar Ecol Prog Ser 222: 63-72

Landsberg JH (2002) The effects of harmful algal blooms on aquatic organisms. Reviews in Fisheries Science 10: 113-390

Lehtimaki J, Sivonen K, Luukkainen R, Niemela S I (1994) The effects of incubation time, temperature, light, salinity, and phosphorus on growth and hepatoxin production by *Nodularia* strains. Arch Hydrobiol 130:269-282

Levitus S, Antonov J, and Boyer T (2005) Warming of the world ocean, 1955-2003. Geophys Res Lett 32:L02604

Liu Z, Häder DP, Sommaruga R (2004) Occurrence of mycosporine-like amino acids (MAAs) in the bloom-forming cyanobacterium *Microcystis aeruginosa.* J Plank Res 26: 963-966

Moore RE (1996) Cyclic peptides and depsipeptides from cyanobacteria: a review. J Ind Microbiol 16: 134-143

Moss B, McKee D, Atkinson D, Collings SE, Eaton JW, Gill AB, Hatton HK, Heyes T and Wilson D (2003) How important is climate? Effects of waming, nutrient addition and fish on phytoplankton in shallow lake microcosms. J Appl Ecol 40: 782-792

Murrell MC, Lores EM (2004) Phytoplankton and zooplankton seasonal dynamics in a subtropical estuary: importance of cyanobacteria. J Plank Res 26: 371-382

Nagle DG, Paul VJ (1998) Chemical defense of a marine cyanobacterial bloom. J. Exp Mar Biol Ecol 225: 29-38

Nagle DG, Paul VJ (1999) Production of secondary metabolites by filamentous tropical marine cyanobacteria: ecological functions of the compounds. J. Phycol 35: 1412-1421

O'Neil J, Dennison WC (2005) *Lyngbya majuscula* in Southeast Queensland waterways. Chapt. 6 In: Abal E, Dennison WC (eds) Healthy Waterways - healthy catchments. South East Queensland Regional Water Quality Strategy, Brisbane City Council. Brisbane, Australia. pp 119-148.

Paerl HW (2006) Nutrient and other environmental controls of harmful cyanobacterial blooms along the freshwater-marine continuum. These proceedings, in press.

Paerl HW, Bland PT, Bowles ND, Haibach ME (1985) Adaptation to high intensity, low wavelength light among surface blooms of the cyanobacterium *Microcystis aeruginosa*. Appl Env Microbiol 49:1046-1052

Pangilinan RF (2000) Effects of light and nutrients on intraspecific secondary metabolite variation in the benthic cyanobacterium *Lyngbya majuscula*. M.S. thesis, University of Guam, 29 p.

Park S, Brett MT, Müller-Solger A, Goldman CR (2004) Climatic forcing and primary productivity in a subalpine lake: Interannual variability as a natural experiment. Limnol Oceaogr 49: 614-619

Paul VJ, Thacker R, Banks K, Golubic S (2005) Benthic cyanobacterial bloom impacts on the reefs of South Florida (Broward County, USA). Coral Reefs 24: 693-697

Paul VJ, Cruz-Rivera E, Thacker RW (2001) Chemical mediation of macroalgal-herbivore interactions: ecological and evolutionary perspectives. In: McClintock J, Baker B (eds) Marine Chemical Ecology, CRC Press, LLC, pp. 227-265

Pennings SC, Pablo SR, Paul VJ (1997) Chemical defenses of the tropical benthic, marine cyanobacterium *Hormothamnion enteromorphoides*: diverse consumers and synergisms. Limnol Oceanogr 42: 911-917

Prospero JM, Lamb PJ (2003) African droughts and dust transport to the Caribbean: climate change implications. Science 302: 1024-1027

Raikow DF, Sarnell O, Wilson AE, Hamilton SK (2004) Dominance of the noxious cyanobacterium *Microcystis aeruginosa* in low-nutrient lakes is associated with exotic zebra mussels. Limnol Oceanogr 49: 482-487

Ramos AG, Martel A, Codd GA, Soler E, Coca J, Rdo A, Morrison LF, Metcalf JS, Ojeda A, Suarez S, Petit M (2005) Bloom of the marine diazotrophic cyanobacterium *Trichodesmium erythraeum* in the Northwest African upwelling. Mar Ecol Prog Ser 301: 303-305

Rapala J, Sivonen K, Lyra C, Niemela S I (1997) Variation of microcystins, cyanobacterial hepatoxins, in *Anabaena* spp. as a function of growth stimuli. Appl Environ Microbiol 63: 2206-2212

Richardson LL, Kuta KG (2003) Ecological physiology of the black band disease cyanobacterium *Phormidium corallyticum*. FEMS Microbiol Ecol 43: 287-298

Robarts RD, Zohary T (1987) Temperature effects on photosynthetic capacity, respiration, and growth rates of bloom-forming cyanobacteria. N Z J Mar Freshw Res 21: 391-399

Rohrlack T, Christoffersen K, Dittmann E, Nogueira I, Vasconcelos V, Börner T (2005) Ingestion of microcystins by *Daphnia*: Intestinal uptake and toxic effects. Limnol Oceanogr 50: 440-448

Rohrlack T, Dittmann E, Börner T, Christoffersen K (2001) Effects of cell-bound microcystins on survival and feeding of *Daphnia* spp. Appl Env Microbiol 67: 3523-3529

Sabine CL, Feely RA, Gruber N, Key RM, Lee K, Bullister JL, Wanninkhof R, Wong CS, Wallace DWR, Tilbrook B, Millero FJ, Peng T-H, Kozyr A, Ono T, Rios AF (2004) The oceanic sink for anthropogenic CO_2. Science 305: 367-371

Scheffer M, Straile D, van Nes EH, Hosper H (2001) Climatic warming causes regime shifts in lake food webs. Limnol Oceanogr 46: 1780-1783

Schopf JW (2000) The fossil record: tracing the roots of the cyanobacterial lineage. In: Whitton BA, Potts M (eds) The ecology of cyanobacteria. Kluwer Academic Publishers, Dordrecht, pp. 13-35

Sellner KG, Olson MM, Kononen K (1994) Copepod grazing in a summer cyanobacteria bloom in the Gulf of Finland Hydrobiologia 293: 249-254

Sheehan PM, Harris MT (2004) Microbialite resurgence after the Late Ordovician extinction. Nature 430: 75-78

Sivonen K (1990) Effects of light, temperature, nitrate orthophosphate, and bacteria on growth of and hepatotoxin production by *Oscillatoria agardhii* strains. Appl Environ Microbiol 56: 2658-2666

Summons RE, Jahnke LL, Hope JM, Logan GA (1999) 2-Methylhopanoids as biomarkers for cyanobacterial oxygenic photosynthesis. Nature 400: 554-557

Thacker RW, Ginsburg DW, Paul VJ (2001) Effects of nutrient enrichment and herbivore exclusion on coral reef macroalgae and cyanobacteria. Coral Reefs 19: 318-329

Third Assessment Report of the Intergovernmental Panel on Climate Change IPCC (WG I) (2001) Climate Change 2001: The Scientific Basis. Cambridge Univ. Press, Cambridge

Tillett D, Dittmann E, Erhard M, von Döhren, Börner T, Neilan BA (2000) Structural organization of microcystin biosynthesis in *Microcystis aeruginosa* PCC7806: an integrated peptide-polyketide synthetase system. Chemistry & Biology 7: 753-764

Tonk L, Visser PM, Christiansen G, Dittmann E, Snelder EOFM, Wiedner C, Mur LR, Huisman J (2005) The microcystin composition of the cyanobacterium *Planktothrix agardhii* changes toward a more toxic variant with increasing light intensity. App Env Microbiol 71: 5177-5181

Treydte KS, Schleser GH, Helle G, Frank DC, Winiger M, Haug GH, Esper J (2006) The twentieth century was the wettest period in northern Pakistan over the past millennium. Nature 440: 1179-1182

Tsuda RT, Kami HT (1973) Algal succession on artificial reefs in a marine lagoon environment in Guam. J Phycol 9: 260-264

Utkilen H, Gjolme N (1992) Toxin production by *Microcystis aeruginosa* as a function of light in continuous cultures and its ecological significance. Appl Environ Microbiol 58: 1321-1325

Verburg P, Hecky RE, Kling H (2003) Ecological consequences of a century of warming in Lake Tanganyika. Science 301: 505-507

Walther G-R, Post E, Convey P, Menzel A, Parmesan C, Beebee TJC, Fromentin J-M, Hoegh-Guldberg O, Bairlein F (2002) Ecological responses to recent climate change. Nature 416: 389-395

Watanabe M F, Oishi S (1985) Effects of environmental factors on toxicity of a cyanobacterium (*Microcystis aeruginosa*) under culture conditions. Appl Environ Microbiol 49: 1342-1344

Watkinson AJ, O'Neil JM, Dennison WC (2005) Ecophysiology of the marine cyanobacterium *Lyngbya majuscula* (Oscillatoriaceae) in Moreton Bay, Australia. Harmful Algae 4: 697-715

Webster PJ, Holland GJ, Curry JA, Chang H-R (2005) Changes in tropical cyclone number, duration, and intensity in a warming environment. Science 309: 1844-1846

Wiedner C, Visser PM, Fastner J, Metcalf JS, Codd GA, Mur LR (2003) Effects of light on the microcystin content of *Microcystis* strain PCC 7806. Appl Env Microbiol 69: 1475-1481

Xie S, Pancost RD, Yin H, Wang H, Evershed RP (2005) Two episodes of microbial change coupled with Permo/Triassic faunal mass extinctions. Nature 434: 494-497

Xue L, Zhang Y, Zhang T, An L, Wang X (2005) Effects of enhanced ultraviolet-B radiation on algae and cyanobacteria. Crit Rev Microbiol 31: 79-89

Chapter 12: Watershed management strategies to prevent and control cyanobacterial harmful algal blooms

Michael F Piehler

The University of North Carolina at Chapel Hill, Institute of Marine Sciences, 3431 Arendell Street, Morehead City, NC 28557, piehler@unc.edu

Abstract

The tenets of watershed management – a focus on the land area linked to the water body, the incorporation of sound scientific information into the decision-making process and stakeholder involvement throughout the process – are well-suited for the management of cyanobacterial harmful algal blooms (C-HABs). The management of C-HABs can be viewed as having two main areas of focus. First, there is mitigation – control and/or removal of the bloom. This type of crisis response is an important component to managing active C-HABs and there are several techniques that have been successfully utilized, including the application of algicides, physical removal of surface scums and the mechanical mixing of the water column. While these methods are valuable because they address the immediate problem, they do not address the conditions that exist in the system that promote and maintain C-HABs. Thus, the second component of a successful C-HAB management strategy would include a focus on prevention. C-HABs require nutrients to fuel their growth and are often favored in longer-residence time systems with vertical stratification of the water column. Consequently, nutrients and hydrology are the two factors most commonly identified as the targets for prevention of C-HABs. Management strategies to control the sources, transformation and delivery of the primary growth-limiting nutrients have been applied with success in many areas. The most effective of these include controlling land use, maintaining the integrity of the landscape and applying best management practices.

In the past, notable successes in managing C-HABs have relied on the reduction of nutrients from point-sources. Because many point sources are now well-managed, current efforts are focused on non-point source nutrient reduction, such as runoff from agricultural and urban areas. Non-point sources present significant challenges due to their diffuse nature. Regardless of which techniques are utilized, effective watershed management programs for decreasing the prevalence of C-HABs will require continuing efforts to integrate science and management activities. Ultimately, it is increased coordination among stakeholders and scientists that will lead to the development of the decision-making tools that managers require to effectively weigh the costs and benefits of these programs.

Introduction

Watershed management programs (Davenport 2002) have been effective tools for addressing ecological problems in bodies of water, including lakes, rivers, estuaries and coastal seas. They are derived from the reasoning that integrative management of the activities within the confines of the catchment of a body of water can affect the amount and transport of pollutants to the water body of concern (Fig. 1) (US EPA 1996). The concept of watershed-based management also includes involving area stakeholders in key decisions – all the way from the establishment of baseline data, to codifying rules to reduce the levels of the pollutants of concern. There are many prominent examples of the application of watershed management throughout the world, including the Chesapeake Bay (Hill and Nelson 1994, Cestti et al. 2003) and Baltic Sea (Elmgren and Larsson 2001). Cyanobacterial harmful algal blooms (C-HABs) occur in waters with sufficient nutrient supply and light levels, factors which are directly affected by human activities in watersheds (Paerl 1997). Like eutrophication, C-HABs are often linked to human modification of nutrient supplies (Watson et al. 1997). This contribution will discuss watershed management activities aimed at preventing C-HABs from occurring and management options designed to remediate water bodies once a C-HAB has occurred.

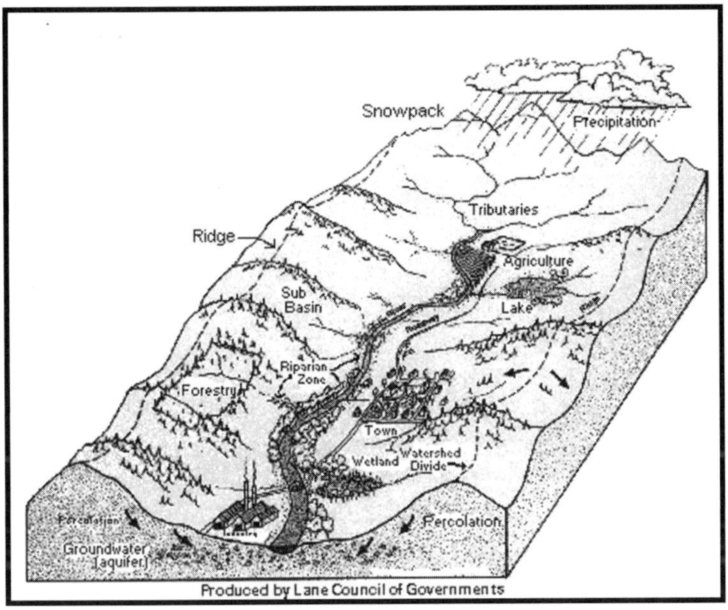

Fig. 1. The geographic, hydrologic and land use properties typical of watersheds. Working within these boundaries is the hallmark of watershed management. (From http://www.epa.gov/owow/watershed)

A summary of the research investigating the relationship between watershed management strategies and the occurrence of C-HABs

In past efforts to control C-HABs, management strategies have been developed to control the factors that affect C-HAB formation and persistence, particularly nutrients and hydrology. Watershed management strategies to reduce the prevalence of C-HABs most often include nutrient controls, specifically for nitrogen and phosphorus. In the past, single nutrient controls were implemented in some systems (Edmondson and Lehman 1981). However, due to the variation in nutrient limitation throughout the freshwater- marine continuum, dual N and P reductions may be required (Paerl et al. 2004). Reduced N to P ratios (Smith 1983) and eutrophic conditions have been shown to result in cyanobacterial dominance in some water bodies (Paerl 1997). Hydrologic manipulations such as increasing base flow, timed increases in discharge and destratification can also be effective when there is sufficient water available (Chorus and Bartram 1999). C-HAB prevalence has likely been enhanced by human activities in watersheds, and management efforts are designed to ameliorate the human impacts.

Human impacts

Growing population and changing land use (Vitousek & Mooney, 1997) often result in significant impacts on the delivery of nutrients to surface waters (Schueler and Holland 2000). Sources of nutrients generally increase as human use of a region increases. Wastewater, agricultural discharge, stormwater and industrial sources of nutrients are among the major contributors. Another change that accompanies human development of the landscape is an alteration of the transport mechanisms of nutrients to waterbodies (Line et al. 2002). Increased imperviousness (amount of land through which water can not infiltrate) within the watershed, ditching, channelization of streams and rivers and removal of the native vegetation contribute to increases in rates and quantities of nutrients transported from the land to the adjacent water bodies (Schueler and Holland 2000). Finally, human activities often lead to significant loss of the natural landscapes that either retain or remove nutrients. Losses of wetlands and landscape alterations at the watershed scale are well-documented and greatly impact both nutrient and water transport from human-dominated regions (Vitousek et al. 1997). There is likely a cumulative impact on nutrient export from human development of the landscape that includes the effects of increased sources, modified transport and altered biological processing of nutrients (Fig. 2).

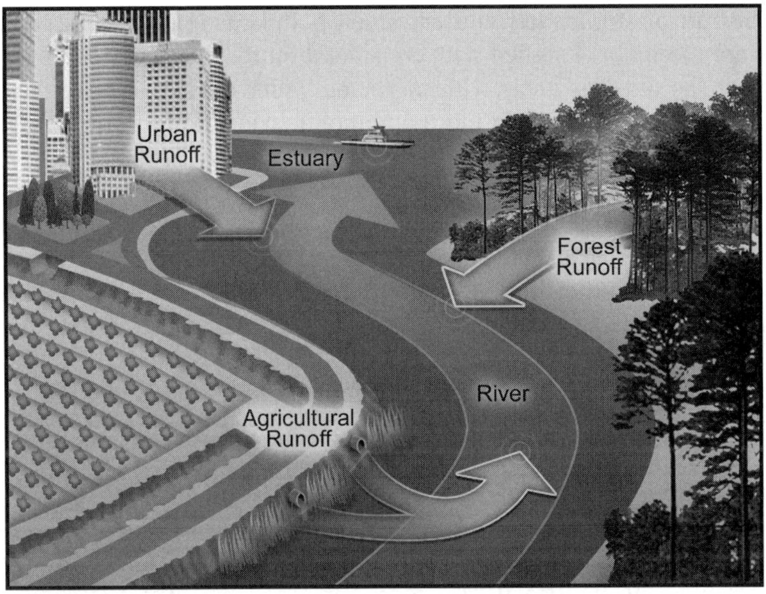

Fig. 2. Changes in land cover and land use resulting from human activities can lead to both changes in sources of pollutants and transport of water from the landscape. Both of these changes affect the success of C-HABS and must be considered in management strategies.

Sources of nutrients that increase with human activity include sewage treatment plant discharges, fertilizers associated with agriculture and other activities (e.g., lawns, gardens and golf courses), increased atmospheric N deposition and increased stormwater carrying myriad non-point sources associated with urbanization (Paerl 1997, Boesch et al. 2001a). In the first stages of human development, the shifts in land use are generally from forest to agriculture and residential. Once an area is populated, a secondary shift in land use away from open space (forest and agriculture) toward residential and industrial often occurs (Beach 2002). Through this sequence of land use alteration, the nutrient sources change and the trend generally results in more nutrients transported to aquatic environments (Line et al. 2002).

Human modification of the transport of water from land to surface waters has occurred in a large proportion of watersheds on a global scale (Vitousek et al. 1997). If sources of nutrients are present, transport of water from the landscape to the aquatic systems usually translates into enhanced nutrient transport. Urbanization leads to increased imperviousness due to increases in the areal coverage of roads and rooftops (Schueler and Holland 2000). However, imperviousness alone is not responsible for changes in transport of stormwater and entrained pollutants (Beach 2002). Urban areas are often not designed with consideration for their impact on nutrient export. The result is areas with more and wider roads that have the additional affect of encouraging more automobile use, and thus creating larger sources of nutrients and other potential pollutants (Trobulak & Frissell 2000). Consideration of the complex interactions development has on water quality during the design and planning process can reduce detrimental effects. An approach that includes strategies to reduce street and parking lot coverage, plan the siting of building lots and conserve natural areas has been demonstrated to reduce the impacts of urbanization on stormwater transport to adjacent waters (Schueler & Holland 2000).

Attenuation of nutrients during transport through the watershed can be significantly affected by human activities. Water transport form the landscape to the aquatic environments is significantly increased by ditching, channelization and increased imperviousness (Williams et al. 1997). The resulting shorter residence times often reduce biological attenuation of the nutrient load from a watershed. Higher volumes of faster moving water are less likely to be effectively filtered in small headwater streams, wetlands and riparian areas. Human activities also lead to direct negative effects on natural systems that attenuate the nutrient load from the landscape. Degradation of headwater streams that often occurs in areas with intensive human use is likely to significantly reduce nitrogen removal and retention (Peterson et al. 2001). In the US, wetland loss is on the order of 50% since the pre-settlement era and despite legal protections, loss is likely to continue to increase (Mitsch & Gosselink 2000). Even with laws in place to protect wetlands, management of wetland resources is a significant challenge (La Peyre et al. 2001). Wetlands are known sinks for nutrients transported to them from watersheds (Reddy & Gale 1994) and they further enhance water quality by transforming N from highly biologically available inorganic forms to less labile organic forms (Craft et al. 1989). Significant biological removal of nitrogen also occurs in riparian areas via denitrification (Jacobs & Gilliam 1985, Spruill 2004). The function of the riparian areas can be compromised by human activities, diminishing the ability of the landscape to retain and remove nutrients (Groffman et al. 2003).

Management techniques

Management activities in watersheds aimed at decreasing the occurrence of C-HABs can be roughly divided into efforts to control environmental factors that promote blooms (e.g., nutrients, water residence time) and other efforts to remove blooms once they have occurred (e.g., algicides, removal of scums and destratifying the water column). These two approaches can be generalized into preventative measures and mitigation measures. Some combination of both approaches will likely have to be included in effective management plans.

Preventative measures are most often the preferred approach to managing C-HAB occurrence. As described above, controlling nutrients and freshwater discharge are the most common management strategies. In order to effectively control these factors, steps must be taken to repair damage to the landscape and water bodies that are contributing to the enhanced load of nutrients. Nutrient management and water management clearly overlap because of the importance of water as a vector for transport of nutrients. Typical water management activities to control C-HABs include minimizing consumptive uses of water by residential, industrial and agricultural activities, among others. There are also examples of C-HAB management plans including the removal of stream obstructions, such as dams and impoundments. Efforts to maintain or restore the natural connectivity between aquatic and terrestrial systems are also effective tools to manage surface flow. Finally, when tenable, management of flow regimes can be an option to reduce C-HABs (Maier et al. 2001).

Nutrient management has been pursued in the context of broader eutrophication control in many aquatic systems throughout the world (NRC 2000). The process for managing nutrients to control C-HABs includes conducting monitoring and experiments to determine the nutrients limiting C-HAB success in the specific system. Phosphorus is generally believed to limit C-HAB growth in freshwater, and nitrogen is thought to limit their growth in salt water. There are however, many site specific factors along the freshwater marine continuum that can make this determination complicated. Once the limiting nutrient is identified, the major sources of that nutrient in the watershed should also be identified. Monitoring, modeling and experimental work must then be undertaken to set an effective and achievable target for nutrient reductions. With the target set, management practices can then be put in place to reduce export of nutrients from the landscape to the aquatic environment. Point sources of nutrients are generally managed first, followed by non-point sources.

Specific nutrient reduction approaches are designed for the type of land use being addressed. Agricultural and urban landscapes are most often management targets and clearly have different sources and transport vectors for nutrients. Sources of nutrients in urban watersheds include automobile exhaust, sundry materials deposited on streets (fertilizer, deicer, vegetation) and un-permitted sewage discharges (Novotony 2002). Urban planning to prevent conditions conducive to nutrient export from urban watershed were discussed above. Other techniques can be utilized in existing urban watersheds including, decreasing impacts of imperviousness (utilizing pervious pavement, decreasing connectivity of impervious areas), increasing surface storage (stormwater treatment and retention structures) and engineering for increased infiltration (Novotony 2002). In agricultural watersheds, best management practices have been designed to minimize sources of nutrients such as fertilizer, and maximize nutrient retention on the landscape. Examples of effective practices include no-till farming, installation of water control structures, maintaining riparian buffers and variable nutrient applications (Lilly 1991).

C-HAB mitigation measures are designed to curtail a bloom once it has started. They are more often applied in drinking waters because of the higher risks of human exposure. Examples of remedial techniques include the application of algicides, oxidants and coagulants (Chorus and Bartram 1999). Caution needs to be taken to avoid the exacerbation of the effects from C-HABs by untimely application of mitigation measures. Active blooms can release their toxins after the application of algicide and create a more dangerous situation (Jones and Orr, 1994, Kenefick et al. 1992). Details of the application of mitigation strategies in drinking waters can be found in Westrick in this volume.

Examples of watershed management strategies reducing the occurrence of C-HABs

Several systems with C-HAB issues have documented significant reduction or elimination of C-HABs resulting from effective watershed management. Among them are Lake Washington in the United States, which in large part to reduction of P inputs from point sources (sewage), experienced a dramatic decline in C-HABs (Edmondson and Lehman 1981), Also, Lake Erie in the US (Likens 1972), and Himmerfjärden in Sweden have shown large declines in C-HABS as a result of a combination of reduction in sewage inputs and some non-point sources including agriculture (Elmgren and Larsson 2002).

There are also examples of successful C-HAB management plans whose primary focus has been on non-point source nutrients. Among those commonly identified are the work of the Murray-Darling Basin Commission (MDBC 1993) and the New South Wales Blue-Green Algal Task Force (NSWBGATF 1993). These successful plans share many attributes. Among the important components of successful plans to prevent and remediate C-HABs are complete public involvement, broad educational efforts across all stakeholders, a sound scientific basis, tangible metrics of success, identification of costs and benefits, mechanisms for adaptation and the inclusion of both preventative and remedial strategies. These programs have not yet documented cause and effect links between the applications of watershed management strategies geared toward non-point source nutrients and C-HAB reductions. However, given the soundness of the plans described above, a good model has certainly developed for effective management plans and documented successes in bloom control is likely to occur soon.

There are other challenges faced by current efforts to control and prevent C-HABs beyond the difficulty of controlling non-point source nutrients. Some management programs lack clear metrics of success, which prevents the establishment of causal links to management activities. To increase the likelihood of demonstrable successes of watershed management in reducing the prevalence of C-HABs, continued and expanded coordination between managers and scientists is required. Scientists must clearly identify which factors will control C-HABs in specific settings and must help managers identify the relevant processes to monitor in order to evaluate success. Management plans must include sufficiently rigorous assessments of the metrics that are being evaluated to permit a statistical assessment of their success. This often means that whole ecosystem monitoring must be a significant component of the programs.

Components of successful watershed management programs to reduce the prevalence of C-HABs

The US Environmental Protection Agency (EPA) model for watershed management includes four major components in its suggestion for how activities should proceed (US EPA 1996). First, the activities should all include broad **stakeholder involvement**. By including stakeholders in the process from planning to codifying laws, programs will be understood and valued by the public at large. Second, the focus of the management strategy is based on the **geographic unit of the watershed**, which often requires cooperation among multiple jurisdictions. There is some deviation from the physical boundaries of the catchment with mobile sources such as atmospheric deposition, but the primary focus is at the watershed level. The watershed management plans must apply sound management principles and this includes **coordinated management activities**. Finally there should be a clear **management schedule**.

The EPA model also calls for the use of sound science to support all levels of decision making including:

- Ecological assessments
- Identifying environmental objectives based on the ecological and societal requirements
- Identifying of priority issues
- Developing of detailed action plans
- Implementing the plans
- Evaluating and adapting the plans as they proceed

The elements of this watershed management framework are applicable and appropriate for preventing and controlling C-HABs. Because of the high profile health and ecological impacts of C-HABs, significant stakeholder involvement is beneficial in many respects: as an educational outlet, in empowering the affected parties and in creating ownership of the plans among the public at large. C-HABs are highly visible and evoke strong emotions among the public. The watershed management process could provide a constructive outlet for the rational concerns that these blooms cause. There are now significant resources available for guiding the development of watershed management programs to restore and protect watershed function (US EPA 2003).

Considerations when applying watershed management strategies to drinking water and recreational waters

In drinking water reservoirs management of the occurrence of C-HABs must be more aggressive because of the enhanced risk to public health (Chorus and Bartram 1999). The risk management context largely determines the nature of the management strategy. In recreational waters, a preventative strategy to reduce nutrient loading and perhaps modify the flow regime would be a viable approach. Recreational waters are more often managed within the context of the functioning of the ecosystem in which they are located. In managing the occurrence of C-HABs in drinking water supplies there must be fast and effective crisis management tools present in the management plan. Some of these approaches include application of algicides and flocculants. Details on drinking water supply management are presented by Westrick in this volume.

Research gaps: Improving watershed management strategies to reduce the occurrence of C-HABs

There is a significant body of literature on the application of watershed management to reduce environmental problems (NRC 2000). Application of watershed management to control C-HABs, particularly in systems with predominantly non-point source nutrient loading has not been as extensive. Among the basic information that remains to be determined are the nutrients limiting C-HAB productivity along the continuum from freshwater to marine environments, the importance of variation in ratios of limiting nutrients and the potential for micronutrient limitation. There are broad data from some systems such as re-oligotrophying lakes (Jeppesen et al 2005), but accurate generalizations can not yet be made with sufficient certainty in all systems. Once a more accurate prediction of which nutrients limit C-HAB success in specific systems is developed, it is critical to examine the impacts of existing watershed management programs with other goals on C-HAB potential. Some programs with more general goals such as eutrophication control may also effectively control C-HABs, while others may in theory promote them (Piehler et al. 2002).

There are also some more specific, but critical pieces of information that would significantly further our ability to prevent C-HABs and our ability to forecast the effectiveness of the management regimes. Assessing the response time of the ecosystem to nutrient management in systems with significant sedimentary nutrient supplies is critical. There are many systems that have years of nutrient supplies stored in the sediments, which clearly affects the timing of results from controls of external nutrient supplies (Marsden 1989, Sondergaard et al. 2003). This information may also lead to the consideration of management options for internal loading that have been applied in some systems (Chorus and Bartram 1999). It is important that we acquire better data on the chemical forms of phosphorus and nitrogen loads from varied land uses and relate the nature of the loads to C-HABs nutritional requirements.

Land use changes are affecting sources, transport and fate of nutrients (Correll et al. 1994) that control C-HAB potential. Some extrinsic factors such as impervious cover have been utilized as indicators of land use change. Studies to develop generalizable intrinsic indicators of land use change that can be applied as predictive tools for C-HAB potential would significantly enhance our ability to effectively manage bloom potential. Changes in freshwater discharge from water use, global climate change and relative sea level rise are affecting the full suite of aquatic C-HAB habitats. Freshwater supplies downstream will be affected by changes in consumptive uses, releases from reservoirs, and changes in precipitation patterns and amounts. Increases in relative sea level due to global climate change will push salt water farther upstream. A better understanding of the cumulative changes in C-HAB habitat that will result from these forcing features would be highly desirable. Because water management is an option currently utilized to control C-HABs in some areas, examining the interactive effects of freshwater flow modification on nutrient supplies, hydrologic properties and temperature would also be highly beneficial.

Requirements for models of costs and benefits of watershed management strategies to control C-HABs

Because watershed management strategies are driven by, and affect a suite of different costs and benefits, models to predict their effectiveness will have to consider several different factors. Clearly, human health effects of C-HABs will be the first consideration. There are myriad costs and benefits associated with the level human health effects associated with C-HABs that are addressed in other chapters throughout this volume. Consideration must also be given to the ecological costs and benefits of management. These costs and benefits must consider the impacts that C-HABs have on the ecology and biogeochemistry of the systems in which they are present. Finally the economic costs and benefits will have to be included in any rigorous modeling effort. Determining the economic implications of watershed management includes assigning an economic value to both human health and ecology, and balancing that against the costs of the management program. There are additional economic considerations in the cost-benefit analysis, including costs associated with the loss of recreational space, indirect costs resulting from modifications of human activities and impacts of C-HABs on tourism and real estate values.

Conclusions

Watershed management is an integral part of efforts to control diffuse inputs of material to bodies of water. It provides a basis to construct the scientific, social, and policy frameworks that are required for a successful management program. Because nutrients are often identified as the driving forces behind the expansion of C-HABs, and the sources of nutrients are often dominated by non-point source forms, application of watershed management is an excellent option for C-HAB control. Reduction of nutrient inputs and maintenance of flow regimes using watershed management tools are likely the best choices for long term success in preventing and controlling C-HAB development. Mitigation techniques that are applied once a bloom has formed are an important part of any management plan, but prevention is the better choice when it can be achieved. Effective C-HAB plans that are currently underway include sound science to achieve the preventative goals described above and they also include significant public participation in the management process. To sustain C-HAB management programs the public must both value, and be integrated into, the effort. Fully incorporating the unique combination of potential impacts of C-HABs (ecological risk, human health risk and economic impacts) into the decision making framework will provide a valuable tool in efforts to prevent and control these blooms.

References

Beach D (2002) Coastal Sprawl: The effects of urban design on aquatic ecosystems in the United States. Pew Oceans Commission, Arlington, VA.

Boesch DF, Burroughs R H, Baker JE, Mason RP, Rowe CL, Seifert RL (2001) Marine Pollution in the United States: Significant accomplishments, future challenges. Pew Oceans Commission, Arlington, VA.

Cestti R, Srivastava J, Jung S (2003) Agriculture Non-Point Source Pollution Control: Good Management Practices – The Chesapeake Bay Experience. 54 pages, World Bank Publications.

Chorus I, Bartram J (1999) Toxic Cyanobacteria in Water. E&F Spon, London, 416 PP.

Correll DL, Jordan TE, Weller DE (1994) The Chesapeake Bay watersheds: Effects of land use and geology on dissolved nitrogen concentrations. Pages 639-648 in Toward a Sustainable Coastal Watershed: The Chesapeake Experiment. Smithsonian Environmental Research Center, Norfolk, VA.

Craft CB, Broome SW, Seneca ED (1989) Exchange of nitrogen, phosphorus, and organic carbon between transplanted marshes and estuarine waters. J. Environ. Qual. 18, 206-211.

Davenport TE (2002) The Watershed Project Management Guide. CRC Press, New York. 296pp.

Edmondson WT, Lehman JT (1981) The effect of changes in the nutrient income and conditions of Lake Washington. Limnol Oceanogr 26:1-29

Elmgren R, Larsson U (2001) Nitrogen and the Baltic Sea: Managing nitrogen in relation to phosphorus. The Scientific World 1(S2): 371-377

Groffman PM, Bain DJ, Band LE, Belt KT, Brush GS, Grove JM, Pouyat RV, Yesilonis IC, Zipperer WC (2003) Down by the riverside: urban riparian ecology. Front. Ecol. Environ. 1(6), 315-321.

Hill P, Nelson S (1995) Toward a Sustainable Coastal Watershed: The Chesapeake Experiment. Proceedings of a Conference 1-3 June 1994. Chesapeake Research Consortium Publication No. 149.

Jacobs TC, Gilliam JW (1985) Riparian losses of nitrate from agricultural drainage waters. J. Environ. Qual. 14, 472-478.

Jeppesen E, Sondergaard M, Jensen JP, Havens KE, Anneville O, Carvalho L, Coveney MF, Deneke R, Dokulil MT, Foy B, Gerdeaux D, Hampton SE, Hilt S, Kangur K, Kohler J, Lammens E, Lauridsen TL, Manca M, Miracle MR, Moss B, Noges P, Persson G, Phillips G, Portielje R, Schelske CL, Straile D, Tatrai I, Willen E, Winder M (2005) Lake responses to reduced nutrient loading - an analysis of contemporary long-term data from 35 case studies. Freshwater Biology 50:1747-1771.

Jones GJ, Orr PT (1994) Release and degradation of microcystin following algicide treatment of a Microcystis aeruginosa bloom in a recreational lake, as determined by HPLC and protein phosphostase inhibition assay. Water Research 28: 871-876.

Kenefick Sl, Hrudey SE, Peterson HG, Prepas EE (1992) Toxin release from Microcystis aeruginosa after chemical treatment. Water Science and Technology 27:433-440.

La Peyre MK., Reams MA, Mendelssohn IA (2001) Linking actions to outcomes in wetland management: An overview of US state wetland management. Wetlands 21, 66-74.

Likens GE (ed) (1972) Nutrients and Eutrophication. American Soc Limnol Oceanogr Special Symp 1

Lilly JP (1991) Best management practices for agricultural nutrients. North Carolina Cooperative Extension Service.

Line DE, White NM, Osmond DL, Jennings GD, Mojonnier CB (2002). Pollutant Export from Various Land Uses in North Carolina. Water Environ. Res. 14, 100-108.

Maier HR, Burch MD, Bormans M (2001) Flow management strategies to control blooms of the cyanobacterium Anabaena circinalis, in the River Murray at Morgan, South Australia. Regulated Rivers-Research and Management. 17:637-650

Marsden MW (1989) Lake restoration by reducing external phosphorus loading – The influence of sediment phosphorus release. Freshwater Biology 21:139-162.

MDBC (1993) Algal management strategy and technical advisory group report. Murray-Darling Basin Commission. Canberra, Australia.

Mitsch WJ, Gosselink JG (2000) Wetlands, 3rd edition. John Wiley and Sons, Inc., New York, NY.

National Research Council (2000) Clean Coastal Waters. National Academy Press, Washington DC.

Novotny V (2002) Water Quality: Diffuse Pollution and Watershed Management 888 pp, John Wiley and Sons.

NSWBGATF (1993) Blue-green algae. First Annual Report of the NSWBGATF, New South Wales Department of Water Resources. Parramatta, Australia.

Paerl HW (1997) Coastal eutrophication and harmful algal blooms: Importance of atmospheric deposition and groundwater as "new" nitrogen and other nutrient sources. Limnol. Oceanogr. 42, 1154-1165.

Paerl HW, Valdes LM, Piehler MF, Lebo ME (2004) Solving problems resulting from solutions: The evolution of a dual nutrient management strategy for the eutrophying Neuse River Estuary, North Carolina, USA. Environmental Science & Technology 38:3068-3073

Peterson BJ, Wollheim BM, Mulholland PJ, Webster JR, Meyer JL, Tank JL, Marti E, Bowden WB, Valett HM, Hershey AE, McDowell WH, Dodds WK, Hamilton SK, Gregory S, Morrall DD (2001) Control of nitrogen export from watersheds by headwater streams. Science 292, 86-89.

Piehler MF, Dyble J, Moisander PH, Pinckney JL, Paerl HW (2002) Effects of modified nutrient concentrations and ratios on the structure and function of the native phytoplankton community in the Neuse River Estuary, North Carolina USA. Aquat. Ecol. 36, 371-385.

Reddy KR, Gale PM (1994) Wetland processes and water quality: A symposium overview. J. Environ. Qual. 23:875-877.

Schueler T, Holland HK (2000) The Practice of Watershed Protection. Center for Watershed Protection, Ellicot City, MD.

Smith VH (1983) Low nitrogen to phosphorus ratios favor dominance by blue green algae in lake phytoplankton. Science 221:669 671

Sondergaard M, Jensen JP, Jeppesen E (2003) Role of sediment and internal loading of phosphorus in shallow lakes. Hydrobiologia 506:135-145.

Spruill TB (2004) Effectiveness of riparian buffers in controlling ground-water discharge of nitrate to streams in selected hydrogeologic settings of the North Carolina Coastal Plain. Water Science & Technology 49:63–70.

Trobulak SC, Frissell CA (2000) Review of ecological effects of roads and terrestrial and aquatic communities. Conservation Biology 14(1) 18-30.

US Environmental Protection Agency (1996) Watershed Framework Approach. EPA Report 840-S-96-001.

US EPA (2003) Watershed Analysis and Management (WAM) Guide for States and Communities. EPA 841-B-03-007.

Vitousek PM, Mooney HA (1997) Estimates of coastal populations. Science 278, 1211-1212.

Vitousek PM, Mooney HA, Lubchenco J, Melillo JM (1997) Human domination of earth's ecosystems. Science 277:494 – 499.

Watson SB, McCauley E, Downing JA (1997) Patterns in phytoplankton taxonomic composition across temperate lakes of differing nutrient status. Limnology and Oceanography 42:487-495.

Williams JE, Wood CA, Dombeck MP (eds) (1997) Watershed Restoration: Principles and Practices. 549 pp, American Fisheries Society.

Chapter 13: Cyanobacterial toxin removal in drinking water treatment processes and recreational waters

Judy A Westrick

Abstract

Although federal drinking water regulations determine the quality of potable water, many specifics influence how each utility chooses to treatment water. Some of the specifics include source water quality, storage capacity, existing unit process, and space. An overview of the US recreational and drinking water regulations were discussed in context of cyanobacterial toxin removal and inactivation by ancillary as well as auxiliary treatment practices. Ancillary practice refers to the removal or inactivation of algal toxins by standard daily operational procedures where auxiliary treatment practice refers to intentional treatment. An example of auxiliary treatment would be the addition of powder activated carbon to remove taste and odor compounds. The implementation of new technologies as such ultraviolet disinfection and membrane filtration, to meet current and purposed regulations, can greatly affect the algal toxin removal and inactivation efficiencies. A discussion on meeting the current regulations by altering chemical disinfection, ozone, chlorine, chloramines and chlorine dioxide included their ancillary effects on the protection against algal toxins. Although much of the research has been on the efficiency of the removal and inactivation of microcystin LR and several microcystin variants, the discussion included other algal toxins: anatoxin–a, saxitoxins, and cyclindrospermopsin.

Introduction

Before the United States Environmental Protection Agency (USEPA) can develop guidelines or regulations concerning the permissible concentrations of algal toxins for drinking water, they must first determine if cyanoalgal toxins can be removed or inactivated through common drinking water treatment practices. In 2000, a group of experts, through a USEPA initiative, created a cyanoalgal toxin priority list that contained the microcystins, anatoxin–a, and cylindrospermopsin. Microcystins are cyclic peptide hepatotoxins with a conserved (2S, 3S, 8S, 9S)–3–amino–9–methoxy–2,6,8–trimethyl–10–phenyl–deca–4,6–dioc acid, frequently abbreviated as Adda, and variable amino acids. Since there are over 70 variants of microcystins, a short list of four variants, LR, YR, LA and RR, referring to the changes in the amino acids, was recommended based on prevalence and toxicity. Cyanoalgal toxins can be either found inside of the cell, "intracellular", or outside the cell, "extracellular". The efficiency of a drinking water unit process with respect to the removal of cyanoalgal toxins depends on the total concentration of the algal toxins, the form of the toxin, and whether it is intracellular or extracellular. Commonly, dissolved contaminants such as extracellular cyanoalgal toxins are most costly to remove because conventional treatment (flocculation, coagulation, sedimentation and filtration) is usually not effective and advanced treatment processes must be implemented unless the contaminant is oxidized through disinfection. This paper reviews the literature on cyanotoxin removal during drinking water treatment with the initial focus on the removal of extracellular toxins followed by the removal of intracellular toxins. Published treatment reviews include Yoo (1995), Chorus (1999), Westrick (2003), Svrcek (2004), and Newcombe (2004).

With more stringent drinking water regulations and decreases in surface water quality, drinking water treatment trains have become more complex with fewer utilities using conventional treatment (coagulation/sediment/filtration/chlorination) alone. New additions have included more traditional unit processes like powered activated carbon (PAC), pretreatment with oxidants other than chlorine, ballasted flocculation, as well as advanced treatment processes such as membrane filtration, ultraviolet photolysis, granular activated carbon (GAC) absorbers and ozone. The traditional oxidant/disinfectant, chlorine, has been replaced with, or supplemented by, ozone, potassium permanganate, chlorine dioxide, or chloramines to eliminate nuisance organisms (e.g., zebra mussels) from the intake, compounds that cause taste and odor problems, and more recently compounds that react with chlorine disinfectant to produce toxic chlorinated by products.

Extracellular Cyanotoxin Inactivation by Chemical Disinfection Processes

Usually the primary focus of the disinfection process is to inactivate pathogenic organisms. The disinfection process can also be used to degrade several nuisance compounds such as those that cause taste and odor problems and a few regulated organic contaminants such as, some pesticides and fungicides, and industrial waste products. However, the disinfectant may also react with organic compounds in the water to produce toxic or carcinogenic byproducts. With health risk concerns about the exposure to chlorinated disinfection byproducts; the traditional chlorination process is being replaced by or supplemented with chloramination, chlorine oxide, ozone, and UV disinfection. This section focuses on the effectiveness of disinfection oxidants at inactivating important cyanotoxins.

Commonly, disinfectant effectiveness is expressed in terms of the product of the disinfectant concentration (C) in mgL^{-1} and the contact time (T) in minutes; hence CT values are in units of $mgL^{-1}min$. The effectiveness of a disinfectant depends primarily on the water pH, the water temperature, the concentration of compounds in the water that can react with the disinfectant, and the target organism itself. Thus, CT values for a particular disinfectant will also vary according to these factors. Several federal regulations incorporate CT values to guide the inactivation of both microorganisms and chemical contaminants. An example is Table 1, which shows the CT values needed with chorine at different water temperatures and pH values to reduce the level of the protozoan pathogen *Giardia lamblia* by 99.9%. As the temperature increases, the CT values for 99.9% inactivation of *G. lamblia* decrease. On the other hand, as pH increases, the CT values also increase. Since many utilities use these tables to determine which disinfectant dose to use, it is relevant to compare CT values needed to control *G. lamblia* with those needed to control the most common cyanotoxin, microcystin.

The CT values for the inactivation of 10 ug/L and 50 ug/L microcystin LR (MCYLR) for a batch reactor are presented in Table 2 (Acero 2005). When comparing Tables 1 and 2, at pH values of 6 and 7, the chlorine CT values needed for 99.9% inactivation of *G. lamblia* would also decrease 50 ug/L and 10 ug/L microcystin below the World Health Organization's drinking water guideline of 1 ug/L MCYLR , except perhaps at warm water temperatures. However, at a pH above 8, several of the CT values associated with the initial doses of 50 ug/L and 10 ug/L are much larger than those for 99.9% inactivation of *G. lamblia*. The larger CT values for both *G. lamblia* and microcystin inactivation at pH 8 and 9 reflect the change of the equilibrium from a strong oxidant, hypochlorous acid, to a weak oxidant, hypochlorite ion. Determining chlorine CT values for the inactivation of MCYLR gives the utilities a familiar tool which can easily be incorporated into treatment practices.

Two recent publications (Acero 2005 and Ho 2005) proposed chlorine inactivation mechanisms for several microcystins. Acero reported that MCYLR, microcystin RR (MCYRR) and microcystin YR (MCYYR) react with chlorine at a similar rate, suggesting that hydroxylation of the conversed Adda moiety is the likely site of deactivation. In contrast, Ho and coworkers reported that the reaction rates with chlorine for the four MCYs they studied differed, where MCYYR>MCYRR>MCYLR>MCYLA, suggesting the oxidation of various amino acids was the most important mechanism of deactivation. Regardless of the mechanism, both reports suggested that chlorination is an effective treatment for destroying the studied microcystins.

Table 1. Chlorine CT values for 99.9% (3-log) inactivation of G. lamblia cysts. Modified version (EPA Guidance Manual, 2003)

pH	**CT values ($mgL^{-1}min$)**			
	10°C	15°C	20°C	25°C
6	87	58	44	29
7	124	93	62	41
8	182	122	91	61
9	265	177	132	88

Table 2. Chlorine CT values for reducing microcystin concentration to 1 ugL-1 for a batch reactor.

pH	**MCYLR**	**CT values ($mgL^{-1}min$)**			
	(ug/L)	10°C	15°C	20°C	25°C
6	50	46.6	40.2	34.8	30.8
	10	27.4	23.6	20.5	17.8
7	50	67.7	58.4	50.6	44.0
	10	39.8	34.4	29.8	25.9
8	50	187.1	161.3	139.8	121.8
	10	110.3	94.9	82.8	71.7
9	50	617.2	526.0	458.6	399.1
	10	363.3	309.6	269.8	234.9

The inactivation of saxitoxins, cylindrospermopsin and anatoxin–a by chlorination has not been studied as thoroughly as the degradation of the microcystins by chlorination. Recent work suggests that the degradation of saxitoxins by chlorine is also pH dependent; however, in contrast with the microcystins, higher pH values increase the inactivation rate of saxitoxin and thus decrease the CT values needed for inactivation. At pH 9, a CT value of 15 $mgL^{-1}min$ (chlorine residual of 0.5 mgL^{-1} for 30 minutes) degraded the five studied saxitoxins by 90% (Nicholson, 2003). The saxitoxins order of reactivity towards chlorine was as follows, with the most toxic saxitoxin being the most susceptible:

GTX5=dcSTX>STX>GTX3=C2>GTX2

Cylindrospermopsin is very susceptible to oxidation by chlorine when the free chlorine residual is above 0.5 mgL-1 with a 30 minute contact time, CT values 15 mg min/L and at a pH above 6 (Senogles 2000), implying that cylindrospermopsin is more susceptible to chlorination than microcystins. These researchers suggested that the more effective degradation of both saxitoxins and cylindrospermopsin at higher pH values is probably due to both the saxitoxins and cyclindrospermopsin being in the unprotonated form. Anatoxin–a is not degraded by chlorination (Carlile 1994). Thus, different cyanotoxins react differently to chlorine. When a cocktail of cyanotoxins is present in a drinking water source, it is important to note that cylindrospermopsin and microcystin are more effectively inactivated at lower pH values, while saxitoxins are more effectively inactivated at higher pH values and anatoxin–a may not be degraded appreciably.

The toxicities of chlorination byproducts of microcystins, nodularin, saxitoxin and cylindrospermopsin have been examined. Microcystin and nodularin byproducts are not toxic by acute toxicity tests such as the mouse bioassay (Nicholson 1994), protein phosphatase inhibition assay, and mutagenicity (Tsuji 1997). Similarly, the byproducts from the chlorination of saxitoxins have been tested by mouse bioassay and have found not to be acutely toxic (Newcombe 2002a and Nicholson 2003). In contrast, the chlorination byproducts of cylindrospermopsin produce a low level of liver damage and genotoxicity (Shaw 2001). An oral 90 and 170 day toxicity study of chlorinated solutions with microcystin, saxitoxin and cylindrospermopsin showed that only cylindrospermopsin byproducts produced ill effects, i.e. fatty vacuolations of the liver in mouse (Senogles–Derham 2003). Chronic testing has not been performed for any of the algal toxins.

Other chlorine disinfection processes such as chloramines and chlorine dioxide have shown little promise towards degrading cyanotoxins. Chloramine doses, as high as 20 mgL^{-1} with 2 day exposures, did not degrade microcystins (Nicholson 1994). Although chlorine dioxide is used in several plants to remove taste and odor compounds, it is not recommended for treatment of microcystins since the dosage would not be cost effective (Kull 2004). Although one might speculate that cylindrospermopsin, anatoxin–a, and saxitoxins are not susceptible to chloramine and chlorine dioxide treatment, studies are needed to validate this statement.

Ozone is an effective disinfectant that oxidizes organic compounds either directly as molecular ozone or indirectly through a hydroxyl radical. Microcystin and anatoxin–a are inactivated at ozone doses where a residual is maintained for several minutes, CT values of 0.1 ugL-1min and 0.3 ugL-1min, respectively. (Keijola 1988, Himberg 1989, Bruchet 1998, Hart 1998, Rositano 1998, 2001, Newcombe 2002b). An acute toxicity screen was negative for microcystin ozone byproducts. The saxitoxin family appears to have low to moderate susceptibility to ozone oxidation, with a CT value of 6.9 mg/L min not being effective (Rositano 2001, Newcombe 2002b). The author of this review speculates that ozone would be an effective treatment to inactivate cylindrospermopsin because ozone is highly reactive toward unsaturated bonds. The confounding factors of ozone treatment for cyanoalgal toxins are 1) oxidation of the dissolved carbon competes with the destruction of algal toxins and 2) sensitive to alkalinity and temperature.

Extracelluar Cyanotoxin Removal by Activated Carbon Adsorption

Two types of activated carbon are used in the drinking water industry: powdered activated carbon (PAC) and granular activated carbon (GAC). Common sources of activated carbon are coal, coconut, and wood. Several studies suggest that mesoporous (i.e., pore diameters between 2–50 nm), wood–based activated carbon is more effective at removing cyanotoxins than other types of activated carbon. PAC is usually used seasonally to remove taste and odor compounds and agricultural chemicals, or as an emergency barrier during industrial/commercial spills. Drinking water utilities can test the relative effectiveness of various PAC types for removing a contaminant(s) of concern by performing jar tests, and then purchase the carbon with the most capacity for their target contaminants. A series of jar tests suggest different microcystins have different adsorption efficiencies:

RR>YR>LR>LA (Cook 2002)

The efficiency of saxitoxins removal by PAC mirrors the order of toxicity, with saxitoxin STX being the most toxic:

STX>GTX>C (Newcombe 2004)

Very little work has been performed on the adsorption of anatoxin–a and cylindrospermopsin to PAC. More systematic studies are needed to determine which PAC type, dosage, and contact time are most appropriate. Although more research is needed to guide drinking water utilities on the effectiveness of PAC for anatoxin–a, cylindrospermopsin and mixtures of cyanotoxins, the author of this review still stresses the importance of site specificity and in–house jar testing to select the most effective activated carbon.

GAC can be used in filter beds along with other filter media or can be used as a stand–alone fixed bed adsorber. GAC filters are used primarily to remove toxic chemicals and occasionally to remove substances in the water that cause taste and odor problems. They lose their organic compound adsorption capabilities usually within a couple of months of use and any organic removal after that comes from biological activity on the filter. Commonly, the effectiveness and degradation of a GAC filter is measured by total organic carbon (TOC) in the filter effluent. When 70–80% of the TOC entering the filter is measured in the filter effluent, usually after at least three months of service, the GAC media is replaced or reactivated. Newcombe and co–workers (2002b) reported microcystin in the GAC effluent at 80% TOC breakthrough suggesting that GAC filters would not be an effective barrier for controlling microcystin, since GAC media is not replaced monthly. However, frequently GAC absorbers are designed to last several months before 80% TOC breakthrough occurs, and since the adsorber media is replace upon 80% TOC breakthrough, GAC adsorbers may be an effective microcystin and other algal toxin barrier. Newcombe and coworkers (2002b) also studied the effect on cyanotoxins of ozonating the water before GAC adsorber entry, using two unit processes as one barrier. This barrier will oxidize microcystin, cylindrospermopsin and anatoxin–a, and the GAC absorber would provide an additional barrier to saxitoxins not oxidized by this treatment. To the best of the author's knowledge, the removal of cylindrospermopsin and anatoxin–a by GAC adsorption has not been investigated.

Extracellular Cyanotoxin Inactivation by Advanced Drinking Water Treatment Processes

The absorbance of ultraviolet (UV) energy can break molecular bonds without chemical addition and is used to inactivate many pathogens in drinking water. Normally, the UV treatment process uses a low to medium pressure lamp with wavelengths between 200 and 300 nm. Commonly, in the water treatment industry, the UV dose is expressed in milli–joules/cm^2 (mJ/cm^2). A dose of 40 mJ/cm^2 for the inactivation of *Crytosporidium parvum* oocysts is considered economically feasible. Much work has been done on the photolytic destruction of microcystins, cylindrospermopsin, and anatoxin–a, but the UV doses that effectively degrade microcystin, cylindrospermopsin, and anatoxin–a range from 1530 to 20,000 mJ/cm2 (Tsuji 1995, Chorus 1999, Senogles 2000), or several orders of magnitude higher than that needed for *Cryptosporidium* oocyst. Thus, UV irradiation is not considered economically feasible for cyanotoxin degradation.

The use of titanium dioxide and UV photolysis together successfully degrades MCYLR and cylindrospermopsin (Lawton 1999, Feitz 1999, Cornish 2000, Senogles 2001, Sherpard 2002). Although this is not a cost efficient treatment process for utilities, it may be viable in point–of–use treatment (e.g., devices used in thc home). The combination of hydrogen peroxide with either ozone or UV irradiation enhances the efficiency of microcystin destruction (Rositano 1998, Qiao 2005, respectively). With hydrogen peroxide at 0.1 mgL^{-1} and ozone at 0.2 mgL^{-1}, 1 mgL^{-1} of MCYLR was completely removed in 30 minutes (CT value of 6 mgL^{-1}min). However, more research on these combinations for the degradation of cyanotoxins is needed to assess their feasibility.

Extracelluar and Intracellular Cyanotoxin Removal by Filtration (other than GAC)

Semi–permeable membranes such as those used in reverse osmosis filters remove many contaminants from water, especially those whose size is larger than the membrane pores. Advances in membrane technology have made this process more versatile, dependable and economically feasible as a drinking water treatment process. Reverse osmosis and nanofiltration membranes remove all pathogens and also large molecules above the membrane pore size (100 Daltons for reverse osmosis, 200 Daltons for nanofilters). However, the membranes are not necessarily foolproof. When concentrations of the removed substance become sufficiently high traces of inorganic and organic compounds may pass through the membrane barrier. A reverse osmosis membrane study using between 10 ugL^{-1} and 130 ugL^{-1} of MCY LR, RR and nodularin removed greater than 95% of the toxin from waters with a range of salinities (Neumann 1998, Vouri 1997). A water with high conductivity such as brackish water may allow slightly greater breakthrough of cyanotoxins than otherwise (Vuori 1997).

Several nanofiltration studies report from 82% to complete microcystin removal (Fawell 1993, Muntisov 1996, Simpson 2002, Smith 2002).
Microfiltration and ultrafiltration use membranes with greater pore sizes than the membranes used in reverse osmosis and nanofiltration. Ultrafiltration is capable of removing intact bacteria (including the cyanobacteria), but not all dissolved organic compounds (Fig. 1). Microfiltration, with its larger pore size, may not remove all intact bacteria and is not effective in removing dissolved organic compounds. Since 50% to 95% of the cyanotoxin is intracellular, ultrafiltration and (to a degree) microfiltration are effective in removing intracellular cyanotoxins in drinking water supplies. Studies (Chow 1997, and Zhou 2001) suggest that both microfiltration and ultrafiltration either as a stand–alone treatment process or as a replacement to conventional filtration are excellent at removing intact cyanobacteria and their intracellular toxins. With regard to conventional water filtration (coagulation, sedimentation, and then filtration), a bench – top jar test apparatus and a full –scale pilot plant resulted in 70% and 99.9% removal, respectively, of *Microcystis* cells (Drikas 2001). This study suggests that conventional treatment is effective in removing intracellular cyanotoxins. These studies reported very little cell breakage during the cell removal process, implying that the process did not cause significant intracellular cyanotoxin release.

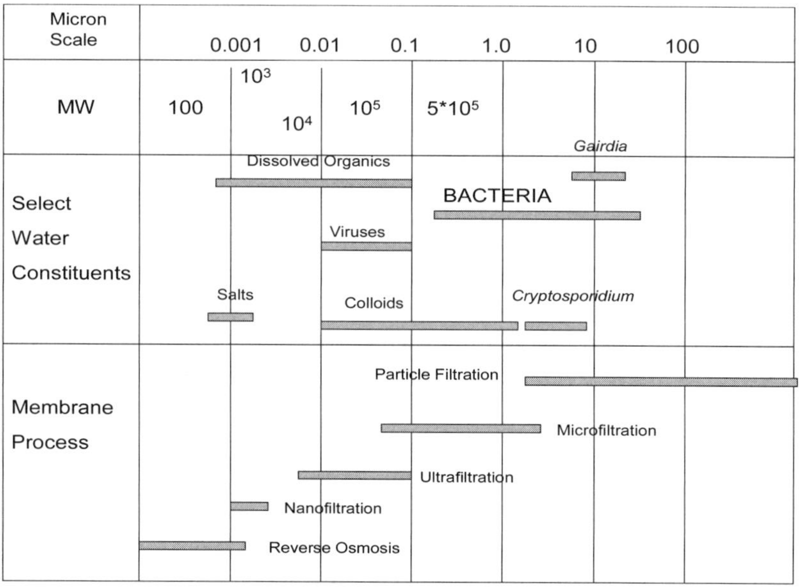

Fig. 1. A summary of membrane processes and their filtration characteristics. Modified version (Tech Brief 1999)

Through Plant Removal and Inactivation of Cyanotoxins: A Multi–barrier Approach

Toxin–producing cyanobacteria are abundant in surface waters and so they impose a degree of risk to drinking water utilities. The toxicity of microcystin LR has prompted the World Health Organization to publish a guideline value of 1.0 ug MCYLR/L for drinking water. Several studies have investigated the concentration of cyanotoxins in drinking water treatment plants. The most recent studies have been done by Carmichael (2001), Karner (2001), Hoeger (2005), and Wood (2005).

During two year study, 1996 to 1997, Carmichael (2001) measured toxins in source and finished (i.e., filtered) waters in 10 utilities in North America. Of the 677 samples tested, 539 (80%) were positive for microcystins when tested using the enzyme linked immunosorbent assay (ELISA). Of the positive samples, 4.3% were higher than the 1 ug/L WHO guideline. Only two finished water samples were above 1 ugL^{-1}, suggesting that during the study many of the utilities were adequately removing or inactivating microcystins.

Karner's (2001) study evaluated source and treated water from five Wisconsin drinking water plants with four plants located on one lake. The reported total concentration of microcystin (intracellular and extracellular toxin) was determined by ELISA. The source waters ranged from tenths of ugL^{-1} to 7 ugL^{-1}. The five plants demonstrated 1-3 log removal of microcystin. This study suggested that pretreatment alone (i.e., before filtration) with potassium permanganate, copper sulfate, and PAC reduced algal toxins as much as 61%. Karner reported the percent of removal, but did not mention the initial microcystin concentrations. There may be several site specific reasons that pretreatment worked for these plants; two of the utilities have at least 5.5 day water detention times before being distributed to the public and another plant used a lower dose of $KMnO_4$. These conditions could have played an important role in achieving the 61% decrease of microcystin. This study stresses the uniqueness of individual treatment trains.

Hoeger's study investigated microcystin removal through two drinking water plants. One plant had pre–ozonation (1.0 mgL^{-1}), rapid sand filtration, intermediate ozonation (0.5 mgL^{-1}), GAC filtration follow by slow sand filtration and the other had conventional treatment followed by chlorination. The first plant was challenged with a microcystin concentration of approximately 8.0 MCY ugL^{-1} where as the conventional plant was challenged with a concentration of 0.2 MCY ugL^{-1}. Both treatment plants removed microcystin below the WHO guideline.

During the summer of 2005, Wood and coworkers (2005) studied cyanobacteria and algal distribution as well as microcystin concentrations in source and finished waters from five utilities in Oklahoma, Vermont, Texas, California and Florida. They found that several of the source waters had low levels of microcystin from 0.2 ugL^{-1} MCY to 0.5 ugL^{-1} MCY and none of the finished waters were above 0.05 ugL^{-1} MCY, as determined by ELISA. This study also showed that typical conventional treatment removed from 2 to 5 log of algal units mL^{-1} (Fig. 2) and finished water commonly contained less than 2 algal units mL^{-1}.

New Technologies

Electrochemical degradation was investigated as a new drinking water technology to degrade extracellular microcystin LR through a hydroxyl radical mechanism similar to other advanced oxidation processes (Feng 2005). At an applied current of 100mA, the degradation rate was 0.219 min^{-1} and the half–life was 2.5 min. This experiment suggested that the electrochemical oxidation may be a promising approach for microcystin degradation.

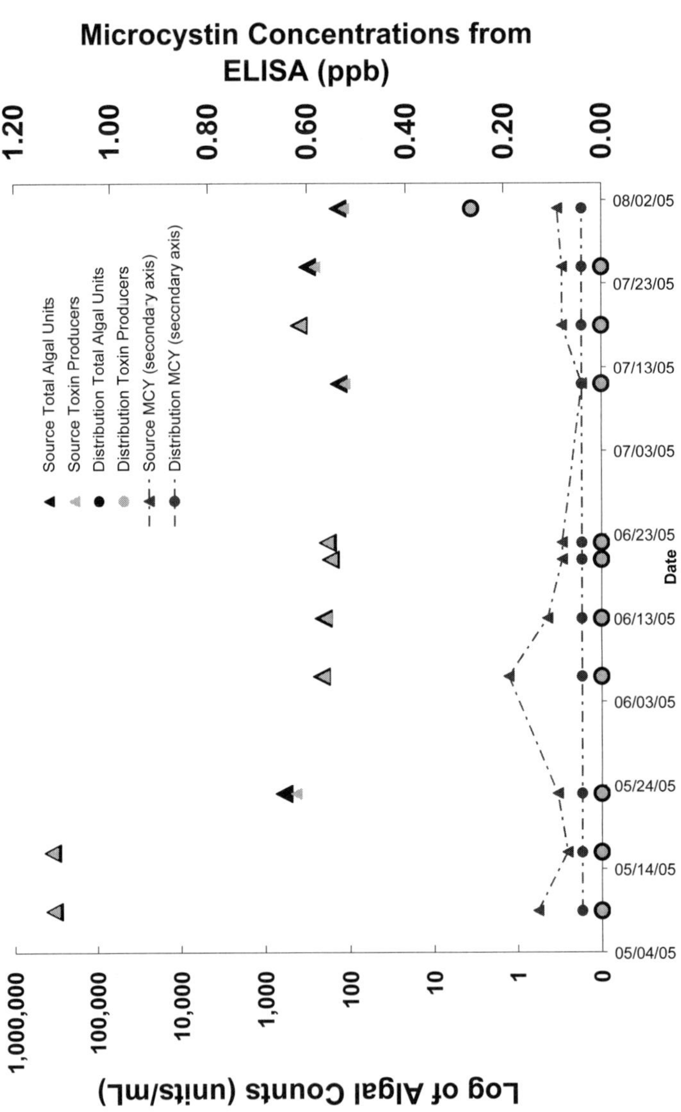

Fig. 2. This graph depicts the removal of intact algal cells and microcystin by conventional treatment.

Conclusion

Drinking water utility managers have several source water and treatment options for removing and inactivating both intracellular and extracellular cyanotoxins. The first consideration should be to optimize the effectiveness of existing processes. Intracellular cyanotoxins can be removed effectively by conventional treatment and membrane filtration. Chloramination and UV treatment are not effective for inactivating extracellular algal toxins. Although several oxidants inactivate some cyanotoxins, it is important to remember that not one oxidant inactivates all cyanotoxins. Chlorine effectively degrades extracellular cyanotoxins both microcystins and cylindrospermopsin between pH 6.0 and 8.0, and saxitoxins at pH values at 9 and higher. Utilities that want to determine the risk of cyanotoxins to their system may use a simple and rapid procedure published by the World Health Organization (Chorus 1999).

References

Acero JL, Rodriguez, E, Meriluoto, J (2005) Kinetics of reaction between chlorine and the cyanobacterial toxin microcystins. Water Research 39:1628–1638.

Bruchet A, Bernazeau F, Baudin I, Peironne P (1998) Algal toxins in surface water analysis and treatment. Water Supply 16 (1–2):619–623.

Carmicheal WW (2001) Assessment of Blue–Green Algal Toxins in Raw and Finished Drinking Water. AWWA Research Foundation

Carlile PR (1994) Further Studies to Investigate Microcystin–LR and Anatoxin–a Removal from Water. Report No. 0458, Foundation of Research, UK.

Chorus I and Bartram J (1999) Toxic Cyanbacteria in Water: A Guide to their Public Health Consequences, Monitoring and Management. London, UK: E & FN Spon.

Chow C, Drikas M, House J, Burch M, Velzeboer R, Gimbel R (1997) A study of membrane filtration for the removal of cyanobacterial cells AQUA 46(6):324–334.

Cook D and Newcombe G (2002) Removal of microcystin variants with powdered activated carbon. Water Science Techonology:Water Supply, 2 (5–6):201–207.

Cornish IT, Lawton LA, Robertson P (2000) Hydrogen peroxide enhanced photocatalytic oxidation of microcystin–LR using titanium dioxide. Applied Catalyst B: Environmental 25:481–485.

Drikas, M, Chow CWK, House J, Burch MD (2001) Using coagulation, flocculation, and settling to remove toxic cyanobacteria. Journal of American Water Works Association 93:2:100.

EPA Guidance Manual (May 2003) Appendix A. Glossary. LT1ESWTR Disinfection Profiling and Benchmarking. 103

Fawell J, Hart J, James H, Parr W (1993) Blue–green algae and their toxins: analysis, treatment, and environmental Control. Water Supply 11(3–4) 243–249.

Feng C, Sugiura N, Masaoka Y, Masaoka T (2005) Electrochemical Degradation of Microcystin–LR. Journal of Environmental Science and Health A40:453–465.

Fritz AJ, Waite TD, Jones GJ, Boyden BH, Orr PT (1999) Photocatalytic degradation of the blue green algal toxin microcystin–LR in a natural organic–aqueous matrix. 33(2):243–249.

Hart J, Fawell J, Croll B (1998) The fate of bath intra and extracellular toxins during drinking water treatment. Water Supply 16(1–2):611–616.

Himberg K, Keijola A, Hiisvirta L, Pyysalo H, Sivonen (1989) The effect of Water Treatment Processes on the removal of hepatotoxins from *Microcystis* and *Oscillatoria* Cyanobacteria: A Laboratory Study. Water Research 23 (8): 979–984

Ho L, Newcombe G, Rinck–Pfeiffer (2005) submitted to Water Research

Hoeger SJ, Hitzfield BC, and Dietrich DR (2005) Occurene and elimination of cyanobacterial toxins in drinking water treatment plants. Toxicology and Applied Pharmacology 203 231–242.

Keijola A, Himberg K, Esala A, Sivonen K, Kiisvirata L (1988) Removal of cyaonbacterail toxins in water treatment processes: laboratory and pilot plant experiments. Toxicological Assessment 3:643–656.

Kull T, Backlund PH, Karlsson KM, Meriluoto JA (2004) Oxidation of the Cyanobacterial Hepatotoxin Microcystin–LR by Chlorine Dioxide: Reaction Kinetics, Characterization, and Toxicity of Reaction Products. Environmental Science Technology 38:6025–6031.

Lawton LA, Robertson, PK (1999) Physio–chemical treatment methods for the removal of microcystins (cyanobacterial hepatotoxins) from potable waters. Chemical Society Review 28(4):217–224.

Muntisov M and Tromboli P (1996) Removal of algal toxins using membrane technology. Water Journal Australian Water Association 23(3):34.

Neumann, U and Wecknesser J (1998) Elimination of microcystin peptide toxins from water by reverse osmosis. Envirnomential Toxicology Water Quality 13(2):143–148.

Newcombe, G and Nicholson BC (2002a) Treatment options for the saxitoxins class of cynatoxins. Water Science and Technology:Water Supply 2(5–6):271–275.

Newcombe G (2002b) Removal of Algal Toxins from Drinking Water Using Ozone and GAC. Denver, Colorado: American Water Works Research Foundation.

Newcombe G and Nicholson B (2004) Water treatment options for dissolved cyanotoxins. Journal of Water Supply and Technology–AQUA 52(4):227–239.

Nicholson BC, Rositano J, Burch MD (1994) Destruction of cyanobacterial peptide heptatotoxins by chlorine and chloramines. Water Research 28:1297–1303.

Nicholson BC, Shaw GR, Morrell J, Senogles PJ, Woods TA, Papageorgiou C, Kapralos C, Wickramasinghe W, Davis BC, Eaglesham GK, Moore MR (2003) Chlorination for degrading saxitoxins (paralytic shellfish poison) in water. Environmental Technology 24:1341–1348.

Qiao RP, Li N, Qi XH, Wang QS, Zhuang YY (2005) Degradation of microcystin–RR by UV radiation in the presence of hydrogen peroxide. Toxin 45(6);745–752.

Rositano J, Nicholson B, and Pieronne P (1998) Destruction of cyanobacterial toxins by ozone. Ozone Science Engineering 20:223–238.

Rositano J, Newcombe G, Nicholson B, Sztajnbok P (2001) Ozonation of NOM and alglal toxins in four treated waters. Water Research 36(1):23–32.

Senogles P, Shaw G, Smith M., Norris R., Chiswell R., Mueller J, Sadler R. Eaglesham (2000) Degradation of the cyanobacterial toxin cylindrospermopsin from Cylindrospermopsis raciborskii, by chlorination. Toxin 38:1203–1213.

Senogles PJ, Smith M, Shaw G (2001) Photocatalytic degradation of the cyanotoxin cylindrospermopsin, using titanium dioxide and UV radiation. Water Research 35(5):1245–1255.

Senogles–Derham PJ, Seawright A, Shaw G, Wickramisingh W, Shahin M (2003) Toxicological aspects of treatment to remove cyanobacterial toxins from drinking water determined using the heterozygous P53 transgenic mouse model. Toxicon 41(8) 979–988.

Shaw GR, Shen X, Wickramasinghe W, Senogles P, Easglesham GK, Lam PKS, Moore MR (2001) Toxicological aspects of byproducts of chlorination of the cyanobacterial toxin, cylindrospermopsin. Toxicology 164(1–3):174.

Shephard MR, Stockenstrom S, de Villiers D, Engelbracht WJ, Wessel GFS (2002) Degradation of microcystin toxins in a falling fim photocatalytic reactor with immobilized titanium dioxide catalyst. Water Research 36(1):140–146.

Simpson MR and MacLeod BW (2002) An integrated approach of algal by–products including bench scale evaluation of nanofiltration for microcystin removal. Proceedings of the AWWA 2002 Water Quality Technology Conference Seattle Washington 10–14 November.

Smith DP, Falls V, Levine AD, McLeod BW, Simpson M, Champlin TL (2002) Nanofiltration to augement conventional treatment for removal of algal toxins, taste and odor compounds, and natural organic matter. Proceedings of the AWWA 2002 Water Quality Technology Conference Seattle Washington 10–14 November.

Svrcek, C. and Smith, D.W. 2004. An Overview of the Cyanobacteria Toxins and the Current State of Knowledge on Water Treatment Options. Journal of Environmental Engineering and Science 3: 155–185.

Tech Brief. (March 1999) National Drinking Water Clearinghouse http://www.nesc.wvu.edu/ndwc/pdf/OT/TB/TB10_membrane.pdf.

Tsuji K, Watanuki T, Kondo F, Watanabe MF, Suzuki SM, Nakazawa H, Suzuki M, Uchida H, Harada KI, (1995) Stability of microcystin from cyanobacterial –II. Effect of UV light on decomposition and isomerization. Toxin 33(12):1619–1631.

Tsuji K, Watanuki T, Kondo F., Watanabe,MF, Nakazawa H, Suzuki, M. Uchida H, Harada KI (1997) Stability of microcystins from cyanbacteria –IV. Effect of chlorination on decomposition. Toxicon 35(7):1033–1041.

Vuori E, Pelander, Himberg K, Waris M, Niinivaara K (1997) Removal of nodularin from brackish water with reverse osmosis or vacuum distillation. Water Research 31(11): 2922–2924.

Westrick JA (2003) Everything a Manager Should Know about Algal Toxins but was Afraid to Ask. Journal of American Water Works Association 95:26–34.

Wood S, Abbott S, Westrick JA (2005) Update on a national preliminary algal toxin occurrence study that monitored source and distribution drinking water (presentation) Kentucky/Tennessee America Water Works Association Conference Covington KY, September 12, 2005.

Yoo RS, Carmichael WW, Hoehn RC, and Hurdey SE (1995) Cyanobacterial (Blue–Green Algal) Toxins: A Resource Guide. Denver, Colorado: American Water Works Research Foundation.

Zhou H and Smith DW (2001) Advanced technologies in water and wastewater treatment Canadian Journal of Civil Engineering 28 (Supplement 1):49–66.

Chapter 14: Causes, Mitigation, and Prevention Workgroup Posters

Application of immobilized titanium dioxide photocatalysis for the treatment of microcystin–LR

Antoniou MG,[1] de la Cruz AA,[2] Dionysiou DD[1]

[1]Department of Civil and Environmental Engineering, University of Cincinnati, 765 Baldwin Hall, Cincinnati, OH 45221–0071; E–mail: dionysios.d.dionysiou@uc.edu. [2]US Environmental Protection Agency, Office of Research and Development, 26 West Martin Luther King Drive, Cincinnati, OH 45268.

Introduction

This research is currently focused on the development of efficient water treatment processes for the hepatotoxin, microcystin–LR (MC–LR). Both conventional and advanced oxidation technologies have been tested against this cyanotoxin. Pilot plant studies have shown that conventional treatment processes such as coagulation, flocculation, and sedimentation result in increased levels of soluble toxin. Chlorination, activated carbon adsorption, and chemical oxidation [ozonation, Fenton reagent (FR)] have been used for the inactivation, physical removal, and degradation of the toxin, respectively. A promising chemical oxidation technology for the treatment of MC–LR is titanium dioxide (TiO_2) photocatalysis. This emerging "green" technology efficiently performs water purification and disinfection. Due to high catalytic surface area and minimized mass transfer limitations, TiO_2 in suspension exhibits high photocatalytic activity.

However, nanosize TiO_2 particles are difficult to handle and remove after their application in MC–LR treatment.

Hypotheses

In this study, immobilized TiO_2 photocatalysis was utilized as a proposed alternative to suspended TiO_2 treatment for MC–LR. Immobilized TiO_2 has higher mass transfer limitations than the suspended TiO_2. Also, fewer active catalytic sites are available to the contaminant resulting in reduced photocatalytic activity. Thus, it is necessary to immobilize TiO_2 particles onto substrates and enhance the photocatalytic and structural properties of TiO_2 material. We are developing novel methods for fabricating porous and non–porous photocatalytic films that possess high photocatalytic activity for a given TiO_2 mass.

Methods

Highly active photocatalytic films were prepared with two methods resulting into porous and non–porous films on glass and stainless steel substrate, respectively. The non–porous films are synthesized by a modified sol–gel method employing a mixture of titania sol and colloidal TiO_2 particles to immobilize commercially available TiO_2 powder nanoparticles. The porous films are prepared with an acetic acid based sol–gel method modified with surfactant templates.

Results

Control experiments demonstrated the high stability of MC–LR. The observed degradation in all the experiments with the films is due to the photocatalytic properties of the titanium dioxide. Both dark reaction and direct photolysis of MC–LR showed no obvious degradation.

Conclusion

The study showed that immobilized titanium dioxide photocatalysis could effectively destroy MC–LR in water, at concentrations up to 5000 ppb.

Environmental conditions, cyanobacteria and microcystin concentrations in potable water supply reservoirs in North Carolina, U.S.A.

Burkholder JM,[1] Touchette BW,[2] Allen EH,[1] Alexander JL,[1] Rublee PA[3]

[1]Center for Applied Aquatic Ecology (CAAE), North Carolina State University, Raleigh, NC 27606. [2]Center for Environmental Studies, Elon University, Elon, NC 27244. [3]Department of Biology, University of North Carolina at Greensboro, Greensboro, NC 27402

Introduction

Run–of–river impoundments or reservoirs, the "lakes" of the Southeast, provide potable water supplies and recreational value for rapidly urbanizing areas. Cyanobacteria blooms and the potential for cyanotoxin contamination of water supplies have not been well studied in these turbid systems. The objective of this ongoing research (summers, 2002–current) is to characterize environmental conditions and cyanobacteria composition, abundance, and microcystin concentrations in potable water supply reservoirs of different age in North Carolina. The reservoirs selected for this work provide potable water to two million people in the western and central areas of the state.

Materials and Methods

Physical/chemical environmental conditions have been assessed using a YSI multiprobe water quality system (model 566MPS). Nutrient concentrations (TP, SRP, TN, NO_3^-+NO_2^-, NH_4^+) and chlorophyll *a* concentrations have been analyzed by the state–certified CAAE water quality laboratory. Cyanobacteria assemblages have been identified and quantified following current keys of Komárek and colleagues (phase contrast microscopy, 600x), supplemented by molecular probes where available. Total microcystins (free and cell–bound fractions) in raw and finished water have been quantified using enzyme–linked immunosorbent assays (ELISA, Envirologix, Inc.), confirmed by high–performance liquid chromatography with mass spectroscopy (LCMS). Here, nutrient and microcystin concentrations were compared in reservoirs of two age groupings: newer reservoirs (20–30 years post–fill, n = 5) and older reservoirs (60–85 years post–fill, n = 6). Differences in the two age groupings in selected physical, chemical, and biological factors were determined by non–parametric one–way ANOVA (Mann–Whitney–Wilcoxon test; SAS Institute, Inc. 1999; $\alpha = 0.05$). Linear regressions were also conducted to examine relationships between biological parameters and physical/chemical factors.

Results

These potable water supply reservoirs are eutrophic, with high nutrient levels, high turbidity, and low alkalinity. High precipitation in turbid, well–flushed systems is not conducive to cyanobacteria blooms, and two of three summers were above–average in precipitation, yet cyanobacteria comprised 60–95% (usually more than 75%) of the total phytoplankton cell number each summer, with densities as high as ~400,000 cells mL–1. Common toxigenic cyanobacteria included Anabaena circinalis, Anabaena flos–aquae, Aphanizomenon flos–aquae, Microcystis aeruginosa, Cylindrospermopsis raciborskii; blooms of Lyngbya wollei and Planktothrix cf agardhii also occurred. Microcystins were detected in most samples, at low concentrations (less than 1 µg L–1). Older reservoirs had significantly higher NO3–+NO2–, TP, and microcystin concentrations, and significantly lower TN:TP ratios than newer reservoirs. Total microcystin levels were positively correlated with TP and TN concentrations in newer reservoirs, and with TP in older reservoirs, indicating potential importance of both TN and TP in blooms.

Conclusion

North Carolina potable water supply reservoirs in both age groups are impacted by nutrient over–enrichment and cyanobacteria blooms, including toxigenic species that may adversely impact the utility of these systems for potable water supplies and recreational activities as the watersheds become increasingly urbanized with associated increases in nutrient inputs.

Removal of microcystins using portable water purification systems

Edwards C, Ramshaw C, Lawton LA

School of Life Sciences, The Robert Gordon University, St Andrew Street, Aberdeen, UK AB25 1HG

Introduction

Blooms of blue–green algae occurring in lakes and rivers are frequently toxic, producing a range of cyclic peptides known as microcystins. Extensive studies on toxicity and several toxicoses have led to a WHO guide line of 1 µg per litre in drinking water supply. Consequently there has been a large amount of research on the efficiency of water treatment processes to remove algal cells and toxins along with a range of management strategies.

Over recent years a number of portable water purification systems have been developed to meet the needs of recreational, military and emergency use. However, limited research on domestic and field purification devices has shown that few systems are able to remove microcystins to below the recommended level. As many lakes produce blooms, and an alternative water source may not be available, it is important to examine the performance of the portable purification systems.

In our study two systems were investigated, an MSR®Miniworks™ unit and a First Need® Deluxe from General Ecology both of which are claimed to exceed EPA requirements for removal of bacteria and protozoa. Both systems were evaluated for the removal of dissolved toxins and cyanobacterial cells.

Methods

Both filters were used and cleaned according to manufacturer's instructions. Micrcystin–LR (MC–LR) and the cell free extract of *M. aeruginosa* PCC 7820 which contained MC–LR, –LY, –LW and –LF were prepared as previously described (Lawton *et al.* 1995) and dissolved in tap and lake water to give realistic concentrations (50 μg L^{-1}) . Microcystins were extracted from water samples using solid phase extraction and analysed by HPLC as described by Lawton, *et al.* 1994. Removal of *Microcystis* cells was quantified using a particle analyser.

Results

Virgin filters removed 100% of MC–LR at a concentration of 50 μg l^{-1} from distilled water and lake water. Experiments were repeated where the water samples were spiked with the extract which contained MC–LR, -LY, –LW and –LF. Although systems performed well, the MSR no longer removed 100% MC–LR. Performance after cleaning was assessed and the MSR unit deteriorated, removing only 66.2–88.8 % of microcystin variants compared to 100% removal by thc First Need unit.

The ability to remove whole cells was investigated, along with determination of intra– and extra–cellular microcystin concentrations. Both filters removed 100 % of cells but high concentrations of the microcystin variants were detected in the MSR filtered water, indicating that cells had been damaged and toxins released.

Conclusion

This short study demonstrated that the First Need filter was highly effective at removing both dissolved microcystins and cells, whereas the performance of the MSR filter rapidly declined, despite the fact that only 6 L of lake water had been passed though it and would therefore not be recommended such an application. More research is needed on continued use of the First Need filter to ensure that performance is maintained and there is no release of microcystins from trapped cells.

References

Lawton L.A., Edwards C., Beattie K.A., Pleasance S., Dear G.J. and Codd G.A. (1995) Natural Toxins. 3, 50–57.

Lawton L.A., Edwards C. and Codd (1994) Analyst 119, 1525–1530.

Multiple Scenarios for Fisheries to Increase Potentially Toxin Producing Cyanobacteria Populations in Selected Oregon Lakes

Eilers JM,[1] St. Amand A[2]

[1] MaxDepth Aquatics, Inc., Bend, OR, j_eilers@maxdepthaq.com; [2] PhycoTech, Inc. St. Joseph, MI, astamand@phycotech.com

Introduction

The dominance of cyanobacteria, many of which produce toxins, in lakes is often associated with external loads of phosphorus from activities in the watersheds. However, we have identified multiple pathways in selected Oregon lakes whereby fisheries management activities play a crucial role in promoting cyanobacteria populations.

Hypotheses

The proliferation of cyanobacteria in freshwater environments is aided by increased availability of phosphorus. Phosphorus availability is increased by alteration of native food–webs, thus changing the trophic dynamics of nutrient cycling in lakes. Fisheries management can alter food webs numerous ways.

Methods

Several Oregon lakes were investigated through the use of paleolimnological techniques. Sediment cores were dated using ^{210}Pb and changes in water quality and cyanobacterial populations were examined using akinetes (resting cells) preserved in the sediments.

Results

Fish populations were shown to alter trophic structure and thus increase phosphorus availability in several ways. In Diamond Lake, the inadvertent introduction of a minnow (tui chub, *Gila bicolor*) native to an adjoining basin led to increased fish biomass and translocation of nutrients from the shallow waters to the pelagic zone. Diamond Lake was treated with rotenone in 1954, thus eliminating all fish and resulting in an immediate drop in cyanobacteria populations. The cyanobacteria densities in Diamond Lake have increased in recent years with the reintroduction of the tui chub and leading to lake closures. In Odell Lake, the State intentionally introduced kokanee (land–locked sockeye salmon, *Oncorhynuchus nerka*) to enhance a native salmonid fishery. The kokanee became the dominant fish and increased nutrient cycling by consuming large quantities of zooplankton in the metalimnion and recycling nutrients back into the photic zone. The native salmonids occupy the hypolimnion during the summer, providing relatively little opportunity for returning phosphorus back to the photic zone in the summer. Devils Lake is a shallow coastal lake with a native salmonid fishery and introduced centrarchids. In an effort to reduce the abundant macrophyte growth throughout much of the lake, the local lake district (with permission of the State) introduced triploid grass carp (*Ctenopharyngoden idella*). The stocking was successful and the grass carp eventually ate all submerged macrophytes in the lake. Once the macrophytes were eliminated, cyanobacterial populations increased dramatically most likely because of increased light availability, reduced competition for nutrients, and increased nutrient supply associated with disturbance of the sediment by the grass carp. Crane Prairie Reservoir is a shallow impoundment in central Oregon and historically was a rainbow trout fishery. It currently has five introduced, non–native fish species and is experiencing major blooms of cyanobacteria.

Conclusions

Fish management activities can increase the likelihood of cyanobacterial blooms through a variety of mechanisms involving translocation of nutrients, increased retention of nutrients in the photic zone, substrate disturbance, and reduction of large cladoceran zooplankton.

Removal of the cyanobacterial toxin microcystin–LR by biofiltration

Eleuterio L, Batista JR

Department of Civil and Environmental Engineering, University of Nevada Las Vegas

Introduction

The occurrence and persistence of the cyanobacterial toxin, microcystin–LR, in natural waters has been reported worldwide, and its risk to public health and animals has been associated with water consumption. The effects of this toxin on humans and animals include total failure of respiratory system, hepatocyte necrosis and tumor promotion in the liver. Conventional water treatment processes such as coagulation, flocculation and filtration, have failed to remove algal toxins to recommended levels required by the World Health Organization (WHO). However, there has been reported effective biological degradation of microcystin–LR in field and laboratory studies using water samples from lakes where cyanobacterial blooms have historically occurred.

Hypothesis

Given that biological degradation of microcystin–LR is very effective, and that filters in water treatment plants can successfully remove naturally organic matter (NOM) via biofiltration, it is expected that microcystin–LR can be degraded by biologically active filters. Biological filters are established in water treatment plants when ozonation is introduced as a disinfectant. Following the approval by the US Environmental Protection Agency legislation to reduce Disinfection By–Products (DBP's) in potable waters, biological filters have become an important water treatment unit to meet the newly established DBP's standards.

Methods

Bench–scale microcystin biodegradation tests were carried out using an enrichment bacterial culture from Lake Mead, Nevada. Three bioreactors were incubated at room temperature for 7 days in the dark to avoid phototrophic growth. In each reactor containing 100 ml of Errington & Powell's medium, 32 mg/L of microcistin–LR was added. In order to evaluate the biodegradability of microcystin in the presence of different amounts of carbon, the amounts of glucose, citric acid, L–glutamic acid and succinic acid from the aforementioned medium were varied to obtain bioreactors containing 100%, 50 % and 0% additional carbon. Sub–samples from the reactors were taken daily for microcystin analysis and evaluation of its degradation rates.

The microcystin–degrading enrichment culture was then used to inoculate two bench–scale biofilters operating with typical design parameters of a drinking water treatment facility. The filters were packed with variable amounts of silica sand (effective size 0.51 mm) and anthracite (effective size 0.9 mm) to provide different empty bed contact times (EBCT). After a biofilm was established on the surface of the filter media, the filters were fed in continuous mode at hydraulic loading rate of 2.5 m/h. Dechlorinated tap water containing readily biodegradable organic matter (i.e. acetate and formate), bentonite and the toxin were added tothe filters. Acetate and formate are typical by–products of the ozonation of natural organic matter (NOM) and they are present in the influent water to the filtration units. Bentonite was used to simulate the particulate matter in surface water. Microcystin–LR concentrations varying from 10–130 g/L were added in the influent water. This range corresponds to dissolved microcystin–LR levels detected in lake waters during algal blooms. The concentration of microcystin–LR in the effluent was monitored every 12 hours and determined by Enzyme Linked Immunosorbent Assay (ELISA).

Results

The results of the biodegradation tests revealed a reduction of approximately 67% in the bioreactor to which no additional carbon source was added. Lower reductions were obtained in the experiments with carbon source addition. Therefore, it appears that microcystin itself can be used as a carbon source by the enrichment bacterial culture. These results are encouraging because concentrations of acetate and formate, byproducts of NOM ozonation, in drinking waters are low. Therefore, the potential to remove microcystin via biofiltration is high.

Water quality and cyanobacterial management in the Ocklawaha Chain–of–Lakes, Florida

Fulton RS, Coveney MF , Godwin WF

St. Johns River Water Management District, Palatka, FL USA

Introduction

The Ocklawaha Chain–of–Lakes are large, shallow water bodies located in central Florida. These surface waters are naturally productive. However, water quality has been severely degraded by nutrient loading, primarily from large agricultural operations. Water quality in the lakes ranges from mesotrophic to hypereutrophic, and the lakes have experienced prolonged severe cyanobacterial blooms.

Restoration program

The program to manage water quality and cyanobacteria in the lakes includes purchase and restoration of wetland habitat in former agricultural areas to reduce external phosphorus loading, operation of a marsh flow–way to remove particulate phosphorus from lake water, harvesting of gizzard shad to reduce recycling of stored phosphorus, and re–establishment of desirable aquatic vegetation.

Lake responses

There are strong relationships between external phosphorus loading, phosphorus concentrations, and cyanobacterial biovolume in the Ocklawaha Chain–of–Lakes. Following external phosphorus load reduction and shad harvesting, Lake Griffin has seen substantial improvements in water quality, including decreases in phosphorus and chlorophyll concentrations, and increases in transparency. Cyanobacterial biovolume has also decreased, and there have been changes in the composition, including a decrease in dominance by *Cylindrospermopsis*. The phytoplankton community has shifted from year–round cyanobacterial dominance to cyanobacterial dominance only during the warm season.

Conclusion

Data indicate that meeting phosphorus targets for the lakes will significantly improve water quality. Cyanobacteria likely will remain seasonally dominant even if phosphorus reduction goals are achieved, although at substantially lower biovolume.

A shift in phytoplankton dominance from cyanobacteria to chlorophytes following algaecide applications

Iannacone LR,[1] Touchette BW[1,2]

[1]Department of Biological Science, Elon University, Elon NC 27244
[2]Center for Environmental Studies, Elon University, Elon NC 27244

Introduction

Cyanobacteria can form massive blooms in nutrient enriched environments. Some species are capable of producing hepatotoxins that are harmful to both humans and animals following ingestion. Chemical algaecides are frequently used to control these blooms in order to minimize their impact to both humans and wildlife. In this study, we compared copper sulfate with sodium carbonate peroxyhdrate (PAK–27^{TM}) on natural cyanobacterial populations. Studies have shown that copper does not readily dissipate from the environment and can thus accumulate over time to levels considered lethal to other organisms – especially invertebrates and fishes. PAK–27^{TM} was developed to be an environmentally friendly algaecide that specifically targets cyanobacteria and decomposes within several weeks into water and oxygen. For this study, three levels of algaecide (0.15, 1.5, and 5.0 ppm; either copper sulfate or PAK–27^{TM}) were used to treat a cyanobacterial bloom (dominated by *Anabaena circinalis*, *Microcystis aeruginosa*, and *Planktolyngbya limnetica)* cultured in 4L microcosms. The treatments were monitored over time, to evaluate changes in species abundance and composition.

Hypotheses

Both algaecides would be effective in controlling algal blooms. There would be treatment (algaecide) level–specific community responses due to chemical applications, with most of the changes in phytoplankton community composition occurring at lower levels.

Methods

A total of 65 microcosms (4L cubitainers) were used during this investigation. Chemical treatments were implemented during a bloom period at peak densities, and consisted of three dose levels of sodium carbonate peroxyhdrate or copper sulfate (0.15 ppm, 1.5 ppm, and 5.0 ppm; n=10). Population densities were monitored on Day–1, Day–5, and Day–15. Plankton counts were performed using a Sedgwick–Rafter counting cell according to APHA methods. During incubation, all microcosms were situated within the lake proper to insure natural temperature and light regimes.

Results & Conclusion

The experiment involved microcosms inoculated with pond water (late September – October) that was experiencing a cyanobacterial bloom comprised primarily of *Anabaena circinalis* (up to 6,000 units mL^{-1}), *Microcystis aeruginosa* (100 units mL^{-1}), and *Planktolyngbya limnetica* (200 units mL^{-1}). Copper treatments were more effective than PAK–27™ on all three cyanobacterial species by Day–15, as illustrated by a two–fold decrease in *A. circinalis*, and greater than a 10–fold decrease in both *M. aeruginosa*, and *P. limnetica*. In contrast, no significant declines were observed in microcosms treated with PAK–27™. One of the more interesting findings from this study was the dominance of green algae following chemical removal of cyanobacteria. In this case, for low and moderate copper, the removal of cyanobacteria resulted in significant increases in *Staurastrum sp.* (up to 500 units mL^{-1}), *Eudorina elegans* (up to 7,000 units mL^{-1}), and *Scenedesmus sp.* (up to 6,000 units mL^{-1}). This shift in community composition, from cyanobacteria to green algae following chemical treatments, may indicate a competitive exclusion of green algae by cyanobacteria. Moreover, this response indicates that cyanobacteria are less tolerant to copper relative to green algae. Nevertheless, high copper levels (at 5 ppm) resulted in substantial declines in all algal taxonomic groups.

Ultrasonically–induced degradation of microcystin LR and RR: Identification of byproducts and effect of environmental factors

Song W,[1] Rein K,[1] de la Cruz A,[2] O'Shea KE[1]

[1]Department of Chemistry and Biochemistry, Florida International University, University Park, Miami, FL 33199. [2]US EPA, National Exposure Research Laboratory, Microbiological & Chemical Exposure Assessment Research Division, 26 W ML King Drive, Cincinnati, OH 45268–1314

Introduction

Microcystins (MCs) are a family of strongly hepatotoxic peptides produced by different species of cyanobacteria commonly found in lakes, water reservoirs, and recreational facilities. The increased eutrophication of fresh water supplies has led to the increase in the incidence of cyanobacteria blooms and concerns over the public health implications of these toxins in the water supply. Conventional water treatment methods are poor at removing low concentrations of the cyanotoxins, and specialized treatment is usually necessary for treatment of contaminated water.

Hypotheses

Advanced oxidation technologies (AOTs) employ photochemical and radical processes for the oxidation of pollutants. While AOTs have shown tremendous promise for the remediation a variety of anthropogenic pollutants, there has been a limited number of reports on the remediation of naturally occur toxins by AOTs. Unlike AOTs involving photochemical process, ultrasonic irradiation can be used for slurries and turbid solutions such as those encountered during cyanobacterial blooms. Ultrasonic treatment is a reagent–free process and does not produce disinfection by–products. We hypothesize that ultrasonic irradiation can be used to effectively destroy MCs.

Methods

Purification of MC–LR from a laboratory culture, the ultrasonic irradiation, HPLC, and LC–MS procedures are available from the literature. Microcystin–RR was purchased from ALEXIS.

Results

Ultrasonic irradiation leads to the rapid degradation of MC–LR and dramatically reduces the PP1 toxicity of the treated solution. Hydroxyl radical is responsible for a significant fraction of the observed degradation, but other processes (hydrolysis/pyrolysis) are also important. The decomposition products of the ultrasonic destruction of microcystin–LR and microcystin RR were analyzed by liquid chromatography–mass spectrometry (LC–MS) and the mechanisms of degradation involve oxidation of the Adda moiety. The effects of pH, Fe^{2+} and H_2O_2 on the ultrasonic degradation were investigated.

Conclusions

The major by–products of ultrasonically induced degradation of MCs result from hydroxyl radical attack on the benzene ring of Adda, substitution and cleavage the Adda conjugated diene structure. The initial rate of MC degradation is strongly pH dependent, in a manner mirrored by the pH dependence of toxin hydrophobicity. While hydroperoxide and organic peroxides are formed during ultrasonically induced irradiation of MCs, addition of Fe^{2+} effectively destroys the peroxides and promotes further oxidation of the MCs. These findings suggest that ultrasonically induced irradiation may be a suitable method for the treatment and detoxification of MCs in drinking water and in response to bioterrorism.

Cultural eutrophication of three midwest urban reservoirs: The role of nitrogen limitation in determining phytoplankton community structure

Pascual DL, Johengen TH, Filippelli GM, Tedesco LP, Moran D

The cultural eutrophication of three Midwest urban reservoirs (Ford Lake, MI; Belleville Lake, MI; and Eagle Creek Reservoir, IN) has resulted in impaired water quality. Nutrient loading to these reservoirs has resulted in the formation of nuisance algal blooms, including possible toxin–producing and/or taste and odor causing, heterocyst–forming blue–green genera such as *Anabaena*, *Aphanizomenon*, and *Cylindrospermopsis*. Analysis of monthly nutrient concentrations (Total P, NO_3^-, NH_4^+) taken from 1998 – 2000 for two southeastern Michigan reservoirs, Ford Lake and Bellville Lake, and weekly nutrient data taken from 1976 – 1996 and bi–weekly data collected in 2003 for Eagle Creek Reservoir, Indiana showed consistent annual trends of NO_3^- + NH_4^+ depletion and P abundance from mid– to late summer, suggesting that phytoplankton growth became seasonally N–limited in these reservoirs. Data from the three reservoirs showed that low N–to–P ratios correlated more strongly with phytoplankton standing stock than N or P alone. Data from Eagle Creek Reservoir showed that low N–to–P ratios preceded an increase in heterocystous *Anabaena* and *Aphanizomenon* concentrations. In 2005, a combination bioassay and high resolution sampling study on Eagle Creek Reservoir began to determine if nuisance algal blooms of these heterocystous blue–green algae were preceded by transition from nitrogen rich (P–limited) to nitrogen poor (N–limited) growth conditions.

Cyanobacteria in eutrophied fresh to brackish lakes in Barataria estuary, Louisiana

Ren L, Mendenhall W, Atilla N, Morrison W, Rabalais NN

Louisiana Universities Marine Consortium, 8125 Hwy. 56, Chauvin, LA 70344

Introduction

Lakes Cataouatche, Salvador and des Allemands in upper Barataria Bay estuary, LA were historically oligotrophic to hypereutrophic fresh and brackish waters. Mississippi River diverted water into Cataouatche and Salvador since late 2003 changes salinity and introduces high nutrient loads. Lac des Allemands does not directly receive diverted water, but receives high nutrient loads from adjacent lands.

Hypotheses

We examined cyanobacteria to detect increases in the extent and duration of cyanobacterial blooms and HABs in response to nutrient additions, particularly the high nitrogen load from the Mississippi River.

Methods

We conduct monthly and biweekly surveys for epifluorescent counts of HABs, HPLC pigments, chlorophyll biomass, and associated water quality. 10-L microcosm experiments of nutrient additions and mixtures of Mississippi River water with lake water have been used to examine nutrient limitation shifts in community structure, and stimulation of potentially toxic HABs. Limited toxin analyses have been conducted.

Results

Common cyanobacteria (and their toxins) include *Anabaena* cf. *circinalis* (saxitoxins and microcystins), *Microcystis* spp. (microcystins), *Cylindrospermopsis raciborskii* (anatoxin), *Anabaenopsis elenkinii*, *Planktonlyngbya* spp., *Raphidiopsis curvata*, and other *Anabaena* spp. (hepatotoxins and/or neurotoxins). In Lakes Cataouatche and Salvador, where diatoms generally dominated phytoplankton communities, cyanobacteria were mostly detected in late spring and summer. The highest concentrations of *A. circinalis* in August had significant increase from 6×10^6 cells L^{-1} in 2003 to 1.1×10^7 in 2004, whereas *R. curvata* remained at the same level of 5×10^5 cells L^{-1}. HPLC pigment analyses and chemical taxonomy results suggest an increase in overall chlorophyll *a* reaching over 70 μg L^{-1} in Lake Salvador in 2003 and 180 μg L^{-1} in spring 2004. *Microcystis* sp. and *Anabaena* spp. were the main contributors to these elevated chlorophyll *a* levels. Lac Des Allemands had higher levels of chlorophyll *a* in comparison to the other lakes due to high abundances of *Anabaena* spp and *Microcystis* sp.. The phytoplankton community was generally dominated by *A. circinalis* during most annual cycles (3.9×10^4 - 6.0×10^7 cells L^{-1}) as indicated by good correlation between its biomass and chlorophyll data. *A. circinalis* counts were high in 2003-2004 in Lac des Allemands, while other cyanobacteria such as *Cylindrospermopsis* spp., *Raphidiopsis* spp. and *Aphanizomenon* sp. increased in 2004 compared to 2003. Dense cyanobacterial blooms occurred in all three lakes in spring and early summer of 2005. Microcosm experiments indicate that phytoplankton growth in the lakes is potentially N limited most of the time. Nutrient additions stimulated the growth of most potentially harmful cyanobacteria. Microcystins were detected in Lake Salvador in April 2005 at levels of 0.3-0.6 μg (protein phosphates inhibition assay; G. Boyer, SUNY-Syracuse, unpubl. data).

Conclusion

The Davis Pond Mississippi River diversion is not yet fully operational, but limited outflows are affecting the salinity and nutrient regimes in Barataria estuary. Nutrient inputs, either from periodic diverted Mississippi River water or adjacent areas, have increased the growth of harmful bacteria and changed the phytoplankton community in three upper basin lakes. Little is known about trophic transfer of cyanobacterial toxins, but recent studies indicate there may be both environmental effects and human health concerns.

Chemical characterization of the algistatic fraction of barley straw (*Hordeum vulgare*) inhibiting *Microcystis aeruginosa*

Ferrier MD,[1] Waybright TJ,[1] Terlizzi DE[2]

[1]Hood College, 410 Rosemont Ave. Frederick, MD, USA 21701
[2]University of Maryland Sea Grant Extension Program, Center of Marine Biotechnology, 701 E. Pratt St. Baltimore, MD, USA 21202

Introduction

The algistatic properties of barley straw (*Hordeum vulgare*) have been observed in laboratory studies and *in situ.* Laboratory studies have produced inhibition as well as stimulation of growth in freshwater and marine species. While taxa known to be inhibited have increased, comparatively little has been done to isolate and classify the compound(s) responsible for the algistatic effect.

Hypotheses

The aim of this study was to confirm and characterize the nature of the toxic component(s) in decomposing barley straw that inhibit the growth of *Microcystis aeruginosa*.

Methods

A microplate assay system was developed using *M. aeruginosa* to help isolate and identify the inhibitory components of barley straw extract. *M. aeruginosa* was selected for the bioassay as it has been consistently inhibited by barley straw extract in studies conducted in our laboratory and by others. The 24–well plate assay utilizes *in vivo* fluorescence monitoring with a TECAN GENios plate reader for determination of chlorophyll–a levels in each 2 mL culture.

Results

Fractionation and partial characterization of inhibitory extracts prepared using several different procedures suggests the inhibitors are polyphenolics with molecular weights between 1000 and 3000. Percolation of the aqueous extract through a Polyamide CC6 resin or through various MW cutoff filters resulted in the loss of algistatic activity, which confirms this assertion. Hydrolysis of the extract resulted in little change in the activity profile. Fractionation by HPLC methods yielded a highly potent multi–compound fraction, which is algicidal at 353 mg L^{-1} and algistatic between 11.1 – 3.53 mg L^{-1}.

Conclusion

The consistent inhibition of *M. aeruginosa* observed by ourselves and several other investigators suggests that the algistatic components of barley straw may be useful in the management of *M. aeruginosa in situ*. In the Chesapeake Bay region recent closures of public beaches in response to *M. aeruginosa* blooms and the presence of microcystin in livers of great blue herons in a recent mortality event emphasizes the need for management methods.

Invertebrate herbivores induce saxitoxin production in *Lyngbya wollei*

Thacker RW, Camacho FA

University of Alabama at Birmingham

Introduction

Most studies of freshwater benthic algal communities have attributed changes in community composition to anthropogenic eutrophication, even though selective herbivory can influence community structure by removing palatable species. The benthic community of Lake Guntersville, AL, USA is dominated by the cyanobacterium *Lyngbya (Plectonema) wollei*, but also includes the green alga *Rhizoclonium hieroglyphicum* and a variety of invertebrate herbivores, such as snails (*Pleurocera annuliferum*) and amphipods (*Hyalella azteca*). The dominance of *L. wollei* in this community may be reinforced by the production of chemical defenses, including saxitoxin (STX).

Hypotheses

1. *L. wollei* is less palatable to snail and amphipod grazers than *R. hieroglyphicum*.
2. Herbivores decrease the growth rates of monocultures of *L. wollei*.
3. Herbivores induce the production of STX.
4. Increased STX concentrations are correlated with decreased *L. wollei* growth rates.

Methods

We used artificial foods to test the palatability of these primary producers to an herbivorous snail, *Pleurocera annuliferum*, and to an omnivorous amphipod, *Hyalella azteca*. For amphipods, we also examined the palatability of crude extracts of *L. wollei*, pure STX, and *L. wollei* sheath material. We grew 1 g monocultures of *L. wollei* to test whether snails, amphipods, and mechanical damage induce STX production. We grew 1, 2, and 3 g of *L. wollei* and *R. hieroglyphicum* in a response surface design to examine how snail herbivory and potential competitors impact STX production.

Results

Both snails and amphipods preferred *R. hieroglyphicum* over *L. wollei*. *L. wollei* crude extracts and pure STX stimulated amphipod feeding; amphipods were deterred by *L. wollei* sheath material. In monocultures, snail herbivory generated strong compensatory growth, while amphipod herbivory decreased *L. wollei* growth rates. Snail herbivory induced high concentrations of STX, but amphipod herbivory did not. In the response surface design, with low N:P ratios, increased STX concentrations were correlated with decreased relative growth rates, suggesting a cost of STX production. However, no trade–offs were observed in monocultures, with higher N:P ratios.

Conclusions

Our results indicate that invertebrate herbivores can strongly influence the composition of freshwater benthic algal communities and the production of cyanobacterial secondary metabolites. Since previous reports of STX production in *L. wollei* have documented a high variability in toxicity among locations, efforts to reduce toxicity should not only focus on reducing nutrient availability, but should also consider interactive effects of palatability, herbivory and competition. Trade–offs between *L. wollei* growth rates and saxitoxin production may depend on the relative supply of nitrogen and phosphorus. The dominance of *L. wollei* in aquatic communities may be maintained by both induced chemical defenses and strong compensatory growth.

A comparison of cyanotoxin release following bloom treatments with copper sulfate or sodium carbonate peroxyhdrate

Touchette BW[1,2], Edwards CT[1], Alexander J[3]

[1]Department of Biological Science, Elon University, Elon NC 27244;
[2]Center for Environmental Studies, Elon University, Elon NC 27244;
[3]Marine Science Program, University of South Carolina, Columbia, SC 29208

Introduction

Nuisance and harmful algal blooms are often controlled by the use of chemical treatments. However, many cyanobacteria retain cyanotoxins within their cell structure, and upon cell lyses (cell death) will release these toxins into the water column. These secondary compounds can be potentially harmful to humans who drink these waters, either from contaminated drinking water reservoirs or from recreational activities (e.g., swimming, water skiing) that promote water ingestion. For example, in the 1970's approximately 150 people (mostly children) in Palm Island, Australia were hospitalized with severe hepatoenteritis and kidney failure. This outbreak was attributed to a cyanotoxin, cylindrospermopsin, which had accumulated in drinking waters following the reservoir's treatment with copper sulfate. In this study, we evaluated both cell-bound and soluble (released following cell lyses) microcystin concentrations, following chemical treatment (with copper sulfate, or sodium carbonate peroxyhydrate; PAK-27™) of a cyanobacterial bloom dominated by *Microcystis aeruginosa*.

Hypotheses

The application of algaecides will increase the level of soluble (free) microcystins and decrease the level of cell bound microcystins. The degree of algaecide treatment will influence the how much cyanotoxin will be released following treatment.

Methods

Bloom samples (composed primarily of *Anabaena* and *Microcystis*) were incubated in 4 L microcosms. The microcosms were dosed with varying levels of algaecides (low ~0.15 ppm, moderate ~1.5 ppm, and high ~5.0 ppm), of either $CuSO_4$ or PAK-27™. All microcosms were situated within the lake to insure natural temperature and light regimes. Water samples were collected (100 mL) initially (Day-1), and 10-, 20-, and 30-days post treatment. Samples were analyzed for microcystin-LR using ELISA (Enzyme-Linked ImmunoSorbent Assay) molecular techniques. The water samples were filtered to separate cell-bound and soluble microcystin. The soluble fraction was concentrated to 2.0 mL using solid phase extraction techniques (C-18 silica cartridges; Waters Sep-Pak Plus). The cell-bound fractions, collected on glass fiber filters, were extracted in 80% methanol. Chlorophyll-*a* levels were determined spectrophotometrically.

Results and Conclusion

In this study, we observed significant declines in cell-bound microcystin-LR by Day-10 (by as much as 0.8 μg L^{-1}) and continued through Day-30 (up to 1.8 μg L^{-1}) in copper-treated microcosms. There were no reductions in cell-bound microcystin-LR observed in PAK-27™ treated microcosms. The declines in cell-bound toxins observed in this study indicate that copper sulfate chemically disrupted cyanobacterial cells, thereby releasing toxins into the water column. This release of toxin was observed in the soluble microcystin-LR fraction by Day-20 (by as much as 1.3 μg L^{-1}) for both PAK-27™ and copper treatments. When considering that the World Health Organization (WHO) placed a provisional guideline of 1.0 μg L^{-1} for potable water, it is critical that cyanobacterial blooms be approached with caution when applying chemical treatments. In this study, for example, the increase of soluble toxin was nearly double the recommended level by the WHO.

Chapter 15: Cyanotoxins Workgroup Report

Workgroup Co-chairs:
Rex A Pegram, *U.S Environmental Protection Agency*, Tonya Nichols, *U.S. Environmental Protection Agency*

Workgroup Members[1]**:** Stacey Etheridge, Andrew Humpage, Susan LeBlanc, Adam Love, Brett Neilan, Stephan Pflugmacher, Maria Runnegar, Robert Thacker

Authors: Rex A Pegram, Andrew R Humpage, Brett A Neilan, Maria T Runnegar, Tonya Nichols, Robert W Thacker, Stephan Pflugmacher, Stacey M Etheridge and Adam H Love

Introduction

The Cyanotoxins Workgroup was charged with the identification and prioritization of research needs associated with: the identification of cyanotoxins; toxicokinetics and toxicodynamics of cyanotoxins; human susceptibility to the toxins; cyanobacterial genetics/omics and factors for inclusion in predictive models of toxin production; and risk reduction from an intentional or accidental release of cyanotoxins. Papers presented for the Cyanotoxins Session of the symposium on toxin types, toxicokinetics, and toxicodyamics (See Humpage this volume), cyanobacterial genetics of toxin production (See Neilan this volume), and parameters related to human risks from cyanobacterial exposure (See Love this volume) set the stage for Cyanotoxins Workgroup discussions.

A consensus was achieved regarding the need to focus on the major identified classes of cyanotoxins. The group expressed the belief that the most significant toxic components of presently occurring harmful algal blooms have been identified, and the knowledge gaps for these most prevalent toxins are great enough to warrant the attention of most of our

---

[1]See workgroup member affiliations in Invited Participants section.

future research. This belief does not negate the need to study mixtures of cyanotoxins and toxin precursors, especially those most likely to occur within a given bloom. Moreover, there is also a significant likelihood that novel cyanobacterial blooms and toxins will continue to emerge, and future identification of unknown bloom-forming species and their toxins will require ongoing diligence.

Charge 1: Identification

> Identify and prioritize research needs concerning the identification and quantification of cyanobacteria, cyanotoxins, and their toxicities.

The public health, environmental health, and economic impacts resulting from harmful algal blooms (HABs) create the need to prioritize research for the detection and identification of cyanotoxins. To comply with the Safe Drinking Water Act (SDWA), the U.S. EPA must identify contaminants that may require regulation in the future and periodically publish the resulting Contaminant Candidate List (CCL). The EPA identifies the data gaps associated with the contaminants on the CCL to prioritize research and data compilation to determine whether regulation of candidate contaminants is warranted.

In 1998 and again in 2005, EPA listed "Cyanobacteria (blue-green algae), other freshwater algae, and their toxins" on the CCL. The Workgroup felt that the CCL should be more specific; for example, listing the individual cyanobacteria and toxins of concern. Many genera and species of cyanobacteria are non-toxic; however, many other genera and species can produce a single known toxin and, conversely, a single species can produce a range of cyanotoxins. More detailed listings may evolve because the CCL is an iterative process that assimilates new data on contaminants as well as recommendations from authorized workgroups. More information can be obtained on the EPA CCL website, http://www.epa.gov/safewater/ccl/index.html.

Research Priority: Protocols for efficient production, certification, and distribution of pure toxins and standards of consistent, high quality.

Rapid and sensitive detection of cyanotoxins is an important goal for mitigating potential risks posed by HABs. Sensitivity and specificity of analytical methods for identifying the different classes of cyanotoxins and for accurate quantification depend on the availability of high quality toxin standards. The workgroup expressed concern that some published research was not as reliable as it might be because of reliance on in-house purification of toxins or use of commercial material that was later shown to be of poor quality. Standardized protocols are needed for the efficient production, certification, and distribution of pure toxins and standards. Commercial standards for cylindrospermopsin, some microcystins, nodularin, anatoxin-a, and saxitoxins are available (Table 1). However, concern regarding the reliability of some of the non-certified commercial standards has been expressed. Larger (mg) quantities of some cyanotoxins are available from certain research labs involved in cyanobacterial research, for example, the Australian Water Quality Centre (http://www.awqc.com.au/). Another concern is that recent regulations of high potency agents and toxins hamper the acquisition and storage of large toxin quantities that are needed for analytical methods development, validation, and, toxicological studies. Although safeguards are needed to prevent illicit uses of cyanotoxins, provisions for safe transportation and use in secure facilities are needed.

Table 1. Currently available cyanotoxin standards and certified reference materials (CRM)

Toxin type	Congener	Supplier	Catalog No.
Paralytic Shell-fish Poisoning	Saxitoxin dihydrochloride	Food and Drug Administration	Reference Standard Saxitoxin
	saxitoxin diacetate	National Research Council Canada	CRM-STXdiAc
	saxitoxin dihydrochloride	(http://imb-ibm.nrc-cnrc.gc.ca/crmp/shellfish/psp_e.php)	CRM-STX-d
	neosaxitoxin	"	CRM-NEO-b
	decarbamoylsaxitoxin	"	CRM-dcSTX
	gonyautoxin 1&4	"	CRM-GTX1&4-b
	gonyautoxin 2&3	"	CRM-GTX2&3-b
	gonyautoxin 5 – aka B1	"	CRM-GTX5-b
	decarbamoylgonyautoxin 2&3	"	CRM-dcGTX2&3
	decarbamoylneosaxitoxin	"	CRM-dcNEO
	N-sulfocarbamoylgonyautoxins C1&2	"	CRM-C1&2
	Saxitoxin in acetic acid	Sigma-Aldrich	BCR663 CRM
	dc-saxitoxin	"	BCR543, BCR542 CRM
	neosaxitoxin	"	SO170
	saxitoxin diacetate salt	"	S1417

Toxin type	Congener	Supplier	Catalog No.
	Radiolabeled saxitoxin (^{3}H-STX)	Currently available at no cost through collaboration between FDA, NOAA and IAEA	
Microcystins	Microcystin LR	Alexis; Calbiochem; NRC Canada	350-012; 475815
	Microcystin LF	Alexis; Calbiochem	350-081; 475814
	Microcystin LW	Alexis; Calbiochem	350-080; 475818
	Microcystin RR	Alexis; Calbiochem; NRC Canada	350-043; 475816
	Microcystin-LR	Sigma-Aldrich	M-2912
	Microcystin LR	Cyano-Biotech	
	Microcystin LF	http://www.cyano-biotech.com	
	Microcystin LW	"	
	Microcystin RR	"	
	Microcystin YR	Alexis; Cyano-Biotech	ALX-350-044
	MC-7-desMethyl LR	NRC Canada	(see website)
	Nodularin	Alexis; NRC Canada	ALX-350-061
Other toxins	(+/-) –Anatoxin-a fumarate	A.G. Scientific	A-1065
	Cylindrospermopsin	National Research Council Canada, in collaboration with the Australian Water Quality Centre	CYN-CRM

This issue has also been addressed by other international organizations. A technical report available from the Organisation for the Prohibition of Chemical Weapons (OPCW) website (http://www.opcw.org/html/global-/s_series/98/s78_98.html) indicates that international access to tritiated saxitoxin is extremely limited, and that most nations do not have access to expensive HPLC systems for monitoring saxitoxin levels. Increased availability and distribution of standards, in addition to inexpensive monitoring and detection systems, are urgently needed to monitor cyanobacterial toxins. The UN Food and Agriculture Organization (FAO) also addressed this issue. "It is possible to measure PSP [paralytic shellfish poisoning] compounds by a number of analytical-chemical methods but they all have some limitations, and they often cannot easily be operated because of the lack of reference materials, although recently some progress has been made in this area. However, they are expensive and mainly available from one source. The efforts undertaken by the European Commission's SMT Programme have led to shellfish reference materials with certified mass fractions of some of the toxicologically most significant PSP toxins. Despite these positive developments, the analytical situation remains difficult and the lack of pure PSP compounds in sufficient quantities for repeated dose toxicity studies is a limiting factor in the development of reliable risk assessment." (Marine Biotoxins, FAO Food and Nutrition Paper Number 80, 2004). The recent Harmful Algal Research and Response plan (HARRNESS 2005) also listed establishment of reference material infrastructure and improved availability and distribution of toxins and their metabolites as top priorities.

The workgroup concluded that reliable, well-characterized standards are needed for as many of the common toxins and congeners as possible so that research results from different laboratories can be reconciled. The lack of pure toxins and standards not only limits monitoring efforts, but the ability of investigators to conduct the chronic, low-dose toxicity studies required for risk assessment. Currently it is inadvisable to have detailed protocols for the production of cyanotoxins widely distributed, as these could be used as recipes by bioterrorists. One option for safely acquiring adequate quantities of standards for research is to designate a few reputable and qualified laboratories as infrastructure resources for producing, certifying, and distributing specific chemical groups on an "at-cost" basis. These services must be provided at a cost that is not prohibitive of their use. In-house production by individual investigators was not the preferred option due to the resource and economic drain that would impede research efforts, and because of the lack of ability to fully characterize the material produced.

Standards for four groups of toxins were designated high priority: microcystins, cylindrospermopsins, anatoxin-a, and saxitoxins (i.e., PSPs). All four toxin groups are relevant to the U.S. EPA because they are highly distributed in the U.S. All four groups can impact drinking water reservoirs, farm irrigation, and recreational activities in freshwater environments.

Dangerous Goods Regulations and the Transport of Cyanotoxins

Cyanotoxins come under Dangerous Goods (DG) regulations for transport of hazardous materials (Metcalf et al. 2006). These regulations are defined under UN and IATA frameworks, but many countries have their own specific regulations as well, particularly for control of import and export. By law, such materials must be packed for transport by people specially trained and licensed for DG handling to ensure compliance with DG transport regulations. Freight companies will not accept these materials unless a licensed DG handler certifies that the package and labeling comply with the regulations. Material Safety Data Sheets specific to the DG must accompany the package. Only a few freight agents routinely transport DGs around the world. Most of them employ Custom's Brokers to facilitate importation into the destination country. It is important to use one of these companies to avoid delays, despite the added cost. It is also important to obtain legal opinion to ensure compliance with the plethora of DG and other relevant regulations, particularly for saxitoxins and microcystins because of the added Chemical Weapons Convention (CWC) regulations that apply to these toxins. Australia's CWC regulations require an export permit and quarterly reports on toxin use from the recipient. Although significant effort is required to develop this capability, it is no longer acceptable to simply mail "research materials" without taking DG and other regulations into account. These requirements will limit the number of labs that find it worthwhile to produce and supply cyanotoxins, again pointing to the need for infrastructure resources for production, certification, and distribution.

Developing Standard Methods to Separate and Identify Toxic Components of Raw Water and Crude Extracts

Prior to identifying research needs concerning the development of analytical methods to identify cyanotoxins, the workgroup recognized several overarching considerations and discussed current methods development projects. Methods are needed to identify the cyanobacterial species that comprise a bloom, as well as to identify individual cyanotoxins within the bloom. Additionally, it is important to develop methods that discriminate between cyanobacterial strains and species that are capable and incapable of producing cyanotoxins that may pose risks in both recreational and drinking waters.

Microscopy or morphology-based monitoring is the method traditionally used to identify potentially toxigenic cyanobacteria. A monitoring or screening process has been instituted in Australia to monitor sources of drinking water, whereby the presence and quantity of cyanobacterial cells are identified based on microscopic observations of water samples. When cyanobacteria are present in significant numbers ("Alert Levels"), additional biochemical and molecular assays further characterize the cyanobacteria present. If toxicity screening assays are positive, additional treatment processes are included to mitigate the likelihood of toxins persisting through to the finished water. This approach relies on cellular morphology to identify species that can potentially produce toxins; however, this method does not determine whether a particular toxin is, in fact, present.

Standard methods exist for the separation and identification of many cyanobacterial toxins (for examples, see Meriluoto and Codd 2005, Moore 1996, WHO 1999 [http://www.who.int/docstore/water_sanitation_health-/toxicyanobact/]). However, there is continuing need to develop assays for newly discovered toxins, for methods refinement, and for validation and certification by organizations such as the AOAC International, a non-profit association of analytical communities. A current project is refining existing methods and developing new methods to support the collection of cyanotoxin occurrence data in cells and surface water. The project, entitled "Determination and Significance of Emerging Algal Toxins (Cyanotoxins)," is sponsored by the American Water Works Research Foundation (AwwaRF Project 2789, http://www.awwarf.org/research/TopicsAndProjects/projectSnapshot.aspx?pn=2789).

As better analytical methods have been developed in the shellfish industry (for example more sophisticated mass spectrometry), more toxin analogs have been detected. However, since toxicological data are unavailable for these compounds, regulators cannot use an evidence-based approach for toxicity classification. It is therefore important that toxicological characterization keeps pace with detection of new compounds. An alternative approach is further development of a structure-activity classification system. Less satisfactory is use of the default assumption that toxicity of an analog is equivalent to that of the most toxic known congener.

Emerging and novel cyanobacteria and their toxins. The workgroup considered the need for field ready, reliable, and inexpensive methods to detect currently known, common cyanobacteria and their toxins to be a higher priority research need than the identification of novel cyanobacteria and cyanotoxins. The workgroup agreed that the major cyanotoxins which pose health and ecological risks have been identified. Emerging toxins were considered of somewhat lesser importance than the need for improved methods to assess known organisms and toxins. Historically, new toxins have been found due to accidental poisoning of wild, farm or domestic animals (or in the case of cylindrospermopsin, humans), rather than through formal surveys. Better links with veterinarians and National Parks staff, for example, would facilitate this process.

Complex mixtures. The identification and prioritization of research needs concerning cyanotoxin identification and quantification represents a unique challenge because cyanobacterial toxins generally occur as mixtures. The toxins usually occur as mixtures of analogs of the same toxin type (for example, there are about eighty known microcystins) and/or mixtures of different toxin types (for example, microcystins mixed with cylindrospermopsin. Toxicity screening assays are required to determine the total toxicity of mixtures present in blooms because all mixture components are not likely to be identified in toxin identification assays.

Synergisms in complex mixtures. Another aspect of the "known toxin" and "new toxin" issue is the observation that toxicity assessments of crude cell extracts usually indicate greater toxicity than can be attributed to the known toxins within the mixture. Similar patterns appear in many studies of chemical ecology (Paul et al. 2001, Pietsch et al. 2001). Although in some cases this situation can be taken as evidence of undiscovered toxins, the possibility has also been raised that there are compounds produced by cyanobacteria that, although not toxic themselves, can modify the effect of the known toxins. These could be non-toxic analogs of known toxins. This issue has received little research attention thus far. Furthermore, the list of non-toxic but biologically active compounds known to be produced by cyanobacteria is continually growing (see Table 1 in Humpage this volume).

Because compounds present in minor concentrations may have additive or synergistic effects with the pure toxins of interest, a clear research goal is to identify compounds in the crude extracts that enhance the toxicity of pure compounds, and to identify whether any compounds in crude extracts can reduce the toxicity of pure compounds. Therefore, assessments of one or several toxins may not be sufficient to characterize the risk posed by the mixture of cyanotoxins in a bloom. This issue is of particular importance to the design of toxicological experiments whose aim is the provision of data to support regulatory limits for the toxins in drinking water. These experiments are universally done using pure toxins, but some consideration should be given to the inclusion of at least one treatment group receiving a crude extract for comparison. However, if such experiments demonstrate these hypothesized effects, then the range of compounds requiring detection may multiply considerably.

Research Priority: Expand the use of toxicity screening assays with a focus on pathology-based and mechanism-based screening

Pathology-based and mechanism-based toxicity screening

Due to the lack of readily available, validated, analytical methods that are capable of detecting the range of cyanotoxins known to exist, and of reference standards for them, the workgroup prioritized the expansion of toxicity screening assays with a focus on pathology-based and mechanism-based screening assays. Several approaches were suggested: 1) *in vitro,* utilizing cell-based assays and molecular techniques; 2) *in vivo,* such as the zebra fish egg test and various crustacean-based assays; 3) developmental screens involving fish or mouse embryos and; 4) assays that account for temporal effects. Screening assays should be as simple and sensitive as possible. Additional considerations are cost per assay and required technical expertise and resources. Screening assays may be used to detect toxins during occurrence surveys, to determine the environmental effects of the toxins, to determine the environmental factors that influence toxin production, and to monitor water bodies used for drinking water. Each of these uses will require the selection of an assay appropriate for the particular use and research need.

As previously mentioned, the basic question of how to deal with mixtures arose as a major complicating factor in developing cyanotoxin screening methods. Other factors to be considered are the effects of degradation rates, pH, photolysis, and total organic carbon (TOC) on the assays. Additionally, effects of toxins may have a temporal aspect that is not easily addressed by short-term *in vitro* assays. The workgroup agreed that the advantage of toxicity screening methods is that they measure toxicity and do not just identify individual toxins. This distinction is important because of the enormous toxicity differences between congeners. Cyanobacteria frequently lose the ability to produce certain congeners during laboratory cultivation. Therefore, it should be anticipated that results from field screening assays may differ from the more advanced laboratory analysis with cultured cyanobacteria.

Based on the Organisation for Economic Co-operation and Development (OECD) Guideline for Testing Chemicals-Direction 210, the zebra fish egg test was proposed as a rapid and sensitive screening tool for evaluating cyanotoxicity with relevance to understanding the environmental effects of these toxins. Ongoing studies are evaluating the effects of cyanotoxins (purified and raw extracts) on zebra fish, and a future assessment of these studies will help in evaluating the utility of this approach as a screening tool for cyanotoxicity. Assays based on transfected mammalian cell-lines that rapidly detect the characteristic biochemical effects of the cyanotoxins are also being developed. It is expected that these will provide a diagnostic capability that will help identify the toxin(s) involved.

Biochemical assays

Biochemical assays have been published for a number of cyanotoxins. Examples include protein phosphatase inhibition assays for microcystins and a protein synthesis inhibition assay for cylindrospermopsin (CYN). Sigma has developed a Protein Tyrosine Phosphatase (PTP) assay kit which can be used to screen for PP1 inhibitors, while Biosense (www.biosense.com) has a test kit in development that uses an immobilized PP2A active site as the receptor in a competitive binding assay micro-plate format.

Antibody-based assays exist for the PSPs (Jellett Rapid Test for PSPs and Ridascreen ELISA for saxitoxin), but little work has been done so far to prove their efficacy for use with the range of congeners produced by freshwater cyanobacteria. Established ELISA kits are available for microcystins. Various products have been evaluated previously (see AwwaRF report 2789). An ELISA for CYN has recently been announced by Abraxis. Biosensors embedded in semiconductor microchips may provide an alternative means of detecting the presence of cyanotoxins, instead of the current reliance on HPLC and ELISA-based assays. Biochips for saxitoxin have already been developed, for example Dill et al. (2001). This publication demonstrates the ability of CombiMarix Corporation's microarray antibody chip to detect saxitoxins (http://www.combimatrix.com/). Aptamers specific for microcystin have been developed but assays based on them do not yet have the required sensitivity and specificity. Assays based on artificially produced antibody fragments are also at the experimental stage. Although the sensitivity of biochips operating in reservoirs and water treatment facilities remains to be determined, the development of these biosensors represents an important area for future research. A drawback of all antibody-based technologies is that affinity for the toxins does not correlate with toxicity, and new analogues may not be detected even though they are toxic (see AwwaRF report 2789). The development of a biosensor to assay only for a particular cyanotoxin would preclude the identification of the contaminating cyanobacteria since multiple genera may produce the same toxin.

Availability of antibodies

There are no readily available non-proprietary antibodies for saxitoxin (STX) or any of the other cyanotoxins, although a number of laboratories around the world have developed or are in the process of developing antibodies for use in their own research or assay development. Proactive steps should be taken to link these laboratories with kit makers so that assay development can be accelerated. The development of reliable assays is also contingent upon a previously identified priority – enhancing the supply of reliable chemical standards.

Research Priority: Develop rapid PCR-based assays for the presence of specific cyanobacteria and the potential for production of specific toxins

Molecular-based monitoring

The *in vitro* molecular approaches based on the polymerase chain reaction (PCR) have advantages as screening tools for the presence of toxigenic species because they have the potential to yield more rapid and more sensitive results (See Charge 5, below). To select appropriate genetic markers, the Workgroup agreed that more genetic information is needed. There are currently no whole genomes available for any toxin-producing cyanobacteria. Although the microcystin B biosynthetic gene cluster (*mycB*) is known, those encoding other cyanotoxins such as PSPs and cylindrospermopsin are not yet known. Molecular methods for assessing potential toxicity are limited by this lack of information.

Generally, the technical expertise and resources required for PCR-based techniques are greater than that of a trained microscopist who identifies cyanobacterial cells, but are less than that of an analytical chemistry lab employing HPLC identification methods and that of a lab conducting toxicity screening assays. Established techniques exist for using the genetic sequence of the 16S ribosomal RNA subunit to identify species of cyanobacteria. Some laboratories are developing assays that use the presence of toxin biosynthetic genes, such as *mycB*, to judge whether a bloom has the potential to produce toxins. Every strain of cyanobacteria with the gene for toxin production is capable of producing toxins, but the conditions for toxin production may not be present. Conversely, if the cyanobacteria lack a gene for toxin production, no toxins will be produced. To identify both toxicity genes and individual cyanobacteria, molecular screening assays should include toxin producing and genera- and species-specific biomarkers. Historical occurrence data, once developed, is invaluable in selecting appropriate markers. However, ecological theory suggests that as a region's environmental conditions change, the organisms found in that region will also change. This process is hypothesized to account for the northward spread of *Cylindrospermopsis raciborskii* that has been observed as global warming continues.

A key development in our ability to predict the occurrence of C-HABs has been the identification of the microcystin gene cluster (*mycB*). This gene can be used as an indicator of potential toxin production, alerting water quality managers to the possibility of a toxic bloom. Reservoirs can be monitored by extracting DNA from whole algal communities, and testing for the presence of the microcystin gene by a PCR-based assay (e.g., Hisbergues et al. 2003). A test for CYN-producing genes has also been published (Fergusson and Saint, 2003). Similar tests would give additional warnings for the potential presence of other toxins, including saxitoxin and anatoxins. However, the biosynthetic genes responsible for the production of these latter toxins are currently unknown. In addition, genes that interact with the toxin gene clusters remain to be fully described in *Microcystis*, and are completely unknown for the other organisms. Whole genome sequencing, followed by bioinformatic analyses, is another approach that can be used to search for toxin gene clusters (see Charge 5 for a detailed discussion).

Research Priority: Develop molecular and toxicological fractionation methods to identify toxic components of raw water

As described earlier in this report, many compounds might act synergistically to augment the effects of cyanotoxins. Crude extracts from blooms often are more toxic than can be accounted for by their content of the purified component toxin(s). This enhanced toxicity may be due to interaction with other (known or unknown) cyanotoxins or because there are other compounds that enhance toxicity. These compounds may include water treatment byproducts, environmental/UV degradation byproducts, other toxic components in raw water, and bacterial metabolites. Because our current knowledge of cyanotoxin interactions with other contaminants is limited, there is clearly a need for continued toxicity testing of any potential byproducts produced by treatments to improve water quality. Similarly, we know very little of the reactive chemistry of cyanotoxins in matrices, including their degradation chemistry and their physical and chemical reactions with sediment and other complex matrices.

The approach of comparing the toxicity of crude cell extracts with that of purified toxins has been undertaken by several labs. Such comparisons performed *in vitro* using cell cultures (preferably of human origin) are needed. One hypothesis is that these unidentified compounds are analogues of known cyanotoxins. This hypothesis should be evaluated in organic chemistry labs with an interest in secondary metabolites. The toxicity of novel analogues should be assessed in isolation and in combination with known analogues. Immunoaffinity is an approach for distinguishing the effect of cyanotoxins from that of other compounds. This approach was applied to microcystin. The toxin was removed by binding to an immunoaffinity column containing antibody to microcystin (Kondo et al.1996), followed by toxicity assays of the separated eluents. This approach could be expanded to cylindrospermopsin-producing cultures or blooms once suitable antibodies are available. An advantage of this method is that the bound toxins can be easily recovered.

Non-mammalian model systems that have provided information of relevance to public health risk assessment include *Xenopus* oocytes and zebrafish. Although these models aid our understanding of effects of cyanotoxins in the environment, any effects must ultimately be confirmed in at least one mammalian model system to assess risks to human health. Since *in vitro* experiments are limited by experimental constraints and difficulty of interpretation, it will be necessary to use mammalian model systems to confirm all hypothesized mechanisms of action.

The above discussion notwithstanding, research prioritization requires some knowledge of the primary risks in a particular country or region. Established detection methods should be implemented to provide local cyanobacteria and cyanotoxin occurrence data. This approach focuses on common analogs, but substantial differences in toxicity or chemistry between the identified and unidentified analogs could lead to inaccurate assessments of risk. A comparison to the toxicity caused by the identified analogues and crude cell extracts can be used to assess the overall risk. Thus, the assessment of risk in local and regional areas may benefit from an on-going prioritization process.

Characterization of unknown toxic compounds

There is a high likelihood that novel cyanobacterial blooms and toxins will continue to emerge. A key research priority will be the identification of unknown bloom-forming species and their toxins. If a new, emerging cyanobacterium (e.g. *Lyngbya* cf. *confervoides* on the coast of Florida), produces unknown toxins, standard chemical and toxicological methods are needed to separate and identify the toxic components. Several labs working on natural products from cyanobacteria have established standard procedures for cyanobacterial culture, activity assay-based fractionation, and LC, MS, and NMR of unknown compounds. These processes have proven fruitful in the past (see Table A.1 in Humpage this volume, and references cited therein). These studies generally provide enough chemical information to enable the development of methods to identify the compounds. Although initial *in vitro* cellular or enzyme inhibition data may be obtained, further toxicological characterization of the compounds is needed to produce meaningful risk assessments.

Recommendation: Develop more local expertise to identify cyanobacteria and their toxins. Several symposium attendees stated that their local water quality management teams do not have the expertise to identify cyanobacteria or cyanotoxins. Education of local water quality agents will be needed to ensure that blooms of toxic organisms will be detected. In addition, simple and fast procedures for toxin detection need to be distributed to local agencies. Although ELISA-based detection kits are already available for microcystins, cylindrospermopsins, and saxitoxins (e.g. from R-Biopharm), some local agencies were not aware of their existence. Outreach and education to local water quality agencies will be needed to implement any findings and recommendations of a national research plan. The development of predictive models for bloom and toxin occurrence is a long-term goal.

Charge 1: Identify and prioritize research needs concerning the identification and quantification of cyanobacteria, cyanotoxins, and their toxicities.

Near-term Research Priorities

- Developing protocols to ensure the availability of reliable reference standards of cyanotoxins as well as antibodies for use in biochemical assays
- Developing local expertise to identify cyanobacteria and their toxins
- Developing PCR-based screening assays to identify cyanobacteria and the potential for toxin production; develop an ELISA assay for anatoxin-a

- Developing ways to increase the rapidity, sensitivity, specificity, and robustness of biological toxicity screening assays
- Obtaining ecological and genetic information on mechanisms of toxin production and toxin regulation.

Long-term Research Priorities

- Developing biologically-based in-line, real-time biosensors for automated water quality monitoring
- Assessing the interactions of cyanotoxins with complex mixtures and complex matrices, including identification of transformation byproducts of toxins resulting from water treatment and environmental or UV degradation
- Developing a predictive model of the occurrence of C-HABs and toxic blooms.

Charge 2: Toxicokinetics

Identify and prioritize toxicokinetic research needed to improve human health risk assessments.

Toxicokinetic research on cyanotoxins to date has been limited to basic studies of tissue distribution/elimination and initial efforts to characterize primary metabolism of the more prevalent toxins. Determinations of the relative contribution of different exposure routes to total exposure, the role of metabolism in toxic responses and detoxification, and particularly, the characteristics of human toxicokinetics, are important endeavors for future research.

Research Priority: Labeled compounds and antibodies are needed for Toxicokinetic studies

Microcystins (MC) have been radioactively labeled in individual laboratories as needed for research. The only exception is ^{3}H- microcystin (Robinson et al. 1989) that was labeled by Amersham for the Pathophysiology Division, United States Army Medical Research Institute of Infectious Diseases, Ft Detrick, Md. Brooks and Codd (1987) grew *Microcystis* in culture with sodium ^{14}C-bicarbonate. This required exposure to radioactivity for long periods, and resulted in a labeled product with low specific activity. The method used by the majority of researchers for labeling MC is reduction of the methyl dehydroalanine residue by NaB^3H_4. The products of reaction were characterized by Meriluoto et al. (1990) as retaining the toxicity of the native MC-LR with somewhat less potency. This labeled compound differs from native MC by not being able to form a covalent bond with protein phosphatase 1 and 2A (PP1 and PP2A): the targets of MC in cells and *in vivo*. Falconer et al. (1986) and Runnegar et al. (1986) labeled the tyrosine of MC-YM enzymatically with iodine 125 and showed that the monoiodinated product of the reaction retained the same toxicity in mice as the native MC-YM while the di-iodinated product had somewhat less potency.

There is only one report of a labeled cylindrospermopsin. This derivative resulted from the growth of the cyanobacterium with sodium ^{14}C-bicarbonate: (Norris et al. (2001). Radiolabeled saxitoxin (^3H-STX) is currently available at no cost through a collaboration between FDA, NOAA and IAEA.

Many antibodies have been prepared commercially that are incorporated into ELISA kits for the detection of MC in water. Research groups have also developed antibodies to MC for use in their laboratories. Commercial MC monoclonal antibodies are marketed by Alexis. Dr. Michael Weller (Technische Universitat Munchen, Institut fur Wasserchemie) has investigated the specificity of these antibodies towards MC-LR and other congeners providing a good basis for their potential scientific use. Nevertheless, the specificity of most of these antibodies needs further characterization. The degree to which antibodies detect MC when it is covalently bound to the protein phosphatases or tightly bound in the inhibitory complex is also unknown. Matrix effects also have not been evaluated. Antibody detection of congeners, metabolites, and degradation products following water treatment also requires study. Metcalf et al. (2002) compared the cross-reactivity of MC-LR, conjugates of three commercial ELISA kits, and an-in house ELISA. Cross-reactivity for the compounds tested depended on the solvent used. Antibodies are used in the ALGAETOX project for the development of biosensor systems to detect and quantify algal blooms and toxins (TU Berlin, Fachgebiet Landsschatokologie, insbesondere Okotoxikologie). With these caveats, antibodies have been used to detect MC in tissue/cells for immunochemical analysis and immunohistochemical localization as well as for some quantitation.

A new ELISA kit for the detection of cylindrospermopsin via antibodies is now available from Biosense Laboratories. The antibody binds to CYN with no cross-reactivity with other non-related toxins or compounds. The assay has a detection range between 0.05 and 2.0 ppb and can be used with a variety of environmental samples such as water and fish tissues. Abraxis recently (Jan 2007) announced production of a CYN ELISA, but it is not known whether the antibodies are available separately.

The production of stable standardized radiolabeled toxins and antibodies with known specificities are important near-term needs for toxicokinetics research.

Research Priority: Classical toxicokinetics (distribution, half-life, elimination, etc.) in laboratory animals

Studies employing radiolabeled microcystins have demonstrated that the liver is the main site of toxin accumulation (after oral, intraperitoneal, or intravenous dosing) and that hepatic toxin concentrations remain relatively constant for up to 6 days after treatment. These findings include: 30-40% of ^{125}I-MC-YM dose in liver at death or 24 hr after i.p. injection (Runnegar et al 1986); 76% of ^{14}C-MC-LR dose in liver at 1 min and 77% in liver at 60 min after i.p. injection (Brooks and Codd 1987); 56% of ^{3}H-MC-LR dose in liver at death or 6 hr after i.p. injection (Robinson et al 1989); 30% of ^{3}H- dihydro MC-LR dose in liver at 45 min after i.v. injection (Meriluoto et al 1990); 67% in liver of mice injected i.v. after1 hr and 6 d (Robinson et al 1991); and 47-65% of ^{3}H- dihydro MC-LR dose in liver at 5 hr after i.v. injection of pigs (Stotts et al 1997). Pretreatment with rifamycin (inhibitor of MC uptake in hepatocytes) decreased the hepatic MC concentration after dosing and protected from MC toxicity if given 5 min before MC dosing (Runnegar et al. 1993).

Small amounts in the kidneys, intestine, and carcass indicated that these differently labeled MCs all concentrated similarly in the liver, and that there was no significant loss of label from MC. Mice had excreted 9 and 15% of an i.v. dose in the urine and feces after 6 days, respectively. Although none of the labeled microcystins are ideal (see Meriluoto et al. 1990 for a discussion), they all have similar organotropism, and subsequent work from many laboratories with animals, perfused livers, and isolated hepatocytes confirmed the findings.

Microcystin antibodies have been used for MC quantitation in tissues of treated mice and fish: (Liu et al. 2000; Guzman et al. 2003). Immunostaining and semi-quantitative analysis in mice were reported by: Yoshida et al. (1998), i.p. mice, liver accumulation; Yoshida et al (1997), oral in mice with liver accumulation; Ito et al. (2000), oral dosing with liver accumulation in mice receiving a lethal dose. Oral dosing showed some additional MC staining in the intestine. Guzman et al. (2003) reported MC staining in the nucleus as well as cytoplasm of hepatocytes in a section of mouse liver. Liver accumulation was also shown in rainbow trout (Fischer et al. 2000).

Table 2. Tissue Distribution and Elimination of Radiolabeled Microcystin-LR in Mice

Species & Route	Dose & Isotope	Time	Tissues	% of Dose	Half Life
Mice i.p.	45 μg/kg Sublethal &100 μg/kg Lethal	6 hrs or death	Liver Kidney Intestine Carcass	>50 1 10 10	
Mice i.p.	70 μg/kg	60 min	Liver Kidney Intestine	60 1 9	29 min
Male Mice i.v.	^{3}H MYC-LR 35 μg/kg Sublethal	1 min	Liver Intestine Kidney Carcass Plasma	23 5 2 30 25	α-phase: 0.8 min β-phase: 6.9 min
Male Mice i.v.	^{3}H MYC-LR 35 μg/kg Sublethal	60 min	Liver Intestine Kidney Carcass Plasma	67 9 1 6 trace	
Male Mice i.v.	^{3}H MYC-LR 35 μg/kg Sublethal	6 days	Urine Feces Liver	9 15 67	Covalently bound in liver cytosol: Day 1 - 83% Day 6 - 42%

Adapted from Robinson et al. (1989, 1991)

^{14}C-labeled cylindrospermopsin was administered i.p. to mice resulting in excretion of CYN mainly in the urine but also in the feces (Norris et al. 2001). The excretion patterns showed substantial inter-individual variability between predominantly fecal or urinary excretion, but excretion patterns were not related in any simple manner to toxicity. The authors also found CYN in the liver: 21% of dose at 6 hrs and 13% at 48 hr. Additional evidence indicated CYN or a metabolite was tightly bound to protein. Analysis by HPLC indicated that this tightly bound CYN was in part a hydrophilic metabolite with a different elution time. This single report shows the complexity of CYN distribution, metabolism and excretion. Much more work is needed to confirm and extend the findings particularly given the intra-mouse variability reported.

Toxicokinetic and toxicodymanic studies have been performed for saxitoxins (STXs) in cats (Andrinolo et al. 2002). Oral doses of gonyautoxins 2/3 were administered to cats to determine how toxins are absorbed in the digestive system, where they are absorbed, the absorption rate, the maximal concentration in the plasma, and the time needed to reach that maximum. Oral uptake was efficient, toxin distributed rapidly throughout the body (including the brain), was cleared by simple glomerular filtration, and there was no evidence of metabolism of the toxins. Urine appears to be a major route of toxin excretion in humans (Gessner et al. 1997). The toxicokinetics of saxitoxinol in rats has also been investigated (Hines et al. 1993; Naseem 1996).

β-Methylamino alanine (BMAA) toxicokinetic studies in rats and monkeys demonstrated rapid and virtually complete uptake via the oral route, and distribution throughout the body (Duncan et al. 1991; 1992). The brain contained only about 0.08% of the administered dose at 48 hrs. There was evidence of active transport across the blood-brain barrier via the large neutral amino acid carrier (Km=2.9mM), but this would not be rapid when BMAA is in competition with normal levels of natural amino acids (Duncan et al. 1992; Jalaludin and Smith 1992). Approximately 1.4% of an oral dose, and 1.8% of an i.v. dose, were excreted unmetabolized in the urine by 48 hrs, whereas approximately 22% could be accounted for in total as either unmetabolized or acid hydrolysable (Duncan et al. 1992; Jalaludin and Smith 1992; Humpage this volume). L-amino acid oxidase has been shown to metabolize BMAA, eventually leading to N-methylglycine, but the rate appears to be quite slow (Hashmi and Anders 1991). The racemate and L-forms of BMAA caused cerebellar dysfunction and degeneration of GABAergic cells, presumably through excitotoxicity, but multiple low doses of racemate did not shown evidence of cumulative toxicity (Seawright et al. 1990).

Transport

Microcystins. Microcystins require transport via specific organic anion transport proteins to cross cell membranes (Runnegar et al.1993; Runnegar et al. 1995a; Fischer et al. 2005). Many studies have shown uptake of MC by active transport in hepatocytes but not in other cells including hepatic nonparenchymal cells. There is a good correlation between MC accumulation and toxicity. Hepatocytes from rats and mice as well as from many aquatic animals have been studied. The organ specificity of MC is due to transporters that predominate in the liver. MCs can inhibit protein phosphatase activity with similar potency in cell/tissue extracts from any eukaryote tested so far. The uptake of MC is inhibited by bile acids and dyes. Perfused livers and whole animals were protected from toxicity if pretreated with these compounds.

Very recently the transporters that mediate MC uptake were identified, opening the possibility for many new studies (Fischer et al. 2005). *Xenopus laevi* oocytes were injected with complementary RNA of the known multi-specific organic anion transporting polypeptides (OATPs), and uptake of ^{3}H-dihydro MC-LR was measured. Three human OATPs were shown to mediate the transport of MC: OATP1B1 (was OATPC), OATP1B3 (was OATP8) and OATP1A2 (was OATPA). The first two are mainly expressed in liver, thus explaining the MC uptake inhibition in hepatocytes by bile salts and other compounds (substrates or inhibitors of OATPs), as well as the protection by predosing with rifamycin *in vivo* (Runnegar et al. 1993). The third transporter is also expressed at the blood-brain barrier. This finding indicates the possibility of MC uptake by the brain. Brain uptake of MC by OATP1A2 cannot be tested directly because the enzyme is of human origin. However, research in several labs on whole body distribution using radioactive MC (four differently labeled isotopes) indicates that brain uptake must be very low. Nevertheless, this issue and the possibility of equivalent transporters in the rat and mouse should be examined. As a first step, mammalian/human cells transfected with the OATP1A2, or any other cell that expresses the transporter, should be tested for MC uptake and intracellular changes such as protein phosphatase activity inhibition and cytoskeletal damage. It is also necessary to confirm that cell lines expressing those OATPs as found in liver can take up MC with activity similar to that found in hepatocytes.

Cylindrospermopsin. It is thought that CYN permeates membranes in all cells, although further experimental demonstration is needed. This can be accomplished only when radiolabeled CYN becomes available. There is some indication that hepatocytes are more sensitive to CYN toxicity than other cells. This could be due to differences in metabolism and/or uptake.

Research Priority: Determine the role of metabolism in toxicity and detoxification

The accepted pathway for microcystin excretion is conjugation to the thiol of glutathione (GSH). This may be excreted as MC-GSH or processed to the gamma glutamyl cysteine conjugate and finally to the cysteine conjugate, and then excreted. These reactions have been shown to occur enzymatically *in vitro* with cell extracts from many sources (i.e., Pflugmacher et al. 1998, Pflugmacher et al. 2001, Takenaka 2001). This is the most likely *in vivo* pathway given the changes in GSH peroxidase and in GSH transferases that follow MC dosing. Evidence of this pathway in intact cells or *in vivo* after MC dosing includes detection and identification of MC metabolites formed *in vivo* in mouse and rat liver (Kondo et al. 1992, Kondo et al. 1996). The GSH adduct (through nucleophilic reaction of the thiol group of GSH to the alpha-beta unsaturated carbonyl of the Mdha moiety (methyl dehydroalanine)) was formed as a small (not specified) percentage of the dose in mice injected i.p. with MC-RR (LD_{50}) between 3 and 24 hr. Sipia et al. (2002) and Karlson et al (2003) reported the presence of the GSH conjugate of nodularin as well as native nodularin in mussels and flounders off the Finnish coast. A nodularin conjugate has also been found in *Artemia salina* (Beattie et al. 2003). Wiegand and Pflugmacher (2005) reported that MC-GSH has a lower affinity for protein phosphatase than the native MC. Although MC excretion in mammals has been proposed to be mainly intestinal, the chemical nature of excreted MC has not been determined. Experimental studies are needed to determine and quantify the metabolism and excretion of MC and metabolites.

There is evidence for metabolic activation of cylindrospermopsin: inhibitors of CYP450s reduce CYN acute toxicity (Runnegar et al. 1994, Froscio et al. 2003), genotoxicity (Humpage et al. 2005), and CYN-dependent inhibition of glutathione synthesis (Runnegar et al. 1995). As noted above, Norris et al. (2001 and 2002) found evidence of CYN/or metabolite binding to protein in the liver, and HPLC analysis indicated that this tightly bound CYN derivative was in part a hydrophilic metabolite with a different elution time. For both MC-LR (and congeners) and CYN, research is needed to characterize the formation, chemical nature, and toxicity of metabolites in animals and humans.

Exposure experiments indicated that fish are susceptible to saxitoxins when administered orally or by i.p. injections, and cytochrome P4501A was induced, suggesting a role for the enzyme in saxitoxin metabolism (Gubbins et al. 2000). Saxitoxin composition and concentrations in mussels and human biological specimens demonstrated that toxin metabolism in humans occurs. Following a human fatality associated with consumption of crabs contaminated with saxitoxins, toxin compositions in the crab and clinical samples (human gut, liver and urine) indicated that toxin metabolism occurred and that toxin conversion began in the victim's gut (Llewellyn et al. 2002).

Charge 2: Identify and prioritize toxicokinetic research needed to improve human health risk assessments.

The recent Harmful Algal Research and Response plan (HARRNESS 2005) indicates top priorities for biosynthesis and metabolism research include: 1) identifying metabolites that contribute to animal or human illness; 2) identifying metabolites that are useful as biomarkers for longer term exposure and; 3) providing toxicological and pharmacokinetic information on HAB toxins and metabolites to help determine toxin uptake, metabolism, and clearance rates. We concur with these research needs and further delineate research priorities as follows:

Near-term Research Priorities

- Provide labeled compounds and better characterized antibodies for toxicokinetic studies
- Develop standardized methods for sample preparation for analysis of toxins/metabolites in complex matrices (tissues)
- Classical toxicokinetics (distribution, half-life, elimination, etc.) in laboratory animals

Long-term Research Priorities

- Determine role of metabolism (if any) in toxicity and detoxification
  - Identify metabolites for priority toxins
  - Mixtures issues: do toxins share common metabolic pathways; are enzymes induced or inhibited that could affect toxicity of other toxins or repeated exposures?
  - Is the promotion of oxidative stress an additional toxic factor of cyanobacterial toxins?
  - Human *in vitro* metabolism (tissue, cell and enzyme specific studies)
- Apply toxicokinetic methods to describe exposure route-dependent and species-specific differences in toxicity
  - Develop physiologically-based toxicokinetic models for priority toxins
- Developmental toxicokinetics
  - Describe transport of priority toxins across placenta
  - Describe the toxicokinetics of saxitoxin in developmental neurotoxicity

Charge 3: Toxicodynamics

Identify and prioritize toxicodynamic research needed to improve human health risk assessments.

Research Priority: Assess dose-response relationships and mechanisms with a focus on low concentration exposures including both *in vitro* and *in vivo* studies as appropriate

Acute Toxicity and Known Mechanisms

Microcystins. Microcystins cause adverse effects by rapidly binding and inhibiting the activity of ser/thr protein phosphatases PP1 and PP2A. The inhibition is irreversible and its consequences are well characterized for acute toxic doses (Ito et al. 2002). MC-LR has a LD_{50} (i.p., mice, 24hr) of 60 μg/kg. The primary acute effect of protein phosphatase inhibition is hyperphosphorylation of many cellular proteins including the hepatocellular cytoskeleton, which causes loss of cell-cell contacts and thus intra-hepatic hemorrhage. Death is due to hypovolemic shock (Runnegar and Falconer 1986; Falconer and Yeung 1992; Runnegar et al. 1993). Other acute effects include altered mitochondrial membrane permeability, generation of reactive oxygen species, and induction of apoptosis (Fladmark et al. 1999; Humpage and Falconer 1999; Ding et al. 2000; Hooser 2000), likely due to a fatal loss of control of regulatory phosphorylation. Morphological changes have been characterized using histopathologic and electron microscopic techniques in rodents, pigs, sheep, and chickens. The Caruaru hemodialysis incident in Brazil extended these findings to humans (see Jochimsen et al 1998; Pouria et al. 1998; Azevedo et al. 2002; Soares et al. 2006; Yuan et al. 2006).

Bulera et al. (2001) showed by toxicogenomic analysis that a non-lethal i.p. injection of MC-LR in male rats resulted in a distinctive pattern of gene expression change when probed by microarray with 1600 rat genes at 3 and 6 hrs after dosing. The pattern differed from that seen with other well characterized liver toxins. The dose of MC (50 μg/kg) did not cause any detectable histological change. This is consistent with the changes in liver function such as receptor-mediated endocytosis that is seen in hepatocytes at doses of MC-LR that cause no significant cytoskeletal changes (Hamm-Alvarez et al. 1996; Runnegar et al. 1997). Although MCs may affect cellular activities independent of the phosphatase inhibition, the impacts of protein phosphatase inhibition may directly or indirectly account for hepatocyte functional alterations, oxidative stress, apoptosis, and changes in gene expression. Protein phosphorylation status plays a critical role in cellular signaling and hence all aspects of cellular function. In addition a number of these effects are also found in cells treated with other chemically distinct protein phosphatase inhibitors such as okadaic acid (see Gehringer 2004 for a review).

Protein phosphatase inhibition by MC is mediated principally by the ADDA group (Goldberg et al. 1995), although most microcystin variants contain dehydroalanine which can covalently link to a cysteinyl sulphur on the phosphatase (Humpage this volume). Congeners of MC-LR with an intact ADDA will also inhibit protein phosphatases. The covalent binding of N-methyl dehydroalanine to cys is secondary in protein phosphorylation, but is important in the metabolism of MC. GSH competes for binding at the N-methyl dehydroalanine to form the GSH conjugate of MC.

Cylindrospermopsin – cytotoxicity. LD_{50} experiments with CYN indicate a delayed toxicity (2.0 mg/kg, i.p. mouse, after 24 hrs but 0.2 mg/kg after 5 days; Ohtani et al. 1992). The pathology caused by CYN has been studied by light and electron microscopy in male mice dosed i.p. or orally. The liver and kidney were the organs most affected (Falconer et al.1999). Acute CYN poisoning results in lipid accumulation in the liver followed by hepatocellular necrosis (Terao et al. 1994; Seawright et al. 1999). Non-hepatic effects include destruction of the proximal tubules of the kidney (Falconer et al. 1999), as well as cytotoxic and thrombotic effects in other tissues. In addition, CYN extract caused varying lesions in other organs as reported earlier (Hawkins et al. 1985). Sub-chronic oral exposure resulted in mainly hepatic and renal effects (Humpage and Falconer 2003). Effects of poisoning in humans included hepatoenteritis and renal insufficiency (Byth 1980).

Native CYN has been shown to be a potent and irreversible inhibitor of protein synthesis at the translation step (Froscio et al. 2001; Runnegar et al. 2002; Froscio et al. 2003). It has also been shown to inhibit GSH synthesis (Runnegar et al. 1995b). However, cell death and the pathology in CYN-dosed animals cannot be explained fully by protein and GSH synthesis inhibition. Further, there is protection from CYN toxicity by cytochrome P450 inhibitors (Runnegar et al. 1995b; Froscio et al. 2003). The best interpretation is that a CYN-derived metabolite formed by a P450-catalyzed reaction is responsible for increased and/or different toxicity compared to that caused by the parent CYN. This is in agreement with the finding that the LD_{50} is lower (more toxicity) at 7 days than after one day in mice (Hawkins et al. 1985; Ohtani et al. 1992; Froscio et al. 2003). Perhaps the early toxicity is due to P450-dependent activation, and the later toxicity is the result of protein synthesis inhibition or a combination of the two effects.

Chemical intermediates in the synthesis of CYN showed that the sulfate ester is not necessary for toxicity since the CYN-DIOL (lacking the sulfate) inhibited protein synthesis in hepatocytes with the same potency of native CYN. The orientation of the C-7 hydroxyl did not matter as the epimer was toxic with similar potency (Runnegar et al. 2002). More recently it was found that synthetic deoxy-CYN (lacking the hydroxyl group at C-7), like natural deoxy-CYN, retains the toxicity of CYN (Looper et al. 2005). Treatment of CYN with chlorine yielded 5-chloro-CYN and cylindrospermic acid. These degradation compounds were shown to have at least 50-fold less toxicity than intact CYN in a mouse bioassay (Banker et al. 2001).

In contrast to MC, more work needs to be done to determine the mechanism of toxicity of CYN both in vitro and in vivo. At present, neither the structural determinants of CYN toxicity nor the cellular targets of CYN have been elucidated. Effects of cylindrospermopsin on non-hepatic tissues have not been well investigated. Renal effects seemed to be important after chronic low dose exposure (Humpage and Falconer 2003), and thrombotic effects have been described by a number of researchers. Preliminary results from reproductive and teratogenic toxicity assessments indicate that larger studies will be required.

Cylindrospermopsin – genotoxicity. Cylindrospermopsin has been shown to be genotoxic in various *in vitro* assay systems. Concentrations of 1 – 10 µg/ml (2.4 – 24 µM CYN) caused DNA fragmentation and loss of whole chromosomes in the cytokenesis-blocked micronucleus assay using the human white blood cell-line WIL2-NS (Humpage et al. 2000a). However, CHO K1 cells did not show this effect when exposed to 1.2 – 2.4 µM CYN (Fessard and Bernard 2003). To investigate the role of metabolism in this discrepancy, genotoxicity was studied in primary mouse hepatocytes (Humpage et al. 2005). Cylindrospermopsin-induced DNA fragmentation, as detected by the COMET assay, was seen at CYN concentrations as low as 0.05 µM, well below the cytotoxicity EC_{50} of 0.5 µM. Genotoxicity was eliminated by application of the CYP450 inhibitors omeprazole (100 µM) or SKF525A (50 µM) (Humpage et al. 2005), indicating that a CYN metabolite probably induces DNA fragmentation. More recently, a range of standard *in vitro* assays have been used to confirm that DNA fragmentation is the major mechanism leading to genotoxicity (Humpage this volume). Research is needed to identify metabolic pathways and toxic metabolites that produce acute toxicity and genotoxicity. There is considerable evidence for involvement of CYP450s, but the specific isoforms have not been identified. Isoform identification is needed to determine whether toxic biotransformation pathways occur in humans and to assess the applicability of animal data to human risk.

In vivo, DNA fragmentation was described in the livers of mice treated with cylindrospermopsin (Shen et al. 2002). Intra-peritoneal injection of a single dose of 0.2 mg CYN/kg induced an increase in hepatic DNA strand breakage within 6 hours that normalized by 72 hours. Evidence from ^{32}P-post-labeling experiments using the livers of CYN-treated mice also suggested that DNA adducts are formed (Shaw et al. 2000).

Saxitoxins. The saxitoxins are potent voltage-gated sodium channel antagonists, causing numbness, paralysis and death by respiratory arrest. Analogue potency varies greatly, with saxitoxin having an LD_{50} (i.p. mouse) of 10 μg/kg, but saxitoxin C1 being at least 160-fold less toxic (Oshima 1995). Toxicological studies to date have assessed acute exposure effects, as would be expected from shellfish poisoning. Repeated, low-level exposure studies are needed to assess effects from drinking water exposure (Humpage this volume). Research is needed to verify and extend a report that tolerance to PSPs develops (Kuiper-Goodman et al. 1999). Mammalian studies are needed to extend a demonstration of neurodevelopmental disturbances in fish (Lefebvre 2002).

Toxicodynamic studies have been performed for gonyautoxins (GTXs) in cats to assess their toxic effects and intoxication illness in mammals when administered orally (Andrinolo et al. 2002). The primary physiological effect was a dramatic decrease in arterial pressure, which eventually resulted in death. Oral doses of 35 μg/kg of GTX 2/3 epimers and a plasma level of 36 ng/ml were the lethal limits for cats. The specific binding of saxitoxin to brain tissue of mammals (Trainer and Baden 1999; Llewellyn et al. 2004; Cianca et al. 2007) and sub-lethal, short term exposure of dissolved saxitoxin to larval fish (Lefebvre et al. 2004) have also been examined. In rats dosed i.p. with 5 or 10 μg saxitoxin/kg body wt, saxitoxin was bound to sodium channels in all brain regions at ppm levels. Larval fish exposed to dissolved saxitoxin exhibited reductions in sensory-motor function after 48 h and paralysis by 4 days. Larval exposure also resulted in reduced growth and survival of the fish several weeks later.

Nodularins. Nodularins are hepatotoxic cyclic peptides of similar structure to the microcystins, containing 5 amino acids rather than 7 (Rinehart and Namikoshi 1994). ADDA is still present but dehydroalanine is replaced by N-methyl-dehydrobutyrine (Rinehart et al. 1988). The smaller ring size prevents this latter moiety from coordinating with the phosphatase cysteine, preventing nodularin from bind covalently (Lanaras et al. 1991;

Craig et al. 1996; Bagu et al. 1997). However, due to the high affinity of ADDA for the active site, this lack of covalent binding does not affect toxin potency, which is similar to that of microcystin-LR (Ki's are of the order 0.1 – 1.5 nM; (Honkanen et al. 1990; MacKintosh et al. 1990; Honkanen et al. 1991)). The lack of covalent binding may allow nodularin to reach other sites in the cell, a mechanism by which nodularins may act as direct carcinogens (Ohta et al. 1994; Bagu et al. 1997; Humpage this volume).

Anatoxin-a/Homoanatoxin-a. These toxins are nicotinic acetylcholine receptor agonists having a LD_{50} of 200 µg/kg (Carmichael et al. 1979; Carmichael 1994). Residence of these toxins at post-synaptic cholinergic receptors results in nerve depolarization (Swanson et al. 1990; Huby et al. 1991; Swanson et al. 1991; Wonnacott et al. 1991). Typical symptoms in mice are loss of muscle coordination, gasping, convulsions, and death within minutes from respiratory arrest (Carmichael et al. 1979). Dog deaths have been attributed to poisoning by these toxins when the animals have licked their coats after swimming (Codd et al. 1992; Edwards et al. 1992; Falconer and Nicholson, pers. comm.). A single human fatality was attributed to anatoxin-a poisoning by a county coroner after the victim swam in a scum-covered pond (Behm 2003). Although anabaena was identified in stool from the victim and another swimmer who became ill but survived, mass spectrometric analyses for anatoxin-a in blood were inconclusive (Carmichael, pers. comm.). The peaks for anatoxin-a and the indigenous amino acid, phenylalanine, occur in the same portion of the spectrum because these compounds are isobaric and elute similarly in reversed phase liquid chromatography (Furey et al. 2005). Anatoxins have not been linked to human poisoning via drinking water (Humpage this volume).

Anatoxin-a(s). Anatoxin-a(s) is a phosphorylated cyclic N-hydroxyguanine, with a structure and action similar to organophosphate pesticides (Mahmood and Carmichael 1986, 1987; Hyde and Carmichael 1991). It is a potent acetylcholinesterase inhibitor with a LD_{50} (i.p., mouse) of 20 µg/kg. The *in vivo* toxic effects are similar to those of anatoxin-a but with the addition of salivation (hence the "s") and lacrimation (Mahmood and Carmichael 1986, Carmichael 1987, Matsunaga et al. 1989). No human illness has been attributed to this toxin (Humpage this volume).

Research Priority: *In vitro* and *in vivo* studies of chronic and repeated exposures

Subchronic and Chronic Toxicity

Microcystins. Reliable *in vitro* data are lacking because hepatocytes lose the transporters for MC with time in culture; thus, repeated/continuous exposure is not possible. This problem can now be overcome in part by using other cell lines that have been transfected to express the newly identified transporters. The effects of low doses of MC have been studied over a period of 65 hrs (Humpage and Falconer 1999). The authors reported a dose dependent (10-100 pM) balance between proliferation and apoptosis of hepatocytes in culture. Lower microcystin doses appeared to affect cell cycle control by suppressing apoptosis and promoting cell division in polyploid hepatocytes *in vitro* (Humpage and Falconer, 1999; Lankoff et al. 2003). Microcystin has also been shown to inhibit DNA repair (Lankoff et al. 2004; 2006a,b). These effects are consistent with the tumor promotion seen in *in vivo* studies, such as enhancement of the growth of hepatic and colonic pre-cancerous lesions in animal models (Fujiki and Suganuma 1993; Ito et al. 1997a; Humpage et al. 2000b; Sueoka et al. 1997). Microcystin exposure has been linked to human liver and colon cancer incidence (Yu 1995; Fleming et al. 2002; Zhou et al. 2002). Using an intestinal crypt cell line, Zhu et al. (2005) explored the mechanism of tumor promotion that is shared by MC and chemically unrelated protein phosphatase inhibitors such as okadaic acid, and showed that MC transformed these cells. Transformation is an important step in the development of carcinogenesis. These cells also had increased proliferation, and the authors went on to show that MC treatment resulted in increased expression and activity of proteins that control cell cycle signaling such as Akt, JNK and MAPK pathways. This study also provides a mechanistic explanation for the hyperplasia in the colon seen in mice initiated with azoxymethane and then exposed to MC in the drinking water for 212 days (Humpage et al. 2000b). To date, findings for MC do not clearly and unequivocally demonstrate a potential for carcinogenicity. The substantial evidence of tumor promotion by MC has recently led the Agency for Research on Cancer (AIRC) to classify MC-LR as "possibly carcinogenic to humans" (Group 2B). A summary of this evaluation has been published (Grosse et al. 2006), and a full report will be published as an AIRC monograph.

A number of rodent studies (and one pig study) with repeated dosing of MC or bloom extracts containing MC have been reported. Some of these studies have been used to calculate reference doses (tolerable daily intakes), LOAELs and NOAELs. These will be addressed by other groups.

We have some indication of the effect of repeated sublethal dosing: daily i.p. dosing with MC-YM for up to 28 days resulted in mild liver damage (Elleman et al. 1978). Bloom extract given over 1 year orally to mice caused liver changes at the higher doses and some indication of increased incidence of tumors (Falconer et al. 1988). Oral consumption of MC containing cells by pigs for 8 weeks resulted in dose dependent liver toxicity (Falconer et al.1994). Miniosmotic pump continuous administration of MC-LR in male rats for 28 days resulted in dose dependent increases in hepatic oxidative stress, in several serum biochemical toxicity indicators, and in histopathology that correlated with protein phosphatase inhibition (Solter et al. 1998; Guzman and Solter 1999). Oral and i.p. dosing of mice with MYC-LR caused dose-dependent liver lesions, but no evidence of developmental toxicity in pregnant females (Fawell et al. 1999). No toxicity was observed in female mice dosed orally with 20 μg/L of MC-LR over 18 months (Ueno et al. 1999).

Cylindrospermopsin. Oral toxicity of low levels of CYN have been investigated by Humpage and Falconer (2003) in mice over 10-11 weeks. Dosing with a *Cylindrospermopsis* culture extract and with CYN indicated that toxic changes were dose dependent with the kidney being most sensitive to the toxin. In addition, a study by Falconer and Humpage (2001) showed the potential for carcinogenicity of CYN-containing culture extracts. Male mice were observed for 30 weeks after three sublethal doses of *C. raciborskii* extract were given orally over a 6 week period followed by treatment with TPA. At the dose used (equivalent of 2.75 to 8.25 mg CYN), the mice exhibited none of the histological changes seen in acute CYN toxicity. However, this treatment appeared to induce the formation of overt cancers in the mice by 30 weeks. Five of 53 treated mice developed neoplastic lesions whereas none of 27 controls did so. The odds ratio for carcinogenesis following CYN treatment was 6.2, but the 95% confidence interval on this estimate was 0.33 to 117, a statistically insignificant result (p=0.16). Nevertheless, this study presents the clearest evidence yet that cylindrospermopsin is not only genotoxic but may also be carcinogenic, which has implications for low-level repeated exposures to CYN. The potential carcinogenicity of CYN needs to be investigated in a larger study. Because this study used *Cylindrospermopsis* culture material, it is not possible to state that CYN alone was the causative agent. The potential carcinogenicity of CYN and of CYN containing extracts should be investigated in a larger number of animals, as well as in a non-rodent species. Cylindrospermopsin has been placed on the 'candidate list' for full toxicological studies by the USEPA and NIEHS. This can only be accomplished if large amounts of CYN and CYN containing extracts are prepared and made available.

Research Priority: Cyanotoxin Mixtures

An important point to reiterate is that single microcystins or single cylindrospermopsins almost never occur in nature. Instead mixtures of toxins are the norm, so there is a need to understand toxin interactions. This will enable the calibration of studies done using MC-LR and the formulation of regulations to be based on toxicity equivalents rather than quantities of individual compounds.

Crude extracts from blooms have been found to be more toxic than the purified component toxin(s). This is probably due to the presence of other congeners or classes of toxins that enhance toxicity. Similar patterns appear in many studies of chemical ecology (Paul et al. 2001). Other compounds present in minor concentrations may have additive or synergistic effects with the pure compounds of interest. A research need is to characterize the compounds in crude extracts that enhance the toxicity of pure compounds, and to identify whether any compounds in crude extracts can reduce the toxicity of pure compounds. An additional research need is to characterize interactions between cyanotoxins and other bioactive compounds in raw and finished water such as metals or disinfection byproducts.

Charge 3: Identify and prioritize toxicodynamic research needed to improve human health risk assessments.

General

- Determine effects from low concentration exposures starting with *in vitro* studies, proceeding to *in vivo* studies as appropriate.
- In contrast to MC, more work needs to be done to determine the mechanism of toxicity of CYN both *in vitro* and *in vivo,* including the structural determinants of CYN toxicity and the cellular targets of CYN.
- Studies are needed to assess the effects of repeated and chronic exposures to individual cyanotoxins and appropriate mixtures.

Mixtures

Near-term Research Priorities

- Mixtures of toxins are the norm, so we need to understand toxin interactions.
  - Priorities are mixtures of microcystins and cylindrospermopsin
  - Use *a priori* information (regional occurrences) to prioritize candidate mixtures

Long-term Research Priorities

- Congener mixtures, especially effects of mixtures of microcystin analogues
- Determine and characterize compounds in the crude extracts that enhance the toxicity of pure compounds, and identify whether any compounds in crude extracts can reduce the toxicity of pure compounds.

Individual Toxins

- *Microcystins: Near-term Research Priorities*
  - Analog mechanisms (including non-toxic ones) in comparison to MCY-LR
  - Comprehensive studies of teratogenicity and reproductive toxicity
  - Chronic animal studies to assess carcinogenicity (ideally these studies should test cyanobacterial extract that may contain modifiers of toxicity (as would be present in water) and with purified MC

- *Microcystins: Long-term Research Priorities*
  - Epidemiological studies into links with human cancer. This requires a biomarker of low dose exposure for which ELISA may be an option (Hilborn et al. 2005).
  - Long term repeated dosing studies are needed in species other than rodents
  - Evaluate effects of modified MCs (and other cyantoxins) produced as byproducts of water treatment

- *Cylindrospermopsin: Near-term Research Priorities*
  - Genotoxicity mechanism including the role of specific CYP450 isoforms in CYN activation and identification of genotoxic metabolites
  - Identification of biomarkers of exposure and effect
  - *In vivo* carcinogenicity trials when adequate pure toxin is available

- Detailed toxicokinetics, including distribution and binding studies are needed to explain the observed delayed toxicity
- Studies of extra-hepatic effects including renal, thrombotic, immunologic, reproductive, and developmental effects
- Human effects – An opportunity exists for follow-up of exposed humans on Palm Island, Australia

- *Anatoxins and Saxitoxins*
  - Chronic low dose exposures/neural developmental toxicity mechanisms *(near-term research priority)*
  - ATXa(s) – studies comparing ATXa(s) and organophosphate pesticide toxicity *(longer-term research priority)*
- *BMAA: Near-term Research Priority*
  - Further assess the association with neurodegenerative disease (Cox and Sacks 2002)
- *BMAA: Long-term Research Priorities*
  - Mechanism of bioaccumulation.
  - Mechanism of toxicity (glutaminergic excitotoxicity or other effects such as disruption of protein structure/function).
  - Trophic studies to determine routes of human exposure.
- *Lower priority*
  - Nodularins – use microcystin as model
  - Lyngbya toxins

Charge 4: Susceptibility

> Identify and prioritize research needed to improve our understanding of human susceptibility factors for adverse effects from cyanotoxin exposure.

Factors affecting susceptibility to cyanotoxins have generally not been investigated for the majority of the toxins. The elucidation of toxic mechanisms as described above within the toxicokinetics and toxicodynamics charges will provide direction and insights for research on susceptibility. Comparative toxicology studies of microcystins have yielded some findings relevant to susceptibility, as summarized below.

In vivo Differences in Responses to Microcystins

There is evidence that nutritional status, sex, age, and strain of laboratory animals can influence the severity of *in vivo* MC toxicity. Mechanistic explanations for these variations in susceptibility are unknown. Fasted rats were more sensitive to MC toxicity (Miura et al. 1991). The LD_{50} for fasted rats dosed i.p. was half that of fed rats (72 μg/kg and 122 μg/kg, respectively). Several studies have indicated that male mice are more sensitive than females to the effects of repeated doses of MC (Falconer et al. 1988). Age can also influence susceptibility to MC; older male mice were more sensitive to oral doses of MC-LR than young male mice (Ito et al. 1997b). Strain differences were demonstrated when Ito et al. (2000) found that Balb/C mice were more sensitive to MC-LR than ICR mice following 12 weeks of repeated oral dosing. It is reasonable to expect that differences in sensitivity to MCs will also be present in human populations.

Researchers have also compared MC toxicity resulting from different dosing modalities. Liver damage from ileal loop dosing was similar to that observed after intravenous dosing (Stotts et al. 1997a, 1997b; Dahlem et al. 1989). Intracheal administration of MC-LR to male mice was as toxic as i.p. doses, with a delay of 60 min (Ito et al. 2001). Inhalation exposure of BALB/C mice to 260–265 mg MC-LR/m3 for 0.5, 1.0, or 2.0 hours/day for 7 days resulted in degeneration and necrosis of the respiratory epithelium but no liver damage (Benson et al Toxicon 45: 691-698, 2005). The lack of hepatotoxicity was attributed to the deposited dose (2.5 mg microcystin/kg in the high-dose group) being far below the no observable adverse effect level (NOAEL) of 200 mg/kg/day for induction of hepatotoxicity.

Charge 4: Identify and prioritize research needed to improve our understanding of human susceptibility factors for adverse effects from cyanotoxin exposure.

The Harmful Algal Research and Response plan (HARRNESS 2005) also prioritized studies to investigate integrated toxin effects and mechanisms of susceptibility with emphasis on characterizing acute and long term effects of HAB toxins, defining mechanisms of susceptibility (e.g. identifying special risk populations), and integrating laboratory animal model data and wildlife exposure information with human exposures and disease. We concur with these research needs and further delineate research priorities below.

Near-term Research Priorities

- Mechanisms as indicated by toxicokinetics/toxicodynamics work
  - P450/GST profile–susceptibility may be determined by polymorphically expressed metabolizing enzymes
    - Focus on links between CYN metabolism by CYP450s and human susceptibility due to variations in expression of human CYP enzymes
    - MC susceptibility may be similarly affected by variability in GSTs

Long-term Research Priorities

- Race, age, sex, and individual health status
- Nutritional status
- Epidemiological studies with an emphasis on dose/exposure characterization
- Identification of special risk populations through integration of laboratory animal model data and wildlife exposure information with human exposures and disease

Charge 5: Genetics/OMICS of Cyanobacterial Toxin Production

Identify and prioritize genetic/OMICS research needed to improve the prediction of cyanobacterial HABs.

Research Priority: Characterization of genomes, transcriptomes, and proteomes

Whole genome sequencing of *Microcystis, Cylindrospermopsis, Lyngbya,* and *Anabaena* and bioinformatic analyses are of high priority and may be more cost-effective than attempting to isolate individual genes that influence the expression of the toxin gene clusters. Whole genome sequencing produces huge amounts of data in a short time and enables the characterization of biosynthetic pathways and the development of tools, such as microarrays, needed to describe synthesis regulation. As sequencing costs are constantly declining this option is becoming more and more feasible for most projects. The Sivonen lab in Helsinki, Finland, is nearing completion of the whole genome sequencing of a hepatotoxin-producing *Anabaena*, and a microcystin-producing *Microcystis* genome is nearly finished at the Pasteur Institute. However, the present capacity and speed of sequencing in USA-based institutes is such that more genomes can be done relatively quickly and cheaply. Determining more whole genome sequences from these harmful cyanobacteria will allow comparisons between, for example, toxic and non-toxic genomes of the same species, speeding up identification of not only toxin genes but also regulatory genes involved in toxin expression. We propose that priority should be given to determining the whole genomes of a second *Microcystis*, toxic and non-toxic strains of both *Cylindrospermopsis* and *Lyngbya*, and two neurotoxic *Anabaena* strains – one saxitoxin producer and one anatoxin-a producer. If whole-genome sequencing is not possible, the toxin gene clusters should at least be identified, thereby enabling the development of PCR-based assays for potential toxin production. In addition to whole genome sequences, global transcriptional and proteomic analyses would reveal mRNA transcripts and proteins correlated with the expression of cyanobacterial toxins. This information would advance our understanding of how toxin production is regulated by cyanobacteria and in response to the ecophysiologic influences of external forces (e.g., light, nutrient availability, etc.).

The complete biosynthetic gene clusters are known for microcystin and nodularin. This information has allowed for detailed studies on the regulation of cyanotoxin production and the development of molecular tools for detecting particularly harmful species in the environment. These genes can be used as an indicator of potential toxin production, alerting water quality managers to the possibility of a toxic bloom. Reservoirs can be monitored by extracting DNA from whole algal communities, and using a PCR-based assay to test for the presence of a gene fragment involved in coding the microcystin synthetase complex (e.g., Hisbergues et al. 2003). Genetic information is also critical for detecting and understanding the production of other cyanobacterial toxins. Only a partial gene cluster analysis has been derived for cylindrospermopsin synthetase. Identification of the complete genetic sequence for cylindrospermopsin synthetase should be given high priority due to the prevalence and distribution of this toxin. Of lower priority are the gene clusters encoding the synthesis of the anatoxins, saxitoxins, and other paralytic shellfish poisons found in freshwater. It is proposed that a strain of C. raciborskii AWT205 (cylindrospermopsin), Oscillatoria or Anabaena spp. (anatoxin-a and anatoxin-a(s)), and either C. raciborskii T3 or a strain of A. circinalis (saxitoxin) be chosen for genome sequencing. Contaminating bacteria in co-culture with these cyanobacteria should be removed prior to DNA extraction and cloning. In addition, genes that interact with the toxin gene clusters remain to be fully described in *Microcystis*, and are completely unknown for the other organisms. The genomic data are needed to apply proteomic and transcriptomic analyses. Global transcriptional and proteomic analyses can reveal mRNA transcripts and proteins correlated with the expression of cyanobacterial toxins.

If whole genome sequencing is not feasible, an alternative approach is to:

- Use global transcriptional and proteomic analyses to locate mRNA transcripts and proteins correlated with the expression of cyanobacterial toxins;
- Use these data to identify the biosynthetic pathways;
- Mutate the pathways to confirm;
- Identify biosynthetic genes for expression;
- Use this information to develop genetic markers for toxin production;
- Communicate: outreach/education to distribute these tools to local water quality agencies.

Research Priority: Identification, characterization, and confirmation of unknown genes for toxin pathways

As noted above, the major cyanobacterial toxin pathways that have not yet been identified or fully characterized are cylindrospermopsin, saxitoxin, anatoxin-a, and anatoxin-a(s). The proposal of a putative biosynthetic pathway has been a successful approach to identifying toxin gene clusters. The first step usually involves feeding experiments with isotope-labeled precursors for the identification of metabolic intermediates. Once these intermediates have been identified, candidate enzymes can be postulated to catalyze these reactions. The next step usually involves searching for genes that encode those putative enzymes in the producing organisms, via degenerate PCRs targeting of conserved domains in these enzyme families, or by screening genomic libraries with labeled probes for putative genes. The most suitable target genes encode unique enzymes that are not involved in primary metabolic pathways. Because cyanobacterial (and other bacterial) toxin genes are organized into clusters, it is usual that the rest of the enzymes involved in the biosynthesis are located adjacent to the identified candidate gene. Following the identification and sequencing of a candidate gene, "genome walking" techniques such as adaptor-mediated and inverse

PCR can be implemented in order to sequence regions around the candidate gene, thereby revealing whether this candidate gene is part of the entire toxin gene cluster.

Confirmation of the role of a putative toxin gene cluster in toxin production is usually accomplished by creating a mutant strain. Confirmation is attained when deletion of a critical part of the toxin gene cluster abolishes toxin production. Mutation methods include gene deletion, insertional mutagenesis, and point mutation. Candidate organisms for anatoxin-a toxin gene cluster identification and confirmation include *Anabaena flos-aquae*, *Oscillatoria* sp., or *Aphanizomenon* sp. The only known producer of anatoxin-a(s) is an *Anabaena* sp. Knockout confirmation of a saxitoxin putative gene cluster could be conducted in a variety of organisms, including *Anabaena, Cylindrospermopsis*, certain heterotrophic bacteria, and several dinoflagellate species. The putative Cylindrospermopsin gene cluster should be mutated in *Cylindrospermopsis raciborskii* or *Anabaena bergii*. *Nodularia spumigena* is suitable for identification and confirmation of a nodularin synthetase gene.

Research Priority: Expression of biosynthetic genes

Cyanobacteria are notoriously difficult to culture in the absence of contaminating bacteria, and methods for genetic manipulation have been described for only a few species. Therefore, verification of putative toxin biosynthesis genes in cyanobacteria by mutational studies is intrinsically difficult. The only cyanobacterial toxin gene cluster that has been characterized by gene disruption and mutant analysis is that for microcystin in *Microcystis*, *Anabaena*, and *Planktothrix*. An alternative approach to gene disruption and mutant analysis is the cloning, heterologous expression, and biochemical characterization of purified recombinant protein. Several cyanobacterial toxin gene clusters have been identified by screening genomic libraries with probes that target candidate biosynthesis genes. Upon identification of candidate gene clusters, the sub-cloning and heterologous expression of candidate genes has led to the characterization of gene clusters for several cyanobacterial metabolites, such as jamaicamide, lyngbyatoxin A, and barbamide. Heterologous expression is not limited to individual genes. Entire gene clusters may also be transferred to metabolically engineered strains that are capable of complex secondary metabolite production, such as *Escherichia coli* BAP1. This approach was successfully used to express the biosynthesis of the siderophore yersiniabactin and a precursor to the antibiotic erythromycin. Once heterologous expression is established in a host strain, which is amenable to genetic manipulation, mutant analysis can be performed with relative ease. In addition, such a system would be essential for the rational design of metabolic products with enhanced pharmacological activities via combinatorial biosynthesis.

Biosynthetic enzymes in *Cylindrospermopsis raciborskii* T3 have been assayed and there was evidence to indicate differences between the metabolism of STX and the C1+2 toxins and high turnover rates of STX biosynthetic enzymes (Pomati et al. 2004). Saxitoxin metabolic pathways have been studied through examinations of the purine degradation pathway (Pomati et al. 2001).

Research Priority: Gene probe development for strain detection and potential for gene transfer

The development and validation of gene probes for phylogenetic classification and toxin-production potential are needed. Further research is required to validate the currently available toxin gene probes for use on a variety of sample matrices such as water and sediment. Gene probes for paralytic shellfish poisons and the anatoxins have not been developed. Probes are also needed for differentiating cyanobacterial potential to produce the different congeners of microcystin and the saxitoxin-related molecules. The acquisition of toxin gene clusters by non-toxic species and the recognition of new toxins is a constant threat. Continued monitoring is required, using a combination of genetic, microbiological and toxicological approaches.

Charge 5: Identify and prioritize genetic/OMICS research needed to improve the prediction of cyanobacterial HABs.

Research Priorities

- **Genomes, transcriptomes, and proteomes**
  - *Near-term Research Priorities*
    - High priority: achieving the completed sequence of cylindrospermopsin synthetase due to the prevalence and distribution of this toxin
    - There is broad consensus regarding the high priority of including the toxic freshwater cyanobacteria in future genome sequencing projects, particularly *Microcystis, Cylindrospermopsis, Lyngbya,* and *Anabaena*

- *Long-term Research Priorities*
  - Of lower priority are the gene clusters encoding synthesis of anatoxins and saxitoxin and other paralytic shellfish poisons found in freshwater
  - Apply proteomic and transcriptomic analyses for the understanding of the ecophysiology of cyanotoxin production

- **Identification and characterization of unknown genes for toxin pathways**
  - *Near-term Research Priorities*
    - The major cyanobacterial toxin pathways not yet identified or fully characterized are cylindrospermopsin, saxitoxin, anatoxins-a, and -a(s)
    - Identify metabolic precursors using isotope- or radiolabeled feeding experiments
    - Search for genes that encode putative biosynthetic enzymes in the producing organisms
    - Sequence candidate genes and regions around the gene

- **Mutation of gene clusters to confirm role in toxin production**
  - *Near-term Research Priorities*
    - A saxitoxin putative gene cluster could be mutated in any of the toxin-producing organisms where the toxin gene sequences have been found
    - The putative Cylindrospermopsin gene cluster should be mutated in *Cylindrospermopsis raciborskii* or *Anabaena bergii*
    - The nodularin synthetase gene cluster should be mutated in *Nodularia spumigena.*

- **Expression of biosynthetic genes**
  - *Longer-term Research Priorities*
    - Cloning, heterologous expression, and biochemical characterization of purified recombinant protein
    - Screening genomic libraries with probes that target candidate biosynthesis genes
    - Sub-cloning and heterologous expression of candidate genes to characterize toxin-coding gene clusters
    - Once heterologous expression is established in a host strain amenable to genetic manipulation, mutant analysis can be performed

- **Gene probe development for strain detection and potential for gene transfer**
  - *Near-term Research Priority*
    - Further research is required to validate the toxin gene probes
  - *Long-term Research Priorities*
    - Gene probes directed to both phylogenetic and toxigenic markers are also required for the detection of cyanobacteria that produce paralytic shellfish poisons and the anatoxins
    - Develop a series of probes for differentiating cyanobacterial potential to produce the different congeners of microcystin and the saxitoxins

Charge 6: Predictive Model Development

> What additional information is needed to develop predictive models for the production and fate of cyanotoxins?

Physical, chemical, genetic, physiological, and ecological information are all needed to develop models to predict the production of cyanotoxins. At present, the genetic mechanisms that regulate toxin production are mostly uncharacterized, although we do know that toxin production can vary in response to ecological factors, including temperature, light availability, nutrient availability, grazing, and competition (Yin et al. 1997, Kearns and Hunter 2000, Thacker et al. 2005). Mechanistic understanding of the interactions between these modulating factors and the expression of biosynthetic genes and cofactors in the target organisms is needed. The capability to predict the fate of cyanotoxins once produced is dependent on knowledge of toxin transport and reactivity/adsorption with various environmental and water treatment matrices.

Genomic and proteomic analyses would advance our knowledge of the cofactors required for toxin expression, including possible promoters of toxin synthesis. These analyses may also reveal how toxins are transported, stored in the cell, or released to the extracellular environment. The natural roles of many cyanotoxins also remain to be described. Although cyanotoxins are often hypothesized to act as defenses against predation, some herbivores preferentially feed on cyanotoxins (Paul et al. 2001, Camacho and Thacker 2006). The growth of sympatric bacteria and algae can be inhibited by cyanotoxins, but secretions from these potential competitors can also regulate toxin production (Kearns and Hunter 2000).

Research Priority: Characterize environmental, biochemical and trophic factors that regulate toxin gene expression

Previous toxin-regulation studies have yielded disparate results, primarily due to a lack of methodological standardization. Different culture techniques (eg. batch cultures, chemostats, turbidistats, etc.) can have a significant impact on experimental outcomes, as can the different methods used for toxin quantification (eg. toxin versus wet or dry cell weight, cell number, total protein content, chlorophyll content, optical density, etc.). Physiological differences between cyanobacterial strains may also affect experimental results and the interpretation of data. Therefore, when investigating the influence of any given environmental parameter on toxin production, the organism's optimal growth requirements must be taken into account before assigning "high/low" values to experimental variables (eg. light intensity, nutrient concentration etc.).

The characterization of natural and engineered non-toxic mutants may be the key to understanding why cyanobacteria produce secondary metabolites. By comparing and contrasting mutant and wild-type responses to different growth conditions, it may be possible to pinpoint the physiological pathways linked to toxin production. The discovery of complete cyanotoxin biosynthesis-gene clusters (eg. anatoxin, saxitoxin, and cylindrospermopsin), and the development of genetic manipulation techniques, is critical to this area of research. Investigating cyanotoxin production at the genetic level, through methods such as RT-PCR and reporter gene analysis may prove to be more efficient and cost-effective compared to traditional end-product analyses. The positive influence of light on the transcription of microcystin biosynthesis genes has been demonstrated. Similar methods may also be employed to investigate the effects of temperature, stress, nitrogen and phosphorus types and levels, and trace metals in other cyanobacteria following the elucidation of their respective toxin biosynthesis-gene clusters. Ultimately, the provision of genome sequences for the model toxin-producing strains will allow the global analysis of genetic transcription, including the interaction between, and regulation of, both secondary and primary metabolism in these bloom-forming microorganisms.

Research on the regulation of toxin production in cyanobacteria has mainly focused on the transcriptional regulation of toxin genes. However, toxin production may also depend on factors such as the availability of cofactors in the producing cells, as well as the stability and post-translational modification of biosynthetic enzymes. One cofactor essential for all polyketide and non-ribosomal peptide producing enzymes is phosphopantetheine, which acts as a covalent attachment site for activated acyl-

and peptidyl chains. Phosphopantetheine is transferred from coenzyme A to a conserved serine residue of acyl- and peptidyl carrier proteins by phosphopantetheinyltransferase. So far, little effort has been made to characterize cyanobacterial phosphopantetheinyltransferases involved in toxin production. Bacteria have also been shown to affect toxin production by algae. Removal of bacteria from saxitoxin-producing dinoflagellate cultures was shown to reduce their toxicity. Similar effects have been observed with domoic acid-producing diatoms. It is believed that trophic interactions between bacteria and algae may be responsible for this effect on toxin production. Research is needed to describe the constituents and interactions in toxin-production pathways.

Research Priority: Mechanisms of toxin transport

Previous toxin-regulation studies have observed an increase in extracellular microcystins under certain growth conditions. The active export of the toxin, by the ABC transporter McyH, has been hypothesized, but not yet proven. The overexpression, purification and reconstitution of cyanotoxin-associated transporters such as McyH into membrane vesicles may enable us to demonstrate transport *in vitro* with radiolabeled toxins. While bioinformatic and mutational data suggest that McyH is responsible for microcystin transport, it is still unclear whether this enzyme transports the toxin to the extracellular environment or to the thylakoid membranes. Intracellular localization of McyH and other cyanotoxin-associated ABC transporters by *in situ* methods may shed light on this area of research. Techniques for observing the intracellular trafficking of all cyanotoxins, including development of specific toxin and transporter antibodies, as well as cyanobacterial imaging methods, are required.

Research Priority: Determine the ecological/physiological role(s) of cyanotoxins

Without question, toxins provide some functional benefit to cyanobacterial. This statement follows from the observation that toxins, or structurally similar compounds, are produced by most of the surveyed cyanobacterial lineages. Equally compelling is the inference that cyanotoxins may play a critical role in cyanobacterial populations or trophic interactions. What remains in question is what, precisely, is that role. Some cyanobacterial toxins may serve as physiological tools to increase survival in hostile environmental conditions, or to enable the colonization of a novel habitat. Others may play a defensive role and act to prohibit the invasion of other strains or species into an occupied niche, limit the advance of neighboring cells, or discourage grazers and predators. An additional role could be as signaling molecules, mediating cell-cell communication or quorum sensing. It is likely that whatever roles cyanotoxins play, these roles change in accordance with the flux of biotic and abiotic components of the environment.

Characterization of the beneficial roles of toxins in cyanobacterial cells or populations requires delineation of the molecular effects of the specific toxins on the physiology and fitness of the producing microorganisms, populations, and ecology of the various species. Knowledge of the genetic basis for cyanotoxin biosynthesis and regulation will enable a functional genomic approach to the study toxin roles. DNA or protein chips, and gene/protein expression profiling tools in general, could offer valuable data on co-transcription of toxic genes/proteins with other genetic networks in experimental sets chosen to mimic changing environmental conditions. Such conditions could also include the stress induced by other competing microorganisms, predators or grazers. A functional association of toxin biosynthetic genes with cellular and metabolic responses to specific stressors will undoubtedly help in elucidating the toxins molecular role. Protein expression studies could provide further information regarding the activation of metabolic pathways (e.g., phosphorylation of signaling proteins) in association with toxin production, and the functional roles of toxins in activities such as cyanobacterial quorum sensing and perception.

Mutagenesis and gene disruption techniques could confirm functional genomic data, and be crucial for investigating the role of toxins in the physiology of producing cyanobacteria. The function of any given cyanotoxin, as well as the conditions that induce variation in both toxin production levels and toxin structure, could be elucidated by studying toxin gene mutants. Broad physiological parameters (e.g., growth) are needed to understand how mutations in toxic genes affect single-cell phenotypes or the structure of an entire population. Both predators and bacteriophage viruses have been shown to play important roles in the ecology and dynamics of cyanobacterial species. The role of protozoan predation on the natural selection of toxin-producing cyanobacterial species is not clear. To date, possible connections between bacteriophage and the production of cyanobacterial toxins have not been explored. These data are needed to produce theoretical and empirical models of the conditions that favor the proliferation of toxin-producing cyanobacteria.

A clearer understanding of interactions between cyanobacterial physiology and physical, chemical, and biological factors such as light availability, nutrient availability, grazers, and competitors is needed to produce models that successfully predict the production and fate of cyanotoxins. Knowledge of mechanisms regulating the expression of biosynthetic genes and cofactors in the target organisms is needed to characterize these interactions. Although these studies can be conducted using global transcriptional and proteomic assays, greater sensitivity and power would be gained if the biosynthetic pathways were known a priori. Integrating these studies with knowledge of the entire genome sequence would provide additional predictive power.

Charge 6: What additional information is needed to develop predictive models for the production and fate of cyanotoxins?

Genetic, physiological, and ecological information are all needed to develop a model to predict the production of cyanotoxins.

Research Priorities

- **Environmental effects on toxin production**
  - *Near-term Research Priority*
    - Methodological standardization for toxin regulation studies - take into account organism's optimal growth requirements before assigning "high/low" values to experimental variables (eg., light intensity, nutrient concentration etc.)
  - *Long-term Research Priorities*
    - Characterize natural and engineered non-toxic mutants to understand why cyanobacteria produce secondary metabolites
    - Compare and contrast mutant and wild-type responses to different growth conditions to pinpoint physiological pathways linked to toxin production
    - Investigate cyanotoxin production at the genetic level, through methods such as RT-PCR and reporter gene analysis to examine the effects of light, temperature, stress, nitrogen source and level, and trace metals in cyanobacteria following the elucidation of their respective toxin biosynthesis-gene clusters
- **Other factors affecting toxin production**
  - *Near-term Research Priorities*
    - Continue characterization of the transcriptional regulation of toxin genes
    - Genomic and proteomic analyses to describe the cofactors required for toxin expression, including possible promoters of toxin synthesis

  - o Describe the dependence of toxin production on factors such as the availability of cofactors in the producing cells, as well as the stability and post-translational modification of biosynthetic enzymes
  - o Characterize cyanobacterial phosphopantetheinyltransferases involved in toxin production
- *Long-term Research Priority*
  - o Determine effects of trophic interactions between bacteria, bacteriophage viruses, and algae on toxin production

- **Physiology**
  - *Long-term Research Priorities*
    - o Describe the mechanisms of interactions with light availability, nutrient availability, grazers, and competitors by examining the expression of biosynthetic genes and cofactors in the target organisms
    - o Integration of these studies with knowledge of the entire genome sequence would provide additional predictive power.

- **Toxin transport/fate**
  - *Near-term Research Priorities*
    - o Describe toxin adsorption in environmental and water treatment martrices
    - o Use overexpression, purification, and reconstitution of cyanotoxin-associated transporters such as McyH into membrane vesicles to enable us to demonstrate transport *in vitro* with radiolabeled toxin
    - o Intracellular localization of McyH and other cyanotoxin-associated ABC transporters by *in situ* methods
  - *Long-term Research Priorities*
    - o Development of techniques for observing the intracellular trafficking of all cyanotoxins, including development of specific toxin and transporter antibodies, as well as cyanobacterial imaging methods, are required

- **Role of toxins in ecology/physiology of toxic cyanobacteria**
  - *Long-term Research Priorities*
    - o Characterize the roles of cyanobacterial toxins within the producing organisms by describing the molecular function of the specific toxin on the physiology and fitness of the producing microorganism, and on its influence on population structure and ecology of the given species

- Use functional genomic investigations to describe associations of toxin biosynthetic genes with cellular and metabolic responses to specific stressors
- Describe the roles of toxins in activities such as cyanobacterial quorum sensing and perception using protein expression studies to provide crucial information regarding the activation of metabolic pathways (e.g., phosphorylation of signaling proteins) associated with toxin production
- Confirm functional genomic data using mutagenesis and gene disruption techniques to elucidate the roles of toxins in the physiology of producing cyanobacteria
- Determine the role of protozoan predation in the natural selection of toxin-producing cyanobacterial species

Charge 7: Intentional/Accidental Release

> Identify and prioritize research needed to reduce risks from an intentional or accidental release of cyanotoxins.

Several cyanotoxins have the potential to cause significant harm to human populations if intentionally or accidentally released through an environmental contamination event. The National Homeland Security Research Center (NHSRC) must now address critical issues that other programs may not have had to consider previously. Recent security interests have placed new and more pressing demands on the assessment of risks from exposure to biological agents through deliberate contamination. Distribution and possession of saxitoxin is already governed by Schedule 1 of the Chemical Weapons Convention as well as the Select Agent Rule. Therefore, saxitoxin and other cyanotoxins need to be evaluated for their risks to human health as well as their potential for intentional contamination. The high toxicity of several cyanotoxins at low concentrations warrants a thorough risk assessment of the most hazardous cyanotoxins. Research on exposure pathways and modes of delivery is needed to identify the cyanotoxins and the environmental conditions that present the greatest health risks.

Research Priority: Scientifically-sound risk assessments are needed to identify toxins with potential for use as terrorist agents

Data are needed to support scientifically-sound assessments of human-health risks from exposure to many cyanotoxins, particularly the risks posed by mixtures of cyanotoxins. Health risks should be based not only on exposure concentration, but also on route-of-exposure. Cyanotoxicity studies have focused primarily on toxicity resulting from oral ingestion. Much less is known about toxicity resulting from dermal and inhalational exposures. Many critical issues concerning mixtures and routes-of-exposure should be addressed, including bioavailability, interactions between toxins, direct modes of action, and indirect effects through immune system activation.

The risk assessment process involves parallel assessments of the hazard and exposure. However, additional parameters such as feasibility and vulnerability are required to assess risks from malevolent releases of biohazardous agents or materials. In the event of a release of a biological agent, the United States Environmental Protection Agency (USEPA) would respond in accordance with the Homeland Security Presidential Directives 5, 7, 9 and 10. The USEPA has mandated roles in decontamination, water infrastructure protection, and risk assessment. The risk associated with a deliberate exposure situation is one of the drivers for decisions regarding evacuation, decontamination, and eventual re-entry into a site or re-use of a water system. A contaminating event with cyanobacteria and/or cyanotoxins would require several layers of investigation. Contamination with saxitoxin (and potentially other cyanotoxins) would require distinguishing between a natural or intentional contaminating event. Risk-based decisions would need to be made concerning decontamination.

Several existing studies describing the chemical nature and characteristics of the saxitoxins are informative for risk assessments. Degradation and biotransformation of the saxitoxins have been investigated *in vitro* (e.g. Sullivan et al. 1983; Laycock et al. 1995), and in response to temperature (e.g. Lawrence et al. 1994), pH (e.g. Hall and Reichardt 1984; Gago-Martinez et al. 2001), combinations of temperature and pH (e.g. Ghazarossian et al. 1976, Indrasena and Gill 1999, Indrasena and Gill 2000). Saxitoxins in drinking water have been degraded using chlorine, and exposure to the treated solution did not result in increased cancer (Senogles-Derham et al. 2003). Removal of saxitoxins from drinking water using granular activated carbon, ozone, and hydrogen peroxide also has been examined (Orr et al. 2004).

The need to control access to potential terrorist agents such as saxitoxins is evident. However, the classification of toxins as potential terrorist agents causes restrictions to be placed on their distribution, thereby limiting research efforts. Efficient means for supplying sufficient quantities of cyanobacteria and cyanotoxins to researchers are needed because much of the data on which to base sound risk assessments are unavailable. Distribution efficiency can by increased by ensuring that requested quantities closely correspond to research (and commercial) needs. For example, laboratories developing reference standard materials may require larger quantities than research labs conducting physical or biochemical analyses. The regulations regarding handling should be clear and manageable to promote compliance. The most current information regarding registration and handling of select agents can be accessed via internet websites (http://www.opcw.org and http://www.selectagents.gov).

Early detection of cyanobacterial HABs and the presence of cyanotoxins in the environment, whether due to natural or intentional causes, is critical for public health protection. Improved techniques are needed to detect and identify HAB organisms, and to determine the potential for toxin production. Techniques also are needed to identify and quantify cyanotoxins, alone and in mixtures, and in various matrices. Molecular, genetic, and biochemical techniques hold promise for the development of techniques that are field-ready, rapid, reliable, and inexpensive. Homeland Security will benefit from methods that enable the presence of HABs and cyanotoxins in the environment to be classified as intentional or natural events.

Additional issues of importance to Homeland Security include the potential for bioaccumulation of cyanotoxins in plants, the aerosolization of cyanotoxins, and the induction of HABs in urban drinking water sources. Research has indicated the potential for cyanotoxin accumulation in agricultural crops when surface water contaminated with cyanobacteria is used for spray irrigation. Colonies and single cells of *Microcystis aeruginosa* and microcystin were retained by salad lettuce after spray irrigation with water containing microcystin-producing cyanobacteria (Codd et al. 1999). The uptake of toxins (pure as well as from cyanobacterial crude extracts) in different crop plants was also shown by Pflugmacher et al. (2006), Peuthert et al (2007 in press) and Järvenpää et al. (2007). Additional research is needed to characterize the risk posed by contaminated spray irrigation. Research is also needed to characterize the potential for aerosolization of cyanotoxins, the stability of cyanotoxins in various environmental matrices, the bioavailability of cyanotoxins bound to various matrices, the bioavailability of inhaled cyanotoxins, and cyanotoxin dermal transfer coefficients. Research is needed to assess the potential for intentional induction of cyanobacterial HABs in surface waters, and to develop emergency risk management procedures.

Charge 7: Identify and prioritize research needed to reduce risks from an intentional or accidental release of cyanotoxins.

Near-term Research Priorities

- Scientifically-sound risk assessments are needed to identify toxins with potential for use as terrorist agents
  - Use appropriate dose-response data to define "significant quantities" below which there is no threat associated with the shipment and possession of cyanotoxins for research purposes
- Detection methods
  - Advances in monitoring for cyanobacteria and cyanotoxins are needed for early detection to prevent exposure
  - Biochemical and molecular techniques that yield detection signatures of cyanotoxins and cyanobacterial blooms are needed to discriminate natural versus unnatural contamination

Long-term Research Priorities

- Stability and degradation studies
  - Characterize cyanotoxin stability in various matrices, including food crops after contaminated spray irrigation
  - Characterize the physical and chemical parameters that enhance bioaffinity of toxin to various substrates
  - Characterize bioaccumulation and other natural concentration mechanisms
- Exposures other than oral
  - Describe changes in cyanotoxin modes of action based on exposure pathway and the presence of other cyanotoxins
  - Describe cyanotoxin aerosolization potential, stability, and bioavailability
  - Characterize the health risk of cyanotoxin inhalation, alone and in mixtures
  - Determine dermal transfer coefficients for cyanotoxins

Summary of Near-Term Research Needs for Highest Priority Cyanotoxins

Table 3 presents a summary of priority near-term research needs for the most prevalent cyanobacterial toxins. "Needs" boxes marked with a check (✓) indicate that little or no data is currently available for a given area of need. "Exist" boxes left blank indicate that the Workgroup was not aware of existing data or research efforts that address the specified area.

Table 3. Summary of Near-Term Research Needs for Highest Priority Cyanotoxins

Higher Research Priorities for the Given Charges	Current	MCY	CYN	STX	ATX
Protocols for efficient production, certification, and distribution of pure toxins and standards of consistent high quality	Need	Congeners	Deoxy	Lyngbya-related	✓
	Exist	LR	AWQC, NRC-IMB	FDA, NRC-IMB	
Study of toxicity from mixture components	Need	✓	✓	✓	✓
	Exist				
Labeled toxins	Need	✓	✓	Have	✓
	Exist	Laboratory Protocols		FDA,NOAA and IICH	
Toxin antibodies	Need	More	More	More	✓
	Exist	Commercial kits and antibodies	ELISA kits	ELISA kits	
Toxicokinetics -Whole animal (exposure route, distribution, metabolism, excretion) -Human in vitro (tissue, cell, enzymes) -PBPK modeling	Need	More	More	More	More
	Exist	Mice	Some	Rats & Cats	Some
Toxicodynamics	Need	More	More	More	More

Higher Research Priorities for the Given Charges	Current	MCY	CYN	STX	ATX
-In vitro screening for dose response effects to low, repeated, continuous exposures -In vivo screening for dose response effects to low, repeated, continuous exposures -Mixtures	Exist	Some	Some	Some	Some
Susceptibility Relies on knowledge from toxicokinetics and toxicodynamics	Need	More	✓	More	✓
	Exist	Age, gender		Some	
-Genomes or toxin gene clusters -Molecular marker for potential toxicity	Need	Have	Cylindrospermopsis	Lyngbya	Anabaena
	Exist	In progress (M.Anabaena)			
-Genetic regulation of toxin production and fate -Utilize for predictive models	Need	More	✓	✓	✓
	Exist	Some			
-Sound risk assessment to determine IF any of these toxins should be considered an intentional threat -Define significant quantities	Need	✓	✓	✓	✓
	Exist	Lots of data available	Limited data	Limited data	Limited Data

AWQC = Australian Water Quality Centre; FDA = US Food and Drug Administration; NRC-IMB = National Research Council Canada, Institute for Marine Biosciences; NOAA = National Oceanographic and Atmospheric Administration; IICH = International Isotope Clearing House

References

Andrinolo D, Iglesias V, Garcia C, and Lagos N (2002) Toxicokinetics and toxicodynamics of gonyautoxins after an oral toxin dose in cats. Toxicon 40: 699-709.

Azevedo S, Carmichael WW, Jochimsen EM, Rinehart KL, Lau S, Shaw GR, Eaglesham GK (2002) Human intoxication by microcystins during renal dialysis treatment in Caruaru-Brazil. Toxicology 181: 441-446.

Bagu JR, Sykes BD, Craig MM, Holmes CFB (1997) A molecular basis for different interactions of marine toxins with protein phosphatase-1. J-Biol-Chem 8: 5087-5097.

Banker R, Carmeli S, Werman M, Teltsch B, Porat R, Sukenik A (2001) Uracil moiety is required for toxicity of the cyanobacterial hepatotoxin cylindrospermopsin. J Tox Environ Health A 62: 281-288.

Beattie KA, Ressler J, Wiegand C, Krause E, Codd GA, Steinberg CEW, Pflugmacher S (2003) Comparative effects and metabolism of two microcystins and nodularin in the brine shrimp Artemia salina. Aquatic Tox 62: 219-226.

Behm D (2003) Coroner cites algae in teen's death. In Milwaukee Journal Sentinal. Milwaukee.

Benson JM, Hutt JA, Rein K, Boggs SE, Barr EB, Fleming LE (2005) The toxicity of microcystin LR in mice following 7 days of inhalation exposure. Toxicon. 45(6):691-8.

Brooks WP, and Codd GA (1987) Distribution of *Microcystis aeruginosa* peptide toxin and interactions with hepatic microsomes in mice. Pharmacology and Toxicology 60: 187-191.

Bulera SJ, Eddy SM, Ferguson E, Jatkoe TA, Reindel JF, Bleavins MR, De La Iglesia FA. (2001) RNA expression in the early characterization of hepatotoxicants in Wistar rats by high-density DNA microarrays. Hepatology. 33(5):1239-58.

Byth S (1980) Palm Island Mystery Disease. Medical Journal of Australia 2: 40-42.

Camacho FA, Thacker RW (2006) Amphipod herbivory on the freshwater cyanobacterium *Lyngbya wollei*: chemical stimulants and morphological defenses. Limnology and Oceanography 51: 1870-1875.

Carmichael WW (1994) The toxins of cyanobacteria. Sci Am 270: 78-86.

Carmichael WW, Biggs DF, Peterson MA (1979) Pharmacology of anatoxin-a produced by the freshwater cyanophyte *Anabaena flos-aquae* NRC-44-1. Toxicon 17: 229-236.

Cianca RC, Pallares MA, Barbosa RD, Adan LV, Martins JM, Gago-Martinez A (2007) Application of precolumn oxidation HPLC method with fluorescence detection to evaluate saxitoxin levels in discrete brain regions of rats. Toxicon 49(1): 89-99.

Codd GA, Edwards C, Beattie KA, Barr WM, Gunn GJ (1992) Fatal attraction to cyanobacteria? *Nature* 359: 110-111.

Codd GA, Metcalf JS, Beattie KA (1999) Retention of Microcystis aeruginosa and microcystin by salad lettuce (Lactuca sativa) after spray irrigation with water containing cyanobacteria. Toxicon. 37: 1181-1185.

Cox PA, Sacks OW (2002) Cycad neurotoxins, consumption of flying foxes, and ALS-PDC disease in Guam. Neurology 58: 956-959.

Craig M, Luu HA, McCready TL, Williams D, Andersen RJ, Holmes CFB (1996) Molecular mechanisms underlying the interaction of motuporin and microycystins with type-1 and type-2a protein phosphatases. Biochemistry & Cell Biology 74: 569-578.

Dahlem AM, Hassan AS, Swanson SP, Carmichael WW, Beasley VR. (1989) A model system for studying the bioavailability of intestinally administered microcystin-LR, a hepatotoxic peptide from the cyanobacterium Microcystis aeruginosa. Pharmacol Toxicol. 64(2):177-81.

Dill K, Montgomery DD, Wang W, Tsai JC (2001) Antigen detection using microelectrode array microchips. Analytica Chimica Acta 444: 69-78.

Ding, WX, Shen, HM, and Ong, CN (2000) Critical role of reactive oxygen species and mitochondrial permeability transition in microcystin-induced rapid apoptosis in rat hepatocytes. Hepatology 32: 547-555.

Duncan MW, Markey SP, Weick BG, Pearson PG, Ziffer H, Hu Y, and Kopin IJ (1992) 2-Amino-3-(methylamino)propanoic acid (BMAA) bioavailability in the primate. Neurobiol Aging 13: 333-337.

Duncan MW, Villacreses NE, Pearson PG, Wyatt L, Rapoport SI, Kopin IJ et al. (1991) 2-amino-3-(methylamino)-propanoic acid (BMAA) pharmacokinetics and blood-brain barrier permeability in the rat. J Pharmacol Exp Ther 258: 27-35.

Elleman TC, Falconer IR, Jackson AR, Runnegar MT (1978) Isolation, characterization and pathology of the toxin from a Microcystis aeruginosa (= Anacystis cyanea) bloom. Aust J Biol Sci 31(3): 209-18.

Edwards C, Beattie KA, Scrimgeour CM, Codd GA (1992) Identification of anatoxin-a in benthic cyanobacteria (blue-green algae) and in associated dog poisonings at Loch Insh, Scotland. Toxicon 30: 1165-1175.

Falconer IR, Buckley T, Runnegar MT (1986) Biological half-life, organ distribution and excretion of 125-I-labelled toxic peptide from the blue-green alga Microcystis aeruginosa. Aust J Biol Sci 39: 17-21.

Falconer IR, Smith JV, Jackson AR, Jones A, Runnegar MT. (1988) Oral toxicity of a bloom of the Cyanobacterium microcystis Aeruginosa administered to mice over periods up to 1 year. J Toxicol Environ Health 24(3): 291-305.

Falconer IR, Yeung DSK (1992) Cytoskeletal changes in hepatocytes induced by *Microcystis* toxins and their relation to hyperphosphorylation of cell proteins. Chem-Biol-Interact 81: 181-196.

Falconer IR, Burch MD, Steffensen DA, Choice M, Coverdale OR (1994) Toxicity of the blue-green alga (cyanobacterium) *Microcystis aeruginosa* in drinking water to growing pigs, as an animal model for human injury and risk assessment. Env Toxicol Water Qual 9:131-139.

Falconer IR, Humpage AR (2001) Preliminary evidence for in vivo tumour initiation by oral administration of extracts of the blue-green alga *Cylindrospermopsis raciborskii* containing the toxin cylindrospermopsin. Environ Toxicol 16: 192-195.

Falconer IR, Hardy SJ, Humpage AR, Froscio SM, Tozer GJ, Hawkins PR (1999) Hepatic and renal toxicity of the blue-green alga (cyanobacterium) *Cylindrospermopsis raciborskii* in male Swiss Albino mice. Environmental Toxicology 14: 143-150.

Fawell JK, Mitchell RE, Everett DJ, Hill RE (1999) The toxicity of cyanobacterial toxins in the mouse: I microcystin-LR. Hum Exp Toxicol 18(3): 162-7.

Fergusson KM, Saint CP (2003) Multiplex PCR assay for Cylindrospermopsis raciborskii and cylindrospermopsin-producing cyanobacteria. Environ Toxicol 18(2):120-125.

Fessard V, Bernard C (2003) Cell alterations but no DNA strand breaks induced in vitro by cylindrospermopsin in CHO K1 cells. Environ Toxicol 18(5): 353-9.

Fischer WJ, Altheimer S, Cattori V, Meier PJ, Dietrich DR, Hagenbuch B (2005) Organic anion transporting polypeptides expressed in liver and brain mediate uptake of microcystin. Toxicol Appl Pharmacol. 203(3):257-63.

Fischer WJ, Hitzfeld BC, Tencalla F, Eriksson JE, Mikhailov A, Dietrich DR. (2000) Microcystin-LR toxicodynamics, induced pathology, and immunohistochemical localization in livers of blue-green algae exposed rainbow trout (oncorhynchus mykiss). Toxicol Sci 54(2): 365-73.

Fladmark, KE, Brustugun, OT, Boe, R, Vintermyr, OK, Howland, R, Gjertsen, BT (1999) Ultrarapid caspase-3 dependent apoptosis induction by serine/threonine phosphatase inhibitors. Cell Death and Differentiation 6: 1099-1108.

Fleming LE, Rivero C, Burns J, Williams C. Bean JA, Shea KA, Stinn J (2002) Blue-green algal (cyanobacterial) toxins, surface drinking water, and liver cancer in Florida. Harmful Algae 1: 157-168.

Froscio SM, Humpage AR, Burcham PC, Falconer, IR (2001) Cell-free protein synthesis inhibition assay for the cyanobacterial toxin cylindrospermopsin. Environ Toxicol 16: 408-412.

Froscio SM, Humpage AR, Burcham PC, Falconer IR (2003) Cylindrospermopsin-induced protein synthesis inhibition and its dissociation from acute toxicity in mouse hepatocytes. Environmental Toxicology 18: 243-251.

Fujiki H, Suganuma M (1993) Tumor promotion by inhibitors of protein phosphatases 1 and 2A: The okadaic acid class of compounds. Advances in Cancer Research 61: 143-194.

Furey A, Crowley J, Hamilton B, Lehane M, James KJ (2005) Strategies to avoid the mis-identification of anatoxin-a using mass spectrometry in the forensic investigation of acute neurotoxic poisoning. J Chromatogr A 1082: 91-97.

Gago-Martinez A, Moscoso SA, Leao Martins JM, Rodriguez Vazquez JA, Niedzwiadek B, Lawrence JF. (2001) Effect of pH on the oxidation of paralytic shellfish poisoning toxins for analysis by liquid chromatography. J Chromatogr A. 905(1-2):351-7.

Gehringer MM (2004) Microcystin-LR and okadaic acid-induced cellular effects: a dualistic response. FEBS Lett 557(1-3): 1-8.

Gessner BD, Bell P, Doucette GJ, Moczydlowski E, Poli MA, Van Dolah F, Hall S. (1997) Hypertension and identification of toxin in human urine and serum following a cluster of mussel-associated paralytic shellfish poisoning outbreaks. Toxicon 35(5): 711-22.

Ghazarossian VE, Schantz EJ, Schnoes HK, Strong FM (1976) A biologically active acid hydrolysis product of saxitoxin. Biochem Biophys Res Commun. 68(3):776-80.

Goldberg J, Huang H, Kwon Y, Greengard P, Nairn AC, Kuriyan, J (1995) Three-dimensional structure of the catalytic subunit of protein serine/threonine phosphatase-1. Nature 376: 745-753.

Grosse Y, Baan R, Straif K, Secretan B, El Ghissassi F, Cogliano V, Cantor KP, Falconer IR, Levallois P, Verger P, Chorus I, Fujiki H, Ohshima H, Shibutani M, Lankoff A, Agudo A, Chan PC, Fan A, Karagas M, Mirvish S, Searles Nielsen S, Runnegar M, Ward MH, Wishnok J, Dietrich D, Junghans T, De Rosa C (2006) Carcinogenicity of nitrate, nitrite, and cyanobacterial peptide toxins. Lancet Oncology 7: 628-629.

Gubbins MJ, Eddy FB, Gallacher S, Stagg RM (2000) Paralytic shellfish poisoning toxins induce xenobiotic metabolising enzymes in Atlantic salmon (Salmo salar). Mar Environ Res 50(1-5): 479-83.

Guzman RE, Solter PF (1999) Hepatic oxidative stress following prolonged sublethal microcystin LR exposure. Toxicol Pathol 27(5): 582-8.

Guzman RE, Solter PF, Runnegar MT (2003) Inhibition of nuclear protein phosphatase activity in mouse hepatocytes by the cyanobacterial toxin microcystin-LR. Toxicon 41(7): 773-81.

Hall S, Reichardt, PB (1984) Cryptic paralytic shellfish toxins. *In* E. P. Ragelis (Ed) Seafood Toxins, pp. 113-124. Washington, DC: American Chemical Society.

Hamm-Alvarez SF, Wei X, Berndt N, Runnegar M (1996) Protein phosphatases independently regulate vesicle movement and microtubule subpopulations in hepatocytes. Am J Physiol 271(3 Pt 1): C929-43.

HARRNESS (2005) Harmful Algal Research and Response: A National Environmental Science Strategy 2005-2015. Ramsdell, J. S., D. M. Anderson, and P. M. Glibert (Eds.), Ecological Society of America, Washington, DC, 96 pp.

Hashmi M, and Anders MW (1991) Enzymatic reaction of beta-N-methylaminoalanine with L-amino acid oxidase. Biochim Biophys Acta 1074: 36-39.

Hawkins PR, Runnegar MTC, Jackson ARB, Falconer IR (1985) Severe hepatotoxicity caused by the tropical cyanobacterium (blue-green alga) *Cylindrospermopsis raciborskii* (Woloszynska) Seenaya and Subba Raju isolated form a domestic supply reservoir. Applied and Environmental Microbiology 50: 1292-1295.

Hilborn ED, Carmichael WW, Yuan M, Azevedo SMFO (2005) A simple colorimetric method to detect biological evidence of human exposure to microcystins. Toxicon 46: 218-221.

Hines HB, Naseem SM, Wannemacher RW Jr. (1993) [3H]-saxitoxinol metabolism and elimination in the rat. Toxicon 31(7): 905-8.

Hisbergues M, Christiansen G, Rouhiainen L, Sivonen K, Borner T. (2003) PCR-based identification of microcystin-producing genotypes of different cyanobacterial genera. Arch Microbiol. 180(6):402-10.

Honkanen RE, Dukelow M, Zwiller J, Moore RE, Khatra BS, Boynton AL (1991) Cyanobacterial nodularin is a potent inhibitor of type 1 and type 2a protein phosphatases. Molec. Pharmacol. 40: 577-583.

Honkanen RE, Zwiller J, Moore RE, Daily SL, Khatra BS, Dukelow M, Boynton, AL (1990) Characterization of microcystin-LR, a potent inhibitor of type 1 and type 2a protein phosphatases. J-Biol-Chem 265: 19401-19404.

Hooser, SB (2000) Fulminant hepatocyte apoptosis in vivo following microcystin-LR administration to rats. Toxicol Pathol 28: 726-733.

Huby NJS, Thompson P, Wonnacott S, Gallagher T (1991) Structural modification of anatoxin-a. Synthesis of model affinity ligands for the nicotinic acetylcholine receptor. *Journal of the Chemical Society, Chemical Communications* **4**: 243-245.

Humpage AR (this volume) Toxin types, toxicokinetics and toxicodynamics.

Humpage AR, Falconer IR (1999) Microcystin-LR and liver tumour promotion: Effects on cytokinesis, ploidy and apoptosis in cultured hepatocytes. Environmental Toxicology 14: 61-75.

Humpage AR, Falconer IR (2003) Oral toxicity of the cyanobacterial toxin cylindrospermopsin in male Swiss albino mice: Determination of no observed adverse effect level for deriving a drinking water guideline value. Environ Toxicol 18: 94-103.

Humpage AR, Fenech M, Thomas P, Falconer IR (2000a) Micronucleus induction and chromosome loss in transformed human white cells indicate clastogenic and aneugenic action of the cyanobacterial toxin, cylindrospermopsin. Mutat Res 472: 155-161.

Humpage AR, Hardy SJ, Moore EJ, Froscio SM, Falconer IR (2000b) Microcystins (cyanobacterial toxins) in drinking water enhance the growth of aberrant crypt foci in the mouse colon. Journal of Toxicology & Environmental Health. Part A 61: 155-165.

Humpage AR, Fontaine F, Froscio S, Burcham P, and Falconer IR (2005) Cylindrospermopsin genotoxicity and cytotoxicity: Role of cytochrome P-450 and oxidative stress. Journal of Toxicology and Environmental Health-Part a-Current Issues 68: 739-753.

Hyde EG, Carmichael WW (1991) Anatoxin-a(s), a naturally occurring organophosphate, is an irreversible active site-directed inhibitor of acetylcholinesterase (EC 3.1.1.7). Journal of Biochemical Toxicology 6, 3: 195-201.

Indrasena WM, Gill TA (1999) Thermal degradation of paralytic shellfish poisoning toxins in scallop digestive glands. Food Research International 32: 49-57.

Indrasena WM, Gill TA (2000) Storage stability of paralytic shellfish poisoning toxins. Food Chemistry 71: 71-77.

Ito E, Takai A, Kondo F, Masui H, Imanishi S, Harada K (2002) Comparison of protein phosphatase inhibitory activity and apparent toxicity of microcystins and related compounds. Toxicon. 40(7):1017-25.

Ito E, Kondo F, Harada K (2001) Intratracheal administration of microcystin-LR, and its distribution.
Toxicon 39(2-3):265-71.

Ito E, Kondo F, Harada K (2000) First report on the distribution of orally administered microcystin-LR in mouse tissue using an immunostaining method. Toxicon 38(1): 37-48.

Ito E, Kondo F, Terao K, Harada KI (1997a) Neoplastic nodular formation in mouse liver induced by repeated intraperitoneal injections of microcystin-LR. Toxicon 35: 1453-1457.

Ito E, Kondo F, Harada K (1997b) Hepatic necrosis in aged mice by oral administration of microcystin-LR.
Toxicon. 1997 Feb;35(2):231-9.

Jalaludin B, and Smith W (1992) Blue-green algae (cyanobacteria). Medical Journal of Australia 156: 744.

Järvenpää S-, Lundberg-Niinistö C, Spoof L, Sjövall O, Tyystjärvi E, Meriluoto J (2007) Effects of microcystins on broccoli and mustard, and analysis of accumulated toxin by liquid chromatography-mass spectrometry. Toxicon (in press).

Jochimsen EM, Carmichael WW, An JS, Cardo DM, Cookson ST, Holmes CE, Antunes MB, de Melo Filho DA, Lyra TM, Barreto VS, Azevedo SM, Jarvis WR (1998) Liver failure and death after exposure to microcystins at a hemodialysis center in Brazil. N Engl J Med 338: 873-876.

Karlsson K, Sipiä V, Krause E, Meriluoto J, Pflugmacher S (2003) Mass spectrometric detection and quantification of nodularin-R in flounder livers. Environ Tox 18: 284-288.

Kearns KD, Hunter MD (2000) Green algal extracellular products regulate antialgal toxin production in a cyanobacterium. Environ Microbiol. 2(3):291-7.

Kondo F, Ikai Y, Oka H, Okumura M, Ishikawa N, Harada K, Matsuura K, Murata H, Suzuki M (1992) Formation, characterization, and toxicity of the glutathione and cysteine conjugates of toxic heptapeptide microcystins. Chem Res Toxicol. 5(5):591-6.

Kondo F, Matsumoto H, Yamada S, Ishikawa N, Ito E, Nagata S (1996) Detection and identification of metabolites of microcystins formed in vivo in mouse and rat livers. Chem-Res-Toxicol 9: 1355-1359. N Engl J Med 338(13): 873-8. Erratum in: N Engl J Med 339(2): 139.

Kuiper-Goodman T, Falconer I, Fitzgerald J (1999) Human health aspects. In Toxic Cyanobacteria In Water. A Guide To Their Public Health Consequences, Monitoring and Management. Chorus, I., and Bartram, J. (eds). London: E & FN Spon on behalf of WHO, pp. 113-153.

Lanaras T, Cook CM, Eriksson J, Meriluoto J, Hotokka, M (1991) Computer modelling of the 3-dimensional structures of the cyanobacterial hepatotoxins microcystin-LR and nodularin. Toxicon 29, 7: 901-906.

Lankoff A, Banasik A, Deperas M, Kuźminski K, Tarczyńska M, Jurczak T, Wojcik A (2003). Effect of microcystin - LR on cell cycle progression, mitotic spindle and apoptosis in CHO-K1 cells. Toxicol Applied Pharmacol 189: 204-213.

Lankoff A, Krzowski L, Glab J, Banasik A, Lisowska H, Kuszewski T, Gozdz S, and Wojcik A (2004). DNA damage and repair in human peripheral blood lymphocytes following treatment with microcystin-LR. Mutation Res. 559: 131-142.

Lankoff A, Bialczyk J, Dziga D, Carmichael WW, Lisowska H, Wojcik A (2006a). Inhibition of nucleotide excision repair (NER) by the PP1 and PP2A inhibitor-microcystin-LR in UV-irradiated CHO-K1 cells. Toxicon, 48: 957-965.

Lankoff A, Bialczyk J, Dziga D, W. Carmichael WW, Gradzka I, Lisowska H, Kuszewski T, Gozdz S, Piorun I, Wojcik A (2006b) The repair of gamma-radiation-induced DNA damage is inhibited by microcystin-LR, the PP1 and PP2A phosphatase inhibitor Mutagenesis 21: 83 - 90.

Lawrence JF, Maher M, Watson-Wright W (1994) Effect of cooking on the concentration of toxins associated with paralytic shellfish poison in lobster hepatopancreas. Toxicon. 32(1):57-64.

Laycock MV, Kralovec J, Richards R (1995) Some in vitro chemical conversions of paralytic shellfish poisoning (PSP) toxins useful in the preparation of analytical standards. J. Mar. Biotech. 3: 121-125.

Lefebvre KA (2002) Sublethal effects of saxitoxin on early development and behavioral performance in fish. In Xth International Conference on Harmful Algae. St. Pete Beach, Florida, US.

Lefebvre KA, Trainer VL, Scholz NL (2004) Morphological abnormalities and sensorimotor deficits in larval fish exposed to dissolved saxitoxin. Aquat Toxicol 66(2): 159-70.

Liu BH, Yu FY, Huang X, Chu FS (2000) Monitoring of microcystin-protein phosphatase adduct formation with immunochemical methods. Toxicon. 38:619-32.

Llewellyn LE, Dodd MJ, Robertson A, Ericson G, do Koning C, and Negri AP (2002) Post-mortem analysis of samples from a human victim of a fatal poisoning caused by the xanthid crab, Zosimus aeneus. Toxicon 40(10): 1463-1469.

Llewellyn L, Negri A, Quilliam M (2004) High affinity for the rat brain sodium channel of newly discovered hydroxybenzoate saxitoxin analogues from the dinoflagellate Gymnodinium catenatum.
Toxicon 43 (1): 101-4.

Looper RE, Runnegar MTC, Williams, RM (2005) Synthesis of the putative structure of 7-deoxycylindrospermopsin: C7 oxygenation is not required for the inhibition of protein synthesis. Angewandte Chemie-International Edition 44: 3879-3881.

Love AH (this volume) Determining important parameters related to cyanobacterial alkaloid toxin exposure.

MacKintosh C, Beattie KA, Klumpp S, Cohen P, Codd GA (1990) Cyanobacterial microcystin-LR is a potent and specific inhibitor of protein phosphatases 1 and 2A from both mammals and higher plants. FEBS-Lett 264: 187-192.

Mahmood NA, Carmichael WW (1986) The pharmacology of anatoxin-a(s), a neurotoxin produced by the freshwater cyanobacterium *Anabaena flos-aquae* NRC 525-17. Toxicon 24, 5: 425-434.

Mahmood NA, Carmichael WW (1987) Anatoxin-a(s), an anticholinesterase from the cyanobacterium *Anabaena flos-aquae* NRC-525-17. Toxicon 25: 1221-1227.

Matsunaga S, Moore RE, Niemczura WP, Carmichael WW (1989) Anatoxin-a(s), a potent anticholinesterase from *Anabaena flos-aquae*. J Amer Chem Society 111: 8021-8023.

Metcalf JS, Beattie KA, Ressler J, Gerbersdorf S, Pflugmacher S, Codd GA (2002) Cross-reactivity and performance assessment of four microcystin immunoassays with detoxication products of the cyanobacterial toxin, microcystin-LR. J Water Supply Res Technol-Aqua 51: 145-151.

Metcalf JS, Meriluoto JA and Codd GA (2006) Legal and security requirements for the air transportation of cyanotoxins and toxigenic cyanobacterial cells for legitimate research and analytical purposes. Toxicol Lett. 163:85-90.

Meriluoto, J and Codd GA, eds. (2005) Toxic Cyanobacterial Monitoring and Cyanobacterial Analysis. Abo , Finland : Abo Akademi University Press.

Meriluoto JA, Nygard SE, Dahlem AM, Eriksson JE (1990) Synthesis, organotropism and hepatocellular uptake of two tritium-labeled epimers of dihydromicrocystin-LR, a cyanobacterial peptide toxin analog. Toxicon 28(12): 1439-46

Miura GA, Robinson NA, Lawrence WB, Pace JG (1991) Hepatotoxicity of microcystin-LR in fed and fasted rats. Toxicon 29(3): 337-46.

Moore RE. (1996) Cyclic Peptides and depsipeptides from cyanobacteria: a review. J Industrial Microbiol Biotechnol 16(2):134-143.

Naseem SM (1996) Toxicokinetics of [3H]saxitoxinol in peripheral and central nervous system of rats. Toxicol Appl Pharmacol 141(1): 49-58.

Neilan BA (this volume) The genetics and genomics of cyanobacterial toxicity.

Norris RL, Seawright AA, Shaw GR, Smith MJ, Chiswell RK, and Moore MR (2001) Distribution of 14C cylindrospermopsin in vivo in the mouse. Environ Toxicol 16: 498-505.

Norris RL, Seawright AA, Shaw GR, Senogles P, Eaglesham GK, Smith MJ, Chiswell RK, Moore MR. (2002) Hepatic xenobiotic metabolism of cylindrospermopsin in vivo in the mouse. Toxicon. 40(4):471-6.

Ohta T, Sueoka E, Iida N, Komori A, Suganuma M, Nishiwaki R (1994) Nodularin, a potent inhibitor of protein phosphatases 1 and 2A, is a new environmental carcinogen in male F344 rat liver. Cancer Res 54: 6402-6406.

Ohtani I, Moore RE, Runnegar MTC (1992) Cylindrospermopsin: A potent hepatotoxin from the blue-green alga *Cylindrospermopsis raciborskii*. J Amer Chem Society 114: 7941-7942.

Orr PT, Jones GJ, Hamilton GR (2004) Removal of saxitoxins from drinking water by granular activated carbon, ozone and hydrogen peroxide--implications for compliance with the Australian drinking water guidelines. Water Res. 38(20):4455-61.

Oshima Y (1995) Postcolumn derivatization liquid chromatographic method for paralytic shellfish toxins. Journal of AOAC International 78: 528-532.

Paul VJ, Cruz-Rivera E, and Thacker RW (2001) Chemical mediation of macroalgal-herbivore interactions: ecological and evolutionary perspectives. In: Marine Chemical Ecology, McClintock JB & Baker BJ, eds., pp. 227-265.

Peuthert A, Chakrabarti S, Pflugmacher S (2007) Uptake of microcystins-LR and –LF (cyanobacterial toxins) in seedlings of several important agricultural plant species and the correlation with cellular damage (lipid peroxidation). Environmental Toxicology (in press)

Pietsch C, Wiegand C, Ame MV, Nicklisch A, Wunderlin D, Pflugmacher S (2001) The effects of a cyanobacterial crude extract on different aquatic organisms: Evidence for cyanobacterial toxin modulating factors. Environ Tox 16: 535-542.

Pflugmacher S, Wiegand C, Oberemm A, Beattie KA, Krause E, Codd GA, Steinberg CE (1998) Identification of an enzymatically formed glutathione conjugate of the cyanobacterial hepatotoxin microcystin-LR: the first step of detoxication. Biochim Biophys Acta 1425(3): 527-33.

Pflugmacher, S (2002) Possible allelopathic effects of cyanotoxins, with reference to microcystin-LR, in aquatic ecosystems. Environ Tox 17: 407-413.

Pflugmacher S, Wiegand C, Beattie KA, Krause E, Steinberg CEW, Codd GA (2001) Uptake, effects, and metabolism of cyanobacterial toxins in the emergent reed plant Phragmites australis (cav.) trin. ex steud. Environ Toxicol Chem 20: 846-852.

Pflugmacher S, Jung K, Lundvall L, Neumann S, Peuthert A (2006) Effects of cyanobacterial toxins and cyanobacterial cell-free crude extract on germination of Alfalfa (*Medicago sativa*) and induction of oxidative stress. Environ Toxicol Chem 25: 2381-2387.

Pomati F, Manarolla G, Rossi O, Vigetti D, Rossetti C (2001) The purine degradation pathway: possible role in paralytic shellfish toxin metabolism in the cyanobacterium Planktothrix sp. FP1.
Environ Int. 27(6):463-70.

Pomati F, Moffitt MC, Cavaliere R, Neilan BA (2004) Evidence for differences in the metabolism of saxitoxin and C1+2 toxins in the freshwater cyanobacterium Cylindrospermopsis raciborskii T3.
Biochim Biophys Acta. 1674(1): 60-7.

Pouria S, de Andrade A, Barbosa J, Cavalcanti RL, Barreto VT, Ward CJ, Preiser W, Poon GK, Neild GH, Codd GA. (1998) Fatal microcystin intoxication in haemodialysis unit in Caruaru, Brazil. Lancet 352(9121): 21-6.

Rinehart KL, Namikoshi M (1994) Structure and biosynthesis of toxins from blue-green algae (cyanobacteria). Journal of Applied Phycology 6: 159-176.

Rinehart KL, Harada K, Namikoshi M, Chen C, Harvis CA (1988) Nodularin, microcystin and the configuration of ADDA. Journal of the American Chemical Society 110: 8557-8558.

Robinson NA, Miura GA, Matson CF, Dinterman RE, and Pace JG (1989) Characterization of chemically tritiated microcystin-LR and its distribution in mice. Toxicon 27: 1035-1042.

Robinson NA, Pace JG, Matson CF, Miura, GA, Lawrence WB (1991) Tissue distribution, excretion and hepatic biotransformation of microcystin-LR in mice. J Pharmacol Exp Ther 256: 176-182.

Runnegar MT, Falconer IR (1986) Effect of toxin from the cyanobacterium *Microcystis aeruginosa* on ultrastructural morphology and actin polymerization in isolated hepatocytes. Toxicon 24: 109-115.

Runnegar MT, Kong S, and Berndt N (1993) Protein phosphatase inhibition and in vivo hepatotoxicity of microcystins. Am J Physiol 265: G224-230.

Runnegar MT, Falconer IR, Buckley T, and Jackson AR (1986) Lethal potency and tissue distribution of 125I-labelled toxic peptides from the blue-green alga *Microcystis aeruginosa*. Toxicon 24: 506-509.

Runnegar MT, Maddatu T, Deleve LD, Berndt N, Govindarajan S (1995a) Differential toxicity of the protein phosphatase inhibitors microcystin and calyculin A. J Pharmacol Exp Ther. 273: 545-53.

Runnegar MT, Kong SM, Zhong YZ, and Lu SC (1995b) Inhibition of reduced glutathione synthesis by cyanobacterial alkaloid cylindrospermopsin in cultured rat hepatocytes. Biochem Pharmacol 49: 219-225.

Runnegar MT, Kong SM, Zhong YZ, Ge JL, and Lu SC (1994) The role of glutathione in the toxicity of a novel cyanobacterial alkaloid cylindrospermopsin in cultured rat hepatocytes. Biochem Biophys Res Commun 201: 235-241.

Runnegar M, Wei X, Berndt N, Hamm-Alvarez SF (1997) Transferrin receptor recycling in rat hepatocytes is regulated by protein phosphatase 2A, possibly through effects on microtubule-dependent transport. Hepatology 26(1): 176-85.

Runnegar MT, Xie CY, Snider BB, Wallace, GA, Weinreb, SM, Kuhlenkamp J (2002) In vitro hepatotoxicity of the cyanobacterial alkaloid cylindrospermopsin and related synthetic analogues. Toxicological Sciences 67: 81-87.

Seawright AA, Brown AW, Nolan CC, and Cavanagh JB (1990) Selective degeneration of cerebellar cortical neurons caused by cycad neurotoxin, L-beta-methylaminoalanine (L-BMAA), in rats. Neuropathol Appl Neurobiol 16: 153-169.

Seawright AA, Nolan CC, Shaw GR, Chiswell RK, Norris RL, Moore MR, Smith MJ (1999) The oral toxicity for mice of the tropical cyanobacterium *Cylindrospermopsis raciborskii* (Woloszynska). Environmental Toxicology 14: 135-142.

Senogles-Derham PJ, Seawright A, Shaw G, Wickramisingh W, Shahin M (2003) Toxicological aspects of treatment to remove cyanobacterial toxins from drinking water determined using the heterozygous P53 transgenic mouse model. Toxicon. 41(8):979-88.

Shaw GR, Seawright AA, Moore MR, Lam PK (2000) Cylindrospermopsin, a cyanobacterial alkaloid: evaluation of its toxicologic activity. Ther Drug Monit 22(1): 89-92.

Shen XY, Lam PKS, Shaw GR, Wickramasinghe W (2002) Genotoxicity investigation of a cyanobacterial toxin, cylindrospermopsin. Toxicon 40: 1499-1501.

Sipia VO, Kankaanpaa HT, Pflugmacher S, Flinkman J, Furey A, James KJ (2002) Bioaccumulation and detoxication of nodularin in tissues of flounder (Platichthys flesus), mussels (Mytilus edulis, Dreissena polymorpha), and clams (Macoma balthica) from the northern Baltic Sea. Ecotoxicol Environ Saf 53(2): 305-11.

Soares RM, Yuan M, Servaites JC, Delgado A, Magalhães VF, Hilborn ED, Carmichael WW,. Azevedo SMFO (2006) Sub-lethal exposure from microcystins to renal insufficiency patients in Rio de Janeiro – Brazil. Environ Toxicol. 21: 95-103.

Solter PF, Wollenberg GK, Huang X, Chu FS, Runnegar MT (1998) Prolonged sublethal exposure to the protein phosphatase inhibitor microcystin-LR results in multiple dose-dependent hepatotoxic effects. Toxicol Sci 44(1): 87-96.

Stotts RR, Twardock AR, Haschek WM, Choi BW, Rinehart KL, Beasley VR (1997) Distribution of tritiated dihydromicrocystin in swine. Toxicon. 35(6): 937-53.

Stotts RR, Twardock AR, Koritz GD, Haschek WM, Manuel RK, Hollis WB, Beasley VR (1997) Toxicokinetics of tritiated dihydromicrocystin-LR in swine. Toxicon. 35(3):455-65.

Sueoka E, Sueoka N, Okabe S, Kozu T, Komori A, Ohta T, Suganuma M, Kim SJ, Lim IK, Fujiki H (1997) Expression of the tumor necrosis factorα gene and early response genes by nodularin, a liver tumor promoter, in primary cultured rat hepatocytes. J Cancer Res and Clin Oncol 123: 413-419.

Sullivan JJ, Iwaoka WT, Liston J (1983) Enzymatic transformation of PSP toxins in the littleneck clam (Protothaca staminea). Biochem Biophys Res Commun. 114(2):465-72.

Swanson KL, Rapoport H, Albuquerque EX, Aronstam RS (1990) Nicotinic acetylcholine receptor function studied with synthetic (+)-anatoxin-a and derivatives. In *Marine toxins. Origin, structure, and molecular pharmacology*. Hall, S., and Strichartz, G. (eds). Washington, DC: American Chemical Society, pp. 107-118.

Swanson KL, Aronstam RS, Wonnacott S, Rapoport H, Albuquerque EX (1991) Nicotinic pharmacology of anatoxin analogs. I. Side chain structure-activity relationships at peripheral agonist and noncompetitive antagonist sites. J Pharmacol Exp Ther 259: 377-386.

Takenaka S (2001) Covalent glutathione conjugation to cyanobacterial hepatotoxin microcystin LR by F344 rat cytosolic and microsomal glutathione S-transferases. Environ Toxicol Pharm 9(4):135-139.

Terao K, Ohmori S, Igarashi K, Ohtani I, Watanabe MF, Harada KI (1994) Electron microscopic studies on experimental poisoning in mice induced by cylindrospermopsin isolated from blue-green alga Umezakia natans. Toxicon 32: 833-843.

Thacker RW, McLeod AM, McLeod SW (2005) Herbivore-induced saxitoxin production in the freshwater cyanobacterium *Lyngbya wollei.* Algological Studies 117: 415-425.

Trainer V, Baden DG (1999) High affinity binding of red tide neurotoxins to marine mammal brain. Aquatic Toxicology 46(2): 139-148.

Ueno Y, Nagata S, Tsutsumi T, Hasegawa A, Watanabe MF, Park HD, Chen GC, Chen G, Yu S-Z (1996) Detection of microcystins, a blue-green algal hepatotoxin, in drinking water sampled in Haimen and Fusui, endemic areas of primary liver cancer in China, by highly sensitive immunoassay. Carcinogenesis 17: 1317-1321.

Ueno Y, Makita Y, Nagata S, Tsutsumi T, Yoshida F, Tamura S-I, Sekijima M, Tashiro F, Harada T, Yoshida T (1999) No chronic oral toxicity of a low dose of microcystin-LR, a cyanobacterial hepatotoxin, in female BALB/c mice. Environ Toxicol 14: 45-55.

Wiegand C, Pflugmacher S (2005) Ecotoxicological effects of selected cyanobacterial secondary metabolites: a short review. Toxicol Appl Pharmacol 203: 201-18.

WHO, 1999). Toxic Cyanobacteria in Water: A guide to their public health consequences, monitoring and management. Chapter 13. Laboratory analysis of cyanotoxins.

Wonnacott S, Jackman S, Swanson KL, Rapoport H, Albuquerque EX (1991) Nicotinic pharmacology of anatoxin analogs. II. Side chain structure-activity relationships at neuronal nicotinic ligand binding sites. *J Pharmacol Exp Ther* 259: 387-391.

Yin Q, Carmichael WW, Evans WR (1997) Factors influencing growth and toxin production by cultures of the freshwater cyanobacterium *Lyngbya wollei* Farlow ex Gomont. J. of Applied Phycology 9:55-63.

Yoshida T, Makita Y, Nagata S, Tsutsumi T, Yoshida F, Sekijima M, Tamura S, Ueno Y (1997) Acute oral toxicity of microcystin-LR, a cyanobacterial hepatotoxin, in mice. Nat Toxins 5(3): 91-5.

Yoshida T, Makita Y, Tsutsumi T, Nagata S, Tashiro F, Yoshida F, Sekijima M, Tamura S, Harada T, Maita K, Ueno Y (1998) Immunohistochemical localization of microcystin-LR in the liver of mice: a study on the pathogenesis of microcystin-LR-induced hepatotoxicity. Toxicol Pathol 26(3): 411-8.

Yu SZ (1995) Primary prevention of hepatocellular carcinoma. Journal of Gastroenterology and Hepatology 10: 674-682.

Yuan M, Carmichael WW, Hilborn ED (2006) Microcystin analysis in human sera and liver from human fatalities in Caruaru, Brazil 1996. Toxicon 48: 627-640.

Zhou L, Yu H, Chen K (2002) Relationship between microcystin in drinking water and colorectal cancer. Biomed-Environ-Sci 15: 166-171.

Zhu Y, Zhong X, Zheng S, Ge Z, Du Q, Zhang S (2005) Transformation of immortalized colorectal crypt cells by microcystin involving constitutive activation of Akt and MAPK cascade. Carcinogenesis 26(7): 1207-14.

Chapter 16: Toxin types, toxicokinetics and toxicodynamics

Andrew Humpage[1,2,3]

[1]Australian Water Quality Centre, PMB 3, Salisbury, Adelaide, SA 5108, Australia. [2]Discipline of Pharmacology, School of Medical Sciences, University of Adelaide, SA 5005, Australia. [3]Cooperative Research Centre for Water Quality and Treatment, PMB 3, Salisbury, Adelaide, SA 5108, Australia.

Introduction

Cyanobacteria produce a wide array of bioactive secondary metabolites (see Table A.1 in Appendix A), some which are toxic (Namikoshi and Rinehart 1996; Skulberg 2000). Those toxic to mammals include the microcystins, cylindrospermopsins, saxitoxins, nodularins, anatoxin-a, homoanatoxin-a, and anatoxin-a(s). It has been recently suggested that β-methylamino alanine (BMAA) may be a new cyanobacterial toxin (Cox et al. 2003; Cox et al. 2005). The public health risks of cyanotoxins in drinking water have recently been reviewed (Falconer and Humpage 2005b). The aim of this paper is to concisely review our current knowledge of their acute toxicity, mechanisms of action, toxicokinetics and toxicodynamics.

Microcystins

Microcystins (MCs) are a group of at least 80 variants based on a cyclic heptapeptide structure (Fig. 1). All toxic microcystin structural variants contain a unique hydrophobic amino acid, 3-amino-9-methoxy-10-phenyl-2,6,8-trimethyl-deca-4(E),6(E)-dienoic acid (ADDA). The prototype-compound is MC-LR, which has leucine and arginine at the two hypervariable positions in the ring structure (X and Y, respectively, in Fig. 1). Substitution of other amino acids at these sites, or methylation of residues at other sites, leads to wide structural variability (Namikoshi et al. 1990; Namikoshi et al. 1992d; Namikoshi et al. 1992b; Namikoshi et al. 1992c; Namikoshi et al. 1992a; Namikoshi et al. 1995; Namikoshi et al. 1998; Sivonen and Jones 1999). These toxins are produced by a wide variety of planktonic cyanobacteria including *Microcystis aeruginosa, M. viridis, M. ichthyoblabe, M. botrys, Planktothrix argardhii, P. rubescens, P. mougeotii, Anabaena flos-aquae, A. circinalis, A. lemmermannii, Nostoc spp.,* and *Snowella lacustris* (Botes et al. 1982; Codd and Carmichael 1982; Botes et al. 1985; Kusumi et al. 1987; Krishnamurthy et al. 1989; Sivonen et al. 1990; Harada et al. 1991; Watanabe et al. 1991; Sivonen et al. 1992; Ueno et al. 1996; Vezie et al. 1998; Marsalek et al. 2000; Fastner et al. 2001). The species most often cited as microcystin producers are *M. aeruginosa* (worldwide) and the *Planktothrix* species (Northern Europe). Microcystin production has also been linked with some benthic species: *Haphalosiphon hibernicus* and *Oscillatoria limnosa* (Prinsep et al. 1992b; Mez et al. 1997). Other benthic species have been implicated, but the difficulty of culturing these species has precluded clear identification of the organisms responsible.

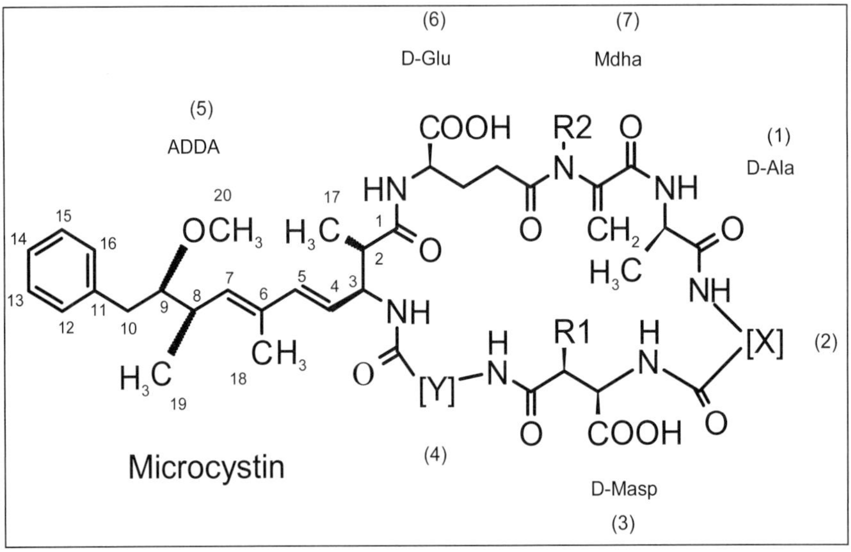

Fig. 1. General structure of the microcystins

The primary site of toxic action of the microcystins is the active site of protein phosphatases 1 and 2A (Eriksson et al. 1990; MacKintosh et al. 1990; Runnegar et al. 1995b). This activity is mediated principally by the

ADDA group (Goldberg et al. 1995) although most microcystin variants contain dehydroalanine, which can undergo covalent linkage to a cysteinyl sulphur on the phosphatase. This makes the inhibition irreversible.

MC-LR has a LD_{50} (ip, mice, 24hr) of 60 μg kg^{-1}. The primary acute effect of protein phosphatase inhibition is hyperphosphorylation of many cellular proteins including the hepatocellular cytoskeleton, which causes loss of cell-cell contacts and intra-hepatic haemorrhage. Death is due to hypovolemic shock (Runnegar and Falconer 1986; Falconer and Yeung 1992; Runnegar et al. 1993). Other acute effects include altered mitochondrial membrane permeability, generation of reactive oxygen species and induction of apoptosis (Fladmark et al. 1999; Humpage and Falconer 1999; Ding et al. 2000; Hooser 2000), most likely due to a fatal loss of control of regulatory phosphorylation. Uptake is via specific organic anion transport proteins (Runnegar et al. 1991; Runnegar et al. 1995a; Fischer et al. 2005); hence MCs exhibit a predominantly hepatic organotropism, although enteric and even dermal effects have been demonstrated in certain circumstances (Falconer and Buckley 1989; Falconer et al. 1992). Studies of tissue distribution using radio-labelled toxin have confirmed the liver as the main site of toxin accumulation (~70% of a sub-lethal iv dose) and that the toxin level in this organ remains constant for up to 6 days post treatment (Falconer et al. 1986; Runnegar et al. 1986; Brooks and Codd 1987; Robinson et al. 1989; Robinson et al. 1990; Robinson et al. 1991). Bile acids and compounds that block bile acid uptake inhibit microcystin hepatic uptake and toxicity (Runnegar et al. 1981; Thompson et al. 1988; Thompson and Pace 1992; Runnegar et al. 1995a). Formation of glutathione metabolites of MC-LR and MC-RR has been demonstrated (Kondo et al. 1996). Toxin is rapidly cleared from the blood, after which time the main albeit slow route of excretion is via the faeces (Robinson et al. 1991). Human acute intoxication via renal dialysis (possibly in combination with cylindrospermopsin) resulted in visual disturbances, nausea, vomiting and death from liver failure (Carmichael et al. 2001; Azevedo et al. 2002), whereas sub-lethal exposure resulted in elevation of liver enzyme activities in the serum (Falconer et al. 1983).

Lower microcystin concentrations (pM) appear to suppress apoptosis and promote cell division in polyploid hepatocytes in vitro (Humpage and Falconer 1999), effects which may be linked to the enhancement of the growth of hepatic and colonic pre-cancerous lesions in animal models (Fujiki and Suganuma 1993; Ito et al. 1997; Humpage et al. 2000b). Microcystin exposure has been linked to human liver and colon cancer incidence (Yu 1995; Fleming et al. 2002; Zhou et al. 2002).

Cylindrospermopsins

The cylindrospermopsins (CYNs, Fig. 2) are alkaloids comprised of a tricyclic guanidino moiety linked via a hydroxylated bridging carbon (C7) to uracil (Ohtani et al. 1992). Structural variants are 7-epi-CYN and 7-deoxy-CYN (Norris et al. 1999; Banker et al. 2000), the latter having slightly lower potency than the 7-hydroxylated variants (Looper et al. 2005). The uracil moiety is required for toxicity (Banker et al. 2001; Runnegar et al. 2002). CYN's are produced by *Cylindrospermopsis raciborskii, Aphanizomenon ovalisporum, Anabaena bergii, Umezakia natans, Raphidiopsis curvata*, and as yet other unidentified species (Hawkins et al. 1985; Harada et al. 1994; Banker et al. 1997; Hawkins et al. 1997; Shaw et al. 1999; Li et al. 2001a; Schembri et al. 2001).

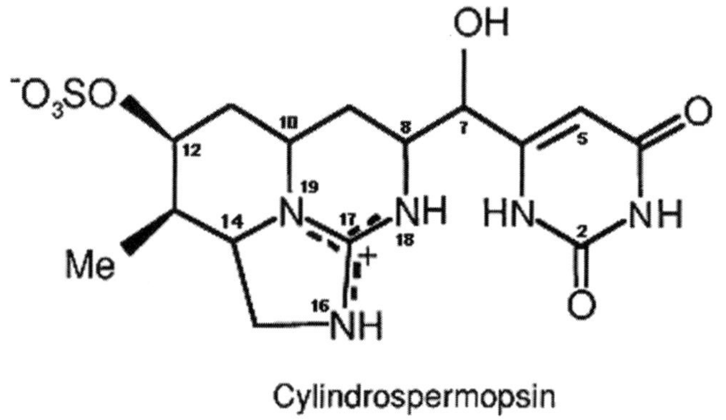

Fig. 2. Structure of cylindrospermopsin

The LD_{50} of CYN indicates a delayed toxicity (2.0 mg kg^{-1}, ip mouse, after 24 hrs but 0.2 mg kg^{-1} after 5 days; Ohtani et al. 1992). The primary toxic effect of the parent compound appears to be irreversible protein synthesis inhibition (Terao et al. 1994; Froscio et al. 2001, 2003; Looper et al. 2005). However, there is also evidence for metabolic activation as inhibitors of CYP450's are able to reduce acute toxicity (Runnegar et al. 1994; Froscio et al. 2003), CYN-dependent inhibition of glutathione synthesis (Runnegar et al. 1995c), and genotoxicity (Humpage et al. 2005). The evidence for CYP450 involvement in the in vivo toxicosis is less clear (Norris et al. 2002). Acute CYN poisoning results in lipid accumulation in the liver followed by hepatocellular necrosis (Terao et al. 1994; Seawright et al.

1999). Non-hepatic effects include destruction of the proximal tubules of the kidney (Falconer et al. 1999), as well as cytotoxic and thrombotic effects in other tissues. Intraperitoneal injection of radio-labelled CYN resulted in predominantly hepatic and, to a lesser extent, renal distribution of the toxin (Norris et al. 2001). There was some evidence for the formation of metabolites, but these were not characterised. Sub-chronic oral exposure resulted in mainly hepatic and renal effects (Humpage and Falconer 2003). Effects of poisoning in humans included hepatoenteritis and renal insufficiency (Byth 1980).

Genotoxic effects of CYN have been demonstrated in vitro using the cytokinesis-blocked micronucleus assay (Humpage et al. 2000a) and the comet assay (Humpage et al. 2005). Strand breakage and loss of whole chromosomes were demonstrated to occur at concentrations below those that caused overt cytotoxicity. Hepatic DNA fragmentation has also been demonstrated in vivo after a single intraperitoneal dose of cylindrospermopsin (Shen et al. 2002). There is some evidence of carcinogenicity in vivo (Falconer and Humpage 2001), but more work is required to confirm this.

Saxitoxins (Paralytic Shellfish Toxins (PSTs))

The saxitoxins (Fig. 3) have been extensively studied due to their involvement in paralytic shellfish poisoning where toxigenic marine dinoflagellates are consumed by shellfish, which concentrate the toxins and can deliver toxic quantities to consumers of the shellfish (Kao 1993). Saxitoxins are alkaloids based on a 3,4,6-trialkyl tetrahydropurine skeleton which can be further carbamylated, sulphated or N-sulphocarbamylated to produce a range of perhaps 30 analogues (Shimizu 2000), some of which are found only in freshwater cyanobacteria (Onodera et al. 1997b; Lagos et al. 1999; Molica et al. 2002). They are produced in the freshwater environment by *Aphanizomenon* spp., *Anabaena circinalis, Cylindrospermopsis raciborskii, Lyngbya wollei, Planktothrix* spp., and other unidentified species (Jackim and Gentile 1968; Ikawa et al. 1982; Humpage et al. 1994; Carmichael et al. 1997; Lagos et al. 1999; Kaas and Henriksen 2000; Pomati et al. 2000; Li et al. 2000; Li et al. 2003).

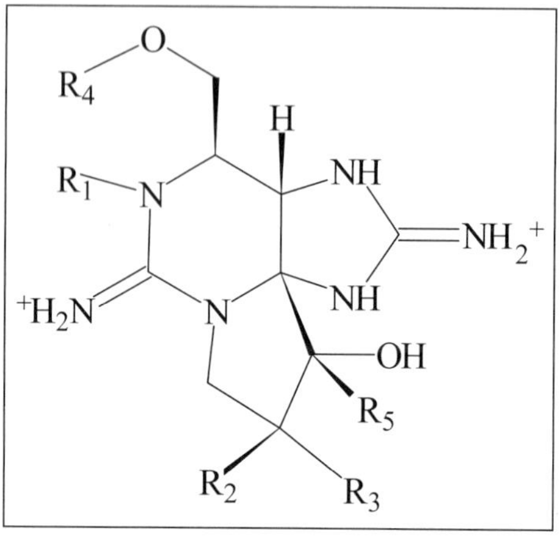

Fig. 3. General structure of the saxitoxins

Toxin	R1	R2	R3	R5	Net Charge	Relative mouse toxicity
R4 = $CONH_2$ (carbamate toxins)						
STX	H	H	H	OH	+2	1.000
neoSTX	OH	H	H	OH	+2	0.924
GTX1	OH	H	OSO_3^-	OH	+1	0.994
GTX2	H	H	OSO_3^-	OH	+1	0.359
GTX3	H	OSO_3^-	H	OH	+1	0.638
GTX4	OH	OSO_3^-	H	OH	+1	0.726
R4 = $CONHSO_3^-$ (n-sulfocarbamoyl (sulfamate) toxins)						
GTX5 (B1)	H	H	H	OH	+1	0.064
GTX6 (B2)	OH	H	H	OH	+1	-
C1	H	H	OSO_3^-	OH	0	0.006
C2	H	OSO_3^-	H	OH	0	0.096
C3	OH	H	OSO_3^-	OH	0	0.013
C4	OH	OSO_3^-	H	OH	0	0.058
R4 = H (decarbamoyl toxins)						
dcSTX	H	H	H	OH	+2	0.513
dcneoSTX	OH	H	H	OH	+2	-
dcGTX1	OH	H	OSO_3^-	OH	+1	-
dcGTX2	H	H	OSO_3^-	OH	+1	0.651
dcGTX3	H	OSO_3^-	H	OH	+1	0.754
dcGTX4	OH	OSO_3^-	H	OH	+1	-
LWTX4	H	H	H	H	+2	<0.004

R4 = $COCH_3$ (Lyngbya wollei toxins)						
LWTX1	H	OSO_3^-	H	H	+1	<0.004
LWTX2	H	OSO_3^-	H	OH	+1	0.072
LWTX3	H	H	OSO_3^-	OH	+1	0.021
LWTX5	H	H	H	OH	+2	0.139
LWTX6	H	H	H	H	+2	<0.004
R4 = COC_6H_4OH (Gymnodinium catenatum toxins)						
GC1	H	H	OSO_3^-	OH	+1	-
GC2	H	OSO_3^-	H	OH	+1	-
GC3	H	H	H	OH	+2	-

Modified from Nicholson and Burch (2001)

Fig. 3 (cont). General structure of the saxitoxins

These toxins are potent voltage-gated sodium channel antagonists, causing numbness, paralysis and death by respiratory arrest. Analogue potency varies greatly, with saxitoxin having an LD_{50} (ip mouse) of 10 μg kg^{-1}, but C1 being at least 160-fold less toxic (Oshima 1995). Toxin uptake and toxicokinetics of a number of analogues have been studied in cats (Andrinolo et al. 1999; Andrinolo et al. 2002b; Andrinolo et al. 2002a). Oral uptake was efficient, and toxin distributed rapidly throughout the body, including the brain. Clearance was via simple glomerular filtration, and there was no evidence of metabolism of the toxins. Toxicological studies to date have assumed the acute exposure paradigm of shellfish poisoning rather than sub-chronic low-dose as might be expected from drinking water. Evidence for development of tolerance to PSTs has been presented (Kuiper-Goodman et al. 1999). Neuro-developmental disturbances have been demonstrated in fish (Lefebvre 2002) but these have not been studied in mammals.

Nodularins

Nodularins (Fig. 4) are hepatotoxic cyclic peptides of similar structure to the microcystins except that they are composed of 5 amino acids rather than 7 (Rinehart and Namikoshi 1994). Variants duc to substitution of arginine with homoarginine or valine (motuporin) have been described (de Silva et al. 1992; Namikoshi et al. 1993; Namikoshi et al. 1994), but these appear to be relatively rare. ADDA is still present but dehydroalanine is replaced by N-methyl-dehydrobutyrine (Rinehart et al. 1988). The smaller ring size prevents this latter moiety from coordinating with the phosphatase cysteine, and so nodularin does not bind covalently (Lanaras et al. 1991; Craig et al. 1996; Bagu et al. 1997). However, due to the high affinity of ADDA for the active site, this lack of covalent binding does not affect toxin potency, which is similar to that of microcystin-LR (Ki's are of the order 0.1 – 1.5 nM; Honkanen et al. 1990; MacKintosh et al. 1990; Honkanen et al. 1991). This lack of covalent binding may allow nodularin to reach other sites in the cell, and this has been suggested as a mechanism by which this toxin might act as a direct carcinogen (Ohta et al. 1994; Bagu et al. 1997). *Nodularia spumigena* appears to be the sole freshwater cyanobacterial source of nodularin (motuporin was isolated from a marine sponge). *N. spumigena* generally prefers brackish waters and so has had only localised impacts on human drinking water sources (for example, in Lake Alexandrina, South Australia in the early 1990's).

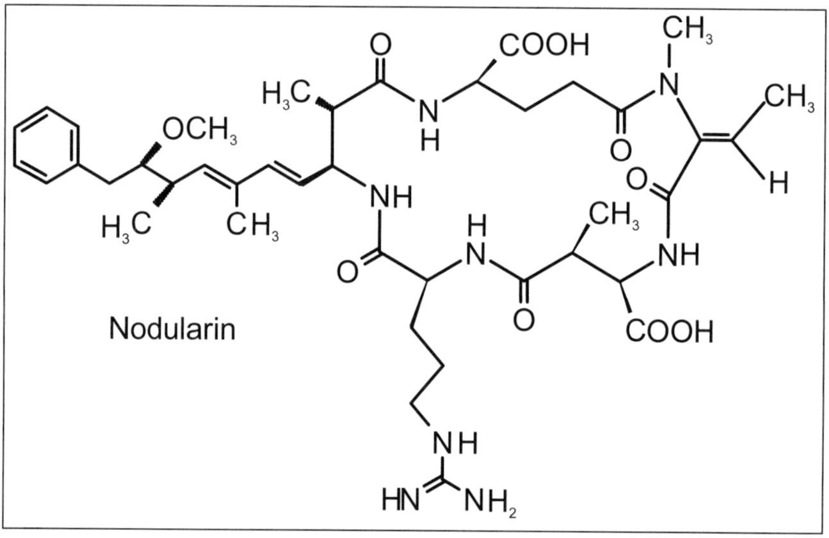

Fig. 4. Structure of nodularin

Anatoxin-a/Homoanatoxin-a

Anatoxin-a (2-acetyl-9-azabicyclo(4-2-1)non-2-ene; (Fig. 5) and/or homoanatoxin-a (propionyl residue replaces acetyl at C2) are produced by *Anabaena flos-aquae, A. planktonica, Aphanizomenon spp., Planktothrix formosa*, and a benthic *Oscillatoria* spp. (Carmichael et al. 1975; Carmichael and Gorham 1978; Sivonen et al. 1989; Edwards et al. 1992; Skulberg et al. 1992; Bruno et al. 1994; Bumke-Vogt et al. 1999). These toxins are nicotinic acetylcholine receptor agonists having a LD_{50} of 200 μg kg^{-1} (Carmichael et al. 1979; Carmichael 1994). Residence of these toxins at post-synaptic cholinergic receptors results in nerve depolarisation (Swanson et al. 1990; Huby et al. 1991; Swanson et al. 1991; Wonnacott et al. 1991). Typical symptoms in mice are loss of muscle coordination, gasping, convulsions and death within minutes from respiratory arrest (Carmichael et al. 1979). Dog deaths have been attributed to poisoning by these toxins when the animals have licked their coats after swimming (Codd et al. 1992; Edwards et al. 1992; Falconer and Nicholson, personal communication). A single human fatality has been attributed to poisoning by anatoxin-a after the victim swam in a scum-covered pond (Behm 2003), but further investigation suggests that this is unlikely to be correct (Carmichael et al. 2004). Anatoxins have not been linked to human poisoning via drinking water, although evidence has been presented that such a risk may exist in Florida (Burns 2005).

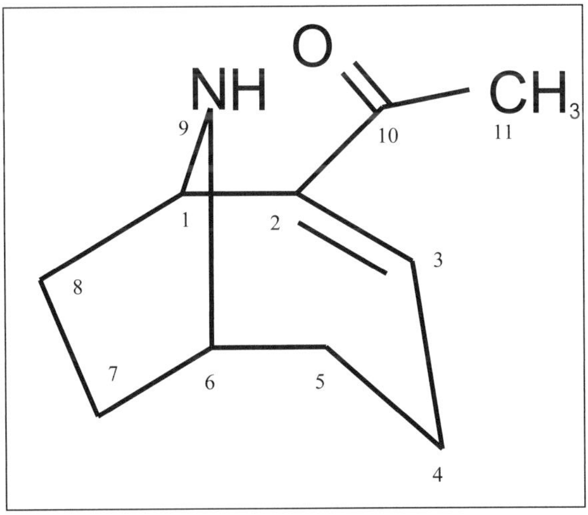

Fig. 5. Structure of anatoxin-a

Anatoxin-a(s)

Anatoxin-a(s) (Fig. 6) is a phosphorylated cyclic N-hydroxyguanine, with a structure and action similar to organophosphate pesticides (Mahmood and Carmichael 1986, 1987; Hyde and Carmichael 1991). It is a potent acetylcholinesterase inhibitor with a LD_{50} (ip, mouse) of 20 μg kg^{-1}. The in vivo toxic effects are similar to those of anatoxin-a but with the addition of salivation (hence the "s") and lacrimation (Mahmood and Carmichael

1986, 1987; Matsunaga et al. 1989). Anatoxin-a(s) is produced by *Anabaena flos-aquae* and *A. lemmermannii* (Matsunaga et al. 1989; Onodera et al. 1997a), and the latter has been implicated in the deaths of water birds in Denmark (Onodera et al. 1997a). No human illness has been attributed to this toxin.

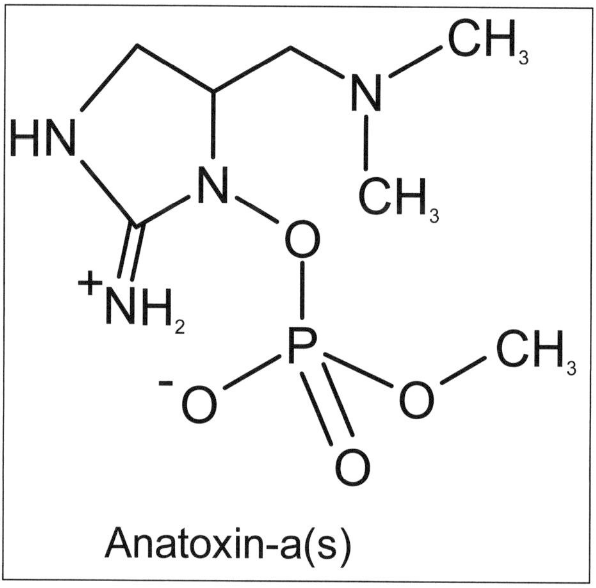

Fig. 6. Structure of anatoxin-a(s)

β-Methylamino alanine

β-Methylamino alanine (BMAA) (Fig. 7) is an old toxin that has recently been found to be of cyanobacterial origin (Cox et al. 2005). Whether cyanobacteria are the only source is not known. BMAA was described in 1967 in extracts of cycad seeds from Guam, and suggested as a possible causative agent of certain neurodegenerative disorders that were prevalent on the island (Vega and Bell 1967). Early studies in monkeys dosed with high levels of BMAA produced effects similar to those seen in humans (Spencer et al. 1987). BMAA was found to be a glutaminergic agonist capable of producing excitotoxicity, but only at relatively high concentrations (EC_{50} in cell-lines of 300 μM; Weiss et al. 1989a; Weiss et al.

1989b). Sodium bicarbonate is required as a cofactor due to the spontaneous formation of the carbamate, turning the monocarboxylic BMAA into a dicarboxylic glutamate mimic (Myers and Nelson 1990). Toxicokinetic studies in rats and monkeys demonstrated rapid and virtually complete uptake via the oral route, and distribution throughout the body, including the brain (Duncan et al. 1991; Duncan et al. 1992), although the latter organ only contained about 0.08% of the administered dose by 48 hrs. There was evidence of active transport across the blood-brain barrier via the large neutral amino acid carrier (Km=2.9mM), but this would not be rapid when BMAA is in competition with normal levels of natural amino acids (Duncan et al. 1992; Jalaludin and Smith 1992). Approximately 1.4% of an oral dose, and 1.8% of an iv dose, were excreted unmetabolised in the urine by 48 hrs, whereas approximately 22% could be accounted for in total (unmetabolised plus acid hydrolysable; Duncan et al. 1992; Jalaludin and Smith 1992). L-amino acid oxidase has been shown to metabolise BMAA, eventually leading to N-methylglycine, but the rate appears to be quite slow (Hashmi and Anders 1991). Multiple sub-lethal doses were shown to be non-cumulative (Seawright et al. 1990). Based on these and other studies, and the likely concentrations of BMAA in cycad seed flour, it was suggested that this toxin was unlikely to be the sole causative agent of the Guam neurodegenerative disease (Duncan et al. 1990). The hypothesis that BMAA might bio-accumulate in cycad seed-consuming flying foxes (Cox and Sacks 2002), and then the demonstration of high levels of the toxin not only in museum exhibits of flying foxes, but also in the brains of patients who died from neurodegenerative disorders in both Guam and Canada, has re-ignited the debate (Banack and Cox 2003; Cox et al. 2003; Murch et al. 2004a; Murch et al. 2004b). Finally, the demonstration that many species of cyanobacteria produce BMAA (Cox et al. 2005) has made this an issue of potential concern for the provision of safe drinking water. Much more work needs to be done before a proper assessment can be made of this "new" cyanotoxin.

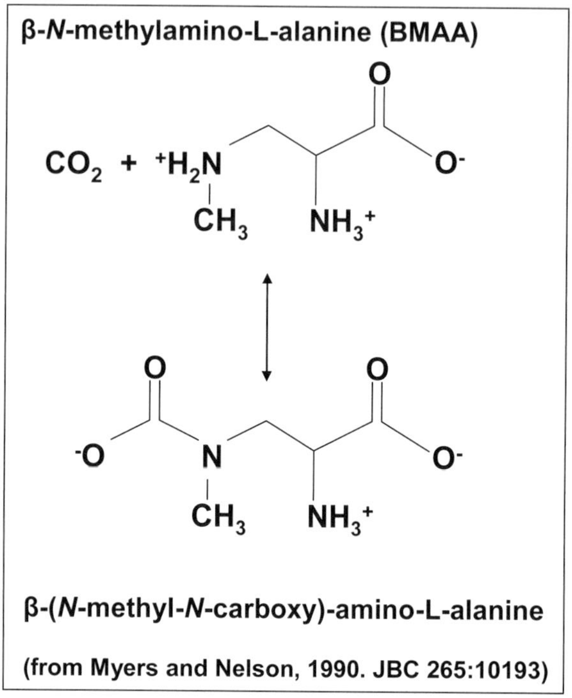

Fig. 7. Structure of BMAA and its reaction with CO_2.

Undiscovered cyanotoxins and other cyanobacterial bioactive compounds

Given the range of bioactive compounds known to be produced by cyanobacteria (see Table A.1 in Appendix A for a non-exhaustive list of "non-cyanotoxin" cyanobacterial bioactive compounds) it is not surprising that a few have turned out to be toxic to mammals. It is unlikely that we have found all of the toxins because unexplained toxicity has been observed, for example, in *C. raciborskii* (Hawkins et al. 1997; Bernard et al. 2003; Fastner et al. 2003; Saker et al. 2003), in *Anabaena* spp. (Baker and Humpage 1994), and in a *Phormidium* spp. (Baker et al. 2001). Furthermore, new toxin analogues continue to be reported (Onodera et al. 1997b; Banker et al. 2000; Molica et al. 2002; Negri et al. 2003). The fact that known toxins are usually found in new locations once people look for them, for example, recent discoveries of CYN in New Zealand (Stirling and Quilliam 2001), Thailand (Li et al. 2001b), Germany (Fastner et al. 2003), Brazil (Carmichael et al. 2001) and Florida (Burns 2005), suggests that the toxigenic species are widespread and that no country should consider itself immune from the risk to public health even from the known toxins. A further point that needs reinforcing is that single microcystins or single cylindrospermopsins almost never occur in nature. Instead mixtures of toxins are the norm, and so we need to understand toxin interactions. This will enable the calibration of studies done using MC-LR and the formulation of regulations to be based on toxicity equivalents rather than quantities of individual compounds.

Research Needs

Research into the toxicology of cyanotoxins is still lacking in a number of important areas:

- Microcystins:
  - Epidemiological studies into links with human cancer. This requires a biomarker of low dose exposure for which ELISA may be an option (Hilborn et al. 2005).
  - Chronic animal studies into links with cancer.
  - Effects of mixtures of microcystin analogues, and of microcystins with cylindrospermopsin.
- Cylindrospermopsin:
  - Human effects – An opportunity exists for follow-up of exposed humans on Palm Island, Australia.
  - Animal studies for Guideline formulation: Toxicokinetics, Chronic exposure, Carcinogenicity, Reproductive toxicity.
  - Effects of mixtures (with microcystins).
  - Cell-based studies: To better understand mechanism(s) of toxic and genotoxic action, leading to identification of biomarkers of exposure & effect (Falconer and Humpage 2005a).
- Neurotoxins:
  - Episodic and chronic low dose exposures – particularly any effect on neural development.
- BMAA:
  - Confirm association with neurodegenerative disease.
  - Mechanism of bioaccumulation.
  - Mechanism of toxicity (glutaminergic excitotoxicity or other effects eg disruption of protein structure/function?).
  - Trophic studies to determine routes of human exposure.

References

Andrinolo D, Michea LF, Lagos N (1999) Toxic effects, pharmacokinetics and clearance of saxitoxin, a component of paralytic shellfish poison (PSP), in cats. Toxicon 37: 447-464

Andrinolo D, Iglesias V, Garcia C, Lagos N (2002a) Toxicokinetics and toxicodynamics of gonyautoxins after an oral toxin dose in cats. Toxicon 40: 699-709

Andrinolo D, Gomes P, Fraga S, Soares-da-Silva P, Lagos N (2002b) Transport of the organic cations gonyautoxin 2/3 epimers, a paralytic shellfish poison toxin, through the human and rat intestinal epitheliums. Toxicon 40: 1389-1397

Azevedo S, Carmichael WW, Jochimsen EM, Rinehart KL, Lau S, Shaw GR, Eaglesham GK (2002) Human intoxication by microcystins during renal dialysis treatment in Caruaru-Brazil. Toxicology 181: 441-446

Bagu JR, Sykes BD, Craig MM, Holmes CFB (1997) A molecular basis for different interactions of marine toxins with protein phosphatase-1. J Biol Chem 8: 5087-5097

Baker PD, Humpage AR (1994) Toxicity associated with commonly occurring cyanobacteria in surface waters of the Murray-Darling Basin, Australia. Australian Journal of Marine and Freshwater Research 45: 773-786

Baker PD, Steffensen DA, Humpage AR, Nicholson BC, Falconer IR, Lanthois B et al. (2001) Preliminary evidence of toxicity associated with the benthic cyanobacterium Phormidium in South Australia. Environ Toxicol 16: 506-511

Banack SA, Cox PA (2003) Biomagnification of cycad neurotoxins in flying foxes - Implications for ALS-PDC in Guam. Neurology 61: 387-389

Banker R, Teltsch B, Sukenik A, Carmeli S (2000) 7-Epicylindrospermopsin, a toxic minor metabolite of the cyanobacterium Aphanizomenon ovalisporum from lake Kinneret, Israel. J Nat Prod 63: 387-389

Banker R, Carmeli S, Hadas O, Teltsch B, Porat R, Sukenik A (1997) Identification of cylindrospermopsin in Aphanizomenon ovalisporum (cyanophyceae) isolated from Lake Kinneret, Israel. Journal of Phycology 33: 613-616

Banker R, Carmeli S, Werman M, Teltsch B, Porat R, Sukenik A (2001) Uracil moiety is required for toxicity of the cyanobacterial hepatotoxin cylindrospermopsin. J Toxicol Environ Health A 62: 281-288

Barchi JJ, Moore RE, Patterson GML (1984) Acutiphycin and 20,21-didehydroacutiphycin, new antineoplastic agents from the cyanophyte Oscillatoria acutissima. Journal of the American Chemical Society 106: 8193-8197

Behm D (2003) Coroner cites algae in teen's death. In Milwaukee Journal Sentinal. Milwaukee

Bernard C, Harvey M, Briand JF, Bire R, Krys S, Fontaine JJ (2003) Toxicological comparison of diverse Cylindrospermopsis raciborskii strains: Evidence of liver damage caused by a French C. raciborskii strain. Environmental Toxicology 18: 176-186

Berry JP, Gantar M, Gawley RE, Wang M, Rein KS (2004) Pharmacology and toxicology of pahayokolide A, a bioactive metabolite from a freshwater spe-

cies of Lyngbya isolated from the Florida Everglades. Comp Biochem Physiol C Toxicol Pharmacol 139: 231-238
Botes DP, Kruger H, Viljoen CC (1982) Isolation and characterisation of four toxins from the blue-green alga, Microcystis aeruginosa. Toxicon 20, 6: 945-954
Botes DP, Wessels PL, Kruger H, Runnegar MTC, Santikarn S, Smith RJ et al. (1985) Structural studies on cyanoginosins-LR, -YR, -YA, and -YM, peptide toxins Microcystis aeruginosa. Journal of the Chemical Society, Perkin Transactions 1: 2747-2748
Brooks WP, Codd GA (1987) Distribution of Microcystis aeruginosa peptide toxin and interactions with hepatic microsomes in mice. Pharmacology and Toxicology 60: 187-191
Bruno M, Barbini DA, Pierdominici E, Serse AP, Ioppolo A (1994) Anatoxin-a and a previously unknown toxin in Anabaena planctonica from blooms found in Lake Mulargia (Italy). Toxicon 32: 369-373
Bumke-Vogt C, Mailahn W, Chorus I (1999) Anatoxin-a and neurotoxic cyanobacteria in German lakes and Reservoirs. Environmental Toxicology 14: 117-125
Burns J (2005) Assessment of Cyanotoxins in Florida's Surface Waters and Associated Drinking Water Resources. Final Report to the Florida Harmful Algal Bloom Task Force, Florida Fish and Wildlife Conservation Commission. St. Johns River Water Management District, St Petersburg, Florida
Byth S (1980) Palm Island Mystery Disease. Medical Journal of Australia 2: 40-42.
Cardelina JH, Moore RE (eds) Malyngic Acid, a New Fatty Acid from Lyngbya majuscula. Tetrahedron Letters 36: 993-996
Cardelina JH, Marner FJ, Moore RE (1979) Structure and Absolute Configuration of Malyngolide, and Antibiotic from the Marine Blue-Green Alga. Journal of Organic Chemistry 44: 4039-4042
Carmeli S, Moore RE, Patterson GML (1990a) Polytoxin and new scytophycins from three species of Scytonema. Journal of Natural Products 53, 6: 1533-1542
Carmeli S, Moore RE, Patterson GML (1990b) Tantazoles: unusual cytotoxic alkaloids from the blue-green alga Scytonema mirabile. Journal of the American Chemical Society 112: 8195-8197
Carmeli S, Moore RE, Patterson GML (1991) Mirabazoles, minor tantazole-related cytotoxins from the terrestrial blue-green alga Scytonema mirabile. Tetrahedron Letters 23: 2593-2596
Carmichael WW (1994) The toxins of cyanobacteria. Sci Am 270: 78-86
Carmichael WW, Gorham PR (1978) Anatoxins from clones of Anabaena flos-aquae isolated from lakes in western Canada. Mitteilungen - Internationale Vereinigung fur Theoretische und Angewandte Limnologie 21: 285-295
Carmichael WW, Biggs DF, Gorham PR (1975) Toxicology and pharmacological action of Anabaena flos-aquae toxin. Science 187: 542-544
Carmichael WW, Biggs DF, Peterson MA (1979) Pharmacology of anatoxin-a produced by the freshwater cyanophyte Anabaena flos-aquae NRC-44-1. Toxicon 17: 229-236

Carmichael WW, Evans WR, Yin QQ, Bell P, Moczydlowski E (1997) Evidence for paralytic shellfish poisons in the freshwater cyanobacterium Lyngbya wollei (Farlow ex Gomont) comb. nov. Appl Environ Microbiol 63: 3104-3110

Carmichael WW, Azevedo SM, An JS, Molica RJ, Jochimsen EM, Lau S et al. (2001) Human fatalities from cyanobacteria: chemical and biological evidence for cyanotoxins. Environ Health Perspect 109: 663-668

Carmichael WW, Yuan M, Friday CF (2004) Human Mortality from Accidental Ingestion of Toxic Cyanobacteria-A Case Re-examined (poster). In 6th Int Conf on Toxic Cyanobacteria Bergen, Norway

Carter DC, Moore RE, Mynderse JS (1984) Structure of Majusculamide C, a Cyclic Depsipeptide from Lyngbya majuscula. Journal of Organic Chemistry 49: 236-241

Codd GA, Carmichael WW (1982) Toxicity of a clonal isolate of the cyanobacterium Microcystis aeruginosa from Great Britain. FEMS Microbiology Letters 13: 409-411

Codd GA, Edwards C, Beattie KA, Barr WM, Gunn GJ (1992) Fatal attraction to cyanobacteria? Nature 359: 110-111

Cox PA, Sacks OW (2002) Cycad neurotoxins, consumption of flying foxes, and ALS-PDC disease in Guam. Neurology 58: 956-959

Cox PA, Banack SA, Murch SJ (2003) Biomagnification of cyanobacterial neurotoxins and neurodegenerative disease among the Chamorro people of Guam. Proceedings of the National Academy of Sciences of the United States of America 100: 13380-13383

Cox PA, Banack SA, Murch SJ, Rasmussen U, Tien G, Bidigare RR et al. (2005) Diverse taxa of cyanobacteria produce beta-N-methylamino-L-alanine, a neurotoxic amino acid. Proceedings of the National Academy of Sciences of the United States of America 102: 5074-5078

Craig M, Luu HA, McCready TL, Williams D, Andersen RJ, Holmes CFB (1996) Molecular mechanisms underlying the interaction of motuporin and microycystins with type-1 and type-2a protein phosphatases. Biochemistry & Cell Biology 74: 569-578

De Mule MCZ, De Caire GZ, De Cano MS, De Halperin DR (1991) Bioactive compounds from Nostoc muscorum (cyanobacteria). Cytobios 66: 169-172

de Silva ED, Williams DE, Andersen RJ, Klix H, Holmes CFB, Allen TM (1992) Motuporin, a potent protein phosphatase inhibitor isolated from the Papua New Guinea sponge Theonella swinhoei Gray. Tetrahedron Letters 33: 1561-1564

Ding WX, Shen HM, Ong CN (2000) Critical role of reactive oxygen species and mitochondrial permeability transition in microcystin-induced rapid apoptosis in rat hepatocytes. Hepatology 32: 547-555

Duncan MW, Steele JC, Kopin IJ, Markey SP (1990) 2-Amino-3-(methylamino)-propanoic acid (BMAA) in cycad flour: an unlikely cause of amyotrophic lateral sclerosis and parkinsonism-dementia of Guam. Neurology 40: 767-772

Duncan MW, Markey SP, Weick BG, Pearson PG, Ziffer H, Hu Y, Kopin IJ (1992) 2-Amino-3-(methylamino)propanoic acid (BMAA) bioavailability in the primate. Neurobiol Aging 13: 333-337

Duncan MW, Villacreses NE, Pearson PG, Wyatt L, Rapoport SI, Kopin IJ et al. (1991) 2-amino-3-(methylamino)-propanoic acid (BMAA) pharmacokinetics and blood-brain barrier permeability in the rat. J Pharmacol Exp Ther 258: 27-35

Edwards C, Beattie KA, Scrimgeour CM, Codd GA (1992) Identification of anatoxin-a in benthic cyanobacteria (blue-green algae) and in associated dog poisonings at Loch Insh, Scotland. Toxicon 30: 1165-1175

Eriksson J, Meriluoto J, Toivola D, Karaki H, Han YG, Hartshorne D (1990) Hepatocyte deformation induced by cyanobacterial toxins reflects inhibitions of protein phosphatases. Biochem-Biophys-Res-Commun 173: 1347-1353

Falconer I, Humpage A (2005a) Cyanobacterial (blue-green algal) toxins in water supplies: cylindrospermopsins. In 12th International Symposium on Toxicity Assessment (ISTA-12). Skiathos, Greece

Falconer IR, Buckley TH (1989) Tumour promotion by Microcystis sp. a blue-green alga occurring in water supplies. Medical Journal of Australia 150: 351

Falconer IR, Yeung DSK (1992) Cytoskeletal changes in hepatocytes induced by Microcystis toxins and their relation to hyperphosphorylation of cell proteins. Chem-Biol-Interact 81: 181-196

Falconer IR, Humpage AR (2001) Preliminary evidence for in vivo tumour initiation by oral administration of extracts of the blue-green alga Cylindrospermopsis raciborskii containing the toxin cylindrospermopsin. Environ Toxicol 16: 192-195

Falconer IR, Humpage AR (2005b) Health risk assessment of cyanobacterial (blue-green algal) toxins in drinking water. International Journal of Environmental Research and Public Health 2: 43-50

Falconer IR, Beresford AM, Runnegar MT (1983) Evidence of liver damage by toxin from a bloom of the blue-green alga, Microcystis aeruginosa. Med-J-Aust 1: 511-514

Falconer IR, Buckley T, Runnegar MT (1986) Biological half-life, organ distribution and excretion of 125-I-labelled toxic peptide from the blue-green alga Microcystis aeruginosa. Aust-J-Biol-Sci 39: 17-21

Falconer IR, Dornbusch M, Moran G, Yeung SK (1992) Effect of the cyanobacterial (blue-green algal) toxins from the Microcystis aeruginosa on isloated enterocytes from the chicken small intestine. Toxicon 30 7: 790-793

Falconer IR, Hardy SJ, Humpage AR, Froscio SM, Tozer GJ, Hawkins PR (1999) Hepatic and renal toxicity of the blue-green alga (cyanobacterium) Cylindrospermopsis raciborskii in male Swiss Albino mice. Environmental Toxicology 14: 143-150

Fastner J, Erhard M, von Dohren H (2001) Determination of oligopeptide diversity within a natural population of Microcystis spp. (Cyanobacteria) by typing single colonies by matrix-assisted laser desorption ionization-time of flight mass spectrometry. Applied and Environmental Microbiology 67: 5069-5076

Fastner J, Heinze R, Humpage AR, Mischke U, Eaglesham GK, Chorus I (2003) Cylindrospermopsin occurrence in two German lakes and preliminary assessment of toxicity and toxin production of Cylindrospermopsis raciborskii (Cyanobacteria) isolates. Toxicon 42: 313-321

Fischer WJ, Altheimer S, Cattori V, Meier PJ, Dietrich DR, Hagenbuch B (2005) Organic anion transporting polypeptides expressed in liver and brain mediate uptake of microcystin. Toxicol-Appl-Pharmacol 205: 257-265

Fladmark KE, Brustugun OT, Boe R, Vintermyr OK, Howland R, Gjertsen BT et al. (1999) Ultrarapid caspase-3 dependent apoptosis induction by serine/threonine phosphatase inhibitors. Cell Death and Differentiation 6: 1099-1108

Fleming LE, Rivero C, Burns J, Williams C, Bean JA, Shea KA, Stinn J (2002) Blue-green algal (cyanobacterial) toxins, surface drinking water, and liver cancer in Florida. Harmful Algae 1: 157-168

Frankmolle WP, Knuble G, Moore RE, Patterson GML (1992) Antifungal Cyclic Peptides from the Terrestrial Blue-Green Alga Anabaene laxa: Structures of Laxapohycin-A Laxaphycin-B, Laxaphycin-D, and Laxaphycin-E. Journal of Antibiotics 45: 1458-1466

Froscio SM, Humpage AR, Burcham PC, Falconer IR (2001) Cell-free protein synthesis inhibition assay for the cyanobacterial toxin cylindrospermopsin. Environ Toxicol 16: 408-412

Froscio SM, Humpage AR, Burcham PC, Falconer IR (2003) Cylindrospermopsin-induced protein synthesis inhibition and its dissociation from acute toxicity in mouse hepatocytes. Environmental Toxicology 18: 243-251

Fujii K, Sivonen K, Kashiwagi T, Hirayama K, Harada K (1999) Nostophycin, a novel cyclic peptide from the toxic cyanobacterium Nostoc sp 152. Journal of Organic Chemistry 64: 5777-5782

Fujiki H, Sugimura T (1987) New classes of tumor promoters: teleocidin, aplysiatoxin and palytoxin. Advances in Cancer Research 49: 223-264

Fujiki H, Suganuma M (1993) Tumor promotion by inhibitors of protein phosphatases 1 and 2A: The okadaic acid class of compounds. Advances in Cancer Research 61: 143-194

Fujiki H, Moore RE, Mori M, Nakayasu M, Terada M, Sugimura T (1981) Indole alkaloids: dihydroteleocidin B, teleocidin, and lyngbyatoxin A as members of a new class of tumor promoters. Proc-Natl-Acad-Sci-U-S-A 78: 3872-3876

Fujiki H, Ikegami K, Hakii H, Suganuma M, Yamaizuma Z, Yamazato K et al. (1985) A blue-green alga from Okinawa contains aplysiatoxins, the third class of tumor promoters. Jpn-J-Cancer-Res 76: 257-259

Gerwick WH, Reyes S, Alvarado B (1987) Two malyngamides from the Caribbean cyanobacterium Lyngbya majuscula. Phytochemistry 26, 6: 1701-1704

Gerwick WH, Mrozek C, Moghaddam MF, Agarwal SK (1989) Novel cytotoxic peptides from the tropical marine cyanobacterium Hormothamnion enteromorphoides. 1. Discovery, isolation and initial chemical and biological characterization of the Hormothamnins from wild and cultured material. Experientia 45, 2: 115-121

Gerwick WH, Jiang ZD, Agarwal SK, Farmer BT (1992) Total Structure of Homothamin A, a Toxic Cyclic Undecapeptide From the Tropical Marine Cyanobacterium Homothamnion entermorphoides. Tetrahedron Letters 48: 2313-2324

Gerwick WH, Lopez A, Van Duyne GD, Clary J, Ortiz W, Baez A (1986) Hormothamnione, a Novel Cytotoxin Strylchromone from the Marine Cyanophyte Hormothamnione enterpomorphoides Grunow. Tetrahedron Letters 17: 1979-1982

Gleason FK (1990) The natural herbicide, cyanobacterin, specifically disrupts thylakoid membrane structure in Euglena gracilis strain Z. FEMS Microbiology Letters 68: 77-82

Goldberg J, Huang H, Kwon Y, Greengard P, Nairn AC, Kuriyan J (1995) Three-dimensional structure of the catalytic subunit of protein serine/threonine phosphatase-1. Nature 376: 745-753

Gromov BV, Vepritskiy AA, Titova NN, Mamkayeva KA, Alexandrova OV (1991) Production of the antibiotic cyanobacterin LU-1 by Nostoc linckia CALU 892 cyanobacterium. Journal of Applied Phycology 3: 55-59

Gross EM, Wolk CP, Juttner F (1992) Fischerellin, a New Allelochemical from the Freshwater Cyanobacterium Fischerella musicola. Phycologia 27: 686-692

Gulavita N, Hori A, Shimazu A (1988) Aphanorphine, a Novel Tricyclic Alkaloid from the Blue-Green Alga Aphanizomenon flos-aquae. Tetrahedron Letters 29: 4381-4384

Harada K, Ohtani I, Iwamoto K, Suzuki M, Watanabe MF, Watanabe M, Terao K (1994) Isolation of cylindrospermopsin from a cyanobacterium Umezakia natans and its screening method. Toxicon 32: 73-84

Harada K, Fujii K, Shimada T, Suzuki M, Sano H, Adachi K, Carmichael WW (1995) Two cyclic peptides, anabaenopeptins, a third group of bioactive compounds from the cyanobacterium Anabaena flos-aquae NRC 525-17. Tetrahedron Letters 36: 1511-1514

Harada K, Suomalainen M, Uchida H, Masui H, Ohmura K, Kiviranta J et al. (2000) Insecticidal compounds against mosquito larvae from Oscillatoria agardhii strain 27. Environmental Toxicology 15: 114-119

Harada K, Ogawa K, Matsuura K, Nagai H, Murata H, Suzuki M et al. (1991) Isolation of two toxic heptapeptide microcystins from an axenic strain of Microcystis aeruginosa, K-139. Toxicon 29, 4/5: 479-489

Hashmi M, Anders MW (1991) Enzymatic reaction of beta-N-methylaminoalanine with L-amino acid oxidase. Biochim Biophys Acta 1074: 36-39

Hawkins PR, Runnegar MTC, Jackson ARB, Falconer IR (1985) Severe hepatotoxicity caused by the tropical cyanobacterium (blue-green alga) Cylindrospermopsis raciborskii (Woloszynska) Seenaya and Subba Raju isolated form a domestic supply reservoir. Applied and Environmental Microbiology 50: 1292-1295

Hawkins PR, Chandrasena NR, Jones GJ, Humpage AR, Falconer, IR (1997) Isolation and toxicity of Cylindrospermopsis raciborskii from an ornamental lake. Toxicon 35: 341-346

Herfindal L, Oftedal L, Selheim F, Wahlsten M, Sivonen K, Doskeland SO (2005) A high proportion of Baltic Sea benthic cyanobacterial isolates contain apoptogens able to induce rapid death of isolated rat hepatocytes. Toxicon 46: 252-260

Hilborn ED, Carmichael WW, Yuan M, Azevedo SMFO (2005) A simple colorimetric method to detect biological evidence of human exposure to microcystins. Toxicon 46: 218-221

Honkanen RE, Dukelow M, Zwiller J, Moore RE, Khatra BS, Boynton AL (1991) Cyanobacterial nodularin is a potent inhibitor of type 1 and type 2a protein phosphatases. Molecular Pharmacology 40: 577-583

Honkanen RE, Zwiller J, Moore RE, Daily SL, Khatra BS, Dukelow M, Boynton AL (1990) Characterization of microcystin-LR, a potent inhibitor of type 1 and type 2a protein phosphatases. J-Biol-Chem 265: 19401-19404

Hooser SB (2000) Fulminant hepatocyte apoptosis in vivo following microcystin-LR administration to rats. Toxicol Pathol 28: 726-733.

Huby NJS, Thompson P, Wonnacott S, Gallagher T (1991) Structural modification of anatoxin-a. Synthesis of model affinity ligands for the nicotinic acetylcholine receptor. Journal of the Chemical Society, Chemical Communications 4: 243-245

Humpage AR, Falconer IR (1999) Microcystin-LR and liver tumour promotion: Effects on cytokinesis, ploidy and apoptosis in cultured hepatocytes. Environmental Toxicology 14: 61-75

Humpage AR, Falconer IR (2003) Oral toxicity of the cyanobacterial toxin cylindrospermopsin in male Swiss albino mice: Determination of no observed adverse effect level for deriving a drinking water guideline value. Environ Toxicol 18: 94-103

Humpage AR, Fenech M, Thomas P, Falconer IR (2000a) Micronucleus induction and chromosome loss in transformed human white cells indicate clastogenic and aneugenic action of the cyanobacterial toxin, cylindrospermopsin. Mutat Res 472: 155-161

Humpage AR, Hardy SJ, Moore EJ, Froscio SM, Falconer IR (2000b) Microcystins (cyanobacterial toxins) in drinking water enhance the growth of aberrant crypt foci in the mouse colon. Journal of Toxicology & Environmental Health Part A 61: 155-165

Humpage AR, Fontaine F, Froscio S, Burcham P, Falconer IR (2005) Cylindrospermopsin genotoxicity and cytotoxicity: Role of cytochrome P-450 and oxidative stress. Journal of Toxicology and Environmental Health-Part a-Current Issues 68: 739-753

Humpage AR, Rositano J, Bretag AH, Brown R, Baker PD, Nicholson BC, Steffensen DA (1994) Paralytic shellfish poisons from Australian cyanobacterial blooms. Australian Journal of Marine and Freshwater Research 45: 761-771

Hyde EG, Carmichael WW (1991) Anatoxin-a(s), a naturally occurring organophosphate, is an irreversible active site-directed inhibitor of acetylcholinesterase (EC 3.1.1.7). Journal of Biochemical Toxicology 6 3: 195-201

Ikawa M, Wegener K, Foxall TL, Sasner JJ (1982) Comparisons of the toxins of the blue-green alga Aphanizomenon flos-aquae with the Gonyaulax toxins. Toxicon 20 4: 747-752

Ishibash M, Moore RE, Patterson GML, Xu CF, Clardy J (1986) Scytphycins, Cyto-Toxic and Antimycotic agents from the Cyanophyte Scytonema pseudo-hofmanni. Journal of Organic Chemistry 51: 5300-5306

Ishida K, Matsuda H, Okita Y, Murakami M (2002) Aeruginoguanidines 98-A-98-C: cytotoxic unusual peptides from the cyanobacterium Microcystis aeruginosa. Tetrahedron 58: 7645-7652

Ishitsuka MO, Kusumi T, Kakisawa H (1990) Microviridin: a novel tricyclic depsipeptide from the toxic cyanobacterium Microcystis viridis. Journal of the American Chemical Society 112: 8180-8182

Ito E, Kondo F, Terao K, Harada KI (1997) Neoplastic nodular formation in mouse liver induced by repeated intraperitoneal injections of microcystin-LR. Toxicon 35: 1453-1457

Jackim E, Gentile J (1968) Toxins of a blue-green alga: similarity to a saxitoxin. Science 162: 915-916

Jakobi C, Oberer L, Quiquerez C, Konig WA, Weckesser J (1995) Cyanopeptolin S, a sulfate-containing depsipeptide from a water bloom of Microcystis sp. FEMS Microbiology Letters 129: 129-133

Jalaludin B, Smith W (1992) Blue-green algae (cyanobacteria). Medical Journal of Australia 156: 744

Kaas H, Henriksen P (2000) Saxitoxins (PSP toxins) in Danish lakes. Water Research 34: 2089-2097

Kao CY (1993) Paralytic Shellfish Poisoning. In Algal Toxins in Seafood and Drinking Water. Falconer I (ed) London: Academic Press Limited, pp 75-86

Kaya K, Sano T, Beattie KA, Codd GA (1996) Nostocyclin, a novel 3-amino-6-hydroxy-2-piperidone-containing cyclic depsipeptide from the cyanobacterium Nostoc sp. Tetrahedron Letters 37: 6725-6728

Knubel G, Larsen LK, Moore RE, Levine IA, Patterson GML (1990) Cytotoxic, antiviral indolocarbazoles from a blue-green alga belonging to the Nostocaceae. Journal of Antibiotics 43 10: 1236-1239

Koehn FE, Longley RE, Reed JK (1992) Microcolins a and b, new immunosuppressive peptides from the blue-green alga Lyngbya majuscula. Journal of Natural Products 55 5: 613-619

Kondo F, Matsumoto H, Yamada S, Ishikawa N, Ito E, Nagata S et al. (1996) Detection and identification of metabolites of microcystins formed in vivo in mouse and rat livers. Chem-Res-Toxicol 9: 1355-1359

Krishnamurthy T, Szafraniec L, Hunt DF, Shabanowitz J, Yates RJ, Hauer CR et al. (1989) Structural characterization of toxic cyclic peptides from blue-green algae by tandem mass spectrometry. Proc-Natl-Acad-Sci-U-S-A 86: 770-774

Kuipcr-Goodman, T, Falconer, I, and Fitzgerald, J (1999) Human health aspects. In Toxic Cyanobacteria In Water. A Guide To Their Public Health Consequences, Monitoring and Management. Chorus I, Bartram J (eds) London: E & FN Spon on behalf of WHO, pp 113-153

Kusumi T, Ooi T, Watanabe MM, Takahsh MM, Kakisawa H (1987) Cyanoviridin-RR, a toxin from the cyanobacterium, (blue-green alga) Microcystis aeruginosa. Tetrahedron Letters 28: 4695-4698

Lagos N, Onodera H, Zagatto PA, Andrinolo D, Azevedo SMFQ, Oshima Y (1999) The first evidence of paralytic shellfish toxins in the freshwater cyanobacterium Cylindrospermopsis raciborskii, isolated from Brazil. Toxicon 37: 1359-1373

Lanaras T, Cook CM, Eriksson J, Meriluoto J, Hotokka M (1991) Computer modelling of the 3-dimensional structures of the cyanobacterial hepatotoxins microcystin-LR and nodularin. Toxicon 29 7: 901-906

Lefebvre KA (2002) Sublethal effects of saxitoxin on early development and behavioural performance in fish. In Xth International Conference on Harmful Algae. St Pete Beach, Florida, US

Li R, Carmichael WW, Liu Y, Watanabe MM (2000) Taxonomic re-evaluation of Aphanizomenon flos-aquae NH-5 based on morphology and 16S rRNA sequences. Hydrobiologia 438: 99-105

Li R, Carmichael WW, Brittain S, Eaglesham G, Shaw G, Liu Y, Watanabe MM (2001a) First report of the cyanotoxins cylindrospermopsin and deoxycylindrospermopsin from Raphidiopsis curvata (Cyanobacteria). Journal of Phycology 37: 1121-1126

Li R, Carmichael WW, Brittain S, Eaglesham GK, Shaw GR, Mahakhant A et al. (2001b) Isolation and identification of the cyanotoxin cylindrospermopsin and deoxy-cylindrospermopsin from a Thailand strain of Cylindrospermopsis raciborskii (Cyanobacteria). Toxicon 39: 973-980

Li R, Carmichael WW, Pereira P (2003) Morphological and 16S rRNA gene evidence for reclassification of the paralytic shellfish toxin producing Aphanizomenon flos-aquae LMECYA 31 as Aphanizomenon issatschenkoi (Cyanophyceae). Journal of Phycology 39: 814-818

Looper RE, Runnegar MTC, Williams RM (2005) Synthesis of the putative structure of 7-deoxycylindrospermopsin: C7 oxygenation is not required for the inhibition of protein synthesis. Angewandte Chemie-International Edition 44: 3879-3881

MacKintosh C, Beattie KA, Klumpp S, Cohen P, Codd GA (1990) Cyanobacterial microcystin-LR is a potent and specific inhibitor of protein phosphatases 1 and 2A from both mammals and higher plants. FEBS-Lett 264: 187-192

Mahmood NA, Carmichael WW (1986) The pharmacology of anatoxin-a(s), a neurotoxin produced by the freshwater cyanobacterium Anabaena flos-aquae NRC 525-17. Toxicon 24, 5: 425-434

Mahmood NA, Carmichael WW (1987) Anatoxin-a(s), an anticholinesterase from the cyanobacterium Anabaena flos-aquae NRC-525-17. Toxicon 25: 1221-1227

Marner FJ, Moore RE (1977) Majusculamides A and B, two Epimeric Lipodipeptides from Lyngbya majuscula Gomont. Journal of Organic Chemistry 42: 2815-2818

Marsalek B, Blaha L, Hindak F (2000) Review of toxicity of cyanobacteria in Slovakia. Biologia 55: 645-652

Martin C, Oberer L, Ino T, Konig WA, Busch M, Weckesser J (1993) Cyanopeptolins, new depsipeptides from the cyanobacterium Microcystis sp. PCC 7806. Journal of Antibiotics 46: 1550-1556

Matern U, Oberer L, Falchetto RA, Erhard M, Konig WA, Herdman M, Weckesser J (2001) Scyptolin A and B, cyclic depsipeptides from axenic cultures of Scytonema hofmanni PCC 7110. Phytochemistry 58: 1087-1095

Matsunaga S, Moore RE, Niemczura WP, Carmichael WW (1989) Anatoxin-a(s), a potent anticholinesterase from Anabaena flos-aquae. Journal of the American Chemical Society 111: 8021-8023

Mez K, Beattie K, Codd G, Hanselmann K, Hauser B, Naegeli H, Preisig H (1997) Identification of a microcystin in benthic cyanobacteria linked to cattle deaths on alpine pastures in Switzerland. Eur J Phycol 32: 111-117

Molica R, Onodera H, Garcia C, Rivas M, Andrinolo D, Nascimento S et al. (2002) Toxins in the freshwater cyanobacterium Cylindrospermopsis raciborskii (Cyanophyceae) isolated from Tabocas reservoir in Caruaru, Brazil, including demonstration of a new saxitoxin analogue. Phycologia 41: 606-611

Moon SS, Chen JL, Moore RE, Patterson GML (1992) Calophycin, a Fungicidal Cyclic Decapeptide from the Terrestrial Blue-Green Alga Calothrix fusca. Journal of Organic Chemistry 57: 1097-1103

Moore RE, Entzeroth M (1988) Majusculamide D and deoxymajusculamide D, two cytotoxins from Lyngbya majuscula. Phytochemistry 27 10: 3101-3103

Moore RE, Cheuk C, Patterson GML (1984) Hapalindoles: new alkaloids from the blue-green alga Hapalosiphon fontinalis. Journal of the American Chemical Society 106: 6456-6457

Moore RE, Yang XQ, Patterson GML (1987) Fontonumide and Anhydrohapaloxindole A, two new Alkaloids from the Blue-Green Alga Haplalosiphon fontinalis. Journal of Organic Chemistry 52: 3733-3777

Moore RE, Bornemann V, Niemczura WP, Gregson JM, Chen JL, Norton TR et al. (1989) Puwainaphycin c, a cardioactive cyclic peptide from the blue-green alga Anabaena BQ-16-1. Use of two dimensional 13 C-13 C and 13 C-15 N correlation spectroscopy in sequencing the amino acid units. Journal of the American Chemical Society 111: 6128-6132

Murakami M, Okita Y, Matsuda H, Okino T, Yamaguchi K (1994) Aeruginosin 298-A, a thrombin and trypsin inhibitor from the blue-green alga Microcystis aeruginosa (NIES-298). Tetrahedron Letters 35: 3129-3132

Murch SJ, Cox PA, Banack SA (2004a) A mechanism for slow release of biomagnified cyanobacterial neurotoxins and neurodegenerative disease in Guam. Proceedings of the National Academy of Sciences of the United States of America 101: 12228-12231

Murch SJ, Cox PA, Banack SA, Steele JC, Sacks OW (2004b) Occurrence of beta-methylamino-L-alanine (BMAA) in ALS/PDC patients from Guam. Acta Neurologica Scandinavica 110: 267-269

Myers TG, Nelson SD (1990) Neuroactive carbamate adducts of beta-N-methylamino-L-alanine and ethylenediamine. Detection and quantitation under physiological conditions by 13C NMR. J Biol Chem 265: 10193-10195

Mynderse JS, Moore RE (1978) Toxins from Blue-Green Algae: Structures of Oscillatoxin A and three Related Bromine-Containing Toxins. Journal of Organic Chemistry 43: 2301-2303

Nagatsu A, Kajitani H, Sakakibara J (1995) Muscoride A: A new oxazole peptide alkaloid from freshwater cyanobacterium Nostoc muscorum. Tetrahedron Letters 36: 4097-4100

Namikoshi M, Rinehart KL (1996) Bioactive compounds produced by cyanobacteria. Journal of Industrial Microbiology & Biotechnology 17: 373-384

Namikoshi M, Rinehart KL, Sakai Y, Sivonen K, Carmichael WW (1990) Structures of three new cyclic heptapeptide hepatotoxins produced by the cyanobacterium (blue-green alga) Nostoc sp. strain 152. Journal of Organic Chemistry 55: 6135-6139

Namikoshi M, Choi BW, Sakai R, Sun F, Rinehart KL (1994) New nodularins: A general method for structure assignment. Journal of Organic Chemistry 59: 2349-2357

Namikoshi M, Sivonen K, Evans WR, Sun F, Carmichael WW, Rinehart KL (1992a) Isolation and structures of microcystins from a cyanobacterial water bloom (Finland). Toxicon 30 11: 1473-1479

Namikoshi M, Choi BW, Sun F, Rinehart KL, Evans WR, Carmichael WW (1993) Chemical characterization and toxicity of dihydro derivatives of nodularin and microcystin-LR, potent cyanobacterial cyclic peptide hepatotoxins. Chem Res Toxicol 6: 151-158

Namikoshi M, Sivonen K, Evans WR, Carmichael WW, Rouhainen L, Luukainen R, Rinehart KL (1992b) Structures of three new Homotyrosine Containing Microcystins and a new Homophenylalanine Variant from Anabaena SP. Strain 66. Chem Res Toxicol 5: 661-666

Namikoshi M, Sun FR, Choi BW, Rinehart KL, Carmichael WW, Evans WR, Beasley VR (1995) Seven more microcystins from homer lake cells - application of the general method for structure assignment of peptides containing alpha,beta-dehydroamino acid unit(s). Journal of Organic Chemistry 60: 3671-3679

Namikoshi M, Sivonen K, Evans WR, Carmichael WW, Sun F, Rouhiainen L et al. (1992c) Two new L-serine variants of microcystins-LR and -RR from Anabaena sp. Strains 202 A1 and 202 A2. Toxicon 30: 1457-1464

Namikoshi M, Rinehart KL, Sakai R, Stotts RR, Dahlem AM, Beasley VR et al. (1992d) Identification of 12 hepatotoxins from a Homer Lake bloom of the cyanobacteria Microcystis aeruginosa, Microcystis viridis and Microcystis wesenbergii : nine new microcystins. Journal of Organic Chemistry 57: 866-872

Namikoshi M, Yuan M, Sivonen K, Carmichael WW, Rinehart KL, Rouhiainen L et al. (1998) Seven new microcystins possessing two L-glutamic acid units, isolated from Anabaena sp. strain 186. Chem Res Toxicol 11: 143-149

Negri A, Stirling D, Quilliam M, Blackburn S, Bolch C, Burton I et al. (2003) Three novel hydroxybenzoate saxitoxin analogues isolated from the dinoflagellate Gymnodinium catenatum. Chem Res Toxicol 16: 1029-1033

Nicholson BC, Burch MD (2001) Evaluation of Analytical Methods for Detection and Quantification of Cyanotoxins in Relation to Australian Drinking Water Guidelines. National Health and Medical Research Council of Australia, the Water Services Association of Australia, and the Cooperative Research Centre for Water Quality and Treatment, Australia, pp 57 Available from http://nhmrc.gov.au/publications/_files/eh22.pdf

Norris RL, Seawright AA, Shaw GR, Smith MJ, Chiswell RK, Moore MR (2001) Distribution of 14C cylindrospermopsin in vivo in the mouse. Environ Toxicol 16: 498-505

Norris RL, Eaglesham GK, Pierens G, Shaw GR, Smith MJ, Chiswell RK et al. (1999) Deoxycylindrospermopsin, an analog of cylindrospermopsin from Cylindrospermopsis raciborskii. Environmental Toxicology 14: 163-165

Norris RL, Seawright AA, Shaw GR, Senogles P, Eaglesham GK, Smith MJ et al. (2002) Hepatic xenobiotic metabolism of cylindrospermopsin in vivo in the mouse. Toxicon 40: 471-476

Ohta T, Sueoka E, Iida N, Komori A, Suganuma M, Nishiwaki R et al. (1994) Nodularin, a potent inhibitor of protein phosphatases 1 and 2A, is a new environmental carcinogen in male F344 rat liver. Cancer Res 54: 6402-6406

Ohtani I, Moore RE, Runnegar MTC (1992) Cylindrospermopsin: A potent hepatotoxin from the blue-green alga Cylindrospermopsis raciborskii. Journal of the American Chemical Society 114: 7941-7942

Okino T, Matsuda H, Murakami M, Yamaguchi K (1993) Microginin, an angiotensin-converting enzyme inhibitor from the blue-green alga Microcystis aeruginosa. Tetrahedron Letters 34: 501-504

Onodera H, Oshima Y, Henriksen P, Yasumoto T (1997a) Confirmation of Anatoxin-a(s), in the cyanobacterium Anabaena lemmermannii, as the cause of bird kills in Danish lakes. Toxicon 35: 1645-1648

Onodera H, Satake M, Oshima Y, Yasumoto T, Carmichael WW (1997b) New saxitoxin analogues from the freshwater filamentous cyanobacterium Lyngbya wollei. Nat Toxins 5: 146-151

Orjala J, Nagle DG, Hsu VL, Gerwick WH (1995) Antillatoxin: An exceptionally ichthyotoxic cyclic lipopeptide from the tropical cyanobacterium Lyngbya majuscula. Journal of the American Chemical Society 117: 8281-8282

Oshima Y (1995) Postcolumn derivatization liquid chromatographic method for paralytic shellfish toxins. Journal of AOAC International 78: 528-532

Park A, Moore RE, Patterson GML (1992) Ischerindole L, a new isonitrile from the terrestrial blue-green alga Fischerella muscicola. Tetrahedron Letters 33 23: 3257-3260

Patterson GML, Carmeli S (1992) Biological effects of tolytoxin (6-hydroxy-7-o-methyl-scytophycin b), a potent bioactive metabolite from cyanobacteria. Archiv fuer Hydrobiologie 157: 406-410

Pomati F, Sacchi S, Rossetti C, Giovannardi S, Onodera H, Oshima Y, Neilan BA (2000) The freshwater cyanobacterium Planktothrix sp FP1: molecular identification and detection of paralytic shellfish poisoning toxins. Journal of Phycology 36: 553-562

Prinsep MR, Moore RE, Levine IA, Patterson GML (1992a) Westiellamide, a Bistratamide-Related Cyclic Peptide from the Blue-Green Alga Westiellopsis prolifica. Journal of Natural Products 55: 140-142

Prinsep MR, Caplan FR, Moore RE, Patterson GML, Honkanen RE, Boynton AL (1992b) Microcystin-LR from a blue-green alga belonging to the stigonematales. Phytochemistry 31 4: 1247-1248

Rinehart KL, Namikoshi M (1994) Structure and biosynthesis of toxins from blue-green algae (cyanobacteria). Journal of Applied Phycology 6: 159-176

Rinehart KL, Harada K, Namikoshi M, Chen C, Harvis CA (1988) Nodularin, microcystin and the configuration of ADDA. Journal of the American Chemical Society 110: 8557-8558

Robinson NA, Miura GA, Matson CF, Dinterman RE, Pace JG (1989) Characterization of chemically tritiated microcystin-LR and its distribution in mice. Toxicon 27: 1035-1042

Robinson NA, Matson CF, Miura GA, Lynch TG, Pace JG (1990) Toxicokinetics of [3 H]microcystin-LR in mice. The FASEB Journal 4: A753,#2823

Robinson NA, Pace JG, Matson CF, Miura GA, Lawrence WB (1991) Tissue distribution, excretion and hepatic biotransformation of microcystin-LR in mice. Journal of Pharmacology and Experimental Therapeutics 256: 176-182

Runnegar M, Berndt N, Kaplowitz N (1995a) Microcystin uptake and inhibition of protein phosphatases: effects of chemoprotectants and self-inhibition in relation to known hepatic transporters. Toxicol Appl Pharmacol 134: 264-272

Runnegar M, Berndt N, Kong SM, Lee EY, Zhang L (1995b) In vivo and in vitro binding of microcystin to protein phosphatases 1 and 2A. Biochem Biophys Res Commun 216: 162-169

Runnegar MT, Falconer IR (1986) Effect of toxin from the cyanobacterium Microcystis aeruginosa on ultrastructural morphology and actin polymerization in isolated hepatocytes. Toxicon 24: 109-115

Runnegar MT, Falconer IR, Silver J (1981) Deformation of isolated rat hepatocytes by a peptide hepatotoxin from the blue-green alga Microcystis aeruginosa. Naunyn-Schmiedebergs-Arch-Pharmacol 317: 268-272

Runnegar MT, Gerdes RG, Falconer IR (1991) The uptake of the cyanobacterial hepatotoxin microcystin by isolated rat hepatocytes. Toxicon 29: 43-51

Runnegar MT, Kong S, Berndt N (1993) Protein phosphatase inhibition and in vivo hepatotoxicity of microcystins. Am-J-Physiol 265: G224-230

Runnegar MT, Falconer IR, Buckley T, Jackson AR (1986) Lethal potency and tissue distribution of 125I-labelled toxic peptides from the blue-green alga Microcystis aeruginosa. Toxicon 24: 506-509

Runnegar MT, Kong SM, Zhong YZ, Lu SC (1995c) Inhibition of reduced glutathione synthesis by cyanobacterial alkaloid cylindrospermopsin in cultured rat hepatocytes. Biochem Pharmacol 49: 219-225

Runnegar MT, Kong SM, Zhong YZ, Ge JL, Lu SC (1994) The role of glutathione in the toxicity of a novel cyanobacterial alkaloid cylindrospermopsin in cultured rat hepatocytes. Biochem Biophys Res Commun 201: 235-241

Runnegar MT, Xie CY, Snider BB, Wallace GA, Weinreb SM, Kuhlenkamp J (2002) In vitro hepatotoxicity of the cyanobacterial alkaloid cylindrospermopsin and related synthetic analogues. Toxicological Sciences 67: 81-87

Saker ML, Nogueira ICG, Vasconcelos VM, Neilan BA, Eaglesham GK, Pereira P (2003) First report and toxicological assessment of the cyanobacterium Cylindrospermopsis raciborskii from Portuguese freshwaters. Ecotoxicology and Environmental Safety 55: 243-250

Sano T, Kaya K (1995) Oscillamide Y, A chymotrypsin inhibitor from toxic Oscillatoria agardhii. Tetrahedron Letters 36: 5933-5936

Schembri MA, Neilan BA, Saint CP (2001) Identification of genes implicated in toxin production in the cyanobacterium Cylindrospermopsis raciborskii. Environ Toxicol 16: 413-421

Seawright AA, Brown AW, Nolan CC, Cavanagh JB (1990) Selective degeneration of cerebellar cortical neurons caused by cycad neurotoxin, L-betamethylaminoalanine (L-BMAA), in rats. Neuropathol Appl Neurobiol 16: 153-169

Seawright AA, Nolan CC, Shaw GR, Chiswell RK, Norris RL, Moore MR, Smith MJ (1999) The oral toxicity for mice of the tropical cyanobacterium Cylindrospermopsis raciborskii (Woloszynska). Environmental Toxicology 14: 135-142

Shaw GR, Sukenik A, Livne A, Chiswell RK, Smith MJ, Seawright AA et al. (1999) Blooms of the cylindrospermopsin containing cyanobacterium, Aphanizomenon ovalisporum (Forti), in newly constructed lakes, Queensland, Australia. Environmental Toxicology 14: 167-177

Shen XY, Lam PKS, Shaw GR, Wickramasinghe W (2002) Genotoxicity investigation of a cyanobacterial toxin, cylindrospermopsin. Toxicon 40: 1499-1501

Shimizu Y (2000) Paralytic Shellfish Poisons: Chemistry and mechanism of action. In Seafood and Freshwater Toxins: Pharmacology, Physiology, and Detection. Botana LM (ed) New York: Marcel Dekker Inc, pp 151-172

Shin HJ, Murakami M, Matsuda H, Ishida K, Yamaguchi K (1995) Oscillapeptin, an elastase and chymotrypsin inhibitor from the cyanobacterium Oscillatoria agardhii (NIES-204). Tetrahedron Letters 36: 5235-5238

Sivonen K, Jones G (1999) Cyanobacterial Toxins. In Toxic Cyanobacteria In Water. A Guide To Their Public Health Consequences, Monitoring and Management. Chorus I, Bartram J (eds) London: E & FN Spon on behalf of WHO, pp 41-111

Sivonen K, Himberg K, Luukkainen R, Niemela SI, Poon GK, Codd GA (1989) Preliminary characterization of neurotoxic cyanobacteria blooms and strains from Finland. Toxicity Assessment 4: 339-352

Sivonen K, Niemela SI, Niemi RM, Lepisto L, Luoma TH, Rasanen LA (1990) Toxic cyanobacteria (blue-green algae) in Finnish fresh and coastal waters. Hydrobiologia 190: 267-275

Sivonen K, Namikoshi M, Evans WR, Gromov BV, Carmichael WW, Rinehart KL (1992) Isolation and structures of five microcystins from a Russian Microcystis aeruginosa strain calu 972. Toxicon 30 11: 1481-1485

Skulberg OM (2000) Microalgae as a source of bioactive molecules - experience from cyanophyte research. Journal of Applied Phycology 12: 341-348

Skulberg OM, Carmichael WW, Andersen RA, Matsunaga S, Moore RE, Skulberg R (1992) Investigations of a neurotoxic oscillatorian strain (cyanophyceae) and its toxin. Isolation and characterization of homoanatoxin-a. Environmental Toxicology and Chemistry 11: 321-329

Spencer P, Nunn PB, Hugon J, Ludolph, A, Ross SM, Roy DN, Robertson RC (1987) Guam amyotropic lateral sclerosis-Parkinsonism-dementia linked to a plant excitant neurotoxin. Science 237: 517-522

Stirling DJ, Quilliam MA (2001) First report of the cyanobacterial toxin cylindrospermopsin in New Zealand. Toxicon 39: 1219-1222

Swanson KL, Rapoport H, Albuquerque EX, Aronstam RS (1990) Nicotinic acetylcholine receptor function studied with synthetic (+)-anatoxin-a and derivatives. In Marine toxins. Origin, structure, and molecular pharmacology. Hall S, Strichartz G (eds) Washington, DC: American Chemical Society, pp 107-118

Swanson KL, Aronstam RS, Wonnacott S, Rapoport H, Albuquerque EX (1991) Nicotinic pharmacology of anatoxin analogs. I. Side chain structure-activity relationships at peripheral agonist and noncompetitive antagonist sites. J Pharmacol Exp Ther 259: 377-386

Terao K, Ohmori S, Igarashi K, Ohtani I, Watanabe MF, Harada KI et al. (1994) Electron microscopic studies on experimental poisoning in mice induced by cylindrospermopsin isolated from blue-green alga Umezakia natans. Toxicon 32: 833-843

Thompson WL, Pace JG (1992) Substances that protect cultured hepatocytes from the toxic effects of microcystin-LR. Toxicology in Vitro 6: 579-587

Thompson WL, Bostian KA, Robinson NA, Pace JG (1988) Protective effects of bile acids on cultured hepatocytes exposed to the hepatotoxin, microcystin. The FASEB Journal 3 3: A372 #846

Tsukamoto S, Painuly P, Young KA, Yang X, Shimizu Y (1993) Microcystilide A: A novel cell-differentiation-promoting depsipeptide from Microcystis aeruginosa NO-15-1840. Journal of the American Chemical Society 115: 11046-11047

Ueno Y, Nagata S, Tsutsumi T, Hasegawa A, Yoshida F, Suttajit M et al. (1996) Survey of Microcystins in Environmental Water by a Highly Sensitive Immunoassay Based on Monoclonal Antibody. Nat-Toxins 4: 271-276

Vega A, Bell EA (1967) α-Amino-β-methyl aminopropionic acid, a new amino acid from seeds of Cycas circinalis. Phytochemistry 6: 759-762

Verpritskii AA, Gromov BV, Titova NN, Mamkaeva KA (1991) Production of the Antibiotic-Algicide Cyanobacterin LU-2 by the Filamentous Cyanobacterium Nostoc SP. Microbiology - English Translation 60: 675-679

Vezie C, Brient L, Sivonen K, Bertru G, Lefeuvre JC, Salkinojasalonen M (1998) Variation of microcystin content of cyanobacterial blooms and isolated strains in lake Gand-lieu (France). Microbial Ecology 35: 126-135

Watanabe MF, Watanabe M, Kato T, Harada KI, Suzuki M (1991) Composition of cyclic peptide toxins among strains of Microcystis aeruginosa (blue-green algae, cyanobacteria). Botanical Magazine, Tokyo 104: 49-57

Weiss JH, Christine CW, Choi DW (1989a) Bicarbonate dependence of glutamate receptor activation by beta-N-methylamino-L-alanine: channel recording and study with related compounds. Neuron 3: 321-326

Weiss JH, Koh JY, Choi DW (1989b) Neurotoxicity of beta-N-methylamino-L-alanine (BMAA) and beta-N-oxalylamino-L-alanine (BOAA) on cultured cortical neurons. Brain Res 497: 64-71

Wonnacott S, Jackman S, Swanson KL, Rapoport H, Albuquerque EX (1991) Nicotinic pharmacology of anatoxin analogs. II. Side chain structure-activity relationships at neuronal nicotinic ligand binding sites. J Pharmacol Exp Ther 259: 387-391

Yu SZ (1995) Primary prevention of hepatocellular carcinoma. Journal of Gastroenterology and Hepatology 10: 674-682

Zhou L, Yu H, Chen K (2002) Relationship between microcystin in drinking water and colorectal cancer. Biomed Environ Sci 15: 166-171

Appendix A

Table A.1. Some of the many bioactive compounds that have been isolated from cyanobacteria

Name	Chemical Class	Activity	Source	Reference
Acutiphycin	-	Antineoplastic	Oscillatoria acutissima	(Barchi et al. 1984)
Aeruginosins	Linear depsipeptides	Protease inhibitors	Microcystis spp., Oscillatoria spp.	(Murakami et al. 1994)
Aeruginoguanidines	Peptide	Cytotoxic	M. aeruginosa	(Ishida et al. 2002)
Anabaenapeptins	Cyclic peptides	Vasodilation	Anabaena flos-aquae, other spp.	(Harada et al. 1995)
Antillatoxin	Cyclic lipopeptide	Ichthyotoxin	Lyngbya majuscula	(Orjala et al. 1995)
Aphanorphine	Alkaloid	-	Aphanizomenon flos-aquae	(Gulavita et al. 1988)
Aplysiatoxins	Phenolic bislactone	Tumour promoters	Lyngbya majuscula	(Fujiki et al. 1985; Fujiki and Sugimura 1987)
Apoptogens	Unknown	Induction of apoptosis	Various benthic species	(Herfindal et al. 2005)
"Bioactive compounds"	-	-	Nostoc muscorum	(De Mule et al. 1991)
Calophycin	Decapeptide	Antifungal	Calothrix fusca	(Moon et al. 1992)
Cyanobacterin	-	Algicide, herbicide, antibiotic	Nostoc linckia	(Gleason 1990; Gromov et al. 1991)

Name	Chemical Class	Activity	Source	Reference
Cyanobacterin	-	Antibiotic, algicide	Nostoc spp.	(Verpritskii et al. 1991)
Cyanopeptolins	Depsipeptide	Protease inhibitors	Microcystis spp.	(Martin et al. 1993; Jakobi et al. 1995)
Fisherellin	-	Allelotoxin	Fischerella musicola	(Gross et al. 1992)
Fontomumide	Alkaloid	-	Hapalosiphon fontinalis	(Moore et al. 1987)
Haplaindoles	Alkaloids	-	Hapalosiphon fontinalis	(Moore et al. 1984)
Hormothamnins	Cyclic undecapeptide	Cytotoxic, antimicrobial, antimycotic	Hormothamnione enteromorphoides	(Gerwick et al. 1989; Gerwick et al. 1992)
Hormothamnione	-	Cytotoxic	Hormothamnione enteromorphoides	(Gerwick et al. 1986)
Indolcarbazoles	-	Cytotoxic, antiviral	Nosctocaceae	(Knubel et al. 1990)
Insecticidal compounds	-	-	Oscillatoria agardhii	(Harada et al. 2000)
Ischerindole	Isonitrile	-	Fischerella musicola	(Park et al. 1992)
Laxaphycin	Cyclic undeca- or dodeca-peptides	Cytotoxic, antimicrobial, antimycotic	Anabaena laxa	(Frankmolle et al. 1992)
Lyngbyatoxins	Cyclic dipeptides	Tumour promoters	Lyngbya majuscula	(Fujiki et al. 1981)
Majusculamides	Heptacyclo-depsipeptides	Cytotoxic, antifungal	Lyngbya majuscula	(Marner and Moore 1977; Carter et al. 1984; Moore and Entzeroth 1988)
Malyngamides	-	-	Lyngbya majuscula	(Gerwick et al. 1987)
Malyngolide	-	Antibiotic	Lyngbya majuscula	(Cardelina et al. 1979)

Name	Chemical Class	Activity	Source	Reference
Malynic acid	Fatty acid	Cytotoxin	Lyngbya majuscula	(Cardelina and Moore 1980)
Microcolins	Peptide	Immuno-suppressant	Lyngbya majuscula	(Koehn et al. 1992)
Microcystilide	Depsipeptide	Cell differentiation promoter	Microcystis aeruginosa	(Tsukamoto et al. 1993)
Microginin	Linear pentapeptide	ACE inhibitor	Microcystis aeruginosa	(Okino et al. 1993)
Microviridins	Depsipeptide	Protease inhibitors	M. viridis	(Ishitsuka et al. 1990)
Mirabazoles	Alkaloids	Cytotoxins	Scytonema mirabile	(Carmeli et al. 1991)
Muscoride	Oxazole peptide alkaloid	-	Nostoc muscorum	(Nagatsu et al. 1995)
Nostocyclin	Depsipeptide	-	Nostoc spp.	(Kaya et al. 1996)
Nostophycin	Cyclic peptide	-	Nostoc spp.	(Fujii et al. 1999)
Oscillamide	Cyclic peptide	Chymotrypsin inhibitor	Oscillatoria agardhii	(Sano and Kaya 1995)
Oscillapeptin	-	Chymotrypsin and elastin inhibitor	Oscillatoria agardhii	(Shin et al. 1995)
Oscillatoxin	-	Toxin	Oscillatoria spp.	(Mynderse and Moore 1978)
Pahayokolide A	-	Antibiotic, cytotoxic	Lyngbya spp. (freshwater)	(Berry et al. 2004)
Polytoxin	-	-	Scytonema spp.	(Carmeli et al. 1990a)
Puwainaphycin	Cyclic peptide	Cardioactive	Anabaena spp.	(Moore et al. 1989)
Scyptolins	Depsipeptides	-	Scytonema hofmanii	(Matern et al. 2001)

Name	Chemical Class	Activity	Source	Reference
Scytophycins, tolytoxin	-	Cytotoxic, antimycotic	Scytonema pseudohofmanni	(Ishibash et al. 1986; Patterson and Carmeli 1992)
Tantazoles	Alkaloids	Cytotoxins	Scytonema mirabile	(Carmeli et al. 1990b)
Westiallamide	Cyclic hexapeptide	Cytotoxic	Westiellopsis prolifica	(Prinsep et al. 1992a)

Chapter 17: The genetics and genomics of cyanobacterial toxicity

Brett A Neilan, Pearson LA, Moffitt MC, Mihali KT, Kaebernick M, Kellmann R, Pomati F

Cyanobacteria and Astrobiology Research Laboratory, School of Biotechnology and Biomolecular Sciences, The University of New South Wales, Sydney 2052, NSW, Australia

Introduction

The past ten years has witnessed major advances in our understanding of natural product biosynthesis, including the genetic basis for toxin production by a number of groups of cyanobacteria. Cyanobacteria produce an unparalleled array of bioactive secondary metabolites; including alkaloids, polyketides and non–ribosomal peptides, some of which are potent toxins. This paper addresses the molecular genetics underlying cyanotoxin production in fresh and brackish water environments. The major toxins that have been investigated include microcystin, cylindrospermopsin, nodularin, the paralytic shellfish poisons (PSP), including saxitoxin, and the anatoxins.

Non–ribosomal peptide synthesis is achieved in prokaryotes and lower eukaryotes via the thiotemplate function of large, modular enzyme complexes, known collectively as peptide synthetases. Most non–ribosomal peptides from microorganisms are classified as secondary metabolites, that is, they rarely have a role in primary metabolism, growth, or reproduction but have evolved to somehow benefit the producing organism. Most cyanobacterial genera have either been shown to produce non–ribosomal peptides or have them encoded within their genomes. Early work on the genetics of cyanobacterial toxicity led to the discovery of one of the first examples of hybrid peptide–polyketide synthetases. This enzyme complex directed the production of the cyclic heptapeptide microcystin and, as one of the largest known bacterial gene clusters, is encoded by more than 55 kb. Orthologs of microcystin synthetase have been found in several strains of *Microcystis* and other genera of toxic cyanobacteria, including *Planktothrix*, *Nostoc*, *Anabaena*, *Nodularia*, *Phormidium*, and *Chroococcus*. The homologous gene cluster in *Nodularia* is predicted to be responsible for the synthesis of the pentapeptide nodularin, providing evidence of genetic recombination and possibly transfer during the evolution of these compounds. Genomic information related to microcystin and nodularin synthesis has also indicated their environmental and cellular regulators, as well as associated transport mechanisms.

More recently, hybrid peptide and polyketide synthetic pathways have been implicated in the production of the alkaloid cylindrospermopsin. Other predicted biosynthetic pathways are also under scrutiny and are being used in the search for candidate gene loci involved in PSP and anatoxin production. The correlation between toxicity and salt tolerance may raise future concerns as these cyanobacteria could compromise the safety of recycled and desalinated drinking water supplies.

The pattern of acquisition of genes responsible for cyanobacterial toxicity is not, on the whole, related to the evolution of potentially toxic species and the global distribution of toxic strains has been the topic of several phylogeographical studies. The exceptions to this include the species *Nodularia spumigena* and Australian strains of *Anabaena circinalis* that produce saxitoxin. Toxin biosynthesis gene cluster–associated transposition and the natural transformability of certain species allude to a broader distribution of toxic taxa. The information gained from the discovery of these toxin biosynthetic pathways has enabled the genetic screening of various environments for drinking water quality management. Understanding the role of these toxins in the producing microorganisms and the environmental regulation of their biosynthesis genes may also suggest the means for controlling toxic bloom events.

Genes involved in the biosynthesis of cyanotoxins

This section of the paper describes the characterization of the biosynthetic gene clusters that have either, by mutation, or by functional prediction been shown to be required for toxin production. Where possible the chemical structures of the toxins will not be revised here unless recent related genetic or enzymology data is available. Other genes encoding cofactors and regulators, for example, are also intimately linked to cyanotoxin production and, where appropriate, these will also be reviewed. Due to the number of toxins to be considered and their distribution across various cyanobacterial genera, each toxin biosynthetic pathway will be discussed separately, in the chronological order of their elucidation.

Microcystin synthetases

The structurally related hepatotoxins microcystin and nodularin are synthesized nonribosomally by the thiotemplate functions of large multifunctional enzyme complexes containing both nonribosomal peptide synthetase (NRPS) and polyketide synthetase (PKS) domains. The gene clusters encoding these biosynthetic enzymes, *mcyS* (microcystin) and *ndaS* (nodularin), have recently been sequenced and partially characterized in several cyanobacterial genera including *Microcystis*, *Anabaena*, *Planktothrix*, and *Nodularia* (Dittmann et al. 1997; Tillett et al. 2000; Christiansen et al. 2003; Moffitt et al. 2004; Rouhiainen et al. 2004). These fundamental studies have afforded insight into the evolution of cyanotoxin biosynthesis, and have provided the groundwork for current PCR–based toxic cyanobacteria detection methods.

The microcystin biosynthesis gene cluster, *mcyS*, was the first complex metabolite gene cluster to be fully sequenced from a cyanobacterium. In *M. aeruginosa* PCC7806, the *mcyS* genomic locus spans 55 kb and comprises 10 genes arranged in two divergently transcribed operons (*mcyA–C* and *mcyD–J*). The larger of the two operons, *mcyD–J*, encodes a modular PKS (McyD), two hybrid enzymes comprising NRPS and PKS modules (McyE and McyG), and enzymes putatively involved in the tailoring (McyJ, F, and I) and transport (McyH) of the toxin. The smaller operon, *mcyA–C* encodes three NRPS (McyA–C). Interestingly, the arrangement of ORFs in the *mcyS* cluster was found to be different in the organisms *Microcystis*, *Anabaena* and *Planktothrix* (Fig. 1). The *Anabaena mcyS* cluster adheres to the 'co–linearity' rule of NRPS pathways that predicts the order of catalytic processes involved in the biosynthesis of a non–ribosomal metabolite is generally the same as the order of the genes which encode their catalytic enzymes (Kleinkauf and von Dohren 1996), however the *Microcystis* and *Plankothrix* clusters do not adhere to this rule.

The formation of Adda (3–amino–9–methoxy–2,6,8,–trimethyl– 10–phenyl–4,6–decandienoic acid) putatively involves enzymes encoded by *mcyD–G* and *mcyJ*, based on bioinformatic analyses and homology to related enzymes. The hybrid NRPS/PKS enzyme, McyG, constitutes the first step in Adda biosynthesis–the putative activation of phenylacetate, catalyzed by the NRPS adenylation domain. The activated phenylacetate is transferred to the 4–phosphopantetheine cofactor of the first carrier domain. Phenylacetate is then extended by several malonyl–CoA elongation steps and subsequently modified by C–methylations, reduction and dehydration, all catalyzed by the PKS modules of McyD, E, and G. The aminotransferase domain of McyE converts the polyketide to a β–amino acid in the final step of Adda biosynthesis. The NRPS module of the second hybrid PKS/NRPS enzyme, McyE, is thought to be involved in the activation and condensation of D–Glu with Adda. No phospopantetheine transferase, that catalyzes the post–translational modification of the holo–NRPS, is associated with the *mcyS* gene cluster.

The *mcyF* ORF was originally predicted to encode a glutamate racemase, responsible for the epimerisation of the L–Glu residue of microcystin (Tillett et al. 2000; Nishizawa et al. 2001). A subsequent study by Sielaf et al. (2003), contested this theory and provided evidence that McyF acts exclusively as an Asp racemase to form the D–*erythro* β–methylaspartate (D–MeAsp) residue. It was proposed that the D–Glu residue is provided by a L–Glu racemase resident external to the *mcyS* gene cluster. However, earlier feeding experiments with isotopic substrates identified that the D–MeAsp residue is formed via a novel pathway (Moore et al. 1991) that would be consistent with McyF acting as a Glu racemase. Mutagenesis experiments in *P. agardhii* have shown that the production of Adda also involved an O–methylation step catalyzed by the putative monofunctional tailoring enzyme, McyJ (Christiansen et al. 2003).

The remaining biosynthetic enzymes in the microcystin biosynthesis pathway (NRPSs) are putatively involved in the specific activation, modification, and condensation of substrate amino acids onto the linear peptide chain that is then cyclized to produce microcystin. In *M. aeruginosa* PCC7806, initially, McyA adds L–Ser to the growing chain, followed by the addition of D–Ala. This step is followed by the addition of L–Leu and D–MeAsp residues (McyB) followed by the addition of L–Arg (McyC), and subsequent thioesterase–dependent cyclization and release of the final peptide product (Tillett et al. 2000).

An additional monofunctional tailoring enzyme, McyI is hypothesized to play a role in the modification of microcystin. McyI shows greatest homology to a group of D–3–phosphoglycerate dehydrogenase (D–3–PGDH) enzymes from various archaeal species. In *E. coli*, D–3–PGDH enzymes are responsible for the first step in the pathway for serine biosynthesis (Sugimoto and Pizer 1968). The role of this enzyme in microcystin synthesis is unknown, however, the MeDha residue in microcystin is produced from L–Ser, therefore McyI may play a role in the production of L–Ser or conversion of L–Ser to MeDha.

An ABC transporter–like gene *mcyH*, is believed to be involved in the transport of microcystin (Pearson et al. 2004). This transporter may be responsible for the thylakoid localization of the toxin (Shi et al. 1995; Young et al. 2005) or for the extrusion of the toxin under certain growth conditions, including exposure to high and red light (Kaebernick et al. 2000).

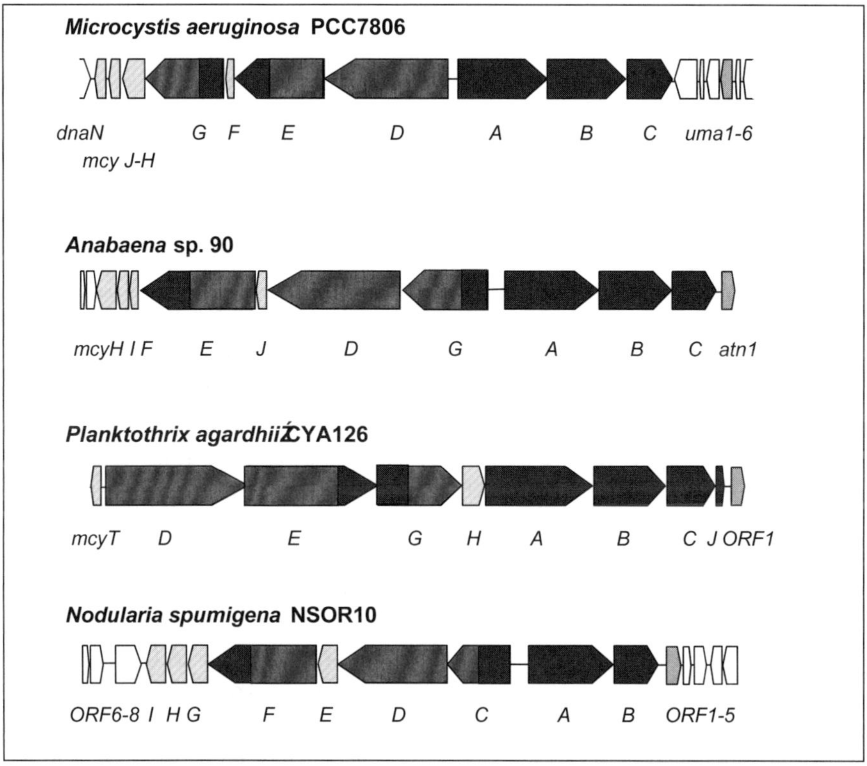

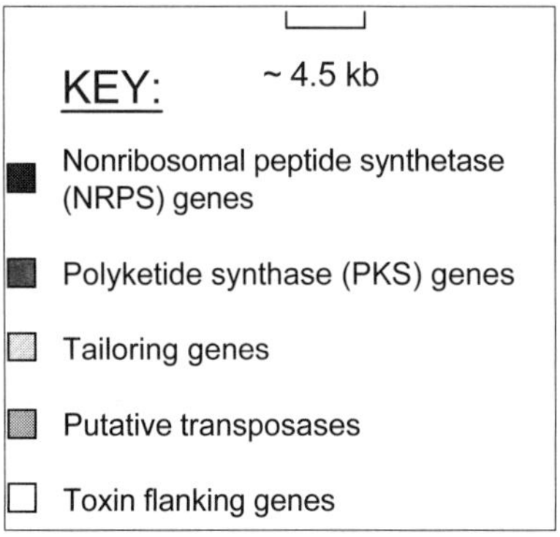

Fig. 1. The microcystin biosynthesis gene clusters and flanking regions found in *M. aeruginosa*, *Anabaena*, and *P. agardhii*, as well as the nodularin biosynthetic genes from *N. spumigena*. The key describes the scheme for polyketide synthase, peptide synthetase, and tailoring enzymes.

Nodularin synthetase

The nodularin biosynthesis gene cluster *ndaS*, from *Nodularia spumigena* NSOR10, has also been recently sequenced and characterized (Moffitt and Neilan 2004). The 48 kb region of the genome consists of nine ORFs (*ndaA–I*) transcribed from a bidirectional regulatory promoter region (Fig. 1). While most of the *ndaS* encoded genes have homologs in the *mcyS* cluster, their arrangement adheres more closely to the 'co–linearity' rule of NRPS pathways that predicts the order of catalytic processes involved in the biosynthesis of a non–ribosomal metabolite is generally the same as the order of the genes which encode their catalytic enzymes (Kleinkauf and von Dohren 1996).

The proposed pathway for nodularin biosynthesis is similar to that proposed for microcystin. Functional assignment of the enzymes was based on bioinformatic analysis and homology to the microcystin synthetase enzymes. The Adda side–chain is produced via a mixed NRP/PKS pathway from a phenylacetate starter unit and several malonyl–CoA extensions (NdaC, D and F). The NRPS module of the hybrid NRP/PKS, NdaF, subsequently adds D–Glu to the growing chain. Two NRPS enzymes, NdaA and B, complete the cyclic pentapeptide by adding the final amino acid residues, L–Thr, D–MeAsp and L–Arg. The NRPS modules responsible for the activation of D–Ala and D–Leu in *mcyS* (McyA and B) are absent from *ndaS* as nodularin lacks these moieties.

The *ndaS* cluster also encodes several putative monofunctional enzymes that may play a role in the modification and transport of nodularin. *ndaE* encodes an O–methyltransferrase, *ndaG* encodes a putative L–Asp/L–Glu racemase, and *ndaI* encodes an ABC transporter. Also encoded within the *ndaS* cluster is a D–3–PGDH homolog, NdaH, that shares 71% identity with McyI. It has been hypothesized that NdaH may therefore catalyse the conversion of MeThr to MeDhb in the final peptide structure (Moffitt and Neilan 2004).

Cylindrospermopsin synthetase

Cylindrospermopsin (Fig. 2) consists of a tricyclic sulfated and methylated guanidinium backbone, which is linked to hydroxymethyluracil (Ohtani et al. 1992). Its biosynthesis has been partially elucidated by labeled precursor feeding studies (Burgoyne et al. 2000). Glycine, five intact acetates, and an S–adenosylmethionine methyl group are confirmed cylindrospermopsin precursors. Biosynthesis of cylindrospermopsin is initiated by the production of guanidinoacetate from glycine and an, as yet unidentified, guanidino donor. The molecular backbone of cylindrospermopsin is then assembled by successive condensations of five intact acetates to the guanidinoacetate starter unit, followed by tailoring reactions, such as C-methylation, ketoreduction, sulfation and cyclizations to complete biosynthesis. Not all parts of the cylindrospermopsin molecule are accounted for, however. For instance, the origin of the guanidino and uracil moieties remains to be investigated (Burgoyne et al. 2000).

The molecular structure and feeding experiments of cylindrospermopsin suggested that it is produced by a mixed non–ribosomal peptide and polyketide pathway. Using a reverse genetic approach, gene fragments of two putative cylindrospermopsin biosynthesis genes were detected in cylindrospermopsin–producing strains of *Cylindrospermopsis raciborskii* and *Anabaena bergii* (Schembri et al. 2001). Phylogenetic screening in this study revealed that the presence of these two genes was directly linked to cylindrospermopsin production. The partial cylindrospermopsin biosynthesis gene cluster from *Aphanizomenon ovalisporum* was sequenced in a later study (Shalev–Alon et al. 2002), and revealed the presence of an amidinotransferase (*aoaA*) adjacent to a hybrid NRPS/PKS (*aoaB*) and a type I PKS (*aoaC*) (Fig. 2b). Enzymes encoded by these three genes are believed to initiate cylindrospermopsin biosynthesis (Shalev–Alon et al. 2002). AoaA is believed to synthesise the guanidinoacetate starter unit from glycine and an unidentified guanidino donor. It provided the highest homology to vertebrate arginine:glycine amidinotransferase. Functional modelling of AoaA revealed that residues corresponding to those involved in arginine and glycine substrate binding in the human amidinotransferase were conserved with regard to the amino acid and the topology in AoaA (Kellmann et al. 2005). Contrary to the study by Burgoyne et al (2000), which could not detect any incorporation of labelled arginine into guanidinoacetate, the data presented strongly suggested that arginine is the natural guanidino donor in cylindrospermopsin biosynthesis. However, this needs to be verified experimentally. Following guanidinoacetate synthesis, the hybrid NRPS/PKS, AoaB, is believed to recruit guanidinoacetate for polyketide extension. Catalytic domains present in AoaB are adenylation domain, acyl carrier protein, β–ketoacylsynthase and acyltransferase. To date, an NRPS with an adenylation domain that activates guanidinoacetate has not been reported. Substrate–binding residues of the AoaB adenylation domain differed from those of other adenylation domains, which may reflect the structure its substrate guanidinoacetate (Kellmann et al. 2005). AoaB may thus activate guanidinoacetate and add the first acetate extender unit. While *aoaC* has only been partially sequenced, it contains a β–ketoacylsynthase, an acyltransferase and a dehydrogenase domain. AoaC may add a further acetate extender unit and also reduce one of the carboxylate carbons (Fig. 2b).

Saxitoxin and the paralytic shellfish toxin biosynthesis

Paralytic shellfish poisoning (PSP) toxins are natural alkaloids. Like the cyclic peptide toxins, PSP toxins are considered secondary metabolites. Almost 30 years after the correct structure of saxitoxin, the parent compound of PSP toxins, was elucidated by X–ray crystallography (Schanz et al. 1975), there has been much speculation as to how this peculiar alkaloid is naturally produced. Nevertheless, very few investigations have contributed to the knowledge of saxitoxin and related compound biosynthesis. The genes encoding enzymes involved in PSP toxin biosynthesis are at present unknown.

In the early 1980s, however, Shimizu and colleagues documented fundamental studies concerning the biosynthesis of the saxitoxin carbon perhydropurine–skeleton (Shimizu et al. 1984; Shimizu 1986a, 1986b). According to feeding experiments carried out with ^{13}C, ^{15}N, and ^{2}H–labelled precursors in the dinoflagellate *Alexandrium tamarense* and the cyanobacterium *Aphanizomenon flos–aquae*, saxitoxin was shown to be derived from arginine and acetate via a Claisen–type condensation between C2 of arginine (or ornithine) and C1 of acetate. This reaction is uncommon, although it characterizes the first step of porphyrin biosynthesis with the condensation of succinate and glycine to form aminolevulinic acid (Shimizu 1996). Subsequently, the amino group of arginine is transformed into a guanidino group by transfer of an amidino moiety from another molecule of arginine. Two guanido groups originating from arginine are incorporated into tricyclic carbon backbone of saxitoxin, however the origin of the carbamate side–chain has not been reported. The side chain C13 originates from the S–adenosylmethionine methyl group. In total, to synthesize one saxitoxin molecule three arginine, one acetate, and one methionine molecule were required. This sequence of biochemical reactions was unprecedented in the literature and it remains unique.

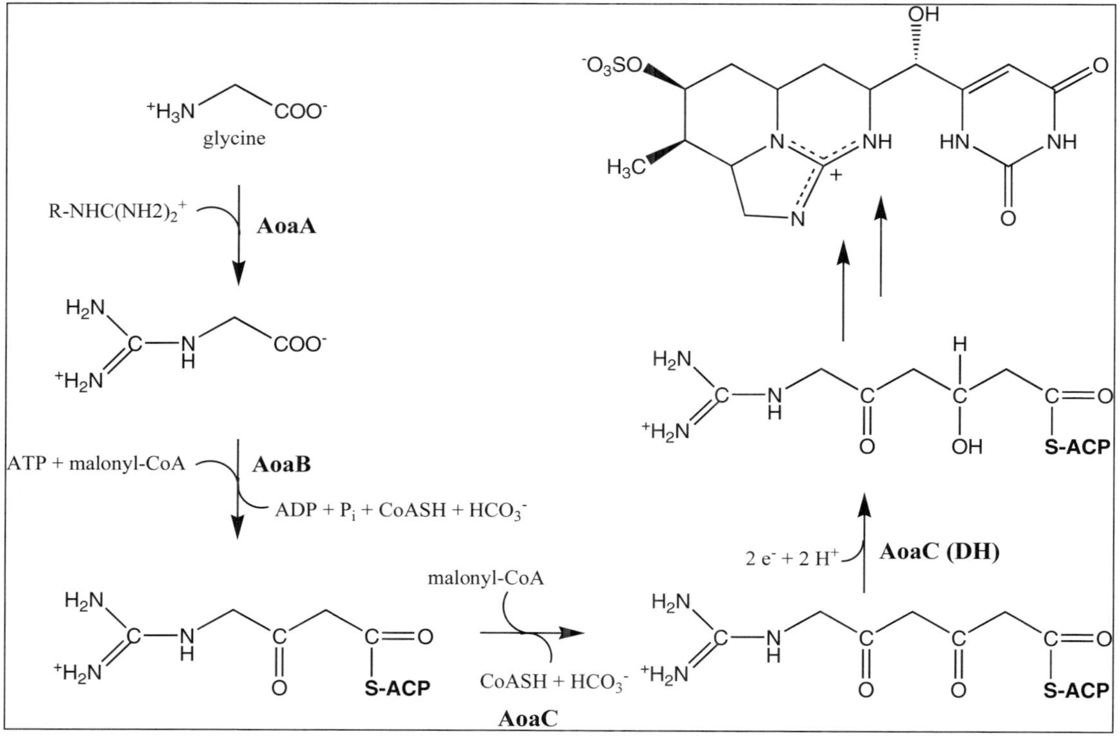

Fig. 2. (a) Biochemical reactions that three putative cylindrospermopsin biosynthesis gene products AoaA, AoaB and AoaC are thought to catalyse

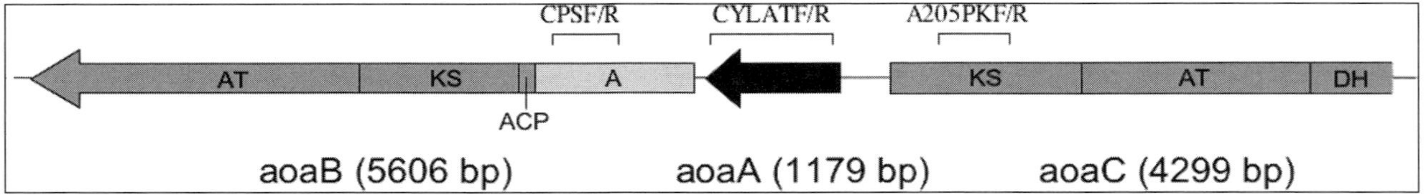

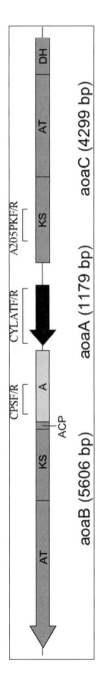

Fig. 2. (b) arrangement of aoaA, aoaB and aoaC in *A. ovalisporum* The light grey shading indicates non–ribosomal peptide synthetase, dark grey polyketide synthase modules. The brackets with primer names indicate the regions that were amplified and sequenced for phylogenetic analysis. A: aminoacyl adenylation; AT: acyl transferase; DH: dehydrogenase, KS: β–ketosynthase, ACP: acyl–carrier protein.

No biosynthetic enzymes have yet been identified and shown to be involved in saxitoxin production. Recent research, however, has identified certain candidate enzymes that may be somehow linked to saxitoxin production. Using differential display, Taroncher–Oldenburg and Anderson (2000) characterized differential gene expression in the toxic dinoflagellate *Alexandrium fundyense* during the early G_1 phase of the cell cycle, coinciding with the onset of toxin production in this organism (Taroncher–Oldemburg et al. 1997). An S–adenosylhomocysteine hydrolase, a methionine aminopeptidase, and a histone–like protein were isolated, although none of these genes show any correlation with the proposed saxitoxin biosynthetic pathway. On the other hand, Sako and co–workers (2001) purified and characterized a N–sulfotransferase from the toxic dinoflagellate *Gymnodinium catenatum* that was uniquely and specifically capable of transferring a sulphate residue from 3'–phosphoadenosine 5'–phosphosulfate (PAPS) to the N–21 carbamoyl group present in saxitoxin and gonyautoxin2+3. This enzyme, a monomeric 60 kDa protein, showed activity with only three substrates, each yielding a distinct product: saxitoxin→gonyautoxin 5, gonyautoxin 2→C–1, and gonyautoxin 3→C–2. The activity of this N–21 sulfotransferase suggested a possible pathway for saxitoxin synthesis in dinoflagellates by which saxitoxin represents the basic structure for the production of all other PSP derivatives. The hypothesized sequence of conversions, saxitoxin→gonyautoxin 2+3→C–toxins, is the reverse of a speculated pathway based on the time–dependent accumulation of various saxitoxin analogues during the synchronised growth of *A. fundyense* (Taroncher–Oldenburg et al. 1997). Another sulfotransferase, highly specific for saxitoxin analogues, has been purified from the PSP–producing dinoflagellate *G. catenatum* (Yoshida et al. 2002). This sulfotransferase converted 11–hydroxy saxitoxin, a synthetic analogue that is believed to be a natural intermediate, into gonyautoxin–2. Sequence information for these two sulfotransferases could not be obtained due to their low yield and instability. It is also possible that the enzymes do not normally play a role in saxitoxin biosynthesis since N–sulfotransferase activity has been detected in a non–toxic strain of *G. catenatum*, and was absent in toxic *A. tamarense* (Oshima 1995).

By exploring the differences between genomes of saxitoxin–producing and non–toxic *A. circinalis* strains, a carbamoyl–phosphate synthase, a S–adenosylmethionine dependent–methyltransferase, a transposase, an acetyltransferase, and several toxic–strain specific hypothetical and regulatory proteins, including the 60 kDa chaperonin GroEL, were identified (Pomati and Neilan 2004; Pomati et al. 2004a). A toxic strain specific gene coding for a putative Na^+ dependent transporter was recently recovered from a saxitoxin–producing strain of *A. circinalis* and was suggested to be involved in Na^{+}–specific pH homeostasis (Pomati et al. 2004a). This gene was successfully applied as a molecular probe in the laboratory for the environmental screening of saxitoxin–producing strains of *A. circinalis*. It has not as yet been possible, however, to demonstrate the direct involvement of any of these genes mentioned above in the biosynthesis of saxitoxin.

Biosynthesis of anatoxins

Anatoxin–a is a neurotoxic alkaloid produced by a number of cyanobacterial species including *Anabaena flos–aquae, Aphanizomenon flos–aquae,* and *Planktothrix* sp. (formerly *Oscillatoria* sp.) (Carmichael 1992). It has previously been proposed (Gallon et al. 1990) that anatoxin–a is synthesized from ornithine via putrescine through the activity of ornithine decarboxylase, similar to the structurally related tropane alkaloid pathways in plants. Subsequent studies (Gallon et al. 1994) suggested the involvement of arginine and Δ^{1}–pyrroline in the same hypothetical pathway but results were not definitive and could be attributed to trans–amination. Later studies (Hemscheidt et al. 1995b) have shown that C1 of the glutamic acid is retained during the formation of the toxin and is not lost by decarboxylation and therefore is incompatible with the hypothesis that the carbon atoms of the pyrrolidine ring are derived in the same way as are the tropane alkaloids. It was further shown that a five–carbon unit is derived intact from an amino acid precursor of the glutamic acid family while the other carbons are derived form acetate extender units.

Recently elucidated toxin biosynthesis pathways in cyanobacteria seem to mainly involve polyketide synthases (PKS) and non–ribosomal peptide synthetases (NRPS) e.g. microcystin, nodularin, jamaicamides and cylindrospermopsin. It is therefore proposed that the anatoxin–a biosynthesis genes would encode a mixed PKS/NRPS system for activating an amino acid of the glutamic acid family followed by the incorporation of 3 acetates in a polyketide manner.

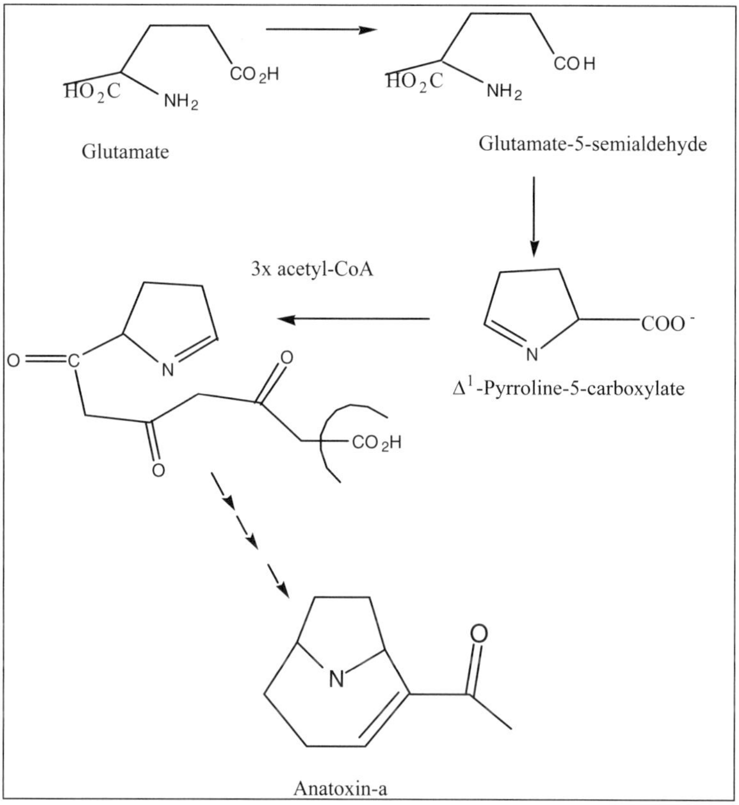

Fig. 3. Putative biosynthetic pathway of anatoxin–a. An amino acid of the glutamic acid family is activated followed by the addition of three acetate extender units in a polyketide manner, further steps may include reductions and a cyclization which are putatively encoded by the corresponding PKS tailoring domains. Note that the first two steps might also be involved in the proline biosynthesis pathway.

Due to the structural similarity of homoanatoxin–a and anatoxin–a it is believed that they are produced via the same pathway (Hemscheidt et al. 1995b), the terminal carbon of homoanatoxin–a seems to be derived from L–methionine via S–adenosyl–methionine (Namikoshi et al. 2004).

For anatoxin–a(s), feeding experiments have shown that the carbons of the triaminopropane backbone and the guanidino group are derived from L–arginine (Moore et al. 1992), whereas (2S,4S)–4–hydroxyarginine is an intermediate (Hemscheidt et al. 1995a).

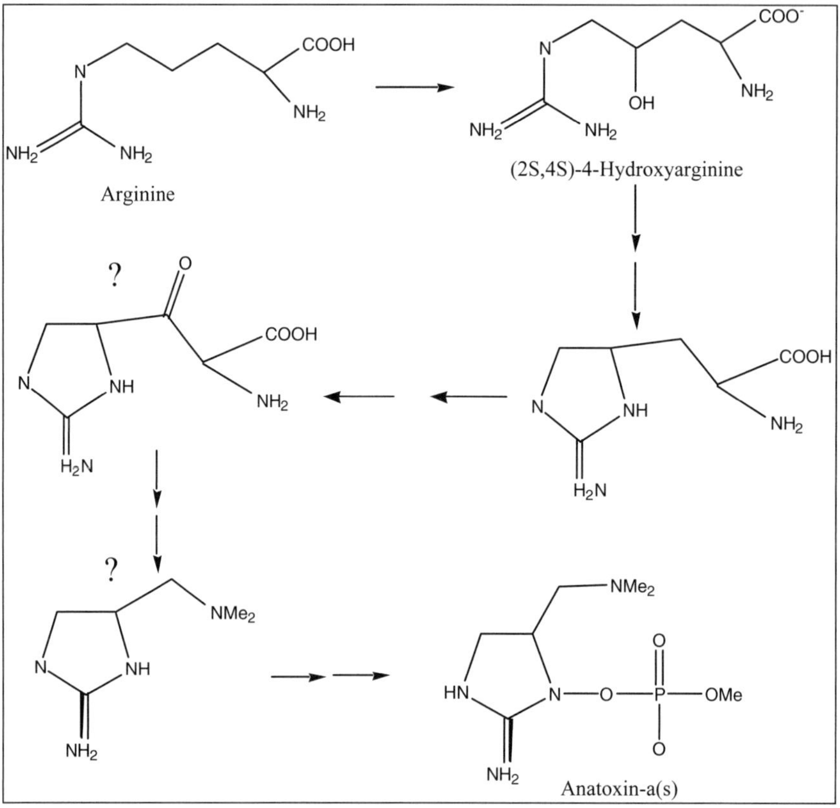

Fig. 4. Putative biosynthetic pathway of anatoxin–a(s). L–Arginine is a precursor and (2S,4S)–4–hydroxyarginine an intermediate, all other intermediates are hypothetical.

Biological and environmental factors influence the genetic expression of cyanotoxins

Most previous work in this area has investigated the end–products of gene expression. Due to the fact that most investigations have not been performed on a single set of strains, the experimental designs have not been standardized and the data not interpreted against the same specific growth controls. For the most part, production of a toxin by cyanobacteria appears to be constitutive. Evidence suggests, however, that the conditions of a bloom environment may alter the levels of toxin produced. Various studies have focused on the typically encountered environmental factors that may influence changes in the production of toxin. The variable ecological parameters that toxic cyanobacteria are exposed to include, temperature, light, nitrogen, iron, phosphate, predators, and other microorganisms.

Microcystin production and microcystin synthetase gene regulation

Hepatotoxin production in cyanobacteria is thought to be influenced by a number of different physical and environmental parameters, including nitrogen, phosphorous, trace metals, growth temperature, light, and pH (Sivonen 1990; Lukac and Aegerter 1993; van der Westhuizen and Eloff 1985; Song et al. 1998). However, due to the fact that most regulatory investigations have not been standardized, the subject of hepatotoxin regulation remains a somewhat contentious issue.

Several batch culture experiments have suggested that high microcystin production in cyanobacteria is correlated with high nitrogen and phosphorus concentrations (Sivonen in 1990; Vezie et al. 2002). Conversely, low iron concentrations have been correlated with increased toxin production (Lukac and Aegerter 1993). While these results suggest that microcystin production is influenced by nutrients and trace metals, the observed toxin fluctuations were probably due to the indirect affects of nitrogen, phosphorous and iron on cell growth rate. Long et al (2000) observed that under nitrogen–limited conditions, fast growing *M. aeruginosa* cells are smaller, of lower mass and contain higher intracellular levels of toxin than slow–growing cells. These results strongly suggest a positive linear relationship between the microcystin content of cells and their specific growth rate. This generalized model for microcystin regulation may also explain the conflicting results yielded from other batch culture investigations where variables such as temperature, light and pH have been tested (Orr and Jones 1998).

Microcystin has been shown to bind iron(III) and other cations and as such it has been considered that microcystin may be produced as a siderophore. Microcystin content was shown to be inversely proportional to the concentration of iron(III) in the media of a *Microcystis* culture (Lukac et al. 1993; Utkilen et al. 1995; Bickel et al. 2000). These studies may indicate microcystin synthesis is strain–specific, or a more complex response to light and iron (III) is associated with other external influences, including light levels and the presence of organic complexes.

Studies into the effect of different light intensities on microcystin production have revealed conflicting results. Whilst some studies indicate that production of microcystin is higher in response to low light intensities, others have indicated that the production (cellular levels) of microcystin is highest under high light conditions (Utkilen et al. 1992; van der Westhuizen et al. 1985; Watanabe et al. 1985; Sivonen et al. 1990; Rapala et al. 1997; Kaebernick et al. 2000). Downing et al. (2005) have recently proposed that most of the differences seen in the published work regarding microcystin production is due to variations in nitrogen uptake and assimilation, possibly in response to carbon fixation. This data then proposes that biologically available nitrogen and phosphorus are the real determinants of microcystin production in the environment, rather than other influences on growth rate, including light and temperature. Interestingly, growth of *Anabaena* at different temperatures has been correlated with the production of certain microcystin isoforms, as discussed later.

While most toxin regulation studies have focused on direct measurements of cellular toxin, the description of the *mcy* gene cluster by Tillett et al. (2000) enabled a closer examination of microcystin regulation at the molecular level. Kaebernick et al. (2000) used the RNase protection assay to measure the transcription of *mcyB* and *mcyD* under a variety of different light conditions. High light intensities and red light were correlated with increased transcription, while blue light led to reduced transcript levels. Interestingly, the authors observed two light thresholds, between dark and low light (0 and16 μmol of photons m^{-2} s^{-1}), and medium and high light (31 and 68 μmol of photons m^{-2} s^{-1}), at which a significant increase in transcription occurred. The same group later discovered that transcription of *mcy* genes occurs via two polycistronic operons, *mcyABC* and *mcyDEFGHIJ,* from a central bidirectional promoter between *mcyA* and *mcyD* (Kaebernick et al. 2002). Interestingly, alternate transcriptional start sites were identified for both operons when cells were cultured under high or low light intensities (ie. 68 or 16 μmol of photons m^{-2} s^{-1} respectively). Interestingly, this central regulatory region of *mcyS* also contains sequence motifs for both Fur and NtcA DNA binding proteins.

Several studies have reported an increase in extracellular microcystin content following exposure of cultures to high light conditions (Rapala et al. 1997; Bottcher et al. 2001; Kaebernick et al. 2000; Wiedner et al. 2003). Wiedner et al. (2003), found that on average, extracellular microcystin concentrations were 20 times higher when cells were cultures at 40 μmol of photons m^{-2} s^{-1} compared to those grown at 10 μmol of photons m^{-2} s^{-1}. However, it is important to note that the extracellular microcystin concentrations at both irradiances accounted for only 2.47 and 0.22% of the total microcystin content, respectively. Kabernick et al. (2001) proposed that microcystin may be constitutively produced under low and medium light intensities, and exported when a higher threshold intensity is reached. The recently identified ABC transporter McyH, may be responsible for this apparent export, however increased cell lysis and leakage of the toxin at higher irradiances can not be ruled out at this stage. Further investigation is clearly required.

An insertional inactivation mutant was produced via the homologous recombination of a chloramphenicol resistance cassette into the *mcyB* of *M. aeruginosa* PCC7806 (Dittmann et al. 1997). Comparative analysis of the proteomes of the wild–type and the mutant cultures resulted in the identification of a protein, MrpA, that was only strongly expressed in the wild–type and not by the non–toxic mutant (Dittmann et al. 2001). This protein and another protein encoded by a gene downstream of *mrpA* were homologous to RhiA and RhiB from *Rhizobium leguminosarum*. The RhiABC proteins are thought to play a role in nodulation and are regulated via a quorum–sensing mechanism. In addition, the *mrpA* transcripts were up–regulated in response to blue light. The results of this work led to the conclusion that microcystin may act as an extracellular signalling molecule and may be associated with a light–sensing mechanism in *Microcystis*.

Regulation of nodularin gene transcription and toxin production

Like *mcyS*, the *ndaS* gene cluster is transcriptionally regulated by a bi–directional promoter region. Analysis of transcription of the *ndaS* cluster found that it is transcribed as two polycistronic mRNA, *ndaAB, ORF1,* and *ORF2,* and *ndaC* (Moffitt 2003). The two genes downstream of *ndaAB*, *ORF1* and *ORF2*, encode a putative transposase and a putative high light–inducible chlorophyll–binding protein, respectively. It is not clear why these proteins are also co–transcribed with the *ndaS* gene cluster. *ORF2* has been identified in all strains of toxic *Nodularia* and the association between *ORF2* and nodularin biosynthesis may suggest a physiological function associated with high–light stress in the cells producing it. A putative heat shock repressor protein, encoded by the gene *ORF3*, was also identified downstream of ORF2, which may be involved in the transcriptional regulation of the *ndaS* genes in response to heat stress.

Growth experiments using *N. spumigena* batch cultures also suggest a link between light quality and nodularin production. The batch cultures preferred higher irradiances for growth (45 to 155 μmol of photons $m^{-2}.s^{-1}$) and these cultures produced the highest concentration of intracellular and extracellular toxin, although there appeared to be no significant difference between these values (Lehtimaki et al. 1994; Lehtimaki et al. 1997). *N. spumigena* BY1 batch cultures have been shown to grow best at temperatures between 25°C and 28°C (Lehtimaki et al. 1997). The level of intracellular and extracellular nodularin was lower in batch cultures grown at temperatures of 19°C and 28°C. Temperature had a similar effect on the production of microcystin by *Microcystis* and *Anabaena* with highest production rates at around 25°C (Rapala et al. 1998; Codd et al. 1988). Similarly, the growth rate of *N. spumigena* and nodularin production is higher in moderate salinities between 0.5% and 2% NaCl when compared to lower (0%) or higher (3%) salinities. The concentration of intracellular nodularin was highest in the cultures with the highest growth rate (0.5% to 2% NaCl), although the extracellular concentration of nodularin was higher in the cultures maintained in media containing 0.5% to 1% NaCl (Blackburn et al. 1996; Lehtimaki et al. 1997).

The presence of ammonium appeared to have a negative effect on the growth of *N. spumigena* BY1 (Lehtimaki et al. 1997). The growth rate of batch cultures was lowest in media supplemented with the highest concentration of nitrate, and the nodularin concentration also decreased. A batch culture study of *N. spumigena* strain GR8b found that the total nodularin concentration was highest at the end of the experiment, under low nitrate levels (Repka et al. 2001). The opposite was found true for the non–nitrogen fixing genera *Oscillatoria* and *Microcystis* which produced higher levels of microcystin when grown in the presence of high levels of nitrogen (Sivonen et al. 1990; Utkilen et al. 1995; Watanabe et al. 1985). Analysis of the effect of phosphate levels on nodularin production indicated that batch cultures of *N. spumigena* BY1 grown at high concentrations of phosphate produced the highest levels of nodularin per cell dry weight. Recent studies of batch and chemostat cultures found that phosphate had no effect on the total nodularin concentration. Analysis of microcystin–producing cultures *Microcystis*, *Anabaena*, and *Oscillatoria* indicated that there is, in general, an increase in the cellular concentration of microcystin when cultures were grown in media containing higher levels of phosphorus (Utkilen et al. 1995; Rapala et al. 1997; Sivonen et al. 1990).

Saxitoxin and paralytic shellfish toxin expression

The regulation of saxitoxin production in cyanobacteria and the metabolic role of PSP toxins within the producing microorganisms is poorly understood. Historically, factors influencing the biosynthesis of saxitoxin and related compounds in cyanobacteria have also been the subject of sub–optimal studies, both in the laboratory and in the environment. Since the genes for saxitoxin production are presently unknown, this section will focus on what influences PSP toxin accumulation.

One of the most detailed reports in the literature concerning the physico–chemical features associated with saxitoxin production, describes the environmental conditions during one of the largest PSP toxin producing blooms on record (Bowling and Baker 1996). The cyanobacterial bloom, dominated by up to half a million cells of *A. circinalis* per millilitre, affected 1000 km of river waters in the Barwon–Darling basin, New South Wales, Australia. Numerous livestock deaths were reported during this occurrence, and the neurotoxicity of *A. circinalis* samples was demonstrated by mouse bioassay. This cyanobacterial bloom was attributed to reduced river flow as a result of intense drought. Under such environmental conditions nutrient concentrations, especially phosphorus, were found to be very high. Alkaline pH (>8.5) and very high ammonia values, in some instances more than 1 mg L^{-1}, also characterized the majority of sampling sites along the river. Additionally, most of the neurotoxic *A. circinalis* samples were collected from water with high electrical conductivity (Bowling and Baker 1996). Similar environmental parameters, in particular elevated pH and high conductivity, have also been described in Australia for another potentially toxic bloom of *A. circinalis* in Lake Cargelligo, New South Wales, during 1991 (Bowling 1994).

In the laboratory, changes in toxicity or in toxin concentration of cyanobacterial isolates can be induced, although such variations range from only two to four–fold. Among the environmental parameters tested on growth and PSP toxin accumulation were culture age, temperature, light, principal nutrients, salinity, and pH. These investigations indicated that cyanobacteria produce PSP toxins under conditions that are most favourable for their growth as has been reported for other cyanobacterial secondary metabolites (Sivonen and Jones 1999).

Cyanobacteria produce more PSP toxins during their late exponential growth phase, and it has been reported that saxitoxin biosynthesis in *Aphanizomenon flos–aquae* NH–5 was inversely proportional to its growth rate (Gomaa 1990). Comparable results were also described for PSP–producing strains of *Lyngbya wollei* and *Cylindrospermosis raciborskii* (Yin et al. 1997; Castro et al. 2004). High light intensity was shown to result in reduced toxicity in *L. wollei* cultures, coupled to an increase in biomass (Yin et al. 1997). On the other hand, temperature seems to have a marked effect on PSP toxin production. Yin and colleagues (1997) documented a net decrease in PSP toxin content of *L. wollei* cultures with increasing temperature. Similarly, *A. circinalis* was found to produce more saxitoxin and gonyautoxin 2/3 per cell at low temperatures (15°C) compared to values above 30°C (Rossetti and Pomati unpublished data). Data consistent with these results have been recently described in *C. raciborskii*, where changes in the extracellular levels of PSP toxins were noted (Castro et al. 2004). Modifications in the toxin profile between 19 and 25°C were also detected. These documented effects have been explained by either a lower stability or higher biodegradation rate of saxitoxins at high temperatures, as well as by significant changes in the rate of cell division (Yin et al. 1997; Castro et al. 2004).

Alkalinity is a feature that often characterizes the environment of toxic cyanobacterial blooms, in particular those dominated by PSP–producing species, such as *A. circinalis* (Bowling and Baker 1996). High salinity, recorded as increased conductivity, has also been correlated with neurotoxic blooms of *A. circinalis* in Australia (Bowling 1994; Bowling and Baker 1996) and *C. raciborskii* blooms in Brazil (S. Azevedo personal communication). Intracellular saxitoxin levels were recently found to respond to the alkalinity and Na^+ content of culture media (Pomati et al. 2004b), and by influencing Na^+ fluxes it was possible to induce a corresponding modulation of saxitoxin accumulation (Pomati et al. 2003; Pomati et al. 2004b). Since maintenance of pH and sodium homeostasis are strictly correlated in cyanobacteria (Horikoshi 1991; Maestri & Joset 2000; Waditee et al. 2001), sodium transport and cycling seem to play a vital role in the regulation of intracellular saxitoxin levels. A Na^+–dependent transporter gene, specific to PSP toxin–producing strains of *A. circinalis,* has also recently been cloned and may somehow be involved in the sodium–coupled transport of saxitoxins (Pomati et al. 2004a).

Certain nutrients, in particular light, nitrogen and phosphorus, are essential for cyanobacterial growth. Phosphorus is usually the limiting factor for autotrophic growth in freshwaters, and hence small changes in this nutrient's concentrations may affect toxin production merely as a result of influencing growth. In *L. wollei*, optimum growth levels for PO_4 also correspond to the optimal conditions for PSP toxin production (Yin et al. 1997). Most microbial secondary metabolites are molecules that are rich in nitrogen. Nitrogen–fixing species, such as PSP toxin producing cyanobacteria, are not dependent on fixed nitrogen in their growth media for toxin production (Rapala et al. 1993; Lehtimäki et al. 1997), although very high concentrations of this element were found to inhibit PSP toxin accumulation in cultures of *L. wollei* (Yin et al. 1997). Rising levels of calcium have been shown to increase PSP toxin synthesis in *L. wollei* (Yin et al. 1997), however, no other trace element has been investigated for their effect on the production of saxitoxin in cyanobacteria.

Regulators of cylindrospermopsin production

Direct studies on the expression of cylindrospermopsin–biosynthesis genes have not been performed, however the growth rates and cylindrospermopsin production rates of *C. raciborskii* in response to different levels and types of nitrogen sources were examined in the study by Saker and Neilan (2001). It was found that cultures grown in the absence of a fixed nitrogen source provided the highest concentration of cylindrospermopsin on a per dry weight basis, with a corresponding lowest growth rate. Cultures supplemented with nitrate were intermediary with regard to growth rate and toxin production, while cultures supplemented with ammonia provided the lowest toxin concentrations, but the highest growth rates.

Regulation of anatoxin production

Different media and culture duration have been shown to effect the anatoxin–a content in *Anabaena flos–aquae* (Gupta et al. 2002). Rapala et al. (1993) found that temperature reduced the levels of anatoxin–a regardless of growth rate. In addition, growth in nitrogen–free media has been correlated to anatoxin–a production, however, results in these studies were strain specific and varied widely in response to the different growth conditions employed.

Multiple genera of cyanobacteria share common biosynthetic processes

Microcystin producers

Comparative studies of the *mcyS* gene clusters from *M. aeruginosa*, *P. agardhii* (Christiansen et al., 2003) and *Anabaena* sp. (Rouhiainen et al. 2004) have noted variations in the arrangement of *mcyS* genes between these different species of cyanobacteria, although the proposed toxin biosynthetic processes are thought to be similar (Fig. 1). The *M. aeruginosa* and *Anabaena* sp. *mcyS* clusters are arranged into two divergently transcribed operons, however, the arrangement of genes within these operons differs between the two species. In *P. agardhii*, the *mcyS* cluster also has a distinctive arrangement and lacks *mcyF* and *mcyI*. Furthermore, the *P. agardhii mcyS* cluster contains an additional gene, *mcyT*, upstream of the central promoter region. This gene is thought to encode a putative type–II thioesterase enzyme, which was proposed to have an editing role by removing mis–primed amino acids from the NRPS and PKS enzymes. The characterization of *mcyS* in *M. aeruginosa*, *P. agardhii*, and *Anabaena* sp. has enabled the study of the origins and evolution of hepatotoxin biosynthesis in cyanobacteria. Identification of transposases associated with the *mcyS* and *ndaS* gene clusters and subsequent phylogenetic analysis has led to the theory that horizontal gene transfer and recombination events are responsible for the sporadic distribution of the *mcyS* gene cluster throughout the cyanobacteria and the various microcystin isoforms that have been identified to date (Tillett et al. 2001; Mikalsen et al. 2003, Tanabe et al. 2004). Recently, a phylogenetic study by Rantala et al. (2004 has contradicted this theory, suggesting that the genetic associated with hepatotoxicity was acquired early in evolution and then lost over time. Collectively, these results suggest that recombination, gene loss, and horizontal gene transfer can explain the distribution and variation regarding microcystin production in different cyanobacterial genera.

Nodularin–producers

Nodularin is produced by all strains of the species *N. spumigena* and no other cyanobacterial species, however, the structural analog motuporin, has been isolated from a sponge. Recent genetic studies suggest that the *ndaS* cluster evolved from the *mcyS* cluster through deletion of two NRPS modules (Mikalsen et al. 2003; Moffitt and Neilan 2004; Rantala et al. 2004).

Cylindropsermopsin producers

The production of cylindrospermopsin has been reported in *Cylindrospermopsis raciborskii*, *Aphanizomenon ovalisporum*, *Anabaena bergii*, *Umezakia natans*, and *Raphidiopsis curvata* (Banker et al. 1997; Harada et al. 1994; Li et al. 2001; Schembri et al. 2001). Apart from *C. raciborskii*, which was classified based on both morphology and 16S rRNA gene sequencing, other cylindrospermopsin–producing cyanobacteria were initially only classified by morphology. Subsequent molecular phylogenetic analyses have revealed that the Israeli and Australian *A. ovalisporum* strains reported to produce cylindrospermopsin may in fact represent a species of the genus *Anabaenopsis*, while the cylindrospermopsin–producing *U. natans* and *A. bergii* were closely related to each other and phylogenetically intermediary between the genera *Anabaenopsis* and *Nodularia* (Kellmann et al. 2005). Interestingly, the phylogenetic distance between cylindrospermopsin–producing cyanobacterial species was greater than between the corresponding cylindrospermopsin biosynthesis genes

(NRPS, PKS, and amidinotransferase), which was strong evidence for horizontal transfer of "toxin genes" between these organisms.

Saxitoxin and other PSP–producers

According to the feeding experiments (Shimizu et al. 1984; Shimizu 1986a, 1986b), both the dinoflagellate *Alexandrium tamarense* and the cyanobacterium *Aphanizomenon flos–aquae* synthesize saxitoxin via an identical sequence of biochemical reactions. It is therefore feasible that all PSP toxin–producing microorganisms share a homologous biosynthetic process. Whether putative biosynthetic enzymes for PSP toxin production are common to different strains of the same species, or if the genes are conserved among phylogenetically distant PSP toxin–producing species, is open to question. Because the molecular basis for saxitoxin production is unproven, the origin and the evolution of PSP toxin production remains a mystery.

The toxin profiles of cyanobacteria are consistent traits that appear to have a genetic basis (Castro et al. 2004; Negri et al. 2003; Negri et al. 1997; Velzeboer et al. 2000). Isolates of the same cyanobacterial species from geographically distant locations provide different toxin profiles (Velzeboer et al. 2000), which remain constant for each strain over many generations. The production of PSP toxins is geographically segregated in *A. circinalis* and *C. raciborskii*. Only Australian isolates of *A. circinalis* (Beltran and Neilan 2000) and Brazilian isolates of *C. raciborskii* produce PSPs (Neilan et al. 2003). Unique analogs of saxitoxin have been found in *L. wollei*, a species that is phylogenetically unrelated to other PSP–producing cyanobacteria (Onodera et al. 1997). PSP toxins have been isolated from a broad range of microorganisms, including dinoflagellates and heterotrophic bacteria. This scattering of saxitoxin–producing microorganisms across two of the three kingdoms of life again suggests the potential for the lateral exchange of saxitoxin biosynthetic genes between dinoflagellates, cyanobacteria and other bacteria, probably via a prokaryotic ancestor. This hypothesis is validated by the fact that the synthetic capability of PSP toxin production is limited to certain strains and is not universal to a species. Evidence has previously been found for possible lateral gene transfer events between toxic strains of *A. circinalis* (Pomati and Neilan 2004). This would further support the fact that the basic molecular and genetic machinery underlying saxitoxin biosynthesis is shared by all PSP–producing micro–organisms.

Preferential expression of one toxin over another in certain cyanobacteria

The biological or ecological controls for the preferential production of a particular toxin by any given cyanobacterial strain are not known. Although a rare occurrence, individual strains of cyanobacteria have been shown to produce more than one type of toxin and, therefore this preferential expression is under–pinned by the genes required to produce these compounds. Contentious in this area are reports that describe multiple toxins isolated from a natural bloom sample that is described as having a single species composition. In this situation, minority members of the bloom may be contributing to the toxin profile but are not detected by standard microscopic methods. This highlights another critical area of research, the need for highly skilled microbiologists that can achieve and maintain culture axenicity for future genetic and chemical analyses.

Expression of multiple congeners of a single type of cyanotoxin

Microcystin isoforms

More than 65 isoforms of microcystin are produced by five genera of cyanobacteria. While current data suggest that *Oscillatoria* strains produce only one major toxin at any time, *Anabaena* spp. and *Microcystis* spp. are capable of producing two to four microcystin isoforms simultaneously (Luukkainen et al. 1993). The structure of microcystin differs primarily at the two L–amino acids, and secondarily on the presence or absence of the methyl groups on D–MeAsp and/or Mdha (Namikoshi et al. 1998), however, substitutions of all moieties within microcystin have been reported (Rinehart et al. 1994; Sivonen et al. 1996; Chorus et al. 1999). Varying levels of toxicity have also been reported for each microcystin isoform analysed (Rinehart et al. 1994).

The large number of microcystin variants existing in nature has been attributed to the relaxed substrate specificity of the adenylation domains in the *mcyS* NRPS modules (Mikalsen et al. 2003). In addition to this, toxin variation is also caused by genetic differences in the microcystin synthetase genes themselves. For example, variation within the *mcyB1* module (proposed to result from recombinations between modules *mcyB1* and *mcyC*), has been correlated with the production of different microcystin isoforms. It has been postulated that variation in *mcyS* genes may be due to several different evolutionary processes, including horizontal gene transfer, lateral recombination and genetic deletions (Mikalsen et al. 2003; Moffitt and Neilan 2004; Rantala et al. 2004; Tanabe et al. 2004). Since the production of microcystin is not restricted to distinct evolutionary clades of species, transposition of the *mcyS* gene cluster across cyanobacterial genera is a strong possibility. Indeed genes encoding transposases have been identified within the flanking regions of *mcyS* (and *ndaS*) clusters, strongly supporting this theory.

While variation at the genetic level appears to be the primary explanation for the plethora of microcystin variants identified, environmental factors may also play a role. Rapala et al. (1997) observed that growth of *Anabaena* cultures at temperatures less than 25°C resulted in the preferential production of microcystin–RR rather than microcystin–LR, which occurred at higher temperatures. Nitrogen added to the growth medium and increasing temperatures also increased the proportion of microcystin variants demethylated in amino acid 3.

Nodularin and related compounds

In contrast, to the wide range of microcystin isoforms present in nature, only seven naturally occurring isoforms of nodularin have been reported. Two of these isoforms, produced by a New Zealand *Nodularia* sp. bloom, have variations within the Adda residue, which reduces or abolishes the toxicity of the compound (Rinehart et al. 1994). The D–Glu residue is essential for the toxicity of nodularin, as esterification of its free carboxyl abolishes toxicity. However, substitutions at position 1 have little effect on toxicity. The other two isoforms, nodularin–Har and motuporin are variable at position 2. Nodularin–Har is produced by the strain *N. harveyana* PCC7804, with the L–Arg, replaced with L–homoarginine (L–Har) (Saito et al. 2001; Beattie et al. 2000). Motuporin has been isolated from the Papua New Guinea sponge *Theonella swinhoei*, and may be synthesized by a commensal cyanobacterium. The L–Arg residue of nodularin is replaced by L–Val in motuporin (deSilva et al. 1992). The L–Val residue is responsible for the additional cytotoxicity of motuporin against cancer cell lines.

The many forms of PSPs

Strains of certain cyanobacterial species, including *L. wollei,* produce a variety of PSP toxin congeners (Onodera et al. 1997). More than thirty different saxitoxin analogs and derivatives are known to date (Sivonen and Jones 1999). There is no indication, however, of a PSP toxin that is preferentially produced by toxic cyanobacterial strains. In addition, each isolate can be characterized by its particular toxin profile (Velzeboer et al. 2000), which is putatively determined by the genetic information carried by its genome. The reason why a particular PSP toxin dominates the profile of a given cyanobacterial strain is unknown. It can be proposed that both the PSP toxin genetics of a cyanobacterium, together with the availability of primary metabolic substrates, is crucial for the production of one PSP toxin over another. Different cyanobacterial strains may also have endogenous enzymes that apply modifications to the basic saxitoxin structure, leading to the production of a range of congeners. Certain environmental conditions are known to favour the stability or the interconversion of individual saxitoxin analogs, although these physical processes mainly occur in the extracellular environment. Transformation reactions may also occur in living cyanobacterial cells, as a result of the aging of both laboratory cultures and natural blooms. Under these conditions, reduction and hydrolysis of carbamate and N–sulphocarbamoyl toxins can yield products of greater toxicity, such as saxitoxin and gonyautoxin, which may become dominant over their original precursors (Jones and Negri 1997). Additionally, high temperatures (> 25ºC) may play a role in the preferential accumulation of decarbamoyl–saxitoxin in *C. raciborskii* (Castro et al. 2004). However, the transformation of saxitoxin into gonyautoxin 2/3 due to a shift to 19ºC from 25ºC has also been reported.

Structures of the anatoxins

Anatoxin–a has one known naturally occurring analogue, homoanatoxin–a. Homoanatoxin–a possesses an extra C–12 methyl group that is probably derived from S–adenosylmethionine and has a similar toxicity to anatoxin–a. So far homoanatoxin-a has only been detected in *Oscillatoria formosa* (Skulberg et al. 1992) and *Raphidiopsis mediterranea* (Namikoshi et al. 2003). The simultaneous production of homoanatoxin–a and anatoxin–a in these species has recently been reported (Namikoshi et al. 2003; Aráoz et al. 2005). As the biosynthetic pathway for both congeners are believed to be almost identical, it is plausible that these organisms posses only a single gene cluster for toxin biosynthesis and the selective action of a tailoring enzyme(s) results in the production of the two variants.

Evolutionary advantages conferred by cyanotoxin production

Toxin biosynthesis by cyanobacteria, is energetically expensive. Toxic species typically commit about 2% of their genomes in order to produce a metabolite that may constitute 2% of the cell's dry mass. In addition, mutation of the genes encoding these pathways does not reduce the viability of cultures in the laboratory and there appears to be a mosaic distribution of toxigenic strains globally in the environment. The ecophysiological basis for cyanotoxin production is a paradox that has directed most of the studies described so far in this paper. Elucidating their role in the life history of the producing organisms is a critical issue in water quality management.

There is evidence that cyanobacteria and their toxins may have an effect on zooplankton (cladoceran and rotiferan) population structure, and that this in turn may guide ecological processes responsible for cyanobacterial success. Cyanobacterial cells are generally a poor food source for zooplankton and are often selectively avoided (DeMott 1991). As a result, zooplankton feed on phytoplankton that are otherwise in competition with cyanobacteria. It would be premature to postulate that toxic cyanobacterial dominance is planned and guided by cyanotoxin production, however, feeding deterrence has been one of the earliest roles suggested for these metabolites. Whether the compounds causing toxicity and deterrence are one and the same has recently been questioned (Rohrlack et al. 1999; Kaebernick et al. 2001; Reinikainen et al. 2001).

Heterotrophic eubacteria, fungi, phytoflagellates and protozoans, commonly associated with cyanobacteria, may also be affected by cyanotoxin production. Many bloom–forming cyanobacteria exhibit optimal growth in the presence of contaminant heterotrophic bacteria. For example, *Pseudomonas aeruginosa* was found to be chemotactically attracted to the heterocysts of *Anabaena oscillarioides*, which subsequently led to the establishment of a mutualistic relationship between the two bacteria sharing the fixed N_2 (Pearl 1984; Pearl and Gallucci 1985). Similarly, *M. aeruginosa* cells exhibit greater cell specific rates of CO_2 fixation when in association with bacteria and protozoan grazers (Pearl and Millie 1996). It has been suggested that the production and secretion of extracellular metabolites, such as cyanotoxins, may play a role in attracting these beneficial hosts while at the same time repelling antagonistic microbes and higher order grazers (Pearl and Millie 1996). An allelopathic function for microcystin has also been suggested as a result of the toxin's growth inhibition of various algal species of the genera *Chlamydomonas*, *Haematococcus*, *Navicula* and *Cryptomonas* (Keating 1978).

Advantages conferred by microcystin biosynthesis

Microcystin's affinity for iron and other cations such as copper, calcium and zinc, indicates the molecule's siderophoric properties (Humble et al. 1994) and suggests a putative role as an iron scavenging molecule (Utkilen and Gjolme 1995). Extracellular siderophores bind Fe^{2+} from the environment for use by the cell under conditions of low iron availability (Lukac and Aegerter 1993). Intracellularly, however such siderophoric properties may have a negative effect on cell functions due to competition with primary metabolites for limited essential iron. Alternatively, under conditions of high intracellular iron, an intracellular siderophore may have a protective function by chelating ferrous ions, forming iron–microcystin complexes and thus keeping the cellular level of free radicals low (Utkilen and Gjolme 1995). Such iron–microcystin complexes may also be stored until low iron conditions allow the release of Fe^{2+} for cellular processes. However, microcystin or complexes thereof have not been identified as stored substances in cellular inclusions, and are more commonly found on the thylakoid and plasma membranes (Shi et al. 1995; Young et al. 2005). It is believed that the Adda moiety of the toxin may bind to the thylakoid, leaving the polar peptide ring to bind metals from the cytoplasm (Orr and Jones 1998). This association with the photosynthetic apparatus of the cell may also indicate a putative function in light harvesting and chromatic adaptation mechanisms exhibited by some cyanobacteria.

Putative roles of cylindrospermopsin

Only one study has examined the effects of cylindrospermopsin on a filter–feeding zooplankter, *Daphnia magna* (Nogueira et al. 2004). Upon exposure to a cylindrospermopsin–producing strain of *Cylindrospermopsis raciborskii* Cylin–A, *D. magna* experienced high mortality, significantly reduced individual body growth and reduced fecundity. A second, and non–cylindrospermopsin producing strain also exhibited toxic effects on *D. magna*, due to an unidentified toxin, however the effects were less severe than those observed for cylindrospermopsin. The response of zooplankton to cyanotoxins is very species–specific. In the case of *D. magna* a bloom of cylindrospermopsin–producing cyanobacteria may reduce the grazing–pressure, and therefore give these cyanobacteria a survival advantage over non–producing strains. However, comprehensive ecological studies are required to verify this hypothesis.

The convergent evolution of saxitoxin and PSPs

PSP toxins have been long considered among the most enigmatic of all microbial natural products, for several reasons. These neurotoxins represent one of the most potent classes of venoms, with a highly specific effect targeting voltage–gated sodium channels in excitable cells. Few stimuli induce or repress saxitoxin production in these microorganisms, and the metabolic role of PSP toxins has historically thought to be related to a possible defense strategy against unknown predators. It is a general and reasonable principle that metabolites of all kinds should play a beneficial role in the producing microorganism. The biosynthesis of toxins requires significant cellular energy and it seems unlikely that evolution would tolerate such wasted metabolism. There is limited evidence, however, to suggest, that PSP toxins confer an evolutionary or survival advantage to the producing microorganism.

As stated, the non–phylogentic distribution of saxitoxin production suggests the lateral acquisition of PSP toxin biosynthetic genes by toxic strains. This is considered to be a relatively rare event in the evolution of microorganisms, especially if it involves the exchange of genes between phylogenetically distant species (Eisen and Fraser 2003) or even kingdoms. This could be fostered under particular or critical growth conditions, as the laterally transferred genes could be maintained by the recipient microorganism if they encoded a function that was essential for cell survival. For example, the production of PSP toxins may confer an advantage under adverse environmental conditions. Recently, evidence was found linking saxitoxin production to the maintenance of cyanobacterial homeostasis under alkaline pH or Na^+ stressed conditions (Pomati et al. 2004b). The blockage of Na^+ uptake by saxitoxin was demonstrated in bacterial and cyanobacterial strains (Pomati et al. 2003b), suggesting that this inhibition could represent a possible mechanism that PSP toxin–producing cyanobacteria employ to cope with conditions of elevated pH and salinity. As mentioned previously, these environmental features are not uncommon in rivers and water–bodies characterised by the seasonal occurrence of saxitoxin–producing cyanobacterial blooms. Other results strongly suggest that the genetic differences between saxitoxin–producing and non–toxic *A. circinalis* strains are due to a distinctive adaptation to specific environmental conditions (Beltran and Neilan 2000; Pomati and Neilan 2004). The genetic heterogeneity of *A. circinalis* is explained, to some extent, by genes associated with the maintenance of Na^+ homeostasis (Pomati et al. 2004a). Taken together, these considerations support the hypothesis that the production of PSP–toxins could represent a potential evolutionary advantage.

The microorganisms producing PSP toxins, and often most of those living in their immediate environment, do not possess nerves or any of the molecular systems that characterize neural transmission, the target of the saxitoxin channel–blocking effect. Ion channels, however, are a common feature that distinguishes all biological membranes. The function of ion channels range from motility or nutrient uptake, in bacteria, to the highest levels of complex neural transmission in the animal central nervous system. While commonly having different structures across phylogenetically unrelated organisms, ion channels often share similar specificity and cellular function. Interactions between saxitoxin and bacterial ion channels have been investigated (Pomati et al. 2003b). The identification and characteristics of the prokaryotic channel(s) inhibited by saxitoxin are yet to be described. The hypothesis, however, of an interaction in the environment between this natural neurotoxin with the ion channels of cyanobacteria, as well as other eukaryotic and prokaryotic planktonic competitors is possible.

Conclusion

While much progress has been made over the past ten years regarding the genetic basis for cyanobacterial toxin biosynthesis, there is now a crucial need to focus investigations on the regulation of toxin production. The toxin biosynthesis pathways have been elucidated for the major cyanobacterial threats to drinking water quality. Microcystin synthetase has been characterized at the genetic and enzyme level, and the public availability of the *M. aeruginosa* genome should be imminent. Together, the knowledge of toxin biosynthesis and the ability to perform global analyses of transcription and translation in *Microcystis* will afford the most complete investigation of the ecophysiological function of microcystin. These systems biological approaches may also be performed on samples in the environment, providing a real–time appraisal of the metabolism of cyanobacteria as they occur in harmful blooms.

However, numerous problems still challenge the success of research into the molecular biology of toxic cyanobacteria. These include the availability of a wide range of quality toxin standards, axenic cultures, methods for genetic transformation and mutagenesis, and the broad use of standardized protocols for reporting toxin levels, production rates, and gene expression. The genomes of other model toxic cyanobacteria are also required if major advances are to be made, including those of the main producers of cylindrospermopsin (*C. raciborskii*) and saxitoxin (*A. circinalis*).

References

Banker R, Carmeli S, Hadas O, Teltsch B, Porat R, Sukenik A (1997) Identification of cylindrospermopsin in Aphanizomenon ovalisporum (Cyanophyceae) isolated from lake Kinneret, Israel. J Phycol 33:613–6

Beltran EC, Neilan BA (2000) Geographical segregation of the neurotoxin–producing cyanobacterium Anabaena circinalis. Appl Environ Microbiol 66:4468–74

Bowling LC (1994) Occurrence and possible causes of a severe cyanobacterial bloom in Lake Cargelligo, New South Wales. Aust J Mar Freshwater Res 45:737–45

Bowling LC, Baker PD (1996) Major cyanobacterial bloom in the Barwon–Darling River, Australia, in 1991, and underlying limnological conditions. Mar Freshwater Res 47:643–57

Burgoyne DL, Hemscheidt TK, Moore RE, Runnegar MTC (2000) Biosynthesis of cylindrospermopsin. J Org Chem 65:152–6

Carmichael WW (1992) Cyanobacteria secondary metabolites– the cyanotoxins. J Appl Bacteriol 72:445–59

Castro D, Vera D, Lagos N, Garcia C, Vasquez M (2004) The effect of temperature on growth and production of paralytic shellfish poisoning toxins by the cyanobacterium Cylindrospermopsis raciborskii C10. Toxicon 44:483–9

Christiansen G, Fastner J, Erhard M, Borner T, Dittmann E (2003) Microcystin biosynthesis in planktothrix: genes, evolution, and manipulation. J Bacteriol 185(2):564–72

Dittmann E, Erhard M, Kaebernick M, Scheler C, Neilan BA, von Dohren H, Borner T (2001) Altered expression of two light–dependent genes in a microcystin–lacking mutant of Microcystis aeruginosa PCC 7806. Microbiol 147(11):3113–9

Dittmann E, Neilan BA, Erhard M, von Dohren H, Borner T (1997) Insertional mutagenesis of a peptide synthetase gene that is responsible for hepatotoxin production in the cyanobacterium Microcystis aeruginosa PCC 7806. Mol Microbiol 26(4):779–87

Downing TG, Sember C, Gehringer MM, Leukes W (2005) Medium N:P ratios and specific growth rate co–modulate microcystin and protein content in Microcystis aeruginosa PCC7806 and M. aeruginosa UV027. Microbial Ecol 49:1–6

Eisen JA Fraser CM (2003) Phylogenomics: intersection of evolution and genomics. Science 300, 1706–1707

Gallon JR, Chit KN, Brown EG (1990) Biosynthesis of the tropane–related cyanobacterial toxin Anatoxin–a: Role of ornithine decarboxylase: Phytochem 29:1107–1111

Gallon JR, Kittakoop P, Brown EG (1994) biosynthesis of anatoxin–a by Anabaena flos–aquae: examination of primary enzymatic steps: Phytochem 35:1195–203

Gupta N, BASB, LR (2002) Growth characteristics and toxin production in batch cultures of Anabaena flos–aquae: effects of culture media and duration. World J Microbiol Biotech 18:29–35

Harada K, Ohtani I, Iwamoto K, Suzuki M, Watanabe MF, Watanabe M, Terao K (1994) Isolation of cylindrospermopsin from a cyanobacterium Umezakia natans and its screening method. Toxicon 32:73–84

Hemscheidt T, Burgoyne DL, Moore RE (1995a) Biosynthesis of anatoxin–a(s). (2S,4S)–4–hydroxyarginine as an intermediate. J Chem Soc, Chem Comm 205–6

Hemscheidt T, Rapala J, Sivonen K, Skulberg OM (1995b) Biosynthesis of anatoxin–a in Anabaena flos–aquae and homoanatoxin–a in Oscillatoria formosa. J Chem Soc, Chem Comm 1361–2

Horikoshi K (1991) Microorganisms in alkaline environments. VCH, New York, pp 110

Kaebernick M, Neilan BA, Borner T, Dittmann E (2000) Light and the transcriptional response of the microcystin biosynthesis gene cluster. Appl Environ Microbiol 66(8):3387–92

Kleinkauf H, Von Dohren H (1996) A nonribosomal system of peptide biosynthesis. Eur J Biochem 236(2):335–51

Jones GJ, Negri AP (1997) Persistence and degradation of cyanobacterial paralytic shellfish poisons (PSPs) in freshwaters. Water Res 31:525–33

Kellmann R, Mills T, Neilan BA (2005) Functional modelling and phylogenetic distribution of putative cylindrospermopsin biosynthesis genes. J Mol Evol 61:1–5

Lehtimäki J, Moisander P, Sivonen K, Kononen K (1997) Growth, nitrogen fixation, and nodularin production by two Baltic Sea cyanobacteria. Appl Environ Microbiol 63:1647–56

Li RH, Carmichael WW, Brittain S, Eaglesham GK, Shaw GR, Liu YD, Watanabe MM (2001) First report of the cyanotoxins cylindrospermopsin and deoxycylindrospermopsin from Raphidiopsis curvata (cyanobacteria). J Phycol 37:1121–6

Long BM, Jones GJ, Orr PT (2001) Cellular microcystin content in N–limited Microcystis aeruginosa can be predicted from growth rate. Appl Environ Microbiol 67(1):278–83

Lukac M, Aegerter R (1993) Influence of trace metals on growth and toxin production of Microcystis aeruginosa. Toxicon 31(3):293–305

Luukkainen R, Sivonen K, Namikoshi M, Fardig M, Rinehart KL, Niemela SI (1995) Isolation and identification of eight microcystins from thirteen Oscillatoria agardhii strains and structure of a new microcystin. Appl Environ Microbiol 59(7):2204–9

Maestri O, Joset F (2000) Regulation by external pH and stationary growth phase of the acetolactate synthase from Synechocystis PCC6803. Mol Microbiol 37:828–38

Mikalsen B, Boison G, Skulberg OM, Fastner J, Davies W, Gabrielsen TM, Rudi K, Jakobsen KS (2003) Natural variation in the microcystin synthetase operon mcyABC and impact on microcystin production in Microcystis strains. J Bacteriol 185(9):2774–85

Moffitt MC (2003) Non–ribosomal biosynthesis of the cyanobacterial toxin nodularin. PhD thesis UNSW

Moffitt MC, Neilan BA (2004) Characterization of the nodularin synthetase gene cluster and proposed theory of the evolution of cyanobacterial hepatotoxins. Appl Environ Microbiol 70(11):6353–62

Moore RE, Chen JL, Moore BS, Patterson GML, Carmichael WW (1991) Biosynthesis of Microcystin–LR. Origin of the carbons in the Adda and Masp units. J Am Chem Soc 113:5083–4

Moore BS, Ohtani I, d KCB, Moore RE, Carmichael WW (1992) Biosynthesis of anatoxin–a(s): origin of the carbons: Tetrahed. Lett 33:6595–8

Namikoshi M, Murakami T, Fujiwara T, Nagai H, Niki T, Harigaya E, Watanabe MF, Oda T, Yamada J, Tsujimura S (2004) Biosynthesis and transformation of homoanatoxin–a in the cyanobacterium Raphidiopsis mediterranea Skuja and structures of three new homologues: Chem Res Toxicol 17:1692–6

Negri A, Llewellyn LE, Doyle J, Webster N, Frampton D, Blackburn S (2003) Paralytic shellfish toxins are restricted to few species among Australia's taxonomic diversity of cultured microalgae. J Phycol 39:663–667

Negri AP, Jones GJ, Blackburn SI, Oshima Y, Onodera H (1997) Effect of culture and bloom development and of sample storage on paralytic shellfish poisons in the cyanobacterium Anabaena circinalis. J Phycol 33:26–35

Neilan BA, Saker ML, Fastner J, Torokne A, Burns BP (2003) Phylogeography of the invasive cyanobacterium Cylindrospermopsis raciborskii. Mol Ecol 12:133–140

Nishizawa T, Asayama M, Shirai M (2001) Cyclic heptapeptide microcystin biosynthesis requires the glutamate racemase gene. Microbiol147(5):1235–41

Nogueira IC, Saker ML, Pflugmacher S, Wiegand C, Vasconcelos VM (2004) Toxicity of the cyanobacterium Cylindrospermopsis raciborskii to Daphnia magna. Environ Toxicol 19:453–9

Onodera H, Satake M, Oshima Y, Yasumoto T, Carmichawel WW (1997) New saxitoxin analogues from the freshwater filamentous cyanobacterium Lyngbya wollei. Nat Toxins 5:146–51

Ohtani I, Moore RE, Runnegar MTC (1992) Cylindrospermopsin – a potent hepatotoxin from the blue–green– alga Cylindrospermopsis raciborskii. J Am Chem Soc 114:7941–7942

Orr PT, Jones GJ (1998) Relationship between microcystin production and cell division rates in nitrogen–limited Microcystis aeruginosa cultures. Limnol Oceanog 43(7):1604–1614

Oshima Y (1995) Chemical and enzymatic transformation of paralytic shellfish toxins in marine organisms. In: Harmful Algal Blooms. Lassus P, Arzul G, Erard E, Gentien P, Marcaillou C (eds), Lavoisier, Paris, pp 475–480

Pearson LA, Hisbergues M, Borner T, Dittmann E, Neilan BA (2004) Inactivation of an ABC transporter gene, mcyH, results in loss of microcystin production in the cyanobacterium Microcystis aeruginosa PCC 7806. Appl Environ Microbiol 70(11):6370–8

Pomati F, Neilan BA (2004) PCR–based positive hybridisations to detect genomic diversity associated with bacterial secondary metabolism. Nucleic Acid Res 32(1):e7

Pomati F, Neilan BA, Manarolla G, Suzuki T, Rossetti C (2003a) Enhancement of intracellular saxitoxin accumulation by lidocaine hydrochloride in the cyanobacterium Cylindrospermopsis raciborskii T3 (Nostocales). J Phycol 39:535–42

Pomati F., B. P. Burns and B. A. Neilan (2004a) Identification of a Na^+ dependent transporter associated with saxitoxin production in the cyanobacterium Anabaena circinalis. Appl Environ Microbiol 70:4711–9

Pomati F, Rossetti C, Manarolla G, Burns BP, Neilan BA (2004b) Interaction between Intracellular Na^+ levels and saxitoxin production in Cylindrospermopsis raciborskii T3. Microbiol 150:455–61

Pomati F, Rossetti C, Calamari D, Neilan BA (2003b) Effects of saxitoxin and veratridine on bacterial Na^+–K^+ fluxes: a prokaryotic–based saxitoxin bioassay. Appl Environ Microbiol 69:7371–6

Rantala A, Fewer DP, Hisbergues M, Rouhiainen L, Vaitomaa J, Borner T, Sivonen K (2004) Phylogenetic evidence for the early evolution of microcystin synthesis. Proc Natl Acad Sci USA 101(2):568–73

Rapala J, Sivonen K, Luukkainen R, Niemelä SI (1993) Anatoxin–a concentration in Anabaena and Aphanizomenon at different environmental conditions and comparison of growth by toxic and non–toxic Anabaena strains, a laboratory study. J App Phycol 5:581–91

Rohrlack T, Dittmann E, Henning M, Borner T, Kohl JG (1999) Role of microcystins in poisoning and food ingestion inhibition of Daphnia galeata caused by the cyanobacterium Microcystis aeruginosa. Appl Environ Microbiol 65(2):737–9

Rouhiainen L, Vakkilainen T, Siemer BL, Buikema W, Haselkorn R, Sivonen K (2004) Genes coding for hepatotoxic heptapeptides (microcystins) in the cyanobacterium Anabaena strain 90. Appl Environ Microbiol 70(2):686–92

Saker ML, Neilan BA (2001) Varied diazotrophies, morphologies, and toxicities of genetically similar isolates of Cylindrospermopsis raciborskii (Nostocales, Cyanophyceae) from northern Australia. Appl Environ Microbiol 67:1839–45

Sako Y, Yoshida T, Uchida A, Arakawa O, Noguchi T, Ishida Y (2001) Purification and characterization of a sulfotransferase specific to N–21 of saxitoxin and gonyautoxin 2+3 from the toxic dinoflagellate Gymnodinium catenatum (Dinophyceae). J Phycol 37:1044–51

Schantz EJ, Ghazarossian VE, Schnoes FM, Strong JP, Springer JD, Pezzanite JD, Clardy J (1975) The structure of saxitoxin. J Am Chem Soc 97:1238–9

Schembri MA, Neilan BA, Saint CP (2001) Identification of genes implicated in toxin production in the cyanobacterium Cylindrospermopsis raciborskii. Environ Toxicol 16:413–21

Shalev–Alon G, Sukenik A, Livnah O, Schwarz R, Kaplan A (2002) A novel gene encoding amidinotransferase in the cylindrospermopsin producing cyanobacterium Aphanizomenon ovalisporum. FEMS Microbiol Lett 209:83–7

Shi L, Carmichael WW, Miller I (1995) Immunogold localization of hepatotoxins in cyanobacterial cells. Arch Microbiol 163(1):7–15

Shimizu Y (1986a) Biosynthesis and biotransformation of marine invertebrate toxins. In: Natural Toxins. Harris JB (ed) Clarendon Press, Oxford UK, pp 115–125

Shimizu Y (1986b) Toxigenesis and biosynthesis of saxitoxin analogues. Pure Appl Chem 58:257–62

Shimizu Y (1996) Microalgal metabolites – a new perspective. Ann Rev Microbiol 50:431–65

Shimizu Y, Norte M, Hori A, Genenah A, Kobayashi M (1984) Biosynthesis of saxitoxin analogues: the unexpected pathway. J Am Chem Soc 106:6433–4

Sielaff H, Dittmann E, Tandeau De Marsac N, Bouchier C, Von Dohren H, Borner T, Schwecke T (2003) The mcyF gene of the microcystin biosynthetic gene cluster from Microcystis aeruginosa encodes an aspartate racemase. Biochem J 373(3):909–16

Sivonen K, Jones G (1999) Cyanobacterial toxins. In: Toxic Cyanobacteria in Water. Chorus I, Bartram J (eds) WHO E & FN Spon Publishers, London, pp 41–111

Sugimoto E, Pizer LI (1968) The mechanism of end product inhibition of serine biosynthesis. I. Purification and kinetics of phosphoglycerate dehydrogenase. J Biol Chem 243(9):2081–9

Tanabe Y, Kaya K, Watanabe MM (2004) Evidence for recombination in the microcystin synthetase (mcy) genes of toxic cyanobacteria Microcystis spp. J Mol Evol 58(6):633–41

Taroncher–Oldenburg G, Anderson DM (2000) Identification and characterization of three differentially expressed genes, encoding S–adenosylhomocysteine hydrolase, methionine aminopeptidase, and a histone–like protein, in the toxic dinoflagellate Alexandrium fundyense. Appl Environ Microbiol 66:2105–12

Taroncher–Oldenburg G, Kulis DM, Anderson DM (1997) Toxin variability during the cell cycle of the dinoflagellate Alexandrium fundyense. Limnol Oceanogr 42:1178–88

Tillett D, Dittmann E, Erhard M, von Dohren H, Borner T, Neilan BA (2000) Structural organization of microcystin biosynthesis in Microcystis aeruginosa PCC7806: an integrated peptide–polyketide synthetase system. Chem Biol 7(10):753–64

Utkilen H, Gjolme N (1995) Iron–stimulated toxin production in Microcystis aeruginosa. Appl Environ Microbiol 61(2):797–800

Velzeboer RMA, Baker PD, Rositano J, Heresztyn T, Codd GA, Raggett SL (2000) Geographical patterns of occurrence and composition of saxitoxins in the cyanobacterial genus Anabaena (Nostocales, Cyanophyta) in Australia. Phycologia 39:395–407

Waditee R, Hibino T, Tanaka Y, Nakamura T, Incharoensakdi A, Takabe T (2001) The halotolerant cyanobacterium Aphanothece halophytica contains an Na^+/H^+ antiporter, homologous to eukaryotic ones, with novel ion specificity affected by C–terminal tail. J Biol Chem 276:36931–8

Yin QQ, Carmichael WW, Evans WR (1997) Factors influencing growth and toxin production by cultures of the freshwater cyanobacterium Lyngbya wollei Farlow ex Gomont. J Appl Phycol 9:55–63

Yoshida T, Sako Y, Uchida A, Kakutani T, Arakawa O, Noguchi T, Ishida Y (2002) Purification and characterization of a sulfotransferase specific to O–22 of 11–hydroxy saxitoxin from the toxic dinoflagellate Gymnodinium catenatum (Dinophyceae). Fisheries Science 68:634–42

Young FM, Thomson C, Metcalf SS, Lucocq JM, Codd GA (2005) Immunogold localization of microcystins in cryosectioned cells of Microcystis. J Struct Biol 151(2):208–14

Chapter 18: Determining important parameters related to cyanobacterial alkaloid toxin exposure

Adam H Love

Forensic Science Center, Lawrence Livermore National Laboratory, P.O. Box 808, L-178, Livermore, CA 94550

Introduction

The United States is faced with having to address critical issues that have not been addressed in the past, as recent security interests have placed new and more pressing demands on the assessment of risk from exposure to potential threats resulting from deliberate contamination. Such assessments are the basis for decisions about future research priorities, actions taken to prevent intentional contaminant releases, and developing detailed plans for response if such an event was to occur. High fidelity assessments are based on robust knowledge of the numerous parameters that can be used to predict the fate, transport, persistence, and toxicity for contaminants under numerous circumstances. Therefore, identification and determination of these parameters are the primary steps for evaluation of potential threats.

Numerous toxic substances exist, but very few have the potential to cause mass casualties from widespread contamination. Conventional chemical and biological weapons have strict laws that control their possession, transport, and use. Military studies of conventional agents have identified and determined many of the parameters necessary for accurate threat assessments. Spanning the threat space between chemical and biological weapons are biotoxins. Biotoxins are chemical toxins that are of natural biological origin. These compounds also have unique physio-chemical properties compared to other chemical and biological agents. Most natural toxins have few controls over possession, transport, and use; thus, these biotoxins have greater potential availability than conventional threat agents. They can range from large proteins (10s to 100s kDa) to small molecular weight molecules (~100-~500 MW). Large biotoxins must be extracted from biological material, but the small molecular weight toxins can be obtained either through biological extraction (Barros et al. 2004) or through methods for direct chemical synthesis published in scientific journals (Kishi et al. 1977; Jacobi et al. 1984; Mansell 1996).

Cyanobacteria produce some of the most potent biotoxins known. Many of these cyanobacterial toxins (cyanotoxins) have low-level acute toxicity that are comparable to the toxicity levels from conventional chemical weapons and therefore warrant consideration as a potential threat (Burrows and Renner, 1999). The alkaloid class of cyanotoxins contains three types of the toxins considered here; saxitoxins (STX), anatoxins (ATX), and anatoxin-a(S). As alkaloids, each of these compounds are small molecular weight organic compounds with basic functional groups and a heterocyclic ring containing at least one nitrogen atom (Fig. 1). These toxins are rapidly acute neutrotoxins, although the toxicological mode of action is different for each type. STX is a tightly regulated biotoxin because of its listing on the Chemical Weapons Convention Schedule 1 list. Possession, transport, and use of ATX and anatoxin-a(S) are unregulated. The increasing occurrence of cyanotoxins in freshwater drinking supplies, as a result of increasing algal bloom containing cyanobacteria, have resulted in "Cyanobacteria and its toxins" listed on the US EPA Drinking Water Candidate Contaminate List 2 in February 2005. Other than acute toxicity levels, our current understanding of alkaloid cyanotoxins is limited and prevents an accurate assessment of whether these toxins pose a legitimate threat.

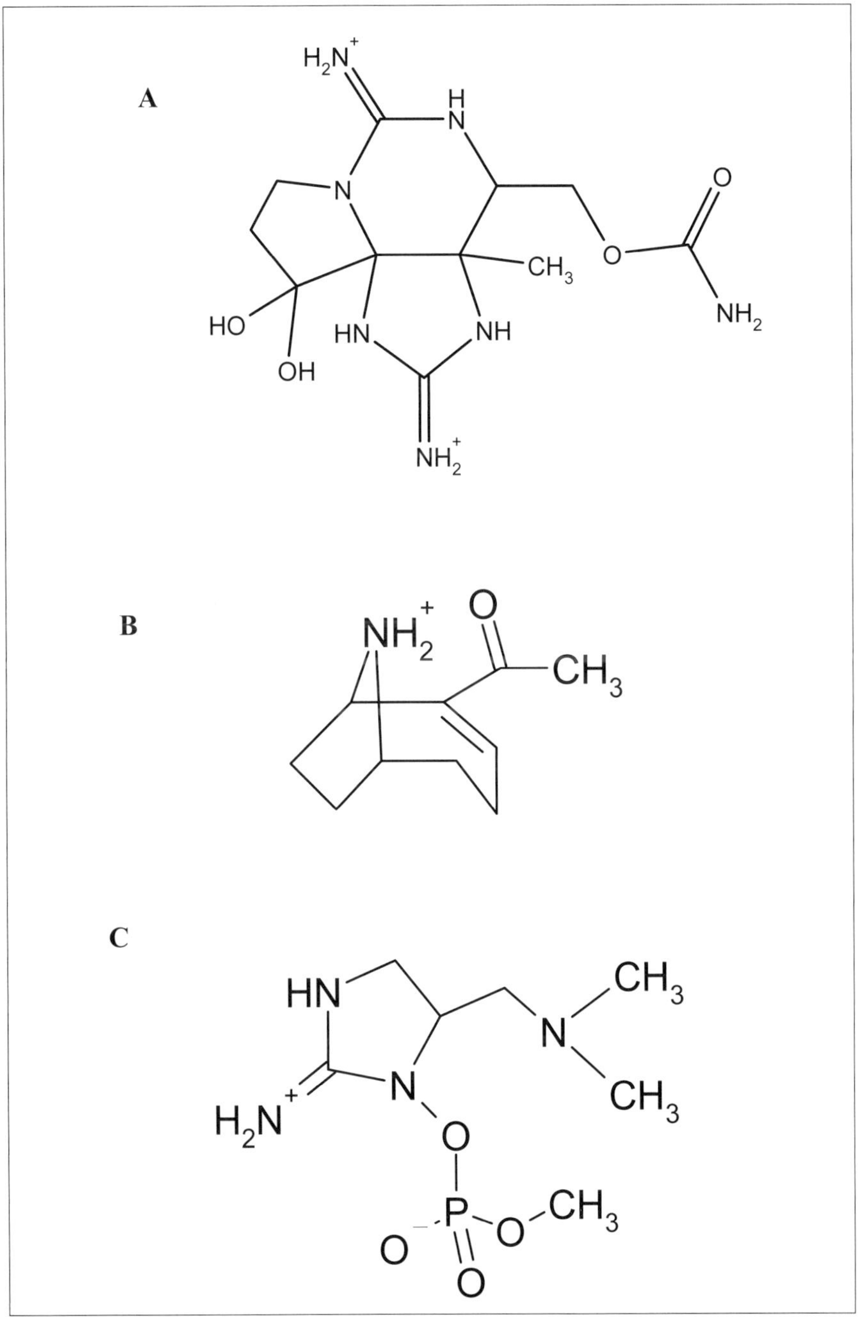

Fig. 1. Structure of alkaloid cyanotoxins; A) saxitoxins B) anatoxins, C) anatoxin-a(S)

Model and Parameters for Assessment of Potential Threats

Since cyanobacteria naturally release alkaloid toxins into freshwater sources, numerous pathways of exposure are being considered in order to assess the information requirements for potential water quality regulations. A schematic model for any general assessment and the links between different components are shown in Fig. 2. Most of the parameters needed for an assessment of a natural cyanotoxin release are also applicable to an intentional cyanotoxin release. A greater number of these model parameters have been determined for microcystin, a cyclic polypeptide liver cyanotoxin, since it is easier to measure quantitatively due to an aromatic ring in its structure that permits UV detection. Recognizing the need for greater protection from these naturally produced toxins, the WHO and several countries have established water quality regulations for microcystin based on conservative assumptions using this greater but still limited scientific understanding. Although assessment of the potential risk from intentional alkaloid cyanotoxin contamination has considerable overlap with natural releases, below is a discussion of some additional pieces of information that would be useful to know in order to address issues specific to intention releases.

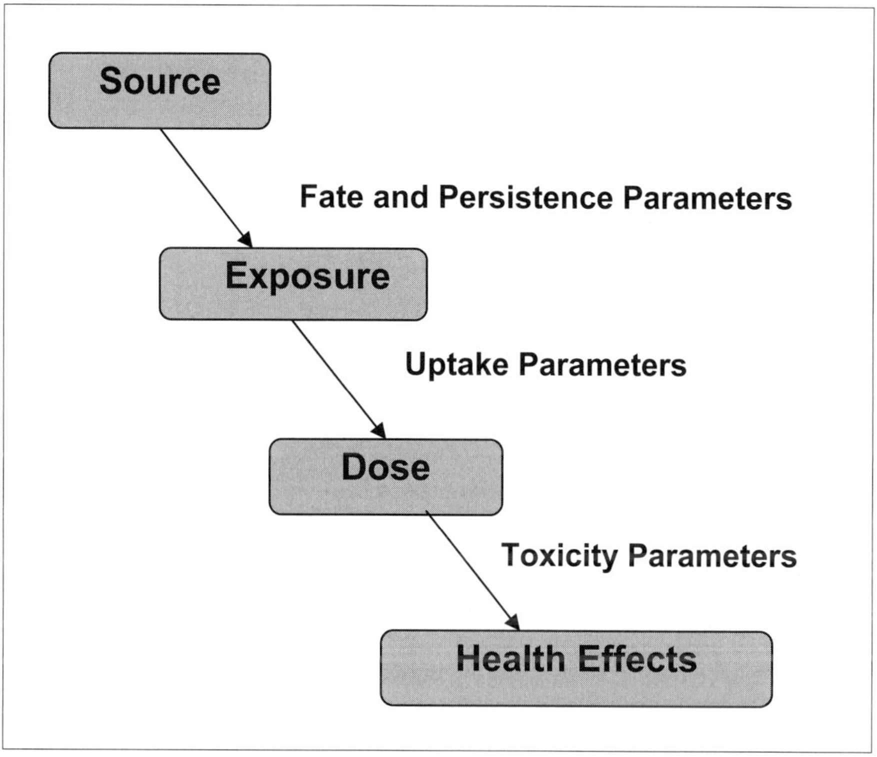

Fig. 2. Schematic Model of Information Needed for Threat Assessment

Source

The starting point for any contaminant model of exposure is characterizing the magnitude and duration of the source of contamination. The information about the source affects all other parts of the exposure model. For natural releases, environmental factors and the biological community that result in cyanotoxin production limit the source characteristics, but intentional releases do not have the same source constraints. Thus, the potential is greater for the source of intentional contamination to reach acute concentrations than under natural conditions. For intentional contamination there is unique information that would be useful to be able to characterize about the source, such as the method of cyanotoxin acquisition and any attribution signatures. Identification of other chemicals associated with the intentional release that indicate if the toxins were synthesized directly or extracted from algal blooms provides important information about the technical abilities of the party involved with an intentional release. Other attribution signatures that can result from related temporal, spatial, physical, chemical, biological, or environmental information which can be used to narrow the identification of the responsible party is also a desirable capability for characterizing the cyanotoxin source.

Fate and Persistence Parameters

Once released, a mechanistic understanding of how cyanotoxins interact with their environment permits some prediction of how long these toxins persist in their environment and in what matrix the toxins are expected to reside. The primary parameters that can be used construct a mechanistic understanding are 1) flow conditions, 2) conditions and rates for degradation, 3) adsorption/complexation reactions and the affects of such associations on degradation and transport, and 4) enhanced degradation mechanism, such as biodegradation and catalysis. Little quantitative research has been performed to determine these parameters, although there have been some previous studies that indicate some general properties of some of these alkaloid cyanotoxins:

- Both STX and ATX are stable under acidic conditions, labile under alkaline conditions, and resistant to oxidation by HOCl (Burrows and Renner 1999).
- STX is thermally labile at normal temperatures (Burrows and Renner 1999; Indrasena and Gill 2000);
- ATX is photolabile on the order of hours to weeks (Stevens and Krieger 1991).
- Little research on fate and persistence has been performed for anatoxin-a(S), but it has structurally similarities to existing organophosphorous pesticides, such as glyphosate, where most of the fate and persistence parameters are well-established (Rueppel et al. 1977).

Thus, models for the assessment of these cyanotoxins require a quantitative and mechanistic understanding of environmental interaction for a quantitative assessment and prediction of the potential threat for conditions not specifically tested in the laboratory.

Exposure Matrices

The exposure matrices for intentionally released cyanotoxins overlap with the matrices currently being considered for natural cyantoxin occurrences.

Surface water bodies that are used either for recreation or for drinking water sources are the primary exposure matrix receiving consideration for natural releases and can represent either an ingestion or dermal exposure (Burrows and Renner 1999). Other natural exposure matrices receiving consideration are 1) aerosols generated during irrigation or showering and 2) contaminated crops, either within plant cells or as surface contaminants, on agricultural products irrigated with contaminated water. Aerosols represent an inhalation exposure, whereas contaminated crops represent an ingestion hazard. Different exposure pathways may have significantly different levels of toxicity and exposure matrices may affect the bioavailability of the toxins. Our current knowledge of important exposure matrices and the difference between different exposure pathways is limited, but is an important component for development an overall assessment of the potential threat of these toxins.

Uptake Parameters

Very little is known about how exposure pathways and uptake dynamics affect the overall toxicity of alkaloid cyanotoxins. This is where the greatest difference exists between natural and intentional cyanotoxin contamination. Natural releases are more likely to pose a long-term exposure/uptake, whereas intentional releases are more likely to pose a short-term exposure/uptake. These fundamental differences are what result in the emphasis of acute toxins over chronic toxins for intentional contamination assessment consideration.

Dose Characterization and Toxicity Parameters

The acute toxicity levels of these alkaloid toxins are well-established for mice and there is information about human toxicity resulting from STX exposure in shellfish. As stated above, the effective dose that results from bioavailability modification of the exposure matrix may have a significant affect of the actual toxicity. Presently, the toxicokinetic and toxicodynamics information is limited and does not allow for a mechanistic understanding of human toxicity. Although no long-term adverse affects have been observed from sub-lethal exposure, there has not been any determination if there is any biological response that may result in the detection of a biomarker or interference with sensitive cellular function, such as prenatal development.

Health Effects

The ultimate endpoint for any threat assessment from an intentional contamination is the human health impact. Such an assessment may be significantly different for natural vs. intentional exposure since magnitude and duration of the different sources may have important differences. Because dose and toxicity information is limited, estimates of deaths, hospitalizations, and any long-term health effects have large uncertainties. A better understanding of dose and toxicity mechanisms would permit better estimates of human health impacts and may allow for the identification of improved methods for diagnosis and treatment. Implementation of the complete assessment model should assist in determining if a set of conditions exist where these toxins pose a legitimate public health threat, and if so, understanding the spatial distribution and total number of health impacts for both natural and intentional contamination scenarios.

Model Impacts for Homeland Security Management

Accurate modeling predictions are needed for a high-fidelity threat assessment, but also the model information is needed as a tool for management of the potential risk. An understanding of the issues related to an intentional release of these toxins is a prerequisite for proper preparation and an appropriate response to such an event. A typical diagram for the management of Homeland Security threats is shown in Fig. 3. Prior to an incident of national security there are "Prevention & Preparation" steps to mitigate the impact of intentional contamination events. During and after a contamination event there is an "Assessment & Response" Phase followed by a "Decontamination and Recovery" Phase. Insight gained from a thorough assessment will result in improved consequence management to such an event.

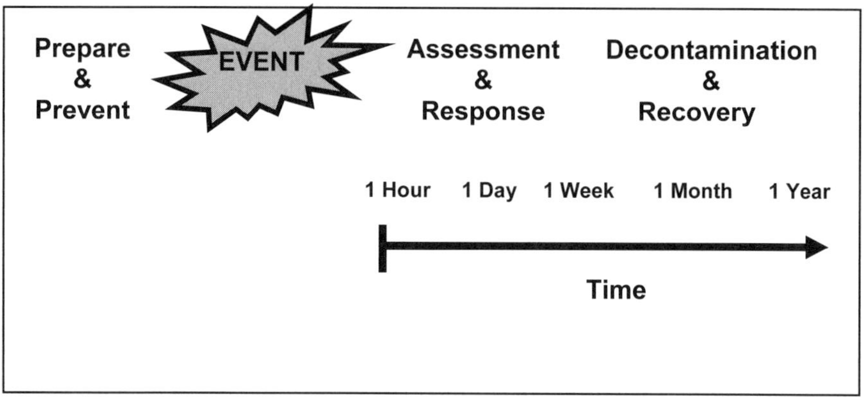

Fig. 3. Simplified Management Perspective of Homeland Security Threats

Prepare & Prevent

Source material management and monitoring are the primary means of mitigating the impact from an intentional biocontamination event. For cyanotoxins that pose a legitimate threat, appropriate quantities necessary for legitimate research purposes (i.e. understanding natural cyanotoxins occurrence and affects or pharmaceutical applications) need to be established without hindering the advancement of science. As with other highly toxic materials, possession of quantities of alkaloid cyanotoxins that the assessment model implicates as a legitimate threat can be tracked and inventoried. Currently methods for detection and quantification of these toxins require highly trained scientists, and sensors for these toxins do not exist. Monitoring using periodic sampling is likely to be more effective for natural cyanotoxin releases, where exposure tend to be chronic. For acute exposure from intentional releases, continuous monitoring is the only way to provide any degree of health protection for short duration exposures. These criteria are part of numerous existing efforts to understand how to best implement Early Warning Systems to the numerous potential threats of intentional contamination. Fate and persistence information on these cyanotoxins is needed for establishing criteria for Early Warning Systems.

Assessment & Response

After an event where alkaloid cyanotoxins were intentionally released, the recognition and identification could result from either monitoring alerts or significant changes in public health status. In either case, identification of these toxins as the contaminate source is critical to effect triage and prevent additional exposures soon after the initial recognition. Clinical toxin identification procedures for legitimate cyanotoxin threats, established through the toxicological parameters of the threat assessment model, should be established for early diagnosis and analytical confirmation methods for important exposure matrices are additionally needed. Once the toxin is identified, a number of responses are necessary to minimize any additional exposure, including identification and elimination of the contaminant source, infrastructure modification to prevent further exposure, forensic sampling and analyses, and public notification. In some cases, the source may not be readily known and using the information collected an event reconstruction can be performed to identify the location and characteristics of the contaminate source. Any event reconstruction will be based on threat assessment model parameters that are then utilized to execute the model backwards (Fig. 1) so exposure and health effects information can be used to determine source characteristics. For such efforts to be successful, high-fidelity data and models are required.

Decontamination & Recovery

Safe reoccupation/utilization is the goal after a biocontamination event. Efforts for recovery of critical infrastructure affected by the intentional contamination may require active decontamination if the agent doesn't naturally degrade over time. Methods for decontamination can utilize information about the degradation, biodegradation, and catalysis of alkaloid cyanotoxins determined for the threat assessment. The toxicological information from the threat assessment could be valuable in establishing public health standards for decontamination also. A basic tenet of Recovery is risk communication which includes educating the stakeholders of the risks associated with the incident, decontamination efforts, and safe reoccupation of decontaminated sites.

Conclusions

Science-based decision making required robust and high-fidelity mechanistic data about the system dynamics and impacts of system changes. Alkaloid cyanotoxins have the characteristics to warrant consideration for their potential threat. Since insuffcent information is available to construct a systems model for the alkaloid cyanotoxins, STX, ATX, and anatoxin-a(S), an accurate assessments of these toxins as a potential threat for use for intentional contamination is not possible. Alkaloid cyanotoxin research that contributed to such a model has numerous areas of overlap for natural and intentional health effects issues that generates dual improvements to the state of the science. The use of sensitivity analyses of systems models can identify parameters that, when determined, result in the greatest impact to the overall system and may help to direct the most efficient use of research funding. This type of modeling-assisted experimentation may allow rapid progress for overall system understanding compared to observational or disciplinary research agendas. Assessment and management of risk from intentional contamination can be performed with greater confidence when mechanisms are known and the relationships between different components are validated. This level of understanding allows high-fidelity assessments that do not hamper legitimate possession of these toxins for research purposes, while preventing intentional contamination that would affect public health. It also allows for appropriate response to an intentional contamination event, even if the specific contamination had not been previous considered. Development of science-based decision making tools will only improve our ability to address the new requirements addressing potential threats to our nation.

References

Barros LPC, Monserrat JM Yunes JS (2004) Determination Of Optimized Protocols For The Extraction Of Anticholinesterasic Compounds In Environmental Samples Containing Cyanobacteria Species. Environmental Toxicology And Chemistry 23 (4): 883-889

Burrows WD, Renner SE (1999) Biological Warfare Agents As Threats To Potable Water. Environmental Health Perspectives 107 (12): 975-984

Indrasena WM, Gill TA (2000) Storage Stability Of Paralytic Shellfish Poisoning Toxins. FOOD CHEMISTRY 71 (1): 71-77

Jacobi PA, Martinelli MJ, Polanc S (1984) Total Synthesis Of (+/-)-Saxitoxin. Journal Of The American Chemical Society 106 (19): 5594-5598

Mansell HL (1996) Synthetic Approaches To Anatoxin-a. TETRAHEDRON 52 (17): 6025-6061

Rueppel Ml, Brightwell Bb, Schaefer J, Marvel JT (1977) Metabolism And Degradation Of Glyphosate In Soil And Water. Journal Of Agricultural And Food Chemistry 25 (3): 517-528

Stevens DK, Krieger RI (1991) Stability Studies On The Cyanobacterial Nicotinic Alkaloid Anatoxin-A. Toxicon 29 (2): 167-179

Tanino H, Nakata T, Kaneko T, Kishi Y (1977) Stereospecific Total Synthesis Of D,L-Saxitoxin. Journal Of The American Chemical Society 99 (8): 2818-2819

Chapter 19: Toxins Workgroup Poster Abstracts

Microginin peptides from *Microcystis aeruginosa*

Drummond AK, Schuster T, Wright JLC

Center for Marine Science, University of North Carolina Wilmington, Wilmington, NC 28409

Introduction

Freshwater cyanobacteria have been shown to produce several classes of unique peptide metabolites. *Microcystis aeruginosa* in particular has been a rich source of many interesting peptides, most notably the heptapeptide microcystins. In addition to the cyclic hepatotoxic microcystins, *M. aeruginosa* also produces microginins, a family of linear peptides composed of 3–6 amino acid residues. Previously characterized microginins show angiotensin converting enzyme (ACE) inhibition as well as leucine aminopeptidase M inhibition. Consequently, these compounds are of interest as lead compounds in the discovery of novel antihypertensive agents as well as treatments for congestive heart failure.

Hypothesis

Cyanobacteria have been extensively studied for toxins that they produce. In addition to these toxins, other unusual cyanopeptides are continually found. This suggests the presence of other complex biosynthetic pathways capable of producing a range of novel peptide metabolites with exotic chemistries and important bioactivities. Thus, we hypothesize that *M. aeruginosa* is likely to be a source of new bioactive secondary metabolites.

Methods

Microcystis aeruginosa cells (UTEX 2385) were grown in B3N media in a 14/10 light–dark cycle. Cells were collected by vacuum filtration and extracted with 80% methanol. The organic extract was dried and applied to a reversed phase column followed by elution with a step gradient of aqueous methanol. The fractions of interest were combined and further purified by size exclusion chromatography. Appropriate fractions were then combined and further subjected to reversed–phase HPLC, yielding microginin 527 (compound **2**, rt = 16.4 min) and microginin 690 (compound **1**, rt = 17.9 min) in addition to several more minor congeners.

Results

Compound 1 was isolated as a white powder. The HRMS and NMR data revealed a molecular formula of $C_{37}H_{46}N_4O_9$ (m/z = 691.334305). The UV data (λ_{max} 224, 276 nm) indicated the presence of an aromatic amino acid. Indeed, the ESI–MS data revealed a strong fragment ion at [M+H –180] and m/z 180 consistent with a C–terminus tyrosine moiety. Another fragment ion at m/z 343 revealed a second tyrosine residue adjacent to the C–terminal tyrosine. The fragment ion at m/z 128 suggested an N–terminus 3–amino–2–hydroxy decanoic acid (Ahda). A moiety corresponding to 161 Da remained unaccounted for. 2D NMR analysis suggested that this unknown component was a fourth amino acid, perhaps a modified tyrosine. Compound **2** was also isolated as a white powder. This compound had the same UV absorbance as compound 1 and shared fragment ions of m/z 128 and m/z 180 but had a molecular weight of 528. It was deduced that this compound shared the N–terminus Ahda residue and a single C–terminus tyrosine.

Conclusion

Cyanobacteria that produce toxins often have the ability to produce other secondary metabolites. We have isolated two previously undiscovered peptides from *Microcystis aeruginosa* as well as some minor derivatives. These compounds may belong to a new family of microginins containing an unknown residue that we believe to be a modified tyrosine. Further NMR analysis is required to fully characterize the nature of this amino acid derivative.

Inactivation of an ABC transporter, *mcyH*, results in loss of microcystin production in the cyanobacterium *Microcystis aeruginosa* PCC 7806

Pearson LA,[1] Hisbergues M,[2] Börner T,[2] Dittmann E,[2] Neilan BA[1]

[1]School of Biotechnology and Biomolecular Sciences, The University of New South Wales, Sydney, Australia 2052, [2]Institute for Biology, Humboldt University, Chausseestr, 117, 10115 Berlin, Germany.

Introduction

The cyanobacterium *Microcystis aeruginosa* is widely known for its production of the potent hepatotoxin microcystin. Microcystin is synthesized nonribosomally by the thiotemplate function of a large, modular enzyme complex encoded within the 55 kb microcystin synthetase (*mcy*) gene cluster. Also encoded within the *mcy* gene cluster is a putative ATP binding cassette (ABC) transporter, McyH. This study details the bioinformatic and mutational analyses of McyH and offers functional predictions for the hypothetical protein.

Hypotheses

It is hypothesized that *mcyH* encodes an ABC–transporter that is responsible for the biosynthesis and/or transport of Microcystin.

Methods

The *mcyH* gene has been characterized bioinformatically via structural, functional and phylogenetic analyses. In addition, an *mcyH* null mutant has been engineered and characterized with respect to its ability to produce and export microcystin. The McyH enzyme has been heterologously expressed in *E. coli,* purified and used to raise anti–McyH antibodies. These antibodies have been used in immunoblotting experiments to investigate McyH expression in mutant and wild–type strains of *M. aeruginosa.*

Results

The McyH transporter is putatively comprised of 2 homodimers, each with an N–terminal hydrophobic domain and a C–terminal ATPase. Phylogenetically, McyH was found to cluster with members of the ABC–A_I subgroup of ABC ATPases, suggesting an export function for the protein. The m*cyH* null mutant strains were unable to produce microcystin. Whilst the *mcyH* deletion had no apparent effect on the transcription of other *mcy* genes, the complete microcystin biosynthesis enzyme complex could not be detected in *mcyH* mutant strains. Expression of McyH was reduced in *mcyA* and *mcyB* mutants and completely absent in the *mcyH* mutant.

Conclusion

By virtue of its association with the *mcy* gene cluster and the bioinformatic and experimental data presented in this study, we predict McyH functions as a microcystin exporter and is, in addition, intimately associated with the microcystin biosynthesis pathway.

Chapter 20: Analytical Methods Workgroup Report

Workgroup Co–chairs: Armah A de la Cruz; Michael T Meyer

Workgroup Members[1]: Kathy Echols, Ambrose Furey, James M Hungerford, Linda Lawton, Rosemonde Mandeville, Jussi AO Meriluoto, Parke Rublee, Kaarina Sivonen, Gerard Stelma, Steven W Wilhelm, Paul V. Zimba

Authors: Armah A de la Cruz, Parke Rublee, James M Hungerford, Paul V Zimba, Steven W Wilhelm, Jussi AO Meriluoto, Kathy Echols, Michael T Meyer, Gerard Stelma, Rosemonde Mandeville, Linda Lawton, Kaarina Sivonen, and Ambrose Furey

The topic of exposure assessment overlaps with other topic areas of this workshop. It includes considerations of establishment of long term monitoring and event response, sampling protocols, development and standardization of organism and toxin assays, funding mechanisms, and public outreach. The development of a coordinated infrastructure (funding, human resources, and facilities, materials and equipment) is key to successfully addressing the threat posed by CHABs.

The establishment of validated standardized protocols to detect cyanobacteria and cyanotoxins is of considerable importance given the increased occurrence of CHABs worldwide. Standardized methods are needed for studies assessing occurrence, monitoring and toxicity studies which are essential aspects of risk assessment and management and the development of guidance and regulation.

---

[1] See workgroup member affiliations in Invited Participants section.

Development of cyanobacteria and cyanotoxin standards for research and monitoring

There is a lack of reliable, quantitative standards for analytical determination of any of the toxins produced by cyanobacteria. Currently, while some of these toxins can be purchased commercially, availability and quantities are unreliable; and the identity and purity of the compounds is not guaranteed. Cases of either false identity or low purity standards have been documented in the scientific literature.

The criteria for selecting which toxins should be produced are:

1. Prevalence in US waters then global prevalence and
2. Documented risk of health effects (primarily irreversible human health effects, but also direct and indirect environmental impacts).

These toxins were discussed extensively during the ISOC HAB meeting. The toxins that need to be produced are microcystins, cylindrospermopsins, anatoxins (anatoxin–a, homoanatoxin–a, anatoxin–a[s]), saxitoxins (many of these are already commercially available at acceptable quality through shellfish monitoring programs), nodularin, and lipopolysaccharides. In addition to these, there are many unknown toxic and bioactive compounds that may become important in the future (Erhard et al.1997; Cox et al. 2003; Berry et al. 2004). One example of this is BMAA (ß–methylamino–L–alanine) which has recently been discovered to be present in many species of cyanobacteria (Cox et al. 2005).

Since microcystins have over 80 variants and congeners that vary in toxicity, it is recommended that several of the most prevalent variants are produced initially. These would be microcystin–LR, –YR, –RR, and –LA and their 3/7–demethylated analogues. Other variants of microcystin would be added to the list as needs arise. Since all of these toxins are derived from cyanobacterial cultures, care should be taken to insure sufficient quantities for monitoring and research purposes. One of the problems to overcome is ensuring cyanobacterial strain purity in order to maintain a consistent level of toxin production. An example of a well–characterized producer of seven microcystins is *Microcystis aeruginosa M.*TN–2 strain maintained in modified Fitzgerald media (Lee and Chou 2000). Algal cultures for production of large volumes of anatoxin–a[s], cylindrospermopsins, homoanatoxin–a must be identified and made available through culture collections. Optimal culturing practices need to be determined, particularly maximal toxin production as a function of temperature, light, and nutrient supply (Downing et al. 2005). Currently, there are no practical and efficient synthetic methods for any of these chemical compounds except for anatoxin–a (Danheiser et al.1985).

Unlabeled and labeled stable isotope standards (for mass spectrometric work) are also needed. The first priority would be the production of standards of known concentration in solution, then neat, pure toxin and spiked matrices (i.e. cyanobacteria, shellfish, finished water, and food supplements). Extraction methodologies for optimal recovery need to be determined, and calibrated between laboratories, particularly for animal tissues during assessment of whole body burden. Standards should be certified for identity, purity and transport/long–term storage stability. Standards should be certified by gravimetric, NMR, and/or chemiluminescence nitrogen detection. For standards usable for biological research, biological activity information in a defined experimental setup should be included.

While there are no certified reference materials (CRMs) for any of these toxins (with the exception of the saxitoxin group), a number of companies are already pursuing this direction. CRMs of these calibration standards would be the ultimate goal. CRMs are used to evaluate the measurement precision and calibration of the laboratories and the analytical instruments used for toxin analysis. The consistency between laboratories and methods can be compared and authenticated by using CRMs.

Standards should be readily available and reasonably priced (non–profit preferred). When developing the structure for distribution of these standards they would be available in small amounts that could not be used for malicious purposes. Therefore, procedure for obtaining these standards should be kept straightforward to reduce paperwork and infrastructure.

The recommendations of the workgroup are to support these activities by pooling government agency, academic and international resources for the development of a reliable source of standards. Without such interactive support, gains in general knowledge, and managing or controlling CHAB will be slowed.

Sampling

To determine the occurrence and assess the risk of cyanobacterial harmful algal blooms (CHAB), it is important to collect samples that reflect the actual site or source conditions. Samples may consist of water, plankton, invertebrates, vertebrates, or sediments. Analyses may include toxins, genomic identification, enzyme or antibody assays, whole organism or tissue specific toxicity assays, or histopathology.

Lakes, reservoirs, estuaries, and streams are all potential sources of blooms. It is important to realize that the spatial and temporal distribution of any CHAB bloom is heterogeneous. Since CHABs growth rates range from 0.25 to 1.0 doublings per day, the field sampling efforts must consider growth rates of the cyanobacteria. Additional, care must be taken during the sampling effort, specifically with regard to altering the natural vertical distribution of cyanobacteria if surface scums are present.

The development of standard sampling procedures must be developed and validated. Aspects to consider include:

1. Safety protocols
2. Sample equipment (including cross contamination issues)
3. Field filtration
4. Sample stabilization (pH, temperature, light, control of degradation, etc.)
5. Sample transportation and storage
6. Sample documentation

Safety concerns during toxin collection must consider both short–term and long–term exposure hazards. A validated standard method for the field and the laboratory is necessary for collecting samples in a CHAB. Standard paraphernalia that should always be worn includes lab coats, gloves, masks, and goggles. Protocols for safe handling of fresh tissue, freeze–dried materials, including cell biomass, sediments, and neat toxins must be developed. General procedures used in Class II (biohazard) laboratories are recommended to minimize exposure risk to personnel.

Sample collection and preservation is dependent on the end use (i.e., toxin analysis, molecular experimentation, culture based, and tissues for histology). For example, culture based approaches require that the samples not be affected by perturbations, whereas, samples for toxin analysis requires different handling procedures. It is imperative that sampling techniques and equipment are used to minimize sample contamination (equipment, human, and cross contamination). For example, for many organic contaminants, glass and Teflon are preferable to plastics due to sorption and can impede cellular growth (if collecting for growth, this is important). Additionally, if the goal is to provide geospatial information (i.e., toxin abundance maps) the use of integrated versus discrete samples and fixed or randomly assigned locations may be used. There is, however, a need for standardized approaches for the analysis of whole water, particulate and dissolved toxins. Filtrate can be derived from a number of different filter sizes (i.e. glass fiber filters of 1.2 or 0.7 microns, versus membranes of 0.45 or 0.2 microns). Tissue toxin analysis requires rapid preservation of samples, the ability to efficiently extract and quantify the toxin in light of differential matrix effects.

The collection of supporting data is critical to relate toxin abundance to physical and chemical causative variables. Physico–chemical parameters linked to cyanobacteria would include: dissolved–organic materials (DOM), pH, macro and micro–nutrients, temperature, turbulence, supporting plankton information including bacteria, algae, and zooplankton present, as well as light quality and quantity.

As with many other sample types, obtaining representative samples representative of the material/area being sampled, is paramount. The use of remote sensing to determine regions of interest may be useful in defining sampling efforts.

Sample processing and detection methods

Toxins produced by CHAB are of concern because of their demonstrated adverse affects to human health and the ecosystem. These compounds include the high priority toxins such as the microcystins (and nodularins), the cylindrospermopsins, and anatoxin–a, and many others including newly emerging toxins. Toxin–producing CHABs have been documented throughout the world and many regions of the United States. Simultaneous cyanotoxin profiling is a challenging area and even though we are not in the position to set limits at this point, current detection methods for each or some group of cyanotoxin, including screening and quantitative methods, hold promise for cyanobacterial toxin detection at the current WHO guidance levels (McElhiney and Lawton, 2005). There is, however, no suitable method to simultaneously extract and detect all the high priority cyanotoxins of interest to the Agency due to their biochemical differences. Routine monitoring is an unmet need in the US. Monitoring needs to be instituted and should address multiple facets including frequency of occurrence, performance of water utilities, transport within ecosystems, phytoplankton profiling, and toxin profiling.

Knowledge gaps in spite of the publication or availability of these methods, for regulation of the toxins, additional work is needed to support validation of sufficiently reliable and rugged methodology. This additional work will include development and field trials to address several important requirements. These include evaluating analyte stability, approaches and requirements for preconcentration (versus various sample matrix interferences) lyses methodologies of cells, and others. Analyte stability issues can be exemplified by the case of anatoxin–a which degrades rapidly at >7.5 pH, hence analytical samples must be maintained in acidic conditions. LC/MS methods detect the degradation products of anatoxin–a and anatoxin–a is easily mis-identified as phenylalanine. As part of sample preparation/sample extraction, differences in intra– and extra–cellular toxin concentrations must be considered. Intracellular concentrations, for example, relate potential toxicity in the case of cell lyses (due to processing or progression of the bloom).

Standard extraction procedures must also be developed. The following should be taken into account:

1. Solvent suitability for target toxins
2. Solvent suitability for multiple toxins determinations, where needed
3. Solvent disposal/safety issues
4. Protein binding (covalent and noncovalent) should be addressed
5. Other sample matrices such as biological tissues and sediment should be considered

Sample preconcentration must be addressed.

1. Choice of sorbent – to favor analyte versus matrix
2. Standardized protocols

Detection methods include:

1. Screening methods (assays such as ELISA, PP2A and other suitable methods)
2. Physicochemical (primarily separation) Methods (i.e., LC–UV, LC–Fluorescence, LC–MS/MS)

Total procedure time and complexity impact ruggedness, cost, and training needs and include:

1. Time to prepare sample (including extraction and cleanup)
2. Number and complexity of procedural steps (impacts difficulty and ruggedness)
3. Automation (column switching, robotics)
4. Data workup (software)
5. Total analysis time – considering above

Simple and reliable field testing methods are also needed. The most commonly used field methods are immunoassay–based tests (i.e., tube, microplate, strip formats). Commercially–available microcystin immunoassay kits are widely used to screen water samples for microcystins, with or without pre–concentration. Field methods are useful for quick screening of samples on–site, so immediate remedial actions can be taken and reduces the number of samples that require further analytical confirmation in the laboratory.

Cost is another important consideration and includes: initial instrument investment, kits and consumables, and labor costs. Safety considerations and restrictions (disposal, radiolabels, etc.) are also important.

A practical guide manual, "Toxic: Cyanobacteria monitoring and cyanotoxin analysis," was just published commissioned and published by the European Community in September 2005 (Meriluoto and Codd 2005). This manual provides a comprehensive method for cyanobacteria and widely–studied cyanotoxins of interest. Methodologies for sampling and analytical methods are defined for many toxins in this publication and it would be useful to update this type of publication every 3–5 years with additional toxin methodologies and state of the art methodologies.

Setting Priorities

A critical component to the understanding of the potential for toxic episodes is an established program for monitoring and event validation. In its most basic format, monitoring must include tier–based approaches which are coupled to rapid and precise methodologies that can quantify toxin occurrence and/or measure biological effect. General field monitoring, incorporating remote sensing platforms (e.g., satellite imagery, deployed sentinel systems) need to be rapidly corroborated by laboratory analyses (i.e., toxin quantification, identification and culture analyses of cloned axenic organisms). The following outlines some identified abilities and areas of future prioritization for exposure assessment of cyanobacterial harmful algal blooms (CHABs).

Sentinel technology for CHABs includes "low" to high technology approaches.

1. Ground based sampling by volunteer groups has been an effective monitoring method in various rivers, lakes, and estuaries (VT, NY, and FL). This can be the first step in obtaining samples for characterization of any CHAB.

2. Remote sensing technology provides a method to identify blooms and appropriate sampling locations to assess CHAB. Remote sensing can use surface platforms: fixed or mobile units (such as Finnish use of ferries to sample the Baltic Sea, also Fig. 1 (see Color Plate 7)) or satellite imagery (e.g., Fig. 2 see Color Plate 7). Decisions regarding fixed sampling locations and use of drogue/physical circulation–driven sampling are essential to answer specific questions.

3. Remote sensing can take advantage of the presence of unique phycobiliprotein and carotenoid photopigments which provide distinctive markers for identification of cyanobacteria. The presence of coccoid cyanobacteria can be specifically identified by the presence of myxoxanthin, whereas aphanizanthin is diagnostic for filamentous cyanobacteria. Cyanobacteria can also be identified using specific absorbance characteristics of the phycobilins–particularly absorbance at 630 nm attributed to the phycocyanins (Fig. 1). The use of reciprocal reflectance data inversion can be used to accentuate spectral properties of interest. Equipment available for cyanobacteria detection includes a variety of hyperspectral sensors (see Ritchie and Zimba 2005, for a review of available sensor equipment and techniques for identification of various pigment signatures). Miniaturized dual spectral radiometers can provide a means of assessing total biomass and specifically cyanobacteria populations. This and other models offer the ability to simultaneously assess available light and algal reflectance, thereby allowing the use of all but highly transitional varying light conditions. One focus is the need to develop specific cyanobacterial reflectance models that are not ratio methods for estimating cyanobacterial biomass. Although these methods can be valuable, one cannot solely rely on satellite imaging data. Sometimes, cyanotoxins are present when there are no visible blooms.

Future Directions

One general goal is the development of a new generation of biosensors. Ideally these biosensors would be low cost, sensitive, reliable and relatively simple to use. Development of more extensive miniaturized biosensors will allow better cyanobacteria or cyanotoxin assessment. For instance, use of submersed hyperspectral radiometers coupled with biochip nanotechnology designed to assess cell surface recognition compounds, antibody coatings, and/or toxin recognition polymers will provide enhanced identification methods. These systems can provide sentinel type monitoring through fixed platforms, floating arrays, or on ships.

Fig. 1. Samples of currently available deployable systems that can be used in situ to sense conditions associated with potential CHAB events. 1). Flow Cam system from Fluid Imaging Technologies can identify cell type in situ. 2). Fluoroprobe system can identify water column phytoplankton based on a combination of 6 different fluorescence signatures. 3). NAS nutrient analyzer can detect in situ biogeochemical shifts that can be linked to pending CHAB events. (See Color Plate 7).

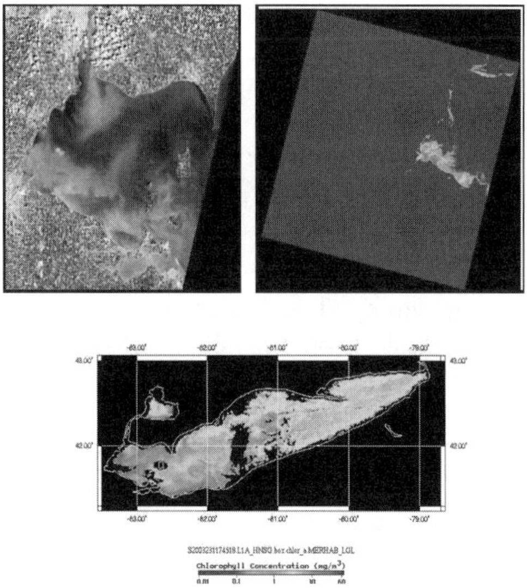

Fig. 2. Sample imagery available from satellites appropriate for monitoring CHAB events. True color imagery from Land Sat 7 (upper left, Rinta–Kanto et al. 2005) can be used to demonstrate potential algal blooms, which appear as green discolorations in the water column. The cyanobacterial–specific pigment phycocyanin can be elucidated from the appropriate applications of other algorithms (upper right, Vincent et al. 2003). Other imagery, such as daily Sea Wifs chlorophyll estimates (bottom) is available more frequently but provides less spatial resolution. (See Color Plate 7).

A critical need is increased knowledge of genetic markers for toxic cyanobacteria and their phenotypic expression:

1. Cyanobacteria identification is essential as a first step to provide firm bases for comparison of algal groups. This would include expanding our knowledge of the diversity of potentially important CHAB organisms and their associated biosynthetic pathways. Current efforts combining phenotypic appearance with molecular approaches will help unify taxonomic identification procedures (Komarek and Anagostidis, 2005).
2. Support of culture collections with access by qualified investigators is one important mechanism for this task, as is support for genetic characterization and "classical" characterization of isolates.
3. The ability to identify toxin forming strains (e.g. *Lyngbya*, *Trichodesmium spp.*).
4. The ability to identify and characterize new toxins (e.g. BMAAs, newly discovered *Trichodesmium* neurotoxin, other bioactive compounds).
5. Toxin biosynthetic pathways need to be elucidated. For example, although microcystin biosynthesis genes have been characterized, information on biosynthetic pathways for other CHAB toxins (i.e., cylindrospermopsin, anatoxin a, saxitoxin, BMAA) is largely unknown.

Proper sentinel deployment is also an essential need. Cyanobacterial blooms may occur on regional scales that are not easily detected by satellite imagery. Moreover, interference from atmospheric events (i.e., cloud cover, solar flares) can impede the ability of satellites to "see" events. Buoys/sentinel devices deployed *in situ* or on mobile platforms can avoid the atmospheric interference problems. Currently technologies are limited to fluorescence–based sensors targeting pigments similar to satellite systems. Future applications must move beyond this, and target the development of applications that can determine both cell type and cellular toxicity. Examples of similar systems include the Environmental Sample Processor (ESP) being developed at the Monterey–based Aquarium Research Institute (MBARI). Incorporation of emerging technologies into sentinel devices will allow for focused responses to events as they occur in real time.

Major future developments in the area of sentinel deployments will involve miniaturization of chemical sensory technology (i.e., hand held/deployable mass spectrometers) and will allow for real–time, on–line detection of toxins. Continued insight into the genetic mechanisms of toxin production, combined with advanced autonomous tools to characterize communities based on molecular markers, will allow for the determination of both the presence (DNA) and activity (RNA) of genetic systems capable of producing toxin production. Linkage of these systems to remotely deployable biosensors that can be incorporated into real time microsensors should allow for accurate characterization of cell abundance, toxin concentrations and toxin activity (Layton et al. 1998; Simpson et al. 2001; Yan et al. 2001; Mioni et al. 2003).

Specific priorities

1. Encourage interaction of CHAB researchers with marine scientists in order to effectively transfer existing remote platform and network technology to ongoing and future CHAB studies
2. Encourage widespread placement of remote sensors on available mobile platforms such as ferries, commercial and government over flights.
3. Encourage outreach programs to educate and recruit non–scientists as stakeholders
4. Support the development of sensor technologies; technologies with great promise include: microarrays; bioreporters, cytotoxicity monitoring, PCR technologies.
5. Encourage incorporation of microfluidics and nanotechnology into sensor development

Overarching considerations

CHAB events and impacts occur within the larger context of ecosystem processes. Therefore, the development of research strategies and activities should include consideration of complimentary and ongoing studies whenever possible. For example, since nutrient inputs influence cyanobacterial activity, site selection should favor areas with adjacent watershed studies or ongoing synoptic sampling and long term monitoring when possible.

CHAB issues fall within the mandate of multiple federal agencies (EPA, DHHS, DI, DC, DOD, DHS) as well as health and environmental agencies from the local to state levels). Therefore, an effective approach should be a coordinated program with funding and administrative support across interested agencies. A valuable component of many existing harmful algal bloom research programs is an outreach component, and this will also be necessary for a well–coordinated cyanobacterial research program. Outreach activities contribute to public awareness and support of research funding. They may also build the capacity of a widely distributed surveillance network to rapidly detect and response to the onset of CHAB events.

References

Berry JP, Gantar M, Gawley RE, Wang M, Rein KS (2004) Pharmacology and toxicology of pahayokolide A, a bioactive metabolite from a freshwater species of *Lyngbya* isolated from the Florida Everglades. Comp Biochem Physiol Part C: Toxicol Pharmacol 139(4):231–238

Cox PA, Banack SA, Murch SJ (2003) Biomagnification of cyanobacterial neurotoxins and neurodegenerative disease among the Chamorro people of Guam. PNAS 100(23):13380–13383

Cox PA, Banack SA, Murch SJ, Rasmussen U, Tien G, Bidigare RR, Metcalf JS, Morrison LF, Codd GA, Bergman B (2005) Diverse taxa of cyanobacteria produce ß–N–methylamino–L–alanine, a neurotoxic amino acid. PNAS 102(14):5074–5078

Danheiser RL, Morin JM Jr, Salaski EJ (1985) Efficient total synthesis of (±)–anatoxin a. J Am Chem Soc 107(26):8066–8073

Erhard M, von Dohren H, Jungblut P (1997) Rapid typing and elucidation of new secondary metabolites of intact cyanobacteria using MALDI–TOF mass spectrometry. Nat Biotechnol 15:906–909

Komárek, J, Anagostidis K (2005) Sübwasserflora von Mitteleuropa. 19(2): Cyanoprokaryota 2: Oscillatoriales. Elsevier, Munich, Germany

McElhiney J, Lawton LA (2005) Detection of the cyanobacterial hepatotoxins microcystins. Toxicol Appl Pharmacol 203(3): 219–230

Layton AC, Muccini M, Ghosh M, Sayler GS (1998) The construction of a bioluminescent reporter strain to detect polychlorinated biphenyls. Appl Environ Microbiol 64(12): 5023–5026

Lee T–H, Chou H–N (2000) Isolation and identification of seven microcystins from a cultured M.TN2 strain of *Microcystis aeruginosa*. Bot Bull Acad Sin 41:197–202

McElhiney J, Lawton LA (2005) Detection of the cyanobacterial hepatotoxins microcystins. Toxicol Appl Pharmacol 203(3):219–230

Meriluoto J, Codd GA (eds) (2005) TOXIC: Cyanobacterial monitoring and cyanotoxin analysis. Abo Akademi University Press, Abo, Finland

Mioni CE, Howard AM, DeBruyn JM, Bright NG, Twiss MR, Applegate BM, Wilhelm SW (2003) Characterization and field trials of a bioluminescent bacterial reporter of iron bioavailability. Mar Chem 83:31–46

Ritchie JC, Zimba PV (2005) Hydrological application of remote sensing water quality, including sediment and algae. In: Anderson M (eds) Encyclopedia of Hydrology John Wiley, Chichester, England

Simpson ML, Sayler GS, Patterson G, Nivens DE, Bolton EK, Rochelle JM, Arnott JC, Applegate BM, Ripp S, Guillorn MA (2001) An integrated CMOS microluminometer for low–level luminescence sensing in the bioluminescent bioreporter integrated circuit. Sens Actuat B: Chem, 72(2):134–140

Yan F, Erdem A, Meric B, Kerman K, Ozsoz M, Sadik OA (2001) Electrochemical DNA biosensor for the detection of specific gene related to *Microcystis* species. Electrochem Commun 3(5):224–228

Chapter 21: Cyanotoxins: sampling, sample processing and toxin uptake

Jussi AO Meriluoto, Lisa EM Spoof

Department of Biochemistry and Pharmacy, Åbo Akademi University, 20520 Turku, Finland

Introduction

There are several cyanobacterial (blue–green algal) toxin groups which have been implicated in human and animal illnesses and mortalities. Sampling and sample processing of cyanotoxins will be discussed as well as toxin uptake in different organisms.

Cyanotoxins

The paper will concentrate on the following commonly occurring cyanotoxins: hepatotoxic microcystins/nodularins (together over 80 analogues), cytotoxic cylindrospermopsin and neurotoxic anatoxin–a (Fig. 1). In addition, there are several other cyanotoxins that deserve attention in local or national monitoring programmes: anatoxin–a(S), saxitoxin family, dermatotoxic alkaloids, lipopolysaccharides, beta–*N*–methylamino–L–alanine (BMAA) etc.

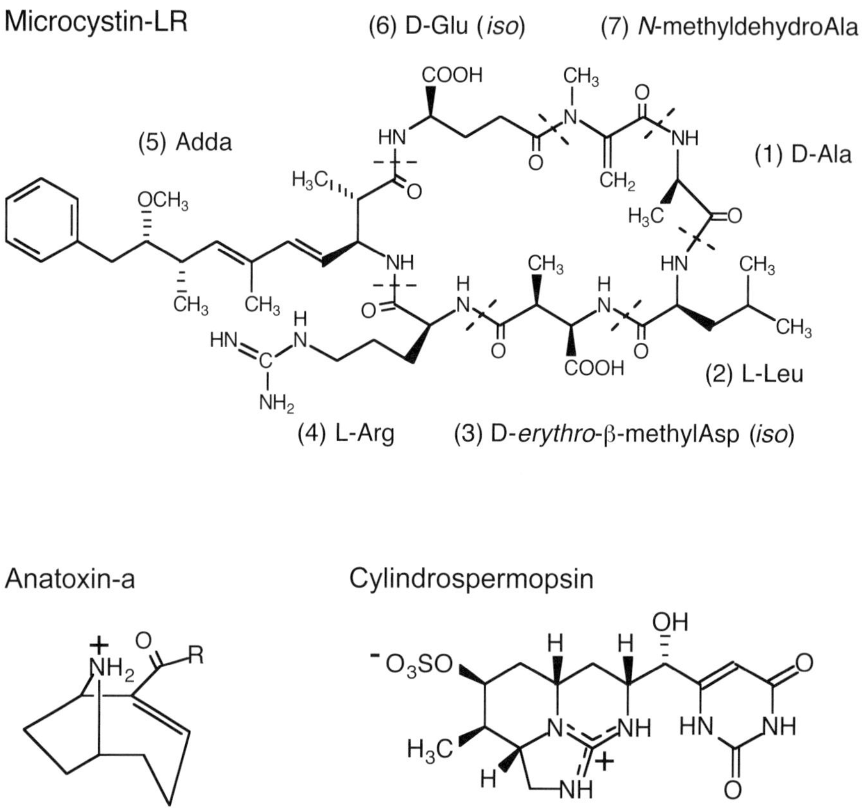

Fig. 1. Chemical structure of microcystin–LR, anatoxin–a (R=CH_3) and cylindrospermopsin.

The commercial availability of the main cyanotoxins, microcystins, anatoxin–a and cylindrospermopsin is limited. The available toxins do not always meet critical norms of purity or guaranteed quantity (Fig. 2). It was recently reported that when three commercial standards of microcystin–RR were tested only one of them actually contained microcystin–RR. The second standard was a mixture of microcystin–RR, and its variant [Dha^7]microcystin–RR, and the third one contained [Dha^7]microcystin–RR only (Kubwabo et al. 2004). When certified reference materials are unavailable, method development related to sample processing and analysis is hampered, and, from a critical viewpoint, the analytical results may not be regarded as truly quantitative.

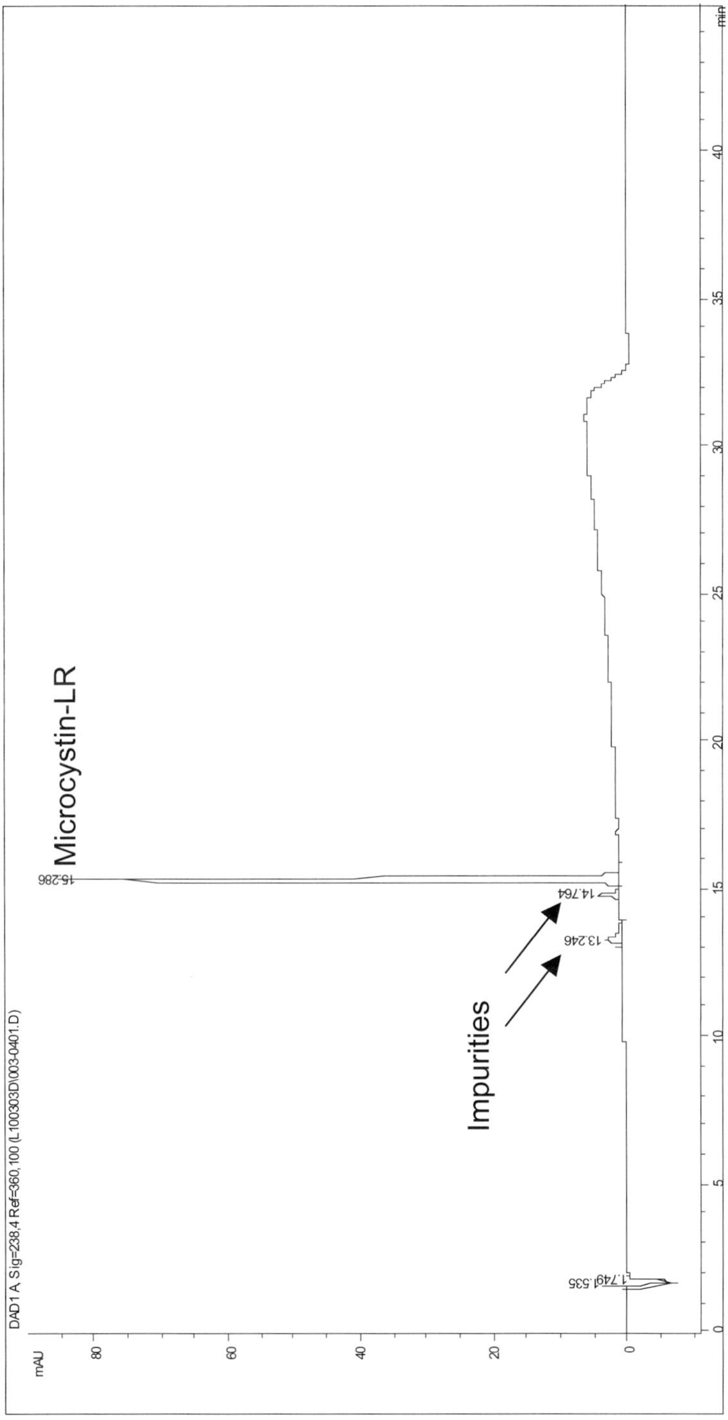

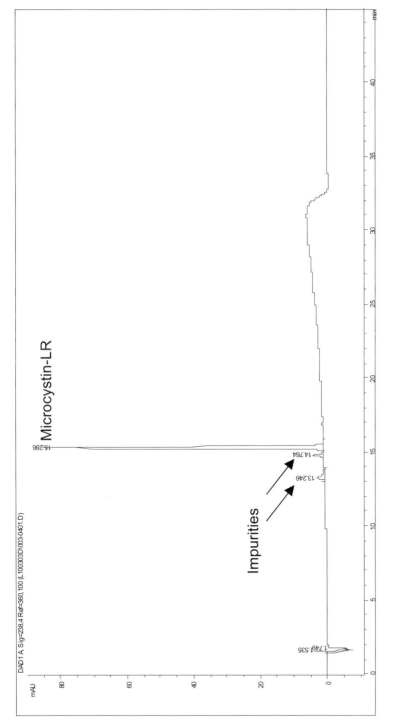

Fig. 2. Trace of commercial microcystin–LR (92%) with impurities (8%, having microcystin–like UV spectra); separation on amide–C16 phase, aqueous trifluoroacetic acid–acetonitrile gradient

Matrices; where, what, when and how to sample

Cyanotoxins are known to occur in a number of matrices and sample types: a) Water – source/recreational waters (fresh, estuarine, brackish & marine) and treated water. Sediments can contain at least stable peptide toxins, microcystins and nodularins. b) Biological materials: phytoplankton including food supplements, zooplankton, shellfish, fish, terrestrial animals including humans, sea birds, aquatic plants, agri– and horticultural products etc. The wide distribution of microcystins and nodularin in food webs is well documented. The main sources of exposure from the human health point of view are drinking water, recreational waters, shellfish and fish, and for some consumers, food supplements. The importance of crop plants as a source of exposure is unclear.

Comprehensive monitoring of lakes and reservoirs (which may be used for both water abstraction and recreational activities) requires extensive resources as the sampling should have coverage in temporal, horizontal and depth dimensions (Fig. 3). In most cases only a fraction of the waterbodies can be monitored in a satisfactory manner (i.e., Finland with its close to 200,000 lakes). In studies performed at the University of Helsinki, Finland, in the 1980s, about 50% of the cyanobacterial blooms tested contained toxins – the majority of them hepatotoxins (microcystins) (Sivonen et al. 1990). Later data from other countries corroborate these findings, and in some countries most of the studied cyanobacterial blooms have been toxic. In our studies on Åland Islands, SW Finland (Spoof et al. 2003), 113 samples taken from 93 different locations in 2001 were analysed for microcystins and nodularin–R. The purpose was to monitor the prevalence of microcystins in a set of non–selected lakes, not only in those where cyanobacterial blooms were observed. Both eutrophic and oligotrophic freshwater lakes and also some brackish waters were included. Intracellular toxins exceeding 0.2 $\mu g\ L^{-1}$ were confirmed by three analytical techniques in samples from 14 locations (15% of the waterbodies). The highest recorded microcystin concentration in the Åland Islands lakes in 2001 was 42.0 $\mu g\ L^{-1}$.

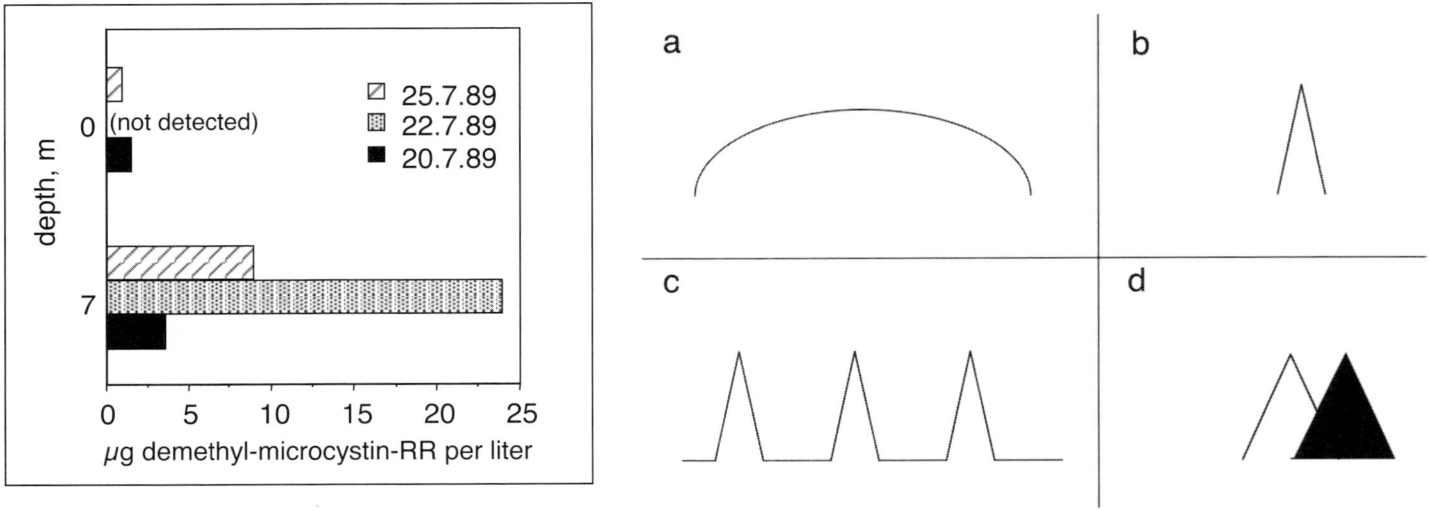

Fig. 3. Left panel: Vertical movement of a *Planktothrix* population in a lake, measured as microcystin at different depths. Right panel: Cyanobacterial blooms can be (a) persistent, (b) intermittent and/or (c) recurrent. Situation (d) represents two toxin profiles at two locations of a waterbody or two populations producing different toxins.

Chlorophyll–a, a measure of phytoplankton in routine water monitoring, can be useful as a first estimation of maximum intracellular microcystin concentration (Fastner et al. 1999, Lindholm and Meriluoto 1991). When microcystin concentrations in German fresh water bodies were studied (Fastner et al. 1999), intracellular microcystin concentrations were below 10 µg l^{-1} in over 70% of the pelagic water samples. The microcystin to chlorophyll–a ratio in the German study varied usually between 0.1–0.5 with maxima of 1–2. The extracellular toxin (for microcystins, <10% of the total toxin in log phase populations, but much higher for cylindrospermopsin) is usually relatively quickly degraded by biotic and abiotic factors, diluted or sedimented in a non–specific manner. Nevertheless, release of cyanotoxins during water treatment, induced by e.g. pre–oxidation, can lead to problems in the production of safe drinking water as extracellular toxins are difficult to remove with conventional treatment technology.

Foodstuffs in contact with cyanobacteria such as shellfish, fish, and agri– and horticultural products irrigated with cyanotoxin–containing water should be monitored for toxins at least occasionally during the bloom periods of cyanobacteria and immediately after blooms. Intensive blooms of known toxin producers require higher levels of vigilance.

Sample extraction and clean–up for microcystins, anatoxin–a and cylindrospermopsin

Effective toxin recovery from cyanobacterial cells is impossible without disruption of the strong cell wall structure. Most procedures use freeze–drying or freeze–thawing for this, and the extraction of toxins is enhanced by ultrasonication in a bath or probe–type sonicator. Toxins can be extracted from filter discs containing harvested cyanobacterial cells or from weighed lyophilized bloom material. Most cyanotoxins can be extracted with (acidified) aqueous methanol. Extraction of tissue samples usually resembles that of phytoplankton samples except for longer extraction times. Adequate toxin recovery has been seldom validated. The complete recovery of microcystins from tissues is practically impossible as microcystins form (non–covalent and) covalent bonds with their target proteins (Ernst et al. 2005).

The sample concentration and clean–up methodology developed for main microcystins/nodularins, anatoxin–a and cylindrospermopsin in phytoplankton and water samples is reasonably robust but still requires great carefullness when naturally occurring concentrations of cyanotoxins (in most cases below 10 µg/L) are analysed. Microcystins and nodularins are usually concentrated on reversed–phase solid–phase extraction cartridges (SPE; C18 or polymeric materials) which have limited clean–up capacity. For microcystins in raw and treated waters, there is an ISO standard (ISO 20179:2005). The ISO method specifies the reversed–phase SPE and HPLC conditions and it was validated for microcystin–RR, –YR, and –LR through an international intercalibration exercise. Immunoaffinity cartridges, now commercially available, can offer superior clean–up for the peptide toxins in water (Rivasseau and Hennion 1999, Lawrence and Menard 2001, Aguete et al. 2003, Aranda-Rodriguez et al. 2003) and more difficult tissue matrices (Harada et al. 2001). Molecular imprinted polymers, still experimental, allow selective molecular recognition of certain known compounds and can be useful for pre–treatment and concentration of microcystins (Chianella et al. 2002, Kubo et al. 2004). Anatoxin–a can be concentrated on either C18 (after pH adjustment to 9.6) or on weak cation exchange sorbents (James and Sherlock 1996). Cylindrospermopsin is very hydrophilic and can be concentrated on graphite carbon SPE columns (Norris et al. 2001, Metcalf et al. 2002).

Many of these sample preparation and analytical methods have been discussed and presented in a detailed manner in the book "TOXIC: Cyanobacterial Monitoring and Cyanotoxin Analysis" published by Åbo Akademi University Press in 2005 (eds. J Meriluoto and GA Codd 149 pp ISBN 951–765–259–3). The book was one of the deliverables of the research project "TOXIC – Barriers against Cyanotoxins in Drinking Water". The TOXIC project was funded by the European Commission under the Fifth Framework Programme (contract number EVK1–CT–2002–00107) in 2002–2005. The TOXIC project involved ten European research groups in nine countries and comprised four programmes focussing on Raw Water Quality, Analysis, Treatment and Exploitation. The manual presents the core methods which have been developed, standardised and used within the Raw Water Quality and Analysis programmes. Contributions to the manual have been made by researchers at Åbo Akademi University (FIN), University of Dundee (UK), University of Lodz (PL) and DVGW Technologiezentrum Wasser (D). The manual contains comprehensive chapters on the identification and sampling of cyanobacteria, and analytical methods for microcystins, anatoxin–a and cylindrospermopsin in phytoplankton and water. Moreover, 22 standard operating procedures (SOPs) for the actual monitoring and analysis work are included. HPLC–diode–array detection (HPLC–DAD) was chosen as the main analytical tool because a) HPLC–DAD (but not HPLC–FL, cf. anatoxin–a, or LC–MS) was available at all partner laboratories b) HPLC can be used as a general method applicable to several toxin classes and c) HPLC can separate individual toxin variants and degradation products which was important because TOXIC aimed at the identification of toxin derivatives formed during water treatment.

Simultaneous SPE of several cyanotoxin classes has been a challenge due to the large differences in analyte structure (Fig. 1) and polarity (Fig. 4). Cylindrospermopsin is very hydrophilic and carries both a positive and a negative charge at neutral pH. Anatoxin–a is hydrophilic and has a positive charge at neutral pH. Microcystins have the Adda residue with a lipophilic side chain, two negatively charged carboxylic acid functions at neutral pH, and arginine–containing microcystins have a positive charge in the guanidino group (pKa > 12). Taken together, cation–exchange could be an alternative for SPE for anatoxin–a and cylindrospermopsin and also for many, but not all, microcystins. The elution of the cation–exchange sorbent should avoid high pH due to the instability of anatoxin–a at basic pH. The trapping of all microcystin variants can be done by reversed–phase materials.

We suggest the further exploration of (polymeric) mixed–mode materials with both reversed–phase and cation–exchange functionalities for the simultaneous SPE of microcystins, nodularins, anatoxin–a and cylindrospermopsin. Our very preliminary experiments consisted of the following steps: 1) activation of a 60 mg Waters Oasis MCX cartridge with 2 ml methanol and 2 ml water, 2) loading of known amounts (a few µg) of microcystin–RR, –YR and –LR, anatoxin–a and cylindrospermopsin in cyanobacterial extracts, either separately or in mixtures (loading solvent was ultrapure water or, for microcystins, up to 20% aqueous methanol), 3) washing step with 0–5 ml water, 4) vacuum suction drying for 1 min, 5) elution with 2 ml methanol – aqueous $CaCl_2$ (0.3 g ml^{-1}) 8:2 (vol:vol), 6) evaporation of solvent at 50 °C with nitrogen or argon gas, 7) reconstitution in water or 30 % aqueous methanol 8) sample clarification by centrifugation and 9) HPLC (according to Fig. 4 or by using HPLC conditions optimised for individual analytes). We received satisfactory recoveries (>75%) for all three analyte types when the SPE loading was performed in a small volume (<2 ml), with minimal breakthrough. However, cylindrospermopsin was lost completely when the loading volume was high (100 ml), indicating weak binding of cylindrospermopsin onto the SPE material. The elution of the cartridge must be carefully optimised as the most cationic of the microcystins studied, microcystin–RR, had the lowest recoveries of the microcystins (the use of a weaker cation exchanger with reversed-phase properties might also be an option).

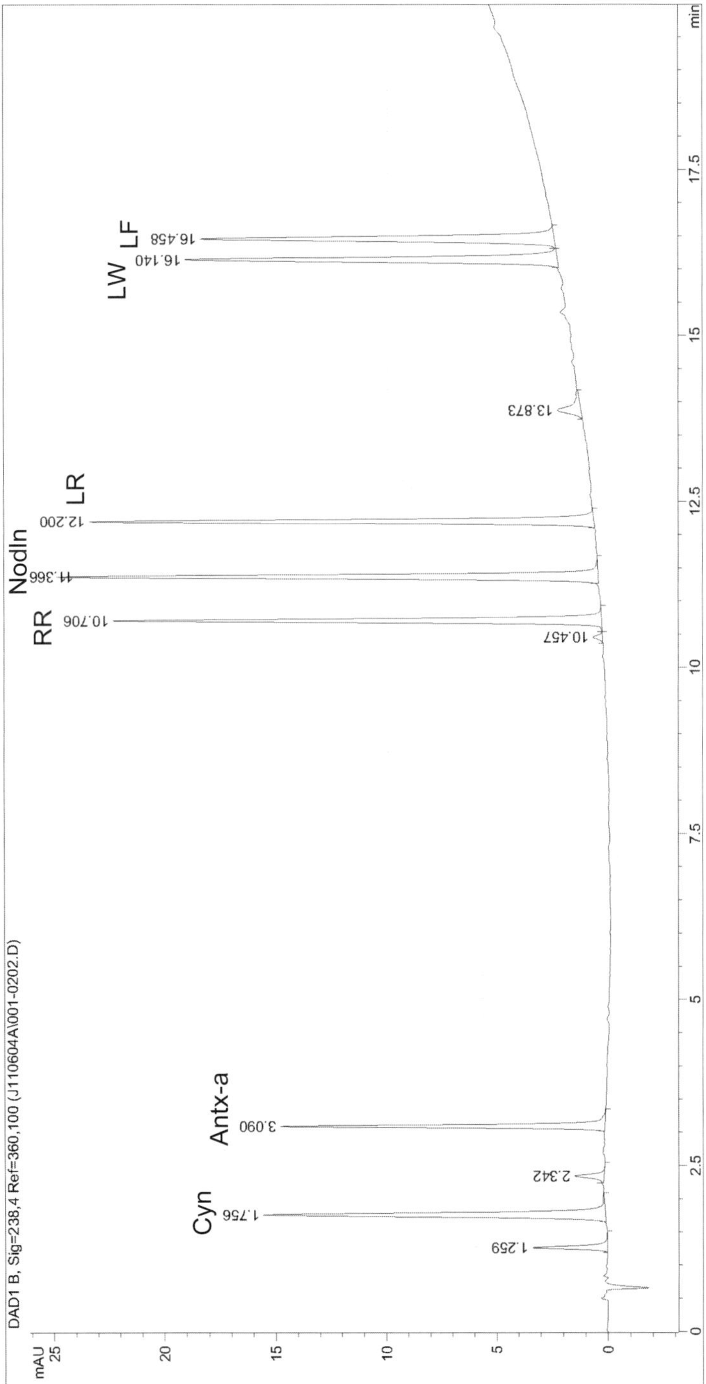

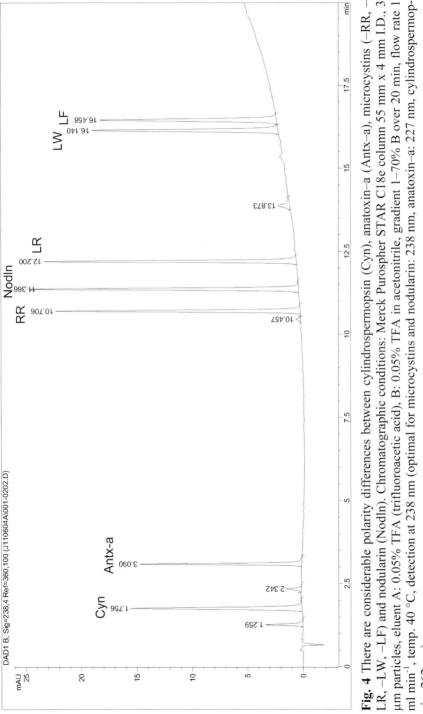

Fig. 4 There are considerable polarity differences between cylindrospermopsin (Cyn), anatoxin–a (Antx–a), microcystins (–RR, –LR, –LW, –LF) and nodularin (Nodln). Chromatographic conditions: Merck Purospher STAR C18e column 55 mm x 4 mm I.D., 3 µm particles, eluent A: 0.05% TFA (trifluoroacetic acid), B: 0.05% TFA in acetonitrile, gradient 1–70% B over 20 min, flow rate 1 ml min^{-1}, temp. 40 °C, detection at 238 nm (optimal for microcystins and nodularin: 238 nm, anatoxin–a: 227 nm, cylindrospermopsin: 262 nm).

Matrix effects

Various matrix components and co–extracted substances make the analyte clean–up, identification and quantitation difficult. These interfering substances can be of either organic (e.g. humic substances, proteins and other biological macromolecules) or inorganic (for microcystins e.g. chlorine, Fe^{3+}, Al^{3+}) nature. It is difficult to get rid of humic substances which co–chromatograph with e.g. microcystins in water samples. Acidic extraction conditions suppress unwanted extraction of many proteins and liquid–liquid extraction of aqueous extracts with non–polar solvents can be used to reduce the concentration of lipophilic substances in relation to microcystins. The processing of tissue samples is facilitated by the use of methods common in forensic science such as tissue digestion with trypsin or pronasc. Cyanotoxins are not affected by these enzymes. The reactivity of chlorineous substances in tap and process waters can be quenched by the addition of sodium thiosulphate. Fe^{3+} and Al^{3+} ions have been suggested to interfere with microcystin detection, e.g. by complex formation or by microcystin destruction (Oliveira et al. 2005).

In LC–MS both severe suppression (more common) and enhancement (less common) of the analyte signal are possible, especially with tissue samples (Dionisio et al. 2004, Karlsson et al. 2005). These effects should be assessed by spiking of the reference matrix with analyte at different stages of sample processing (Karlsson et al. 2005), and corrected for. In order to achieve reasonably low LOD values the use of solid–phase extraction is always recommended for tissue samples.

Cyanotoxins in shellfish and fish

Microcystins/nodularins have been observed in laboratory and field experiments in many aquatic animals including mussels, clams, fish species, crab larvae, prawns, crayfish and zooplankton. There are severe methodological difficulties in the bioaccumulation studies and, because of unclear extraction efficiency and matrix effects in the analytical methods, the reported amounts should be considered as minimum amounts in many cases, especially in the older literature. There are no enforced methods for the tissue analysis of cyanotoxins with the exception of the saxitoxin family (saxitoxin derived from e.g. marine dinoflagellates is regulated in seafood).

Immunoaffinity purification is a very promising form of clean–up for tissue samples (Harada et al. 2001).

Nodularin does not bind covalently to protein phosphatases (Craig et al. 1996, Bagu et al. 1997) like most microcystins do and it may therefore be easier to extract from tissue material than microcystins. Typical levels of extractable microcystins and nodularins detected in tissue materials of aquatic organisms have ranged from undetectable (below 10 ng g^{-1} tissue dry weight) to several tens of μg/g tissue dry weight. The highest levels have been recorded in the hepatic (or hepatopancreatic) tissues of fish and mussels while concentrations in fish muscle have usually been substantially lower or undetectable. It was proposed that a large proportion of the microcystin is bound covalently and irreversibly to the tissue matrix, and cannot be fully extracted from the tissue with methanol (Williams et al. 1997a, Williams et al. 1997b).

Shellfish are readily contaminated by many cyanotoxins and other phycotoxins as they can filter large volume of toxin–containing water without being affected by many substances harmful for higher animals. Mussels exposed to cyanobacterial blooms consisting of over 20000 cells ml^{-1} have been shown to accumulate toxins in sufficient amounts to be unsafe for human consumption (Van Buynder et al. 2001). There are considerable differences in the toxin accumulation characteristics of different bivalve species (Yokoyama and Park 2002) and also in the experimental settings in the reported research making comparisons and predictions difficult. Freshwater mussels *Anodonta cygnea* were demonstrated to accumulate microcystin (70–280 μg per mussel) when they were kept in a culture of *Planktothrix agardhii* (40–60 μg intracellular demethyl–microcystin–RR per litre) (Eriksson et al. 1989). The mussels did not suffer from any obvious toxic symptoms. *Mytilus edulis* mussels were recommended not to be be collected for human consumption during a water bloom of *Nodularia* (Falconer et al. 1992). The gut and associated tissues showed hepatotoxicity to mice. Fragments of *Nodularia* were also detected in the gut of the mussels confirming ingestion of the cyanobacteria. *Anodonta grandis simpsoniana* accumulated microcystins by grazing on toxic phytoplankton and only minimally via uptake of the dissolved toxin (Prepas et al. 1997). In the Baltic Sea food web blue mussels accumulate nodularin present in the cyanobacterium *Nodularia spumigena*. The toxin accumulated in the mussels (recorded maximum concentration in mussels 2150 μg kg^{-1}, (Sipiä et al. 2001)) can be transferred to other organisms consuming mussels, e.g. flounders (recorded maximum concentration in the liver 399 μg kg^{-1} dry weight in the same paper, (Sipiä et al. 2001)) and eiders (sea birds, recorded maximum concentration in the liver 180 μg kg^{-1} dry weight, (Sipiä et al. 2004)).

The derived health alert levels for microcystin and nodularin for adults were 250 μg kg^{-1} for fish, 1100 μg kg^{-1} for prawns and 1500 μg kg^{-1} for mussels (Van Buynder et al. 2001). Boiling of the seafood can redistribute toxins between viscera and flesh as reported for nodularin in prawns (Van Buynder et al. 2001). The depuration mechanism of microcystins when bivalves are placed in pure water have been shown to be biphasic (Prepas et al. 1997), a initial fast decline in toxin concentration is followed by a very slow elimination (duration weeks–months). The depuration of toxins by bivalves may be temperature dependent, being more effective in higher temperatures (Yokoyama and Park 2003).

Bivalves are also resistant to toxic effects of high concentrations of cylindrospermopsin. Cylindrospermopsin from *C. raciborskii* accumulated in exposed *Anodonta cygnea* with the highest concentration in the haemolymph (Saker et al. 2004). The derived health alert levels for cylindrospermopsin were 158 μg kg^{-1} for fish, 720 μg kg^{-1} for prawns and 933 μg kg^{-1} for mussels (Saker et al. 2004).

A coarse generalisation, with many exceptions and reservations, of microcystin, nodularin and cylindrospermopsin levels found in shellfish within 2–3 weeks of exposure: Biomass–bound toxin level in water, N μg L^{-1}, equals roughly to the typical maximum toxin level in shellfish soft tissue, N μg g^{-1} dry weight. Much inter–species variation is expected. The highest toxin levels have been found in the hepatopancreas/viscera (microcystin/nodularin) and haemolymph (cylindrospermopsin), less in the muscle.

Fish flesh is usually safe for consumption (there has been some exceptions) but stomach/intestinal content and internal organs, especially the liver, can contain considerable amounts of cyanotoxins. Phytoplankton–eating fish are exposed to toxins in cyanobacterial biomass. *Tilapia rendalli* was collected from a Brazilian coastal lagoon (phytoplankton dominated by the genus *Microcystis*, average microcystins 46 μg L^{-1}, highest 257 μg L^{-1}) (Freitas de Magalhaes et al. 2001). Microcystins were detected in 57% of the fish liver (concentration of microcystins 0–15 μg g^{-1}) and viscera samples (0–72 μg g^{-1}). In muscle tissue the microcystin concentration peaked at 337 ng g^{-1} (the values concern an intensive bloom period of cyanobacteria in 1999).

In conclusion, the overall message found in the literature is that the consumption of fish muscle tissue cannot be considered a major hazard to human health but there are fish species and/or fish organs which may contain considerable amounts of cyanotoxins. Consumption of mussels and clams collected during cyanobacterial blooms or immediately after blooms should definitely be avoided.

Cyanotoxins in plants

Toxin accumulation in plants could either reduce crops or cause health problems for plant consumers. The best documented cyanotoxin group in this context is microcystins which are known to be taken up and affect a number of terrestrial and aquatic plants (MacKintosh et al. 1990, Kurki-Helasmo and Meriluoto 1998, Pflugmacher et al. 1999, McElhiney et al. 2001, Chen et al. 2004, Mitrovic et al. 2005). Seedling growth is usually inhibited at micromolar (µg mL^{-1}) microcystin concentrations which are seldom found in nature. At micromolar toxin exposure the recorded toxin concentrations in the plant seedlings have been up to a few µg kg^{-1}. Plant exposure to toxins is more probable in hydroponic cultures than in soil–grown cultures. It is currently unclear whether the main crop plants grown in natural conditions can contain high enough concentrations of cyanotoxins to cause a health hazard for consumers. It is possible that the only scenario which can lead to serious toxin contamination of terrestrial plants is spray irrigation with a heavy bloom of toxic cyanobacteria. Especially the contamination of leafy plants where the aerial part is eaten, e.g. lettuce, is possible (Codd et al. 1999).

Plant–based biotests have been suggested for the detection of microcystins (Kós et al. 1995) and cylindrospermopsin (Vasas et al. 2002). Microcystins can either enter the plants through roots (Kurki-Helasmo and Meriluoto 1998) or remain on (in) the leaves after spray irrigation with toxin–containing water (Codd et al. 1999, Siegl et al. 1990, Abe et al. 1996). Translocation of microcystin within the plant is possible (Kurki-Helasmo and Meriluoto 1998). Cylindrospermopsin, a protein–synthesis inhibitor, has been suggested to affect the pollen germination if high enough concentrations of the toxin are present in spray irrigation water (Metcalf et al. 2004).

Biomagnification of cyanobacterial toxins in the cyanobacteria–plant–animal food web was demonstrated in the case of beta–*N*–methylamino–L–alanine (BMAA), a neurotoxic amino acid (Cox et al. 2003, Cox et al. 2005). BMAA was originally discovered in cycads and later in their cyanobacterial root symbiont, the cyanobacterium *Nostoc*. BMAA is suggested to be the cause of neurological disease in Chamorro people in Guam, the diet of whom include cycad seeds and flying fox bats. Nearly a 100–fold increase in BMAA for each trophic level, cyanobacteria – cycad/cyanobacteria symbiosis – flying foxes was observed. BMAA production in cyanobacteria was demonstrated to be ubiqitous, comprising both symbiotic and free–living cyanobacteria (Cox et al. 2005). This finding warrants a global risk assessment for human health. Indeed BMAA has been found in the frontal cortex of Canadian Alzheimer's patients (Cox et al. 2003).

Microcystin/nodularin conjugates

Conjugation of xenobiotics is a common detoxification strategy for different organisms. There is some direct and indirect (elevation of glutathione *S*–transferase activity) evidence of glutathione and cystein conjugates of microcystins and nodularin in animals and plants (Karlsson et al. 2005, Kondo et al. 1992, Pflugmacher et al. 2001, Pflugmacher 2004, Sipiä et al. 2002).

Cyanotoxins in food supplements

Most blue–green algal food supplements collected from the North American market contained microcystins (Gilroy et al. 2000, Lawrence et al. 2001) with concentrations up to 35 $\mu g\ g^{-1}$. Tolerable daily intake of microcystin–LR (0.04 $\mu g\ kg^{-1}$ body weight) is easily exceeded by the consumption of contaminated food supplements (Dietrich and Hoeger 2005).

Cyanotoxins in sediments

Some aquatic animals, including different invertebrates and fish, live in or feed on sediment materials and become in contact with sedimented cyanotoxins and cyanobacterial cells. It was shown that 1 ml of autoclaved (sterilised) sediment could adsorb 13–24 μg of microcystins and 50–82 μg of anatoxin–a (Rapala et al. 1994). Several sediment samples fom Japanese lakes were successfully studied using the ozonolytic MMPB method combined with GC–MS detection (Harada et al. 2001, Tsuji et al. 2001). This strategy was chosen instead of conventional HPLC approach with intact microcystins. Especially the hydrophilic microcystins such as microcystin–RR were strongly adsorbed on the sediment and were found difficult to extract. Several extraction methods were tested. The best recovery of intact microcystins (60% for MC-LR) was obtained by extraction with 5% acetic acid containing 0.1%TFA under ultrasonication. The extract was filtered, evaporated, redissolved in water, applied on C18 SPE and eluted with 90% methanol. The extract still contained several interfering impurities in HPLC–UV. Microcystins were found to interact with humic and fulvic substances, with suspended particulate matter and sediments, however preferably remaining in the aqueous phase (Rivasseau et al. 1998). In river water spiked with microcystin–LR, –YR and –RR, 5 µg L^{-1}, only 9–13% of microcystin was adsorbed on particles and 7–8% on sandy sediment within three days. Nodularin is present in the Baltic Sea sediments (Kankaanpää et al., Finnish Institute of Marine Research).

Conclusion

The frequent presence of cyanotoxins in phytoplankton and aquatic animal specimens is well documented. There is also increasing evidence of toxin contamination of aquatic and terrestrial plants. Sample processing of most toxins in phytoplankton is reasonable robust and at least partially validated in a few intercalibration exercises whereas sample processing of tissue samples, animal and plant, still requires much standardization. The lack of certified reference materials (CRMs) of most cyanotoxins calls for immediate attention as the toxin analyses cannot be considered truly quantitative without access to CRMs.

Acknowledgements

The authors acknowledge travel support from the ISOC–HAB (Research Triangle Park, NC, 2005) organisers, and research funding from the Academy of Finland (project 108947), the European Commission (contract EVK1–CT–2002–00107) and the Tor, Joe and Pentti Borg Foundation. Prof GA Codd and Dr JS Metcalf, University of Dundee, are thanked for samples of cylindrospermopsin and anatoxin–a. Pia Vesterkvist is acknowledged for contributing to Fig. 4.

References

Abe T, Lawson T, Weyers JDB, Codd GA (1996) New Phytol 133 651

Aguete EC, Gago–Martinez A, Leao JM, Rodriquez–Vazquez JA, Menard C, Lawrence JF (2003) Talanta 59 697

Aranda–Rodriguez R, Kubwabo C, Benoit FM (2003) Toxicon 42 587

Bagu JR, Sykes BD, Craig MM, Holmes CFB (1997) J Biol Chem. 272 5087

Chen J, Song L, Dai J, Gan N, Liu Z (2004) Toxicon 43 393

Chianella I, Lotierzo M, Piletsky SA, Tothill IE, Chen BN, Karim K, Turner APF (2002) Anal Chem 74 1288

Codd GA, Metcalf JS, Beattie KA (1999) Toxicon 37 1181

Cox PA, Banack SA, Murch SJ (2003) Proc Natl Acad Sci USA 100 13380
Cox PA, Banack SA, Murch SJ, Rasmussen U, Tien G, Bidigare RR, Metcalf JS, Morrison LF, Codd GA, Bergman B (2005) Proc Natl Acad Sci USA 102 5074
Craig M, Luu HA, McCready TL, Williams D, Andersen RJ, Holmes CFB (1996) Biochem Cell Biol 74 569
Dietrich D, Hoeger S (2005) Toxicol Appl Pharmacol 203 273
Dionisio Pires LM, Karlsson KM, Meriluoto JAO, Kardinaal E, Visser PM, Siewertsen K, Van Donk E, Ibelings BW (2004) Aquat Toxicol 69 385
Eriksson JE, Meriluoto JAO, Lindholm T (1989) Hydrobiologia 183 211
Ernst B, Dietz L, Hoeger SJ, Dietrich DR (2005) Environ Toxicol 20 449
Falconer IR, Choice A, Hosja W (1992) Environ Toxicol Water Qual 7 119
Fastner J, Neumann U, Wirsing B, Weckesser J, Wiedner C, Nixdorf B, Chorus I (1999) Environ Toxicol 14 13
Freitas de Magalhaes V, Moraes Soares R, Azevedo SMFO (2001) Toxicon 39 1077
Gilroy DJ, Kauffman KW, Hall RA, Huang X, Chu FS (2000) Environ Health Perspect 108 435
Harada KI, Kondo F, Tsuji K (2001) J Assoc Off Anal Chem 84 1636
James KJ, Sherlock IR (1996) Biomed Chromatogr 10 46
Karlsson KM, Spoof LEM, Meriluoto JAO (2005) Environ Toxicol 20 281
Kondo F, Ikai Y, Oka H, Okumura M, Ishikawa N, Harada KI, Matsuura K, Murata H, Suzuki M (1992) Chem Res Toxicol 5 591
Kós P, Gorzó G, Surányi G, Borbély G (1995) Anal Biochem 225 49
Kubo T, Hosoya K, Watabe Y, Tanaka N, Sano T, Kaya K (2004) J Sep Sci 27 316
Kubwabo C, Vais N, Benoit FM (2004) J AOAC Int 87 1028
Kurki–Helasmo K, Meriluoto J (1998) Toxicon 36 1921
Lawrence JF, Menard C (2001) J Chromatogr A 922 111
Lawrence JF, Niedzwiadek B, Menard C, Lau BPY, Lewis D, Kuiper–Goodman T, Carbone S, Holmes C (2001) J AOAC Int 84 1035
Lindholm T, Meriluoto JAO (1991) Can J Fish Aquat Sci 48 1629
MacKintosh C, Beattie KA, Klumpp S, Cohen P, Codd GA (1990) FEBS Lett 264 187
McElhiney J, Lawton LA, Leifert C (2001) Toxicon 39 (2001) 1411
Metcalf JS, Beattie KA, Saker ML, Codd GA (2002) FEMS Microbiol Lett 216 159
Metcalf JS, Barakate A, Codd GA (2004) FEMS Microbiol Lett 235 125
Mitrovic SM, Allis O, Furey A, James KJ (2005) Ecotoxicol Environ Safety 61 345
Norris RLG, Eaglesham GK, Shaw GR, Senogles P, Chiswell RK, Smith MJ, Davis BC, Seawright AA, Moore MR (2001) Environ Toxicol 16 391
Oliveira ACP, Magalhães VF, Soares RM, Azevedo SMFO (2005) Environ Toxicol 20 126
Pflugmacher S (2004) Aquat Toxicol 70 169
Pflugmacher S, Codd GA, Steinberg CEW (1999) Environ Toxicol 14 111

Pflugmacher S, Wiegand C, Beattie KA, Krause E, Steinberg CEW, Codd GA (2001) Environ Toxicol Chem 20 846
Prepas EE, Kotak BG, Campbell LM, Evans JC, Hrudey SE, Holmes CFB (1997) Can J Fish Aquat Sci 54 41
Rapala J, Lahti K, Sivonen K, Niemelä SI (1994) Lett Appl Microbiol 19 423
Rivasseau C, Hennion MC (1999) Anal Chim Acta 399 75
Rivasseau C, Martins S, Hennion MC (1998) J Chromatogr A 799 155
Saker ML, Metcalf JS, Codd GA, Vasconselos VM (2004) Toxicon 43 185
Siegl G, MacKintosh C, Stitt M (1990) FEBS Lett 270 198
Sipiä VO, Kankaanpää HT, Flinkman J, Lahti K, Meriluoto JAO (2001) Environ Toxicol 16 330
Sipiä VO, Kankaanpää HT, Pflugmacher S, Flinkman J, Furey A, James KJ (2002) Ecotoxicol Environ Safety 53 305
Sipiä VO, Karlsson KM, Meriluoto JAO, Kankaanpää HT (2004) Environ Toxicol Chem 23 1256
Sivonen K, Niemelä SI, Niemi RM, Lepistö L, Luoma TH, Räsänen LA (1990) Hydrobiologia 190 267
Spoof L, Vesterkvist P, Lindholm T, Meriluoto J (2003) J Chromatogr A 1020 105
Tsuji K, Masui H, Uemura H, Mori Y, Harada KI (2001) Toxicon 39 687
Van Buynder PG, Oughtred T, Kirkby B, Phillips S, Eaglesham G, Thomas K, Burch M (2001) Environ Toxicol 16 468
Vasas G, Gáspar A, Surányi G, Batta G, Gyémánt G, M–Havas M, Grigorszky I, Molnár E, Borbély G (2002) Anal Biochem 302 95
Williams DE, Craig M, Dawe SC, Kent ML, Holmes CFB, Andersen RJ (1997) Chem Res Toxicol 10 463
Williams DE, Craig M, Dawe SC, Kent ML, Andersen RJ, Holmes CFB (1997) Toxicon 35 985
Yokoyama A, Park HD (2002) Environ Toxicol 17 424
Yokoyama A, Park HD (2003) Environ Toxicol 18 61

Chapter 22: Field methods in the study of toxic cyanobacterial blooms: results and insights from Lake Erie Research

Steven W Wilhelm

Department of Microbiology, The University of Tennessee, Knoxville, TN 37996

Abstract

Sound field methodologies are an essential prerequisite in the development of a basic understanding of toxic cyanobacteria blooms. Sample collection, on–site processing, storage and transportation, and subsequent analysis and documentation are all critically dependent on a sound field program that allows the researcher to construct, with minimal uncertainty, linkages between bloom events and cyanotoxin production with the ecology of the studied system. Since 1999, we have collected samples in Lake Erie as part of the MELEE (Microbial Ecology of the Lake Erie Ecosystem) and MERHAB–LGL (Monitoring Event Responses for Harmful Algal Blooms in the Lower Great Lakes) research programs to develop appropriate tools and refine methods necessary to characterize the ecology of the reoccurring cyanobacterial blooms in the systems. Satellite imagery, large ship expeditions, classical and novel molecular tools have been combined to provide insight into both the cyanobacteria responsible for these events as well as into some of the environmental cues that may facilitate the formation of toxic blooms. This information, as well new directions in cyano–specific monitoring will be presented to highlight needs for field program monitoring and/or researching toxic freshwater cyanobacteria.

Introduction

Responses to toxic cyanobacterial events in freshwater systems require, as a first step, the identification of the event in question. Advanced field methods, designed to identify potential events and provide subsequent confirmation, are a critical need in cyanobacterial harmful algal bloom (CHAB) management. This need begets the development of sound approaches and tools for field applications, which will provide a systematic, tier–based response to events (i.e., a series of response ranging from casual observation to event validation and management action). This paper reviews the presently available technologies for field use to identify, confirm and characterize CHABs.

Given the global scale of CHAB events, an appropriate tier–based structure is required that allows for primary identification of a potential event, localized confirmation of this event, and subsequent characterization of the diversity and activity of the community members associated with the event. Combined with an identification of essential future need items, the goal of this paper is to provide the reader with insight into the current state of the available monitoring systems.

Early identification systems: The application of satellites and sentinel warning systems.

Although CHAB events occur on regional scales of meters to kilometers, a broad–based approach to monitoring freshwater systems for potential CHABs is ideal. While not necessarily applicable for small bodies of water (such as rivers and ponds), large bodies of water (e.g., the Laurentian Great Lakes) can be effectively monitored using available satellite imagery. Application of satellite imagery to monitoring large scale oceanic events is well established, with proxies for plankton biomass, sea surface temperature, sea surface height, etc. (Field et al. 1998). Application of these tools as sentinels for potential CHAB events in freshwater environments is enticing, as large scale imagery can be used to provide basic insight into phytoplankton distribution on time–scales associated with satellite flyover. As an example, current imagery for Lake Erie (which has been plagued by reoccurring blooms of toxic *Microcystis aeruginosa* since 1995, (Brittain et al. 2000)) includes true color satellite data (Fig. 1a) as well as inferred products such as phycocyanin (Vincent et al. 2004). While potentially limited by cloud albedo (Fig. 1b, August 23, 2005), this imagery nonetheless provides a simple, global approach to bloom monitoring.

One limitation of these imagery products is that they are currently based solely on the presence of inferred pigments. While blooms of toxic cyanobacteria do lead to significant concentrations of water column chlorophyll a (Rinta–Kanto et al. 2005), other non–toxic phytoplankton can similarly bloom and result in "false alarms" (Fig. 1c, May 3, 2005, see Color Plate 8). As such, any potential events identified by these tools will require more specific ground based confirmation.

Sentinel systems, incorporated into existing or newly deployed buoys, *in situ* monitoring devices, etc. also provide valuable insight into water column conditions that can be used as a first–alert to CHAB events. Predicated upon similar analytical methods as the satellites, currently available early monitoring systems have examined chlorophyll and phycocyanin as proxies for algal biomass. An expanded distribution of these sensors, combined with networking to a single source hub, is an ideal first step in the development of a global monitoring system for North American freshwater environments. The ongoing development of the IOOS (Integrated Ocean Observing System, see http://ocean.us/) and linkage to region nodes such as the proposed GLOS (Great Lakes Observing System, see http://www.glc.org/glos/) should provide stakeholders and end users with the potential to rapidly assess water conditions. Expansion of these sensors to mobile platforms (e.g., Coast Guard vessels, regional car and passenger ferries, public and privately managed docks, etc.) can be expected to further enhance coverage. These integrated networks represent the ideal ground based system for the incorporation of emerging CHAB sensors (discussed later). Moreover, many of the available sentinels can be deployed in small water bodies (e.g., ponds, rivers, reservoirs) that are typically below the resolution of satellites. As with any remote data collection method though, these systems require field sampling and confirmation prior to any action to deal with the bloom event.

Response to potential CHAB events, sample collection and processing.

The collection of a potentially toxigenic water sample that will be used to make management decisions of public access and use of a water resource requires appropriate and timely handling. In studies involving toxic samples particular attention needs to be paid to safety. While no prescription for preventative measures is given here (due to the diversity of sample and toxin types covered in this general overview) it is strongly recommended that newer personnel seek guidance from senior scientists and err on the side of caution concerning exposure risk.

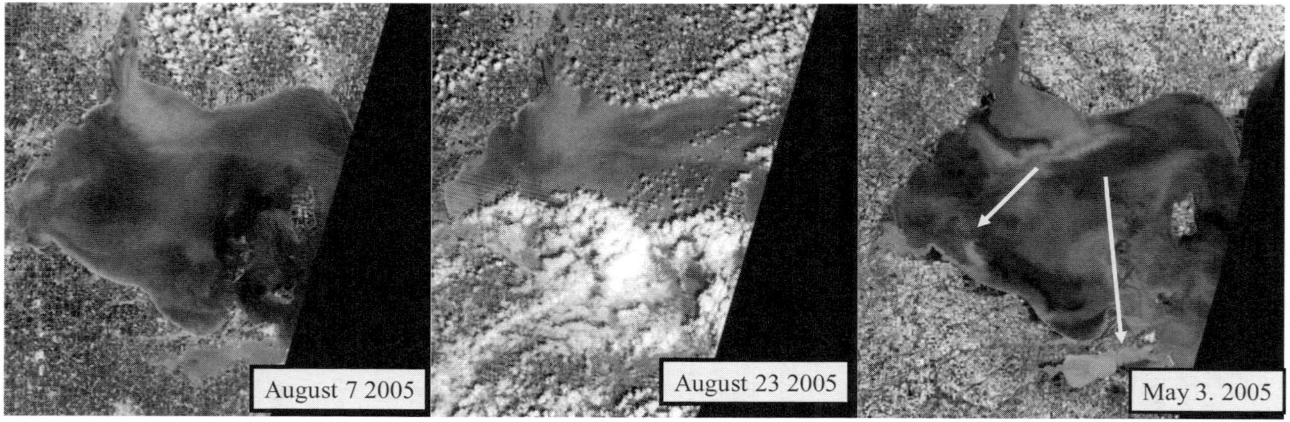

Fig. 1. LandSat 7 image of the western basin of Lake Erie for August 7, 2005. Greenish coloration of the water column in Sandusky Bay and at the mouth of the Maumee River suggests the onset of seasonal cyanobacterial blooms. (See Color Plate 8).

Color Plate 1

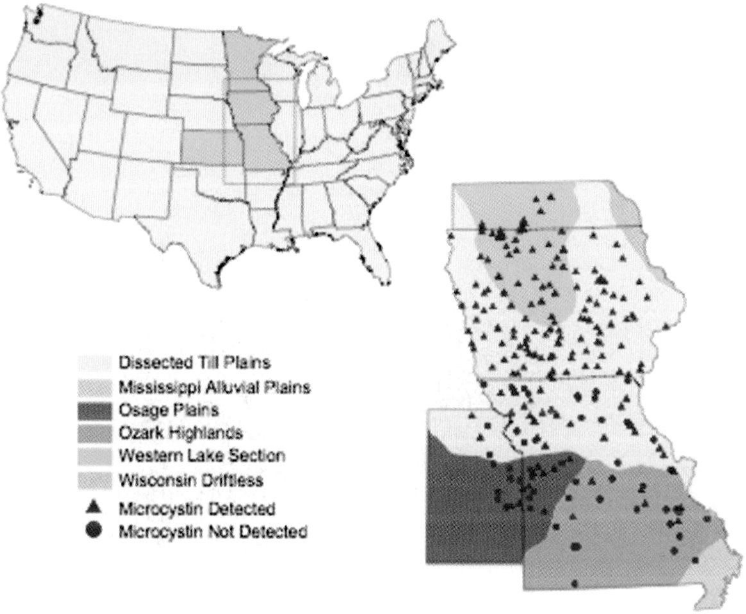

Fig. 3 (Chapter 3). Trends in Microcystin Occurrence in Midwestern Lakes. Taken from Graham et al. 2004. (See page 54).

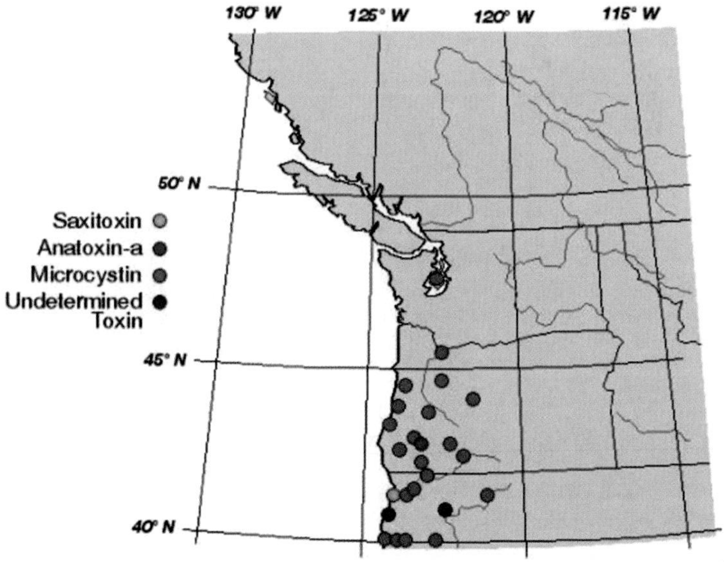

Fig. 4 (Chapter 3). Cyanotoxin Events in the Pacific Northwest (2001–2005). Taken from Carmichael 2006. (See page 54).

Color Plate 2

Fig. 12 (Chapter 3). Anabaena bloom on Lake Pontchartrain (Photo courtesy of John Burns. See page 82).

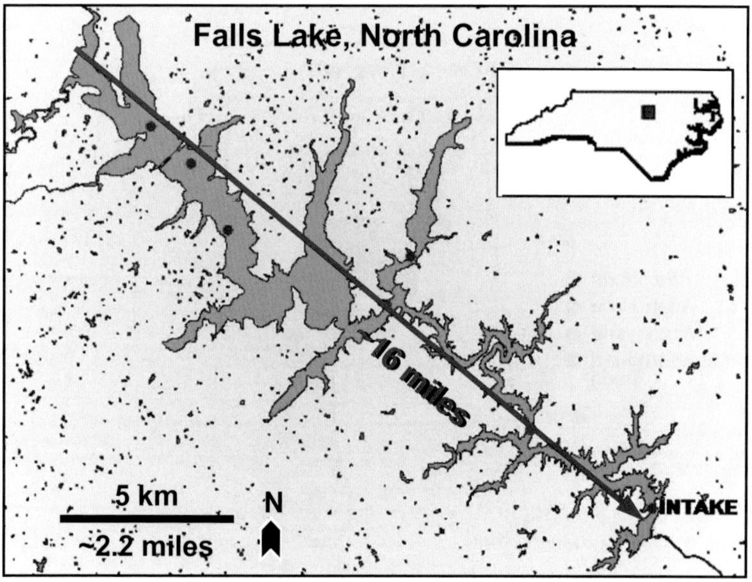

Fig. 13 (Chapter 3). Map of Falls Lake, NC, indicating six sampling stations and the location of the intake for the water treatment plant of the capital city, Raleigh. *The watershed is sustaining rapid human population growth and associated increased nutrient runoff into receiving waters. From Burkholder (2006b).* (See page 82).

Color Plate 3

Fig. 1 (Chapter 5). Mycrocystis Bloom—St. Johns River mid-channel south of the Buckman Bridge. (See page 133).

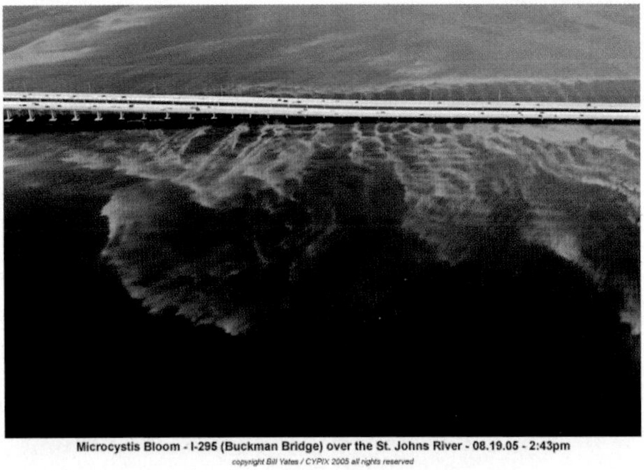

Fig. 2 (Chapter 5). Microcystis Bloom—I295 (Buckman Bridge) over the St. Johns River. (See page 134).

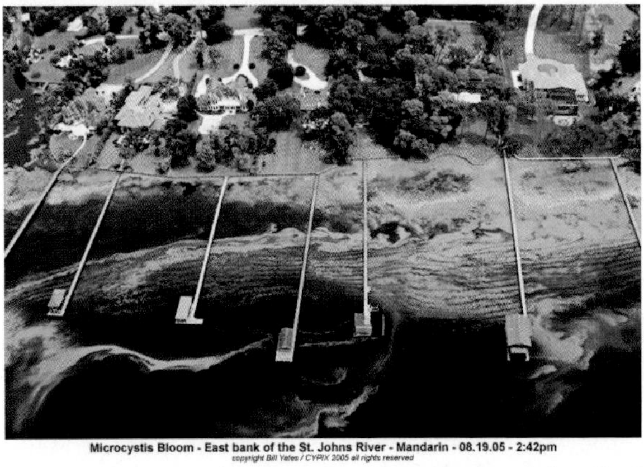

Fig. 3 (Chapter 5). Microcystis Bloom – East bank of the St. Johns River – Mandarin. (See page 134).

Color Plate 4

Fig. 3 (Chapter 6). Posting of warning signs at lake beaches and boat ramps. (See page 150).

Fig. 1 (Chapter 10). Harmful cyanobacterial blooms in a range of nutrient–enriched aquatic ecosystems. Upper left. A bloom of the non–N_2 fixing genera *Microcystis aeruginosa* and *Oscillatoria* sp. in the Neuse River, NC (Photo, H. Paerl). Upper right. A mixed *Microcystis* sp. and *Anabaena* spp. (N_2 fixers) bloom in the St. Johns River, Florida (Photo, J. Burns). Lower left. A bloom of the benthic filamentous N_2 fixer *Lyngbya wollei* in Ichetucknee Springs, Florida (Photo H. Paerl). Lower right. A massive bloom of *Microcystis* sp. and *Anabaena* spp. in Lake Ponchartrain, Louisiana (Photo J. Burns). (See page 219).

Color Plate 5

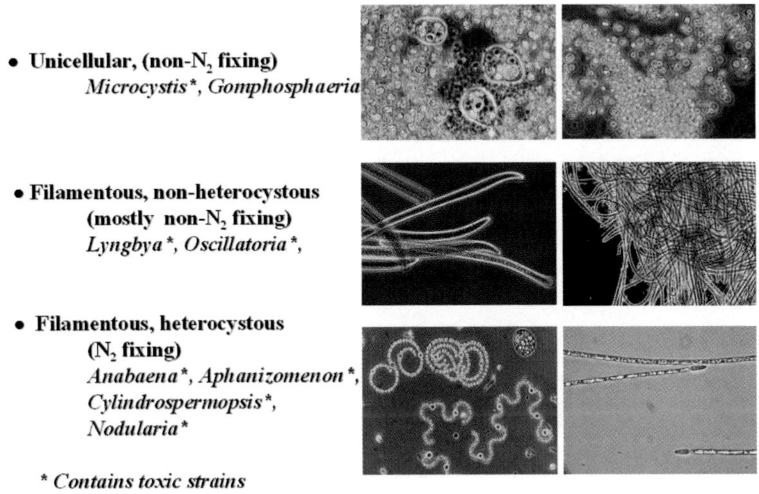

Fig. 2 (Chapter 10). Photomicrographs of genera representing the three major CHAB morphological groups, including coccoid, filamentous non–heterocystous and filamentous heterocystous types. (See page 220).

Fig. 3 (Chapter 10). Painting of Haarlemmermeer, a shallow, eutrophic lake in the Netherlands. Jan van Gooyen, ca. 1650. Note the surface scums characterizing the lake. (See page 220).

Color Plate 6

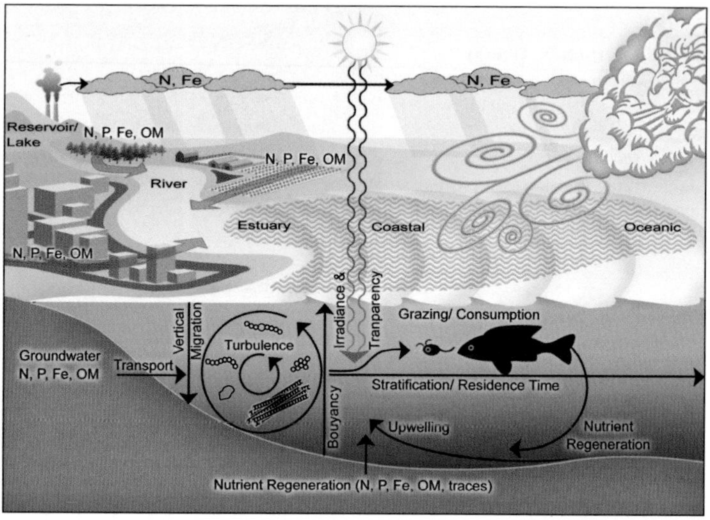

Fig. 4 (Chapter 10). Conceptual diagram, showing the interactive physical, chemical and biotic controls of cyanobacterial blooms along the freshwater–marine continuum. (See page 221).

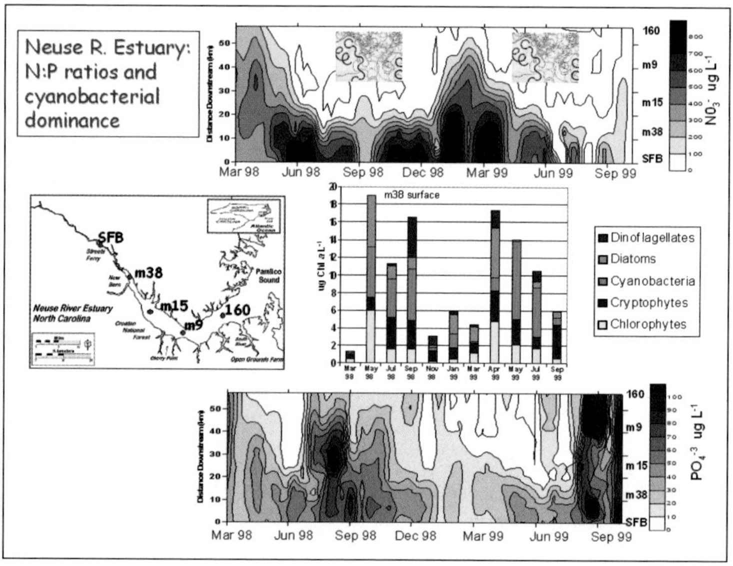

Fig. 5 (Chapter 10). Relationship, in space and time, of nitrate and phosphate concentrations, and relative dominance by cyanobacteria in the Neuse River Estuary, NC. Phytoplankton composition along a transect of 5 locations ranging from the upper oligohaline to lower mesohaline segments of the estuary was determined using high performance liquid chromatographic (HPLC) analysis of diagnostic (for major algal groups) photopigments (see Paerl et al. 2003). The period during which N_2 fixing cyanobacteria were present is indicated by the photomicrograph in the upper frame. (See page 223).

Color Plate 7

Fig. 1 (Chapter 20). Samples of currently available deployable systems that can be used in situ to sense conditions associated with potential CHAB events. 1). Flow Cam system from Fluid Imaging Technologies can identify cell type in situ. 2). Fluoroprobe system can identify water column phytoplankton based on a combination of 6 different fluorescence signatures. 3). NAS nutrient analyzer can detect in situ biogeochemical shifts that can be linked to pending CHAB events. (See page 477).

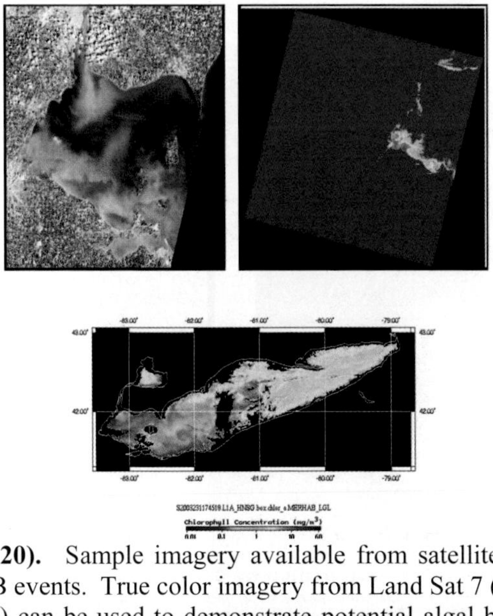

Fig. 2 (Chapter 20). Sample imagery available from satellites appropriate for monitoring CHAB events. True color imagery from Land Sat 7 (upper left, Rinta–Kanto et al. 2005) can be used to demonstrate potential algal blooms, which appear as green discolorations in the water column. The cyanobacterial–specific pigment phycocyanin can be elucidated from the appropriate applications of other algorithms (upper right, Vincent et al. 2003). Other imagery, such as daily Sea Wifs chlorophyll estimates (bottom) is available more frequently but provides less spatial resolution. (See page 477).

Color Plate 8

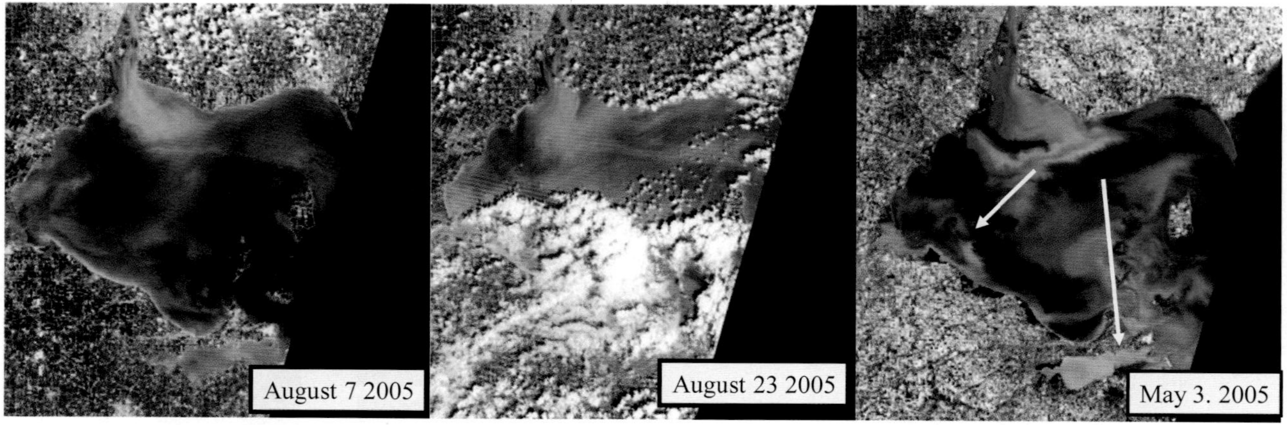

Fig. 1 (Chapter 22). LandSat 7 image of the western basin of Lake Erie for August 7, 2005. Greenish coloration of the water column in Sandusky Bay and at the mouth of the Maumee River suggests the onset of seasonal cyanobacterial blooms. (See page 504).

In field studies, the collection of water samples is commonly dependent upon the sampling site (riverine, pond, large lake) and end use of the sample (e.g., measurement of toxin molecular markers, measurement of toxin concentrations, processing for experiment or culture conditions). For samples to be processed for cellular genetics or toxin content, water collection onto appropriate filters (e.g., 0.2–µm polycarbonate filters) needs to be completed in a manner that ensures trace levels of contamination between samples does not occur. For the analysis of molecular markers (by PCR, microarray, etc.) sample storage is a critical issue as nucleic acids decay. If samples are processed in the field this issue is moot, but if they are to be transported to a laboratory then considerations of buffering and refrigeration, appropriate for the planned analyses must be considered. Several companies now provide stabilizing solutions for RNA and DNA that can be used to reduce degradation during transport. Preservation of samples for analytical measures of toxin concentration are being dealt with in companion papers (Lawton this volume, Meriluoto this volume).

Toxic bloom confirmation – a first level response to a potential threat.

As stated above, sentinel systems to autonomously monitor aquatic bodies for CHAB events require confirmation of any potential positive results. As a first step, a rapid, easy to use and cost–effective diagnostic can be employed to quickly ascertain whether an event is occurring. These tools are now commonplace and can take many forms: strip tests, reactive tube assays, etc. They are used in widely from the detection of pathogenic bacteria (like *E. coli* O157:H7) to home pregnancy tests. Compact and easy to use, these assays typically lack the sensitivity of more refined laboratory techniques, but provide a simply binary (yes/no) output and require little training to employ or interpret. Currently the commercial market consists of slightly more complicated tools (e.g., multiwell ELISA type assays), but it is anticipated that significant demand will drive the development of simpler tools in the near future.

Bloom confirmation and event characterization

Once a bloom occurrence has been identified, it is important for several reasons for full characterization to be carried out. Management of blooms and toxic events as well as public awareness campaigns regarding risks associated with exposure are dependent upon information concerning the responsible CHAB species, the toxin produced, and environmental conditions associated with the event. Collection and archiving of information regarding events remains one of the best scientific tools that may, in the future, allow for a better understanding of factors associated with bloom initiation.

Obviously, characterization of toxins produced during these events is a critical factor. Techniques to characterize and quantify toxins vary almost as widely as the toxins themselves. Moreover, ranges in sensitivity, cross–reactivity, and equipment/technical expertise are also broad. While an important consideration for the analysis of field samples, it should be noted that most of these current tools (like those of the molecular biologist, below) require transport of the sample to a research laboratory as equipment is often too cumbersome for field use. For more details on approaches to measuring toxins, see papers by Lawton and Meriluoto (this issue).

If available, microscopy provides important morphological information that can be used to provide a preliminary characterization of the CHAB associated community. The fact that species of potentially toxic cyanobacteria can be rapidly distinguished from others by morphology remains one of the best tools of the cyanobacteriologist (Chorus and Bartram 1999). However, while this gross morphology allows for distinction to be made between different taxa (a dying art in the scientific world !), morphology cannot be used to distinguish between toxic *vs* non–toxic cells (Ouellette and Wilhelm 2003). Indeed, one consideration in the development of new techniques for the analysis of toxic blooms is that they must be able to make these distinctions: as such the introduction of molecular tools to the CHAB research community holds great promise.

Given the limitations of microscopy, researchers have globally turned toward molecular tools to identify cyanobacteria as well as characterize their ability to produce toxins. Phylogenetically, cyanobacteria have been characterized by several researchers based on the sequence of their 16S rDNA (e.g., (Urbach et al. 1992; Nelissen et al. 1992; Nubel et al. 1997)). This allows researchers to place cyanobacteria into a general phylogeny (*vis a vis* (Woese 2000)). As shown in Fig. 2, this information can be useful as cells will cluster, for the most part, based on their taxonomy. Indeed, one strength of the molecular approach is how it often can corroborate existing taxonomies based on classical (i.e., morphological) approaches (Castenholz 1992; Komárek 2003). As such, sequences from taxonomically unidentified isolates or natural populations can be quickly characterized and a determination as to their potential phylogeny provided.

A set of problems however arise when using the standard "universal marker" (16S rDNA) for phylogeny:

1. The 16S rDNA marker in itself does not allow the user to determine the toxigenic potential of the organism in question, as phylogeny based on this marker has no relation to the potential of cells to produce any of the diverse cyanotoxins.

2. Identification of any organism is based on the quality of the information in the databases (i.e., GenBank, EMBL, the Ribosomal Database Project, *etc.*). In some cases the information in the databases may be misleading, out of date or even incorrect.

3. Application of such a general tool to natural systems is often confounded by an overabundance of non–target organisms. For example, in studies conducted during *Microcystis* blooms in the western basin of Lake Erie, PCR amplification of cyanobacterial 16S rDNA have yielded only *Synechococcus*–like sequences, even though *Microcystis* were abundant during the sample periods (Ouellette et al. 2005).

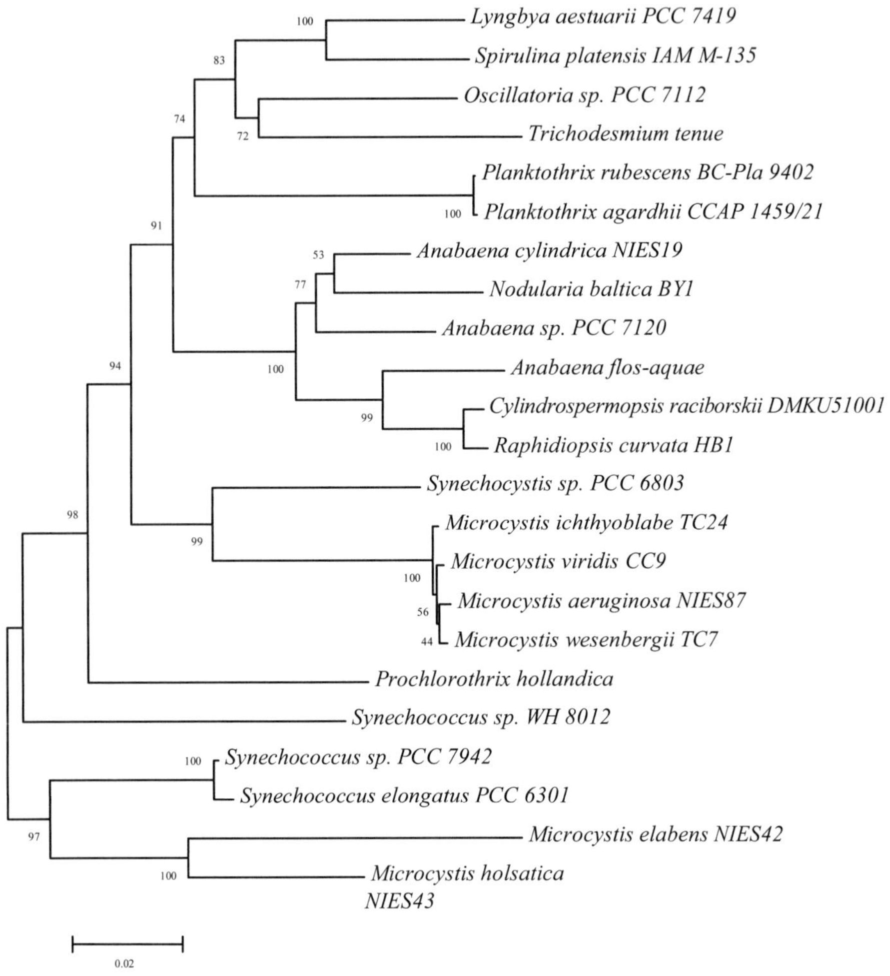

Fig. 2. Phylogenetic analysis of 16s rDNA sequencings of common cyanobacteria. Sequences were acquired from the ribosomal database project, aligned and edited in BioEdit (Hall TA 1999) and dendogram created using the neighbor-joining approach with Mega3 Kumar et al. 1994).

As such, while generic makers of microbial phylogeny are useful with organisms in culture, more specific molecular makers are desired for studies with mixed communities. Functional genes, such as the *nifH* (nitrogen fixation pathway), and conserved regions such as the intergenic *cpcBA–IGS* (between *cpcB* and *cpcA*, which encode for phyobilisome subunits) have been shown to be effective at discriminating specific groups of potentially toxigenic cyanobacteria (Dyble et al. 2002). More specifically, a number of studies have genetic elements directly associated with the production of specific toxins (e.g., (Nonneman and Zimba 2002; Kaebernick et al. 2000; Neilan et al. 2003)). Identification of these genes is of interest as they can be used in populations, alone or with other makers, to not only determine which type of cells may be present, but also whether or not the specific cell line is potentially toxigenic (Ouellette and Wilhelm 2003). While holding great promise for the future, the specificity of these genetic markers also delimits one of their flaws: the specificity of PCR primers for certain toxin genes means that related by different genetic elements are easily missed by these approaches. Moreover, it means that the development of probes is limited to only those strains that we have genetic information for – as such the identification of the > 70 variants of microcystin is a current significant issue.

Once developed though, the movement of these tools from binary (presence/absence) data collection to quantitative data collection is now relatively easy. Quantitative PCR (qPCR, sometimes also known as real–time PCR) is an approach that allows an investigator to estimate the abundance of copies of a target gene of choice in an original sample. Several research groups have now developed and published information using a variety of genetic elements as targets, and have been able to quantify *Anabaena* and *Microcystis* in natural samples (Vaitomaa et al. 2003; Rinta–Kanto et al. 2005; Kurmayer and Kutzenberger 2003). This breakthrough has allowed researchers to not only accurately quantify toxigenic cyanobacteria in natural environments, but has allowed them to do it effectively amongst a background of other cyanobacterial populations.

Future directions for field studies in toxic cyanobacterial blooms

Rapid advancement in monitoring approaches, a better understanding of the causes and phylogenetic diversity, and the development of a national reporting system for CHABs should all be priorities for federal agencies, scientists and system managers. A great deal of excitement and anticipation surround large scale monitoring plans (such as the GLOS described above). However, these approaches need to be tempered with insight into the development of tools (e.g., bioreporters, cytotoxicity monitors, etc.) that can be incorporated into a systems of deployable sentinels that can monitor smaller, regional areas at risk. Expanded distribution of sensors to other platforms (e.g., commercial ferries, government vessels and coastal docks, etc.) will in part offset "gaps" in the system.

Along with enhanced vigilance, the continued insight into the genetic mechanisms of toxin production and the organisms that are capable of these biochemistries, coupled with the development of advanced autonomous tools to characterize communities based on molecular markers, will allow for the determination of both the presence (DNA) and activity (protein or product) of genetic systems capable of producing toxin production. Linkage of these systems to remotely deployable biosensors (e.g., bioluminescent bacterial bioreporter systems, (Layton et al. 1998; Mioni et al. 2003) that can be incorporated into real time microsensors (Simpson et al. 1999) should allow for accurate characterization of cell abundance, toxin concentrations and toxin activity. When coupled to deployable sentinels, and supported by other technologies (e.g., fluorescence based sensing of pigments, etc.) these tools will provide a powerful first alert to CHABs. Moreover, the continued advancement of our understanding of the environmental conditions that lead to CHAB events will provide new avenues to target sensors.

Perhaps a final consideration that needs to be made is that there currently remains no reporting system within North America (an important point as many blooms occur in waters that cross national boundaries) where information can be archived. Modern computer software and nearly universal internet access amongst researchers and managers provide the necessary framework for the development of a national repository for this information. Information on system chemistry, microbial diversity, toxins and physiochemical parameters should all be collected that could, through data mining approaches, provide insight into the causes of CHABs and assist in future management decisions.

Acknowledgements

I thank Dr Greg Boyer, Dr Anthony Ouellette, Dr Renhui Li, and Johanna Rinta–Kanto for their help in formulating some of these thoughts over the past few years and Dr Paul Zimba for providing comments. I would also like to thank the members of the Exposure Assessment panel for insight provided during the ISCOHAB conference. This article includes research supported by funds from the American Water Works Association (AWWA #2818), the National Oceanic and Atmospheric Administration Coastal Ocean Program (MERHAB–LGL, #NA 160 P2788) and the National Science Foundation (DEB–0129118).

References

Brittain SM, Wang J, Babcock–Jackson L, Carmichael WW, Rinehart KL, Culver DA (2000) Isolation and characterization of microcystins, cyclic heptapeptide hepatotoxins from a Lake Erie strain of *Microcystis aeruginosa*. Journal of Great Lakes Research 26:241–249

Castenholz R (1992) Species usage, concept and evolution in the cyanobacteria (blue–green algae). Journal of Phycology 28:737–745

Chorus I, Bartram J (1999) Toxic cyanobacteria in water; a quide to their public health consequences, monitoring and management. E & FN Spon, London

Dyble J, Paerl HW, Neilan BA (2002) Genetic Characterization of *Cylindrospermopsis raciborskii* (Cyanobacteria) Isolates from Diverse Geographic Origins Based on nifH and cpcBA–IGS Nucleotide Sequence Analysis. Applied and Environmental Microbiology 68: 2567–2571

Field C, Behrenfeld M, Randerson J, Falkowski P (1998) Primary production of the biosphere: integrating terrestrial and oceanic components. Science 281: 237–240

Hall TA (1999) BioEdit: a user–friendly biological sequence alignment editor and analysis program for Windows 95/98/NT. (http://www.mbio.ncsu.edu/BioEdit/bioedit.html). Nucl Acids Symp Ser 41: 95–98

Kaebernick M, Neilan BA, Borner T, Dittman E (2000) Light and the transcriptional response of the microcystin biosynthesis gene cluster. Applied and Environmental Microbiology 66: 3387–3392

Komárek J (1999) Coccoid and colonial cyanobacteria. In: J.D. Wehrm and R.G. Sheath (eds), Freshwater algae of North America: Ecology and Classification. Academic Press, New York, pp 59–116

Kumar S, Tamura K, Nei M (1994) MEGA: Molecular Evolutionary Genetics Analysis software for microcomputers.http://www.megasoftware.net/. Comput Appl Biosci 10: 189–191

Kurmayer R, Kutzenberger T (2003) Application of real–time PCR for quantification of microcystin genotypes in a population of the toxic cyanobacterium *Microcystis sp* Applied and Environmental Microbiology 69: 6723—6730

Layton AC, Muccini M, Ghosh MM, Sayler GS (1998) Construction of a bioluminescent reporter strain to detect polychlorinated biphenyls. Applied and Environmental Microbiology 64(12) 5023–5026 Notes: English Article

Mioni CE, Howard AM, DeBruyn JM, Bright NG, Twiss MR, Applegate BM, Wilhelm SW (2003) Characterization and field trials of a bioluminescent bacterial reporter of iron bioavailability. Marine Chemistry 83: 31–46

Neilan BA, Saker ML, Fastner J, Torokne A, Burns BP (2003) Phylogeography of the invasive cyanobacterium *Cylindrospermopsis raciborskii*. Molecular Ecology 12:133–140

Nelissen B, Wilmotte A, Debaere R, Haes F, Vandepeer Y, Neefs JM, Dewachter R (1992) Phylogenetic study of cyanobacteria on the basis of 16s ribosomal RNA sequences. Belg J Bot 125:210–213

Nonneman D, Zimba PA (2002) A PCR–based test to assess the potential for microcystin occurrence in channelcatfish production ponds. Journal of Phycology 38: 230–233

Nubel U, GarciaPichel F, Muyzer G (1997) PCR primers to amplify 16S rRNA genes from cyanobacteria. Applied and Environmental Microbiology 63(8) 3327–3332

Ouellette AJA, Handy SM, Wilhelm SW (2005) Toxic *Microcystis* is widespread in Lake Erie: PCR detection of toxin genes and molecular characterization of associated microbial communities. Microbial Ecology in press

Ouellette AJA, Wilhelm SW (2003) Toxic cyanobacteria: the evolving molecular toolbox. Frontiers in Ecology and the Environment 7:359–366

Rinta–Kanto JM, Ouellette AJA, Twiss MR, Boyer GL, Bridgeman TB, Wilhelm SW (2005) Quantification of toxic *Microcystis spp* during the 2003 and 2004 blooms in Western Lake Erie. Environmental Science and Technology 39: 4198–4205

Simpson ML, Sayler GS, Ripp S, Nivens DE, Applegate BM, Paulus MJ, Jellison GE Jr (1999) Bioluminescent bioreporter integrated circuits form novel whole–cell biosensors. Trends in Biotechnology 16: 332–338

Urbach E, Robertson DL, Chisholm SW (1992) Multiple evolutionary origins of prochlorophytes within the cyanobacterial radiation. Nature 355:267–270

Vaitomaa J, Rantala A, Halinen K, Rouhiainen L, Tallberg P, Mokelke L, Sivonen K (2003) Quantitative real–rime PCR for determination of Microcystin Synthetase E copy numbers for *Microcystis* and *Anabaena* in Lakes. Applied and Environmental Microbiology 69: 7289–7297

Vincent RK, Qin X, McKay RML, Miner J, Czajkowski K, Savino J, Bridgeman T (2004) Phycocyanin detection from LANDSAT TM data for mapping cyanobacterial blooms in Lake Erie. Remote Sensing of Environment 89:381–392

Woese CR (2000) Interpreting the universal phylogenetic tree. Proceedings of the National Academy of Sciences USA 97: 8392–8396

Chapter 23: Conventional laboratory methods for cyanotoxins

Linda A Lawton, Edwards C

School of Life Sciences, The Robert Gordon University, St Andrew Street, Aberdeen, AB25 1HG, UK. Email:L.Lawton@rgu.ac.uk

Introduction

Over recent years it has become apparent that toxic cyanobacterial blooms are on the increase, presenting a hazard to animal and human health (Appendix A, Table A.1). The importance of algal toxins is reflected in their inclusion of EPA recognised contaminants in water (Richardson and Ternes 2005). Microcystins have been extensively studied and reported over recent years. Despite the number of microcystin variants and lack of standards, a large number of biological and chemical methods have been optimised for a variety of matrices, usually cells, water and tissue. Data on chronic and acute toxicity have led to the WHO to set a guideline maximum of 1 μg per litre in drinking water. Methods developed for microcystins are suitable for the pentapeptide nodularins, although these cyanotoxins usually occur in brackish water.

In contrast, relatively little work has been done on methods detection of other known toxins, anatoxins, cylindrospermopsins, BMAA and aplysiatoxins. Saxitoxins being the exception, as they occur widely in the marine environment and many methods have been developed for their detection in shellfish. However, there has been only limited application of these methods to freshwater samples. There are many challenges in assessing and selecting suitable methods since blooms can not only be composed of co–occurring species but it is also known that some species produce multiple classes of toxins.

This paper reviews methods presented in the literature, many of which are currently used for routine monitoring and in research. We discuss the application, validation, cost and practicability of a range of techniques. Priorities, future needs and challenges are addressed.

Analysis of microcystins

Microcystins are the most commonly reported cyanobacterial toxins and this is reflected by the large number of methods for their detection and analysis summarised in Table A.2. Although nodularins are less of a problem in freshwater, most methods developed for microcystins are suitable for nodularins. By far the greatest challenge in analysing microcystins is the fact that there are in excess of 65 variants characterised to date and most likely others yet to be identified. It is essential that any method used has the ability to detect all variants, regardless of availability of standards. Equally important, extraction and separation procedures must be suitable for the chemical range of variants in order to obtain accurate qualitative and quantitative data.

HPLC methods

There are many liquid chromatography based methods in the literature, utilising a range of stationary phases, mobile phases and detectors for both isocratic and gradient separations (Meriluoto 1997). However, reversed–phase chromatography with diode array detection (HPLC–PDA) has been the most widely used approach over the last two decades, as it enables detection of all microcystins based on their characteristic UV spectra (Lawton et al. 1994). Use of a gradient helps to ensures microcystins variants will be separated and despite lack of standards or certified reference materials, quantification of approximate total microcystin content is possible based on purified MC–LR to give MC–LR equivalence. Inter–laboratory validation data supports this approach combined with concentration and clean–up on SPE (Isolute C18) for the monitoring of intra and extra cellular microcystins in water samples as recommended in a "Blue Book" publication in the UK (Environment Agency 1998). Limits of quantification reported are 1–10 ng on column (achieving sub–µg per litre). A recent inter–laboratory trial highlighted the need for certified reference materials as commercial material that is currently available is essentially a laboratory reagent not a standard. When this material is used as a standard it results in varying responses for the same samples in different laboratories (Fastner et al. 2002).This study also highlighted that despite variation in material which is used as standards, a variety of analytical systems and methods yielded similar responses, extraction procedures used for real samples was more problematic, emphasising the need for complete method optimisation. Detection limits have been improved by the use of immunoaffinity SPE for concentration, however, there are still limitations on binding capacity and the volumes loaded which must be overcome if this is to be a practical solution (Lawrence and Menard 2001, Aranda–Rodrigues et al. 2003). Recent advances, using recombinant antibody fragments, have demonstrated potential for the development of cost effective, robust and reproducible immunoaffinity cartridges (McElhiney et al. 2002).

As technology has evolved, LC–ESI–MS or LC–ESI–MS/MS is becoming the preferred technique as it offers greater selectivity and sensitivity than diode array detection. Good sensitivity was achieved using a single quadrupole (LC–ESI–MS), LOD of 11, 72, 21 and 6 pg for MC–LR, MC–RR, MC–YR and nodularin respectively on column (1 mm I.D.) using selected ion monitoring (SIM (Barco et al. 2002)). However, most methods published in the literature use tandem MS, which enables noise reduction and thus greater sensitivity, multiple reaction monitoring, and the removal of the need for complete separation of analytes. This approach enabled the development of a high through–put method which analysed ten microcystins in 2.8 minutes, without the need for complete resolution (Meriluoto et al. 2004). However, although the potential of LC–MS/MS is unequivocal, much work is still needed since most methods have been developed with a limited number of microcystins and there is no way to guarantee detection of unknown microcystins as fragmentation patterns vary considerably with conditions and microcystin chemistry itself. Fig. 1 illustrates the diversity of ionisation under typical reversed phase conditions. Microcystins containing no arginine are more susceptible to the formation of sodium and potassium adducts which is far from ideal in a quantitative application. Therefore, for a robust LC–MS/MS method, there is a requirement for ionisation optimisation and a thorough study on the effects of a wider variety of sample matrices, their effects and overcoming/understanding them. For suppression of sodium and potassium adducts, Yuan et al. demonstrated that the addition of oxalic acid biased the formation of the molecular ion thus increasing the sensitivity although this is seldom used and adduct ions are regularly monitored (Yuan et al. 1999). This work also showed that storage led to increases in adducts.

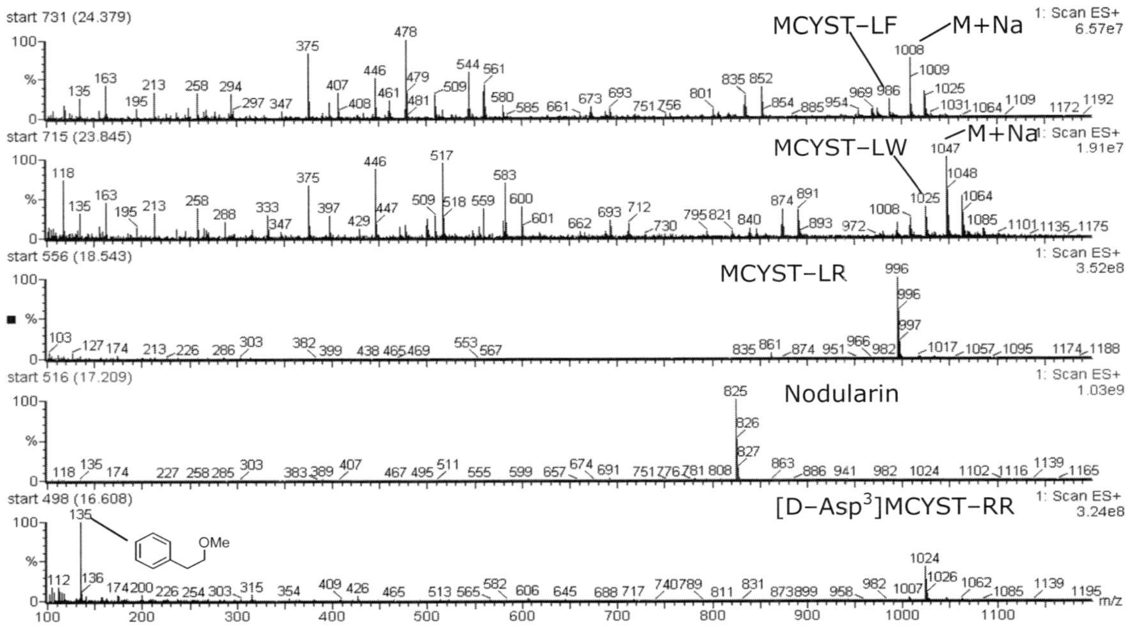

Fig. 1. Mass spectral of microcystin–LF, –LW, –LR, –D–Asp3 RR and nodularin performed using a Waters Micromass LC–ESI–MS. Separation was performed on a Symmetry C18 column using a water/acetonitrile plus 0.05%TFA gradient. Eluent was monitored by diode array (200–400 nm) and positive ion electrospray (100–1200 amu; cone voltage of 80 ev).

Maximizing individual microcystin sensitivity can be achieved by complex methods utilising time scheduled selected reaction monitoring conditions as demonstrated by Bogialli et al (2005). Several reports have examined matrix effects on analysis of tissue samples, illustrating the importance of the inclusion of this work for any method under development and in subsequent validation. Ruiz et al. demonstrated a 15% over estimation of MC–RR in extract from kidney compared to a 37% decrease in detection in liver (Ruiz et al. 2005). From these findings they recommended the use of matrix matched standards for use when quantifying unknown samples.

Matrix–assisted laser desorption (MALDI) has been used in conjunction with TOF analysers for the detection of microcystins and unknown variants in small samples (Welker et al. 2002). Characteristic fragmentation was achieved by post–source decay, which results in destruction of the peptide bonds. Whilst rapid, this offline technique requires some extraction to eliminate matrix/sample interferences, but, as improved matrices are developed, there is future potential for an approach eliminating time consuming sample preparation and chromatography. This is illustrated in a recent publication, describing the use of MALDI linked to a triple quadruple for the qualitative and quantitative determination of spirolide toxins (Sleno and Volmer 2005). The combination of this ionisation technique with sensitive multiple reaction monitoring, proved to be precise and accurate without the need for extensive sample preparation.

Another exciting approach which is rapid and eliminates time consuming SPE, where the microcystins were captured on a hydrophobic chip and subsequently ionised by surface–enhanced laser desorption ionisation–time–of flight MS (SELDI–TOF–MS) enabled determination of 2.5 pg MC–LR in 2 µl (1.2 µg L^{-1}) water (Yuan and Charmichael 2004). However, severe matrix effects were experienced when more complex samples were analysed, and it was not possible to monitor the characteristic m/z 135 due to background interference. Future chip developments could present the way forward although may prove costly.

In–vitro bioassays

To compliment the large number of physico–chemical methods there exists a significant number of bioassays for detection of microcystins. Microcystins and nodularins are strong inhibitors of protein phosphatases, PP–1, PP–2A and PP–3, PP–2A being the most sensitive. This functionality has been exploited to develop assays which provide a direct measure of toxicity. A range of substrates have been used but the most commonly used are *p*–nitrophenol phosphate (*P*–NPP), 4–methylumbelliferyl phosphate (MUP) and 6,8–difluoro–4–methylumbelliferyl phosphate (DiFMUP). The latter has been successfully validated against HPLC and mouse bioassay for the detection of okadaic acid in shellfish (González et al. 2002). This approach has been adapted to a rapid microplate assay for screening microcystins in drinking water without the need for pre–concentration, achieving a detection limit of 0.1 μg L^{-1}, which is well below the provisional guideline value (Bouaïcha et al. 2002). This assay provides a useful pre– or post analytical screen for bioactivity although false positives may be obtained from other phosphatase inhibitors, which may occur in environmental samples. Many researchers have reported good correlation of data obtained by protein phosphatase inhibition assay and HPLC–PDA (Ward et al. 1997, Wirsing et al. 1999). The necessary components are available commercially although there is batch variation in enzyme activity.

Immunoassays

Immunoassays, exploiting polyclonal, monoclonal antibodies and recombinant antibody fragments, are widely used as screening tools for microcystins and nodularins and are well reviewed elsewhere (McElhiney and Lawton 2005, Metcalf and Codd 2003). Several kits are commercially available, in microtitre plate or tube format. Many of the assays/kits use antibodies raised against MC–LR and subsequently may have limited cross reactivity (EnviroLogix Inc, Portland, ME, USA), whereas kits using antibodies raised against ADDA provide improved sensitivity and excellent cross–reactivity (Abraxis LLC, PA, USA: Biosense Laboratories AS, Bergen, Norway). However, the behaviour of non–toxic degradation products including free ADDA is as yet unknown These ELISA kits are supplied in a 96–well microplate format with ready to use reagents enabling screening of up to 96 samples in 2.5 hours with a consumable cost of $400.00. All of these commercial kits are simple to use, rapid and economical for screening. As with phosphatase inhibition assays, immunoassays can be used for detection of microcystins below the WHO guideline without the need for sample pre–concentration.

Other useful methods

A cost effective, rapid, thin layer chromatography (TLC) method has also been developed which enables detection of microcystins to meet the WHO 1 μg L^{-1} guideline. This method relied on visualisation of the microcystins on the developed TLC plate using N,N–dimethyl—1,4–phenylendiamonium dichloride (N,N,–DPDD) and good correlation was achieved compared to protein phosphatase and ELISA assay (Pelander et al. 2000). However, without sophisticated spotting and scanning devices, this is not quantitative, but would serve as a useful screen for known microcystins, although it does require improvements in sample concentration to remove interfering contaminants.

A method for determination of total microcystins relies on oxidation to produce 2–methyl–3–methoxy–4–phenyl–butyric acid (MMPB) from ADDA, which is detected by GC–MS (Kayo and Sano 1999), HPLC–Fl (Sano et al. 1992) or HPLC–TSP. Whilst this method has been demonstrated to be useful for complex samples such as sediments, the need for oxidation, and the fact that only total microcystin is determined, make it a complex, time consuming and expensive screen. Despite these disadvantages, this could be a useful confirmatory method and can be used with a wide range of instrumentation without the need for microcystin standards. Most methods described determine free microcystins, this method will also detect bound microcystin, thus providing a complete picture in metabolism studies.

Capillary electrophoresis based methods exploit high efficiency columns to separate variants often problematic in LC separations such as MC–LR and [D–Asp (Lawton et al. 1994)] MC–LR providing a useful complimentary technique (Bateman et al. 1995). Issues such as sensitivity and interfering compounds have been overcome by improved online and offline sample clean–up.

Combined methods

HPLC–UV/PDA has been shown to be a powerful tool in combination with protein phosphatase inhibition or ELISA assay. HPLC–PP2A was first reported in 1991 as a highly sensitive bioscreen for okadaic acid along with related polyether toxins (Holmes 1991) and later applied for the detection of microcystins in freshwater environments (Boland et al. 1993). These approaches are still used, often along side mass spectrometry to determine complete structure/activity profiles of unknown samples (Ortea et al. 2004).

Fractionation into 96 well plates was used to increase automation and extending the assay to include an immunoassay providing LC–UV/ELISA /PP2A data, achieving detection limits 1000 x more sensitive than UV (Zeck et al. 2001). This paper also compared the response of the same sample to PP2A, and ELISA, using three commercially available antibodies, highlighting huge variation in cross reactivity. Several groups have reported the use of ELISA alongside PP2A inhibition, providing a measure of total microcystins and toxicity, however, the most elegant use of these techniques is the immunophosphatase assay.

Analysis of saxitoxins

Saxitoxins (also known as paralytic shellfish poisons, PSPs) are another complex group of compounds which have presented a challenge over the last two decades. Until June 2005, the only validated method available was the mouse bioassay, routinely used for screening shellfish and phytoplankton. However, there has been much progress in development of methods as summarised in Table A.3 of Appendix A, reflecting their importance in the shellfish industry and the fact that many countries have rigorous guidelines and monitoring requirements.

HPLC analysis

In June 2005 an HPLC method relying on fluorescence detection of the oxidised saxitoxins was approved by AOAC after inter–laboratory validation (Lawrence et al. 2004). Whilst this method is robust, the sample processing is complex and two pre–column oxidation reactions/separations may be needed for quantification of the complete range of saxitoxins. A further problem is that oxidation of some GTXs, dcGTXs, dcSTX and dcNEO results in the production of two fluorescent compounds, thus requiring a broad range of standards. Despite the reported robustness, this is a time consuming and therefore expensive method. Automation of the derivatization procedure would reduce manual processing, however it must be noted the fluorescent products are not stable after a few hours.

An alternative approach using post–column derivatization has been preferred in many labs as it benefits from simple automation. However, three, more recently two, separations are needed to accurately quantify all toxins. This method is sensitive to changes in flow rate, reagent age and temperature. With both pre– and post column derivatization methods, it is ideal to run a sample without oxidisation to confirm peaks are not interfering contaminants.

Several methods using capillary electrophoresis have been reported although, they are not widely used and suffer from low sensitivity due to the low volume injected and the requirement for a very clean sample in order to obtain reproducible chromatography. It is reported that LOD is an order of magnitude greater than HPLC–FL/MS.

A recent publication described a single gradient separation for all saxitoxins with MS/MS detection for qualitative analysis and future optimisation of quantitation provides a promising alternative analytical method (Dell'Aversano et al. 2005). This will provide a simpler, although more expensive method, without the need for oxidation.

Many assays have been described which exploit the functionality of the saxitoxins, i.e. sodium channel blocking activity. Most of these rely on the use of cultured cell lines and specialist techniques/facilities, thus not practical for routine monitoring purposes and out with the scope of this review.

Immunoassays

An immunoassay kit, RIDASCREEN®, is available from R–Biopharm AG (Darmstadt, Germany), which is used widely by commercial organisations for screening shellfish. This is a sensitive (LOD of 50 ppb), quantitative, plate based kit, which requires a microtitre plate reader (450 nm). Each 48 plate allows analysis of up to 42 samples providing results after a one–hour incubation. This is generally used as a rapid screen, providing a yes/no response, providing good correlation with the mouse bioassay for the detection of saxitoxins in shellfish (Inami et al. 2004). This kit has a lot of potential for screening saxitoxins in water, cells and tissues, but the only published report was analysis of crude cyanobacterial cell extracts (Teneva et al. 2005). It must be remembered that there is poor cross reactivity with related compounds, e.g. 12% with neosaxitoxin which is often a major component produced by cyanobacteria.

One of the most promising, commercially available screens, is the Jellet Rapid Test (JRPT: formerly MIST Alert) which is a lateral flow immuno–chromatographic test approach based on antibodies raised to multiple, structurally diverse saxitoxins, providing good cross reactivity and therefore accuracy (Jellet et al. 2002). The JRPT functions in a manner similar to a pregnancy testing kit, providing a yes/no answer within twenty minutes. This has been widely tested across the world in parallel with the mouse bioassay and HPLC, and in many areas now serves as the primary screening tool. Potential use of this system for monitoring saxitoxins in freshwater has yet to be investigated, although, it must be remembered that the level of detection is aimed at the shellfish and some modification for freshwater application would be necessary or a sample concentration step added.

Analysis of cylindrospermopsins

Compared to microcystins and saxitoxins, relatively few methods have been developed for detection of cylindrospermopsins (Table A.4 in Appendix A). This may be due to the fact this is a more recently discovered toxin which was easily detected by HPLC–PDA/MS and/or that events have been limited. HPLC–PDA is good for detection of cylindrospermopsins and its analogues as they have characteristic UV spectra (λ max at 262 nm) however, sample cleanup is necessary to remove co–eluting contaminants (Welker et al. 2002) . HPLC–PDA was used by five out of six laboratories during a recent inter–laboratory comparison of cylindrospermopsin analysis (Törökné et al. 2004). Cylindrospermopsin was extracted from freeze–dried cells by a variety of procedures followed by HPLC analysis to determine method suitability. Whilst all methods were successful for crude extraction/analysis of cylindrospermopsin, further refinements would be necessary if any of these was to be used for monitoring purposes. LC–MS/MS is currently the most favoured method of analysis, providing structural confirmation and sensitive quantification by monitoring the transition from M+H ion (m/z of 416) to the major fragment m/z of 194, achieving a range of 1–600 $\mu g\ L^{-1}$ without sample concentration (Eaglesham et al. 1999). Although cell and invertebrate assays have been used to detect cylindrospermopsin, these are non–specific and insensitive. The development of a sensitive, selective rapid screen for monitoring is essential. However, it is important to remember with cylindrospermopsins in water samples, that these compounds are excreted from the cyanobacterial cell during growth, thus necessitating robust sampling protocols and analysis of extra– and intracellular toxin.

Analysis of anatoxin–a

Apart from the mouse bioassay, all reported methods of detection of anatoxin–a are based on chromatography, with or without derivatization as summarised in Appendix D. LC–UV has been widely used but suffers from limitations such as sensitivity and interferences in complex sample matrices. In recent years sensitive, qualitative and quantitative methods which rely on some form of derivatization procedure included GC–MS, GC–ECD and HPLC with fluorescence detection have been the preferred methods (Himberg 1989, Stevens and Krieger 1988, James et al. 1998). As with most applications, improvements in LC–MS and LC–MS/MS technology have led to increasing use for detection of anatoxin–a and its analogues, eliminating the need for derivatization (Furey et al. 2005, James et al. 2005). However, LC–MS was the sole method used to confirm the presence of anatoxin–a as the most likely cause of a young man's death in 2002, but as it transpired the compound was in fact phenylalanine, but due to the fact that the two compounds are isobaric and have similar retention characteristics, LC–MS alone was insufficient to distinguish between them (Furey et al. 2005). This case, illustrates the need for multiple, robust and complimentary methods and /or detectors. A diode array detector in series would have shown the difference in UV spectra of anatoxin–a and phenylalanine, having maximum absorption at 227 nm and 257 nm respectively. A similar approach, using LC–PDA and LC–MS was recently used to unequivocally identify anatoxin–a associated with a dog poisoning in France (Gugger et al. 2005).

In the short term, the only option for a low cost, rapid screen, could be the TLC method where the anatoxin is reacted with the diazonium reagent, Fast Black K salt, to form an orange–red product (Ojanperä et al. 1991). Although this method is sufficiently sensitive for determination of anatoxin in algal cells (10 $\mu g\ g^{-1}$), pre–concentration of water samples would be necessary. This method should also be suitable for detection of anatoxin analogues.

Analysis of anatoxin–a(s)

The occurrence of this alkaloid cholinesterase inhibitor is rare, as is reflected by the number of methods published. Despite its rarity, anatoxin–a(s) is highly toxic (LD_{50} in mice is 50 $\mu g\ kg^{-1}$ body weight) and has been responsible for several livestock and bird poisonings thus necessitating reliable methods of detection. Lack of a chromophore, limits the use of conventional HPLC methods although mass spectrometry would be an ideal means of detection. Colorimetric bioassays based on acetylcholinesterase inhibition have been the most reliable methods to date, although false positives can be obtained from organo–phosphorus insecticides (Ellman et al. 1961). This assay is a rapid and sensitive laboratory screen, with all necessary enzymes and reagents available from general laboratory suppliers. Biosensors, incorporating enzymes of different sensitivities, have been developed which facilitate specific detection of anatoxin–a(s) below $\mu\ L^{-1}$ level (Devic et al. 2002). A similar biosensor used oxime reactivation of the enzyme to differentiate between anatoxin–a(s) and insecticide inhibition (Villatte et al. 2002). Refinement and commercialisation of these biosensors would be an ideal screen for anatoxin–a(s), being rapid, inexpensive and simple.

Analysis of ß–N–methylamino–L–alanine (BMAA)

A recent publication indicated that this neurotoxic amino acid is produced by a diverse range of cyanobacteria (Cox et al. 2005), a potential hazard, obviating the need for further investigation. Several HPLC methods have been reported including derivatization with 6–aminoquinolyl–*N*–hydroxylsuccinimidyl carbamate followed by RP–HPLC with fluorescence detection with a limit of quantitation reported as 1.2 μg L^{-1}. MS detection of this derivative was also used for additional confirmation. GC–MS has also been used to detect BMAA in cycad seeds as an N–ethoxy carbonyl ethyl ester derivative (Pan et al. 1997). Although these methods have been used to detect BMAA in cycads, flying foxes and brain tissue, further work is needed to provide robust methods, encompassing extraction, concentration/clean–up and quantitative/qualitative analysis to support necessary research and monitoring programs.

Conclusions and Summary

It is clear from the literature that numerous methods are available for most cyanotoxins, although many publications on monitoring data indicate that the favored approach is the use of proven, robust methods for individual toxins. The most effective approach is the utilization of a robust rapid screen, where positive samples are followed up by qualitative and quantitative analysis to provide the essential decision making data needed for successful management strategies (Fig. 2). Currently, rapid screens are available for microcystins, saxitoxins and anatoxin–a(s), whilst optimisation and validation is needed, many publications report good correlation with the mouse bioassay and HPLC.

There is an urgent need for rapid, simple, and inexpensive assays for cylindrospermopsins, anatoxin–a and BMAA. Although methods exist for analysis of BMAA, the fact that a recent study showed 95% of cyanobacteria producing this, some at levels >6,000 μg g^{-1} dry wt, is of concern and rapid screening followed by robust analysis is needed.

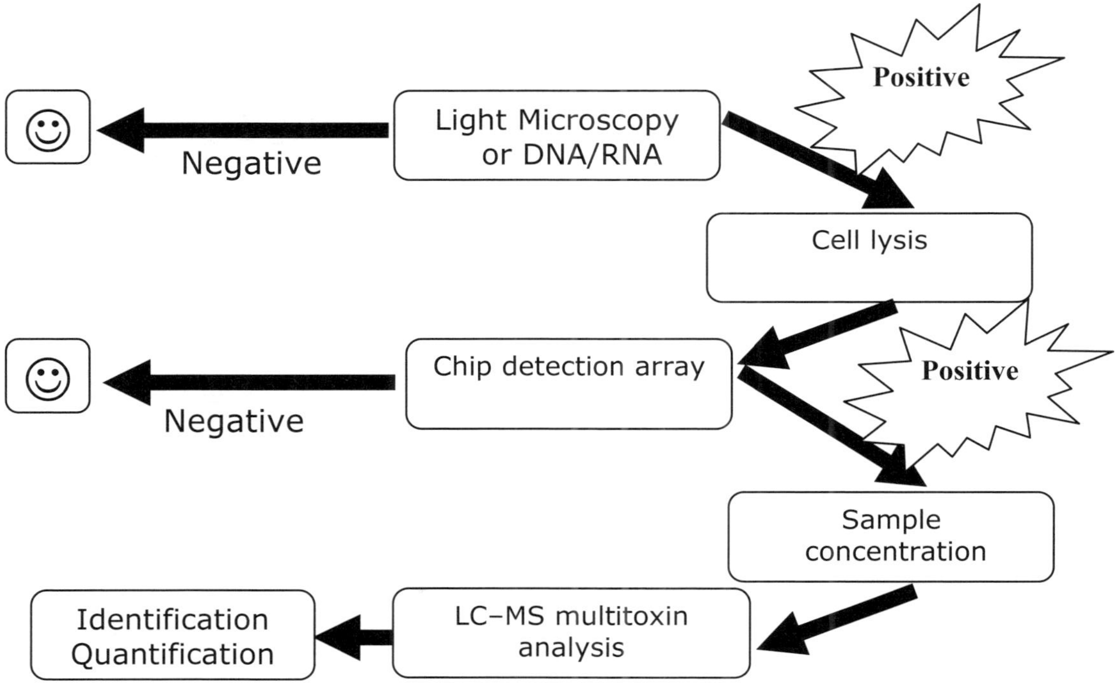

Fig. 2. Future strategy for cyanotoxin monitoring including initial screen to identify the presence of harmful cyanobacteria either by established light microscopy or novel molecular techniques, followed by rapid multitoxin array with the ability to highlight the presence of any of the known classes of cyanotoxins above a pre–determined threshold and where necessary analytical confirmation by LC–MS.

An ideal approach would be a single method capable of extracting and detecting all cyanotoxins. Several publications describe such approaches using LC–MS, but as expected from a group of compounds with diverse chemistry, there are obvious limitations in recoveries during sample processing, chromatographic performance and sensitivity (Dahlmann et al. 2003, Dell'Aversano et al. 2004, Pietsch et al. 2001).

Selection of methods must be based on the application requirements, equipment available and cost. For many organisations it may be more cost effective to out–source the occasional analysis. However, as the incidence of blooms appears to be increasing, the need for more rigorous monitoring is needed, sensible investment is needed to meet recommended guidelines. Most of the methods discussed in this paper are suitable for achieving this goal, although clean–up and concentration is usually necessary for physicochemical methods.

References

Aranda–Rodriguez R, Kubwabo C, Benoit FM (2003) Toxicon 42 587–599

Barco M, Rivera J, Caixach J (2002) J Chromatogr A 959 103 –111

Bateman KP, Thibault P, Douglas DJ, White RL (1995) J Chromatogr A 712 253-268

Bogialli S, Bruno M, Curini R, Di Corcia A, Lagana A, Mari B (2005) J Agric Food Chem 53 6586–6592

Boland MP, Smillie MA, Chen DZX, Holmes CFB (1993) Toxicon 31 1393–1405

Bouaïcha N, Maatouk I, Vincent G, Levi Y (2002) Food Chem Toxicol 40 1677–1683

Chianella I, Lotierzo M, Piletsky SA, Tothill IE, Chen B, Karim K, Turner APF (2002) Anal Chem 74 1288–1293

Cox PA, Banack SA, Murch SJ, Rasmussen U, Tien G, Bidigare RR, Metcalf JS, Morrison LF, Codd GA, Bergman B (2005) PNAS. 102 No 14 5074–5078

Dahlmann J, Budakowski WR, Luckas B (2003) J Chromatogr A 994 45 –57

Dell'Aversano C, Eaglesham GK, Quilliam MA (2004) J Chromatogr A 1028 155–164

Dell'Aversano C, Hess P, Quilliam MA (2005) J Chromatogr A 1081 190 –201

Devic E, Li D, Dauta A, Henriksen P, Codd GA, Marty JL, Fournier D (2002) Appl Environ Microbiol 68 No 8 4102–4106

Eaglesham GK, Norris RL, Shaw GR, Smith MJ, Chiswell RK, Davis BC, Neville GR, Seawright AA, Moore MR (1999) Environ Toxicol 14 151–154

Ellman GL, Courtney KD, Andres V, Featherstone RM (1961) Biochem Pharmacol 7 88–95

Environment Agency (1998) Methods for the examination of Waters and Associated Materials. The Environment Agency, Bristol UK

Fang X, Fan X, Tang Y, Chen J, Lu J (2004) J Chromatogr 1036 233–237

Fastner J, Codd GA, Metcalf JS, Woitke P, Wiedner C, Utkilen H (2002) Anal Bioanal Chem 374 437–444
Fastner J, Heinze R, Humpage AR, Mischke U, Eaglesham GK, Chorus I (2003) Toxicon 42 313–321
Furey A, Crowley J, Hamilton B, Lehane M, James KJ (2005) J Chromatogr A 1082 91– 97
González JC, Leira F, Fontal OI, Vieytes MR, Arévalo FF, Vieites JM, Bermúdez–Puente M, Muniz S, Salgado C, Yasumoto T, Botana LM (2002) Anal Chim Acta 466 233–246
Gugger M, Lenoir S, Berger C, Ledreux A, Druart JC, Humbert JF, Guette C, Bernard C (2005) Toxicon 45 919–928
Harada KI, Nagai H, Kimura Y, Suzuki M, Park HD, Watanabe MF, Luuklkainen R, Sivonen K, Carmichael WW (1993) Tetrahedron 49 9251–9260
Himberg K (1989) J Chromatogr A 481 358–362
Hoeger SJ, Shaw G, Hitzfeld BC, Dietrich DR (2004) Toxicon 43 639–649
Holmes CFB (1991) Toxicon 29 469–477
Inami GB, Crandall C, Csuti D, Oshiro M, Brendan RA (2004) J AOAC Int 87 1133–1142
James HA, James CP (1991) Report No Ro224. Foundation for Water Research, Allen House, The Listons, Liston Road, Marlow UK
James KJ, Furey A, Sherlock IR, Stack MA, Twohig M, Caudwell FB, Skulberg OM (1998) J Chromatogr A 798 147–157
James KJ, Crowley J, Hamilton B, Lehane M, Skulberg OM, Furey A (2005) Rapid Commun Mass Spectrom 19 1167–1175
Jellet JF, Roberts RL, Laycock MV, Quilliam MA, Barrett RE (2002) Toxicon 40 1407–1425
Kayo K, Sano T (1999) Anal Chim Acta 386 107–112
Kim YM, Oh SW, Jeong SY, Pyo DJ, Choi EY (2003) Environ Sci Technol 37, 1899–1904
Kondo F, Ikai Y, Oka H, Matsumoto H, Yamada S, Ishikawa N, Tsuji K, Harada KI, Shimada T, Oshikata M, Suzuki M (1995) Nat Toxins 3 41–49
Lawrence JF, Menard C (2001) J Chromatogr A 922 111–117
Lawrence JF, Niedzwiadek B, Menard C (2004) J AOAC Int 87 No 1 83–100
Lawton LA, Edwards C, Codd GA (1994) Analyst 119 1525–1530
Locke SJ, Thibault P (1994) Anal Chem 66 3436–3446
Manger RL, Leja LS, Lee SY, Hungerford JM, Kirkpatrick MA, Yasumoto T, Wekell MM (2003) J AOAC Int 86 540–543
McElhiney J, Lawton LA (2005) Toxicol Appl Pharmacol 203 219–230
McElhiney J, Lawton LA, Edwards C, Gallacher S (1998) Toxicon 36 417–420
McElhiney J, Drever M, Lawton LA, Porter AJ (2002) Appl Environ Microbiol 68 5288–5295
Meriluoto J (1997) Anal Chim Acta 352 277–298
Meriluoto J, Karlsson K, Spoof L (2004) Chromatographia 59 291–298
Metcalf JS, Codd GA (2003) Chem Res Toxicol 16 103–112
Metcalf JS, Bell SG, Codd GA (2001) Appl Environ Microbiol 67 904–909

Metcalf JS, Lindsay J, Beattie KA, Birmingham S, Saker ML, Törökné AK, Codd GA (2002) Toxicon 40 1115–1120
Mirocha CJ, Cheong W, Mirza U, Kim YB (1992) Rapid Commun Mass Spectrom 6 128–134
Ojanperä I, Vuori E, Himberg K, Waris M, Niinivaara K (1991) Analyst 116 265–267
Ortea PM, Allis O, Healy BM, Lehane M, Shuilleabháin AN, Furey A, James KJ (2004) Chemosphere 55 1395 – 1402
Oshima Y (1995) In: .M Hallegraeff, DM Anderson and AD Cambella [eds] Manual on Harmful marin Algae, IOC Manuals and Guides, 33 81–94
Pan M, Mabry TJ, Cao P, Moini M (1997) J Chromatogr A 787 288–294
Pelander A, Ojanperä I, Lahti K, Niinivaara K, Vuori E (2000) Water Res 34 2643–2652
Pietsch J, Fichtner S, Imhof L, Schmidt W, Brauch HJ (2001) Chromatographia 54 339–344
Richardson SD, Ternes TA (2005) Anal Chem 77 3807–3838
Ruiz MJ, Cameán AM, Moreno IM, Picó Y (2005) J Chromatogr A 1073 257–262
Sano T, Nohara K, Shiraishi F, Kaya K (1992) Int J Environ Anal Chem 49 163–170
Seawright AA, Nolan CC, Shaw GR, Chiswell RK, Norris RL, Moore MR, Smith MJ (1999) Environ Toxicol 14 135–142
Sleno, L, Volmer DA (2005) Anal Chem 77 1509–1517
Stevens DK, Krieger RI (1988) J Anal Toxicol 12 126–131
Takino M, Daishima S, Yamaguchi K (1999) J Chromatogr 862 191–197
Teneva I, Dzhambazov B, Koleva L, Mladenov R, Schirmer K (2005) Toxicon 45 711–725
Törökné A, Asztalos M, Bánkiné M, Bickel H, Borbély G, Carmeli S, Codd GA, Fastner J, Huang Q, Humpage A, Metcalf JS, Rábai E, Sukenik A, Surányi G, Vasas G, Weiszfeiler V (2004) Anal Biochem 332 280–284
Villatte F, Schulze H, Schmid RD, Bachmann TT (2002) Anal Bioanal Chem 372 322–326
Ward CJ, Beattie KA, Lee EYC, Codd GA (1997) FEMS Microbiol Lett 153 465—473
Welker M, Bickel H, Fastner J (2002) Water Res 36 4659–4663
Wirsing B, Flury T, Wiedner C, Neumann U, Weckesser J Environ Toxicol 14 23–29
Yuan M, Carmichael WW (2004) Toxicon 44 561–570
Yuan M, Namikoshi M, Otsuki A, Watanabe MF, Rinehart KL (1999) J Am Mass Spectrom 10 1138–1151
Zeck A, Weller MG, Niessner R (2001) Anal Chem 73 5509–5517
Zhang L, Ping X, Yang Z (2004) Talanta 62 193–200
Zotou A, Jeffries TM, Brough PA, Gallagher T (1993) Analyst 118 753–758

Appendix A

Table A.1 Toxin producing cyanobacteria and their relative human–health risk.

Cyanotoxin	Cyanobacteria	Health Risk	Occurrence
Microcystins	*Microcystis, Anabaena, Planktothrix, Nostoc, Anabaenopsis, Snowella, Woronichinia*	Acute: Liver toxin Chronic: Tumour promoter	Reported globally Very common in Freshwater blooms
Nodularins	*Nodularia* Cyanobacterial symbiont of marine sponge (?)	Acute: Liver toxin Chronic: Tumour promoter and possible carcinogen	Brackish waters Freshwater thermal spring
Saxitoxins	*Anabaena, Planktothrix, Aphanizomenon, Lyngbya, Cylindrospermopsis*	Acute: Neurotoxin	Freshwater reported in North and South America, Australia and Europe
Anatoxin–a	*Anabaena, Planktothrix, Aphanizomenon, Cylindrospermum, Raphidiopsis*	Acute: Neurotoxin	Freshwater, low incidence of reports
Anatoxin–a(s)	*Anabaena*	Acute: Neurotoxin	Freshwater, very low incidence of reports
Lipopolysaccharide	All cyanobacteria	Acute: Inflammation May reduce the bodies detoxification mechanisms	Global and present in all cyanobacterial blooms
Cylindrospermopsin	*Cylindrospermopsis, Anabaena, Umezakia, Aphanizomenon, Raphidiopsis*	Acute: Organ and tissue damage particularly liver Chronic: Possible carcinogen	Increasingly reported especially in warmer climates

Cyanotoxin	Cyanobacteria	Health Risk	Occurrence
Aplysiatoxins	*Lyngbya, Schizothrix, Planktothrix*	Acute: Inflammation affecting skin and GI tract Chronic: tumour promoter	Benthic marine cyanobacteria
Lyngbyatoxin–a	*Lyngbya*	Acute: Inflammation affects skin and GI tract	Benthic marine cyanobacteria
B–methylamino–L–alanine (BMAA)	Present in c. 95% of cyanobacteria tested although the amount of BMAA varies by several orders of magnitude.	Chronic: Neurotoxin (Parkinson–like and dementia)	Global

Table A.2. Methods for the detection of microcystins and nodularins. LoQ – limit of quantitation given as amount on the system and/or concentration; Cost/sample–quoted contract as opposed to "in–house"; Time/sample – run time excluding sample concentration/preparation.

Method	LoQ	Cost/ Sample	Time/ sample	Multi–variant	Comments
HPLC–UV (Meriluoto 1997)	1–10 ng	n/a	30 min	Limited	Very limited identification, dependent on retention time of known standards.
HPLC–PDA (Lawton et al. 1994)	1–10 ng 250 g/l	$300	30–60 min	Yes	Utilizes characteristic spectra to identify microcystins and nodularins. Validated and widely used.
LC–MS (Barco et al. 2002)	50 pg	$400	30 min	Yes	SIM allows low level detection. TIC required for the detection and identification of multiple variants.
LC–MS/MS (Zhang et al. 2004)	2.6 ng/l	$400	30 min	Yes	Fragment analysis provides 'finger print' for individual variants including those with identical masses. MRM enhanced selectivity and sensitivity.
MALDI–TOF (Welker et al. 2002)	1–100 µg/l	n/a	< 1 min	Yes	Identification and structural elucidation. Problems with matrix interference.
SELDI–TOF (Yuan and Carmichael 2004)	2.5 pg/2 µl water	n/a	300 /day	Yes	New technology with a lot of potential
Frit–FAB (Kondo et al. 1995)	20–50 ng	n/a		Yes	Low sensitivity now superseded
MMPB with LC–MS or GC–MS or LC–Fl (Kayo and Sano 1999, Sano et al. 1992)	0.02 µg/l	n/a		Yes	Detects oxidation product of microcystin, no identification of individual variants. Good for complex matrices and bound toxin.

Method	LoQ	Cost/ Sample	Time/ sample	Multi–variant	Comments
TLC (Pelander et al. 2000)	10 ng	n/a	30 min	Yes	Requires automation for reliable quantification. Limited by number of standards available.
CE–UV (Bateman et al. 1995)	3 µg m/l	n/a	12 min	Yes	Limited sample loading volume results may poor LoQ. This can be improved with inline concentration
CE–MS	4 pg 0.2 µg m/l	n/a	12 min	Yes	High speed separation with identification from mass data
Mouse	50–500 µg/kg	$500	24 hr	Yes	Non–specific, specialized facilities and license required
ELISA–adda (McElhiney and Lawton 2005)	20 pg/ml	$250	2.5 hr per plate	Yes	No pre–concentration required. Good sensitivity and cross reactivity. Commercially available.
ELISA (McElhiney and Lawton 2005)	20–3000 pg/ml	$200	3 hr per plate	Yes	Several kits commercially available. No pre–concentration required.
Dipstick (Kim et al. 2003)	95 pg/ml	n/a	12 min	some	Not tested with non–arginine containing microcystins.
Inhibitor immunoassay (Metcalf et al. 2001)	IC_{50} 45 ng	n/a	3 hr	Yes	Good combination of structural detection and bioactivity

Method	LoQ	Cost/ Sample	Time/ sample	Multi–variant	Comments
Protein phosphatase (Ward et al. 1997, Wirsing et al. 1999)	0.1–0.25 µg/L	n/a	1 hr per plate	yes	Provides total toxicity. Components commercially available. No sample concentration needed.
Molecular Imprinted polymers (Chianella et al. 2002)	0.2 µg/L	n/a	25 min	No	Different polymers need for each variant. Very stable with great potential for biosensor.

Table A.3 Methods for the detection of saxitoxins. LoQ – limit of quantitation given as amount on the system and/or concentration; Cost/sample – quoted contract as opposed to "in–house"; Time/sample – run time excluding sample concentration/preparation.

Method	LoQ	Cost/ sample	Time/ sample	Multi–variant	Comments
HPLC–Fluorescence, pre–column derivatization (Lawrence et al. 2004)	80 µg/kg	$400	40 min	Yes	Relatively reliable but two oxidation/separations needed. Complex sample processing would benefit from automation. Validated.
FAB–MS (Mirocha et al. 1992)	200 pg		10 min	Not shown yet	Potential interference issues
LC–QTOF–MS (Fang et al. 2004)	0.1 µg/g	$400	25 min	DcSTX and STX only	Confirmation and quantitation Simple sample treatment and good recoveries.
HILIC–MS/MS (Dell'Aversano et al. 2005)	50–1000 nM (single quad)	$400	30 min	Yes	Confirmation and quantitation Improved sensitivity using SIM (5–30 nM)
HPLC– Fluorescence, pre–column derivatization (Oshima 1995)	65 µg/100 g	$400	1.5 hr	Yes	Three separation methods needed. Some users focus on species dominant group of compounds.
CE–UV/MS (Lock and Thibault 1994)	0.2–1 µg/ml	$400	40 min	Yes	Good sample clean–up essential as it is sensitive to contaminants.
Mouse	0.2 µg/ml	$500	0.5 –2 hr	Yes	Specialized facilities and license required.
Sodium channel receptor (Manger et al. 2003)	2 µg/100 g	$500	6 hr	Yes	Cells commercially available. Requires cell culture facilities

Method	LoQ	Cost/ sample	Time/ sample	Multi–variant	Comments
MIST ALERT/ JRPT (Jellet et al. 2002)	40 µg/100g	$150	20 min	Yes	Rapid and multi–variant Optimised for field use and rapid response for shellfish. LoQ based on regulatory guidelines for shellfish.
ELISA	50 ppb		42 samples/1.5 hr	Yes	Limited cross reactivity. Commercially available.
Locust (McElhiney et al. 1998)	2.5 µg/ml	$30	90 min	Yes	Robust and simple. Locust is a controlled species in many countries, bush cricket is suitable alternative.

Table A.4 Methods for the detection of cylindrospermopsins. LoQ – limit of quantitation given as amount on the system and/or concentration; Cost/sample – quoted contract as opposed to "in–house"; Time/sample – run time excluding sample concentration/preparation.

Method	LoQ	Cost/ sample	Time/ sample	Multi–variant	Comments
HPLC–UV (Törökné et al. 2004)			10–65 min	Not discussed	Limited use
HPLC–PDA (Welker et al. 2002)	1–300 ng	$400	20 min	Yes	Potential matrix effects from extraction procedure
LC–MS/MS (Eaglesham et al. 1999)	1–600 μg/L	$400	10 min	Yes	Minimal processing. Characterization and quantitation obtained. Used for crawfish samples.
Mouse (Seawright et al. 1999, Fastner et al. 2003)	0.2 mg/kg (IP LD_{50})	$500	24 hr	Yes	Specialized facilities and license required. Lengthy assay time.
Brine shrimp (Metcalf et al. 2002)	8.1 μg /ml	$50	24 hr	Not reported	Insensitive and lengthy assay time. Sensitivity increases with incubation time.

Table A.5 Methods for the detection of anatoxin–a. LoQ – limit of quantitation given as amount on the system and/or concentration; Cost/sample – quoted contract as opposed to "in–house"; Time/sample – run time excluding sample concentration/preparation.

Method	LoQ	Cost/ sample	Time/ sample	Multi–variant	Comments
HPLC–UV (Zotou et al. 1993)	1 μg/ l	$300	20	Not discussed	Limited data – isocratic analysis
GC–MS (Himberg 1989)	0.1 ng	$400	10 min	Yes	Derivatization to N–acetyl anatoxin
HPLC–Fluorescence (James et al. 1998)	10 ng/l	$300	35	Yes	Derivatization necessary. Good complementary confirmation technique
LC–MS (Takino et al. 1999)	15 ng/l	$400	25 min	Yes	No derivatization needed. Qualitative and quantitative
LC–QIT–MS (James et al. 2005)	25–1000 μg/l	$400	20 min	Yes	Characterisation and quantification without loss of sensitivity.
GC–ECD (Stevens and Krieger 1988)	2 pg 3 ng/ l	$300	10 – 30 min	Yes	Derivatized with trichloroacetic anhydride. Sensitive but no structural confirmation
LC–TSP–MS (Harada et al. 1993)	1–40 ng	$400	10 min	Yes	Isocratic – not ideal for complex samples without sample specific clean–up. SIM needed. Superseded by ESI and APCI
TLC (Ojanperä et al. 1991)	10 μg/g cells	$20	1 hr multiple samples	Yes	Useful as a rapid screen. Sample concentration needed for water samples.

Chapter 24: Emerging high throughput analyses of cyanobacterial toxins and toxic cyanobacteria

Kaarina Sivonen

Department of Applied Chemistry and Microbiology, P.O.Box 56, Viikki Biocenter, FI–00014 Helsinki University, Finland, e–mail: kaarina.sivonen@helsinki.fi

Introduction

Toxic cyanobacterial mass occurrences (blooms) are commonly found in fresh, brackish, and marine waters (Sivonen and Jones 1999). Cyanobacteria growing in benthic environments have also been shown to contain toxins (Sivonen and Jones 1999; Edwards et al. 1992; Mez et al. 1997; Surakka et al. 2005). The most common cyanobacterial toxins (cyanotoxins) are hepatotoxins (microcystins and nodularins), neurotoxins [anatoxin–a, anatoxin–a(S) and saxitoxins], cytotoxins (cylindrospermopsins), and dermatotoxins (aplysiatoxin and debromoaplysiatoxins) (Sivonen and Jones 1999). Microcystins in freshwaters are most frequently produced by *Microcystis*, *Planktothrix* (formerly *Oscillatoria*) and *Anabaena* (Sivonen and Jones 1999). Microcystin production has been also proven for *Nostoc* (isolates from water and terrestrial environment; Sivonen et al. 1992; Oksanen et al. 2004), terrestrial isolates of *Hapalosiphon* (Prinsep et al. 1992) and *Phormidium* (Izaguirre and Neilan 2004). In brackish waters such as the Baltic Sea or saline lakes and estuaries in Australia and New Zealand, nodularin producing cyanobacterium, *Nodularia spumigena,* frequently occur (Rinehart et al. 1988; Sivonen et al. 1989; Heresztyn and Nicholson 1997). Neurotoxins are commonly produced by *Anabaena* and less frequently by *Aphanizomenon, Lyngbya and Oscillatoria* (Sivonen and Jones 1999). *Cylindrospermum raciborskii, Aphanizomenon, Umezakia, Anabaena* and *Radhidiopsis* produce cylindrospermopsins (Fergusson and Saint 2003), whereas *Lyngbya, Schizothrix* and *Oscillatoria* are primarily implicated as main producers of dermatotoxins (Sivonen and Jones 1999;

see also current listings of toxins and toxin producers in this volume; e.g., Humpage). Cyanobacterial toxins have caused hundreds of animal poisoning cases worldwide (Ressom et al. 1994) and are threat to human health (Kuiper–Goodman et al. 1999). It has become evident that efficient methods to detect cyanobacterial toxins as well as toxic cyanobacteria are needed. This paper will evaluate the potential of the current and emerging high throughput methods for analysis of cyanobacterial toxins and toxin producing organisms.

Detection of toxins

To protect water users from poisoning and exposure to the toxins, it is important to know the identity and quantity of the toxins. Such cases include drinking water, dietary supplements, important areas for recreation, animal–poisoning cases, etc. Several papers in the current issue deal with toxin analyses in detail (Meriluoto and Spoof 2006; Lawton 2006) thus, only the potential for developing high throughput methods are considered here.

The structures of the most wide spread cyanotoxins are known and that has made possible the development of high throughput analysis methods such as enzyme–linked immunosorbent assay (ELISA), protein phosphatase inhibition assay (PPIA), high–performance liquid chromatography (HPLC) and liquid chromatography – mass spectrometry (LC/MS). ELISA is available for cyanobacterial hepatotoxins and saxitoxins whereas, PPIA is used to detect microcystins and nodularins. Both methods have been evaluated to be sensitive and fast screening methods (Harada et al. 1999; Rapala et al. 2002; Hilborn et al. 2005). Both of these analyses can be carried out in multi–well plate formats. The ELISA analyses can be fully automated by using new robotic technologies. The PPIA analysis involves the use of enzyme which readily inactivates without precautions. Both of these methods give total toxin concentrations but do not differentiate the compounds. Matrix–assisted laser desorption/ionisation time–of–flight mass spectrometry (MALDI–TOF) has been proven as a fast screening method for the detection of cyanobacterial peptide toxins and bioactive compounds (Erhard et al. 1997; Fastner et al. 2001; Welker et al. 2004).

To identify individual toxins, separation is needed and identification methods such as HPLC combined with UV, FT or mass detection are required. There are good applications for many toxins and they can be accurately determined (Harada et al. 1999; Sivonen 2000; Meriluoto et al. 2004; Meriluoto and Codd 2005). LC/MS is an excellent method to identify the individual toxins. Improvement of the instruments has also made the method very sensitive. All chromatographic methods are currently rather fast. The analysis times are in minutes and the equipment with auto samplers can be run continuously on nights and weekends. Thus, the limiting step still is sampling and sample preparation/extraction. The availability of standards of certain toxins is still a problem. New developments in this field leading to on–line measurements and fast and easy to use kit detection systems are likely to improve the rapid detection of cyanotoxins in the future (in this monograph, see Lawton 2006, Wilhelm 2006).

Detection of cyanobacteria and genes involved in toxin production

In order to implement any rational mitigation scheme, further research is needed on the biology, ecology and proliferation of cyanobacteria producing toxins. The major toxin producers have been identified by isolating the organisms and showing their toxin production capability (Sivonen and Jones 1999 and the cases thereafter, in this monograph). This has created invaluable culture collections, which contain toxic and non–toxic planktonic cyanobacteria as well as benthic strains. Such culture collections have been valuable sources for physiological studies on toxin production, taxonomic/phylogenetic analyses of toxin producers as well as studies on the biosynthesis of toxins. We cannot identify toxin producers by microscopy since toxic and non–toxic strains of the same species are known to occur (Ohtake et al. 1989; Vezie et al. 1998). The gene clusters involved in microcystin and nodularin production are known. Microcystins (nodularins) are produced non–ribosomally by the multi–enzyme complex consisting of polyketide synthases and peptide synthetases, which was verified by Dittmann et al. (1997) by gene knockout experiments. These biosynthetic gene clusters have been fully sequenced from *Microcystis* (Nishizawa et al. 1999 and 2000; Tillet et al. 2000), *Planktothrix* (Christiansen et al. 2003), Anabaena (Rouhiainen et al. 2004) and *Nodularia* (Moffitt and Neilan 2004). This has made possible to develop molecular methods to identify these toxin producers in samples. However, the biosyntheses of neurotoxins and other cyanobacterial toxins remains to be elucidated and verified by gene knockout experiments before molecular analysis can be developed with high certainty. Toxin analyses will not reveal the producers in most cases. The only exception is nodularin, which is produced thus far, only by Nodularia spumigena. Detection of toxic cyanobacteria by conventional or real–time PCR can serve as an early warning method since PCR detects the presence of toxin producers in very low concentrations.

These and microarray (DNA chip) technologies are the only methods currently available to study toxin producing cyanobacteria in situ.

Conventional PCR

Conventional PCR is a fast and cheap method to detect potentially toxin–producing strains in samples. This method requires only PCR and gel documentation facilities, which are common nowadays in laboratories. Like the methods for toxin analysis, the most time consuming task using molecular methods is the environmental sample collection, DNA isolation and purification. Conventional PCR can be used to detect microcystin/nodularin producers since the gene clusters are now known and several primers are thus, available (Table 1). These gene clusters are large (55 kb for microcystins and 47 kb for nodularin) and offer a wealth of information and possibilities for probe and primer design. In the recent years, it has become evident that cyanobacteria often produce plenty of non–ribosomal cyclic or linear peptides other than microcystins (Rouhiainen et al. 2000; Fastner et al. 2001; Welker et al. 2004) and nodularins (Fujii et al. 1997) and this should be taken into consideration in primer design.

Microcystis aeruginosa is found worldwide, the most frequently occurring microcystin–producing cyanobacterial species and its biosynthetic genes have been known for the longest time. For this reason, most studies have focused on the detection of toxic *Microcystis* strains (Table 1) but more recently extended to include other producers such as *Planktothrix* and *Anabaena*. Simultaneous occurrence of several potential microcystin producers in a lake is not uncommon (Vaitomaa et al. 2003). There are few general primers designed to detect several different producers of microcystins (Hisbergues et al. 2003; Rantala et al. 2004) and to produce sequence information to design genus specific primers (Vaitomaa et al. 2003). In the development of genus–specific primers the testing with all possible microcystin–producing cyanobacteria is required in addition to *in silico* analyses. The other option, to detect specific microcystin producers, was presented by Hisbergues et al. (2003). PCR products generated by general primers were digested with restriction enzymes and producer genera were identified based on the differences in resulting bands. The gene region chosen in that study was *mcy*A, a region missing from the *Nodularia,* thus those primers were not suitable for detection of nodularin producers. Vaitomaa et al. (2003) and Rantala et al. (2004) successfully used *mcy*E primers to get PCR product from all known microcystin and nodularin producers. This gene region is involved in the construction of Adda and activation and condensation of glutamate. Adda and glutamic acid are present in both microcystins and nodularins and are shown to be the most important determinants (excluding the cyclic nature of the compounds) for toxicity of these compounds (Harada et al. 1990; Rinehart et al. 1994; Goldberg et al. 1995). Structural variations of Adda and glutamic acid in microcystins and nodularins are less frequent than other parts of these molecules (L– amino acids in microcystins see Sivonen and Jones 1999), which makes this gene region attractive for primer and probe design. The microcystin and nodularin synthetase sequence information has also been used to study the evolution of these genes (Rantala et al. 2004). The microcystin synthetase genes were ancient and the non–toxic strains have lost these genes during evolution. This also implies that individual strains of cyanobacteria may have retained these genes and new toxin producers among the cyanobacteria currently regarded as non–producing species may still be found.

Conventional PCR is prone to typical PCR based method problems (Wintzingerode et al. 1997) and requires some understanding of basic molecular biology. The primers are always designed and tested only with a subset of strains and thus, may not be readily applicable to new strains/species without further analysis. Environmental samples may also contain PCR inhibitors and this should be controlled especially in the case of negative results. Multiplex–PCR when developed may further improve this method and make it faster.

DNA–based detection methods are only able to identify potential toxin producers. The analysis of high number of *Microcystis* strains and also environmental samples have found only few cases where microcystin synthetase genes were detected but the organism was unable to produce microcystins possibly due to mutations in the large gene cluster (Kaebernick et al. 2001; Tillett et al. 2001; Mikalsen et al. 2003). The study of Kurmayer et al. (2004) showed that a rather high frequency of *Planktothrix* strains/filaments contained the genes but were not producing microcystins in Alpine lakes of Austria (see Table 1).

Molecular methods to detect producers of cylindrospermopsins have been developed (Schembri et al. 2001; Fergusson and Saint 2003). These detection methods are based on polyketide synthase (PKS) and peptide synthetase (PS) gene clusters present only in cylindrospermopsin producing strains, but verification of the involvement of these genes in cylindrospermopsin biosynthesis by knockout is yet to be done. Ferqusson and Saint (2003) developed multiplex PCR assay for *Cylindrospermopsis raciborskii*, *Aphanizomenon ovalisporum* and *Anabaena bergii* based on the PKS and PS genes which correlated well with the actual production of cylindrospermopsin.

Lack of information on the biosynthesis and biosynthetic genes of other toxins such as the cyanobacterial neurotoxins limits the use of molecular methods to detect the producers of these toxins.

Table 1. Conventional PCR analyses of microcystin and nodularin producing cyanobacteria.

Target organisms	Gene	Application and results	Reference
Several cyanobacteria common in cyanobacterial blooms	General primers for peptide synthetases and microcystin synthetase	Several strain of Anabaena, Aphanizomenon, Cylindrospermopsis, Lyngbya, Microcystis, Nodularia, Nostoc, Planktothrix (Oscillatoria), Plektonema, Pseudanabaena and Synechococcus were tested (isolates from Australia, Europe and Japan) – generally strains producing microcystins gave positive results with microcystin synthetase primers	Neilan et al. 1999
Microcystis	NMT region of gene mcyA	Tested with 38 microcystin producing and non–producing strains of Microcystis mostly from Australia and North America	Tillett et al. 2001
Microcystis	mcyB three primer pairs	Tested with whole cells of 30 strains of Microcystis, 4 strains of Planktothrix/(Oscillatoria) 2 Anabaena and 1 Aphanizomenon as well as 200 samples from Chinese lakes. One of the primers pairs (TOX2P/TOX2M) gave consistent results with the toxins analyses.	Pan et al. 2002
Microcystis	mcyA	Some bloom samples were mcyA positive	Baker et al. 2002
Microcystis	mcyB	Applied to water samples collected from 476 channel catfish production ponds in USA, 31% gave positive results	Nonneman & Zimba 2002
Microcystis	mcyB – A1 domain	Single colonies from L. Wannsee (Berlin Germany), 75% of M. aeruginosa, 16% of M. ichtyoblabe and 0% of M. wesenbergii gave positive results. Restriction analysis of PCR products gave 7 restriction types, which differed also by nucleotide sequence. The largest colonies contained highest proportion of the microcystin producing genotypes	Kurmayer et al. 2002, 2003

Target organisms	Gene	Application and results	Reference
Anabaena, Microcystis	*mcyE*	Tested with 13 strains of *Microcystis*, 14 strains of *Anabaena*, 8 strains of *Planktothrix* and a strains of *Nostoc* (isolates mostly from Finland). Primers were genus–specific.	Vaitomaa et al. 2003
Anabaena, Microcystis, Planktothrix (Nostoc)	*mcyA* + RLFP analysis of PCR products	Primers tested with 24 strains of *Microcystis*, 8 *Anabaena*, 11 *Planktothrix*, 2 *Nostoc* and 7 *Nodularia* strains and a lake sample. Consistent results with the microcystin analyses – no product with *Nodularia*. Identification of producer by restriction profile.	Hisbergues et al. 2003
Microcystis	*mcyA* (condensation domain), *mcyB* A1 domain	A total of 244 *Microcystis* colonies from 9 different European lakes were analyzed: 75% of *M. aeruginosa* and *M. botrys*, less than 20% of *M. ichtyoblabe* and *M. viridis* contained mcy genes. *M. wesenbergii* did not contain mcy genes. The maximum proportion of mcy–PCR positive colonies was found among the largest colony group.	Via–Ordorika et al. 2004
Anabaena, Microcystis, Planktothrix, Nostoc, Nodularia	General primers *mcyE, mcyD*	Observed amplification of all tested producers of microcystins and nodularin. Microcystin synthetase genes are ancient. Nodularin biosynthetic genes evolved from microcystin synthetase genes.	Rantala et al. 2004
Planktothrix	*mcyA*	All 49 strains of red–pigmented *P. rubescens* contained mcyA and 23 strains of green–pigmented *P. agardhii* were either with or without mcyA. One strain of *P. agardhii* and 8 strains of *P. rubescens* had *mcyA* genes but were unable to produce microcystins. The population of inactive microcystin genotypes was 5% in Irrsee and 21 % in Mondsee.	Kurmayer et al. 2004

Target organisms	Gene	Application and results	Reference
Planktothrix	*mcyA, B* and *C*	Adenylation domains of 21 strains of *P. agardhii* and *P. rubescens*, were compared to produced microcystin variants. *McyAAd1* with NMT had N–methyl–dehydroalanine and strains without NMT had dehydrobutyrine as amino acid no 7. *McyBAd1* genotype had homotyrosine and other genotype arginine in position 4.	Kurmayer et al. 2005
Microcystis	*mcyA, B, C, D, E, G*	Nine *Microcystis* strains (Spain, USA, Australia, Canada, South–Africa, Morocco) and 8 field colonies (Spain) were tested. Simultaneous detection of several genes using whole cells gave expected results	Ouahid et al. 2005
Planktothrix	*mcyT, TD, A, EG, B, E, CJ, HA*	Primers were tested with 47 non–toxic and toxic *Planktothrix* strains (European isolates), primers targeted to *mcyE* gave the most reliable results	Mbedi et al. 2005
Lake samples – tested with primers designed for *Microcystis*	*mcyB* and *mcyD*	Lake Oneida, NY, USA. Mcy genotypes were present in the water from mid–June to October. 88% of samples were positive for *mcyB* and 79% for *mcyD*	Hotto et al. 2005

Quantitative real–time PCR

The real–time PCR yields quantitative information and answers the question regarding which organism is a major producer of the toxins in a sample. The real–time PCR needs specific instrument that are more expensive than the standard PCR machines and have higher running costs. However, the benefits of this technology being quantitative are obvious compared with the conventional PCR. The current real–time PCR machines are designed for high throughput analyses since they use multi–well plates or carry cells with multiple capillaries. Currently, only few applications of this method for detection of cyanotoxin producers are found in literature (Table 2). Two major methodologies, SyberGreen and TaqMan methods are available; the latter is regarded to be more specific. The optimization of real–time PCR is often difficult and time consuming. The primers should preferably target amplification of short sequences (100 –200 bp) to be efficient. The prerequisite for this method is that the biosynthetic genes are known which restricts the current use of this method to microcystin and nodularin producers.

The quantitative real time PCR method is likely to produce valuable information on the toxin producers. Since it is quantitative, it can be associated with other environmental parameters possibly giving new insights to what factors promote the selection of toxic cyanobacteria and why especially toxic blooms occur. In case where several producers are found in environmental samples, it will reveal which organism is the major producer. It may have great importance when mitigation schemes are designed – *Microcystis* and nitrogen fixing cyanobacteria such as *Anabaena* or low light adapted *Planktothrix* or *Cylindrospermopsis* require different approaches.

Table 2. Quantitative real –time PCR analyses of toxic cyanobacteria

Target/Gene	Method	Major findings	Reference
Microcystis mcyA (122pb) f (mcyB (850pb))/	TaqMan	Primers were specific to *Microcystis* and three copies of the target genes per sample was detected within 2 hours	Foulds et al. 2002
Microcystis/mcyB Intergenic spacer region of phycocyanin (PC)	TaqMan	The proportion of mcy genotypes ranged from 1 to 38% of the all *Microcystis* genotypes (determined by PC genotypes) in Lake Wannsee (Germany).	Kurmayer and Kutzenberger 2003
Anabaena and *Microcystis/mcyE* genus specific primers	SyberGreen	Both potentially toxic *Anabaena* and *Microcystis* co–occurred in both lakes studied (Finland). *Microcystis* was dominant toxin producer in L. Tuusulanjärvi in summer 1999	Vaitomaa et al. 2003
Microcystis/mcyD	TaqMan	Lake Erie: the results indicated presence of other toxin producers in the lake in addition to *Microcystis*	Rinta–Kanto et al. 2005

DNA chips

DNA chips (microarrays) offer new insights into cyanobacterial populations in natural environmental settings. This methodology is very new and has not been widely used in environmental analysis, yet. It is an attractive method for monitoring since large amount of data can be created fast and the method can be automated (the data consist of hybridization results which are analyzed by computer). Current monitoring of cyanobacteria in lakes is based on microscopic identification and cell/colony/filament counting. This microscopic method is prone to pitfalls: the identification of organism is subjective, it requires extensive training, and in addition, microscopic counting is very time consuming and tedious. DNA chip technology can identify all cyanobacteria that are present in a sample accurately and identify the toxin producers depending on the designed probes. It should be emphasized that toxic cyanobacteria cannot be identified by microscopy, thus this new technology is superior when it comes to the identification of potentially toxic cyanobacteria in samples. At present, applications of DNA chip technology for the detection of cyanobacteria are scarce. Rudi et al. (2000) developed a microarray based on 16S rRNA genes for a few groups of cyanobacteria (*Phormidium, Microcystis*, *Planktothrix*, *Anabaena, Aphanizomenon, Nostoc +Anab., Aph.),* Eubacteria (+ chloroplasts). The array was based on membrane bound probes. The environmental DNA was isolated, amplified by PCR, labeled and hybridized with the complementary probes on the membrane. The method was used to detect cyanobacteria in eight Norwegian lakes.

Another recent example of development of microarray to detect cyanobacteria comes from the European Union project MIDI–CHIP (http://www.cip.ugl.ac.be/midichip/). The technology used in the MIDI–CHIP project was so called the universal microarray method (Gerry et al. 1999). This method combines hybridization with ligation detection reaction (LDR), which improves identification. In this method, two probes, a discrimination and a common probe, are needed. The universal array method avoids the limitation of hybridization. It is very difficult to establish standard conditions for different polymorphic DNA targets to be analyzed at the same time. The method consists of oligos (ZipCodes), which have similar thermodynamic behavior and are unrelated to probes. This makes array very flexible and new probes easy to add. The prototype microarray was based on 16S rRNA genes (Castiglioni et al. 2004). Sequences from data bases (281) and strains isolated in MIDI–CHIP project were aligned and phylogenetic groupings were based on Neighbour joining three made in ARB–program (Ludwig et al. 2004). The arrays detected 19 major cyanobacteria groups and were tested with strains and environmental samples and found to be specific. These arrays have been developed further to include a few more important phylogenetic groups as well as the detection of microcystin/nodularin producing cyanobacteria (MIDI–CHIP project, unpublished results).

Genome projects

In recent years, the whole genome sequencing projects have also been extended to cyanobacteria. The first cyanobacterial genome sequenced was *Synechocystis* PCC 6803, which is a model organism to study photosynthesis, as well as stress or high light acclimation processes (Kaneko and Tabata 1997; Marin et al. 2003; Hihara et al. 2001). This strain does not produce any of the known cyanobacterial toxins. Since then, tens of cyanobacterial genomes have been fully or partly sequenced (e.g. http://www.kazusa.or.jp/cyano/; http://genome.jgi–psf.org/; http://www.moore.org/microgenome/). Most recently, toxin producing cyanobacteria genome sequencing projects have been started: two microcystin producing cyanobacteria *Microcystis aeruginosa* (Pasteur Institute) and *Anabaena* strain 90 genome (at the authors' laboratory in collaboration with Beijing Genomics Institute, China) as well as nodularin producing *Nodularia spumigena* from the Baltic Sea (Moore Foundation). Detailed annotation of these genomes is likely to reveal genes associated with toxin production and regulation. Comparison of these genomes will yield information on the metabolic versatility of these organisms and the differences between the various producers of toxins. Knowledge of genome sequences makes it possible to design expression arrays to study the gene expression of these organisms in various conditions as was carried out with *Synechocystis* PCC 6803 or *Anabaena* PCC 7120 strains (Marin et al. 2003, Hihara et al. 2001; Katoch et al. 2004). Gene knockout experiments with planktonic cyanobacteria which will give important information about the function of the yet uncharacterized genes, have been difficult to accomplish. The whole genome sequence of an organism provides wealth of information, which can also be utilized in proteome research. The whole set of proteins of an organism can be separated by 2D gel electrophoresis and identified by mass spectrometry (Simon et al. 2002). Genome sequencing, as well as gene expression and proteome analyses are all high throughput methods. Combination of all these methodologies in case of toxin producing, bloom–forming cyanobacteria is likely to reveal key aspects of the biology of these important organisms. However, several ecologically relevant, mass–occurrence forming and toxin producing cyanobacteria such as

Planktothrix, *Cylindrospermopsis* and all of the neurotoxin producers still wait for ushering into the genome sequencing programs.

Conclusion and Summary

The common occurrence of toxic cyanobacteria causes problems for health of animals and human beings. More research and good monitoring systems are needed to protect water users. It is important to have rapid, reliable and accurate analysis i.e. high throughput methods to identify the toxins as well as toxin producers in the environment. Excellent methods, such as ELISA already exist to analyse cyanobacterial hepatotoxins and saxitoxins, and PPIA for microcystins and nodularins. The LC/MS method can be fast in identifying the toxicants in the samples. Further development of this area should resolve the problems with sampling and sample preparation, which still are the bottlenecks of rapid analyses. In addition, the availability of reliable reference materials and standards should be resolved.

Molecular detection methods are now routine in clinical and criminal laboratories and may also become important in environmental diagnostics. One prerequisite for the development of molecular analysis is that pure cultures of the producer organisms are available for identification of the biosynthetic genes responsible for toxin production and for proper testing of the diagnostic methods. Good methods are already available for the microcystin and nodularin–producing cyanobacteria such as conventional PCR, quantitative real–time PCR and microarrays/DNA chips. The DNA–chip technology offers an attractive monitoring system for toxic and non–toxic cyanobacteria. Only with these new technologies (PCR + DNA–chips) will we be able to study toxic cyanobacteria populations *in situ* and the effects of environmental factors on the occurrence and proliferation of especially toxic cyanobacteria. This is likely to yield important information for mitigation purposes. Further development of these methods should include all cyanobacterial biodiversity, including all toxin producers and primers/probes to detect producers of neurotoxins, cylindrospermopsins etc. (genes are unknown). The on–going genome projects concerning toxin producing cyanobacteria combined with future gene expression and gene knockout experiments as well as proteome research will yield a wealth of information on the biology and metabolic regulation of these organisms in near future.

Acknowledgements

Ms. Anne Rantala and Dr. Leo Rouhiainen are thanked for critically reading the manuscript. The grants from the Academy of Finland and collaborators in EU–project MIDI–CHIP and PEPCY are gratefully acknowledged.

References

Baker JA, Entsch B, Neilan BA, McKay DB (2002) Monitoring changing toxigenicity of a cyanobacterial bloom by molecular methods. Appl Environ Microbiol 68:6070–6076

Castiglioni B, Rizzi E, Frosini A, Sivonen K, Rajaniemi P, Rantala A, Mugnai MA, Ventura S, Wilmotte A, Boutte C, Grubisic S, Balthasart P, Consolandi C, Bordoni R, Mezzelani A, Battaglia C, De Bellis G (2004) Development of a universal microarray based on the ligation detection reaction and 16S rRNA gene polymorphism to target diversity of cyanobacteria. Appl Environ Microbiol 70(12):7161–7172

Christiansen G, Fastner J, Erhard M, Börner T, Dittmann E (2003) Microcystin biosynthesis in *Planktothrix*: genes, evolution, and manipulation. J Bacteriol 185:564–572

Dittmann E, Neilan B, Erhard M, Von Döhren H, Börner T (1997) Insertional mutagenesis of peptide synthetase gene that is responsible for hepatotoxin production in the cyanobacterium *Microcystis aeruginosa* PCC 7806. Mol Microbiol 26:779–787

Edwards C, Beattie KA, Scrimgeour CM, Codd GA (1992) Identification of anatoxin–a in benthic cyanobacteria (blue–green algae) and in associated dog poisonings at Loch Insh, Scotland Toxicon 30:1165–1175

Erhard M, von Döhren H, Jungblut P (1997) Rapid typing and elucidation of new secondary metabolites of intact cyanobacteria using MALDI–TOF mass spectrometry. Nature Biotechnol 15:906–909

Fastner J, Erhard M, von Döhren H (2001) Determination of oligopeptide diversity within a natural population of *Microcystis spp.* (cyanobacteria) by typing single colonies by matrix–assisted laser desorption ionization–time of flight mass spectrometry. Appl Environ Microbiol 67:5069–5076

Fergusson KM, Saint CP (2003) Multiplex PCR assay for *Cylindrospermopsis raciborskii* and cylindrospermopsin–producing cyanobacteria. Environmental Toxicology 18:120–125

Foulds IV, Granacki A, Xiao C, Krull UJ, Castle A, Horgen PA (2002) Quantitation of microcystin–producing cyanobacteria and E. coli in water by 5'–nuclease PCR. J Appl Microbiol 93:825–834

Fujii K, Sivonen K, Adachi K, Noguchi K, Sano H, Hirayama K, Suzuki M, Harada KI (1997) Comparative study of toxic and non–toxic cyanobacterial products: novel peptides from toxic *Nodularia spumigena* AV1. Tetrahedron Letters 38:5525–5528

Gerry NP, Witowski NE, Day J, Hammer RP, Barany G, Barany F (1999) Universal DNA microarray method for multiplex detection of low abundance point mutations. J Mol Biol 292: 251–262

Goldberg J, Huang HB, Kwon YG, Greengard P, Nairn AC, Kuriyan J (1995) Three–dimensional structure of the catalytic subunit of protein serine/threonine phosphatase–1. Nature 376:745–753

Harada KI, Matsuura K, Suzuki M, Watanabe MF, Oishi S, Dahlem AM, Beasley VR, Carmichael WW (1990) Isolation and characterization of the minor components associated with microcystins–LR and –RR in the cyanobacterium (blue–green algae). Toxicon 28:55–64

Harada KI, Kondo F, Lawton F, Lawton LA (1999) Laboratory analysis of cyanotoxins. In: Toxic Cyanobacteria in Water: a Guide to Public Health Significance, Monitoring and Management. I. Chorus and J. Bertram (Edn). The World Health Organization. ISBN 0–419–23930–8. E and FN Spon, London, UK, pp 369–405

Heresztyn T, Nicholson BC (1997) Nodularin concentrations in Lakes Alexandrina and Albert, South Australia, during bloom of the cyanobacterium (blue–green alga) *Nodularia spumigena* and degradation of the toxin. Environ Toxicol Water Qual 12:273–281

Hihara Y, Kamei A, Kanehisa M, Kaplan A, Ikeuchiv M (2001) DNA microarray analysis of cyanobacterial gene expression during acclimation to high light. The Plant Cell 13:793–806

Hilborn DE, Carmichael WW, Yuan M, Azevedo SMFO (2005) A simple colorimertic method to detect biological evidence of human exposure to microcystins. Toxicon 46:218–221

Hisbergues M, Christiansen G, Rouhiainen L, Sivonen K, Börner T (2003) PCR–based identification of microcystin–producing genotypes of different cyanobacterial genera. Arch Microbiol 180:402–410

Hotto A, Satchwell M, Boyer G (2005) Seasonal production and molecular characterization of microcystins in Oneida Lake, New York, USA. Environ Toxicol 20:243–248

Izaguirre G, Neilan BA (2004) Benthic *Phormidium* species that produces microcystin–LR, isolated from three reservoirs in Southern California. Sixth International Conference on Toxic Cyanobacteria, Abstract book pp 50

Kaebernick M, Rohrlack T, Christofferssen K, Neilan BA (2001) A spontaneous mutant of microcystin biosynthesis: genetic characterization and effect on Daphnia. Environ Microbiol 3:669–679

Kaneko T, Tabata S (1997) Complete genome structure of the unicellular cyanobacterium *Synechocystis sp* PCC6803. Plant and Cell Physiology 38:1171–1176

Katoh H, Asthana RK, Ohmori M (2004) Gene expression in the cyanobacterium *Anabaena sp* PCC7120 under desiccation. Microbial Ecol 47:164–174

Kuiper–Goodman T, Falconer I, Fitzgerald J (1999) Human health aspects. In: Toxic cyanobacteria in water. A guide to their Public Health Consequences,

Monitoring and Management. I. Chorus and J. Bartram (edn) The World Health Organization. ISBN 0–419–23930–8 E and FN Spoon, London, pp 113–153

Kurmayer R, Kutzenberger T (2003) Application of real–time PCR for quantification of microcystin genotypes in a population of the toxic cyanobacterium Microcystis sp Appl Environ Microbiol 69:6723–6730

Kurmayer R, Dittmann E, Fastner J, Chorus I (2002) Diversity of microcystin genes within a population of the toxic cyanobacterium *Microcystis spp* In Lake Wannsee (Berlin, Germany). Microb Ecol 43:107–118

Kurmayer R, Christiansen G, Chorus I (2003) The abundance of microcystin–producing genotypes correlates positively with colony size in *Microcystis sp* and determines its microcystin net production in Lake Wannsee. Appl Environ Microbiol 69:787–795

Kurmayer R, Christiansen G, Fastner J, Börner T (2004) Abundance of active and inactive microcystin genotypes in populations of the toxic cyanobacterium *Planktothrix spp* Environ Microbiol 6:831–841

Kurmayer R, Christiansen G, Gumpenberger M, Fastner J (2005) Genetic identification of microcystin ecotypes in toxic cyanobacteria of the genus *Planktothrix*. Microbiology 151:1525–1533

Ludwig W, Strunk O, Westram R, Richter L, Meier H, Yadhukumar, Buchner A, Lai T, Steppi S, Jobb G, Forster W, Brettske I, Gerber S, Ginhart AW, Gross O, Grumann S, Hermann S, Jost R, Konig A, Liss T, Lussmann R, May M, Nonhoff B, Reichel B, Strehlow R, Stamatakis A, Stuckmann N, Vilbig A, Lenke M, Ludwig T, Bode A, Schleifer KH (2004) ARB: a software environment for sequence data. Nucleic Acids Res 32:1363–1371

Marin K, Suzuki I, Yamaguchi K, Ribbeck K, Yamamoto H, Kanesaki Y, Hagemann M, Murata N (2003) Identification of histidine kinases that act as sensors in the perception of salt stress in *Synechocystis sp* PCC 6803. Academy of Sciences of the United States of America 100:9061–9066

Mbedi S, Welker M, Fastner J, Wiedner C (2005) Variability of the microcystin synthetase gene cluster in the genus *Planktothrix* (Oscillatoriales, Cyanobacteria). FEMS Microbiol Lett 245:299–306

Meriluoto J, Codd GA (edn) (2005) TOXIC: Cyanobacterial Monitoring and Cyanotoxin Analysis. Åbo Akademi University Press (Turku), ISBN 951–765–259–3, pp 149

Meriluoto J, Karlsson K, Spoof L (2004) High–throughput screening of ten microcystins and nodularins, cyanobacterial peptide hepatotoxins, by reversed–phase liquid chromatography–electrospray ionisation mass spectrometry. Chromatographia 59:291–298

Mez K, Beattie KA, Codd GA, Hanselmann K, Hauser B, Naegeli H, Preisig HR (1997) Identification of a microcystin in benthic cyanobacteria linked to cattle deaths on alpine pastures in Switzerland. Eur J Phycol 32:111–117

Mikalsen B, Boison G, Skulberg OM, Fastner J, Davies W, Gabrielsen TM, Rudi K, Jakobsen KS (2003) Natural variation in the microcystin synthetase operon *mcyABC* and impact on microcystin production in *Microcystis* strains. J Bact 185:2774–2785

Moffitt MC, Neilan BA (2004) Characterization of the nodularin synthetase gene cluster and proposed theory of the evolution of cyanobacterial hepatotoxins. Appl Environ Microbiol 70:6353–6362

Neilan BA, Dittmann E, Rouhiainen L, Bass RA, Schaub V, Sivonen K, Börner T (1999) Nonribosomal peptide synthesis and toxigenicity of cyanobacteria. J Bact 181:4089–4097

Nishizawa T, Asayama M, Fujii K, Harada KI, Shirai M (1999) Genetic analysis of the peptide synthetase genes for a cyclic heptapeptide microcystin in *Microcystis spp* J Biochem 126:520–529

Nishizawa T, Ueda A, Asayama M, Fujii K, Harada KI, Ochi K, Shirai M (2000) Polyketide synthase gene coupled to the peptide synthetase module involved in the biosynthesis of the cyclic heptapeptide microcystin. J Biochem 127:779–789

Nonneman D, Zimba PV (2002) A PCR–based test to assess the potential for microcystin occurrence in channel catfish production ponds. J Phycol 38:230–233

Ohtake A, Shirai M, Aida T, Mori N, Harada KI, Matsuura K, Suzuki M, Nakano M (1989) Toxicity of *Microcystis* species isolated from natural blooms and purification of the toxin. Appl Environ Microbiol 55:3202–3207

Oksanen I, Jokela J, Fewer D, Wahlsten M, Rikkinen J, Sivonen K (2004) Discovery of rare and highly toxic microcystins from lichen associated cyanobacterium *Nostoc sp* strain IO–102–I. Appl Environ Microbiol 70:5756–5763

Ouahid Y, Pérez–Silva G, del Campo FF (2005) Identification of potentially toxic environmental *Microcystis* by individual and multiple PCR amplifications of specific microcystin synthetase gene regions. Environ Toxicol 20:235–242

Pan H, Song L, Liu Y, Börner T (2002) Detection of hepatotoxic *Microcystis* strain by PCR with intact cells from both culture and environmental samples. Arch Microbiol 178:421–427

Prinsep MR, Caplan FR, Moore RE, Patterson GML, Honkanen RE, Boynton AL (1992) Microcystin–LA from a blue–green alga belonging to the Stigonematales. Phytochemistry 31:1247–1248

Rantala A, Fewer DP, Hisbergues M, Rouhiainen L, Vaitomaa J, Börner T, Sivonen K (2004) Phylogenetic evidence for the early evolution of microcystin synthesis. Proceedings of the National Academy of Sciences of the United States of America 101(2):568–573

Rapala J, Erkomaa K, Kukkonen J, Sivonen K, Lahti K (2002) Detection of microcystins with protein phosphatase inhibition assay, high–performance liquid chromatography–UV detection and enzyme–linked immunosorbent assay. Comparison of methods. Analytica Chimica Acta 466:213–231

Ressom R, Soong FS, Fitzgerald J, Turczynowicz L, El Saadi O, Roder D, Maynard T, Falconer I (1994) Health effects of toxic cyanobacteria (blue–green algae). National Health and Medical Research Council, Australian Government Publishing Service, Canberra, Australia, pp 108

Rinehart KL, Harada KI, Namikoshi M, Chen C, Harvis CA, Munro MHG, Blunt JW, Mulligan PE, Beasley VR, Dahlem AM, Carmichael WW (1988) Nodu-

larin, microcystin, and the configuration of Adda. J Am Chem Soc 110:8557–8558
Rinehart KL, Namikoshi M, Choi BW (1994) Structure and biosynthesis of toxins from blue–green alga (cyanobacteria). J Appl Phycol 6:159–176
Rinta–Kanto JM, Ouellette AJA, Boyer GL, Twiss MR, Bridgeman TB, Wilhelm SW (2005) Quantification of toxic *Microcystis spp* during the 2003 and 2004 blooms in Western Lake Erie using quantitative real–time PCR. Environ Sci Technol 39:4198–4205
Rouhiainen L, Paulin L, Suomalainen S, Hyytiäinen H, Buikema W, Haselkorn R, Sivonen K (2000) Genes encoding synthetases of cyclic depsipeptides, anabaenopeptilides, in *Anabaena* strain 90. Molecular Microbiol 37(1):156–167
Rouhiainen L, Vakkilainen T, Lumbye B, Siemer, Buikema W, Haselkorn R, Sivonen K (2004) Genes coding for hepatotoxic heptapeptides (microcystins) in the cyanobacterium *Anabaena* strain 90. Appl Environ Microbiol 70(2):686–692
Rudi K, Skulberg OM, Kulberg R, Jakobsen KS (2000) Application of sequence–specific labelled 16S rRNA gene oligonucleotide probes for genetic profiling of cyanobacterial abundance and diversity by array hybridisation. Appl Environ Microbiol 66:4004–4011
Schembri MA, Neilan BA, Saint CP (2001) Identification of genes implicated in toxin production in the cyanobacterium *Cylindrospermopsis raciborskii*. Environ Toxicol 16:413–421
Simon WJ, Hall JJ, Suzuki I, Murata N, Slabas AR (2002) Proteomic study of the soluble proteins from the unicellular cyanobacterium *Synechocystis sp*. PCC6803. using automated matrix–assisted laser desorption/ionisation–time of flight peptide mass fingerprinting. Proteomics 2:1735–1742
Sivonen K (2000) Chapter 26, Freshwater cyanobacterial neurotoxins: ecobiology, chemistry and detection. In: Seafood and Freshwater Toxins. L. M. Botana (Edn) Marcel Dekker, Inc, New York, USA, pp 567–582
Sivonen K, Jones G (1999) Cyanobacterial toxins. In: Toxic Cyanobacteria in Water: a Guide to Public Health Significance, Monitoring and Management. I Chorus and J Bertram (Edn) The World Health Organization. ISBN 0–419–23930–8. E and FN Spon, London, UK, pp 41–111
Sivonen K, Kononen K, Carmichael WW, Dahlem AM, Rinehart K, Kiviranta J, Niemelä SI (1989) Occurrence of the hepatotoxic cyanobacterium *Nodularia spumigena* in the Baltic Sea and the structure of the toxin. Appl Environ Microbiol 55(8):1990–1995
Sivonen K, Namikoshi M, Evans WR, Färdig M, Carmichael WW, Rinehart KL (1992) Three new microcystins, cyclic heptapeptide hepatotoxins, from *Nostoc sp* strain 152. Chem Res Toxicol 5(4):464–469
Surakka A, Sihvonen LM, Lehtimäki JM, Wahlsten M, Vuorela P, Sivonen K (2005) Benthic cyanobacteria from the Baltic Sea contain cytotoxic *Anabaena, Nodularia* and *Nostoc* strains and an apoptosis inducing *Phormidium* strain. Environ Toxicol 20:285–292
Tillett D, Dittmann E, Erhard M, von Döhren H, Börner T, Neilan BA (2000) Structural organization of microcystin biosynthesis in *Microcystis aeruginosa*

PCC 7806: an integrated peptide–polyketide synthase system. Chem Biol 7:753–764

Tillett D, Parker DL, Neilan BA (2001) Detection of toxigenity by a probe for the microcystin synthetase A gene (*mcyA*) of the cyanobacterial genus Microcystis: comparison of toxicities with 16S rRNA and phycocyanin operon (phycocyanin intergenic spacer) phylogenies. Appl Environ Microbiol 67:2810–2818

Vaitomaa J, Rantala A, Halinen K, Rouhiainen L, Tallberg P, Mokelke L, Sivonen K (2003) Quantitative real–time–PCR for determination of microcystin synthetase E copy numbers for *Microcystis* and *Anabaena* in lakes. Appl Environ Microbiol 69(12):7289–7297

Vezie C, Brient L, Sivonen K, Betru G, Lefeuvre GC, Salkinoja–Salonen M (1998) Variation of microcystin content of cyanobacterial blooms and isolated strains in Lake Grand–Lieu (France). Microbial Ecology 35:126–135

Via–Ordorika L, Fastner J, Kurmayer R, Hisbergues M, Dittmann E, Komarek J, Erhard M, Chorus I (2004) Distribution of microcystin–producing and non–microcystin–producing *Microcystis sp* in European freshwater bodies: detection of microcystins and microcystin genes in individual colonies. System Appl Microbiol 27:592–602

Welker M, Christiansen G, von Döhren H (2004) Diversity of coexisting *Planktothrix* (cyanobacteria) chemotypes deduced by mass spectral analysis of microystins and other oligopeptides. Arch Microbiol 182(4):288–98

Wintzingerode VF, Göbel UB, Stackebrandt E (1997) Determination of microbial diversity in environmental samples: pitfalls of PCR–based rRNA analysis. FEMS Microbiol Rev 21:213–229

Chapter 25: Analytical Methods Workgroup Poster Abstracts

Early warning of actual and potential cyanotoxin production

Metcalf JS, Morrison LF, Reilly M, Young FM, Codd GA

School of Life Sciences, University of Dundee, Dundee DD1 4HN UK

Introduction

The cyanobacteria which develop into mass populations in aquatic environments commonly include species and strains which produce potent toxins, alongside phylogenetically– or phenotypically similar strains which do not. The diverse range of low molecular weight cyanotoxins which can be produced present health hazards ranging from severe to mild in potable and recreational water resources.

Hypothesis

Risk management of these problems is aided if it is known whether the organisms present have the potential to produce cyanotoxins and whether they actually do so.

Methods

Understanding of the production and abundance of the toxins themselves is being advanced through the use of physicochemical and antibody methods. PCR is finding increasing application for the detection and quantification of cyanotoxin genes, whilst fluorescent *in situ* hybridisation (FISH) is amenable for the localisation of cyanotoxin genes in mixed phytoplankton populations.

Results

Using such methods as an early warning system, we are quantifying microcystins in single filaments and single colonies of cyanobacteria using antibody–based procedures (CQ–ELISA) and measuring the genetic potential for microcystin and cylindrospermopsin production in single filaments and colonies through the use of PCR and FISH. Detection of DNA sequences for cyanotoxin peptide synthetases and polyketide synthases is thereby feasible in *Microcystis*, *Planktothrix*, *Anabaena*, *Nostoc*, *Nodularia*, *Aphanizomenon* and *Cylindrospermopsis*. Some examples are given for the United Kingdom.

Discussion

For these methods to be integrated into effective early warning systems, it is necessary that good systems are also developed for the collection and delivery of samples for analysis and for the rapid reporting of results. For early warning of potential or actual cyanotoxin production to be useful in cyanotoxin risk management, it is also necessary that data interpretation is available and that contingency measures by water utility managers and operatives are already in place.

Conclusion

Early warning of actual or potential cyanotoxin production is possible using a range of methods. Rapid sample processing times in a prepared laboratory can enable results to be obtained within as little as 1–3 hours, and typically on the day of sample receipt.

Detecting toxic cyanobacterial strains in the Great Lakes, USA

Dyble J,[1] Tester PA,[1] Litaker RW,[1] Fahnenstiel GL,[2] Millie DF[3]

[1]NOAA, Center for Coastal Fisheries and Habitat Research, 101 Pivers Island Rd, Beaufort, NC 28516. [2]NOAA, GLERL, Lake Michigan Field Station, 1431 Beach St., Muskegon, MI 49441. [3]Florida Institute of Oceanography, University of South Florida, 100 Eighth Ave. SE, St. Petersburg, FL 33701

Introduction

Some regions in the Great Lakes have been experiencing a resurgence of the cyanobacterial harmful algal bloom (HAB) genera *Microcystis*. Blooms of *Microcystis spp.* that produce the toxin microcystin have detrimental impacts on multiple levels, from disruption of zooplankton grazing to illness and mortality in animals and humans. Thus, it is of great concern that microcystin concentrations above the World Health Organization's recommended limit for drinking water (1 μg/L) have been measured in parts of Lake Huron and western Lake Erie, with particularly high concentrations in wind–accumulated scums. However, not all *Microcystis* strains produce toxins and traditional microscopic analyses are insufficient for discerning whether a bloom is composed of toxic strains. Instead, genetic analyses based on the *mcyB* gene, which is involved in cellular microcystin production, were used to differentiate toxic vs. non–toxic strains and specifically detect the presence of toxic strains of *Microcystis* in environmental samples.

Hypothesis

The use of *mcyB* gene can differentiate toxic and non–toxic *Microcystis* spp. that can potentially produce microcystin toxins.

Methods

Genetic analyses based on the *mcyB* gene, which is involved in cellular microcystin production, were used to differentiate toxic vs. non–toxic strains and specifically detect the presence of toxic strains of *Microcystis* in environmental samples.

Results

DNA sequence analysis of the *mcyB* gene revealed a genetically variable population of *Microcystis* in Saginaw Bay (Lake Huron) and western Lake Erie, with areas containing a greater proportion of toxic *Microcystis* strains also having higher microcystin concentrations, suggesting that changes in bloom toxicity may be the result of shifts in community composition. Another cyanobacterial HAB species, *Cylindrospermopsis raciborskii,* has also recently been detected in the Great Lakes and studies of its distribution and toxicity in this system are on–going.

Conclusion

The application of these methods to monitoring and modeling efforts will be important to protect human and ecosystem health in the Great Lakes region.

A progressive comparison of cyanobacterial populations with raw and finished water microcystin levels in Falls Lake Reservoir

Ehrlich LC,[1] Gholizadeh A,[2] Wolfinger ED,[3] McMillan L[4]

[1]Spirogyra Diversified Environmental Services, 2232 Holland Avenue, Burlington, North Carolina, USA, Tel: 336–570–2520, Email: spiro-gyra@juno.com [2]Laboratory Corporation of America, Burlington, North Carolina, USA. [3]Meredith College, SMB, 3800 Hillsborough Street, Raleigh, North Carolina, USA. [4]City of Raleigh Dept. of Public Utilities, P.O. Box 590, Raleigh, North Carolina, 27602 USA

Introduction

Cyanobacteria and algal toxins have been placed on USEPA's drinking water Contaminant Candidate List (CCL2). Microcystin (MCYST), being the most frequently detected cyanobacterial toxin in water, is of high importance for study in potable water supply reservoirs. Processed water samples from numerous surface water supplies in the USA have been found positive for MCYST. MCYST has been detected in raw and finished water from Falls Lake, Raleigh, NC. Conventional drinking water treatment processes are only partially effective in removing cyanotoxins. To assess the risk of cyanotoxins in surface water supplies, USEPA is evaluating the use of cyanobacterial genera identification and enumeration. This study evaluated a known testing protocol and investigated the occurrence of MCYST in raw and finished water from Falls Lake Reservoir, as well as the relationship of toxin concentration to cyanobacterial populations.

Hypotheses

1. Applying standard sample extraction and concentration procedures increases the sensitivity of the Competitive–Binding ELISA assay.
2. Source water MCYST concentrations are directly related to cyanobacterial cell densities.

3. Water treatment processes used to treat Falls Lake raw water (pre–oxidation > coagulation/flocculation/PAC > sedimentation > filtration > chloramination) effectively remove MCYST from raw water.

Methods

From May 28–July 8, 2003, we collected raw and finished water samples from Falls Lake and Johnson WTP to assess the levels of microcystins. We applied standard sample preparation procedures, including: freeze–thaw/sonication; concentration via lyophilization; methanol–water extraction; and solid phase extraction. We then analyzed the concentrated samples using a commercially–available Competitive–Binding ELISA assay kit. Other aliquots of the water samples were preserved and analyzed by standard taxonomic and direct cell counting techniques.

Results

1. MCYSTs were detected above the assay limit of quantitation (LOQ)(0.160 ppb) in raw water concentrates.
2. MCYSTs in finished water concentrates were significantly lower, at or below the LOQ.
3. MCYSTs and the grouped densities of *Anabaena* and *Aphanizomenon* were weakly correlated ($R^2 = 0.11$).

Conclusions

1. For raw water samples, the sensitivity of the ELISA can be increased by using either an alternate low level protocol (LOQ 0.05 ppb) or pre–concentration techniques.
2. The low level protocol should be employed for finished water samples.
3. Conventional treatment processes removed 60–100% of MCYST.
4. *Anabaena* and *Aphanizomenon* densities may be useful predictors of MCYST levels.

Liquid chromatography using ion–trap mass spectrometry with wideband activation for the determination of microcystins in water

Allis O, Lehane M, Muniz–Ortea P, O'Brien I, Furey A, James KJ

PROTEOBIO, Mass Spectrometry Centre for Proteomics and Biotoxin Research, Cork Institute of Technology, Department of Chemistry, Cork, Ireland

Introduction

Microcystins are a chemically diverse group of heptapeptide toxins that are produced by cyanobacteria (blue–green algae) and over 65 have been characterised, to–date. Microcystins are specific inhibitors of protein phosphatases, are potent tumour promoters, and strict requlatory control is required to comply with the World Health Organisation guideline limit of 1 µg/L. Electrospray ion–trap mass spectrometry (MS) was applied to the determination of microcystins in cyanobacteria and water samples.

Hypothesis

Electrospray ion–trap mass spectrometry (MS) can detect and differentiate various congeners and analogues of microcystins in cyanobacteria and water samples.

Methods

Sample preparation involved C–18 solid phase extraction but large variations in extraction efficiencies were observed for individual microcystins, MC–LR, MC–YR, MC–RR and MC–LA. Both C–18 and amide columns were used for the separation of microcystins using liquid chromatography (LC) and both collision induced dissociation (CID) and MS/MS studies can been carried out simultaneously, using electrospray interfacing.

Results

Microcystins have a unique C–20 β–amino acid side chain (adda) and the cleavage of part of this moiety gives rise to a characteristic fragment ion at *m/z* 135 that allows the MS detection of unknown microcystins using source CID. MS studies revealed that the loss of a water molecule is typical of microcystins but to obtain abundant characteristic fragment ions from microcystins, WideBand activation together with a high collision energy was used. In this mode, both the parent ion, $[M+H]^+$ and the $[M+H–H_2O]^+$ ions were trapped and fragmentation of the latter produced spectral data that were characteristic of individual microcystins.

Conclusion

Liquid chromatography using ion–trap mass spectrometry with wideband activation can characterize individual microcystins in water.

Anatoxin–a elicits an increase in peroxidase and glutathione S–transferase activity in aquatic plants

Mitrovic SM,[1] Stephan Pflugmacher S[2], James KJ[1], Furey A[1]

[1]PROTEOBIO, Department of Chemistry, Mass Spectrometry Centre for Proteomics and Biotoxin Research, Cork Institute of Technology, Bishopstown, Cork, Ireland. [2]Leibniz Institute of Freshwater Ecology and Inland Fisheries, FG Biogeochemical Regulation, Müggelseedamm 301, 12587 Berlin, Germany

Introduction

Although the toxic effects of cyanotoxins on animals have been examined extensively, little research has focused on their effects on macrophytes and macroalgae. To date, only microcystins have been found to be detrimental to aquatic plants.

Hypothesis

Anatoxin–a elicits an increase in peroxidase and glutathione S–transferase activity in aquatic plants.

Methods

The peroxidase activity of the free floating aquatic plant *Lemna minor* and the filamentous macroalga *Chladophora fracta* was measured after exposure to several concentrations of the cyanotoxin, anatoxin–a. The effects of various concentrations of anatoxin–a on the detoxication enzyme, glutathione S–transferase (GST) in *L. minor* were also investigated.

Results

Peroxidase activity (POD) was significantly ($P < 0.05$) increased after 4 days of exposure to an anatoxin–a concentration of 25 μgmL^{-1} for both *L. minor* and *C. fracta*. Peroxidase activity was not significantly increased at test concentrations of 15 μgmL^{-1} or lower. In another experiment, the effects of various concentrations of anatoxin–a on the detoxication enzyme, glutathione S–transferase (GST) in *L. minor* were investigated. GST activity was significantly elevated at anatoxin–a concentrations of 5 and 20 μgmL^{-1}. Photosynthetic oxygen production by *L. minor* was also found to be reduced at these concentrations.

Conclusion

This is the first report to our knowledge of the cyanotoxin anatoxin–a being harmful to aquatic plants.

The mis–identification of anatoxin–a using mass spectrometry in the forensic investigation of acute neurotoxic poisoning

James KJ, Crowley J, Hamilton B, Lehane M, Furey A

PROTEOBIO, Mass Spectrometry Centre for Proteomics and Biotoxin Research, Cork Institute of Technology, Department of Chemistry, Cork, Ireland.

Introduction

Anatoxin–a (AN) is a potent neurotoxin, produced by a number of cyanobacterial species. Forensic investigations of suspected AN poisonings are frequently hampered by difficulties in detecting this toxin in biological matrices due to its rapid decay. Further impediments are the lack of availability of AN analogues and their degradation products. Possible confusion can also occur in identifying AN as the causative agent in both human and animal fatalities due to the presence of the amino acid Phenylalanine (Phe).

Hypothesis

Nano–electrospray hybrid quadrupole time–of–flight (nano ESI QqTOF) MS can accurately differentiate AN and Phe.

Methods

In July 2002 a suspected human intoxication in the USA, that relied on liquid chromatography–single quadrupole MS (LC–MS), confused Phe and AN, since both have similar masses. We previously developed a quadrupole ion–trap (QIT) MS for the determination of AN in cyanobacteria and drinking water. Liquid chromatography–multiple tandem mass spectrometry (LC–MS^n) was employed to study the fragmentation pathway of Phe, in positive mode, to identify characteristic product ions and fragmentation processes.

Results

Reversed–phase LC, using a C_{18} Luna column gave similar retention times and on certain C_{18} columns can co–elute. The molecular related species $[M+H]^+$ *m/z* 166 was used as the precursor ion for LC–MS^n experiments. MS^2–MS^4 spectra displayed major characteristic product ions for Phe. Fragmentation of other adduct ions $[M+Na]^+$ and $[M+NH_3]$ were examined in order to identify distinctive product ions. A comparison of the QIT MS^n data for AN and Phe can prevent misidentification. Nano–electrospray hybrid quadrupole time–of–flight (nano ESI QqTOF) MS was then used to confirm formulae assignments of the product ions using high mass accuracy data and to identify ions in the lower mass range.

Conclusion

Nano–electrospray hybrid quadrupole time–of–flight (nano ESI QqTOF) MS can confirm the formulae assignments of the product ions using high mass accuracy data and to identify ions in the lower mass range.

Cyanobacterial toxins and the AOAC marine and freshwater toxins task force

Hungerford JM

Seafood Products Research Center, USFDA, 22201 23rd Dr SE, Bothell, WA, USA

Introduction

Cyanobacterial toxins have a significant economic and human health impact. Although there is a strong and global need for improved testing methods for these toxins, the demand for new, officially validated methods has not been met. Similarly, marine toxins require extensive monitoring programs and yet officially validated methodology is scarce or often based on outdated mouse bioassays. The AOAC Task Force on Marine and Freshwater toxins addresses this need by focusing efforts, setting priorities, and identifying economic and intellectual resources. The Task Force is an international group of experts on marine and freshwater toxins, and stakeholders who have a strong and practical interest in the development and validation of methods for detection of these toxins. The group establishes methods priorities, determines fitness for purpose, identifies and reviews available methodology, recommends methodology for validation, and identifies complementary analytical tools. Once appropriate analytical methodology has been identified or developed, the Task Force identifies financial and technical resources necessary to validate the methodology.

Since its first meeting in May 2004, the group has grown to include members from Europe, Asia, Africa, North and South America, and also members from Australia and New Zealand. As of September 2005, the group now totals 150 scientists, officials, and others. The toxins Task Force also has several members from the US state health agencies and federal agencies such as the Department of Health and Human Services (Food and Drug Administration, CFSAN Office of Seafoods, CFSAN/OC Shellfish Program, CVM), Department of Defense, Department of Commerce (National Oceanic and Atmospheric Administration), and Environmental Protection Agency.

New official method of analysis

The first product of the new AOAC Task Force is a new Official Method of Analysis for paralytic shellfish poisoning toxins (OMA 2005.6) which for the first time in over 45 years of shellfish monitoring, (saxitoxins) has been approved allowing an alternative to animal testing that will have a worldwide impact. The new method, developed by Health Canada and based on precolumn– oxidation HPLC and fluorescence detection, has also found some application to saxitoxins–producing cyanobacteria.

Cyanobacterial toxins

Although the first year emphasized the marine toxins, efforts addressing cyanobacterial toxins are rapidly increasing with the appointment of many experts and stakeholders in this field. Many oral and poster presentations addressing the cyanobacteria and associated toxins were presented, along with marine toxin presentations, at the group's first major conference, "Marine and Freshwater Toxins Analysis: 1st Joint Symposium and AOAC Task Force Meeting" in Baiona, Spain. At this unique meeting, participants discussed test kits for microcystins and validation needs for the cyanobacterial toxins, in general. The high level of interest recently led to the formation of a new cyanobacterial toxin subgroup.

New subgroup to address cyanobacterial toxins

The new cyanobacterial subgroup, chaired by Task Force member Hans van Egmond (co–chairs to be appointed) will meet for the first time at AOAC's Annual meeting in Orlando, Florida on Sept. 2005, as will several marine toxin subgroups. Also, included in the extensive toxins program for Orlando are two symposia addressing marine and freshwater toxins including presentations on topics in cyanobacterial toxin detection. The Task Force intends to interact extensively with the US EPA and Dr. Armah de la Cruz will also present in Orlando a brief overview of methods discussions from ISOC–HAB. Also planned are Task Force relevant presentations on cyanobacterial toxins and marine toxins at symposia to be held at Pacifichem 2005 in Honolulu, Hawaii, Dec. 2005.

Detection of Toxic Cyanobacteria Using the PDS® Biosensor

Allain B, Xiao C, Martineau A, Mandeville R

Biophage Pharma Inc., 6100 Royalmount, Montreal, QC, H4P 2R2, Canada.

Introduction

Cyanobacteria can produce molecules hazardous to human health (i.e. hepatotoxins and neurotoxins). The ubiquity of cyanobacteria in terrestrial, as well as freshwater and marine environments, suggest a potential for widespread human exposure. We have developed a Pathogen Detection System (PDS®) biosensor to monitor the presence of toxic cyanobacteria in freshwater. This biosensor allows the detection of live bacteria in water and biological fluids, as well as, the detection of cytotoxic compounds on mammalian cells.

Hypotheses

Biophage Pharma Inc., in collaboration with the NRC/BRI, has developed a patented PDS® biosensor based on impedance (for more details visit www.biophagepharma.com). The PDS® biosensor allows the detection and quantification of living pathogens in water and biological fluids. It can also assess cytotoxicity on mammalian cells (normal and cancer cells). As toxic cyanobacteria emerge as potentially hazardous microorganisms for human, animal and marine health, the PDS biosensor can be used to monitor the presence of cyanobacterial cells in drinking waters. In addition, cyanobacterial toxins should be monitored in cyanobacteria–positive samples as well as in invertebrates, fish or grazing animals used for human consumption.

Methods

Samples are mixed directly with broth media, added in the PDS® wells and then monitored for up to 24 hrs.

Results

The PDS® biosensor can detect and quantify with great precision a small number of bacteria (about 5 bacteria/ml) without any pre–amplification step. With the addition of a small pre–amplification step, this limit could be lowered without modifying the total time from sample collection to detection. At very low concentrations, the total time for detection varies between 2 h for fast growing bacteria, to 24 h in very slow growing bacteria, which is at least two times faster than conventional culture techniques. In addition, detection is monitored in real time on a computer screen allowing for immediate action as soon as detection occurs.

Conclusions

The PDS® biosensor allows the detection of a large number of organisms two times faster than conventional culture techniques. The samples do not require preprocessing and can detect very low number of bacterial cells (5 cells/ml). In addition, detection can be monitored in real time on a computer screen allowing for immediate action as soon as detection occurs. The PDS® biosensor can also detect cytotoxicity on mammalian cells.

Development of microarrays for rapid detection of toxigenic cyanobacteria taxa in water supply reservoirs

Rublee PA,[1] Henrich VC,[1] Marshall MM,[1] Burkholder JM[2]

[1]University of North Carolina at Greensboro, Greensboro, NC 27402.
[2]Center for Applied Aquatic Ecology, NC State University, Raleigh, NC 27606

Introduction

Reservoirs are the principal supplies of drinking water for many urban areas and also serve as major recreational areas. In North Carolina, reservoirs are generally characterized by high nutrient levels, high turbidity, high total organic carbon, and low alkalinity. They also commonly develop cyanobacterial blooms which can comprise more than 90% of the phytoplankton cell number at times. Dominant cyanobacteria taxa include *Cylindrospermopsis* spp. (including *Cylindrospermopsis raciborskii*), *Anabaenopsis, Planktolyngbya limnetica, Aphanocapsa* sp., *Aphanizomenon gracile, Oscillatoria, Anabaena, Microcystis,* and *Aphanizomenon*, and microcystins have been found in raw water from these reservoirs.

Materials and Methods

We are developing microarray detection methods for assessment of cyanobacterial taxa in these freshwater systems. In previous work, our prototype arrays were able to detect the DNA of heterotrophic bacteria, eukaryotic protists and cyanobacteria in samples of genomic DNA extracted from three different lakewater samples with a high degree of sensitivity and specificity. To expand the prototype array for cyanobacteria our first step has been to identify cyanobacterial taxa of interest through the literature and field data. Second, cyanobacterial gene libraries were constructed from four representative NC reservoirs using PCR with primers that target cyanobacterial small subunit ribosomal DNA (SSU rDNA) genes. Third, taxon–specific (generally species or clone–specific) PCR primers and 50–mer oligonucleotide probes are designed to complement unique sequences in the variable regions of the SSU rDNA. Fourth, the oligonucleotide probes are printed onto glass slides to form the microarray. Fifth, using both the microarray and real–time PCR, probes are tested for specificity and sensitivity. Finally, the microarray will be used to assess the presence and relative abundance of targeted cyanobacterial taxa over a three year period in NC reservoirs. Microarray results will be compared to direct microscopic count assessments. An additional aspect is that we are also developing oligonucleotide probes for the array that target known toxin biosynthesis genes (e.g. microsytin and cylindrospermopsin synthesis genes).

Conclusion

At present, we are testing our first group of primers and oligonucleotide probes targeting 25 known cyanobacterial taxa (at the species or sub–species level) and 17 novel cyanobacterial clones derived from the NC reservoirs prior to spotting on arrays and field testing. We anticipate that microarrays will be a powerful tool with the potential to assess the abundance of cyanobacterial in near real–time. Such data will aid in management decisions to prevent or mitigate the effects of cyanobacterial blooms.

ARS research on harmful algal blooms in SE USA aquaculture impoundments

Zimba PV

United States Department of Agriculture, Agricultural Research Service, Catfish Genetics Research Unit, P.O. Box 38, Stoneville, MS 38776

Introduction

In the United States, catfish aquaculture accounts for around 70% of the total freshwater revenue (currently around $1 billion annually). During the 15–18 month stocker–sized fry to fillet turnover, effusive algal growth results from the high stocking and feeding rates. Approximately 1.5 mg/L N in the form of unassimilated feed and fish excreta is added to ponds on a daily basis. Pond management must contend with maintaining algal blooms to process this nitrogen while preventing bloom collapse and hypoxic to anoxic dissolved oxygen conditions. In 1999, an Agricultural Research Service (ARS) research unit was formed to assess optimal pond management scenarios for fish production. This program can be divided into two major components: 1) prediction of harmful algal bloom events with the goal of identifying forcing variables leading to bloom events and 2) development of rapid assessment technologies for pond management.

Methods

Microcystin, anatoxin–a, euglenophycin, and prymnesin toxins have been previously reported from this research unit. Fish mortalities have been documented from these blooms in five southeastern states (AR, LA, MS, NC, SC, and TX). In 2000, a synoptic survey of 3% of the total production ponds was conducted in the southeastern 4–state catfish production area after documentation of microcystin–fish mortalities. Water samples were collected from 485 production ponds during a 10–day period, with analyses of pigments, off–flavor, and microcystin toxins (using HPLC/MS). A PCR method using myc b gene was developed to identify algae capable of microcystin synthesis.

Detection of algal blooms is difficult in freshwater, principally due to the alteration of reflectance from algae by suspended solids and color. At present, no modeling software is designed for use in Case II waters. Zimba and Gitelson (in review) have proposed tuning model properties to water column conditions to better estimate standing stock biomass (chl a) and applied this model to aquaculture ponds. The conceptual three–band model [R–1(11)–R–1(12)]×R(13) and its special case, the band–ratio model R(13)/R(11), were spectrally tuned in accord with optical properties of the media and optimal spectral bands (11,12, and 13) for accurate chl *a* estimation were determined.

Results

Myxoxanthophyll, a cyanobacterial carotenoid biomarker, was strongly correlated with microcystin content in the synoptic survey (R= 0.92). Microcystin was detected in over 50% of all ponds sampled, with WHO limits exceeded in <1% of surveyed ponds. Myxoxanthophyll is present only in coccoid cyanobacteria and is a useful first approximation of potential toxic episodes. The PCR method had a sensitivity resolution of *ca.* 10 cells and was able to detect toxic algae at microcystin concentrations >0.25 ng/mL.

The same technique of model tuning (11 for phycocyanin, 12 for chl) is being used for modeling cyanobacterial biomass, rather than indirect ratio methods currently used. Application of Case1 water chl model to freshwater pond data resulted in poor model fit (40–60% explained variance). This new method improved model accuracy by 14%.

Conclusions

Toxic cyanobacteria blooms occur in over 50% of aquaculture impoundments. Development of spectral models for Case 2 waters will serve as a management tool for assessment of cyanobacterial blooms.

Chapter 26: Human Health Effects Workgroup Report

Workgroup Co–Chairs: Elizabeth D Hilborn, John W Fournie

Workgroup Members[1]**:** Sandra MFO Azevedo, Neil Chernoff, Ian R Falconer, Michelle J Hooth, Karl Jensen, Robert MacPhail, Ian Stewart

Authors: Sandra MFO Azevedo, Neil Chernoff, Ian R Falconer, Michael Gage, Elizabeth D Hilborn, Michelle J Hooth, Karl Jensen, Robert MacPhail, Ellen Rogers, Glen R Shaw, Ian Stewart

Introduction

Two types of approaches may be used to evaluate the toxicity of environmental agents:

- Observational population studies may be used to investigate humans or animals in contact with the agent of interest, either prospectively or retrospectively.
- Experimental studies may be performed under controlled conditions on animals, living tissues or cells. These results may be extrapolated to human and animal populations.

Observational epidemiological studies of human and animal populations have the advantage of investigating the effects of environmentally relevant exposures to naturally–occurring mixtures of toxins. Health effects may be identified in the target species of concern. These studies are difficult to implement however, as monitoring of cyanobacteria toxin occurrence in ambient water is essential to document exposure status, and specific associations between exposure and effect are difficult to establish in free–living populations.

---

[1] See workgroup member affiliations in Invited Participants section.

Health studies of standard test species in controlled environments address some of the difficulties described above, but they often involve the use of species other than those of primary interest. Laboratory studies may not closely approximate the toxin, dose or duration of exposure relevant to environmental conditions. The effects of mixtures of natural materials and toxins may be difficult to reproduce during repeated experiments. The advantages and disadvantages of these two strategies reveal that each may complement and enhance the other. Neither approach is complete in itself. Both are essential to the understanding of the potential effects of cyanobacterial toxins on human and animal populations.

The use of laboratory animals in toxicology is based on the premise that the data obtained during controlled exposures may be extrapolated with some degree of confidence to other species including our own. Cyanobacterial toxins are a heterogeneous group of compounds with unrelated structures, toxic endpoints, and mechanisms of action. Cyanobacterial toxins include hepatotoxins, renal toxins, immunotoxins, neurotoxins, skin irritants and sensitizers, although mechanisms of pathogenesis are not fully understood. Future cyanobacterial toxicity studies must consider the following components:

1. Selection of test species. The science of toxicology is replete with examples of highly significant inter–species variability. Differences in susceptibility to acute toxicity (e.g. TCDD), carcinogenicity (e.g. aflatoxin), and teratogenicity (e.g. thalidomide) are well documented. The response of different test species to a given dose of cyanobacterial toxin will depend onspecies–related metabolic/ toxicokinetic characteristics and differences in sensitivity to the agent studied (toxicodynamics). The majority of studies have used the mouse because it is the smallest rodent test species and therefore requires the least amount of toxin. Rodents are commonly used test animals but other animals must be studied in order to address the occurrence of specific toxicity endpoints across species. Determination of the "appropriate" test species will therefore depend on the information known about the site(s) of action and metabolism of the study agent. Although these factors are critical to understanding the effects of cyanobacterial toxins, there are major gaps in our knowledge of these factors for most of the toxins. Given these data gaps, the generally accepted testing strategy is to utilize multiple test species and to compare the results obtained in each species with those known to occur in the target species of concern.
2. Duration of exposure. Studies need to be conducted that approximate environmental exposures to cyanobacterial toxins. This may involve

multiple study designs as free–living populations may be exposed to cyanobacterial toxins chronically at low doses, and/or may experience episodic short–term exposures to high doses. These different exposure scenarios may, in turn, result in different spectrums of toxicity. For example, the data from cylindrospermopsin studies (Hawkins et al. 1997) have shown that the acute LD50 is 2000 ug/kg, contrasted with the 5–day LD50 of only 200 ug/kg. This suggests that the mortality seen in the two exposure groups of animals may be the result of different types of injury, one type being acute, and the other cumulative. Adequate characterization of cyanobacterial–induced toxicity should include a range of exposure scenarios.

3. Route of exposure. Oral, inhalation, and cutaneous exposures to cyanobacterial toxins may be the most likely routes of exposure among human and animal populations. Two episodes of human microcystin exposure by the intravenous route have been documented among patients undergoing dialysis (Jochimsen et al. 1998, Soares et al. 2006). In contrast, most animal toxicology studies have reported health effects associated with intraperitoneal (i.p.) exposure to cyanobacterial toxins. Expense and lack of sufficient test material are the primary reasons for most cyanobacterial toxin studies having been performed using the i.p. exposure route. The oral route of administration has been prohibitively expensive for extended studies since the toxins are far less toxic orally than by the i.p. route (generally by factors of 10 or more). Future laboratory animal studies should include oral, inhalation, and cutaneous exposures to evaluate effects that may be associated with these environmentally relevant exposure routes.

4. Characterization of cyanobacterial toxins and effects of mixtures. Multiple cyanobacteria genera produce the same toxin, and some produce multiple toxins. Many cyanobacterial toxins have yet to be identified and/or characterized. The production of cyanobacterial toxins can be highly variable, and factors associated with toxin gene expression are poorly understood. *Cylindrospermopsis raciborskii*, for example, produces widely varying amounts of cylindrospermopsin depending upon its geographical location and environmental conditions. A recently described cyanobacterial toxin is β–N–methylamino–L–alanine (BMAA). This compound may be associated with adverse neurological effects in humans. It is possible that BMAA occurs in multiple species of cyanobacteria including some that occur in fresh water (Cox et al. 2005). Microcystins may vary temporally, spatially and chemically within a water body, and

> microcystin variants differ significantly in their potential to induce mammalian toxicity (Sivonen and Jones 1999). Experimental evidence also indicates that any given bloom may produce more than one cyanobacterial toxin and the resultant toxicity may be the result of additive and/or synergistic effects among these agents. Studies on cyanobacterial cellular extracts containing either anatoxin–a or cylindrospermopsin have shown that the resulting toxicity was not directly proportional to the amount of toxin in the test material itself (Stevens and Krieger 1991; Falconer et al. 1999). In both reports, the authors hypothesized the existence of other, as yet unidentified toxins that were contributing to the observed adverse effects.

Studies designed to assess the effects of purified cyanobacterial toxins such as microcystin–LR, anatoxin–a, and cylindrospermopsin, are needed to: characterize toxicokinetics and toxicodynamics, develop biomarkers of exposure and effect, and study specific health effects. However, this approach is hindered by a lack of toxin standards, and the fact that the study of individual toxins does not approximate environmentally relevant exposures. The use of cellular extracts from cyanobacterial clonal cultures addresses these limitations. Cultures can be characterized, batched and used for multiple studies among laboratories. The use of these cultures would allow effects research to proceed in the absence of toxin standards, and may be more representative of the spectrum and mix of toxins to which human and animal populations are exposed.

Cyanobacteria occurrence is generally increasing as a result of eutrophication and warming of surface waters. When human and domestic animal populations congregate around surface water sources their waste, and nutrients such as nitrogen and phosphorous, enters the water. Local populations may depend on surface waters to provide drinking water, irrigation, food fish, and recreation. However, as population densities increase, the occurrence of blooms increases, thus heightening the risk of human and animal exposure to cyanobacteria toxins. Given the potency, heterogeneity, and documented occurrence of cyanobacterial toxins in US surface waters, they will increasingly be recognized as a serious public and environmental health concern.

Charge 1

What materials do we need to perform health effects research?

Ideally, investigators need well characterized, pure cyanobacterial toxins to perform toxicity and carcinogenicity studies. The results of studies using pure toxins are easier to interpret, and have fewer confounding variables. However, extended toxicity studies require significant quantities of toxin to dose sufficient numbers of animals by relevant routes of exposure (dermal, oral, and inhalation). These studies are necessary for risk assessment and the development of guidance values.

Currently the supply of pure toxin is limited. Only two oral dosing toxicity studies using pure microcystin (Fawell et al. 1994) and cylindrospermopsin (Humpage and Falconer 2003) have been implemented to date. However, the impact of these types of studies is significant. The Fawell study was used by the World Health Organization (WHO) as the basis for a Guideline Value determination for microcystins. The cylindrospermopsin study is currently under review by WHO. Despite the importance of these types of studies, few are conducted because sufficient quantities of purified toxin are not currently available. There are published chemical synthesis methods for microcystins, including production of the 3–amino–9–methoxy–2,6,8–trimethyl–10–phenyl–4,6–decadienoic acid (ADDA) residue which has been successfully synthesized (Humphrey et al. 1996; Candy et al.1999). However, no commercial production of pure cylindrospermopsin in bulk has been reported. Although research groups have reported chemical methods to synthesize cylindrospermopsin (Xie et al. 2000), this method is prohibitively expensive due to the 20 steps required for synthesis.

Few toxicology studies using pure toxins have been conducted in species other than mice. Because of the limited supply of pure toxin, mice are preferred as an animal model because of their size. However, they may not be the best model for every target species or system of interest. Alternatively, studies with thoroughly characterized toxic extracts of cyanobacteria have yielded valuable data for domestic animals, including a study with growing pigs which was also used by WHO in the derivation of the human drinking water guideline (Falconer et al. 1994).

Pure toxin isolated from live cultures of cyanobacteria is available privately and commercially in limited amounts allowing the conduct of mechanistic studies which require less material. Some investigators have used *C. raciborskii* cultures (Hawkins et al. 1997; Chiswell et al. 1999; Shaw et al. 2000). Cylindrospermopsin has been purified from *C. raciborskii* cultures using high performance liquid chromatography (HPLC) (Chiswell et al. 1999). Other investigators have conducted studies with highly purified cylindrospermopsin isolated from a bloom of *Umezakia natans* collected from Lake Mikata, Fukui, Japan (Harada et al. 1994; Terao et al. 1994). Likewise, studies with approximately 95% pure microcystin–LR have been conducted with toxin produced from *Microcystis aeruginosa*, laboratory strain 7820, and purified by HPLC (Hooser et al.

1990). Limited quantities of characterized microcystin variants LA, LR, RR, and YR, cylindrospermopsin and nodularin are available commercially from Sigma–Aldrich (St. Louis, MO, US) and microcystin variants LF, LR, LW, and RR from EMD Biosciences (San Diego, CA, US). Although the production method is not disclosed, these materials are suspected to have been isolated from live cultures. Regardless of source, it is always important to verify the identity and purity of these toxins, as investigators have reported inconsistencies in research results among different lots of toxins.

Freeze–dried natural bloom material is more widely available in larger quantities for research studies. Dr. Ian Falconer and his colleagues have conducted long–term microcystin toxicity and cancer promotion studies in mice with extracts from dried natural bloom material that was from a batch collected at Lake Mokoan in Victoria, Australia and from Malpas Dam in New South Wales, Australia (Falconer and Humpage 1996). Both sources of material have been characterized and used in previous research (Falconer et al. 1988, 1994). Ten peaks exhibiting the characteristic absorbance spectrum of microcystins were detected using HPLC analysis. However, none of these peaks corresponded to microcystins for which standards were available (microcystin–LR, –YR, and –RR). Bloom materials offer the advantage of reflecting real–world exposure scenarios. However, these materials have to be carefully evaluated for use in studies that may be included in risk assessments. Bloom materials may not be completely characterized and the toxicity of the mixture may be greater than that of the identified individual components indicating that: 1) the mixture may contain additional unidentified toxins; 2) that there may be a synergistic effect among bioactive components of the bloom material.

Since the lack of availability of pure cyanobacterial toxins is currently limiting the ability to perform health effects research, this workgroup strongly recommends the development and maintenance of bulk clonal cyanobacteria cultures shared among laboratories to facilitate screening, bioassay development and long–term exposure studies. Extensive chemical and bioassay characterization of the test materials is needed to ensure appropriate interpretation of study results. Due to the hazardous nature of these materials, clarification and standardization of the import/export regulations is needed to facilitate the use of shared cyanobacterial test materials among laboratories (Metcalf et al. 2006).

Charge 1: What materials do we need to perform health effects research?

Near-term Research Priorities

- Develop and maintain bulk clonal cyanobacteria cultures that can be shared among laboratories to facilitate screening, bioassay development and long–term exposure studies.

Long-term Research Priorities

- Develop lower cost methods of chemical synthesis to produce sufficient quantities of standard cyanobacterial toxins.

Charge 2

What are the health effects associated with chronic and episodic exposures?

Risk assessments of the effects of cyanobacterial toxins on human and animal health require information produced during studies of chronic and episodic exposures, and are insufficient if based solely on studies of acute exposures and lethal endpoints. However, much of our knowledge of toxicity rests upon acute exposures: poisoning of animals in the natural environment, and laboratory animal studies of lethal endpoints. Considerable advances have been made, however, in our understanding of the sub-lethal effects of cyanobacterial toxins, and this must become the focus of research in order to insure continued progress in the understanding of the risks associated with cyanobacterial toxin exposures.

There are strong experimental studies that demonstrate the tumor promotion activity of microcystins and nodularin, and well–designed genotoxicity studies that show cylindrospermopsin to be a potential carcinogen (Falconer 2005). To perform effective risk assessment it is essential to carry out carcinogenicity studies on both groups of cyanobacterial toxins. The present limitation to both National Toxicology Program (NTP) protocol studies is the lack of available purified toxin. It is preferable to use highly characterized toxic extracts of cultures rather than to indefinitely defer these essential carcinogenicity trials, as humans may be currently exposed to cylindrospermopsin via drinking and recreational water exposure.

The highest priority work is to implement carcinogenicity trials with cylindrospermopsin because of the structure of the toxin, preliminary evidence of carcinogenicity, and the current strength of evidence of genotoxicity in a human lymphoblastoid cell line (Humpage et al. 2000a). An *in vivo* study conducted by Falconer and Humpage provided preliminary evidence of the carcinogenicity of cylindrospermopsin in mice (Falconer and Humpage 2001).

Neurodevelopmental studies are required for all groups of cyanobacterial toxins. Currently little is known about neurodevelopmental toxicity. Early work with chronic *Microcystis* extract exposure to male and female mice throughout pregnancy showed cytotoxicity in the hippocampus of neonatal offspring (Falconer et al. 1988). These studies are especially relevant as there is the potential for neonates to be exposed via drinking water in formula during a period of their rapid neurologic development.

Immunomodulation and immunosuppression after exposure to microcystins has been documented (Shen et al. 2003; Shi et al. 2004; Chen et al. 2005). Adverse effects on human leukocytes *in vitro* have been reported at very low doses (Hernandez et al. 2000). The potential immunotoxic effects of microcystin in human and animal populations are uncharacterized. To our knowledge, no studies of the potential immunotoxic effects of cylindrospermopsin have been reported.

There is no strong evidence to date for the teratogenicity of microcystin and anatoxin–a (Chernoff et al. 2002, Rogers et al. 2005, MacPhail et al. 2005). Teratogenicity studies of cylindrospermopsin are underway. Multigenerational studies are needed for most toxins although the implementation of these will need to be deferred until sufficient toxin or characterized clonal bloom material is available.

Charge 2: What are the health effects associated with chronic and episodic exposures?

Near-term Research Priorities

- Conduct carcinogenicity studies with cylindrospermopsin.
- Conduct neurodevelopmental studies for all groups of cyanobacterial toxins.
- Conduct immunotoxicology studies for all groups of cyanobacterial toxins.

- Conduct studies of the effects of acute, episodic and chronic exposures at sublethal concentrations.
- Implement toxicological studies using oral, inhalation, and cutaneous exposure routes.

Long-term Research Priorities

- Implement multigenerational studies for all groups of cyanobacterial toxins.

Charge 3

What are the health effects associated with environmental mixtures of toxins?

In naturally occurring cyanobacterial blooms, a mixture of toxins is present. Most commonly, these are mixtures of microcystins when the bloom is *Microcystis* or *Planktothrix* or of cylindrospermopsins when the bloom is *Cylindrospermopsis, Aphanizomenon, Raphidiopsis* or *Umezakia*. Anatoxin–a and anatoxin–a(s) are uncommonly found occurring with microcystins or cylindrospermopsins in the environment, but when this mixture does occur, synergistic effects are possible. Bloom toxicity therefore represents the combined toxicities of the constituent toxin variants present, and can best be expressed as toxicity equivalents (Falconer 2005).

Under natural circumstances eutrophic water bodies form a sequence of toxic blooms that appear to be determined primarily by water temperature. An observed sequence in Australia is neurotoxic *Anabaena*, followed by hepatotoxic *Microcystis*, followed by cytotoxic *Cylindrospermopsis* in late summer (Bowling 1994). Populations can then potentially be exposed to saxitoxins, followed by successive hepatotoxins with different mechanisms of action.

Effective risk assessment of natural toxic bloom events requires a systematic assessment of mixtures of toxins for which there is currently no data. We have little information about the health effects of exposure to concurrent or sequential mixtures. Monitoring of cyanobacteria toxin occurrence in environmental samples is essential to detecting and understanding potentially associated health effects of exposure to mixtures of toxins.

Risk assessment of populations exposed to natural blooms needs to take into consideration a range of pathological effects generated by different cyanobacterial toxins, in sequence and in combination. These may include measures of hepatic, immune system, and kidney impairment (Falconer 2005). In human populations, neurological impairment can be measured by the study of the neurodevelopment of young children and with a variety of standard tests among adults. Cancer registries may be used to assess potential carcinogenetic effects of cyanobacterial toxins if large populations are found to be exposed.

Charge 3: What are the health effects associated with environmental mixtures of toxins?

Near-term Research Priorities

- Implement toxicology studies of animal models exposed to mixtures of cyanobacterial toxins to evaluate toxin synergism.
- Implement epidemiologic studies of exposed populations to assess a variety of cyanobacterial toxin–associated health effects.
- Perform risk assessments of populations exposed to natural blooms and multiple cyanobacterial toxins, in sequence and in combination.

Charge 4

What research is needed to better understand the public health effects of exposures to cyanobacteria?

A well–designed study of the health effects associated with cyanobacteria exposure in human populations requires comprehensive exposure assessment. Currently, the United States does not routinely monitor drinking or recreational waters for the presence of cyanobacterial toxins despite the reported presence in these waters of potentially toxic blooms. Therefore, the magnitude of the risk of cyanobacteria toxin exposure and associated effects to the US population is unknown. Human exposures to cyanobacterial toxins may be broadly categorized for the purpose of public health investigation into acute or chronic (episodic) exposures. Study methods and approaches will vary depending upon the size of the population and the characteristics of exposure.

Acute exposures to cyanobacteria and their toxins may occur via the oral, dermal, inhalational or intravenous exposure routes. The most common exposures are believed to occur during recreational and occupational contact with cyanobacteria in lakes, rivers and marine waters (Osborne et al. 2001; Stewart et al. 2006). Acute, chronic, and episodic exposures may arise from drinking water (Annadotter et al. 2001; Ressom et al. 1994; Kuiper–Goodman et al. 1999; Duy et al. 2000; Falconer 2005), dietary intake via consumption of cyanobacterial toxin–contaminated foods (Negri and Jones 1995; Nagai et al. 1996; Codd et al. 1999; Saker and Eaglesham 1999; de Magalhães et al. 2001), and dietary supplements (Schaeffer et al. 1999; Gilroy et al. 2000; Lawrence et al. 2001; Saker et al. 2005). Acute human exposure has been documented via contamination of dialysate in hemodialysis clinics (Hindman et al. 1975; Carmichael et al. 2001; Azevedo et al. 2002; Soares et al. 2006). Outcomes may range from no apparent effect to serious morbidity or death associated with toxicosis. Non life–threatening allergic and allergic–like reactions (rhinitis, asthma, eczema, conjunctivitis) have been reported but are poorly quantified, as are acute illnesses (e.g. flu–like reactions, skin rashes) in (presumably) non–allergic individuals (Stewart et al. 2006).

Reports of acute illness have been received from workers sampling potentially toxic cyanobacteria blooms (Reich 2005). These ranged from anaphylactic–like reactions, to dermatitis, gastroenteritis, and respiratory irritation. However, published reports and descriptions of exposed and ill individuals (cases) and case series are needed as the literature is sparse (Turner et al. 1990). This information is needed to develop more specific case definitions, to describe the spectrum of cyanobacteria–associated health effects, and to identify potential susceptibility factors.

Chronic or episodic exposure to cyanobacterial toxins may occur when people are exposed occupationally or by their drinking water. The principal concern is that some cyanobacterial toxins, such as the hepatotoxins, may have carcinogenic potential. Some epidemiological investigations have suggested that hepatocellular carcinoma (HCC) and colorectal cancer rates were significantly higher in regions of China where consumption of untreated pond or ditch water was common when compared to rates in populations drinking deep well water (Ressom et al. 1994; Falconer 2005). However, the epidemiological evidence is contradictory; a recent retrospective study failed to identify a relationship between HCC and consumption of ditch water (Yu et al. 2002). China is a high–risk area for HCC, accounting for some 45% of worldwide mortality and HCC is associated with other risk factors such as chronic viral hepatitis and exposure to aflatoxins (Ming et al. 2002; Yu and Yuan 2004; Shi et al. 2005). Therefore, cyanobacterial toxin exposure assessment is essential to the successful design and implementation of studies to investigate the risk of cancer associated with repeated exposure to these toxins.

Because of the lack of currently identified cyanobacteria toxin–specific health effects, we urgently need to develop specific biomarkers of exposure and effect to use in health studies. Biomarkers of exposure include the toxin itself, metabolites of the parent compound, and DNA or protein adducts specific to the toxin. Recently, the use of a simple colorimetric assay was proposed to screen human serum samples for the presence of microcystins (Hilborn et al. 2005). Work is in progress to identify cylindrospermopsin metabolites and possible adducts that can be used as human biomarkers of exposure (Humpage and Falconer 2005). Biomarkers of effect include: biochemical markers such as serum concentrations of liver transaminases, and quantification of micronuclei in leukocytes (Falconer et al. 1983; Humpage et al. 2000a). The use of biomarkers will strengthen the specificity of association between exposure to cyanobacterial toxins and health endpoints. The carcinogenicity of aflatoxin is now much better understood through the ability to measure aflatoxin B_1, its metabolites and DNA–adducts in urine (Ross et al. 1992; Qian et al. 1994); the epidemiology of cyanobacterial toxins awaits similar advances. Potential future areas for biomarker development include the use of DNA mutations and adducts as markers of genotoxicity. Targeted studies using genomic and proteomic techniques may be useful, however are frequently difficult to interpret. Currently, these approaches are best guided by and aligned with the results of traditional toxicologic studies.

Epidemiologic studies are needed to evaluate the effects of exposures to cyanobacteria and their toxins on human and animal health. Population–based observational study designs that assess cyanobacteria exposure retrospectively are probably the most cost–effective approaches to investigate cyanobacteria bloom–associated health effects at this time. Of particular importance are recent event–triggered investigations. Both acute and chronic health effects may be investigated. However, assigning exposure histories to individuals may be problematic, particularly to those individuals or populations with chronic or episodic exposures. Retrospective studies may lack the information to examine the temporal relationship between exposure and effect, and one design, the case–control study, requires specific case definitions which are not well defined at present.

Prospective population–based observational studies have a greater ability to associate exposure and effect due to improved exposure assessment, which is done before health effects occur. However this approach requires large numbers of exposed persons, is expensive, and time consuming. The dynamic nature of cyanobacteria blooms and toxin production presents significant challenges for exposure characterization, study planning and implementation.

An experimental study design can provide a higher degree of the certainty of associations between exposure and effect. The principal advantage of a large, randomized, controlled exposure study is that random assignment to a study group minimizes the differences between exposure groups that may affect the interpretation of the study results. However, the specific details of any proposed randomized exposure trial would need to be closely reviewed for its scientific merits and for the protection of human subjects. Within human ethical guidelines, an experimental design could, in theory, be applied to the study of certain acute, low risk exposures to cyanobacteria. However, randomized, controlled exposure studies are expensive and labor intensive, and therefore are not recommended as a research priority at this time. Resources should instead be directed to the implementation of well–designed observational studies to investigate the effects of human exposures to ambient concentrations of cyanobacteria and their toxins.

Charge 4: What research is needed to better understand the public health effects of exposures to cyanobacteria?

Near-term Research Priorities

- Develop specific biomarkers of exposure and effect.
- Implement retrospective population–based observational studies.
- Systematically collect descriptions of exposed and ill individuals (cases) and case series.

Long-term Research Priorities

- Implement prospective population–based observational studies in those communities with recurrent exposure events.

Charge 5

What are the determinants of host susceptibility?

Human beings are a diverse lot, varying in age, sex, genetics, nutrition, exposure history and health status. Accurate prediction of the human health risks of cyanobacterial toxins must be based on recognition that each one of these variables, either alone or in combination, could affect an individual's susceptibility to the effects of these toxins. However, little information is available at this time regarding the importance of these (and other) factors in determining host susceptibility. This lack of knowledge is due primarily to the sporadic nature of poisoning episodes in humans and in other animals (i.e. wildlife, livestock, domestic animals), and the absence of a centralized mechanism for collecting and summarizing data from these types of events. Laboratory studies have focused largely on toxicity produced by acute exposures in relatively homogenous laboratory test species, such as mice. Some useful data exists on the effects of cyanobacterial toxins in some domestic animals (Falconer 2005). However, there is far less data on the effects of toxins following sublethal acute, episodic or chronic exposures. As a result, attempts to predict the human health risks of cyanobacterial toxins are riddled with uncertainty.

Despite the paucity of data, it seems likely that several factors could influence the severity of outcome in humans following exposure to cyanobacterial toxins. For example, liver failure would likely be more common following exposure to hepatotoxins in people with prior hepatic disease or insufficiency, and irritation caused by exposure to air–borne toxins may be greater in people with underlying respiratory disease.

Studies on laboratory animals are potentially more informative in identifying host susceptibility factors. The most common practice, referred to as the mouse bioassay, acutely exposes mice by the i.p. route to samples suspected to contain cyanobacterial toxins (Sullivan 1993). This approach has been widely used to determine the safety of seafood samples, as well as to gauge potency during the extraction and purification of toxins from environmental samples. A premium has therefore been placed on standardized conditions of testing among samples (Fernandez and Cembella 1995). As a consequence, the potential importance of most susceptibility factors that may be prevalent in the human population is unknown.

Two episodes of human poisonings have focused attention on age as a potential risk factor for adverse health effects. For example, a large scale outbreak of toxicoses among renal dialysis patients in Brazil found that mortality was significantly higher among older (47 vs. 35 year old) patients (Jochimsen et al. 1998). However, these patients were ill, and exposed to cyanobacterial toxins by the intravenous route; this finding may not be applicable to people in general. Another large–scale poisoning episode involved exposure to *Cylindrospermopsis* via drinking water in Australia (Hawkins et. al 1985; Falconer 2005). This outbreak of hepato–enteritis affected considerably more children that adults. However, in contrast to children, very few adults in the community drank from the local water supply. Therefore, the higher rate of intoxication observed among children in the population may have been due to exposure rather than to age–related susceptibility. The role of human genetic susceptibility to effects associated with cyanobacterial toxins is unknown.

Some laboratory animal data indicate that age at exposure may influence susceptibility to cyanobacterial toxins. For example, the acute LD_{50} in rats of an extract prepared from Alaskan butter clams containing paralytic shellfish poisoning (PSP) toxins increased with age; it is has been surmised that the principal toxin in this extract was saxitoxin (WHO 1984). Compared to newborn rats, adult rats were about 10 times less sensitive following oral exposure, and about 2 times less sensitive following i.p. exposure to the extract (Watts et al. 1966). Regardless of age, however, LD_{50} values were considerably lower for the i.p. than the oral route of dosing (Wiberg and Stephenson 1960). On the other hand, in mice receiving a reported i.p. LD_{50} of purified microcystin LR, the time to death *decreased* substantially with age (from 6 to 36 weeks). A similar age–related decrease in time to death was also reported following an oral LD_{50} exposure of mice to a microcystin–containing extract (Rao et al. 2005). Recent studies found no developmental or long–term effects in mice following repeated prenatal exposure to sublethal doses of microcystin LR, or anatoxin–a (Chernoff et al. 2002; Rogers et al. 2005; MacPhail and Jarema 2005). These latter studies highlight the importance of systematic investigations into the effects of sublethal exposures involving acute, episodic and chronic exposures, in order to make accurate estimations of the health hazards associated with cyanobacterial toxins.

Charge 5: What are the determinants of host susceptibility?

Near-term Research Priorities

- Systematically examine and report susceptibility among multiple test species of animals exposed to cyanobacterial toxins in toxicology studies.
- Collect and report information about susceptibility during case reports and observational studies of humans and animals exposed to cyanobacterial toxins.
- Implement studies of the effects of sublethal acute, episodic and chronic exposures in healthy animals and in animal models of susceptibility.

Charge 6

What are the research needs after exposure/intoxication has occurred?

Little is known about the potential for human health effects that may persist or develop after intoxication with cyanobacterial toxins, although long term effects of microcystin poisoning in sheep have been reported (Carbis et al. 1995). A better understanding of post intoxication effects will depend upon longer–term toxicological studies and upon comprehensive epidemiological evidence, including longitudinal studies initiated after recognition of human and animal exposures.

Timely recognition of exposure events is needed. In global terms, human exposures to cyanobacterial toxins in drinking water supplies may occur sporadically as a consequence of treatment failures, deficiencies in drinking water systems operations, or from the addition of copper sulfate to reservoirs to terminate blooms of cyanobacteria. Epidemiological evidence is useful to describe the relationship between cyanobacterial toxin exposure and human health outcomes. However, comprehensive information is difficult to collect due to the lack of knowledge of the risk of human exposure to cyanobacterial toxins by water quality managers. When potentially toxic cyanobacterial blooms occur, they may be unrecognized as a health threat and public health authorities may not be involved early in the process. Therefore, the majority of outbreaks of human cyanobacterial toxicoses have been studied retrospectively and complete epidemiological and environmental data has rarely been available.

A similar lack of knowledge may preclude identification of potential adverse health effects associated with other potential routes of cyanobacterial toxin exposure such as recreational water activities, consumption of contaminated fish, or dietary supplements contaminated by toxic cyanobacteria. Ideally, for health effects research, these risks would be recognized during the period of exposure to toxic cyanobacteria, and measures taken to protect human health. This timely identification of exposure would allow better collection of epidemiological evidence and consequently improve the analysis of associated health effects and recovery.

Although there are no known antidotes to treat poisonings associated with cyanobacterial toxins, people and animals may be treated by various supportive measures such as intravenous fluid and electrolyte replacement, corticosteroid therapy, assisted ventilation, and maintenance of acid–base balance. Methods to reduce absorption such as gastric lavage, activated charcoal or cholestyramine may also be used.

Potentially, if hepatotoxin poisoning is recognized soon after exposure, survival may be improved using recent advances in acute liver support therapy. Therapeutic use of the bioartificial liver, the molecular adsorbents recirculating system, portal vein arterialisation and human hepatocyte transplantation may supplement the standard treatment for fulminant liver failure – liver transplantation (Park and Lee 2005; Hay 2004; Nguyen et al. 2005; van de Kerkhove et al. 2005; Nardo et al. 2005; Baccarani et al. 2005; Tissières et al. 2005).

Maintenance of tissue oxygenation by intermittent positive pressure ventilation may benefit victims of neurotoxin poisoning. Some have investigated this technique for anatoxin–a poisoning, with mixed results (Carmichael et al. 1975; Carmichael et al. 1977; Beasley et al. 1989; Valentine et al. 1991). Artificial ventilation in conjunction with the potassium channel blocker 4–aminopyridine has shown promise in reversing experimental saxitoxin poisoning (Chang et al. 1996).

However, emergency interventions on human and animal victims of cyanobacterial toxicosis can only make a limited contribution to the understanding of post–intoxication intervention, as typically little is known about the total absorbed dose. There is a need for experimental work designed specifically to investigate the efficacy of various therapeutic approaches.

Although a large number of animal poisonings and animal toxicity tests have been reported, few have measured the effects of chronic or sub–chronic exposure (Falconer et al. 1988; Guzman and Solter 1992; Falconer et al. 1994; Fawell et al. 1994; Ito et al. 1997; Humpage et al. 2000b; Falconer and Humpage 2001; Humpage and Falconer 2003). Therefore, the long–term risk associated with chronic, low–level exposure is less well understood. However, this type of exposure scenario may be the most common among human populations. The relative lack of animal data makes epidemiological studies difficult due to the lack of case definitions and biologic methods to confirm evidence of chronic or subchronic exposure to low levels of cyanobacterial toxins (Kuiper–Goodman et al. 1999).

There is a great disparity of information among cyanobacterial toxins. Microcystins are the group of toxins most frequently studied, and there is more information to support the understanding of the intoxication and detoxification process. However, even in the case of microcystins, research into long–term exposures is needed to investigate reproductive toxicity, teratogenicity and carcinogenicity effects in mammals. For other cyanobacterial toxins, such as cylindrospermopsin and the neurotoxins, anatoxins and saxitoxins, there are large knowledge gaps related to chronic toxicity and to the intoxication and detoxification process.

Charge 6: What are the research needs after exposure/ intoxication has occurred?

Near Term Research Priorities

- Initiate regular monitoring of water bodies at risk for toxic cyanobacteria blooms to enable remediation and timely health effects studies.
- Implement toxicologic studies that encompass multiple health endpoints at frequent time intervals to examine the relationship between initial intoxication, health effects and recovery at multiple life stages.
- Conduct studies to investigate the effectiveness of therapeutic approaches.
- Monitor food and supplements at risk for contamination with cyanobacterial toxins to enable public health intervention and timely health effects studies.

Charge 7

Where are we in the development of predictive models?

Predicting human health outcomes associated with cyanobacterial toxins requires linking exposure estimates with dose–response models. Estimates of toxin occurrence may be potentially derived from the models of bloom dynamics, toxin production, and the fate and transport of the toxins. Mechanistic models have been developed that can simulate a variety of bloom characteristics and they have been useful in determining the significance of a variety of environmental variables that influence bloom dynamics (Bonnet and Poulin 2002; Thébault and Rabouille 2003; Håkanson et al. 2003; Robson and Hamilton 2004; Arhonditsis and Brett 2005; Prokopkin et al. 2006) but their predictive capacity remains limited. In contrast, inductive models, such as those employing artificial neural networks, appear to have the capacity to predict significant aspects of bloom dynamics (Recknagel et al. 1997; Maier et al. 1998; Jeong et al. 2003).

A key step in using predictive models of bloom dynamics to predict exposures will be consideration of toxin fate and transport within aquatic ecosystems. This will help to determine the relative importance of different routes of human exposure to cyanobacterial toxins. For example, there is a potential for oral, inhalational and dermal exposures during recreational water use. Microcosm studies may provide an efficient means to develop and evaluate the models that integrate bloom dynamics, toxin production, fate and transport to estimate such multi–route exposures. Reliable models predicting exposure via drinking water may be more difficult to link to bloom dynamics given the uncertainties regarding the fate of toxins in water processing and the potential for by–products resulting from the interaction of disinfection agents with cyanobacterial toxins.

Mechanistic models of human health effects based on stepwise links from exposure to effect require toxicokinetic and toxicodynamic information that is largely unavailable for the majority of cyanobacterial toxins. Furthermore, dose–response information is only currently available for a small number of toxins for limited types of effects, suitable for use in more basic modeling efforts. Alternatively, more 'intelligent' inductive models may be used to predict human health outcomes from mixtures of cyanobacterial toxins based on data derived from experimental studies that employ characterized bloom extracts rather than pure toxin. While such extracts more accurately represent individual real–world blooms, their complex and unique characteristics may limit the inferences that can be made until large numbers of blooms are characterized and their effects assessed.

The development of predictive models may also benefit from a more extensive characterization of the influence of temporal aspects of exposure on various types of health outcomes. Different health outcomes are likely to predominate as a result of acute, subchronic, chronic and episodic exposures. Only minimal data are currently available that are suitable for discriminating the impact of these temporal aspects of exposure on the variety of potential human health outcomes. Predictive models are needed that may be used by risk managers during decisions related to the costs and benefits of surface water use and exposure.

To the extent that future experimental studies implicate a wider variety of health outcomes as being influenced by cyanobacterial toxin exposure, models will need to be developed to address more comprehensive predictions that include a variety of health outcomes in the human population. Evaluation of such predictions will require corresponding biomarkers of effect suitable for use in human populations to evaluate the validity of model predictions.

While the primary linkage between bloom occurrence and human health effects is presumed to be exposure to cyanobacterial toxins, the effects of cyanobacterial blooms on ecosystem services may also influence human health. While several international efforts are modeling the impact of cyanobacterial blooms on ecosystems, assessing the impact of ecosystem services on human health is emerging as an important area of research. At this time it is difficult to assess the significance of ecosystem services for modeling the effect of cyanobacteria blooms on human health.

Charge 7: Where are we in the development of predictive models?

Near-term Research Priorities

- Implement toxicology studies to determine dose–response relationships using oral, inhalational and dermal exposure routes.
- Implement studies of cyanobacterial toxin fate and transport in environmental media.
- Characterize the toxicokinetics/toxicodynamics of parent toxin and metabolites.

Long-term Research Priorities

- Develop linkages between predictive models of bloom dynamics to dose–response models of human health and ecologic effects outcomes.
- Investigate how cyanobacterial blooms reduce ecosystem services, and the resulting effects on human and animal health.

Conclusion

In summary, we have identified multiple health effects research needs associated with exposure to cyanobacteria and cyanobacterial toxins.

- Affordable toxin standards are needed. However, in the short term, research may be conducted with characterized clonal cyanobacteria cultures that can be shared among laboratories.
- Studies of the health effects of chronic, episodic, and low–dose exposures by environmentally—relevant routes are needed. Carcinogenicity, neurodevelopmental, neurotoxicology, and immunotoxicology studies are of immediate importance.
- Experimental studies of laboratory animals and observational studies of populations exposed to mixtures of cyanobacterial toxins are needed. A goal is to develop risk assessments of populations exposed to natural blooms and mixtures of cyanobacterial toxins.
- The public health consequences of exposure to cyanobacteria and cyanobacterial toxins are poorly characterized. Biomarkers of exposure and effect are needed to use in human studies. Case reports and observational epidemiologic studies are needed.
- Host susceptibility is poorly defined and should be systematically examined among individuals involved in outbreaks of toxicoses, and during experimental and observational studies.
- Timely and accurate exposure assessment is critical to detecting and understanding cyanobacterial toxin—associated health effects. Systematic toxin occurrence monitoring is needed in water and food at risk for contamination with potentially toxic cyanobacteria.
- Critical data gaps remain before predictive modeling of cyanobacteria–associated health effects in populations is possible. Information about the environmental fate and transport of toxins, the effects of cyanobacteria blooms on ecosystem services and population dynamics, the

toxicokinetics/toxicodynamics of parent compounds and metabolites, and dose–response data are needed.

References

Annadotter H, Cronberg G, Lawton L, Hansson HB, Göthe U, Skulberg O (2001) An extensive outbreak of gastroenteritis associated with the toxic cyanobacterium Planktothrix agardhii (Oscillatoriales, Cyanophyceae) in Scania, south Sweden. In: Cyanotoxins – occurrence, causes, consequences. Edited by Chorus I. Berlin: Springer–Verlag; 200–208

Arhonditsis GB, Brett MT (2005) Eutrophication model for Lake Washington (USA) Part I. Model description and sensitivity analysis. Ecol Modell 187(2–3):140–178

Azevedo SM, Carmichael WW, Jochimsen EM, Rinehart KL, Lau S, Shaw GR, Eaglesham GK (2002) Human intoxication by microcystins during renal dialysis treatment in Caruaru–Brazil. Toxicology 181–182:441–446

Baccarani U, Adani GL, Sainz M, Donini A, Risaliti A, Bresadola F (2005) Human hepatocyte transplantation for acute liver failure: state of the art and analysis of cell sources. Transplant Proc 37(6):2702–2704

Beasley VR, Dahlem AM, Cook WO, Valentine WM, Lovell RA, Hooser SB, Harada K, Suzuki M, Carmichael WW (1989) Diagnostic and clinically important aspects of cyanobacterial (blue–green algae) toxicoses. J Vet Diagn Invest 1(4):359–365

Bonnet MP, Poulin M (2002) Numerical modelling of the planktonic succession in a nutrient–rich reservoir: environmental and physiological factors leading to Microcystis aeruginosa dominance. Ecol Modell 156(2–3):93–112

Bowling L (1994) Occurrence and possible causes of a severe cyanobacterial bloom in Lake Cargelligo, New South Wales. Aust J Mar Freshwater Res 45(5):737–745

Candy DJ, Donohue AC, McCarthy TD (1999) An asymmetric synthesis of ADDA and ADDA–glycine dipeptide using the beta–lactam synthon method. J Chem Soc Perkin Trans I 5:559–567

Carbis CR, Waldron DL, Mitchell GF, Anderson JW, McCauley I (1995) Recovery of hepatic function and latent mortalities in sheep exposed to the blue–green alga Microcystis aeruginosa Vet Rec 137(1):12–15

Carmichael WW, Azevedo SM, An JS, Molica RJR, Jochimsen EM, Lau S, Rinehart KL, Shaw GR, Eaglesham GK (2001) Human fatalities from cyanobacteria: chemical and biological evidence for cyanotoxins. Environ Health Perspect 109(7):663–668

Carmichael WW, Biggs DF, Gorham PR (1975) Toxicology and pharmacological action of Anabaena flos–aquae toxin. Science 187(4176):542–544

Carmichael WW, Gorham PR, Biggs DF (1977) Two laboratory case studies on the oral toxicity to calves of the freshwater cyanophyte (blue–green algae) Anabaena flos–aquae NRC–44–1. Can Vet J 18(3):71–75

Chang FCT, Bauer RM, Benton BJ, Keller SA, Capacio BR (1996) 4–aminopyridine antagonizes saxitoxin– and tetrodotoxin–induced cardiorespiratory depression. Toxicon 34(6):671–690

Chen T, Shen P, Zhang J, Hua Z (2005) Effects of microcystin–LR on patterns of iNOS and cytokine mRNA expression in macrophages in vitro. Environ Toxicol 20(1):85–91

Chernoff N, Hunter ES 3rd, Hall LL, Rosen MB, Brownie CF, Malarkey D, Marr M, Herkovits J (2002) Lack of teratogenicity of microcystin–LR in the mouse and toad. J Appl Toxicol 22(1):13–17

Chiswell RK, Shaw GR, Eaglesham G, Smith MJ, Norris RL, Seawright AA, Moore MR (1999) Stability of cylindrospermopsin, the toxin from the cyanobacterium, Cylindrospermopsis raciborskii: Effect of pH, temperature, and sunlight on decomposition. Environ Toxicol 14(1): 155–161

Codd GA, Metcalf JS, Beattie KA (1999) Retention of Microcystis aeruginosa and microcystin by salad lettuce (Lactuca sativa) after spray irrigation with water containing cyanobacteria. Toxicon 37(8):1181–1185

Cox PA, Banack SA, Murch SJ, Rasmussen U, Tien G, Bidigare RB, Metcalf JS, Morrison LF, Codd GA, Bergman B (2005) Diverse taxa of cyanobacteria produce β–N–methylamino–L–alanine, a neurotoxic amino acid. Proc Natl Acad Sci U S A 102(14):5074–5078

de Magalhães VF, Soares RM, Azevedo SMFO (2001) Microcystin contamination in fish from the Jacarepaguá Lagoon (Rio de Janeiro, Brazil): ecological implication and human health risk. Toxicon 39(7):1077–1085

Duy TN, Lam PKS, Shaw GR, Connell DW (2000) Toxicology and risk assessment of freshwater cyanobacterial (blue–green algal) toxins in water. Rev Environ Contam Toxicol 163:113–185

Falconer IR (2005) Cyanobacterial toxins of drinking water supplies: cylindrospermopsins and microcystins. Boca Raton: CRC Press. 279pp.

Falconer IR, Beresford AM, Runnegar MT (1983) Evidence of liver damage by toxin from a bloom of the blue–green alga Microcystis aeruginosa. Med J Aust 1(11):511–514

Falconer IR, Burch MD, Steffensen DA, Choice M, and Coverdale OR (1994) Toxicity of the blue–green alga (cyanobacterium) Microcystis aeruginosa in drinking water to growing pigs, as an animal model for human injury and risk assessment. Environ Toxicol Water Qual 9(2):131–139.

Falconer IR, Hardy SJ, Humpage AR, Froscio SM, Tozer GJ, Hawkins PR (1999) Hepatic and renal toxicity of the blue–green alga (cyanobacterium) Cylindrospermopsis raciborskii in male Swiss albino mice. Environ Toxicol 14(1): 143–150

Falconer IR, Humpage AR (1996) Tumour promotion by cyanobacterial toxins. Phycologia 35(6 Suppl):74–79

Falconer IR, Humpage AR (2001) Preliminary evidence for in vivo tumour initiation by oral administration of extracts of the blue–green alga Cylindrospermopsis raciborskii containing the toxin cylindrospermopsin. Environ Toxicol 16(2):192–195

Falconer IR, Smith JV, Jackson AR, Jones A, Runnegar, MT (1988) Oral toxicity of a bloom of the cyanobacterium Microcystis aeruginosa administered to mice over periods up to one year. J Toxicol Environ Health 24(3):291–305.

Fawell JK, James CP, James HA (1994) Toxins from blue–green algae: Toxicological assessment of microcystin–LR and a method for its determination in water. Water Research Centre, Medenham, England Report number FR 0359/2/DoE 3358/2

Fernandez M, Cembella AD (1995) Mammalian bioassays. In: Manual on harmful marine microalgae, IOC Manuals and Guides No. 33. Edited by Hallegraeff GM, Anderson DM, Cembella AD. Paris: UNESCO 213-224

Gilroy DJ, Kauffman KW, Hall RA, Huang X, Chu FS (2000) Assessing potential health risks from microcystin toxins in blue–green algae dietary supplements. Environ Health Perspect 108(5):435–439

Guzman RE, Solter PF (2002) Characterization of sublethal microcystin–LR exposure in mice Vet Pathol 39(1):17–26

Håkanson L, Malmaeus JM, Bodemer U, Gerhardt V (2003) Coefficients of variation for chlorophyll, green algae, diatoms, cryptophytes and blue–greens in rivers as a basis for predictive modelling and aquatic management. Ecol Modell 169(1):179–196

Harada KI, Ohtani I, Iwamoto K, Suzuki M, Watanbe MF, Watanabe M, Terao K (1994) Isolation of cylindrospermopsin from a cyanobacterium Umezakia natans and its screening method. Toxicon 32(1):73–84

Hawkins PR, Chandrasena NR, Jones GJ, Humpage AR, Falconer IR (1997) Isolation and toxicity of Cylindrospermopsis raciborskii from an ornamental lake. Toxicon 35(3):341–346

Hawkins PR, Runnegar MTC, Jackson ARB, Falconer IR (1985) Severe hepatotoxicity caused by the tropical cyanobacterium (blue–green alga) Cylindrospermopsis raciborskii (Woloszynska) Seenaya and Subba Raju isolated from a domestic supply reservoir. Appl Environ Microbiol 50(5):1292–1295

Hay JE (2004) Acute liver failure. Curr Treat Options Gastroenterol 7(6):459–468

Hernandez M, Macia M, Padilla C, Del Campo FF (2000) Modulation of human polymorphonuclear leukocyte adherence by cyanopeptide toxins. Environ Res 84(1):64–68.

Hilborn ED, Carmichael WW, Yuan M, Azevedo SMFO (2005) A simple colorimetric method to detect biological evidence of human exposure to microcystins. Toxicon 46(2):218–221

Hindman SH, Favero MS, Carson LA, Petersen NJ, Schonberger LB, Solano JT (1975) Pyrogenic reactions during haemodialysis caused by extramural endotoxin. Lancet 2(7938):732–734

Hooser SB, Beasley VR, Basgall EJ, Carmichael WW, Haschek WM (1990) Microcystin–LR–induced ultrastructural changes in rats. Vet Pathol 27(1):9–15

Humpage AR, Falconer IR (2003) Oral toxicity of the cyanobacterial toxin cylindrospermopsin in male Swiss albino mice: determination of no observed adverse effect level for deriving a drinking water guideline value. Environ Toxicol 18(2):94–103

Humpage AR, Falconer IR (Nov. 2005) Personal communication

Humpage AR, Fenech M, Thomas P, Falconer IR (2000a) Micronucleus induction and chromosome loss in transformed human white cells indicate clastogenic and aneugenic action of the cyanobacterial toxin, cylindrospermopsin. Mutat Res 472(1–2):155–161

Humpage AR, Hardy SJ, Moore EJ, Froscio SM, Falconer IR (2000b) Microcystins (cyanobacterial toxins) in drinking water enhance the growth of aberrant crypt foci in the mouse colon. J Toxicol Environ Health A 61(3):155–165

Humphrey JM, Aggen JB, Chamberlin AR (1996) Total synthesis of the serine–threonine phosphatase inhibitor microcystin–LA. J Am Chem Soc 118(47): 11759–11770.

Ito E, Kondo F, Terao K, Harada K (1997) Neoplastic nodular formation in mouse liver induced by repeated intraperitoneal injections of microcystin–LR Toxicon 35(9):1453–1457

Jeong KS, Kim DK, Whigham P, Joo GJ (2003) Modelling Microcystis aeruginosa bloom dynamics in the Nakdong River by means of evolutionary computation and statistical approach Ecol Modell 161(1–2):67–78

Jochimsen EM, Carmichael WW, An J, Cardo DM, Cookson ST, Holmes CEM, Antunes MB, de Melo Filho DA, Lyra TM, Barreto VST, Azevedo SMFO, Jarvis WR (1998) Liver failure and death after exposure to microcystins at a hemodialysis center in Brazil. N Engl J Med 338(13):873–878

Kuiper–Goodman T, Falconer I, Fitzgerald J (1999) Human health aspects. In: Toxic cyanobacteria in water: a guide to their public health consequences, monitoring and management. Edited by Chorus I, Bartram J. London: E & FN Spon on behalf of the World Health Organization 113–153

Lawrence JF, Niedzwiadek B, Menard C, Lau BPY, Lewis D, Kuper–Goodman T, Carbone S, Holmes C (2001) Comparison of liquid chromatography/mass spectrometry, ELISA, and phosphatase assay for the determination of microcystins in blue–green algae products. J AOAC Int 84(4):1035–1044

MacPhail RC, Farmer JD, Jarema KA, Chernoff N (2005) Nicotine effects on the activity of mice exposed prenatally to the nicotinic agonist anatoxin–a. Neurotoxicol Teratol 27(4):593–598

MacPhail RC, Jarema, KA (2005) Prospects on behavioral studies of marine and freshwater toxins. Neurotoxicol Teratol 27(5):695–699

Maier HR, Dandy GC, Burch MD (1998) Use of artificial neural networks for modelling cyanobacteria Anabaena spp. in the River Murray, South Australia. Ecol Modell 105(2–3):257–272

Metcalf JS, Meriluoto JA, Codd GA (2006) Legal and security requirements for the air transportation of cyanotoxins and toxigenic cyanobacterial cells for legitimate research and analytical purposes. Toxicol Lett 163(2):85–90

Ming L, Thorgeirsson SS, Gail MH, Lu P, Harris CC, Wang N, Shao Y, Wu Z, Liu G, Wang X, Sun Z (2002) Dominant role of hepatitis B virus and cofactor role of aflatoxin in hepatocarcinogenesis in Qidong, China. Hepatology 36(5):1214–1220

Nagai H, Yasumoto T, Hokama Y (1996) Aplysiatoxin and debromoaplysiatoxin as the causative agents of a red alga Gracilaria coronopifolia poisoning in Hawaii. Toxicon 34(7):753–761

Nardo B, Caraceni P, Montalti R, Puviani L, Bertelli R, Beltempo P, Pacilè V, Rossi C, Gaiani S, Grigioni W, Bernardi M, Martinelli G, Cavallari A (2005) Portal vein arterialization: a new surgical option against acute liver failure? Transplant Proc 37(6):2544–2546

Negri AP, Jones GJ (1995) Bioaccumulation of paralytic shellfish poisoning (PSP) toxins from the cyanobacterium Anabaena circinalis by the freshwater mussel Alathyria condola. Toxicon 33(5):667–678

Nguyen TH, Mai G, Villiger P, Oberholzer J, Salmon P, Morel P, Bühler L, Trono D (2005) Treatment of acetaminophen–induced acute liver failure in the mouse with conditionally immortalized human hepatocytes. J Hepatol 43(6):1031–1037

Osborne NJT, Webb PM, Shaw GR (2001) The toxins of Lyngbya majuscula and their human and ecological effects. Environ Int 27(5):381–392

Park JK, Lee DH (2005) Bioartificial liver systems: current status and future perspective. J Biosci Bioeng 99(4):311–319

Prokopkin IG, Gubanov VG, Gladyshev MI (2006) Modelling the effect of planktivorous fish removal in a reservoir on the biomass of cyanobacteria. Ecol Modell 190(3–4):419–431

Qian GS, Ross RK, Yu MC, Gao YT, Henderson BE, Wogan GN, Groopman JD (1994) A follow–up study of urinary markers of aflatoxin exposure and liver cancer risk in Shanghai, People's Republic of China. Cancer Epidemiol Biomarkers Prev 3(1):3–10

Rao PVL, Gupta N, Jayaraj R, Bhaskar ASB, Jatav PC (2005) Age–dependent effects on biochemical variables and toxicity induced by cyclic peptide toxin microcystin–LR in mice. Comp Biochem Physiol C Toxicol Pharmacol 140(1):11–19

Recknagel F, French M, Harkonen P, Yabunaka KI (1997) Artificial neural network approach for modelling and prediction of algal blooms. Ecol Modell 96(1–3):11–28

Reich A (Nov. 2005) Personal communication

Ressom R, Soong FS, Fitzgerald J, Turczynowicz L, El Saadi O, Roder D, Maynard T, Falconer I (1994) Health effects of toxic cyanobacteria (blue–green algae). Canberra: National Health and Medical Research Council & Australian Government Publishing Service

Robson BJ, Hamilton DP (2004) Three–dimensional modelling of a Microcystis bloom event in the Swan River estuary, Western Australia. Ecol Modell 174(1–2):203–222

Rogers EH, Hunter III ES, Moser VC, Phillips PM, Herkovits J, Munoz L, Hall LL, Chernoff N (2005) Potential developmental toxicity of anatoxin–a, a cyanobacterial toxin. J Appl Toxicol 25(6):527–534

Ross RK, Yuan JM, Yu MC, Wogan GN, Qian GS, Tu JT, Groopman JD, Gao YT, Henderson BE (1992) Urinary aflatoxin biomarkers and risk of hepatocellular carcinoma. Lancet 339(8799):943–946

Saker ML, Eaglesham GK (1999) The accumulation of cylindrospermopsin from the cyanobacterium Cylindrospermopsis raciborskii in tissues of the Redclaw crayfish Cherax quadricarinatus. Toxicon 37(7):1065–1077

Saker ML, Jungblut AD, Neilan BA, Rawn DFK, Vasconcelos VM (2005) Detection of microcystin synthetase genes in health food supplements containing the freshwater cyanobacterium Aphanizomenon flos–aquae. Toxicon 46(5):555–562

Schaeffer DJ, Malpas PB, Barton LL (1999) Risk assessment of microcystin in dietary Aphanizomenon flos–aquae. Ecotoxicol Environ Saf 44(1):73–80

Shaw GR, Seawright AA, Moore MR, Lam PKS (2000) Cylindrospermopsin, a cyanobacterial alkaloid: evaluation of its toxicologic activity. Ther Drug Monit 22(1):89–92

Shen PP, Zhao SW, Zheng WJ, Hua ZC, Shi Q, Liu ZT (2003) Effects of cyanobacteria bloom extract on some parameters of immune function in mice. Toxicol Lett 143(1):27–36

Shi J, Zhu L, Liu S, Xie WF (2005) A meta–analysis of case–control studies on the combined effect of hepatitis B and C infections in causing hepatocellular carcinoma in China. Br J Cancer. 92(3):607–612

Shi Q, Cui J, Zhang J, Kong FX, Hua ZC, Shen PP (2004) Expression modulation of multiple cytokines in vivo by cyanobacteria blooms extract from Taihu Lake, China. Toxicon 44(8):871–879

Sivonen K, Jones G (1999) Cyanobacterial toxins. In: Toxic cyanobacteria in water: a guide to their public health consequences, monitoring and management. Edited by Chorus I, Bartram J. London: E & FN Spon on behalf of the World Health Organization 41–111

Soares RM, Yuan M, Servaites JC, Delgado A, Magalhães VF, Hilborn ED, Carmichael WW, Azevedo SMFO (2006) Sublethal exposure from microcystins to renal insufficiency patients in Rio de Janeiro, Brazil. Environ Toxicol 21(2):95–103

Stevens DK, Krieger RI (1991) Effect of route of exposure and repeated doses on the acute toxicity in mice of cyanobacterial nicotinic alkaloid anatoxin–a. Toxicon 29(1):134–138

Stewart I, Webb PM, Schluter PJ, Shaw GR (2006) Recreational and occupational field exposure to freshwater cyanobacteria – a review of anecdotal and case reports, epidemiological studies and the challenges for epidemiologic assessment. Environ Health 5(1):6

Sullivan JJ (1993) Methods of analysis for cyanobacterial toxins: dinoflagellate and diatom toxins. In: Cyanobacterial toxins in seafood and drinking water. Edited by Falconer IR. London: Academic Press 29-48

Terao K, Ohmori S, Igarashi K., Ohtani I, Watanabe MF, Harada KI, Ito E, Watanabe M (1994) Electron microscopic studies on experimental poisoning in mice induced by cylindrospermopsin isolated from blue–green alga Umezakia natans. Toxicon 32(7):833–843

Thébault, JM, Rabouille, S (2003) Comparison between two mathematical formulations of the phytoplankton specific growth rate as a function of light and

temperature, in two simulation models (ASTER & YOYO) Ecol Modell 163(1–2):145–151
Tissières P, Sasbón JS, Devictor D (2005) Liver support for fulminant hepatic failure: is it time to use the molecular adsorbents recycling system in children? Pediatr Crit Care Med 6(5):585–591
Turner PC, Gammie AJ, Hollinrake K, Codd GA (1990) Pneumonia associated with contact with cyanobacteria. BMJ 300(6737):1440–1441
Valentine WM, Schaeffer DJ, Beasley VR (1991) Electromyographic assessment of the neuromuscular blockade produced in vivo by anatoxin–a in the rat. Toxicon 29(3):347–357
Van de Kerkhove AP, Poyck PPC, Deurholt T, Hoekstra R, Chamuleau RAFM, van Gulik TM (2005) Liver support therapy: an overview of the AMC–bioartificial liver research. Dig Surg 22(4):254–264
Watts JS, Reilly J, DaCosta FM, Krop S (1966) Acute toxicity of paralytic shellfish poison in rats of different ages. Toxicol Appl Pharmacol 8(2):286–294
WHO, International Programme on Chemical Safety (1984) Aquatic (marine and freshwater) biotoxins. Environmental Health Criteria 37. Geneva: World Health Organization. 95pp.
Wiberg GS, Stephenson NR (1960) Toxicologic studies on paralytic shellfish poison. Toxicol Appl Pharmacol 2:607–615
Xie C, Runnegar MTC, Snider BB (2000) Total synthesis of (±)-cylindrospermopsin. J Am Chem Soc 122(21):5017–5024
Yu MC, Yuan JM (2004) Environmental factors and risk for hepatocellular carcinoma. Gastroenterol 127(Suppl 1):S72–78
Yu SZ, Huang XE, Koide T, Cheng G, Chen GC, Harada K, Ueno Y, Sueoka E, Oda H, Tashiro F, Mizokami M, Ohno T, Xiang J, Tokudome S (2002) Hepatitis B and C viruses infection, lifestyle and genetic polymorphisms as risk factors for hepatocellular carcinoma in Haimen, China. Jpn J Cancer Res 93(12):1287–1292

Chapter 27: Health effects associated with controlled exposures to cyanobacterial toxins

Ian R Falconer

Pharmacology, University of Adelaide Medical School, Adelaide, South Australia and Cooperative Research Centre for Water Quality and Treatment, Salisbury, South Australia.

Abstract

The cyanobacterial toxins of concern as potential human health hazards are those known to occur widely in drinking water sources, and therefore may be present in water for human use. The toxins include a diverse range of chemical compounds, with equally diverse toxic effects. These toxins are not limited to individual cyanobacterial species or genera, and all of the toxins of concern to human health are produced by multiple cyanobacterial species.

Microcystins

The acute effects of microcystins have been investigated mainly by intraperitoneal (i.p.) dosing or oral dosing of experimental animals. Limited information is available from inhalation or intranasal dosing of rodents. No controlled dosing trials have been undertaken with human subjects.

It is well established that the main target for toxic effects of microcystins in mammals is the liver, with adverse effects also seen in the small intestine and kidney. In large animals hemorrhagic responses have also been seen after oral dosing of toxic extracts. Hepatotoxicity is a primary response because microcystins are transported actively into hepatocytes via organic anion transporters which concentrate the toxin in target cells. The toxicity of microcystins results from the inhibition of the catalytic subunit of protein phosphatases 1 and 2A. These enzymes play a vital role in intracellular regulation, in balancing the phosphorylation and dephosphorylation of regulatory and structural proteins, hence altering enzyme activities and structural integrity of the hepatocyte. As a consequence hepatocytes are highly sensitive to microcystin toxicity; the concentration of microcystin causing 50% rat hepatocyte deformity during *in vitro* incubation for 30 min is 30nM. This initial rapid response of hepatocytes to microcystin exposure is cell deformation caused by disruption of the cytoskeleton through hyperphosphorylation of intermediate filament proteins. *In vivo* this results in disintegration of the hepatic cellular architecture, with extensive bleeding into the liver resulting in rapid death of the animal. The i.p. LD_{50} over 24h for rodents for most microcystin variants is 50-100μg/kg bodyweight. The microcystin variants having two arginine residues in the molecule (MCYST-RR variants) have lower toxicity, with an approximate LD_{50} of 300μg/kg. The earliest time-to–death after a lethal intraperitoneal dose in mice is approximately 20-30 minutes. Enterocytes lining the small intestine also actively take up microcystins, and show the same deformation responses *in vitro* to microcystins as do hepatocytes. This damage to enterocytes is likely to be the cause of some of the observed gastrointestinal responses to oral microcystin poisoning.

Intranasal exposure to microcystin-LR is equally toxic as i.p. exposure, with extensive damage to nasal mucosa. Oral toxicity is 5-10 fold lower than i.p. toxicity, and in mice is affected by age, with older mice being more sensitive. Male mice are appreciably more sensitive than female mice, which require at least a two-fold higher microcystin dose for the same adverse response. Time-to-death from a lethal oral dose in rodents is variable, around 6-20 hours. In sheep acute lethal effects are seen from 18h after intra-ruminal dosing, with major liver damage.

Oral sub-chronic dosing trials with microcystins have been undertaken in mice and pigs, with the aim of determining a maximum No Observed Adverse Effect Level (NOAEL). In mice this was carried out by gavage with pure microcystin-LR, and a NOAEL for male mice was determined at 40μg/kg/day, based primarily on histopathological evidence of liver injury and serum analysis for liver function enzymes. In pigs the microcystins supplied though the drinking water were a thoroughly characterised *Microcystis aeruginosa* extract containing several microcystins. This trial resulted in determination of a Lowest Observed Adverse Effect Level (LOAEL) of 100μg/kg/day, also by histopathological examination of the liver. Chronic oral toxicity trials in mice using extracts of *Microcystis* supplied in the drinking water for upto one year showed increased general mortality from infections, and chronic active liver injury at higher doses.

Exposure to drinking water containing *Microcystis* extracts gave evidence of increased tumours in these mice.

Laboratory investigation of tumour promotion in rodents by microcystin has given strong evidence of non-phorbol ester type promotion. *In vivo* exposure to microcystin-LR has shown increased foci of pre-neoplastic cells in livers of mice initiated with carcinogen. Similarly supply of microcystin in drinking water to mice given azoxymethane as a colon tumour initiator, showed increased growth of hyperplastic colon crypts. Mice treated on the skin with the carcinogen dimethylbenzanthracine showed marked increases in papilloma growth when given *Microcystis* extract in drinking water.

The genotoxic effects of microcystin are still under discussion, with evidence from different investigations of both genetic damage and of lack of genotoxicity. There is also evidence of liver carcinogenesis by high repeated intraperitoneal doses of microcystin in mice. There are no non-rodent long term studies of chronic toxicity or carcinogenicity at present.

Reproductive toxicity studies have given conflicting results, possibly due to toxic mixtures of *Microcystis* extracts being injected intraperitoneally in mice. In studies using pure microcystin-LR, or by oral exposure of mice to microcystins in extracts, no clear reproductive or developmental injuries were reported.

Human toxicity is relevant to this discussion, though all evidence of human injury from microcystins is necessarily obtained from retrospective investigation with actual exposures not readily quantified. An epidemiological study of human liver function in populations exposed and unexposed to drinking water drawn from a reservoir carrying a large *Microcystis* bloom (treated with copper sulphate) showed a response characteristic of hepatotoxicity only in the exposed population, and only at the time of the *Microcystis* bloom. In a Brazilian clinic more than 50 dialysis patients died after perfusion with water containing the cyanobacterial toxins microcystin and cylindrospermopsin, with clear evidence of major liver damage.

In southern China there are areas of very high rates of primary liver cancer, and epidemiological studies have linked this to drinking of surface water containing microcystins, as well as to aflatoxin in the diet and infection by viral hepatitis. In another area in China, colon cancer rates have shown a correlation with concentration of microcystins in the various water sources.

Recent evaluation of carcinogenesis from microcystin exposure by the International Agency for Research in Cancer has determined that microcystin-LR and other microcystins are **possible carcinogens**, based substantially on their demonstrable tumour promoting effects in laboratory studies with rodents.

Cylindrospermopsins

These alkaloids have been investigated by i.p. and oral dosing of rodents, but not yet of larger mammals due to limited supply and high cost of the toxins. They are general cytotoxins, causing damage to liver, kidneys, gastrointestinal tract, endocrine organs, the immune system, vascular system and muscle. There appear to be two toxic responses, the rapid toxicity probably linked to a toxic metabolite formed by oxidation in the liver, as there is protection from toxicity by some cyt-P450 inhibitors. The other slower toxic response is due to protein synthesis inhibition, which is a longer-term toxicity showing much higher sensitivity to cylindrospermopsin. As a result, the i.p. LD_{50} in mice at 24h was 2,100 μg/kg bodyweight and the LD_{50} at 5-6 days was 200μg/kg. The acute oral LD_{50} is not yet clearly established in experimental animals, but appears to be about 5,000μg/kg.

Sub-chronic oral toxicity trials by gavage with purified toxin determined a NOAEL in male mice of 30μg/kg/day. This was based on adverse effects on kidney function, which appeared more sensitive than liver injury to cylindrospermopsin toxicity. A long-term oral toxicity study in male and female mice which used hematology and erythrocyte morphology as the most sensitive indicators of adverse effects, determined a Lowest Observed Adverse Effect Level (LOAEL) of 20μg/kg/day in both sexes of mice. Derivation of a provisional Guideline Value for cylindrospermopsin in drinking water from these data resulted in values of approximately 1μg/l., using the standard safety factors for subchronic toxicity in rodents.

In-vitro genotoxicity and mutagenicity testing of cylindrospermopsin in hepatocytes, lymphocytes and several transformed cell lines has demonstrated that it was genotoxic and mutagenic.

Preliminary results for oral carcinogenicity in mice indicate that it may be carcinogenic in a range of tissues.

Data are lacking for the detailed mechanism of action of cylindrospermopsin, though it has been shown that protein synthesis inhibition is exerted at the ribosome level. This is presently under investigation. Current unpublished studies of teratogenicity and developmental toxicity should be available shortly. Chronic oral carcinogenicity studies on cylindrospermopsin are urgently needed for both rodents and non-rodent species.

A human population on an island were exposed to drinking water drawn from a reservoir carrying a heavy bloom of *Cylindrospermopsis,* suffered acute hepato/gastroenteritis after the reservoir was treated with copper sulphate. About 150 children and adults were treated in the hospital, with 85 cases considered severe enough to be airlifted to a more advanced hospital with intensive care facilities. None died. Cylindrospermopsin was subsequently isolated from cultured *Cylindrospermopsis raciborskii* obtained from this reservoir.

Anatoxins

Anatoxin-a and homo-anatoxin-a are small neurotoxic alkaloids, which act as agonists at the neuromuscular junction, causing spontaneous firing and eventually death by respiratory failure. The acute i.p. LD_{50} is 375µg/kg in mice, the intravenous (i.v.) LD_{50} is less than 100µg/kg, with an intranasal LD_{50} of 2000µg/kg and no lethality observed at a 5000µg/kg oral dose. Repeated i.p. injection did not elicit resistance to toxicity. Data are available for sub-chronic oral toxicity. Anatoxin-a by gavage at 15,000µg/kg killed mice within 3 minutes, however 3 of 4 mice receiving 7,500µg/kg/day for 4 weeks survived with no post-mortem pathological changes. Nineteen of 20 mice receiving 3,000µg/kg/day for 4 weeks showed no effects. There is no evidence for reproductive, teratogenic or carcinogenic effects of anatoxin-a.

Anatoxin-a(s) is an organophosphate anticholinesterase causing salivation, muscle weakness, convulsions, and death by respiratory paralysis. The i.p. LD_{50} is 20µg/kg in mice, there are no oral toxicity data.

Dog deaths due to consumption of anatoxin-a and homo-anatoxin-a from cyanobacterial sources have been reported from several countries.

Though a recent accidental death of a teenager in the USA was attributed to consumption of anatoxin-a during swimming, there is doubt over the analytical results. No population-level adverse health effects have been reported.

Saxitoxins

These alkaloids block sodium channels in nerve axons, causing loss of sensation and paralysis and are highly toxic. Saxitoxins are best known as paralytic shellfish poisons, which have caused many human fatalities. The i.p. LD_{50} in mice is 8-10µg/kg, the i.v. LD_{50} is 3.4µg/kg and the oral LD_{50} is 260µg/kg for saxitoxin. Other saxitoxin variants have lower toxicity. Young rats are more susceptible than adults, and prior exposure appears to reduce susceptibility. There are no experimental data for subchronic exposure, reproductive, teratogenic or carcinogenic effects of saxitoxins.

Much human health data for saxitoxin poisoning is available as a result of the population and individual poisonings reported over the last 300 years. National regulatory agencies have set maximum limits for saxitoxins in shellfish for human consumption, and commercial and recreational harvesting is prohibited when shellfish samples contain saxitoxins above the regulatory limit. There does not appear to be evidence for lasting effects of poisoning by saxitoxin, unlike the effects seen following poisoning by other neurotoxic marine shellfish poisons.

Reading

A detailed current account of the health effects related to exposure to cylindrospermopsins and microcystins may be found in

Falconer IR (2005) Cyanobacterial Toxins of Drinking Water Supplies; Cylindrospermopsins and Microcystins. CRC Press, Boca Raton, FL, pp 279

A broad review of cyanobacterial toxins, including ecology and monitoring, can be found in

Chorus I , Bartram J (1999) Toxic Cyanobacteria in Water: A Guide to their Public Health Consequences, Monitoring and Management. E & FN Spon, London, pp 416

A review of the entire field of marine and freshwater algal and cyanobacterial toxins can be found in

Falconer IR (Ed) (1993) Algal toxins in Seafood and Drinking Water. Academic Press, London, pp 224

Chapter 28: Cyanobacterial poisoning in livestock, wild mammals and birds – an overview

Ian Stewart[1,2], Alan A. Seawright[1], Glen R. Shaw[2,3]

[1]: National Research Centre for Environmental Toxicology, University of Queensland, Brisbane, Australia; [2]: Cooperative Research Centre for Water Quality and Treatment, Adelaide, Australia; [3]: School of Public Health, Griffith University, Brisbane, Australia

Abstract

Poisoning of livestock by toxic cyanobacteria was first reported in the 19th century, and throughout the 20th century cyanobacteria–related poisonings of livestock and wildlife in all continents have been described. Some mass mortality events involving unrelated fauna in prehistoric times have also been attributed to cyanotoxin poisoning; if correct, this serves as a reminder that toxic cyanobacteria blooms predate anthropogenic manipulation of the environment, though there is probably general agreement that human intervention has led to increases in the frequency and extent of cyanobacteria blooms. Many of the early reports of cyanobacteria poisoning were anecdotal and circumstantial, albeit with good descriptions of the appearance and behaviour of cyanobacteria blooms that preceded or coincided with illness and death in exposed animals. Early necropsy findings of hepatotoxicity were subsequently confirmed by experimental investigations. More recent reports supplement clinical and post–mortem findings with investigative chemistry techniques to identify cyanotoxins in stomach contents and tissue fluids.

Introduction

Planktonic blooms and surface scums appear to have been recognised for over two thousand years, and indigenous peoples of North America, Africa and Australia may have been aware of the poisonous nature of cyanobacteria (Codd et al. 2005). Detailed studies – including microscopy – of the behaviour of planktonic cyanobacteria blooms began in the late 19th century. Surface scums that were concentrated by wind activity were observed, having a sudden onset and dissipation (Phillips, 1884). The world's first scientific report of the toxic nature of cyanobacteria was written by George Francis, who described in 1878 the rapid death of stock animals at Lake Alexandrina, a freshwater lake at the mouth of the Murray River in South Australia (Francis 1878). Francis described a wind–blown surface scum of the brackish water cyanobacterium *Nodularia spumigena*, incidental consumption of which resulted in the death of sheep, horses, dogs and pigs within periods of 1 to 24 hours. Francis also experimentally dosed a sheep with fresh bloom material. Necropsy examination showed a significant pericardial effusion, "two pints of yellow serum" (i.e. most likely ascitic fluid) in the abdominal cavity, and normal–looking liver, lungs and kidneys (Francis 1878). Codd et al (1994) note that the observations of Francis are entirely in agreement with the modern understanding of *Nodularia*–associated intoxication through its associated cyanotoxin, nodularin. Beasley et al (1989a, 1989b) suggest that another cyanotoxin may have been involved, given that Francis did not report any hepatic pathology characteristic of nodularin intoxication, however the liver lesion caused by nodularin may not have been apparent or obvious to the inexperienced observer. The description of ascitic fluid in this context is strongly suggestive of acute liver damage. Beyond Francis's (likely) observation of ascitic fluid, and the necropsy findings of (possibly) a single animal, it is probably unwise to attempt a reinterpretation of Francis's findings (Francis did report opening "many sheep that died", but it is not clear from the report whether his examinations of those field deaths extended beyond his obvious interest in the apparent absence of bloom material from the stomachs) (Francis 1878).

Since Francis's report in the last quarter of the 19th century there have been numerous descriptions of mammal and bird mortalities associated with exposure to cyanobacteria. The relevant literature throughout the 20th century represents all inhabited continents. Anecdotal and case reports have been collated and/or reviewed by many authors, e.g. (Schwimmer and Schwimmer 1955, 1968; Schwimmer and Schwimmer 1964; Hammer 1968; Codd and Beattie 1991; Carmichael and Falconer 1993; Ressom et al. 1994; Duy et al. 2000; Briand et al. 2003; Codd et al. 2005; Falconer 2005). Many of the earlier reports are circumstantial, noting a temporal association between the presence of cyanobacterial blooms in a waterbody and otherwise unexplained mortalities. Early necropsy findings of hepatotoxicity were subsequently confirmed by experimental toxicology studies into isolated and purified cyanotoxins. More recent reports supplement clinical and post–mortem findings with investigative chemistry techniques to identify cyanotoxins in stomach contents and organs.

Discussion of wild and domestic animal poisonings can be found in most reviews of toxic cyanobacteria, and mention of field deaths often served to introduce and contextualise reports of experimental toxicological investigations into cyanobacterial suspensions, extracts or isolates. As the number of primary anecdotal and case reports and reviews into cyanobacteria–related mortality of birds and terrestrial mammals is extensive, a systematic, comprehensive review has not been conducted in recent times. Such a task is also beyond the scope of this review; therefore we will present some selected discussion of early reports, including palaeontologic findings, then we will discuss more recent investigations that incorporate modern investigative chemistry techniques to supplement clinical and post–mortem findings in order to diagnose cyanotoxin poisoning.

Early reports (pre–1960s)

Lake Alexandrina in South Australia, as mentioned in the Introduction, was a trouble–spot for cyanobacteria–related poisonings, with several hundred deaths associated with a number of bloom events in the latter part of the 19th century. The local Aboriginal people were apparently aware of poisonings in the lake at around 1850 (Codd et al. 1994).

Another early locus was South Africa, where Steyn (1945) suggested that "many thousands" of cattle and sheep mortalities over the preceding 25–30 years were attributable to cyanobacterial poisoning; horses, mules, donkeys, dogs, hares, poultry and waterbirds found dead near affected waterbodies were also thought to be similarly affected. A toxic bloom on the Vaal Dam in 1942 covered an estimated 98 per cent of the reservoir, this being at the time 120km x 24km (at its broadest point) with a 1 teralitre capacity (Stephens 1949). *Microcystis* sp. were identified in the water; fulminant hepatic disease was observed in some animals, as were subacute cases with death ensuing after some two weeks and chronic cases with weeks or months of morbidity preceding either death or recovery. Necropsy findings in fulminant cases included pulmonary haemorrhage, haemoperitoneum, hepatomegaly (with livers dark red or black in colour, and friable), splenomegaly, and occasional "bloody patches…on the mucous membranes of [the stomach and intestines]." Findings in subacute cases were haemo–serous exudate in the pericardium, thorax and peritoneal cavity, the liver being soft or friable and yellow–coloured. Experimental dosing of animals with dam water produced identical clinical signs and post–mortem lesions. Secondary photosensitivity was described in detail: painful inflammation, fissuring of affected epithelium, purulent discharge, sloughing of skin. Unpigmented areas are mainly affected – muzzle, udder and teats, ears (Steyn 1943).

Cyanobacteria–related poisonings were reported in North America also from the latter half of the 19th century – see citations in Codd et al (2005). Howard and Berry (1933) discussed *Anabaena*–related cattle and other animal deaths at an Ontario lake in 1924. Storm Lake in Iowa experienced dramatic bloom events in 1952: associated with *Anabaena flos–aquae* blooms were estimated deaths of 5–7,000 gulls, 560 ducks, 400 coots, 200 pheasants, 50 squirrels, 18 muskrats, 15 dogs, 4 cats, 2 hogs, 2 hawks, 1 skunk, 1 mink, plus "numerous" songbirds. These figures were taken from the recorded number of animals buried by a field worker at the time, so were most probably underestimates of the total number of deaths. Signs of neurotoxicity were seen: prostration and convulsions preceded death; milder cases displayed restlessness, weakness, dyspnoea and tonic spasms. 57 weak and partially paralysed mallards recovered following gastric lavage. A dog drinking the water when sick birds were being gathered then died near the lake shore when trying to escape from workers. Botulism was excluded from the diagnoses, and necropsy findings were unremarkable. Experimental oral or parenteral dosing of bloom filtrate to laboratory rodents and chickens resulted in rapid death – within minutes, though delayed mortalities were reported: one chicken lived for 48 hours, and guinea pigs exposed to bloom filtrate orally or by i.p. injection died "within 24 hours" (Firkins 1953, Rose 1953).

While some of the aforementioned reports were dramatic in terms of the number of animals affected by specific bloom events, equally compelling were early descriptions of the rapidity with which large animals succumbed after exposure to (presumably) neurotoxic cyanobacteria. McLeod and Bondar (1952) noted that a horse, "several" calves, two pigs and a cat all died within an hour of consuming cyanobacteria–affected water at Lake Dauphin, Manitoba in 1945. At the same site in 1951, nine dogs and a horse died within one hour; the horse showed signs of progressive muscular weakness and paralysis, with profuse sweating. Experimental dosing of laboratory rodents was conducted using bloom material (99% *Aphanizomenon flos–aquae*) collected from the lake six days after the last reported poisoning event. Interestingly, oral or i.p. injections of lake water filtrate were not toxic, whereas unfiltered lake water, various preparations of cyanobacterial residue or filtrate from disrupted cellular material resulted in deaths ranging from under 12 hours to around 65 hours. Clinical signs were those of "loss of equilibrium followed by progressive paralysis." Later signs were clonic muscular spasms and respiratory dysfunction leading to pronounced cyanosis. Post–mortem findings from these experimental animals were not reported (McLeod and Bondar 1952).

Prehistoric animal mortalities

The first detailed descriptions of the behaviour of cyanobacteria blooms and cyanobacterial toxicity were written in the latter part of the 19th century (see Introduction), but there is no reason to suppose that both phenomena were not present at earlier stages of our history. Codd et al (1994) suggest that Pliny the Elder may have described a cyanobacterial or algal scum on the River Dnieper in AD77, and Höger (2003) implies that an Old Testament reference may be a description of a piscicidal *Planktothrix* bloom in the River Nile. These contemporaneous descriptions reach modern researchers through the device of written language. However, two reports from the palaeontology literature suggest that cyanobacterial toxicity may be a prehistoric phenomenon. Braun and Pfeiffer (2002) present their hypothesis that at Neumark–Nord in Germany, a Pleistocene (1.8 million – 11,000 years ago) lake assemblage of >70 deer, as well as forest elephant, rhinoceros, auroch (ox) and cave lion skeletons may represent a cyanotoxin–related mass mortality event. The preservation features and antler development suggest rapid, catastrophic death in autumn. Calcified layers in the sediment are thought to represent decomposed cyanobacteria, and absorption spectra of cyanobacteria–specific carotenoids were identified in sediment extracts. The authors used HPLC with UV detection to determine the presence of microcystins in methanol extracts of sediment; a similar (but not identical) UV spectrum to that obtained from an extract of a *Microcystis aeruginosa* strain with an unspecified microcystin congener profile was seen. Other possible explanations for the deaths at Neumark–Nord – butchery by early humans, or a gas eruption – are discussed by the authors but thought to be unlikely (Braun and Pfeiffer 2002). Considering further the cyanotoxin hypothesis, it would be interesting to see further confirmatory investigations for microcystins in the sediment extracts using a second method such as mass spectrometry. Looking beyond the search for microcystins in the sediment, it would seem that sudden, catastrophic death due to ingestion of a cyanobacterial neurotoxin could equally explain these events.

An older mammal and bird assemblage from the Middle Eocene epoch (49 – 37 million years ago) also has features suggestive of a mass poisoning event. The Messel oil shale pit in Germany was an ancient freshwater lake. While the sediment layers are more compacted than those of Neumark–Nord, and chemical and physical transformation of sediments by diagenesis is likely, Koenigswald et al (2004) suggest that there are marked similarities across both sites, and thin layers of siderite at Messel probably represent the decomposed remains of sunken cyanobacteria blooms. Well–preserved skeletons of horses, turtles, bats and birds are found; most were in good health, and stomach contents show that they were well nourished. Apparent features of seasonal death are seen: of some 50 dawn horse skeletons, five are of pregnant mares. The foetuses are all in late stages of development, which again suggests seasonal mortality, as predator pressure on newborn ungulates drives seasonal parturition, usually in late spring. These skeletons were found in different layers, so the picture is that of seasonal deaths occurring in different years. The frequent presence of winged animal skeletons at Messel, which is unusual in lake deposits, also supports a toxic waterbloom hypothesis (Koenigswald et al. 2004). The authors also discuss a finding of aquatic turtle specimens in which two animals are preserved as pairs in close proximity to each other. "At least" five slabs of shale featuring such pairs of turtles have been found; photographs of two pairs are presented in the paper. The authors cite another worker's interpretation of this finding as representing sudden death during mating, which would support a toxic cyanobacteria bloom hypothesis (Koenigswald et al. 2004). However the photographs show two pairs of turtles apparently fused posteriorly but facing away from each other, which is not the normal copulatory position adopted by turtles. The male mounts the female, though infrequent reports of "belly–to–belly" position can be found in the literature (Goode 1967, Kuchling 1999).

If these reports do indeed describe ancient cyanobacteria–related intoxications, they serve as a reminder that toxic cyanobacteria blooms are natural events; they predate recent anthropogenic manipulation of the environment, and may indeed predate human arrival. However, most cyanobacteriologists would acknowledge that the problem of harmful cyanobacteria is exacerbated in modern times due to cultural eutrophication, and may worsen in future with the additional impact of human activity–related climate warming and other population–related pressures.

Recent developments

While many early reports and reviews of cyanobacteria–related poisonings are impressive because of their comprehensive nature and empirical observations supplemented by simple experimental techniques, more recent reports of mammal and bird poisonings reveal the advances in analytical techniques that allow definitive identification of well–characterised cyanotoxins. The following section will present discussion of recent case studies that investigated animal poisonings caused – or likely to have been caused – by cyanobacteria that produce toxins from each of the main cyanotoxin groups that have been characterised to date. Clinical presentation, necropsy findings that reflect the current understanding of the mechanisms of toxicity, and modern investigative chemistry procedures that confirm the diagnosis are discussed. These examples are by no means exhaustive, as different analytical approaches may be applied, and factors such as dose, time period of exposure and ingestion of more than one type of cyanotoxin may complicate the diagnosis.

The experimental toxicology of these isolated cyanotoxins has been well documented, and many excellent reviews are available, e.g. (Ressom et al. 1994, Codd et al. 1999, Carmichael 2001, Falconer 2005). Peracute microcystin and nodularin poisoning presents principally with fulminant haemorrhagic liver injury; pathologic changes can also be seen in the gut, kidney, heart and lungs. Cylindrospermopsin intoxication is characterised by fatty degeneration of the liver, cholestasis, injury to renal tubules and thymic atrophy. Experimental dosing with cylindrospermopsin–producing cyanobacterial extracts results in pathological effects on other tissues and organ systems, including the gastrointestinal tract and spleen.

Cyanobacterial neurotoxins cause death by respiratory failure through various mechanisms: anatoxin–a by a depolarising block at the neuromuscular junction; anatoxin–a(s) is an anticholinesterase, thus preventing the enzymatic degradation of acetylcholine which results in uncontrolled hyperstimulation of muscles. Hypersalivation and lacrimation, which are signs of parasympathetic stimulation, also characterise anatoxin–a(s) poisoning. The saxitoxin group are sodium channel blocking agents, thus interfering with axonal conduction. Death is very rapid – within minutes – when purified cyanobacterial neurotoxins are administered parenterally, particularly from anatoxin–a or saxitoxins. Post–mortem and histology findings are generally unremarkable.

Microcystins

Case study: duck deaths in Japan, 1995 (Matsunaga et al. 1999)

Shin–ike pond in Hyogo Prefecture had become eutrophic from the entry of untreated sewage following an earthquake in January 1995, when a sewage treatment plant was damaged. A bloom of *M. aeruginosa* was evident. In September of that year, some 20 ducks died at the site; Oo–ike pond, 1km distant, also had a cyanobacteria bloom, but no unusual bird deaths were reported from that site. Water samples were collected from both ponds. Necropsy of one of the affected ducks showed a liver that was necrotic and "severely jaundiced…" Preliminary toxicity testing with sonicated cell suspensions using a mouse bioassay resulted in unspecified signs of *Microcystis* toxicity and death within two hours from the Shin–ike pond material, whereas that from Oo–ike pond did not produce signs of acute toxicity. Quantification of microcystins by HPLC revealed that Shin–ike pond lyophilised bloom material contained 318µg/g MC–RR, and 161µg/g MC–LR. Oo–ike pond cyanobacteria contained 29µg/g MC–RR and no detectable MC–LR (Matsunaga et al. 1999). While tissues from affected birds were not analysed for microcystins, the combination of necropsy, bioassay and microcystin quantification, along with differential findings between the suspect pond and a nearby "control" site lend strong support to the presumptive diagnosis of acute microcystin intoxication in this case.

Nodularin

Case study: dog death in South Africa, 1994 (Harding et al. 1995).

In March 1994 a bull terrier was admitted to a Cape Town veterinary hospital with a history of lethargy, vomiting and loss of appetite after drinking from an urban lake the previous day. The lake is known to be eutrophic and the phytoplankton is normally dominated by *M. aeruginosa*. Antibiotic, corticosteroid, antiemetic and choleretic pharmacotherapy was administered, as well as vitamin B complex and intravenous fluids. Serum alanine transaminase was markedly elevated, indicating acute hepatic dysfunction. The dog died on the fifth day after admission; necropsy was not conducted, but liver samples were collected and prepared for histopathological examination. Water and scum samples from the lake were collected two days after the dog drank from it. Microscopic examination revealed a bloom of 95% *Nodularia spumigena*, with the remainder being *M. aeruginosa*. Scum material was acutely toxic in a mouse bioassay, with liver damage suggestive of cyanobacterial hepatotoxin poisoning. Methanol extracts of bloom material analysed by HPLC with photodiode array and UV detection showed a major peak that increased when spiked with nodularin standard, and a UV spectrum that matched that of nodularin. Lyophilised cyanobacteria contained 3.5mg/g nodularin; no evidence for the presence of microcystins in the bloom material was found. Histology findings suggested toxic liver injury, with hepatocyte degeneration and indications of cholestasis (Harding et al. 1995). Again, while tissues were not examined for the presence of cyanotoxins, the combination of clinical presentation, diagnostic tools – liver function tests, histopathology, mouse bioassay – and investigative chemistry strongly supported the diagnosis of nodularin poisoning.

Cylindrospermopsin

Case study: cattle deaths in Australia, 2001 (Shaw et al. 2004).

Two separate poisoning incidents were investigated in Central and Northwest Queensland involving a total of 55 cows. Affected animals were lethargic and recumbent for three or four days ante–mortem. Post–mortem findings were typical of cylindrospermopsin intoxication, with pallid livers and cholecystomegaly. Histopathology showed hepatocyte degeneration and necrosis, nephrosis and multifocal cardiomyopathy. Farm water samples, rumen contents, liver, kidney and muscle were analysed by HPLC-tandem mass spectrometry for cylindrospermopsin, which was found in all samples except muscle. Water and rumen samples contained cylindrospermopsin concentrations in excess of 1mg/L (Shaw et al. 2004). The principal cylindrospermopsin–producing cyanobacterium in tropical and subtropical Queensland is *Cylindrospermopsis raciborskii*. Associated stock animal deaths and human water supply–related acute and chronic poisonings have long been suspected there (Hayman 1992, Thomas et al. 1998, Griffiths and Saker 2003), and livestock industry publications seek to inform farmers, veterinarians and government workers of the risks from this cyanotoxin (Berry 2001).

Anatoxin–a

Case study: dog deaths in France, 2003 (Gugger et al. 2005).

Two dogs died in separate incidents in September 2003 shortly after drinking from the shore of a river in the Jura region. Clinical signs were vomiting, hind limb paresis and respiratory failure preceding death. The smaller dog (2.5kg) sickened and died shortly after emerging from the water, whereas the larger dog (25kg) had a delayed onset of signs and died within five hours. Stomach contents, intestinal contents and liver were sampled as well as water column and benthic biofilm from the river; stomach contents and field samples were examined for phytoplankton identification. An initial screen for three potential cyanotoxins was conducted on biofilm samples: protein phosphatase 2A inhibition assay for microcystins, mouse neuroblastoma *in vitro* assay for saxitoxins, and HPLC with photodiode array for anatoxin–a. No microcystins or saxitoxins were detected, but a compound with similar chromatographic characteristics to anatoxin–a was seen, so subsequent investigations were directed to that end. A total of seven Oscillatorealean cyanobacterial species and genera were identified across field and gastric samples; three of these were isolated and one isolate, *Phormidium favosum*, was found to produce anatoxin–a. This benthic species was found in the biofilm covering river sediments, stones and macrophytes, as well as from the stomach contents of both dogs – *P. favosum* being the dominant cyanobacterium found in the smaller dog's stomach. Identification of anatoxin–a in biofilm, liver and gastro–intestinal contents was confirmed by tandem mass spectrometry; the authors draw attention to the necessity for confirmation of identity by fragmentation ion spectral assignments to allow definitive identification of anatoxin–a, as the amino acid phenylalanine has an identical mass spectrum peak to that of anatoxin–a when single ion monitoring is used. This was also the first report of anatoxin–a in French waters, and the first identification of an anatoxin–a–producing *Phormidium* species. The authors suggest that neurotoxic signs in 37 dogs (with 26 deaths) in 2002 and 2003 in the south of France may also have been associated with anatoxin–a intoxication (Gugger et al. 2005). This comprehensive investigation utilised well–equipped facilities, use of anatoxin–a and phenylalanine standards, and combined expertise in the fields of investigative chemistry, phytoplankton identification and isolation, so it probably represents the current state of the art for the diagnosis of anatoxin–a poisoning.

Anatoxin–a(s)

Case study: waterbird deaths in Denmark, 1993 (Henriksen et al. 1997, Onodera et al. 1997).

Cyanobacteria–associated animal deaths have been reported at Lake Knud sø since 1981. In 1993, two grebes and a coot that died when a cyanobacterial bloom was evident were collected and frozen. Stomach contents were examined microscopically to identify cyanobacteria; *Anabaena lemmermannii* were found in all three birds. This material was lyophilised for further toxin analysis. Four neurotoxic strains of *A. lemmermannii* were isolated and cultured from bloom material collected from the lake. Bloom material, cyanobacterial cultures and stomach contents were analysed for anticholinesterase activity with a colorimetric assay and for microcystins by ELISA. Only low levels of microcystins (ng/kg range) were found in bird gastric contents. Neither anatoxin–a nor saxitoxins were detected by HPLC. Intraperitoneal injection of bloom material into mice produced signs of acute neurotoxicity, including the muscarinic signs of salivation, lacrimation and urinary incontinence that typify anatoxin–a(s) intoxication. Anticholinesterase activity was significant in bloom samples (2.3mg anatoxin–a(s) equivalents/g); *A. lemmermannii* isolates and stomach contents also showed anticholinesterase activity (29–743µg/g, and 2–90µg/kg body weight respectively). Subsequent work using mass spectrometry, nuclear magnetic resonance and circular dichroism definitively identified anatoxin–a(s) in bloom material (Henriksen et al. 1997, Onodera et al. 1997).

Saxitoxins

Case study: sheep deaths in Australia, 1994 (Negri et al. 1995).

Thirteen ewes and one ram died next to or within 150m of a farm dam. Observed signs were trembling, recumbency and crawling. The ram, which was ataxic, was caught by the owner and "immediately died in his arms." Necropsy and histopathology findings were unremarkable. The dam was found to contain a bloom of *Anabaena circinalis*. Mouse bioassay of bloom material caused death within 4–11 minutes, with clinical signs of staggering, gasping, leaping and respiratory failure. High levels of saxitoxins were found in bloom extracts and intestinal contents of one sheep by HPLC with fluorescence detection; principal components were C–toxins and gonyautoxins, with low levels of saxitoxin and decarbamoyl saxitoxin. Neither anatoxin–a nor microcystins were found using GC/MS and HPLC respectively; trace amounts of microcystin–like activity were found with a protein phosphatase assay (Negri et al. 1995). Saxitoxin–producing *Anabaena circinalis* is a particularly problematic cyanobacterium in Australia (Humpage et al. 1994). A 1000 km-long riverine bloom in the spring and summer of 1991 was associated with significant stock losses; one farmer alone reportedly lost over 1,100 sheep during the six–week period of the bloom (NSW Blue–Green Algae Task Force 1992, Smith 2000).

Lesional neurological disease and cyanobacteria?

The case studies summarised above represent investigations into animal deaths associated with the well–characterised cyanotoxins. However, other less well–understood cyanotoxins may be capable of killing animals, and are being actively researched. Wilde et al. (2005) and Williams et al (2007) report on avian vacuolar myelinopathy (AVM), a recently–described wildlife disease. AVM was reportedly responsible for more than 100 bald eagle deaths, and "untold numbers" of waterbirds. Clinical signs of the disease are loss of coordination on land and in water, and difficulty in flying. Necropsy reveals a diffuse vacuolation of myelinated neuronal tissue. Epidemiological investigations suggest that a neurotoxin is responsible, with exposure being seasonal, linked to dietary intake by herbivorous waterfowl from AVM–positive waterbodies, and moving up the food chain (i.e., from waterfowl prey to eagle predator). Current evidence suggests that AVM is not contagious. A common factor at AVM–positive sites – and absent or rare at AVM–negative sites – appears to be an epiphytic cyanobacterium in the order Stigonematales. The authors demonstrate the potential of veterinary epidemiology to advance the understanding of cyanobacteria–related toxicity by use of experimental techniques that are not available in the field of human epidemiology. A mallard duck sentinel trial was conducted, where 20 farm–raised birds were tagged for identification, had wing feathers clipped and were released into an AVM–suspect pond. The site was regularly monitored, symptomatic birds were captured and sacrificed, and at the end of the six–week trial, remaining sentinels were captured and sacrificed. Five symptomatic ducks and ten birds captured at the end of the trial – these 15 being all birds able to be re-captured from the pond – all showed AVM lesions in brain tissues (Wilde et al. 2005). Williams et al (2007) have isolated and cultured the suspect cyanobacterium from environmental samples, and have developed PCR assays to aid detection and identification. Further epidemiological field work using suitable control sites would be of great interest.

"Unusual" animal deaths and cyanobacteria

The previous discussion has concentrated on cyanotoxin–related deaths in common domestic, stock and wild animals. Some other animals reportedly poisoned by cyanobacteria are:

- **Flamingos:** Cyanobacteria–related mortalities have been reported in three flamingo species, both wild and captive (see summary by Codd et al (2003)). Lesser Flamingos in Kenyan soda lakes have been the subject of research interest, with four microcystin congeners and anatoxin–a found in cyanobacterial mats and stomach contents of dead birds at Lake Bogoria. Microcystins and anatoxin–a were also detected in faecal pellets collected from lake shorelines (Krienitz et al. 2003). The same cyanotoxins have been found in feathers taken from poisoned flamingos, with a dietary origin most likely (Metcalf et al. 2006). Mass die–offs of tens of thousands of birds have been reported in these crater lakes, with implications for management and regional and national economies, as flamingos are a significant tourist attraction (Krienitz et al. 2003, Ndetei and Muhandiki 2005). Isolated strains of *Arthrospira fusiformis* from two crater lakes have been shown to produce both microcystin–YR and anatoxin–a, while an *A. fusiformis* strain from a third lake produces anatoxin–a (Ballot et al. 2004, Ballot et al. 2005). Similar Lesser Flamingo poisonings have been reported from alkaline lakes in Tanzania, with toxic *A. fusiformis* implicated (Lugomela et al. 2006). The implications of these findings are significant for several reasons, as A*rthrospira* sp. (also known as *Spirulina*) are the principal food source of Lesser Flamingos, and *Spirulina* spp. are used as a dietary supplement by humans and as a feed additive for livestock. There is a significant body of literature that suggests that consumption of *Spirulina* is not harmful (Ciferri 1983, Belay et al. 1993, Hayashi et al. 1994, Qureshi et al. 1996, Salazar et al. 1998, Abdulquader et al. 2000, Al–Batshan et al. 2001), though presumably these studies refer to non-toxic strains of *Arthrospira* used for both commercial mass production and from wild harvesting. Investigations to determine the relative contribution of *A. fusiformis* to the production of cyanotoxins in Kenyan soda lakes would be of great interest. Some of these lakes are periodically dominated by more well–known toxigenic cyanobacteria such as *Anabaena* and *Microcystis* spp. (Ndetei and Muhandiki 2005), so it will be important to estimate the production capacity of cyanotoxins by various cyanobacteria in field situations when harmful levels of cyanotoxins are present.

- **Insectivorous bats:** Staff at a campground in Alberta, Canada, in the summer of 1985, counted 500 dead bats and estimated over 1,000 deaths on the leeward side of a lake. At least 24 dead mallards were also reported. The lake area was covered with a "thick white scum" that had a blue–green sheen. Examined animals were covered with a green slime; necropsy did not reveal any abnormalities. An alkaloid was extracted from the material covering the carcases. This alkaloid was identified by GC/MS and found to be anatoxin–a (known at the time as Very Fast Death Factor) (Pybus et al. 1986).
- **Rhinoceros:** Four white rhinoceroses were introduced to a South African game reserve in May, 1979. Two months later, two of them were found dead but were unable to be examined. Approximately one week later, another rhino was found dead after being seen to be active the previous day. Macroscopic and microscopic findings were typical of acute hepatotoxicity: hepatomegaly, ascitic fluid, coagulopathy seen in various tissues, severe hepatic necrosis and loss of hepatic architecture. At the time of death, a severe bloom of *M. aeruginosa* covered the park dam, with a surface scum of 4–12cm (Soll and Williams 1985).
- **Honeybees:** In the summer of 1971, "almost total" mortality of bees from 84 hives was associated with the insects watering on the leeward edge of a lake in New South Wales, Australia. That area of the lake was affected by a windborne scum of *A. circinalis*; an apiary on the windward shore was unaffected (May and McBarron 1973).

Exposure routes, avoidance behaviour

Exposure to cyanotoxins in the field is presumably either through contaminated drinking water or by direct consumption of benthic cyanobacteria. Codd et al (2005) also note suggestions that cyanotoxins may move along the food chain to poison wild animals. This would seem to be a plausible interpretation: there is a significant section of the literature devoted to experimental and observational studies of cyanotoxin bioaccumulation and trophic transfer. Much of the literature is concerned with potential human exposure to cyanotoxins through consumption of molluscs and fish; short-term (days, weeks) accumulation of hazardous levels of cyanotoxins in aquatic prey is certainly feasible, and predator animals may be more at risk than humans in some circumstances, since they are more likely to consume the viscera in which cyanotoxins are found in highest concentrations. Several bioaccumulation studies are inconclusive; for some positive examples see: (Negri and Jones 1995, Saker and Eaglesham 1999, Magalhães et al 2001, Sipiä et al 2002, Saker et al 2004, Sipiä et al 2004, Smith and Haney 2006).

Experimental work has shown that the inhalation route may be an efficient method for microcystin–LR to access the circulation (Creasia 1990, Fitzgeorge 1994, Ito 2001), but there is no evidence that cyanotoxins can be effectively aerosolised to toxic concentrations under field conditions. The topic of inhalational exposure to cyanotoxins is clearly under–researched at present.

Codd et al (1992) outlined the premise that exposure to toxic cyanobacteria may not always be coincidental; dogs may actively seek out and consume benthic cyanobacteria. Some descriptions in the report of Hamill (2001) also suggest that dogs may actively consume toxic cyanobacteria. The work of Gugger et al (2005) also appears to support this concept: while descriptions from the field were that two dogs died "soon after drinking water from the shoreline of the…river…" analysis of phytoplankton in the dogs' stomachs revealed the presence of biofilm–associated cyanobacteria that were not seen in water column samples. This thick biofilm reportedly covered sediment, stones and macrophytes at the water's edge (Gugger et al. 2005); it is at least conceivable that those animals were also actively consuming toxic cyanobacteria for their final meal.

As for planktonic cyanobacteria, there are suggestions that wild and domestic animals will avoid drinking from cyanobacterial scums if less–affected water is accessible. Falconer (2005 p.80) notes that detailed investigations of cyanobacteria–related stock animal deaths often reveal that cleaner water was unavailable because of fence lines, thus forcing the animals to consume toxic water. Other workers have reported avoidance behaviour in cattle (Codd 1983) and flamingos (Ndetei and Muhandiki 2005), but Carbis et al (1994) report that sheep did not discriminate between hepatotoxic cyanobacteria–contaminated water and a readily available alternate supply. Lopez Rodas and Costas (1999) report observing ungulates, birds, wasps and unspecified wildlife failing to demonstrate avoidance behaviour, consuming concentrated *M. aeruginosa* scum in Spanish reservoirs. Their subsequent experiments showed that laboratory mice demonstrated an ultimately fatal preference for dense cultures of microcystin–producing *M. aeruginosa* over tap water, and they did not discriminate between toxic and non–toxic strains of *M. aeruginosa* in their drinking water. Steyn (1943) noted that pregnant animals and dairy cows displayed a preference for consuming *Microcystis* scums, suggesting that a relative nutritional deficiency may explain higher mortalities than seen in dry cows.

Treatment of affected animals; research needs

As no complete pharmacological antagonists against cyanotoxins are available, treatment of affected animals is largely supportive. Purgatives and enemas may be helpful; unguents and restriction to shaded areas should be considered for animals suffering secondary photosensitization (Steyn 1943). Other supportive therapies include emesis, activated charcoal and bathing to remove bloom material from the fur (Corkill et al. 1989, Roder 2004). Blood transfusion and correction of electrolyte imbalance has been recommended (Beasley et al. 1989a). Aggressive intervention with fluid replacement and steroids has been suggested for cases of cyanobacterial hepatotoxin poisoning (Roder 2004). While such an approach was unsuccessful in the case (discussed above) of the bull terrier presumed to have died from nodularin intoxication (Harding et al. 1995), there is clearly a need for controlled experiments to evaluate interventional therapies. Pre–treatment with pharmacologic doses of hydrocortisone prevented acute and delayed MC–LR–related mortality in mice (Adams et al. 1985), and dexamethasone and indomethacin pre–treatment blocked renal toxicity caused by MC–LR (Nobre et al. 2001). Pre–treatment with steroids is obviously not a practical therapeutic intervention, but these findings suggest that there is much more to be learned about the immunotoxicology of the cyanobacterial hepatotoxins.

Artificial ventilation has been suggested in cases of cyanobacterial neurotoxin poisoning (Beasley et al. 1989a, Beasley et al. 1989b, Carmichael and Falconer 1993, Roder 2004). While the practical application of this technique in the field would present significant challenges, experimental work to evaluate the benefits of intermittent positive pressure ventilation and other intensive therapy may have implications for the field of human medical care as well as for economically valuable equine and livestock animals. Work conducted in the 1970s on calves and rodents dosed with anatoxin–a–producing cyanobacteria suggested that artificial ventilation is unlikely to be a practical intervention, with one calf failing to achieve effective spontaneous respiration after being ventilated for 30 hours, and rats similarly afflicted after maintenance ventilation of up to eight hours (Carmichael et al. 1975, Carmichael et al. 1977). However, anatoxin–a doses producing up to 95% neuromuscular blockade in rats were reportedly reversible with respiratory support (Valentine et al. 1991). Artificial ventilation was conducted on three rats during experimental administration of anatoxin–a(s); they survived a greater than 4–fold lethal dose in the short term (less than one hour), but the study was not designed to follow through to recovery (Cook et al. 1990). Interventional strategies to treat saxitoxin poisoning using antitoxin or the potassium–channel blocker 4–aminopyridine and artificial ventilation have shown promise, though the toxicodynamics of saxitoxin are complex and the combined neurological, cardiovascular and respiratory effects are incompletely understood (Chang et al. 1993, Benton et al. 1994, Chang et al. 1996, Chang et al. 1997, Benton et al. 1998).

Management of seizures, presumably anticonvulsant pharmacotherapy, has been suggested in cases of cyanobacterial neurotoxin poisoning (Roder 2004), though this may be a moot point if, as suggested at least for anatoxin–a, seizures result from hypoxia–related cerebral ischaemia (Carmichael 1994). Cyanobacterial neurotoxin–related seizures are not well described in the literature, mostly being referred to as “convulsions”, though there are references to clonic spasms (McLeod and Bondar 1952), clonic seizures (Cook et al. 1989, Cook et al. 1998, Carmichael 2001), tonic spasm (Firkins 1953) and tonic seizures (Beasley et al. 1989a, Beasley et al. 1989b). The electrophysiology of these events has not been investigated. Opisthotonus is often described (Carmichael et al. 1975, Gorham and Carmichael 1988, Cook et al. 1989, Carmichael 2001, Codd et al. 2003); this increase in tone of the musculature of the neck and shoulders has some similarity to the clinical manifestation of a tonic seizure, though again the priority from a management perspective would appear to be restoration of oxygen to the tissues. Anticonvulsants have also been administered for management of seizures in cases of cyanobacterial hepatotoxin poisoning, but with little apparent benefit (Corkill et al. 1989).

Further empirical studies are warranted to assess various emergency interventions, including pharmacotherapies, for cases of cyanotoxin poisoning. However, these are potent, rapidly acting toxins, so prevention will likely remain the most effective intervention for the foreseeable future. In the short term this involves restricting access to cyanotoxin–affected drinking water, with longer–term strategies of nutrient reduction to mitigate blooms. Dissemination of information to farmers and the livestock industry is always important; beyond the academic research literature there are some industry and government partnerships that provide helpful resources, e.g. Australia’s Animal Health Surveillance Report (see for example Berry 2001, Elliott 2001), and the Animal Health Expositor from Canada (now Animal Health Perspectives) (Yong 2000a, Yong 2000b). Comprehensive economic analyses of the costs of cyanotoxin–related livestock and wild animal poisonings to the agricultural and tourist industries would be valuable.

Concluding remarks

Cyanotoxins in the environment continue to be hazardous to livestock and wild animals throughout the world. While much is known about the most potent, acutely toxic neurotoxins and hepatotoxins, there is still more to learn about the toxicokinetics and toxicodynamics of even the well–characterised cyanotoxins. Cyanobacteria are rich sources of biologically active compounds, and it is likely that some yet–to–be–described cyanobacterial products will be found to be harmful. Current investigations into the subacute disease avian vacuolar myelinopathy hint at this possibility. While controlled dosing with purified cyanotoxins under laboratory conditions has been an essential endeavour that allows these materials to be characterised and understood, exposures in the field may be much more complex. Subacute or chronic exposures, particularly to hepatotoxic cyanobacteria, may result in clinical presentations characterised by concomitant degenerative and regenerative changes. Expert veterinary clinicians and pathologists, as well as specialist investigative chemistry laboratory services, are invaluable resources for the diagnosis and management of these complex poisonings.

Acknowledgements

Thanks to Wasa Wickramasinghe for helpful discussion. This work was supported by the Cooperative Research Centre for Water Quality and Treatment.

The National Research Centre for Environmental Toxicology is co–funded by Queensland Health, The University of Queensland, Griffith University and Queensland University of Technology.

References

Abdulqader G, Barsanti L, Tredici MR (2000) Harvest of Arthrospira platensis from Lake Kossorom (Chad) and its household use among the Kanembu. J Appl Phycol 12(3–5):493–498

Adams WH, Stoner RD, Adams DG, Slatkin DN, Siegelman HW (1985) Pathophysiologic effects of a toxic peptide from Microcystis aeruginosa. Toxicon 23(3):441–447

Al–Batshan HA, Al–Mufarrej SI, Al–Homaidan AA, Qureshi MA (2001) Enhancement of chicken macrophage phagocytic function and nitrite production by dietary Spirulina platensis. Immunopharmacol Immunotoxicol 23(2):281–289

Ballot A, Krienitz L, Kotut K, Wiegand C, Metcalf JS, Codd GA, Pflugmacher S (2004) Cyanobacteria and cyanobacterial toxins in three alkaline Rift Valley lakes of Kenya—Lakes Bogoria, Nakuru and Elmenteita. J Plankton Res 26(8):925–935

Ballot A, Krienitz L, Kotut K, Wiegand C, Pflugmacher S (2005) Cyanobacteria and cyanobacterial toxins in the alkaline crater lakes Sonachi and Simbi, Kenya. Harmful Algae 4(1):139–150

Beasley VR, Cook WO, Dahlem AM, Hooser SB, Lovell RA, Valentine WM (1989a) Algae intoxication in livestock and waterfowl. Vet Clin North Am Food Anim Pract 5(2):345–361

Beasley VR, Dahlem AM, Cook WO, Valentine WM, Lovell RA, Hooser SB, Harada K, Suzuki M, Carmichael WW (1989b) Diagnostic and clinically important aspects of cyanobacterial (blue–green algae) toxicoses. J Vet Diagn Invest 1(4):359–365

Belay A, Ota Y, Miyakawa K, Shimamatsu H (1993) Current knowledge on potential health benefits of Spirulina. J Appl Phycol 5(2):235–241

Benton BJ, Keller SA, Spriggs DL, Capacio BR, Chang FCT (1998) Recovery from the lethal effects of saxitoxin: a therapeutic window for 4–aminopyridine (4–AP). Toxicon 36(4):571–588

Benton BJ, Rivera VR, Hewetson JF, Chang FCT (1994) Reversal of saxitoxin induced cardiorespiratory failure by a burro–raised alpha–STX antibody and oxygen therapy. Toxicol Appl Pharmacol 124(1):39–51

Berry J (2001) Blue green algae. Animal Health Surveillance Quarterly Report 6(3):11 http://www.animalhealthaustralia.com.au/shadomx/apps/fms/fmsdownload.cfm?file_uuid=2FC528AC-013A-817F-983F-2932E47F83C5&siteName=aahc

Braun A, Pfeiffer T (2002) Cyanobacterial blooms as the cause of a Pleistocene large mammal assemblage. Paleobiology 28(1):139–154

Briand JF, Jacquet S, Bernard C, Humbert JF (2003) Health hazards for terrestrial vertebrates from toxic cyanobacteria in surface water ecosystems. Vet Res 34(4):361–377

Carbis CR, Simons JA, Mitchell GF, Anderson JW, McCauley I (1994) A biochemical profile for predicting the chronic exposure of sheep to Microcystis aeruginosa, an hepatotoxic species of blue–green alga. Res Vet Sci 57(3):310–316

Carmichael WW (1994) The toxins of cyanobacteria. Sci Am 270(1):64–72

Carmichael WW (2001) Health effects of toxin–producing cyanobacteria: "The CyanoHABs". Hum Ecol Risk Assess 7(5):1393–1407

Carmichael WW, Biggs DF, Gorham PR (1975) Toxicology and pharmacological action of Anabaena flos–aquae toxin. Science 187(4176):542–544
Carmichael WW, Falconer IR (1993) Diseases related to freshwater blue–green algal toxins, and control measures. In: Algal toxins in seafood and drinking water. Edited by Falconer IR. London: Academic Press 187–209
Carmichael WW, Gorham PR, Biggs DF (1977) Two laboratory case studies on the oral toxicity to calves of the freshwater cyanophyte (blue–green algae) Anabaena flos–aquae NRC–44–1. Can Vet J 18(3):71–75
Chang FCT, Bauer RM, Benton BJ, Keller SA, Capacio BR (1996) 4–aminopyridine antagonizes saxitoxin– and tetrodotoxin–induced cardiorespiratory depression. Toxicon 34(6):671–690
Chang FCT, Benton BJ, Lenz RA, Capacio BR (1993) Central and peripheral cardio–respiratory effects of saxitoxin (STX) in urethane–anesthetized guinea–pigs. Toxicon 31(5):645–664
Chang FCT, Spriggs DL, Benton BJ, Keller SA, Capacio BR (1997) 4–aminopyridine reverses saxitoxin (STX)– and tetrodotoxin (TTX)–induced cardiorespiratory depression in chronically instrumented guinea pigs. Fundam Appl Toxicol 38(1):75–88
Ciferri O (1983) Spirulina, the edible microorganism. Microbiol Rev 47(4):551–578
Codd GA (1983) Cyanobacterial poisoning hazard in British freshwaters. Vet Rec 113(10):223–224
Codd GA, Beattie KA (1991) Cyanobacteria (blue–green algae) and their toxins: awareness and action in the United Kingdom. PHLS Microbiol Dig 8(3):82–86
Codd GA, Bell SG, Kaya K, Ward CJ, Beattie KA, Metcalf JS (1999) Cyanobacterial toxins, exposure routes and human health. Eur J Phycol 34(4):405–415
Codd GA, Edwards C, Beattie KA, Barr WM, Gunn GJ (1992) Fatal attraction to cyanobacteria? Nature 359(6391):110–111
Codd GA, Lindsay J, Young FM, Morrison LF, Metcalf JS (2005) Harmful cyanobacteria: from mass mortalities to management measures. In: Harmful cyanobacteria. Edited by Huisman J, Matthijs HCP, Visser PM. Dordrecht: Springer 1–23
Codd GA, Metcalf JS, Morrison LF, Krienitz L, Ballot A, Pflugmacher S, Wiegand C, Kotut K (2003) Susceptibility of flamingos to cyanobacterial toxins via feeding. Vet Rec 152(23):722–723
Codd GA, Steffensen DA, Burch MD, Baker PD (1994) Toxic blooms of cyanobacteria in Lake Alexandrina, South Australia—learning from history. Aust J Mar Freshwater Res 45(5):731–736
Cook WO, Beasley VR, Lovell RA (1989) Consistent inhibition of peripheral cholinesterases by neurotoxins from the freshwater cyanobacterium Anabaena flos–aquae: studies of ducks, swine, mice, and a steer. Environ Toxicol Chem 8(10):915–922
Cook WO, Iwamoto GA, Schaeffer DJ, Carmichael WW, Beasley VR (1990) Pathophysiologic effects of anatoxin–a(s) in anaesthetized rats: the influence of atropine and artificial respiration. Pharmacol Toxicol 67(2):151–155

Cook WO, Beasley VR, Dahlem AM, Dellinger JA, Harlin KS, Carmichael WW (1998) Comparison of effects of anatoxin–a(s) and paraoxon, physostigmine and pyridostigmine on mouse brain cholinesterase activity. Toxicon 26(8):750–753
Corkill N, Smith R, Seckington M, Pontefract R (1989) Poisoning at Rutland Water. Vet Rec 125(13):356
Creasia DA (1990) Acute inhalation toxicity of microcystin–LR with mice. Toxicon 28(6):605
Duy TN, Lam PKS, Shaw GR, Connell DW (2000) Toxicology and risk assessment of freshwater cyanobacterial (blue–green algal) toxins in water. Rev Environ Contam Toxicol 163:113–185
Elliott J (2001) Cyanobacterial toxicity. Animal Health Surveillance Quarterly Report 6(1):13 http://www.animalhealthaustralia.com.au/shadomx/apps/fms/fmsdownload.cfm?file_uuid=2FC47F6C-9D1A-FA75-465E-DE8DCA75BBD4&siteName=aahc
Falconer IR (2005) Cyanobacterial toxins of drinking water supplies: cylindrospermopsins and microcystins. Boca Raton: CRC Press
Firkins GS (1953) Toxic algae poisoning. Iowa State Coll Vet 15(3):151–153
Fitzgeorge RB, Clark SA, Keevil CW (1994) Routes of intoxication. In: Detection methods for cyanobacterial toxins. Edited by Codd GA, Jefferies TM, Keevil CW, Potter E. Cambridge: The Royal Society of Chemistry 69–74
Francis G (1878) Poisonous Australian lake. Nature 18:11–12
Goode J (1967) Freshwater tortoises of Australia and New Guinea (in the Family Chelidae). Melbourne: Lansdowne Press
Gorham PR, Carmichael WW (1988) Hazards of freshwater blue–green algae (cyanobacteria). In: Algae and human affairs. Edited by Lembi CA, Waaland JR. Cambridge: Cambridge University Press 403–431
Griffiths DJ, Saker ML (2003) The Palm Island mystery disease 20 years on: a review of research on the cyanotoxin cylindrospermopsin. Environ Toxicol 18(2):78–93
Gugger M, Lenoir S, Berger C, Ledreux A, Druart JC, Humbert JF, Guette C, Bernard C (2005) First report in a river in France of the benthic cyanobacterium Phormidium favosum producing anatoxin–a associated with dog neurotoxicosis. Toxicon 45(7):919–928
Hamill KD (2001) Toxicity in benthic freshwater cyanobacteria (blue–green algae): first observations in New Zealand. N Z J Mar Freshwater Res 35(5):1057–1059
Hammer UT (1968) Toxic blue–green algae in Saskatchewan. Can Vet J 9(10):221–229
Harding WR, Rowe N, Wessels JC, Beattie KA, Codd GA (1995) Death of a dog attributed to the cyanobacterial (blue–green algal) hepatotoxin nodularin in South Africa. J S Afr Vet Assoc 66(4):256–259
Hayashi O, Katoh T, Okuwaki Y (1994) Enhancement of antibody production in mice by dietary Spirulina platensis. J Nutr Sci Vitaminol (Tokyo) 40(5):431–441

Hayman J (1992) Beyond the Barcoo – probable human tropical cyanobacterial poisoning in outback Australia. Med J Aust 157(11–12):794–796

Henriksen P, Carmichael WW, An J, Moestrup O (1997) Detection of an anatoxin–a(s)–like anticholinesterase in natural blooms and cultures of cyanobacteria/blue–green algae from Danish lakes and in the stomach contents of poisoned birds. Toxicon 35(6):901–913

Höger SJ (2003) Problems during drinking water treatment of cyanobacterial loaded surface waters: consequences for human health. Doctoral Thesis Constance: Universität Konstanz http://www.ub.uni-konstanz.de/v13/volltexte/2003/1071//pdf/SJHoeger_Thesis.pdf

Howard NJ, Berry AE (1933) Algal nuisances in surface waters. Can Public Health J 24:377–384

Humpage AR, Rositano J, Bretag AH, Brown R, Baker PD, Nicholson BC, Steffensen DA (1994) Paralytic shellfish poisons from Australian cyanobacterial blooms. Aust J Mar Freshwater Res 45(5):761–771

Ito E, Kondo F, Harada K (2001) Intratracheal administration of microcystin–LR, and its distribution. Toxicon 39(2–3):265–271

Koenigswald Wv, Braun A, Pfeiffer T (2004) Cyanobacteria and seasonal death: a new taphonomic model for the Eocene Messel lake. Paläontol Z 78(2):417–424

Krienitz L, Ballot A, Kotut K, Wiegand C, Pütz S, Metcalf JS, Codd GA, Pflugmacher S (2003) Contribution of hot spring cyanobacteria to the mysterious deaths of Lesser Flamingos at Lake Bogoria, Kenya. FEMS Microbiol Ecol 43(2):141–148

Kuchling G (1999) The reproductive biology of the Chelonia. Berlin: Springer Verlag

Lopez Rodas V, Costas E (1999) Preference of mice to consume Microcystis aeruginosa (toxin–producing cyanobacteria): a possible explanation for numerous fatalities of livestock and wildlife. Res Vet Sci 67(1):107–110

Lugomela C, Pratap HB, Mgaya YD (2006) Cyanobacteria blooms–a possible cause of mass mortality of Lesser Flamingos in Lake Manyara and Lake Big Momela, Tanzania. Harmful Algae 5(5):534–541

Magalhães VF, Soares RM, Azevedo SMFO (2001) Microcystin contamination in fish from the Jacarepaguá Lagoon (Rio de Janeiro, Brazil): ecological implication and human health risk. Toxicon 39(7):1077-1085

Matsunaga H, Harada KI, Senma M, Ito Y, Yasuda N, Ushida S, Kimura Y (1999) Possible cause of unnatural mass death of wild birds in a pond in Nishinomiya, Japan: sudden appearance of toxic cyanobacteria. Nat Toxins 7(2):81–84

May V, McBarron EJ (1973) Occurrence of the blue–green alga, Anabaena circinalis Rabenh., in New South Wales and toxicity to mice and honey bees. J Aust Inst Agric Sci 39(4):264–266

McLeod JA, Bondar GF (1952) A case of suspected algal poisoning in Manitoba. Can J Public Health 43(8):347–350

Metcalf JS, Morrison LF, Krienitz L, Ballot A, Krause E, Kotut K, Pütz S, Wiegand C, Pflugmacher S, Codd GA (2006) Analysis of the cyanotoxins anatoxin-a and microcystins in Lesser Flamingo feathers. Toxicol Environ Chem 88(1):159-167

Ndetei R, Muhandiki VS (2005) Mortalities of lesser flamingos in Kenyan Rift Valley saline lakes and the implications for sustainable management of the lakes. Lakes Reserv Res Manage 10(1):51–58

Negri AP, Jones GJ (1995) Bioaccumulation of paralytic shellfish poisoning (PSP) toxins from the cyanobacterium Anabaena circinalis by the freshwater mussel Alathryia condola. Toxicon 33(5):667-678

Negri AP, Jones GJ, Hindmarsh M (1995) Sheep mortality associated with paralytic shellfish poisons from the cyanobacterium Anabaena circinalis. Toxicon 33(10):1321–1329

Nobre AC, Coelho GR, Coutinho MC, Silva MM, Angelim EV, Menezes DB, Fonteles MC, Monteiro HS (2001) The role of phospholipase A(2) and cyclooxygenase in renal toxicity induced by microcystin–LR. Toxicon 39(5):721–724

NSW Blue–Green Algae Task Force (1992) Final report of the NSW Blue–Green Algae Task Force. Parramatta: NSW Department of Water Resources

Onodera H, Oshima Y, Henriksen P, Yasumoto T (1997) Confirmation of anatoxin–a(s), in the cyanobacterium Anabaena lemmermannii, as the cause of bird kills in Danish lakes. Toxicon 35(11):1645–1648

Phillips W (1884) The breaking of the Shropshire meres. Trans Shropshire Archaeol Nat Hist Soc 7:277–300

Pybus MJ, Hobson DP, Onderka DK (1986) Mass mortality of bats due to probable blue–green algal toxicity. J Wildl Dis 22(3):449–450

Qureshi MA, Garlich JD, Kidd MT (1996) Dietary Spirulina platensis enhances humoral and cell–mediated immune functions in chickens. Immunopharmacol Immunotoxicol 18(3):465–476

Ressom R, Soong FS, Fitzgerald J, Turczynowicz L, El Saadi O, Roder D, Maynard T, Falconer I (1994) Health effects of toxic cyanobacteria (blue–green algae). Canberra: National Health and Medical Research Council & Australian Government Publishing Service

Roder JD (2004) Blue–green algae. In: Clinical veterinary toxicology. Edited by Plumlee KH. St Louis: Mosby 100–101

Rose EF (1953) Toxic algae in Iowa lakes. Proc Iowa Acad Sci 60:738–745

Saker ML, Eaglesham GK (1999) The accumulation of cylindrospermopsin from the cyanobacterium Cylindrospermopsis raciborskii in tissues of the Redclaw crayfish Cherax quadricarinatus. Toxicon 37(7):1065-1077

Saker ML, Metcalf JS, Codd GA, Vasconcelos VM (2004) Accumulation and depuration of the cyanobacterial toxin cylindrospermopsin in the freshwater mussel Anodonta cygnea. Toxicon 43(2):185-194

Salazar M, Martínez E, Madrigal E, Ruiz LE, Chamorro GA (1998) Subchronic toxicity study in mice fed Spirulina maxima. J Ethnopharmacol 62(3):235–241

Schwimmer D, Schwimmer M (1964) Algae and medicine. In: Algae and man. Edited by Jackson DF. New York: Plenum Press 368–412

Schwimmer M, Schwimmer D (1955) The role of algae and plankton in medicine. New York: Grune & Stratton
Schwimmer M, Schwimmer D (1968) Medical aspects of phycology. In: Algae, man, and the environment. Edited by Jackson DF. Syracuse: Syracuse University Press 279–358
Shaw GR, McKenzie RA, Wickramasinghe WA, Seawright AA, Eaglesham GK, Moore MR (2004) Comparative toxicity of the cyanobacterial toxin cylindrospermopsin between mice and cattle: human implications. In: Harmful Algae 2002. Edited by Steidinger KA, Landsberg JH, Tomas CR, Vargo GA. St Petersburg, Florida, USA: Florida Fish and Wildlife Conservation Commission, Florida Institute of Oceanography, and Intergovernmental Oceanographic Commission of UNESCO 465–467
Sipiä VO, Kankaapää HT, Pflugmacher S, Flinkman J, Furey A, James KJ (2002) Bioaccumulation and detoxication of nodularin in tissues of flounder (Platichthys flesus), mussels (Mytilus edulis, Dreissena polymorpha), and clams (Macoma balthica) from the northern Baltic Sea. Ecotoxicol Environ Saf 53(2):305-311
Sipiä VO, Karlsson KM, Meriluoto JA, Kankaapää HT (2004) Eiders (Somateria mollissima) obtain nodularin, a cyanobacterial hepatotoxin, in Baltic Sea food web. Environ Toxicol Chem 23(5):1256-1260
Smith JL, Haney JF (2006) Foodweb transfer, accumulation, and depuration of microcystins, a cyanobacterial toxin, in pumpkinseed sunfish (Lepomis gibbosus). Toxicon 48(5):580-589
Smith PT (2000) Freshwater neurotoxins: mechanisms of action, pharmacology, toxicology, and impacts on aquaculture. In: Seafood and freshwater toxins: pharmacology, physiology, and detection. Edited by Botana LM. New York: Marcel Dekker 583–602
Soll MD, Williams MC (1985) Mortality of a White Rhinoceros (Ceratotherium simium) suspected to be associated with the blue–green alga Microcystis aeruginosa. J S Afr Vet Assoc 56(1):49–51
Stephens EL (1949) Microcystis toxica sp. Nov.: a poisonous alga from the Transvaal and Orange Free State. Trans R Soc S Afr 32(1):105–112
Steyn DG (1943) Poisoning of animals by algae on dams and pans. Farming S Afr 18:489–492, 510
Steyn DG (1945) Poisoning of animals and human beings by algae. S Afr J Sci 1945, 41:243–244
Thomas AD, Saker ML, Norton JH, Olsen RD (1998) Cyanobacterium Cylindrospermopsis raciborskii as a probable cause of death in cattle in northern Queensland. Aust Vet J 76(9):592–594
Valentine WM, Schaeffer DJ, Beasley VR (1991) Electromyographic assessment of the neuromuscular blockade produced in vivo by anatoxin–a in the rat. Toxicon 29(3):347–357
Wilde SB, Murphy TM, Hope CP, Habrun SK, Kempton J, Birrenkott A, Wiley F, Bowerman WW, Lewitus AJ (2005) Avian vacuolar myelinopathy linked to exotic aquatic plants and a novel cyanobacterial species. Environ Toxicol 20(3):348–353

Williams SK, Kempton J, Wilde SB, Lewitus A (2007) A novel epiphytic cyanobacterium associated with reservoirs affected by avian vacuolar myelinopathy. Harmful Algae 6(3):343-353

Yong C (2000a) Cyanobacteria (blue-green algae) poisoning. Animal Health Expositor 2(3):2 http://www.usask.ca/pds/Information/Sept%20%2700.pdf

Yong C (2000b) Blue–green algae poisoning in two dogs. Animal Health Expositor 2(3):3 http://www.usask.ca/pds/Information/Sept%20%2700.pdf

Chapter 29: Epidemiology of cyanobacteria and their toxins

Louis S Pilotto

Faculty of Medicine, University of New South Wales, Australia

Introduction

Epidemiology is defined as the study of the distribution and determinants of health–related states or events in specified populations, and the application of this study to the control of health problems (Last 2001). In this context, "study" includes observation, hypothesis testing, analytic research, and experiments. In turn, each of these methods has an increasing level of sophistication that provides results with differing strength of evidence linking human exposure and health outcome. The World Health Organisation and other agencies, including the National Health and Medical Research Council (NHMRC) in Australia, have developed a classification system for these levels of evidence based on rigor, quality and the minimisation of bias. The NHMRC's new pilot classification allows for studies about aetiology. The classification is tiered and extends from the strongest Level I, that obtained from a systematic review of all studies using a prospective cohort design (Level II evidence) to Level IV, that obtained from case series and cross–sectional studies (Table 1 right hand column). Such aetiological studies are different than intervention studies where the investigators have the opportunity to determine exposure. Intervention studies vary from case–series where the outcomes prior to and after some prescribed intervention are measured and compared, to randomised controlled trials and their systematic reviews (Table 1 left hand column). Of the intervention studies, randomised controlled trials are the gold standard, which by their very nature, are designed to minimise or eliminate bias and allow inferences about exposure and health related effects. However, little of the current evidence about the health effects of cyanobacterial exposure has been obtained from intervention studies. This pilot classification is useful as it acknowledges that evidence in some instances, by the potentially hazard effects of some agents, cannot be derived from intervention studies, and provides a classification for non–experimental studies that supply evidence about aetiology. In this context, this paper examines the epidemiological evidence for cyanobacteria in the aetiology of adverse human health effects (essentially the right hand column in the table). The literature cited is not comprehensive, but has been selected to highlight the well–documented events and to sort the evidence they provide into their relative strengths.

Table 1 Designations of levels of evidence* according to type of research question

Level	Intervention	Aetiology †††
I *	A systematic review of level II studies	A systematic review of level II studies
II	A randomised controlled trial	A prospective cohort study
III–1	A pseudorandomised controlled trial (i.e. alternate allocation or some other method)	All or none §§§
III–2	A comparative study with concurrent controls: • Non–randomised, experimental trial • Cohort study • Case–control study • Interrupted time series with a control group	A retrospective cohort study
III–3	A comparative study without concurrent controls: • Historical control study • Two or more single arm study • Interrupted time series without a parallel control group	A case–control study
IV	Case series with either post–test or pre–test/post–test outcomes	A cross–sectional study

††† If it is possible and/or ethical to determine a causal relationship using experimental evidence, then the 'Intervention' hierarchy of evidence should be utilised. If it is only possible and/or ethical to determine a causal relationship using observational evidence (i.e., cannot allocate groups to a potential harmful exposure, such as nuclear radiation), then the 'Aetiology' hierarchy of evidence should be utilised.

* A systematic review will only be assigned a level of evidence as high as the studies it contains, excepting where those studies are of level II evidence.

§§§ All or none of the people with the risk factor(s) experience the outcome. For example, no smallpox develops in the absence of the specific virus; and clear proof of the causal link has come from the disappearance of small pox after large–scale vaccination.
(Adapted from NHMRC additional levels of evidence and grades for recommendations for developers of guidelines. PILOT PROGRAM. 2005. http://www.nhmrc.gov.au/consult/index.htm)

Case series (Level IV evidence)

Case series provide descriptive information about illness events that raise hypotheses about causation. However, they are not designed to test such hypotheses, and at best provide clues about causation to be further investigated using analytic study designs. There are a number of case series reports in the literature linking cyanobacterial exposure and adverse effects in humans.

Outbreaks of gastroenteritis were reported in Charleston, West Virginia, and in towns along the Ohio River after a period of low rainfall and increased bloom formation (Veldee 1931; Tisdale 1931). Approximately 15% of the 60000 residents of Charlestown were affected. It is likely that people were exposed to the bloom through contaminated drinking water.

In 1959, thirteen people became ill after swimming in a Canadian lake containing cyanobacteria that had recently been linked to a number of livestock deaths. *Microcystis* spp and *Anabaena circinalis* were identified in the stools of a doctor, who had accidentally ingested some of the water (Dillenberg et al. 1960).

On Palm Island just off north–east Australia, an outbreak of a hepatitis–like illness including gastroenteritis affected 138 children and 10 adults. The cause was initially considered a mystery (Byth 1980). It was later revealed that copper sulphate was used to remove an algal bloom from the drinking water reservoir, and the illnesses occurred a week later. It is likely that the copper sulphate lysed the cyanobacterial cells, releasing dissolved cyanotoxins that were not removed from the water by the usual treatment processes, leading to illness over the ensuing week.

While the above reports implicate ingestion as the principal mode of exposure, a British Medical Journal report described two severe pneumonia deaths and 16 cases of gastrointestinal symptoms in health military recruits after canoe training on water containing a bloom of *Microcystis aeruginosa* (Turner et al. 1990). Training involved carrying full packs and taking part in canoe rolls. This report implicates inhalation and/or aspiration as an effective exposure route for adverse effects in humans.

Anabaena and *Microcystis* blooms were found in the newly constructed Itaparica Dam's reservoir that supplied drinking water to a community in Brazil in 1988. Two thousand people developed gastroenteritis leading to 88 deaths over 42 days (Teixeira et al.1993). Investigations implicated the water impounded by the dam as the source of the outbreak with toxins from cyanobacteria considered to be responsible.

While it is highly probable and we are all likely to agree that cyanobacteria and their toxins were responsible for these events, in the absence of proper control groups, quantification of ingestion and inhalation, demonstrated dose–response effects and the like, as epidemiological evidence, this is not considered strong proof of aetiology. Certainly this evidence does not meet epidemiological criteria for causality, does not lend itself to dose–response investigation and is not helpful in assisting in the development of safety guidelines for exposure.

Cross–sectional/ecological studies (Level IV evidence)

A cross–sectional study measures disease occurrence and exposure status at the same time in a given population. Sometimes the exposure status of individuals prior to or even at the time of the onset of the disease is not necessarily known (ecological study). In these studies, exposure to an agent is assumed, based for example, on known contamination of a water supply, but the individuals are not individually asked if they actually consumed the water. Lack of specific individual exposure information prior to the onset of illness limits causality inferences. However, such studies are certainly suggestive and raise issues and hypotheses for further analytic investigations.

An important such study implicating cyanobacteria in the development of liver damage in humans was carried out in Armidale, Australia. Significantly increased levels of the liver enzyme gamma glutamyl transferase (GGT) were found in blood samples from a population of people supplied with a drinking water source from the Malpas dam containing a bloom of *Microcystis* and treated with copper sulphate. These GGT concentrations were significantly greater than in blood samples from people supplied with drinking water from a different source. (Falconer et al.1983). The raised GGT coincided with the bloom which makes it highly suggestive that it was responsible (Fig. 1). The other liver enzymes were not reported to be significantly different between samples from people with access to the different drinking water sources.

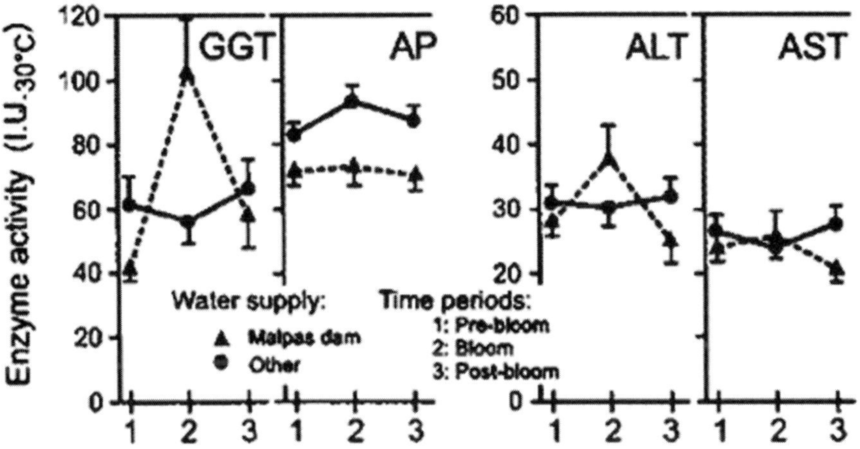

Fig. 1 Liver enzyme levels in blood samples from a pathology laboratory in Armidale, Australia, during a bloom of Microcystis aeruginosa. (GGT – gamma glutamyl transferase; ALT – alanine aminotransferase; AST – aspartate aminotransferase; AP – alkaline phosphatase)

One hundred and fifty–six communities in South–Eastern Australia, providing 32,700 singleton live newborns during 1992–94 were studied to examine the link between potential cyanobacterial exposure through drinking water and birth outcomes. This study linked birth outcomes with reported bloom occurrences in water bodies providing drinking water to the communities involved. Cyanobacterial occurrence and cell density (alert level) in drinking water sources during the first and last trimesters and for the whole gestational period for each birth were used as estimates of exposure. There were statistically significant differences between the proportion of time during the first trimester with cyanobacterial occurrence and the percentage of births that were low birth weight (LBW) and very low birth weight. Significant differences were also found among various categories of first trimester exposure based on average cell density and LBW, prematurity and congenital defects. However, the pattern of these results does not suggest a causal link to cyanobacteria. There were no clear dose–response relationships even with such a large cohort. Analyses based on exposure during the last trimester and total gestation also showed no significant dose–response effects. The results of this study provide no clear evidence for an association between cyanobacterial contamination of drinking water sources and adverse pregnancy outcomes (Pilotto et al.1999).

Yu (1995) reported that six large epidemiological studies have confirmed that populations receiving drinking water from pond–ditch water experience higher hepatocellular carcinoma mortality rates than populations receiving deep–well water. The findings from a study of water sources supported the hypothesis that microcystin in the drinking water of ponds–ditches and rivers, as opposed to deep–well water, was one of the risk factors for the high incidence of primary liver cancer in China (Ueno et al. 1996).

As before, although these studies suggest a link (or not) between cyanobacterial exposure and adverse health effects, they do not take individual exposure into account. The studies work on the assumption that the individuals involved did consume water from the affected water sources but we cannot be sure of this at an individual level. They were not individually asked, or if they were asked, contamination of the water by cyanobacteria (toxins) at the time of the study might not have been examined. Hence, while these studies are highly suggestive, they do not provide epidemiological proof of causality. They certainly raise the hypotheses for further investigation in humans.

Retrospective case–control study (Level III–3 evidence)

A case–control study is an analytic study that compares a group of people with disease to a similar group of people without that disease. Participants are selected into the study on the basis of having the disease. Controls are then selected based on certain inclusion criteria such as age matching. The levels of exposure to some agent each group had before appearance of the disease is then compared, and provides information about that agent being the likely cause for the outcome. Importantly, the exposure of the individuals involved in the study is known, and this information is usually obtained by questionnaire or other recorded information. Often, however, a question is raised about the accuracy of exposure information that relies on subjective recall. This may lead to recall bias, which has been identified as a major consideration to be taken into account when considering the results of case–control studies.

Along 8 Murray River towns in Australia, the risk of gastrointestinal symptoms was significantly associated with drinking chlorinated river water (RR 2.37; 95% CI 1.25, 4.49) during a period of raised cyanobacterial cell counts compared to rain water, and the risk of gastrointestinal (RR

5.20; 95% CI 1.13, 23.95) and skin symptoms (RR 5.21; 95% CI 1.01, 26.80) was associated with using untreated river water rather than rain water for domestic purposes (el Saadi et al. 1995). The width of the confidence intervals does raise doubt about the robustness of these findings. Also, the weekly mean log cyanobacterial count, although weakly correlated ($r = 0.52$) with the weekly proportions of patients presenting to medical practitioners with skin symptoms, this correlation was not statistically significant. This study was based on reports from general practitioners who supplied the researchers with the type of occurrence of illness, and the principal source of drinking and domestic use water, as well as recreational exposure to water. *Anaebaena* (some toxic), *Aphanizomenon* and *Oscillatoria* were the most common species present.

Prospective cohort study (Level II evidence)

A cohort study is an analytic study that involves identification of a cohort of people in which at least two groups can be identified, one that did receive the exposure of interest, and one that did not, and following these groups forward and comparing them for the outcome of interest. It is important in these studies that participants do not have the outcome of interest at the start of the study. This ensures that the risk factor occurs before illness, adding temporality in support of causality.

A prospective cohort study was conducted to investigate health effects of exposure to cyanobacteria as a result of recreational water activities. Participants aged 6 years and over, were interviewed at water recreation sites in South Australia, New South Wales, and Victoria on selected Sundays during January and February 1995. Telephone follow up was conducted 2 and 7 days later to record any subsequent diarrhoea, vomiting, flu–like symptoms, skin rashes, mouth ulcers, fevers, eye or ear irritations. On the Sundays of interview, water samples from the sites were collected for cyanobacterial cell counts and toxin analysis. There were 852 participants, of whom 75 did not have water contact on the day of interview and were considered unexposed. The 777 who had water contact were considered exposed. No significant differences in overall symptoms were found between the unexposed and exposed after two days. At seven days, there was a significant trend to increasing symptom occurrence between the unexposed, those exposed for up to 60 minutes and those exposed for more than 60 minutes ($p = 0.03$). A significant trend to increasing symptom occurrence was also found between unexposed subjects, and those exposed to water with cell counts of less than 5,000; 5,000 to 20,000; 20,000 to

80,000; and greater than 80,000 cells per mL ($p = 0.04$). Participants exposed to more than 5,000 cells per mL for more than one hour experienced a significantly higher symptom occurrence rate than the unexposed. *Microcystis* spp, *Anaebaena* spp, *Aphanizomenon* spp and *Nodularia* spp, some toxic, were identified. However, symptoms were not correlated with the presence of hepatotoxins, but might be due to direct contact with the lipopolysaccharide endotoxins on the surface of cyanobacteria. These results suggest increasing symptom occurrence was associated with increasing duration of contact with water containing cyanobacteria, and with increasing cyanobacterial cell density. The findings suggested that the currently accepted threshold for exposure of 20,000 cells per mL might be too high (Pilotto et al. 1997).

Another cohort study was conducted following an outbreak of liver failure at a dialysis centre in Caruaru, Brazil in the first half of 1996. One hundred and sixteen (89%) patients became ill (50 died from acute liver failure). Symptoms included visual disturbances, nausea and vomiting, headache, muscle weakness, epigastric pain, confusion, bleeding, fever and seizures. To examine risk factors for acute liver failure and death a case definition was established to allow comparison of risk factors for patients receiving dialysis at the city's two dialysis centres. A case was defined as any patient who had dialysis in either centre during February 1996 and who had acute liver failure. Results confirmed that all 101 case patients came from the same dialysis centre, and that centre had its water ("unfinished") supplied by truck from the municipal water–treatment plant as it was not linked to the water distribution system at that time. This water was not filtered or chlorinated. The nearby dialysis centre that recorded no cases received water from the same plant, but the water was sand filtered and chlorinated prior to being distributed to the water distribution system to which it was attached. Microcystins were found in the water reservoir, the delivery truck, the water holding tank and carbon and ion resin water treatment devices at the affected centre, and in the serum and liver tissue of case patients. No microcystins were found at the other dialysis centre. Unfortunately water samples from the time of likely exposure were not available, so it was not possible to quantify individual exposure of case patients (Jochimsen et al. 1998). So while this study provides good evidence for the acute systemic effects of microcystin exposure, it does not allow for an examination of dose response relationships that would contribute to safety guidelines.

Randomised controlled trial (RCT) (Level II evidence)

A RCT is an experimental study in which investigators randomly (usually using computer generated numbers) assign an intervention to a cohort of participants to form two groups, a treatment group and a control group, among which health outcomes can subsequently be compared. In some situations, such as with skin patch testing, it is possible to conduct a modified experimental study where patients become their own controls by randomly allocating a toxic agent, along with positive and negative control agents, to the skin of the same person. While not a true RCT where different individuals are randomised to receive different doses of toxin, both the participant and the person recording skin reactions remains blind to the position of the toxic agent.

A study using this approach on human volunteers was used to assess the skin irritant potential of a range of laboratory grown cyanobacterial species. Cell suspensions and extracts of cyanobacterial cultures of *Microcystis aeruginosa* (non–toxic strain), *Anabaena circinalis* and *Nodularia spumigena* were applied to 64 volunteers in one trial, and *Microcystis aeruginosa* (toxic strain), *Apanocapsa incerta* and *Cylindrospermopsis raciborskii* were applied to 50 volunteers in a second trial. Six cell concentrations of each organism in the range from less than 5000 to greater than 200,000 cells/mL were applied in random order using adhesive skin patches (Finn Chambers®). In addition, the applications included two treatments of each cyanobacterial species, involving whole and lysed cells, and positive (sodium lauryl sulphate) and negative (culture media) controls. Patches were removed after 24 hours and assessment of erythema was made by a dermatologist blinded to the species, cell type and concentration. On average, between 20% and 24% of individuals with 95% confidence interval ±8% reacted across the concentration range tested for these cyanobacterial species. The reaction rates were lower (11% to 15%) among the subset of subjects not reacting to negative controls. The reaction was mostly mild, and in all cases was resolved without treatment. This was the case for both whole and lysed cells with little difference in reaction rates between these two treatments. There was also no dose–response across the concentration range for any of the cyanobacterial species tested. Similar patterns of reaction were observed for atopic and non–atopic individuals. This study provides evidence that a small proportion of healthy people (around 20%) may develop a skin reaction to cyanobacteria in the course of normal water recreation, but the reaction is likely to be mild and resolve without treatment (Pilotto et al. 2004). From these results, skin irritation is not readily translated into a quantitative guideline for recreational water activities.

Conclusion

The evidence clearly links cyanobacteria (toxins) to adverse heath effects, particularly gastrointestinal illness, liver disease, neurological effects, skin reactions and cancer in humans. Exposure is through ingestion, inhalation and/or aspiration and dermal contact. Unfortunately most of the research related to these health effects provides lower levels of evidence, which do not take into account individual exposure and/or exposure levels and lack associated control groups. Individual exposure information, coupled with other potential confounding information, allows an examination of the risks for such exposures to cause adverse effects. Case–control, cohort or randomised controlled trails are required to meet these criteria. However, the very toxic nature of cyanobacteria means they are not readily available for investigation at such an analytic level in humans. This then makes the current body of evidence very limited in its capacity to assist in the development of safety guidelines, especially for ingestion and inhalation. The adverse effects of such toxins essentially restrict their use in experimental studies (RCTs) of ingestion and inhalation in humans.

Opportunities do exist to further refine our knowledge in relation to dermal contact exposure. It would be possible to develop a protocol based on recreational exposure using a RCT design. Appropriate recreational sites could be identified and, with their consent, people attending these sites could be randomised into an intervention group that engages in a prescribed water–related activity and a control group that is provided with an alternate form of recreation. Ideally the control group activity should be water related but not in the water containing cyanobacteria. Thought would need to be given to the range of microorganisms that would need to be measured, other confounding factors that would need to be accounted for, the outcome variables to be recorded, and the sample size required to have confidence in the findings. Ethical considerations require that study sites would be restricted to those believed to be low risk for the occurrence of adverse human health effects. Such a study would overcome many of the biases inherent in cohort and case–control studies would allow an accurate estimation of individual personal exposure and would provide a stronger level of evidence of the effects of recreational exposure to cyanobacteria than currently exists.

References

Byth S (1980) Palm Island mystery disease. Med J Aust 2: 40–42

Dillenberg HO, Dehnel MK (1960) Toxic water bloom in Saskatchewan, 1959. Can Med Assoc J 83:1151–1154

El Saadi O, Esterman AJ, Cameron S, Roder D (1995) Murray River water, raised cyanobacterial cell counts, and gastrointestinal and dermatological symptoms. Med J Aust 162: 122–125

Falconer IR, Beresford AM, Runnegar MTC (1983) Evidence of liver damage by toxin from a bloom of the blue–green alga, Microcystis aeruginosa. Med J Aust 1: 511–514

Jochimsen EM, Carmichael WW, An J, et al (1998) Liver failure and death after exposure to microcystins at a hemodialysis centre in Brazil. N Engl J Med 338: 873–878

Last J (Ed) (2001) A Dictionary of Epidemiology. Fourth Edition

Pilotto LS, Burch MD, Douglas RM, et al (1997) Health effects of exposure to cyanobacteria (blue green algae) during recreational water–related activities. Aust NZ J Public Health 21(6): 562–566

Pilotto LS, Hobson P, Burch M, Ranmuthugala G, Attewell R (2004) Acute skin irritant effects of cyanobacteria (blue green algae) in healthy volunteers. Aust NZ J Public Health 28: 220–224

Pilotto LS, Kliewer EV, Burch MD, Attewell RG, Davies RD (1999) Cyanobacterial contamination in drinking water and perinatal outcomes. Aust NZ J Public Health 23: 154–158

Teixeira Mda G, Costa Mda C, de Carvalho VL, Pereira Mdos S, Hage E (1993) Gastroenteritis epidemic in the area of the Itaparica Dam, Bahia, Brazil. Bull Pan Am Health Organ 27(3):244–53

Tisdale E (1931) Epidemic of intestinal disorders in Charleston, WVa, occurring simultaneously with unprecedented water supply conditions. Am J Public Health 21: 198–200

Turner PC, Gammie AJ, Hollinrake K, Codd GA (1990) Pneumonia associated with cyanobacteria. Br Med J 300:1440–1441

Ueno Y, Nagata S, Tsutsumi T, Hasegawa A, et al (1996) Detection of microcystins, a blue–green algal hepatotoxin, in drinking water sampled in Haimen and Fusui, endemic areas of primary liver cancer in China, by highly sensitive immunoassay. Carcinogenesis 17(6): 1317–1321

Veldee MV (1931) An epidemiological study of suspected water–borne gastroenteritis. Am J Public Health 21(9):1227–1235

Yu SZ (1995) Primary prevention of hepatocellular carcinoma. J. Gastroenterol Hepatol 10(6): 674–682

Chapter 30: Human Health Effects Workgroup Poster Abstracts

Serologic evaluation of human microcystin exposure

Hilborn ED,[1] Carmichael WW,[2] Yuan M,[2] Soares RM,[3] Servaites JC,[2] Barton HA,[4] Azevedo, SMFO[3]

[1]United States Environmental Protection Agency, Office of Research and Development, National Health and Environmental Effects Research Laboratory, Research Triangle Park, North Carolina [2]Department of Biological Sciences, Wright State University, Dayton, Ohio [3]Laboratory of Ecophysiology and Toxicology of Cyanobacteria, Carlos Chagas Filho Biophysics Institute, Federal University of Rio de Janeiro, Rio de Janeiro, Brazil. [4]United States Environmental Protection Agency, Office of Research and Development, National Center for Computational Toxicology, Research Triangle Park, North Carolina

Introduction

Microcystins (MCYST) are among the most commonly detected toxins associated with cyanobacteria blooms worldwide. Biological evidence of human exposure is needed in order to evaluate potential MCYST-associated health effects. MCYST are detectable in free and bound forms in human serum. We will provide an overview of selected methods to detect biological evidence of exposure in humans, and will identify some uncertainties associated with interpretation of results.

Methods

We analyzed serum samples collected from MCYST-exposed patients after exposure events at Brazilian dialysis clinics during 1996 and 2001. We used a commercially available enzyme linked immunoassay (ELISA) method to detect free MCYST, liquid chromatography/mass spectrometry (LC/MS) to detect free MCYST, and gas chromatography/mass spectrometry (GC/MS) to detect 2-methyl-3-methoxy-4-phenylbutyric acid (MMPB). MMPB is derived from both free and protein-bound MCYST by chemical oxidation, so it appears to represent total MCYST present in serum.

Results

Exposed patients provided blood samples for analysis after exposure. In a subset of 10 serum samples we found similar concentrations of free MCYST between the ELISA and LC/MS methods (Spearman $r=0.96$, $p<0.0001$). ELISA measurement of free MCYST was consistently lower than MMPB quantification of total MCYST. ELISA measured free MCYST as 8 – 51 % of total MCYST. Among the larger exposed population, we found evidence of free MCYST in patient serum for more than 50 days after the last date that documented MCYST exposure occurred.

Conclusion

MCYST are present in serum in free and protein-bound forms, though the nature of protein bound forms is uncertain. Analysis of serum samples for the presence of free MCYST may be performed in a cost-effective manner using screening assays such as the ELISA, but they underestimate total circulating concentrations. The relationship between free or total MCYST and absorbed dose is unknown due to limited knowledge of distribution and clearance. We found that free MCYST concentrations in patient serum may be detected for more than 50 days after the last documented exposure occurred. However, it is possible that patients experienced continued MCYST exposure by some route that was undetected during this study. Research is urgently needed to elucidate the human toxicokinetics of MCYST, in part to determine how measured serum levels can be used to estimate MCYST exposure.

The views expressed in this abstract are those of the individual authors and do not necessarily reflect the views and policies of the U.S. Environmental Protection Agency.

Characterization of chronic human illness associated with exposure to cyanobacterial harmful algal blooms predominated by *Microcystis*

Shoemaker RC, House D

Center for Research on Biotoxin Associated Illness, Pocomoke, Md

Introduction

Health effects from exposure to surface waters in the USA experiencing blooms of toxigenic cyanobacteria have not been well characterized. We initially evaluated seven cases of chronic illness following exposure to Lake Griffin, a member of the St. John's chain of lakes in Florida, during a bloom of *Microcystis* that was reported by the St. John's Water Management District. All seven people complained of multiple–system symptoms and demonstrated deficits in visual contrast sensitivity (VCS). Differential diagnoses based on medical histories, physical examinations, complete blood counts, comprehensive metabolic profiles, and assessments of both potentially confounding factors and toxic exposures indicated that exposure to the *Microcystis* bloom was the likely cause of illness. Patient reevaluations after 2 weeks of cholestyramine (CSM) therapy to bind and eliminate toxins demonstrated a statistically significant decrease in the number of symptoms and increase in VCS. The evidence indicated that exposure to the *Microcystis* bloom caused a biotoxin–associated illness similar to those previously reported in association with exposures to waters with high levels of toxigenic dinoflagellates and with exposures to water–damaged indoor environments exhibiting microbial amplification, including toxigenic fungi. We currently report a cohort of 10 patients exposed to *Microcystis* blooms who were evaluated before and after CSM therapy.

Hypotheses

Exposures to *Microcystis* blooms are associated with: (1) chronic illness characterized by multiple–system symptoms and VCS deficits; 2) increased blood levels of leptin and MMP9; 3) decreased blood levels of aMSH, ADH/osmolality, VEGF and ACTH/cortisol, and; 4) symptom resolution, and normalization of VCS and all biomarkers following CSM therapy.

Methods

Ten cases of chronic illness following exposure to *Microcystis* blooms were evaluated using the methods described above. Exposures to *Microcystis* blooms were determined to be the likely cause of illness. Three cases were exposures to blooms predominated by M*icrocystis* and reported by the St. John's Water Management District, whereas six cases were exposed to *Microcystis* blooms reported by the Maryland Department of Natural Resources. The number of symptoms, VCS, and blood levels of leptin, cortisol, osmolality, MMP9, VEGF, aMSH, and ACTH, were measured before and after CSM therapy, and HLA DR genotypes were identified. Repeated measurements of C3a, C4a, interleukin–10, and interferon alpha were also obtained from 3 cases. All measures were compared to those previously obtained from 239 unexposed well patients.

Results

The mean number of symptoms reported by patients was 19.7 out of 37 assessed before CSM therapy, and 3.2 following therapy. VCS increased by about 40% after therapy. Blood levels of blood levels of leptin and MMP9 were significantly higher than controls prior to therapy, whereas aMSH, ADH/osmolality, VEGF and ACTH/cortisol were low. All biomarkers normalized after 2 weeks of therapy except for aMSH. Two HLA DR haplotypes were significantly overrepresented in the cohort. All three cases for whom C3a, C4a, interleukin–10, and interferon alpha were measured showed elevated levels prior to therapy and normal levels following therapy.

Conclusion

The evidence indicated that exposures to *Microcystis* blooms may cause a form of chronic, biotoxin–associated illness that is characterized by abnormalities in symptoms, VCS and multiple biomarkers that resolves with CSM therapy. A randomized, double–blind, placebo–controlled, clinical trial and methods to measure cyanotoxins in blood is needed to confirm this hypothesis.

Chapter 31: Ecosystem Effects Workgroup Report

Workgroup Co–chairs:
John W Fournie, Elizabeth D Hilborn

Workgroup Members[1]:
Geoffrey A Codd, Michael Coveney, Juli Dyble, Karl Havens, Bas W Ibelings, Jan Landsberg, Wayne Litaker

Authors:
Bas W Ibelings, Karl Havens, Geoffrey A Codd, Juli Dyble, Jan Landsberg, Michael Coveney, John W Fournie, Elizabeth D Hilborn

Introduction

Harmful cyanobacterial blooms represent one of the most serious ecological stressors in lakes, rivers, estuaries and marine environments. When there are persistent or frequent blooms with high biomass of cyanobacterial cells, colonies or filaments in the water, a wide range of impacts on the ecosystem may occur. These are well established in the scientific literature and are summarized in Paerl et al. (2001). Blooms may shade the water and thereby inhibit growth of other primary producers including phytoplankton, benthic algae and vascular plants and may elevate pH, particularly in poorly buffered waters. High population densities of large cyanobacteria interfere with food collection by filter–feeding zooplankton. The senescence and subsequent microbial decomposition of blooms may impact benthic macro–invertebrate community structure, as well as fish and other biota, due to increased organic loading and resulting anoxia of sediments, accumulation of NH4 in the water and accompanying increases in pH. Blooms of toxic cyanobacteria have been implicated in mass mortalities of birds and fish (e.g., Matsunaga et al. 1999; Rodger et al. 1994), but

[1] See workgroup member affiliations in Invited Participants section.

the importance of cyanotoxins relative to the other stressors that accompany blooms remains unknown. With persistent blooms, there are substantial declines in biodiversity at all levels ranging from phytoplankton and zooplankton to birds. Changes in nutrient cycling and disruptions of carbon and energy flow in pelagic and benthic food webs are observed (Paerl et al 1998). Where blooms become severe in shallow lakes, a positive feedback loop develops through various biological mechanisms related to the presence of cyanobacteria and fish that maintains a turbid water state (Scheffer and Carpenter 2003).

A major uncertainty regarding the effects of cyanobacterial blooms is the role that cyanotoxins play in contributing to the various biological responses listed above. There are three reasons for this uncertainty: (a) most research to examine cyanobacterial bloom effects at the ecosystem level has focused on factors not associated with toxins but with the mere presence of cyanobacteria; (b) no experimental studies have been done at the whole community level to examine effects of blooms in the presence of vs. absence of cyanotoxins; and (c) experimental studies dealing with cyanotoxins have largely involved exposure of a single species to a single toxin under ideal conditions in the laboratory. Studies have not examined synergistic effects with other natural stressors, nor have they adequately investigated how multiple toxins of natural and anthropogenic origin might affect the biota. Thus, laboratory results are not readily transferable to the field.

The objective of this report is to identify major knowledge gaps regarding the impacts of cyanobacterial blooms on biota in lakes, rivers and estuaries from the individual to ecosystem level. The text is organized around six charges given to the Ecologic Effects Working Group. All of the identified research components are considered by the Working Group to be a high priority. Careful consideration was given to information already available in the primary literature in determining research needs to avoid duplicity of effort. A simple conceptual model illustrates the interrelationship among the research and modeling work discussed in the subsequent sections of this paper (Fig. 1).

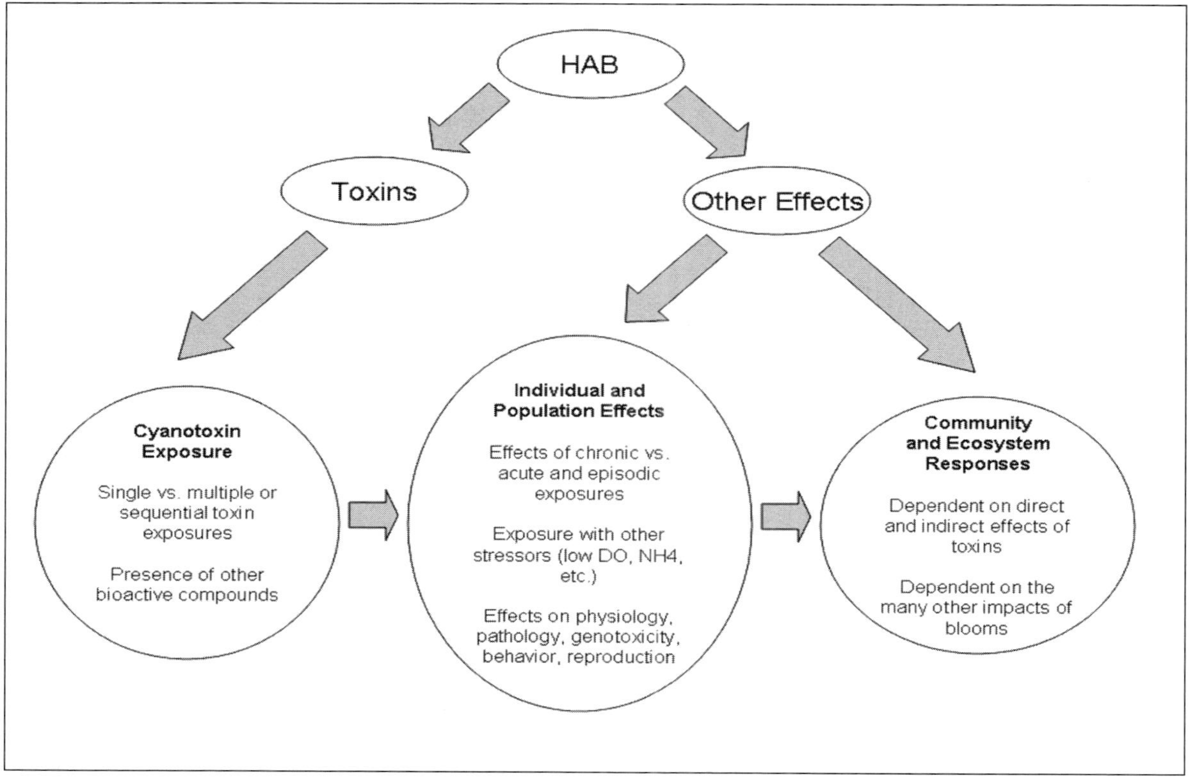

Fig. 1. Conceptual model of the ecosystem effects of cyanobacteria and cyanotoxins

There is a logical order in which the research topics noted here might be addressed, starting with the species level work and then scaling up to the community level with environmentally–relevant experiments based on findings from previous work. Development of community bioaccumulation models may occur in concert with controlled and observational research, so that at any given time, modeling tools may become available for application with clearly identified levels of uncertainty and defined boundaries of applicability.

Charge 1

Identify research needed to quantify effects of cyanotoxins under environmentally relevant conditions.

To understand the effects of cyanotoxins on aquatic biota and ecosystems, it is critical that environmentally–relevant exposure conditions be identified and evaluated. Almost all experimental studies of exposure have been of single species exposed to individual cyanotoxins under optimal conditions (see Ibelings and Havens, this issue for an overview). In addition, there are a limited number of studies that have assessed the distribution of cyanotoxins in lake food webs (Kotak et al. 1996; Ibelings et al. 2005). These studies have furthered our understanding of potential ecological effects, but field studies alone are insufficient to identify associations between exposures and ecological effects during periods when cyanobacterial blooms predominate in aquatic communities.

It is well established in the toxicological literature that stressors may have antagonistic, additive or synergistic effects (Taylor et al. 2005). Hence there is a need for studies that determine how exposure to cyanotoxins alone or in combination with other physiologically stressful conditions (e.g., low dissolved oxygen, high ammonia (NH_4), high pH, poor food quality, high and low temperature, salinity, etc.) affect the fitness of aquatic biota. Often these sub–optimal or stressful environmental conditions coincide with the presence of cyanotoxins in the water, and the relative contribution of each exposure is poorly understood. Bury et al. (1995) demonstrated that NH_4, like dissolved microcystin–LR, impeded fish growth. Additionally, interactions may occur between different classes of the cyanotoxins themselves. Indeed, lipopolysaccharide endotoxins can inhibit glutathione *S*–transferases *in vivo*, thereby reducing the capacity of glutathione *S*–transferases to detoxify microcystins (Best et al. 2002).

Most controlled experiments have examined the biotic effects of microcystin–LR, and to a lesser extent nodularin, with relatively few studies looking at the effects of other cyanotoxins (Table 1). Although some cyanobacteria produce saxitoxins, the ecological effects of these toxins have been studied primarily in marine systems where they are produced by dinoflagellates (Landsberg 2002). Toxins that have been less frequently studied are cylindrospermopsin and its derivatives, and lyngbyatoxins. Given the likelihood that these toxins are present in US waters (Carmichael et al. 1997; Burgess 2001), their effects must be quantified if we are to make predictions about the ecological effects of toxic cyanobacterial blooms with a reasonable level of certainty.

Table 1. Number of peer reviewed papers on ecological effects of cyanotoxins by class of toxins and group of aquatic organisms. Results from October 2005 search (key words: toxin plus organism as listed in the column head) of ISI Web of Knowledge (Thomson Scientific and Healthcare, Stamford, Connecticut).

Class of toxins	Zooplankton	Bivalves	Fish	Waterfowl
Microcystin	73	38	87	13
Nodularin	17	17	16	2
Anatoxin–a and a(s)	11	0	7	6
Cylindrospermopsin	5	2	7	0
Lyngbyatoxin	5	1	9	0
Microviridin	4	0	0	0
Saxitoxin (freshwater)	2	2	6	0

It is becoming increasingly clear that the numerous studies of microcystin–LR may not be representative of the complexities of the interactions between other cyanotoxins, cyanobacterial blooms, and other biota. Complexity arises because: (a) other microcystin variants can be abundant and may be ecologically more relevant, such as the less toxic microcystin–RR which may be taken up preferentially into biota (Xie et al. 2005); (b) cyanotoxins other than microcystins occur widely and have documented adverse ecological effects, such as the association between anatoxin–a(s), anatoxin–a and mass avian mortality events (Henriksen et al. 1997; Krienitz et al. 2003); (c) there is an array of potentially harmful bioactive compounds produced by cyanobacteria which have not been well–studied. For example, microviridin–J has detrimental effects on molting in *Daphnia*, but its effects on other biota are not well understood (Rohrlack et al. 2004). There have been some examples of the toxicity of crude cell–extracts exceeding the expected toxicity of the component cyanotoxins, suggesting that unidentified compounds or synergistic effects are associated with observed toxicity (Lürling 2003).

It also is critical that exposure studies use relevant organisms where possible. For example, there is considerable variation in intraspecies susceptibility to cyanotoxins among fish and birds, but only a relatively small number of species have been studied (Carmichael and Biggs 1978; Fischer and Dietrich 2000). Few experimental studies have been done using waterfowl (Carmichael and Biggs 1978). This has largely necessitated the use of oral toxicity data obtained from the study of other animal groups to estimate the risk of avian toxicity (Krienitz et al. 2003).

Charge 1: Identify research needed to quantify effects of cyanotoxins under environmentally relevant conditions

Near-term Research Priorities

- Laboratory studies exposing key aquatic biota to cyanotoxins under simulated natural conditions, including low dissolved oxygen, elevated pH, elevated ammonia, and other stressors associated with cyanobacteria growth and senescence.
- Field and/or mesocosm studies of the ecologic effects of cyanotoxins under varying environmental conditions.

Charge 2

Identify research needed to quantify the physiological, pathological and behavioral effects of acute, chronic, and episodic exposures to cyanotoxins.

The duration of exposure to cyanobacterial blooms and toxins can range from days to years, yet most studies have investigated the effects of short–term exposures. Research is needed on the effects of long–term exposure and adaptive responses of populations that may employ both existing phenotypic plasticity and adaptive evolution. Although mortality has been examined as a common endpoint, sub–lethal effects require further study. With the exception of *Daphnia*, few studies have examined behavioral responses and their significance to the affected species. A small number of studies have demonstrated that fish behavior is affected by exposure to cyanotoxins in water (Best et al. 2003; Baganz et al. 1998). Further research needs include controlled experiments to examine effects of cyanotoxins and toxin mixtures on: (a) behavior, especially as it relates to escape from predators and ability to acquire resources; (b) reproduction; (c) neurologic function, and (d) genotoxicity. Some work has been performed to evaluate the role of enzyme inhibition and oxidative stress on genotoxicity, however, the relevance of oxidative stress under field conditions is unknown. Inhibition of protein phosphatases is the classic mode of action of microcystins and nodularins, resulting in hyper–phosphorylation of cytoskeletal proteins and the disruption of numerous other phosphorylation–regulated cell processes. Oxidative stress occurs during the detoxification process. Detoxification produces glutathione–microcystin conjugates, a depletion of the cellular glutathione pool and an imbalance in reactive oxygen species (Pflugmacher 2004).

The effects of chronic exposure to cyanotoxins are rarely investigated in aquatic animals, despite their widespread geographic distribution and potential lifelong exposure to toxic cyanobacterial blooms. Some cyanotoxins have been reported to be tumor promoters, and the risk of tumorigenesis increases with chronic exposure. Microcystins, nodularins, cylindrospermopsins, aplysiatoxins, debromoaplysiatoxin, and lyngbyatoxin–a have all been demonstrated to be tumorigenic, but these properties have only been experimentally demonstrated in small mammals or cell assays (Fujiki et al. 1984; Falconer and Humpage 1996). Recent studies have investigated the association between tumor–promoting cyanotoxins and an increased prevalence of fibropapillomas in seaturtles (Landsberg 2002; Arthur et al. 2005).

It is important to consider individual susceptibility as influenced by factors including age, disease, nutrition, and gender. Species–specific susceptibilities include those related to differences in detoxification and metabolism of cyanotoxins. More detailed knowledge is needed, both in the field of toxicodynamics and toxicokinetics among species. There have been a few sub–chronic exposure studies on the accumulation and depuration of cyanotoxins in a limited number of animals. This type of research seems to have focused on bivalves and, to a lesser extent, fish. The bivalve studies showed a biphasic depuration of microcystin. Fluctuating microcystin concentrations during depuration were speculated to be the result of an ongoing process of covalent binding and release of microcystins (Amorim and Vasconcelos 1999). Overall much is unknown about the fate of cyanobacterial cells and cyanotoxins after ingestion. Reports suggest that only a small percentage of the toxins that are ingested with the food end up in the blood and organs of the organisms: 2.7 % in *Daphnia*, and 1.7 % in rainbow trout (Rohrlack et al. 2005; Tencalla and Dietrich 1997). There are barriers to microcystin uptake at various levels. Even if taken up, aquatic organisms have the capacity for detoxification, which in fish is followed by rapid excretion via the biliary excretion system. Studies that examine chronic exposure of biota over ecologically relevant time scales of months – years have, to our knowledge, not been conducted. In addition, very little is known about patterns in the accumulation and depuration of cyanotoxins other than microcystin. Ultimately, good quantitative data is required for a number of well defined endpoints for a range of toxin classes and aquatic biota (Table 2).

Table 2. List of Definitions

Term	Definition
Bioaccumulation:	The process which causes an increased chemical concentration in an aquatic organism compared to the water, due to uptake by all exposure routes (Gray 2002).
Bioconcentration:	Uptake directly from the water, and results in the chemical concentration being greater in an aquatic organism than in the water (Gray 2002).
Biomagnification:	Transfer of a chemical from food to an organism, resulting in a higher concentration in the organism than in its diet. The result may be a concentration of the chemical as it moves up the food chain (Gray 2002).
Biodilution:	Decreased toxin levels are observed at each increase in trophic level in the food web.
Endpoint:	An observable or measurable biological event or chemical concentration (e.g., metabolite concentration in a target tissue) used as an index of an effect of a chemical exposure.
Exposure	
Acute:	Resulting in adverse effects from a single dose or exposure to a substance
Chronic:	Continuous or repeated exposure to a substance over a long period of time, typically the greater part of the total life–span in animals or plants
Subchronic:	Exposure for period typically involving a time period in between acute and chronic
No–Observed–Adverse–Effect Level (NOAEL):	The highest dose at which there are no biologically significant increases in the frequency or severity of adverse effects between the exposed population and its appropriate control; some effects may be produced at this level, but they are not considered adverse or precursors of adverse effects. In comparison, see LOEL.
Lowest Observable Effect Level (LOEL):	The lowest dose which produces an observable effect.
Toxicodynamics:	The determination and quantification of the sequence of events at the cellular and molecular levels leading to a toxic response to an environmental agent.
Toxicokinetics:	The determination and quantification of the time course of absorption, distribution, biotransformation, and excretion of chemicals.

Charge 2: Identify research needed to quantify the physiological, pathological and behavioral effects of acute, chronic, and episodic exposures to cyanotoxins

Near-term Research Priorities

- Investigate the behavioral effects of cyanotoxin exposures
- Investigate sub–lethal effects of chronic exposures of key taxa of aquatic biota including invertebrates, fish, bivalves and others.

Charge 3

Identify research needed to quantify biological effects of exposure to multiple toxicants

An increasing number of reports describe the co–occurrence of different cyanotoxins in aquatic systems, and there is an emerging catalogue of bio–active materials associated with cyanobacterial blooms. For example, co–occurrence has been observed for microcystin and anatoxin–a, microcystin and cylindrospermopsin, and there is an assumed universal co–occurrence of lipopolysaccharides with all other known cyanotoxins (Codd et al. 2005). Research on the death of lesser flamingos in Kenya's Rift Valley lakes demonstrated that both microcystin and anatoxin–a were present in the cyanobacteria on which the birds were feeding (Krienitz et al. 2003). The relative contribution of these cyanotoxins, which differ greatly in their mode of action, to the mass bird mortalities remains unclear, as is the role of co–occurring anthropogenic pollutants like heavy metals and organic pesticides. Thus, research is needed to examine effects of simultaneous and sequential exposure to multiple toxins (Codd et al. 2005). The Working Group considers the highest priority for multiple–exposure studies of effects in US lakes to be the evaluation of the combination of microcystin and cylindrospermopsin. Research also is needed to examine effects of other bio–active compounds, including non–microcystin cyclic peptides and lipopeptides.

Charge 3: Identify research needed to quantify biological effects of exposure to multiple toxicants

Near-term Research Priority

- Investigate the effects of simultaneous and sequential exposure to multiple toxins, particularly the combination of microcystins and cylindrospermopsin, and microcystins and anatoxins

Charge 4

Identify research needed to quantify effects of cyanotoxins at whole community level

Community level effects are a function of: (a) the direct effects of cyanotoxins; (b) the direct effects of cyanobacteria blooms; (c) the indirect effects associated with altered competitive and predatory interactions; and (d) changes in nutrient cycling. Effects may occur in all biota from bacteria to birds and mammals and are integrally linked with a loss of biodiversity in aquatic systems. Key research needs include studies involving complete natural communities and studies with simple food chains to examine the effects of exposure to toxic vs. non–toxic strains of cyanobacteria.

At this time we do not adequately understand the relative importance of different uptake routes of toxins from the environment. Exposure to cyanotoxins can be through direct ingestion of cells, uptake of toxins that are present in the environment, or by the transfer of toxins through the food web. Vectorial transport of cyanotoxins has been demonstrated in a few experiments involving dissolved toxins. Fish may be negatively affected by dissolved toxin uptake through the gills, but other biota are not very sensitive to the toxins once they are extracellular (Lürling & Van der Grinten 2003; Zurawell et al. 1999). There is little information about the relative importance of this exposure route vs. exposure by direct ingestion of toxic cells. In one experiment, pike larvae were exposed to zooplankton that had accumulated dissolved nodularin. There was a strong inhibition of larval feeding rate despite the fact that only 0.03 % of the toxin that was present in the zooplankton was actually taken up by the larvae (Karjalainen et al. 2005). The remainder of the toxin was either metabolized or excreted. If this is a representative result, vectorial transport of only a small amount of the cyanotoxin that is produced at the base of the food web may have significant ecological effects at higher trophic levels. Substantiation of the relevance of vectorial transport and the effects of different classes of cyanotoxins throughout the food web is of prime importance to understand ecological effects of these toxins.

Much research is needed to understand the degree of bioaccumulation that occurs in communities. Bioaccumulation may vary considerably between species, but this has been studied only with a small number of organisms. Most bioaccumulation studies have focused on microcystin. In future research and modeling, it is critical that we distinguish between bioaccumulation, bioconcentration and biomagnification (Table 2). In most studies, the term bioaccumulation is used in a loose way, simply meaning that toxins are present in biota. Ibelings et al. (2005) have argued that biomagnification is the most relevant process to study in food webs since most of the transfer and uptake of cyanotoxins appears to be via food. An increase in microcystin concentration, as it moves up the food chain was not found by these authors, and was not expected due to the low octanol–to–water partition coefficient of microcystin–LR (De Maagd et al. 1999). However it is known that the octanol–to–water coefficient varies widely according to the microcystin variant and correlates with in vivo toxicity to Tetrahymena (Ward and Codd 1999). Cyanobacterial toxins other than microcystin may behave very differently, as indicated by the distribution of the neurotoxin β–N–methylamino–L–alanine (BMAA) in the terrestrial food web on the island of Guam. Biomagnification of this toxin may have accounted for the exposure of the indigenous Chamorro people to high concentrations of BMAA via their consumption of flying foxes (Cox et al. 2003). More knowledge is required on the potential for bioaccumulation, bioconcentration and biomagnification of different cyanotoxins and other cyanobacterial bioactive compounds in the food web.

An important shortcoming of all but a few studies is the absence of data on covalently–bound microcystin in biota. Microcystins are routinely extracted using aqueous methanol, but this does not extract quantities of the methyldehydroalanine–containing microcystins which are covalently bound to protein phosphatases in the cell. Lemieux oxidation does extract these covalently bound forms. Studies that have compared standard aqueous methanol extraction to extraction after Lemieux oxidation have demonstrated that a large part of the total microcystin pool in biota is indeed covalently bound (Table 3). Most of the literature therefore severely underestimates the concentration of total microcystin (free and bound forms). Covalent binding of microcystins may reduce the transfer of free, unbound microcystin along the food chains, and potentially contribute to biodilution of microcystin (Karjalainen et al. 2005). A relevant, but as yet unanswered question, concerns the toxicity and bioavailability of the covalently–bound microcystins (Ibelings et al. 2005).

Table 3. Comparison of standard aqueous methanol extraction and Lemieux oxidation among organisms

Organism	MeOH extraction as % of Lemieux oxidation	Reference
Dungeness Crab (larvae)	0.01	Williams et al., 1997a
Salmon (liver)	24	Williams et al., 1997a
Blue mussel	0.1	Williams et al., 1997b
Zebra mussel	62	Dionisio Pires et al., 2004

Charge 4: Identify research needed to quantify effects of cyanotoxins at whole community level

Near-term Research Priorities

- Investigate the effects of exposure to toxic vs. non–toxic strains of cyanobacteria in natural communities and simple food chains.
- Examine the potential for bioaccumulation, bioconcentration and biomagnifications of different cyanotoxins and other cyanobacterial bioactive compounds in the food web.
- Determine the toxicity and bioavailability of covalently–bound microcystins.

Charge 5

Identify research needed to determine the relative importance of the effects of cyanotoxins vs. the effects of cyanobacteria at the ecosystem level

Cyanobacterial bloom development, maturation and senescence can all result in adverse environmental conditions that affect biota independent of the effects of cyanotoxins. Key research questions at the ecosystem level of inquiry include: (a) how important are the effects of cyanotoxins vs. the effects of cyanobacteria; (b) does the presence of high concentrations of cyanotoxins, for instance through interference with zooplankton grazing, contribute to the stability of the turbid water state in shallow eutrophic lakes; (c) does the increasing occurrence of cyanotoxins in shallow lakes undergoing eutrophication contribute to the shift from the clear to turbid state? Effects of toxins on benthic communities and benthic processes are not well understood, yet those processes play a key role in aquatic food webs and nutrient cycling (Palmer et al. 2000). The relevance of these processes is demonstrated by the consequences of the invasion by zebra mussels (Dreissena polymorpha) in the Laurentian Great Lakes of North America. It has been hypothesized that selective filter feeding by these mussels has been instrumental in the return of Microcystis blooms to Lake Erie (Vanderploeg et al. 2001). However, in the Netherlands, Microcystis is efficiently grazed by Dreissena, resulting in a low concentration of cyanobacteria in areas where the mussels are abundant (Dionisio–Pires et al. 2004). This paradox is not fully understood, although emerging explanations include variation in cyanotoxin concentrations of the Microcystis strains involved and the relevance of nutrient recycling in lakes of widely varying trophic status (Raikow et al. 2004). The pseudofeces of Dreissena are rich in cyanobacteria and they may transfer toxins to the benthic food web, where benthic feeders are potentially exposed to the toxins (Babcock–Jackson et al. 2002). To further address these issues and gain a deeper understanding of interactions between toxins, cyanobacteria and other biota at the ecosystem level, we propose research at a high level of integration, including the use of static and flowing mesocosms under controlled conditions.

Charge 5: Identify research needed to determine the relative importance of the effects of cyanotoxins vs. the effects of cyanobacteria at the ecosystem level

Near-term Research Priorities

- Investigate the importance of the effects of cyanotoxins vs. cyanobacteria at the community level.
- Determine if the presence of high concentrations of cyanotoxins contributes to the stability of the turbid water state in shallow eutrophic lakes.
- Determine if the increasing occurrence of cyanotoxins in shallow lakes undergoing eutrophication contributes to the shift from the clear to turbid state.

Charge 6

Identify how modeling can contribute to a predictive understanding of HAB bloom and cyanotoxin effects.

Basic models relating cyanobacterial growth to nutrient inputs and other environmental conditions are readily available, and their development is not a priority research area. Various factors including phosphorous, nitrogen, iron and light have been studied in the laboratory and are known to have an effect on cyanotoxin concentrations (Wiedner et al. 2003).

However, there is a need for models that relate environmental conditions to cyanotoxin types, concentrations and compartmentation (soluble vs. particulate pools) in blooms and water bodies containing benthic cyanotoxins. Evidence is emerging that cyanotoxin concentrations increase in direct response to exposure to grazers like zooplankton and fish (Jang et al. 2003, Jang et al. 2004). This would strengthen the idea that cyanobacteria produce these energetically costly toxins as a grazer–deterrent, but whether this is the sole or primary purpose for toxin production is an important question that demands further research. While many of these factors impacting toxin production have been studied individually, modeling the interactive effects of both the bottom–up and top–down factors could provide further insight into the potential toxicity of a bloom given a set of environmental conditions.

Models are needed to describe the fate of cyanotoxins in water, sediment and food webs. This is not a trivial undertaking since toxins at every level in the food web are potentially subject to covalent binding, metabolism (detoxification) and excretion, and thus the amount of bioavailable cyanotoxin that is transferred to the next trophic level is a complex issue. As noted, modeling can be an ongoing activity, with predictive certainty and general applicability increasing as ongoing research at the population, community and ecosystem levels provides additional information for model parameterization, calibration and verification.

Conclusions

The authors of this report have identified near term priorities for research on ecological and ecosystem effects of harmful cyanobacterial blooms. Although the negative impact of cyanobacterial blooms on many ecosystems is well known, the specific contribution of cyanobacterial toxins to the harmful effects is hard to distinguish. Most research has involved exposure of a single species to a single toxin under controlled laboratory conditions. More research is needed at the whole community level. The research priorities ordered from the species level and scaling up to the community and ecosystem level that have been identified are:

- To study the effects of cyanotoxins under environmentally relevant conditions, including other environmental stressors; additive or synergistic effects of combinations of cyanobacterial toxins or bioactive compounds produced by cyanobacteria.
- To use relevant, naturally co–occurring organisms in exposure studies; more knowledge is needed on species–specific toxicokinetics and toxicodynamics.
- To obtain good quantitative data for a number of well–defined endpoints for a range of toxin classes and biota.
- To study the effects of long term exposures and adaptive responses of aquatic organisms.
- To identify community level effects of cyanotoxins. Key research needs include studies of simple food chains and natural communities exposed to toxic cyanobacteria.
- To identify the relative importance of different uptake routes from the environment and the extent to which vectorial transport of toxins in the food web takes place.
- To understand the potential for bioaccumulation, bioconcentration and biomagnification of cyanobacterial toxins in aquatic food webs.
- To distinguish between the ecosystem effects of cyanobacterial toxins and the harmful effects of cyanobacterial blooms in general (toxic vs. non toxic blooms).
- To build and test models that relate environmental conditions to cyanotoxin types, concentrations and compartmentalization and models that describe the fate of cyanotoxins in water, sediment and food webs.

References

Amorim A, Vasconcelos V (1999) Dynamics of microcystins in the mussel Mytilus galloprovincialis. Toxicon 37:1041–1052

Babcock–Jackson L, Carmichael WW, Culver DA (2002) Dreissenid musselsincrease exposure of benthic and pelagic organisms to toxic microcystins. Verh Internat Verein Limnol 28:1082–1085

Baganz D, Staaks G, Steinberg C (1998) Impact of the cyanobacterial toxin microcystin–LR on behaviour of zebrafish, Danio rerio. Water Res 32:948–952

Best JH, Pflugmacher S, Wiegand C, Eddy FB, Metcalf JS, Codd GA (2002). Effects of enteric bacterial and cyanobacterial lipopolysaccharides , and of microcystin–LR on glutathione S–tranfersase activities in zebra fish (Danio rerio). Aquatic Toxicology 60: 223–231.

Best JH, Eddy FB, Codd GA (2003). Effects of Microcystis cells, cell extracts and lipopolysaccharide on drinking and liver function in rainbow trout Oncorhynchus mykiss Walbaum. Aquatic Toxicology 64:419–426.

Burgess C (2001). A wave of momentum for toxic algae study. Environ Health Perspect 109:160–1.

Bury NR, Eddy FB, Codd GA (1995) The effects of the cyanobacterium Microcystis–aeruginosa, the cyanobacterial hepatotoxin microcystin–LR and ammonia on growth–rate and ionic regulation of brown trout. J Fish Biol 46:1042–1054

Carmichael WW, Biggs D F (1978). Muscle sensitivity differences in two avain species to anatoxin–a produced by the freshwater cyanophyte Anabaena flos–aquae. Canadian Journal of Zoology 56:510–512.

Carmichael WW, Evans WR, Yin QQ, Bell P, Moczydlowski E (1997). Evidence for paralytic shellfish poisons in the freshwater cyanobacterium Lyngbya wollei (Farlow ex Gomont) comb. nov. Appl Environ Microbiol 63:3104–3110.

Codd GA, Lindsay J, Young FM, Morrison LF, Metcalf JS (2005). Harmful cyanobacteria: From mass mortalities to management measures. In: Harmful Cyanobacteria, Eds. J. Huisman, H.C.P. Matthijs and P.M. Visser, Springer, Dordrecht, The Netherlands, pp. 1–23.

Cox PA, Banack SA, Murch SJ (2003). Biomagnification of cyanobacterial neurotoxins and neurodegenerative disease among the Chamorro people of Guam. PNAS 100: 13380–13383

De Maagd PGJ, Hendriks AJ, Seinen W, Sijm D (1999) pH–dependent hydrophobicity of the cyanobacterial toxin microcystin–LR. Water Res 33:677–680

Dionisio Pires LM, Karlsson KM, Meriluoto JAO, Visser PM, Siewertsen K, Van Donk E, Ibelings BW Assimilation and depuration of microcystin–LR by the zebra mussel, Dreissena polymorpha. Aquat Toxicol. 2004;69:385–396.

Falconer IR, Humpage AR. 1996. Tumor promotion by cyanobacterial toxins. Phycologia 35:74–79.

Fischer WJ, Dietrich DR (2000). Pathological and biochemical characterization of microcystin–induced hepatopancreas and kidney damage in carp (Cyprinus carpio). Toxicol Appl Pharmacol 164:73–81

Fujiki H, Suganuma M, Hakii H, Bartolini G, Moore RE, Takegama S, Sugimura T. (1984). A two–stage mouse skin carcinogenesis study of lyngbyatoxin A. J Cancer Res Clin Oncol 108: 174–176.

Gray JS (2002). Biomagnification in marine systems: the perspective of an ecologist. Mar Poll Bull 45: 46–52

Henriksen P, Carmichael WW, An JS, Moestrop O (1997). Detection of an anatoxin–a(s)–like anticholinesterase in natural blooms and cultures of Cyanobacteria/blue–green algae from Danish lakes and in the stomach contents of poisoned birds. Toxicon 35:901–913

Ibelings BW, Bruning K, de Jonge J, Wolfstein K, Pires LMD, Postma J, Burger T (2005). Distribution of microcystins in a lake food web: No evidence for biomagnification Microb Ecol 49:487–500

Jang MH, Ha K, Joo GJ, Takamura N (2003)Toxin production of cyanobacteria is increased by exposure to zooplankton Freshw Biol 48:1540–1550

Jang MH, Ha K, Lucas MC, Joo GJ, Takamura N (2004). Changes in microcystin production by Microcystis aeruginosa exposed to phytoplanktivorous and omnivorous fish. Aquatic Toxicol 68: 51–59.

Lürling M (2003) Daphnia growth on microcystin–producing and microcystin–free Microcystis aeruginosa in different mixtures with the green alga Scenedesmus obliquus. Limnol Oceanogr 48:2214–2220.

Lürling M, van der Grinten E (2003) Life–history characteristics of Daphnia exposed to dissolved microcystin–LR and to the cyanobacterium Microcystis aeruginosa with and without microcystins. Environ Toxicol Chem 22:1281–1287.

Karjalainen M, Reinikainen M, Spoof L, Miriluoto JAO, Sivonen K, Viitasalo M (2005). Trophic transfer of cyanobacterial toxins from zooplankton to planktivores: Consequences for pike larvae and mysid shrimps. Environ Toxicol 20:354–362

Krienitz L, Ballot A, Kotut K, Wiegand C, Putz S, Metcalf JS, Codd GA, Pflugmacher S (2003) Contribution of hot spring cyanobacteria to the mysterious deaths of lesser flamingos at Lake Bogoria, Kenya. FEMS Microbiol Ecol 43:141–148

Kotak BG, Zurawell RW, Prepas EE, Holmes CFB (1996) Microcystin–LR concentration in aquatic food web compartments from lakes of varying trophic status. Can J Fish Aquat Sci 53:1974–1985

Landsberg JH (2002) The effects of harmful algal blooms on aquatic organisms. Rev Fish Sci 10:113–390

Matsunaga H, Harada KI, Senma M, Ito Y, Yasuda N, Ushida S, Kimura Y (1999). Possible cause of unnatural mass death of wild birds in a pond in Nishinomiya, Japan: Sudden appearance of toxic cyanobacteria. Nat Toxins 7:81–88

Paerl HW, Pinckney JL, Fear JM and Peierls BM (1998). Ecosystem responses to internal and watershed organic matter loading: consequences for hypoxia in

the eutrophying Neuse River Estuary, North Carolina, USA. Marine Ecology Progress Series 166:17–25.
Paerl HW, Fulton III RS, Moisander PH and Dyble J (2001). Harmful algal blooms with an emphasis on cyanobacteria. TheScientificWorld Journal 1:76–113.
Palmer MA, Covich AP, Lake S, Biro P, Brooks, JJ, Cole J, Dahm C, Gibert J, Goedkoop W, Martens K, Verhoeven J and van de Bund WJ (2000). Linkages between aquatic sediment biota and life above sediments as potential drivers of biodiversity and ecological processes. BioScience 50:1062–1075.
Pflugmacher S (2004). Promotion of oxidative stress in the aquatic macrophyte Ceratophyllum demersum during biotransformation of the cyanobacterial toxin microcystin–LR. Aquat Toxicol 70:169–178
Raikow DF, Sarnelle O, Wilson, AE, Hamilton, SK (2004) Dominance of the noxious cyanobacterium Microcystis aeruginosa in low–nutrient lakes is associated with exotic zebra mussels. Limnol Oceanogr 49:482–487
Rodger HD, Turnbull T, Edwards C, Codd GA (1994) Cyanobacterial (blue–green–algal) bloom associated pathology in brown trout, Salmo–trutta L, in Loch Leven, Scotland. J Fish Dis 17:177–181
Rohrlack T, Christoffersen K, Kaebernick M, Neilan BA (2004). Cyanobacterial protease inhibitor microviridin J causes a lethal molting disruption in Daphnia pulicaria. Appl Environ Microbiol 70: 5047–5050
Rohrlack T, Christoffersen K, Dittmann E, Nogueira I, Vasconcelos V, Börner T (2005). Ingestion of microcystins by Daphnia: Intestinal uptake and toxic effects. Limnol Oceanogr 50: 440–448
Scheffer M, Carpenter SR (2003). Catastrophic regime shifts in ecosystems: linking theory to observation TREE 18: 648–656
Taylor RL, Caldwell GS, Bentley MG (2005). Toxicity of algal–derived aldehydes to two invertebrate species: Do heavy metal pollutants have a synergistic effect? Aquatic Toxicol 74: 20–31
Tencalla F, Dietrich D (1997). Biochemical characterization of microcystin toxicity in rainbow trout (Oncorhynchus mykiss). Toxicon 35: 583–595 APR 1997
Vanderploeg HA, Liebig JR, Carmichael WW, Agy MA, Johengen TH, Fahnenstiel GL, Nalepa TF (2001) Zebra mussel (Dreissena polymorpha) selective filtration promoted toxic Microcystis blooms in Saginaw Bay (Lake Huron) and Lake Erie. Can J Fish Aquat Sci 58:1208–1221
Ward CJ, Codd GA (1999). Comparative toxicity of four microcystins of different hydrophobicities to the protozoan, Tetrahymena pyriformis. J Appl Microbiol 86:874–882
Wiedner C, Visser PM, Fastner J, (2003). Effects of light on the microcystin content of Microcystis strain PCC 7806. Appl Environm Microbiol 69:1475–148
Williams DE, Craig M, Dawe SC, Kent ML, Holmes CFB, Andersen RJ (1997) Evidence for a covalently bound form of microcystin–LR in salmon liver and dungeness crab larvae. Chem Res Toxicol 10:463–469
Williams DE, Dawe SC, Kent ML, Andersen RJ, Craig M, Holmes CFB (1997) Bioaccumulation and clearance of microcystins from salt water, mussels,

Mytilus edulis, and in vivo evidence for covalently bound microcystins in mussel tissues. Toxicon 35:1617–1625

Xie LQ, Xie P, Guo LG, Li L, Miyabara Y, Park HD (2005). Organ distribution and bioaccumulation of microcystins in freshwater fish at different trophic levels from the eutrophic Lake Chaohu, China. Environ Toxicol 20:293–300

Zurawell RW, Kotak BG, Prepas EE (1999). Influence of lake trophic status on the occurrence of microcystin–LR in the tissue of pulmonate snails. Freshw Biol 42:707–718

Chapter 32: Cyanobacterial toxins: a qualitative meta–analysis of concentrations, dosage and effects in freshwater, estuarine and marine biota

Bas W Ibelings[1], Karl E Havens[2]

[1]Netherlands Institute of Ecology (NIOO–KNAW) – Centre for Limnology, Rijksstraatweg 6, 3631 AC, Nieuwersluis, The Netherlands; e–mail b.ibelings@nioo.knaw.nl; [2]University of Florida, Department of Fisheries and Aquatic Sciences, Gainesville, FL 33458, USA

Abstract

This paper reviews the rapidly expanding literature on the ecological effects of cyanobacterial toxins. The study employs a qualitative meta–analysis from the literature examining results from a large number of independent studies and extracts general patterns from the literature or signals contradictions. The meta–analysis is set up by putting together two large tables – embodying a large and representative part of the literature (see Appendix A). The first table (Table A.1) reviews the presence (concentrations) of different cyanobacterial toxins in the tissues of various groups of aquatic biota after exposure via different routes, experimentally in the lab or via natural routes in the environment. The second table (Table A.2) reviews the dose dependent effect of toxins on biota. The great majority of studies deal with the presence and effects of microcystin, especially of the MC–LR congener. Although this may partly be justified – MC–LR is an abundant and highly toxic protein – our review also emphasizes what is known about (i) other MC congeners (a number of studies showed a preferred accumulation of the less toxic variant MC–RR in animal tissues), (ii) nodularin (data on a range of biota from studies on the Baltic Sea), (iii) neurotoxins like anatoxin–a(s), which are conspicuously often present at times when mass mortalities of birds occur, (iv) a few studies on the presence and effects of cylindrospermposin, as well as (v) the first examples of ecological effects of newly identified bioactive compounds, like microviridin–J. Data were reorganized to assess to what extent bioconcentration (uptake and concentration of toxins from the water) or biomagnification (uptake and concentration via the food) of cyanobacterial toxins occurs in ecosystems. There is little support for the occurrence of biomagnification, and this reduces the risk for biota at higher trophic levels. Rather than biomagnification biodilution seems to occur in the foodweb with toxins being subject to degradation and excretion at every level. Nevertheless toxins were present at all tropic levels, indicating that some vectorial transport must take place, and in sufficient quantities for effects to possibly occur. Feeding seemed to be the most important route for exposure of aquatic biota to cyanobacterial toxins. A fair number of studies focus on dissolved toxins, but in those studies purified toxin typically is used, and biota do not appear very sensitive to this form of exposure. More effects are found when crude cyanobacterial cell lysates are used, indicating that there may be synergistic effects between different bioactive compounds. Aquatic biota are by no means defenseless against toxic cyanobacteria. Several studies indicate that those species that are most frequently exposed to toxins in their natural environment are also the most tolerant. Protection includes behavioral mechanisms, detoxication of MC and NODLN by conjugation with glutathione, and fairly rapid depuration and excretion. A common theme in much of the ecological studies is that of modulating factors. Effects are seldom straightforward, but are dependent on factors like the (feeding) condition of the animals, environmental conditions and the history of exposure (acclimation and adaptation to toxic cyanobacteria). This makes it harder to generalize on what is known about ecological effects of cyanobacterial toxins. The paper concludes by summarizing the risks for birds, fish, macroinvertebrates and zooplankton. Although acute (lethal) effects are mentioned in the literature, mass mortalities of – especially – fish are more likely to be the result of multiple stress factors that co–occur during cyanobacterial blooms. Bivalves appear remarkably resistant, whilst the harmful effects of cyanobacteria on zooplankton vary widely and the specific contribution of toxins is hard to evaluate.

List of abbreviations.

Abbreviation	Definition
AchE	Acetyl Choline Esterase
ANA(a)(as)	Anatoxin–a; anatoxin–a(s)
BCF	Bioconcentration Facor (concentration of toxic compound in an organism as % of that in water)
BMF	Biomagnification Factor (concentration of toxic compound in an organism as % of that in its diet)
CAT	Catalase (one of the antioxidative enzymes)
CYN	Cylindrospermopsin
GI	Gastrointestinal tract
GPx	Glutathione Peroxidase (one of the antioxidative enzymes)
GR	Glutathione Reductase (one of the antioxidative enzymes)
GSH	Glutathione
GST	Glutathione–S–Ttransferase (catalyst of the formation of MC–GSH conjugates in detoxication)
H_2O_2	Hydrogen peroxide (one of the ROS formed during oxidative stress)
HP	Hepatopancreas
IP	Intraperitoneal injection
LC_{50}	Concentration at which 50 % of the test animals die from exposure to the toxin
LOEC	Lowest Observable Effect Concentration
LPO	Lipid Peroxidation (outcome of oxidative stress)
LPS	Lipopolysacharides
MC	Microcystin
NODLN	Nodularin
PP	Protein phosphatases (inhibition of PP by MC results in hyperphosphorilation of proteins)
PST	Paralytic Shellfish Toxin
ROS	Reactive Oxygen Species (formed during oxidative stress)
SAX	Saxitoxin
SOD	Super Oxide Dismutase (one of the antioxidative enzymes)
TDI	Tolerable Daily Intake (of a toxin like MC)
TEH	Total Extractable Hepatotoxins (sum of toxins and their biotransformation products)
TOSC	Total Oxygen Scavenging Capacity

Introduction

Until just a few years ago statements like "traditionally research has focused on the acute toxicity of microcystin–LR to laboratory mammals" and "there is a general lack of research involving aquatic organisms which may be exposed to toxin producing cyanobacterial blooms in their natural environment" (Zurawell et al. 1999) were fully justified. In contrast, there now exists a large literature from observational and experimental studies dealing with cyanobacterial toxins in aquatic systems. Yet major information gaps remain, and our ability to understand effects is limited by certain attributes of those studies. The aim of this paper is to review the extant literature, identify general patterns in results, and identify key areas where additional research is warranted.

This assessment is complex because in addition to microcystin–LR there are many other toxins produced by cyanobacteria. Some of these toxins, such as nodularin, are closely related to microcystin, while others are quite different (e.g., the neurotoxins anatoxin–a and a(s) and saxitoxin and the protease inhibitor cylindrospermopsin). There also exists an ever increasing list of bioactive compounds produced by cyanobacteria, some of which have been shown to be toxic to selected aquatic biota (like microviridin–J for *Daphnia* (Rohrlack et al. 2004), but many of which have not been studied in any detail. Furthermore, chronic and sub–chronic effects (see Havens et al elsewhere in this volume for definitions) may be more relevant to study than acute lethal effects. Exposure of biota in lakes supporting cyanobacterial blooms is likely to be repetitive and over much of an organism's lifespan. Although cyanobacterial toxins have been claimed to play a role in acute events like mass mortalities of fish and birds, there is usually insufficient evidence to link fish and bird kills directly to these toxins. This does not mean that there are no important sub–lethal effects resulting from chronic exposure in the aquatic ecosystem.

This study employs qualitative meta–analysis of data from the literature, examining results from a large number of independent studies and synthesizing summaries and conclusions addressing the issue of toxic effects. Meta–analysis aims to utilize the increased power of pooled data to clarify the state of knowledge on that issue, and may include quantitative statistical analyses when the data are consistent with that approach – this is not the case here. We set up the meta–analysis by putting together two large tables (see Appendix A). The first table (Table A.1) reviews the presence (concentrations) of cyanobacterial toxins in the tissues of aquatic biota after exposure via different routes, be it experimental or via natural routes in the field. The second table (Table A.2) reviews the dose dependent effects of toxins on biota. While Tables A.1 and A.2 do not include 100% of the published papers on these subjects, they do embody a large and representative part of the literature.

Methodology

The literature was queried using the ISI–Web of Science. Results of the literature search are given in Havens et al (this volume). To assemble Table A.1 – concentrations in biota – the following data were extracted from the literature: (i) biota involved, four groups are distinguished: birds, fish, macroinvertebrates and zooplankton); (ii) type of cyanobacterial toxin studied; (iii) exposure route; (iv) toxin concentrations in biota; (v) concentration in the source (i.e. this could be dissolved purified toxin in an experimental setting, cells from cultures of toxic cyanobacteria or natural seston containing toxic cyanobacteria); and (vi) analytical analysis method that was used to quantify the toxin. In the discussion, data from Table A.1 will be organized in such a way that biomagnification (accumulation of the toxins in biota via consumption of food that contains the toxins) can be quantified. Bioaccumulation of cyanobacterial toxins is often speculated to increase the risk of exposure for aquatic biota (especially at higher trophic levels), but bioaccumulation has seldom been analyzed correctly (see Havens et al, this volume for definitions). Especially for MC it is well known that the analytical method may have a marked effect on the concentration that is measured. This is even more so when MC is measured in biota, rather than in the toxin producing cyanobacteria. Standard MeOH extraction does not include covalently bound MC and analysis using ELISA suffers from cross reactivity between MC or NODLN and their GSH conjugates formed in detoxication. The consequences of this for interpretation of the concentrations given in Table A.1 are discussed.

For Table A.2 the following data were compiled: (i) biota involved; (ii) exposure route; and (iii) dose and effect. In many studies it is not possible to establish true dose–effect relationships because organisms are exposed to only one or two different dosages, and Table A.2 will indicate in which cases sufficient data have been gathered to establish a relationship. Ideally the unit for dose would be units of toxin administered per unit of body weight and per unit of time (for comparison TDI for humans equals a dose of 0.04 $\mu g\ kg\ bw^{-1}$). It is more common however to find toxin contents of the source expressed in $\mu g\ L^{-1}$. Differences in units hamper interpretation across different studies. What will also emerge from Table A.2 is that the exposure route has a strong influence on biological effects. In fish IP injection of MC is often fatal however oral dosage hardly ever results in mortality. As will be discussed, fish–kills in lakes seem to be a consequence of multi–stress factors during blooms of toxic cyanobacteria rather than of direct intoxication. Tables A.1 and A.2 are included as an appendix of this paper.

Results and Discussion

The last 5 years have included a steady increase in the number of papers investigating cyanobacterial toxins in aquatic biota. Whether this increasingly large body of literature is sufficiently broad (in terms of the toxins / bioactive compounds and aquatic biota covered by the studies) and deep (in terms of yielding a detailed understanding of the effects these toxins have on aquatic biota) is another matter. Below we formulate an answer to this question – which is central to this paper – by analyzing the assembled literature presented in the tables.

Toxins in biota

Toxin producing species and their toxins

This review takes a somewhat unusual perspective in that the focus is not on toxin producing cyanobacteria but rather on the biota in the aquatic ecosystem that may be affected by the toxins. Cyanobacteria that (frequently) appear in the tables are the MC producing genera *Microcystis* (species *M. aeruginosa, M. flos aquae* and *M. viridis*) and *Planktothrix* (*P. agardhii* and *P. rubescens*), the main NODLN producing species *Nodularia spumigena* and the anatoxin–a and a(s) producing genus *Anabaena.* Although species within a genus may differ greatly in their ecology (*P. agardhii* for instance is commonly found in hypertrophic, turbid shallow lakes, whereas *P. rubescens* is typically found in clear, deep alpine lakes), we do not stress differences between toxin producers at the species level since often we know little about differences in toxin production within a genus like for instance *Microcystis*.

If there is one thing immediately striking about the tables it is the prominence of MC, especially of MC–LR (not all studies specify which MC congeners were present, although most studies express total MC as MC–LR equivalents since often MC–LR is the only standard used in HPLC analysis). This focus on the presence and effects of MC is probably (partly) justified because surveys of cyanobacterial toxins in several countries have indeed shown that microcystins are prominently present (e.g. in Denmark – Henriksen et al. 1997). The spotlight on MC–LR (one variety in a family of around 70 different MC congeners) may be more biased, and seems influenced by the laboratory work on mammals. MC–LR is relatively toxic, its LD_{50} in mice is 50 µg kg^{-1}, considerably lower than for instance the LD_{50} of MC–RR (600 µg kg^{-1}) (Spoof 2005). Yet as the data in Table A.1 indicate there are a number of studies demonstrating that especially the less toxic MC–RR is taken up into tissues of aquatic biota, for instance in silver carp (Jang et al. 2004); or in freshwater snails (only MC–RR present in foot of *B.aeruginosa* (Chen and Xie 2005) and hepatopancreas of *S.histrica* (Ozawa et al. 2003). On the basis of their observations on the ratio of MC–LR:MC–RR in different tissues and organs, Xie et al. (2004) suggested that MC–LR may be actively degraded during digestion, whereas MC–RR is transported across the intestines and embedded into body tissues. At the same time Xie suggested that MC–RR is not acutely toxic to the carp, since no mortality was observed despite the uptake of MC–RR into various organs. In a study using the *Thamnocephalus* bioassay LC_{50} of MC–RR (and of MC–YR) was actually very close to the LC_{50} of MC–LR. The real difference in this invertebrate was a much gentler slope of MC–RR compared to MC–LR (and YR) when LC_{10} and LC_{90} values were included (Blom et al. 2001).

A further relevant distinction within the microcystins is that between the methyldehydroalanine–containing microcystins which covalently bind to PP in the cell and MC that contains dehydrobutyirine and – like NODLN – that do not bind covalently to PP. For NODLN it has been suggested that because NODLN does not bind covalently its transfer in the food web is facilitated (Kankaanpaa et al. 2001). The same would be true for the dehydrobutyirine containing microcystins, but since concentrations of these have not been analyzed in biota there is no evidence for this. Overall it is clear that studies on MC–LR cover just a small part of the total complexity of interactions between microcystin producing cyanobacteria and aquatic biota, so that the bias indicated by Tables A.1 and A.2 is unjustified. More research is needed on the ecological effects of the whole spectrum of bioactive – potentially harmful – compounds produced by cyanobacteria, including MC congeners other than LR.

Some toxins other than MC have been analyzed in biota. NODLN features fairly prominently in the tables, due to considerable work that has been done on the Baltic sea. This is one of the few regions in the world where there are sufficient data to actually follow concentrations of NODLN and its effects throughout much of the food web (more in discussion on 'bioaccumulation'). Neurotoxins are not a notable group in the table. Interestingly anatoxins (anatoxin–a and/or anatoxin–a(s)) often seem to play a role when toxic cyanobacteria have been implicated in the death of waterfowl (Henriksen et al. 1997; Krienitz et al 2003). Neurotoxicosis can be seen as convulsed extremities and arched back necks (Codd et al. 2005). The death of the lesser flamingos in Africa's Rift valley lakes is an interesting example where multiple cyanobactertial toxins (neurotoxins and hepatoxins) may play a simultaneous role and have synergistic effects (a more accurate description of the interaction between different toxins or toxins and other stress factors may be additive rather than synergistic effects). This is another area where very little is known today.

Biota involved and organs affected

Tables A.1 and A.2 distinguish four groups of aquatic organisms: waterfowl, fish, macroinvertebrates (i.a. bivalves, crabs, prawns, snails) and zooplankton. There are only a handful of studies involving birds (see above). There are many more studies on fish, the other group of aquatic vertebrates where mass mortalities have been attributed to blooms of toxic cyanobacteria. Toxins in fish have been analyzed after exposure via different routes, but the majority of work actually involves studies of fish caught in lake or sea, i.e. fish that has been exposed to toxins via natural routes (exposure through the food web or to dissolved toxins after lysis of blooms). In the ecosystem the feeding guild of a fish species appears to be a primary determinant of exposure to toxins. Phytoplanktivorous fish like silver carp (e.g. Jang et al. 2004) directly consume cyanobacteria. Zooplanktivorous fish like sticklebacks and smelt (Ibelings et al. 2005) feed on zooplankton that directly consume cyanobacteria. Likewise a species like flounder predates on filter feeding blue–mussels. Piscivorous fish that prey on zooplanktivorous fish are one step further removed from the toxin producing cyanobacteria. It may be expected that in the absence of biomagnification (but see discussion below) MC concentrations decrease in the order phytoplanktivorous > zooplanktivorous > piscivorous fish . Omnvirous fish may fit in anywhere. Indeed fish caught in the IJsselmeer, The Netherlands showed an increase in MC in fish–liver moving from larger perch (predatory) to ruffe (benthic) and zooplanktivorous smelt (Ibelings et al. 2005). In contrast Xie et al. (2005) found that MC in various tissues and organs varied as carnivorous > omnvivorous > phytoplanktivorous fish. Fischer and Dietrich (2000) explain that there are several differences in GI tract between a carnivorous fish like rainbow trout and planktivorous and herbivorous cyprinids like carp. Cyprinids (as well as cichlids) possess a much longer ileum with larger surface area and higher resorption capacities, so that carnivorous fish would accumulate less MC, i.e. the opposite of what was found in Chinese lakes. In contrast

Carbis et al. (1997) explain that the neutral or slightly basic conditions in the GI of carp limit absorption of MC, since cells of cyanobacteria would only be digested in an acid environment.

Overall no relationship between feeding guild and toxin concentrations in fish can be pulled out from the data in Table A.1. Data from different studies are hard to compare because toxin concentrations, exposure routes and a host of other biotic and abiotic factors differ between sites and studies. What is clear, and this is found across studies on fish, is that concentrations of MC are mainly present in the gut and liver, to a somewhat lesser extend in kidneys and gonads, and much less in muscle tissue (e.g. Kankaanpaa et al. 2005a; Li et al. 2004; Soares et al. 2004; Malbrouck et al. 2003; Xie et al. 2005). Microcystins are also found in fish faeces in substantial amounts (Jang et al. 2004; Xie et al. 2005) and in pseudofaeces of *Dreissena* (Babcock–Jackson et al. 2002; Pires et al. 2004). This could expose the benthic community to cyanobacterial toxins produced in the pelagic zone.

There is a surprisingly large number of studies on the presence of toxins in macroinvertebrates, especially in bivalves (mussels and clams). In contrast there are very few studies that describe effects of cyanobacterial toxins on these animals, perhaps because generally they seem insensitive (e.g. Saker et al. 2004). A common theme in the studies on macroinvertebrates is the analysis of time courses for accumulation and depuration of toxins. Accumulation is often found to be time dependent and proceeds in an orderly manner. Time lagged acccumulation occurred at deeper sites in the Baltic sea. Mussels at deep sites are primarily exposed to toxic *Nodularia* towards the end of the bloom period, when filaments sink to the sediment, where they over winter. Although mussels at deeper sites contained much less toxin, they did accumulate some NODLN (Sipia et al. 2002). Depuration from mussels is almost always found to be biphasic (Ozawa et al. 2003; Sipia et al. 2001b; Vasconcelos et al. 1995), sometimes concentrations of toxin even increase in the first phase of depuration (Amorim and Vasconcelos 1999). It has been suggested that this is a consequence of dynamics in production and degradation of PP to which the MC are bound (Vasconcelos et al. 2001), but more research is needed (Ozawa et al. 2003). Although depuration is commonly judged to be rapid (e.g. Sipia et al. 2002; Kankaanpaa et al. 2005b; Pereira et al. 2004) it is equally clear that depuration is incomplete even after a considerable period of time. Depuration is temperature dependent and slows down in winter, so that toxins may even be carried on to the next spring (Ozawa et al. 2003). In for instance Lake IJsselmeer this has consequences for thousands of diving ducks that arrive in autumn. Although summer *Microcystis* blooms have dispersed, the mussels (food for the ducks) still contain traces of toxins.

Thus the mussels may be considered a vector that prolongs the time when toxins are able to exert negative effects in that lake ecosystem. In macroinvertebrates hepatotoxins, but also CYN, were mainly found in the haemolymph and hepatopancreas, to a lesser extend also in gonads and muscle tissue (foot).

With respect to zooplankton we see the opposite of the macroinvertebrates papers: there are a large number of studies on effects of cyanobacterial toxins on zooplankton but much less on concentrations of toxins in the animals. The available results suggest that concentrations are relatively high in indiscriminately filter feeding taxa such as *Daphnia* (Ibelings et al. 2005; Kotak et al. 1996a; Thostrup and Christoffersen 1999), and perhaps lower in copepods, but again there is a general lack of data. Toxins seem to be taken up into the body of zooplankton, concentrations of toxin cannot be explained solely by the presence of toxic cyanobacteria in the gut.

Analytical analysis methods

The standard method for analysis of hepatotoxins (MC and NODLN) is HPLC, coupled to diode array UV detection (see Table A.1). ELISA is frequently used because of its high sensitivity (see Spoof 2005) for pros and cons of different methods. A drawback of ELISA is cross reactivity with detoxication metabolites, like the conjugates of MC and GSH. These conjugates have been shown to have a much lower toxicity (Metcalf et al. 2000). Conjugates can be detected using LC–MS, but this is still rarely undertaken (however see Karlsson et al. 2003 and Sipia et al. 2002). Because ELISA suffers from cross reativity some studies on biota in the Baltic Sea have introduced the term TEH – total extractable hepatotoxins, which includes the biotransformation products. TEH almost invariably exceeds the concentrations of untransformed hepatotoxins in biota – see for instance the comparison of NODLN (analyzed on LC–MS) and TEH (ELISA) in Kankaanpaa et al. (2005a), which differ by an order of magnitude, or Lehtonen et al. (2003) where NODLN was < 5 % of TEH in Baltic clams. Another important analytical issue is that of extraction of the toxins. A large number of MC congeners – those that contain methyldehydroalanine – covalently bind to PP in plant and animal cells; these covalently bound MC are not extracted using standard MeOH extraction. The handful of studies that have used Lemieux oxidation – a method that does extract covalently bound MC – demonstrated that a large part of the MC in biota is covalently bound (Williams et al. 1997; Pires et al. 2004) (see Table 2 in Havens et al, this issue). This means that almost all of the concentrations given in Table A.1 seriously underestimate the total amount of MC present in biota. What is unknown – and this is important – is whether all these studies also underestimate the bioavailability and toxicity of MC. Is covalently bound MC in *Daphnia* still (equally) toxic to the fish that swallows cladocerans?

Bioaccumulation

Many studies have suggested that cyanobacterial toxins bioaccumulate in aquatic biota and that this may enhance the risk of exposure of biota higher up in the food web (e.g. Li et al. 2004; Sipia et al. 2001a; Negri and Jones 1995). Xie et al. (2005) present data that demonstrate that MC has a general tendency to accumulate up the food chain, with concentrations being highest in carnivorous and lowest in herbivorous fish species. PST concentrations in *Daphnia magna* grazing on *Aphanizomenon* exceeded those in the cyanobacterium (bioaccumulation factor > 1). Bioaccumulation in most papers however use a loose definition and usually it just means that toxins are present in biota. When a more formal – and informative – definition of bioaccumulation, and the related processes of bioconcentration and biomagnification are used (see Havens et al, this issue for definitions) there is very little evidence to support the notion of bioaccumulation of MC and NODLN in aquatic food webs. Rather the opposite, i.e. biodilution of hepatotoxins in the food web, is supported by the data (Karjalainen et al. 2005). Data in Table A.1 do show however that bioconcentration of NODLN may take place. In an experimental setting two copepods and a ciliate took up dissolved NODLN and accumulated this to concentrations far higher than in the water (BCF ranged from 12–22). Predators of these zooplankters and protozoa would be exposed to substantial concentrations of toxin in their food and may suffer consequences like decreased ingestion rates, as was shown for pike larvae and mysid shrimps feeding on the zooplankton (Karjalainen et al. 2005).

Biomagnification factors express the concentration of a toxin in biota as a percentage of that in their diet. BMF for the Baltic Sea and IJsselmeer are shown in Table 1.1. BMF in the Baltic biota and in *Daphnia* and *Dreissena* from the IJsselmeer are well below 100 %, indicating that the concentration was much below the concentration in the seston. Biomagnification obviously is absent in these cases. BMF in the Baltic of Copepods and clams are exceptionally low; only a very small part of NODLN which is present in the cyanobacteria is taken up by these grazers. The calculation in Table 1 of BMF for grazers of the phytoplankton is very sensitive to the toxin content of the seston used in the calculation, and this content may be highly variable. Values of BMF in the table should be taken as indicative rather than absolute. The BMF of ruffe and especially smelt in the IJsselmeer are > 100 and seem to indicate that biomagnification of MC has taken place. In this case however concentrations of MC in a whole organism (like *Daphnia*) are compared to values for a selected organ where the toxin specifically accumulates (the liver), and this gives a skewed representation of biomagnification (Gray 2002). The difference in BMF between freshwater mussels in the IJsselmeer and their marine counterparts in the Baltic is striking. The very low concentration of MC in *Dreissena* led Ibelings et al. (2005) to the conclusion that the food web linked to filter feeding mussels is hardly exposed to toxins. In contrast Kankaanpaa et al. (2005a) concluded that in the Baltic food webs involving mussels are especially exposed to hepatotoxins. The tenfold difference in BMF supports these apparently opposing conclusions.

Table 1. Biomagnification factors Baltic Sea and IJsselmeer (The Netherlands). BMF were calculated as NODLN (Baltic) or MC (IJsselmeer) content in biota as a percentage of toxin in their diet (e.g. in eiders as a % of that in mussels). For comparison BMF is also calculated as percentage of the concentration in the seston (although this would only qualify as biomagnification for organisms that actually feed on seston, like Daphnia and the mussels). Data compiled from (Engstrom-Ost et al. 2002; Kankaanpaa et al. 2005a; Karjalainen et al in press; Lehtonen et al. 2003; Sipia et al. 2001b; Sipia et al. 2002; Sipia et al. 2004) for the Baltic sea. BMF for the IJsselmeer have been modified from Ibelings et al. (2005). Data on which calculation of BMF are based are taken from Table A.1.

Baltic biota	BMF seston	BMF diet	IJsselmeer biota	BMF seston	BMF diet
Copepods	0.3	0.3	Daphnia galeata	20	20
Blue mussel (Mytilus edulis)	8.9	8.9	Zebra mussel (Dreissena polymorpha)	0.9	0.9
Baltic clam (Macoma baltica)	0.6	0.6	Perch (Percia fluviatilis)	5.9	11
Mysid shrimp (Mysis relicta)	0.3	100	Ruffe (Gymnocephalus cernuus)	13.2	120
Pike larvae (Esox lucieus)	0.2	59	Smelt (Osmerus eperlanus)	53.5	286
Sticklebacks (Gasterosteus aculeatus)	0.05	24			
Flounder (Platichthys flesus)	1.6	19			
Eider (Somateria mollissima)	0.8	8			

Exposure routes

Laboratory studies

Early studies on fish (Tencalla et al. 1994; Kotak et al. 1996b) primarily used the method which is preferred for exposure of mammals in laboratory studies – intra–peritoneal injection. Direct injection of toxins like MC proved to be highly toxic to fish. The effects are comparable to those seen in mammals, but differences are seen as well. Whereas mammals die from haemorrhagic shock following hepatocyte insult, fish die from direct liver failure, necrosis (e.g. Malbrouck et al. 2003, Li et al 2004). The LC_{50} for MC–LR in perch (1500 μg g DW^{-1}, Ibelings et al unpublished data) is well above the LC_{50} for mice, indicating that these fish species are less sensitive to the toxin than warm blooded animals. Nevertheless Sipia et al. (2001a) notes that salmon hepatocytes seem more sensitive to algal toxins than rat hepatocytes. When MC was administered orally (up to 1150 μg MC kg^{-1} bw given by gavage 8 times over 96 h (a total dose of 9200 μg MC kg^{-1} bw) to perch from the IJsselmeer no mortality was seen, although histopathology of the livers showed that MC were having severely detrimental effects. Similar differences between IP injection and gavage can be found in Table A.2 (e.g. Tencalla et al. 1994).

Directly from the water

In ecotoxicology there is a general assumption that uptake from the water is a common route for aquatic vertebrates to accumulate xenobiotic substances (Karjalainen et al. 2003). Indeed some of the studies have demonstrated direct uptake of cyanobacterial toxins from the water, even to the extent that the concentration in biota exceeds those in the water. However, most studies where biota are exposed to dissolved toxins have been in the laboratory (see Table A.2) using purified toxins. These studies have proven valuable in finding the mechanisms through which biota are affected by cyanobacterial toxins but are less informative about the importance of uptake of dissolved toxins in the ecosystem. A specific effect of dissolved MC is inhibition of ATP–ase activity of Na^+ K^+ pumps in the gills of fish and crabs, resulting in ion imbalance (Best et al. 2003; Vinagre et al. 2003; Zambrano and Canelo 1996). Concentrations of dissolved MC are much increased when surface blooms of floating cyanobacteria lyse. When cyanobacteria float to the surface they are exposed to extreme conditions, in particular an increase in irradiance, potentially to damaging levels. Photoprotective mechanisms that may protect the cells from photooxidation are hampered by the co–occurrence of light stress and other stress factors, notably an increase in temperature, desiccation and depletion of inorganic carbon (Ibelings and Maberly 1998). Lehtonen et al. (2003) suggested that the major fate of cyanobacterial blooms in the Baltic is to disintegrate in the water column so that very little reaches the bottom. If this were the case exposure to dissolved toxins would be a major event. Lysis of surface blooms is not unlikely, but we maintain that exposure of biota to high concentrations of dissolved toxin are the exception rather than the rule because processes like mixing, adsorption to clay particles, photolysis and bacterial degradation rapidly reduce the availability of dissolved toxins (Ozawa et al. 2003).

Moreover it has been shown by several authors that some aquatic biota are not sensitive to dissolved cyanobacterial toxins – e.g., brown trout (Best et al. 2001), pike–larvae (Karjalainen et al. 2005) and *Daphnia magna* (Lurling and van der Grinten 2003). Microcystins tend to be quite water soluble and polar, and do not readily pass the lipid bilayer of membranes. It is important to note however that whenever effects of purified dissolved toxins are compared to whole cell extracts biological effects tend to be much enhanced for the latter (Palikova et al. 1998; Oberemm et al. 1999), a possible indication of synergistic effects between MC and other bioactive compounds in cyanobacterial cells. There are exceptions, however. For example, in *A. salina* purified CYN showed a lower LC_{50} than crude *Cylindrospermopsis* extracts, and this may indicate that unidentified compounds in the cyanobacterial cell extracts lowered the bioavailability of the toxin (Metcalf et al. 2002).

Via food (vectorial transport)

Feeding seems to be the most important route for exposure of aquatic biota to cyanobacterial toxins. This seems natural for organisms that directly feed on seston that includes cyanobacteria. Zooplankton, filter feeding bivalves and phytoplanktivorous fish would be among the organisms that are directly exposed to toxins in their food (unless they manage to avoid toxic cyanobacteria – see below 'protective mechanisms'). For those biota that do not feed directly on cyanobacteria, toxins must reach them via the food web. The risk of being exposed to toxins via the food web is much increased if biomagnification takes place. This is commonly found for lipophilic toxicants like PCB, but is less likely for hydrophilic compounds like MC–LR. This congener has a very low octanol to water partition coefficient, but as demonstrated by Ward and Codd (1999) other variants may have higher coefficients, and toxicity to *Tetrahymena* has been shown to vary accordingly. As discussed above biomagnification of MC and NODLN is unlikely and not substantiated by data from the field. Of the amount of toxin ingested with the food very little is actually taken up into the body (e.g. 2.7 % in *Daphnia* in Rohrlack et al. 2005). And even the little toxin that is actually taken up into the blood of *Daphnia* and transported to its organs is subject to detoxication (see later in this paper) and excretion. These processes that dilute toxin concentration act at every step in a food chain. Rather than biomagnification MC and other toxins may be subject to biodilution in the foodweb. Nevertheless toxins are found at higher trophic levels, so there must be some vectorial transport, and as will be discussed below in sufficient quantities to have harmful effects. The presence of toxins in grazers like the zooplankton indicate that cyanobacteria are indeed ingested, despite their reputation of being hard to handle because of their large size. Indeed Work and Havens (2003) found cyanobacteria in the gut of all crustacean zooplankton in a large subtropical lake, including taxa known to produce toxins such as *Anabaena*.

A special case of exposure via food is coprophagy, described in a study on blue mussels (Svensen et al. 2005). A fair number of studies have analyzed toxins in the faeces of various species. Concentrations may be relatively high compared to concentrations in organs and tissues, and the faeces laden with toxins provide a medium for further transport of toxins in aquatic systems, especially towards the benthic community. Examples in Tables 1 and 2 include the faeces of silver carp and *C. gibelio* (Jang et al. 2004), *M. galloprovincialis* (Amorim and Vasconcelos 1999) as well as *M. edulis* (Svensen et al. 2005), and faecal pellets of calanoid copepods (Lehtiniemi et al. 2002).

Effects on biota

Acute vs. chronic effects

Acute effects are those that result from a single exposure to a toxin. This is conceivable under laboratory settings or after large scale lysis of a surface bloom. Biota in the field however will mainly be exposed repeatedly to toxins over a long period of time. This is sub–chronic and chronic exposure (definitions in Havens et al., this issue). An example of a study of acute exposure is that by Kankaanpaa et al. (2002) on sea trout. The fish were exposed to a single bolus of toxic *Nodularia* and time dependent accumulation / depuration of NODLN was coupled to the analysis of damage and recovery of the liver. An example of a sub–chronic exposure study (i.e. on a time scale intermediate between acute and chronic) is that by Pinho et al. (2003) where estuarine crabs were exposed daily for 4–7 d to cell extracts from toxic *Microcystis* or the exposure of carp to *Microcystis* during 28d (Li et al. 2004). Experimental chronic exposure studies where biota are exposed to toxins for the greater part of their lifespan have – necessarily – been restricted to organisms with short generation times, especially zooplankton. There are a fair number of studies in which the effects of cyanobacterial toxins on the life–history of *Daphnia* have been studied. Examples are the studies by (Lurling 2003) and (Hietala et al. 1997). Some of the bivalve studies (accumulation / depuration) lasted for several weeks (e.g., Pires et al. 2004 and Bury et al. 1996) exposed brown trout to MC–LR for a period of 63d, but the great majority of data on chronic exposure to toxic cyanobacteria come from field studies where animals are exposed to toxic cyanobacteria via natural routes (many examples in Tables A.1 and A.2) during extended periods of time.

Overall Table A.2 indicates wide ranging effects of different cyanobacterial toxins on various aquatic organisms. Effects vary from mortality to subtle changes in behavior. Effects in fish include changes in liver enzymology, liver damage and ionic imbalance. Effects of cyanobacterial toxins on the embryonic development of fishes have been studied in two species: zebra fish and loach. Whereas immersion of zebra fish embryos in a solution of purified MC did not result in morphological changes except at the very highest concentration (Oberemm et al 1999), embryonic development of loach was affected by exposure to MC (Liu et al. 2002). In the study on zebra fish it was seen – like in studies on other biota – that crude cell extracts had much stronger effects, resulting in malformations of the fishes.

Effect studies are especially rich in the zooplankton literature. In Table A.2 it can be seen that effects on zooplankton vary from feeding inhibition to reduced reproduction, growth and mortality. Feeding inhibition may actually serve as a protective mechanism, and there is some evidence that especially species that are highly susceptible to MC may protect themselves by strong inhibition of the intake of cyanobacteria (Demott 1999). Studies also have shown that zooplankton is relatively insensitive to dissolved toxins (Demott et al. 1991; Lurling and van der Grinten 2003) so that feeding inhibition may indeed be very effective in preventing harmful exposure to the toxins. A complicating factor in zooplankton studies is that also 'non–toxic' cyanobacteria induce effects like reduced growth and reproduction. Cyanobacteria are generally believed to be food of low quality to zooplankton, especially *Daphnia*, so that direct toxic effects can not always be separated from the effect of insufficient food of good quality (LaurenMaatta et al. 1997). Experimental tests in which toxicity effects were separated from food effects – by adding a sufficient amount of high quality food like the green alga *Scenedesmus* – clearly demonstrate however that nutritional insufficiency of *Microcystis* cannot be solely responsible for the effects on *Daphnia*. Negative effects on survival, growth and population development persisted even when green algae were added. Moreover this was true even when a *Microcystis* mutant was used that no longer produces MC (Lurling 2003). Hence the author concluded that harmful effects by *Microcystis* cannot be the result of MC only. Feeding inhibition (starvation) and unknown bioactive compounds must also play a part.

The studies by Rohrlack and co–workers (Rohrlack et al. 1999b; Rohrlack et al. 2004; Rohrlack et al. 2005) enable direct insight into the effects of MC on *Daphnia* because the wild type *Microcystis* and its mutant only differ in their capacity to produce MC. Several interesting observations were made. Although MC was not responsible for feeding inhibition – the mutant had an equally strong effect – clearly MC had direct toxic effects. Visible first symptoms of MC poisoning included an inhibition of movements of thoracic legs, mandibles, foregut, second antennae, as well as stimulation of gut muscles leading to a permanent contraction of the midgut. These effects became apparent as soon as MC was taken up into the blood. Contraction of the midgut interferes with digestion, nutrient assimilation and uptake of ions. Eventually MC resulted in a breakdown of *Daphnia* metabolism, exhaustion and eventually death. In nature intake of microcystins will be modulated by various factors that were not considered in the study by Rohrlack et al. (2005) like *Microcystis* colony size, presence of alternative food, temperature or condition of the animal (see section below on 'modulating factors').

Dose effect relationships

In studies on Baltic Sea flounder as well as on fish from the IJsselmeer no relationship could be detected between liver histopathology and toxin concentrations (Ibelings et al 2005.; Kankaanpaa et al. 2005a). The lesions that are seen in fish livers caught from systems supporting dense blooms of cyanobacteria may be attributed to hepatotoxin exposure, but other factors like liver parasites and anthropogenic pollutants will also play a part. Kankaanpaa et al. (2005a) concluded that liver histopathology can not be used as a reliable bioindicator of exposure to cyanobacterial toxins. A complicating factor is the dynamic nature of liver damage and recovery. The acute exposure study mentioned earlier where flounders were given a single dose of toxin (Kankaanpaa et al. 2005a) demonstrated that damage is transient, and recovery from liver damage is rapid (on the order of days). Many studies have failed to relate effects to concentrations of toxin. Egg production of *Daphnia* in the IJsselmeer had no relationship with toxin content of the cyanobacteria in the lake (Ibelings et al. 2005). According to

Rohrlack et al. (1999) it may not be the presence of toxins in the seston but the actual intake of toxins that matters. *Daphnia* species that were presumed to differ in susceptibility to MC may actually be equally susceptible – where they actually differ may be in their ingestion rate of toxic *Microcystis* cells. Rohrlack et al. (2004) established a clear relationship between MC ingestion rate and LT_{50} (survival time) of *Daphnia*. Despite all the complicating factors, significant dose–effect relationships have been found and are included in Table A.2. Another example is the dose (and time) dependent mortality in brine shrimp exposed to CYN and MC (Metcalf et al. 2002).

Protective mechanisms

Aquatic biota are by no means defenseless against toxic cyanobacteria. Blooms of toxic *Nodularia* have been around for at least 7000 years in the Baltic, giving other biota sufficient time to adapt to these nuisance cyanobacteria (Bianchi et al. 2000). Several studies indicate that species which are most frequently exposed to the toxins have the highest physiological tolerance. Baltic shrimp are less sensitive than fish larvae, but these larvae only feed on phytoplankton during the first stages of their life. Baltic copepods feed upon and ingest toxic *Nodularia* and they survive and reproduce without apparent harmful effects of the toxins (Engstrom et al. 2000). Where *Thamnocephalus* exhibited reduced survival after grazing upon *Planktothrix* filaments, other zooplankton – naturally co–existing with toxic cyanobacteria– were unaffected (Kurmayer and Juttner 1999).

Sessile organisms like mussels cannot move away from cyanobacteria, but zooplankton and fish may migrate to parts of the system where concentrations of cyanobacteria are low, as has been suggested for fish in the Baltic (Karjalainen et al. 2005). Moreover zooplankton, mussels and fish may temporarily stop feeding when toxic cyanobacteria are present and avoid ingestion in this way. If toxic cyanobacteria can not be avoided and cells are indeed ingested, very little of the toxin present may actually be taken up into the body. Mucoid cyanobacteria like *Microcystis* are resistant to digestion, and there are barriers for the uptake of MC across the gut epithelium into the blood (Fischer and Dietrich 2000). Rohrlack et al. (2005) however showed that presence of *Microcystis* in the midgut of *Daphnia* caused the epithelium to loose cohesion. Cells loose contact with each other and this may facilitate the uptake of MC into the blood. Microcystin was transported by the blood to various organs, where beat rates were slowed down, until finally *Daphnia* died. Although both the MC producing wild type and the mutant (that no longer is capable of MC production) affected cohesion of the epithelium, beat rates were only affected by the MC producing strain, a clear demonstration that MC – if taken up into the blood – is indeed highly toxic to *Daphnia*.

Feeding inhibition in the presence of toxic cyanobacteria is an efficient means to prevent ingestion of the toxins. However the avoidance of toxins must be balanced with the risk of starvation. Demott (1999) found that in *Dahnia magna* exposure to toxic *Microcystis* in a mixture with *Scenedesmus* resulted in a rapid feeding inhibition, but feeding recovered when exposure was continued. DeMott concluded that this pattern of inhibition and recovery may balance the benefits of reduced ingestion of toxin with the disadvantage of a reduced food intake. In an environment with a patchy occurrence of toxic cyanobacteria feeding inhibition would be adaptive if the environment could be sensed correctly (chemical cues) and animals are able to recover quickly from inhibition in the absence of toxic strains.

Detoxication and oxidative stress

Another important process is detoxication of MC, which now has been documented in many aquatic biota, including several animals and macrophytes (Pflugmacher 2004). Metabolic breakdown of MC results in conjugate formation, amongst others with GSH. The formation of these conjugates is catalyzed by the enzyme GST, which has a microsomal and a cytosolic fraction. The activity of cGST has been demonstrated to increase after exposure to MC in zebra fish (Wiegand et al. 1999) and brine shrimp (Beattie et al. 2003), although exceptions have also been described, where activity of the enzyme remained unchanged, e.g. in goldfish (Malbrouck et al. 2004) and carp (Li et al. 2003). GST activity (cGST and mGST) also increased in *Daphnia* after exposure to CYN and an unidentified hepatotoxin (Nogueira et al. 2004). The MC–GSH conjugates have a much reduced toxicity and may be subject to enhanced excretion. Detoxication thus is a significant mechanism that protects biota to acute toxic effects of MC, as long as the capacity for detoxication is not exceeded. As a result of detoxication the cellular GSH pool is depleted and this exposes cells to oxidative stress through the formation of ROS like hydrogen peroxide. Organisms have a wide range of protective mechanisms to oxidative stress, including enzymes like SOD, CAT and GSH–reductase. The latter enzyme needs GSH as a co–substrate and its activity may be reduced when GSH is depleted through its conjugation with MC. Thus exposure to MC may have damaging effects in direct (inhibition of PP) and indirect ways (disbalance in ROS). Jos et al. (2005) showed that crushed cyanobacterial cells (MC released) resulted in enhanced oxidative stress resulting in lipid peroxidation, despite the fact that also levels of defensive enzymes were enhanced.

Studies by several authors in Table A.2 (i.e. Best et al. 2002) indicate that LPS (which are present on the cell–surface of cyanobacteria and on bacteria associated with cyanobacterial blooms) interfere with the detoxication process. In the study by Best and others GST activity in zebra fish was reduced when MC and LPS were offered in combination. Since LPS from different bacterial sources are always present in the aquatic environment (although not all LPS from different sources equally disturbed detoxication when tested) it would be rewarding to study the process of detoxication in the field. Another study (Best et al. 2003) demonstrated that LPS stimulate drinking in fish, the increased volume of water in the gut potentially increases the opportunity for uptake of toxins (including MC) from the water and promotes osmoregulatory imbalance.

Modulating factors

A common theme in much of what has been discussed in this paper is that of modulating factors. The effects that these cyanobacterial toxins have on aquatic biota are seldom straightforward but are modulated by factors in the environment or the status of biota themselves. Examples of modulating factors include condition of the animals, temperature and pre–acclimation /adaptation to cyanobacterial toxins. Hepatotoxicity of MC–LR has been shown to increase in fasted compared to fed animals (Malbrouck et al. 2004). Fasted goldfish showed a more severe and rapid inhibition of PP, and this may be related to differences in the glycogen content of the livers and the rate of MC removal from the body via the bilary excretion system. The tolerance of *Daphnia pulex* to toxic *Microcystis* was shown to be temperature dependent (Hietala et al. 1997) and decreased with higher temperatures. Adaptation to toxic cyanobacteria may play an important role too. *Daphnia* from locations where it is repeatedly exposed to toxic blooms would develop a higher tolerance to the toxins (Gustafson and Hansson 2004). Whether this is truly an adaptive evolutionary response remains to be tested since adaptation during 4–6 generations in the experiments by Gustafson and Hansson (2004) seem insufficient (although adaptation in *Daphnia* has indeed been shown to be a rapid process (e.g. Ebert et al. 2000). The essential message from their work is clear however: whenever the ecological effects of cyanobacterial toxins on biota are considered it is important to understand the history of the species involved. Modulating factors are an important reason why it is so hard to generalize the effects toxic cyanobacteria have on the biota in their environment.

Knowledge gaps

Throughout this paper remarks have been made about knowledge gaps that limit our understanding of the 'true' ecological effects of cyanobacterial toxins. At this point there is no need however to list those gaps extensively, since this is the subject of the paper by Havens et al, this issue. To summarize their main findings, Havens et al recommend further study on the following subjects:

- Studies at the whole community level in the presence vs. absence of cyanobacterial toxins;
- Studies that mitigate the bias towards microcystin, especially MC–LR, i.e. more knowledge is needed about ecological effects of toxins like CYN;
- Studies into synergistic effects of combinations of cyanobacterial toxins and of cyanobacterial toxins and other bioactive compounds from cyanobacterial cells;
- Studies in which biota are exposed to toxins under environmentally relevant conditions (synergistic effects with other stressors like temperature, low oxygen etc);
- More emphasis should be placed into (sub)chronic studies having sub–lethal effects, including those on behavior or genotoxicty; these may be more relevant than acute lethal effects, and more knowledge is needed here;
- What is the fate of toxins produced by cyanobacteria in the ecosystem?; what is for instance the role of detoxication and covalent binding of MC on transfer of toxins in the foodweb?
- Effects of toxins on benthic communities are not well understood;
- In which way and to what extend does toxicity of cyanobacteria interfere with lake restoration?

Conclusions

The qualitative meta–analysis identifies the following general patterns for major groups of aquatic biota (birds, fish, macroinvertebrates and zooplankton).

Birds. On basis of the limited number of studies on the role of toxic cyanobacteria on waterfowl we conclude that aquatic birds are at risk of cyanobacterial toxicosis. Anatoxins seem to play a relatively large role, they are often present when dead birds are found and the symptoms in diseased birds indicate a neurotoxin. Birds may be at high risk because they may directly feed on floating scum of cyanobacteria (personal observation) and are warm blooded animals, like the mammals which have been shown to be sensitive to cyanobacterial toxins in laboratory studies. A disease that must be mentioned here is avian vacuolar myelinopathy (AVM), which is a neurologic disorder primarily affecting bald eagles (*Haliaeetus leucocephalus*) and American coots (*Fulica americana*). The agent of this disease is an uncharacterized neurotoxin produced by a novel cyanobacterial epiphyte of the order *Stigonematales* (Wilde et al. 2005).

Fish. On basis of their study of common carp exposed to *Microcystis*, Li et al. (2005) conclude that fish kills during blooms of cyanobacteria can be assumed to result from extensive liver damage. Zambrano and Canelo (1996) on the other hand state that blockage of the gill activity could be the cause of mass mortalities during blooms of *Microcystis*. Both papers have in common that they put forward that cyanobacterial toxicosis can directly be responsible for the death of fish. We maintain that this is unlikely. Studying the collected data in Table 2 it seems doubtful whether naturally occurring concentrations of cyanobacterial toxins (either dissolved in the water or contained in the cell) are sufficiently high to be directly lethal. Again, generalizations are difficult because there appear to be important differences between fish species. Fischer and Dietrich (2000) related the capacity for uptake of toxins to the morphology of the GI. Perhaps combinations of stress factors that co–occur during blooms of toxic cyanobacteria (high temperature and pH, enhanced levels of ammonia, low oxygen in addition to cyanobacterial toxins) are more likely to cause fish mortality. Important sub–lethal effects of cyanobacterial toxins in fish are more than probable, however. Harmful effects have been seen on embryonic development, on growth of juvenile fish and on adult species. Several organs may be affected (e.g., kidney, heart, gonads), but the liver is the main target. Several studies have shown that around 50 % of fish caught from lakes or estuaries that support cyanobacterial blooms show hepatic lesions that could – partially – be the result of exposure to cyanobacterial toxins.

Macroinvertebrates. Most of the studies on bivalves agree that these animals are quite resistant to different cyanobacterial toxins. This has been shown for freshwater and marine mussels and clams and has been found for hepatotoxins, neurotoxins and CYN. More attention is given to the potential accumulation of toxins in mussels, and especially the risk of vectorial transport to predators (including man). However depuration studies have shown that mussels clear toxins fairly rapidly, so that there is little retention. Nevertheless depuration is seldom complete, and low concentrations may even be carried through to the start of the next cyanobacterial growing season.

Zooplankton. Whenever the ecological significance of cyanobacterial toxins is discussed the primary suggestion is often that they deter grazing by zooplankton. Highly selective grazers like copepods would exert a stronger selection pressure than less selective grazers like *Daphnhia,* but the study by Kurmayer and Juttner (1999) shows that *Daphnia* may play a persisting role in the evolution of MC production (see also studies by Jang et al. 2004) who demonstrated that MC concentrations increased up to five–fold when *Microcystis* was exposed to filtered zooplankton growth medium (*Daphnia* and *Moina* spp). The literature concerning the effects of toxic cyanobacteria on zooplankton is extensive (Table A.2 only shows a selection) but there appear to be many contradictions. This is not surprising since there are numerous complicating factors. Furthermore it is now well established that not all toxic effects can be traced back to the well known cyanobacterial toxins like MC. Although work by Rohrlack et al. (2005) has proven decisively that microcystins are toxic, the same work has shown that also the mutant incapable of producing MC has negative effects like inhibition of feeding. Effects of cyanobacterial blooms also exert effects at the community level. Zooplankton community composition may change towards dominance of smaller cladocerans which have a lower grazing pressure. In this way cyanobacterial blooms may stabilize the turbid state on which some of the cyanobacteria like *Planktothrix agardhii* depend (Scheffer et al. 1997) and interfere with lake restoration. The specific contribution of toxins – as opposed to general negative effects of cyanobacteria – at this high level of integration is unclear however.

Acknowledgments

The authors wish to acknowledge Calvin Walker and Mike Coveney for their comments on the paper.

References

Agrawal MK, Bagchi D, Bagchi SN (2005) Cysteine and serine protease–mediated proteolysis in body homogenate of a zooplankter, Moina macrocopa, is inhibited by the toxic cyanobacterium, Microcystis aeruginosa PCC7806. Comp Biochem Physiol B–Biochem Mol Biol 141(1): 33–41

Amorim A, Vasconcelos V (1999) Dynamics of microcystins in the mussel Mytilus galloprovincialis. Toxicon 37(7): 1041–1052

Babcock–Jackson L, Carmichael WW, Culver DA (2002) Dreissenid mussels increase exposure of benthic and pelagic organisms to toxic microcystins. VerhInternatVereinLimnol 28: 1082–1085

Baganz D, Staaks G, Steinberg C (1998) Impact of the cyanobacteria toxin, microcystin–LR on behaviour of zebrafish, Danio rerio. Water Res 32(3): 948–952

Ballot A, Krienitz L, Kotut K, Wiegand C, Metcalf JS et al. (2004) Cyanobacteria and cyanobacterial toxins in three alkaline rift valley lakes of Kenya – Lakes Bogoria, Nakuru and Elmenteita. J Plankton Res 26(8): 925–935

Beattie KA, Ressler J, Wiegand C, Krause E, Codd GA et al. (2003) Comparative effects and metabolism of two microcystins and nodularin in the brine shrimp Artemia salina. Aquat Toxicol 62(3): 219–226

Best JH, Pflugmacher S, Wiegand C, Eddy FB, Metcalf JS et al. (2002) Effects of enteric bacterial and cyanobacterial lipopolysaccharides, and of microcystin–LR, on glutathione S–transferase activities in zebra fish (Danio rerio). Aquat Toxicol 60(3–4): 223–231

Best JH, Eddy FB, Codd GA (2003) Effects of Microcystis cells, cell extracts and lipopolysaccharide on drinking and liver function in rainbow trout Oncorhynchus mykiss Walbaum. Aquat Toxicol 64(4): 419–426

Best JH, Eddy FB, Codd GA (2001) Effects of purified microcystin–LR and cell extracts of Microcystis strains PCC 7813 and CYA 43 on cardiac function in brown trout (Salmo trutta) alevins. Fish Physiol Biochem 24(3): 171–178

Bianchi TS, Engelhaupt E, Westman P, Andren T, Rolff C et al. (2000) Cyanobacterial blooms in the Baltic Sea: Natural or human–induced? Limnol Oceanogr 45(3): 716–726

Blom JF, Robinson JA, Juttner F (2001) High grazer toxicity of [D–Asp(3) (E)–Dhb(7)]microcystin–RR of Planktothrix rubescens as compared to different microcystins. Toxicon 39(12): 1923–1932

Bury NR, Eddy FB, Codd GA (1996) Stress responses of brown trout, Salmo trutta L, to the cyanobacterium, Microcystis aeruginosa. Environ Toxicol Water Quality 11(3): 187–193

Bury NR, Eddy FB, Codd GA (1995) The Effects of the Cyanobacterium Microcystis–Aeruginosa, the Cyanobacterial Hepatotoxin Microcystin–Lr, and Ammonia On Growth–Rate and Ionic Regulation of Brown Trout. J Fish Biol 46(6): 1042–1054

Carbis CR, Rawlin GT, Grant P, Mitchell GF, Anderson JW et al. (1997) A study of feral carp, Cyprinus carpio L, exposed to Microcystis aeruginosa at Lake

Mokoan, Australia, and possible implications for fish health. J Fish Dis 20(2): 81–91

Chen J, Xie P (2005) Tissue distributions and seasonal dynamics of the hepatotoxic microcystins–LR and –RR in two freshwater shrimps, Palaemon modestus and Macrobrachium nipponensis, from a large shallow, eutrophic lake of the subtropical China. Toxicon 45(5): 615–625

Codd GA, Lindsay J, Young FM, Morrison LF & Metcalf J (2005). Harmful cyanobacteria. From mass mortalities to magament measures. Huisman J, Matthijs HCP, Visser PM. Harmful cyanobacteria. Springer (Dordrecht)

de Magalhaes VF, Soares RM, Azevedo S (2001) Microcystin contamination in fish from the Jacarepagua Lagoon (Rio de Janeiro, Brazil): ecological implication and human health risk. Toxicon 39(7): 1077–1085

Demott WR (1999) Foraging strategies and growth inhibition in five daphnids feeding on mixtures of a toxic cyanobacterium and a green alga. Freshw Biol 42(2): 263–274

Demott WR, Zhang QX, Carmichael WW (1991) Effects of Toxic Cyanobacteria and Purified Toxins On the Survival and Feeding of a Copepod and 3 Species of Daphnia. Limnol Oceanogr 36(7): 1346–1357

Ebert D, Lipsitch M, Mangin KL (2000) The effect of parasites on host population density and extinction: Experimental epidemiology with Daphnia and six microparasites. Am Nat 156(5): 459–477

Engstrom–Ost J, Lehtiniemi M, Green S, Kozlowsky–Suzuki B, Viitasalo M (2002) Does cyanobacterial toxin accumulate in mysid shrimps and fish via copepods? J Exp Mar Biol Ecol 276(1–2): 95–107

Engstrom J, Koski M, Viitasalo M, Reinikainen M, Repka S et al. (2000) Feeding interactions of the copepods Eurytemora affinis and Acartia bifilosa with the cyanobacteria Nodularia sp. J Plankton Res 22(7): 1403–1409

Ferrao–Filho AD, Azevedo S (2003) Effects of unicellular and colonial forms of toxic Microcystis aeruginosa from laboratory cultures and natural populations on tropical cladocerans. Aquat Ecol 37(1): 23–35

Ferrao AS, Azevedo S, DeMott WR (2000) Effects of toxic and non–toxic cyanobacteria on the life history of tropical and temperate cladocerans. Freshw Biol 45(1): 1–19

Fischer WJ, Dietrich DR (2000) Pathological and biochemical characterization of microcystin– induced hepatopancreas and kidney damage in carp (Cyprinus carpio). Toxicol Appl Pharmacol 164(1): 73–81

Gray JS (2002) Biomagnification in marine systems: the perspective of an ecologist. Mar Pollut Bull 45(1–12): 46–52

Gustafson S, Hansson LA (2004) Development of tolerance against toxic cyanobacteria in *Daphnia*. Aquat Ecol 38: 37–44

Henriksen P, Carmichael WW, An JS, Moestrup O (1997) Detection of an anatoxin–a(s)–like anticholinesterase in natural blooms and cultures of Cyanobacteria/blue–green algae from Danish lakes and in the stomach contents of poisoned birds. Toxicon 35(6): 901–913

Hietala J, LaurenMaatta C, Walls M (1997) Life history responses of Daphnia clones to toxic Microcystis at different food levels. J Plankton Res 19(7): 917–926

Ibelings BW, Maberly SC (1998) Photoinhibition and the availability of inorganic carbon restrict photosynthesis by surface blooms of cyanobacteria. Limnol Oceanogr 43(3): 408–419

Ibelings BW, Bruning K, de Jonge J, Wolfstein K, Pires LMD et al. (2005) Distribution of microcystins in a lake foodweb: No evidence for biomagnification. Microb Ecol 49(4): 487–500

Jang MH, Ha K, Lucas MC, Joo GJ, Takamura N (2004) Changes in microcystin production by Microcystis aeruginosa exposed to phytoplanktivorous and omnivorous fish. Aquat Toxicol 68(1): 51–59

Jos A, Pichardo S, Prieto AI, Repetto G, Vazquez CM et al. (2005) Toxic cyanobacterial cells containing microcystins induce oxidative stress in exposed tilapia fish (Oreochromis sp.) under laboratory conditions. Aquat Toxicol 72(3): 261–271

Kankaanpaa H, Vuorinen PJ, Sipia V, Keinanen M (2002) Acute effects and bioaccumulation of nodularin in sea trout (Salmo trutta m. trutta L.) exposed orally to Nodularia spumigena under laboratory conditions. Aquat Toxicol 61(3–4): 155–168

Kankaanpaa H, Turunen AK, Karlsson K, Bylund G, Meriluoto J et al. (2005a) Heterogeneity of nodularin bioaccumulation in northern Baltic Sea flounders in 2002. Chemosphere 59(8): 1091–1097

Kankaanpaa HT, Sipia VO, Kuparinen JS, Ott JL, Carmichael WW (2001) Nodularin analyses and toxicity of a Nodularia spumigena (Nostocales, Cyanobacteria) water–bloom in the western Gulf of Finland, Baltic Sea, in August 1999. Phycologia 40(3): 268–274

Kankaanpaa HT, Holliday J, Schroder H, Goddard TJ, von Fister R et al. (2005b) Cyanobacteria and prawn fanning in northern New South Wales, Australia – a case study on cyanobacteria diversity and hepatotoxin bioaccumulation. Toxicol Appl Pharmacol 203(3): 243–256

Karjalainen M, Kozlowsky–Suzuki B, Lehtiniemi M, Engström–Ost J, Kankaanpää H, Viitasalo M (in press). Nodularin accumulation during cyanobacterial blooms asnd experimental depuration in zooplankton

Karjalainen M, Reinikainen M, Lindvall F, Spoof L, Meriluoto JAO (2003) Uptake and accumulation of dissolved, radiolabeled nodularin in Baltic Sea Zooplankton. Environ Toxicol 18(1): 52–60

Karjalainen M, Reinikainen M, Spoof L, Meriluoto JAO, Sivonen K et al. (2005) Trophic transfer of cyanobacterial toxins from zooplankton to planktivores: Consequences for pike larvae and mysid shrimps. Environ Toxicol 20(3): 354–362

Karlsson K, Sipia V, Kankaanpaa H, Meriluoto J (2003) Mass spectrometric detection of nodularin and desmethylnodularin in mussels and flounders. J Chromatogr B 784(2): 243–253

Koski M, Schmidt K, Engstrom–Ost J, Viitasalo M, Jonasdottir S et al. (2002) Calanoid copepods feed and produce eggs in the presence of toxic cyanobacteria Nodularia spumigena. Limnol Oceanogr 47(3): 878–885

Kotak BG, Zurawell RW, Prepas EE, Holmes CFB (1996a) Microcystin–LR concentration in aquatic food web compartments from lakes of varying trophic status. Can J Fish Aquat Sci 53(9): 1974–1985

Kotak BG, Semalulu S, Fritz DL, Prepas EE, Hrudey SE et al. (1996b) Hepatic and renal pathology of intraperitoneally administered microcystin–LR in rainbow trout (Oncorhynchus mykiss). Toxicon 34(5): 517–525

Krienitz L, Ballot A, Kotut K, Wiegand C, Putz S et al. (2003) Contribution of hot spring cyanobacteria to the mysterious deaths of Lesser Flamingos at Lake Bogoria, Kenya. FEMS Microbiol Ecol 43(2): 141–148

Kurmayer R, Juttner F (1999) Strategies for the co–existence of zooplankton with the toxic cyanobacterium Planktothrix rubescens in Lake Zurich. J Plankton Res 21(4): 659–683

LaurenMaatta C, Hietala J, Walls M (1997) Responses of Daphnia pulex populations to toxic cyanobacteria. Freshw Biol 37(3): 635–647

Lehtiniemi M, Engstrom–Ost J, Karjalainen M, Kozlowsky–Suzuki B, Viitasalo M (2002) Fate of cyanobacterial toxins in the pelagic food web: transfer to copepods or to faecal pellets? Mar Ecol–Prog Ser 241: 13–21

Lehtonen KK, Kankaanpaa H, Leinio S, Sipia VO, Pflugmacher S et al. (2003) Accumulation of nodularin–like compounds from the cyanobacterium Nodularia spumigena and changes in acetylcholinesterase activity in the clam Macoma balthica during short–term laboratory exposure. Aquat Toxicol 64(4): 461–476

Li XY, Liu YD, Song LR, Liu HT (2003) Responses of antioxidant systems in the hepatocytes of common carp (Cyprinus carpio L.) to the toxicity of microcystin–LR. Toxicon 42(1): 85–89

Li L, Xie P, Chen J (2005) In vivo studies on toxin accumulation in liver and ultrastructural changes of hepatocytes of the phytoplanktivorous bighead carp i.p.–injected with extracted microcystins. Toxicon 46(5): 533–545

Li XY, Chung IK, Kim JI, Lee JA (2004) Subchronic oral toxicity of microcystin in common carp (Cyprinus carpio L.) exposed to Microcystis under laboratory conditions. Toxicon 44(8): 821–827

Liras V, Lindberg M, Nystrom P, Annadotter H, Lawton LA et al. (1998) Can ingested cyanobacteria be harmful to the signal crayfish (Pacifastacus leniusculus)? Freshw Biol 39(2): 233–242

Liu YD, Song LR, Li XY, Liu TM (2002) The toxic effects of microcystin–LR on embryo–larval and juvenile development of loach, Misguruns mizolepis Gunthe. Toxicon 40(4): 395–399

Lurling M (2003) Effects of microcystin–free and Microcystin containing strains of the cyanobacterium Microcystis aeruginosa on growth of the grazer Daphnia magna. Environ Toxicol 18(3): 202–210

Lurling M, van der Grinten E (2003) Life–history characteristics of Daphnia exposed to dissolved microcystin–LR and to the cyanobacterium Microcystis

aeruginosa with and without microcystins. Environ Toxicol Chem 22(6): 1281–1287
Magalhaes VF, Marinho MM, Domingos P, Oliveira AC, Costa SM et al. (2003) Microcystins (cyanobacteria hepatotoxins) bioaccumulation in fish and crustaceans from Sepetiba Bay (Brasil, RJ). Toxicon 42(3): 289–295
Malbrouck C, Trausch G, Devos P, Kestemont P (2003) Hepatic accumulation and effects of microcystin–LR on juvenile goldfish Carassius auratus L. Comp Biochem Physiol C–Toxicol Pharmacol 135(1): 39–48
Malbrouck C, Trausch G, Devos P, Kestemont P (2004) Effect of microcystin–LR on protein phosphatase activity and glycogen content in isolated hepatocytes of fed and fasted juvenile goldfish Carassius auratus L. Toxicon 44(8): 927–932
Metcalf JS, Beattie KA, Pflugmacher S, Codd GA (2000) Immuno–crossreactivity and toxicity assessment of conjugation products of the cyanobacterial toxin, microcystin–LR. FEMS Microbiol Lett 189(2): 155–158
Metcalf JS, Lindsay J, Beattie KA, Birmingham S, Saker ML et al. (2002) Toxicity of cylindrospermopsin to the brine shrimp Artemia salina: comparisons with protein synthesis inhibitors and microcystins. Toxicon 40(8): 1115–1120
Mohamed ZA, Carmichael WW, Hussein AA (2003) Estimation of microcystins in the freshwater fish Oreochromis niloticus in an Egyptian fish farm containing a Microcystis bloom. Environ Toxicol 18(2): 137–141
Negri AP, Bunter O, Jones B, Llewellyn L (2004) Effects of the bloom–forming alga Trichodesmium erythraeum on the pearl oyster Pinctada maxima. Aquaculture 232(1–4): 91–102
Negri AP, Jones GJ (1995) Bioaccumulation of Paralytic Shellfish Poisoning (Psp) Toxins from the Cyanobacterium Anabaena–Circinalis by the Fresh–Water Mussel Alathyria–Condola. Toxicon 33(5): 667–678
Nogueira ICG, Saker ML, Pflugmacher S, Wiegand C, Vasconcelos VM (2004) Toxicity of the cyanobacterium Cylindrospermopsis radborskii to Daphnia magna. Environ Toxicol 19(5): 453–459
Nogueira ICG, Pereira P, Dias E, Pflugmacher S, Wiegand C et al. (2004) Accumulation of Paralytic Shellfish Toxins (PST) from the cyanobacterium Aphanizomenon issatschenkoi by the cladoceran Daphnia magna. Toxicon 44(7): 773–780
Oberemm A, Becker J, Codd GA, Steinberg C (1999) Effects of cyanobacterial toxins and aqueous crude extracts of cyanobacteria on the development of fish and amphibians. Environ Toxicol 14(1): 77–88
Ozawa K, Yokoyama A, Ishikawa K, Kumagai M, Watanabe MF et al. (2003) Accumulation and depuration of microcystin produced by the cyanobacterium Microcystis in a freshwater snail. Limnology 4(3): 131–138
Palikova M, Kovaru F, Navratil S, Kubala L, Pesak S et al. (1998) The effects of pure microcystin LR and biomass of blue–green algae on selected immunological indices of carp (Cyprinus carpio L.) and silver carp (Hypophthalmichthys molitrix Val.). Acta Vet BRNO 67(4): 265–272

Pereira P, Dias E, Franca S, Pereira E, Carolino M et al. (2004) Accumulation and depuration of cyanobacterial paralytic shellfish toxins by the freshwater mussel Anodonta cygnea. Aquat Toxicol 68(4): 339–350

Pflugmacher S (2004) Promotion of oxidative stress in the aquatic macrophyte Ceratophyllum demersum during biotransformation of the cyanobacterial toxin microcystin–LR. Aquat Toxicol 70(3): 169–178

Pinho GLL, da Rosa CM, Yunes JS, Luquet CM, Bianchini A et al. (2003) Toxic effects of microcystins in the hepatopancreas of the estuarine crab Chasmagnathus granulatus (Decapoda, Grapsidae). Comp Biochem Physiol C–Toxicol Pharmacol 135(4): 459–468

Pires LMD, Karlsson KM, Meriluoto JAO, Kardinaal E, Visser PM et al. (2004) Assimilation and depuration of microcystin–LR by the zebra mussel, Dreissena polymorpha. Aquat Toxicol 69(4): 385–396

Prepas EE, Kotak BG, Campbell LM, Evans JC, Hrudey SE et al. (1997) Accumulation and elimination of cyanobacterial hepatotoxins by the freshwater clam Anodonta grandis simpsoniana. Can J Fish Aquat Sci 54(1): 41–46

Reinikainen M, Hietala J, Walls M (1999) Reproductive allocation in Daphnia exposed to toxic cyanobacteria. J Plankton Res 21(8): 1553–1564

Reinikainen M, Lindvall F, Meriluoto JAO, Repka S, Sivonen K et al. (2002) Effects of dissolved cyanobacterial toxins on the survival and egg hatching of estuarine calanoid copepods. Mar Biol 140(3): 577–583

Rohrlack T, Christoffersen K, Dittmann E, Nogueira I, Vasconcelos V et al. (2005) Ingestion of microcystins by Daphnia: Intestinal uptake and toxic effects. Limnol Oceanogr 50(2): 440–448

Rohrlack T, Henning M, Kohl JG (1999a) Does the toxic effect of Microcystis aeruginosa on Daphnia galeata depend on microcystin ingestion rate? Arch Hydrobiol 146(4): 385–395

Rohrlack T, Dittmann E, Henning M, Borner T, Kohl JG (1999b) Role of microcystins in poisoning and food ingestion inhibition of Daphnia galeata caused by the cyanobacterium Microcystis aeruginosa. Appl Environ Microbiol 65(2): 737–739

Rohrlack T, Dittmann E, Borner T, Christoffersen K (2001) Effects of cell–bound microcystins on survival and feeding of Daphnia spp. Appl Environ Microbiol 67(8): 3523–3529

Rohrlack T, Christoffersen K, Kaebernick M, Neilan BA (2004) Cyanobacterial protease inhibitor microviridin J causes a lethal molting disruption in Daphnia pulicaria. Appl Environ Microbiol 70(8): 5047–5050

Saker ML, Metcalf JS, Codd GA, Vasconcelos VM (2004) Accumulation and depuration of the cyanobacterial toxin cylindrospermopsin in the freshwater mussel Anodonta cygnea. Toxicon 43(2): 185–194

Scheffer M, Rinaldi S, Gragnani A, Mur LR, vanNes EH (1997) On the dominance of filamentous cyanobacteria in shallow, turbid lakes. Ecology 78(1): 272–282

Sipia VO, Kankaanpaa HT, Pflugmacher S, Flinkman J, Furey A et al. (2002) Bioaccumulation and detoxication of nodularin in tissues of flounder (Platichthys flesus), mussels (Mytilus edulis, Dreissena polymorpha), and clams

(Macoma balthica) from the northern Baltic Sea. Ecotox Environ Safe 53(2): 305–311

Sipia V, Kankaanpaa H, Lahti K, Carmichael WW, Meriluoto J (2001a) Detection of Nodularin in flounders and cod from the Baltic Sea. Environ Toxicol 16(2): 121–126

Sipia VO, Kankaanpaa HT, Flinkman J, Lahti K, Meriluoto JAO (2001b) Time–dependent accumulation of cyanobacterial hepatotoxins in flounders (Platichthys flesus) and mussels (Mytilus edulis) from the northern Baltic Sea. Environ Toxicol 16(4): 330–336

Sipia VO, Karlsson KA, Meriluoto JAO, Kankaanpaa HT (2004) Eiders (Somateria mollissima) obtain nodularin, a cyanobacterial hepatotoxin, in Baltic Sea food web. Environ Toxicol Chem 23(5): 1256–1260

Soares RA, Magalhaes VF, Azevedo S (2004) Accumulation and depuration of microcystins (cyanobacteria hepatotoxins) in Tilapia rendalli (Cichlidae) under laboratory conditions. Aquat Toxicol 70(1): 1–10

Spoof L (2005). Miocrocystins and nodularins. Meriluoto J and Codd GA. Cyanobacterial monitoring and cyanotoxin analysis. Åbo Akademi University Press

Svensen C, Strogyloudi E, Riser CW, Dahmann J, Legrand C et al. (2005) Reduction of cyanobacterial toxins through coprophagy in Mytilus edulis. Harmful Algae 4(2): 329–336

Tencalla FG, Dietrich DR, Schlatter C (1994) Toxicity of Microcystis–Aeruginosa Peptide Toxin to Yearling Rainbow–Trout (Oncorhynchus–Mykiss). Aquat Toxicol 30(3): 215–224

Thostrup L, Christoffersen K (1999) Accumulation of microcystin in Daphnia magna feeding on toxic Microcystis. Arch Hydrobiol 145(4): 447–467

Vasconcelos VM, Sivonen K, Evans WR, Carmichael WW, Namikoshi M (1995) Isolation and Characterization of Microcystins (Heptapeptide Hepatotoxins) From Portuguese Strains of Microcystis–Aeruginosa Kutz Emend Elekin. Arch Hydrobiol 134(3): 295–305

Vasconcelos V, Oliveira S, Teles FO (2001) Impact of a toxic and a non–toxic strain of Microcystis aeruginosa on the crayfish Procambarus clarkii. Toxicon 39(10): 1461–1470

Vinagre TM, Alciati JC, Regoli F, Bocchetti R, Yunes JS et al. (2003) Effect of microcystin on ion regulation and antioxidant system in gills of the estuarine crab Chasmagnathus granulatus (Decapoda, Grapsidae). Comp Biochem Physiol C–Toxicol Pharmacol 135(1): 67–75

Ward CJ, Codd GA (1999). Comparitive toxicity of four microcystins of different hydrophobicities to the protozoan Tetrahymena pyriformis. J. Appl. Microbiol. 86: 874–882

Wiegand C, Pflugmacher S, Oberemm A, Meems N, Beattie KA et al. (1999) Uptake and effects of microcystin–LR on detoxication enzymes of early life stages of the zebra fish (Danio rerio). Environ Toxicol 14(1): 89–95

Wilde SB, Murphy TM, Hope CP, Habrun SK, Kempton J et al. (2005) Avian vacuolar myelinopathy linked to exotic aquatic plants and a novel cyanobacterial species. Environ Toxicol 20(3): 348–353

Williams DE, Craig M, Dawe SC, Kent ML, Holmes CFB et al. (1997) Evidence for a covalently bound form of microcystin–LR in salmon liver and dungeness crab larvae. Chem Res Toxicol 10(4): 463–469

Work KA, Havens KE (2003) Zooplankton grazing on bacteria and cyanobacteria in a eutrophic lake. J Plankton Res 25(10): 1301–1306

Xie LQ, Xie P, Ozawa K, Honma T, Yokoyama A et al. (2004) Dynamics of microcystins–LR and –RR in the phytoplanktivorous silver carp in a sub–chronic toxicity experiment. Environ Pollut 127(3): 431–439

Xie LQ, Xie P, Guo LG, Li L, Miyabara Y et al. (2005) Organ distribution and bioaccumulation of microcystins in freshwater fish at different trophic levels from the eutrophic Lake Chaohu, China. Environ Toxicol 20(3): 293–300

Yokoyama A, Park HD (2003) Depuration kinetics and persistence of the cyanobacterial toxin microcystin–LR in the freshwater bivalve Unio douglasiae. Environ Toxicol 18(1): 61–67

Zambrano F, Canelo E (1996) Effects of microcystin–LR on the partial reactions of the Na+– K+ pump or the gill of carp (Cyprinus carpio linneo). Toxicon 34(4): 451–458

Zurawell RW, Kotak BG, Prepas EE (1999) Influence of lake trophic status on the occurrence of microcystin–LR in the tissue of pulmonate snails. Freshw Biol 42(4): 707–718

Appendix A

Table A.1. Concentrations of various cyanobacterial toxins in a range of aquatic biota. Where possible concentrations are expressed as μg g DW-1. Conversion of wet weight to dry weight in animals using a factor of 0.1 and conversion of C to DW requires multiplication by 1/0.526 {Winberg, 1971). Ash content of DW is neglected. MC is commonly expressed as MC–LR eq, since the majority of studies only used MC–LR as a standard in HPLC. Concentrations are usually presented as a range (lowest – highest values found). Concentration in the 'source' applies to either purified toxin (experiments), cultured cells of cyanobacteria, dissolved toxins in natural waters or cyanobacteria in the seston. For abbreviatons see 'list of abbreviations'.

Organism	Toxin	Exposure route	Conc. in organism (μg g^{-1} DW)	Conc. in source (μg g^{-1} DW)	Analysis method	Remark	Reference
BIRDS							
(i) coots (Fulica atra) (ii) grebes (Podiceps nigricollis, P. cristatus)	ANA(a-s), MC	Natural routes: feeding in foodweb from lake with Anabaena bloom	(i) ANA - Coot: 0.021 Grebe: 0.90 (ii) MC - 0.0005-0.001	ANA: 4-3300 MC: 0.1-0.9	ELISA (MC) AChE assay (ANA)	(i) Anabaena and its toxin found in stomach birds: cyanobacterial toxicosis? (ii) ratio ANAeq : MCeq 40 to > 1000	Henriksen et al, 1997
Lesser flamingo (Phoeniconaias minor)	MC-LR, RR, LF and YR, ANA(a)	Feeding on mats of cyanobacteria, as well as on planktonic Arthrospira	(i) stomach: MC 1.96 (all 4 variants); ANA 43.4 (ii) intestines: MC 0.36 (MC-LR only); ANA 7.62 (iii) faeces: MC 0.48 (all but MC-LF); ANA 2.45	(i) mats - MC: 221-845 ANA: 10-18 (ii) plankton – MC: max 4600 ANA: max 223	HPLC, MALDITOF-MS	(i) intoxication birds likely but lack of relevant data on susceptibility (ii) multiple stress factors	Krienitz et al, 2003; Ballot et al, 2004

Eider (Somateria mollissima)	NODLN	Feeding on mussels in Baltic sea	(i) 0.003-0.18 (ii) 0.1-5.8 μg per liver		ELISA and LC-MS		Sipiä et al, 2004
FISH							
Round goby (Neogobius melanostomus)	MC	Natural routes in lake	Goby liver: 0.6-3.0	91-820			Babcock-Jackson et al, 2002
Sticklebacks (Gasterosteus aculeatus)	NODLN	Feeding on copepods pre-exposed to cyanobacteria	(i) 0.15 (ELISA) (ii) 0.8 (PPase)		ELISA and PPase assay		Ensgtröm-Öst et al, 2002
(i) perch (Percia fluviatilis) (ii) ruffe (Gymnocephalus cernuus) (iii) smelt (Osmerus eperlanus)	MC	Natural routes in lake foodweb	(i) perch liver: 17-51 (ii) ruffe liver: 9-194 (iii) smelt liver: 59-874	7-3912	HPLC	Apparent biomagnification based upon skewed representation	Ibelings et al, 2005
(i) silver carp (Hypophthalmichthys molitrix), (ii) Carassius gibelio	MC-LR; MC-RR	Feeding on Microcystis cells	(i) homogenized tissues: < 5 (H. molitrix) < 0.8 (C. gibelio) (ii) faeces: 11 - 46 (H. molitrix) 21 (C. gibelio)			MC-RR levels in tissues >> MC-LR	Jang et al, 2004
Sea trout (Salmo trutta)	NODLN	Gavage with single dose of Nodularia	(i) liver: 0.019-1.2; max 1.6 (ii) muscle: max. 0.125		ELISA and HPLC	Single oral dose: loss liver architecture 1-2d, partial recovery 4-8 d, complete after 8 d: damage reversible	Kankaapää et al, 2002
Flounder (Platichthys flesus)	NODLN, MC	Feeding on mussels in Baltic sea	(i) 0.02-0.1 TEH prebloom exposure (ii) 0.02-2.23 TEH (ELISA) (iii) nd-0.47 (LC-MS)		ELISA and LC-MS	(i) NODLN variable between individuals: peak conc. in sub-populations (ii) 50 % livers small scale necrosis	Kankaapää et al, 2005a

Flounder (Platichthys flesus)	NODLN	Natural routes in sea	Liver: 0.82-6.37		LC-MS	No biotransformation products (e.g. glutathione adduct) found	Karlsson et al, 2003
(i) northern pike (Esox lucieus (ii) white sucker (Catostomus comersonii)	MC	Feeding on MC containing prey (i.a. gastropods)	Liver: not detected	1.2–6.1 $\mu g\ L^{-1}$	HPLC		Kotak et al, 1996a
Carp (Cyprinus carpio)	MC	Feeding on Microcystis scum in tanks	(i) hepatopancreas: 2.6 (ii) muscle: 0.4				Li et al, 2004
Tilapia rendalli	MC	Natural routes in lake	(i) viscera: 0 – 67.8 (ii) liver: 0 – 31.1 (iii) muscle: 0.003 – 0.026	max. 980 $\mu g\ L^{-1}$	HPLC, ELISA	MC found in 75 % fish samples	Magalhães et al, 2001
Fish (unspecified)	MC	Natural routes in lake	Muscle: 0.01 – 0.4	0.12–0.78 $\mu g\ L^{-1}$	ELISA		Magalhães et al, 2003
Goldfish (Carassius auratus)	MC-LR	IP injection	Liver: 0.5 - 3		PPase	Time course accumulation (first 48 h), depuration (48-96 h); patterns not affected by fastening	Malbrouck et al, 2003, 2004
Tilpapia (Oreochromis niloticus)	MC	On fish-farm to Microcystis bloom	(i) gut: 1.8 – 8.3 (ii) liver: 4.1 – 5.3 (iii) kidney: 3.6 - 4 (iv) muscle: 0.4 – 1.0	1120	ELISA	At times MC in liver > gut	Mohamed et al, 2003
Flounder (Platichthys flesus)	NODLN and MC	Natural routes in Baltic sea	Max 0.4 (TEH)	150-8700 THE; < 2400 NODLN	ELISA and MALDI-TOF-MS		Sipiä et al, 2001, 2002
Tilapia rendalli	MC	(i) Feeding on Microcystis cells + fish food; one exp cells disrupted prior feeding	(i) liver: 2.8 (ii) muscle: 0.08 (iii) faeces: 0.07 (all max values)	14.6	ELISA	(i) 15 or 42 d exposure (ii) accumulation MC less in presence alternative food (iii) depuration phase: only small percentage MC removed	Soares et al, 2004

Silver carp (Hypophthalmichthys molitrix)	MC-LR and RR	Feeding on Microcystis bloom in tanks	(i) faeces: 44.5; MC-LR:RR = 0.75 (ii) intestines: 49 – 115; MC-LR:RR = 0.17 (iii) blood (MC-RR): 0.4 – 50 (iv) liver (MC-RR): 8 - 18 (v) muscle (MC-RR): 0.5 – 1.4	MC: 286–866; MC-LR:RR = 0.57	PPase	(i) in the various tissues and organs always MC-RR found, rarely MC-LR (ii) active degradation of MC-LR during digestion (?)	Xie et al, 2004
Various fish species: phytoplanktivorous (Hypophthalalmichthys molitrix), herbivorous (Parabramis pekinensis), omnivorous (Carassius auratus), carnivorous (Culter ilishaeformis)	MC-LR and RR	Natural routes in lake	(i) intestines: 22 (26 % LR) (ii) blood:14.5 (45 % LR) (iii) liver: 7.8 (iv) bile 6.3 (48 % LR) (v) kidney: 5.8 (30 % LR) (vi) muscle: 1.8 (18 % LR)	240	HPLC	(i) MC in carnivorous > omnivorous > phytoplanktivorous fish (ii) fish at top of foodweb most at risk; general tendency MC to accumulate up the foodchain	Xie et al, 2005
Salmon (Salmo salar)	MC-LR	IP injection	(i) 2.6 – 263 (MeOH) (ii) 138 – 1181 (Lemieux)		(i) PPase after MeoH extract. (ii) Lemieux (GC-MS)	Covalently bound MC made up ~ 74 % of total	Williams et al, 1997a
MACROINVERTEBRATES							
Mytilus galloprovincialis	MC	Grazing on Microcystis cells	(i) mussel: 10.7 during accumulation rising to max of 16 on day 2 depuration (ii) faeces: 140	3.4 µg / 10^7 cells	ELISA	(i) no mussel mortality (ii) during depuration initial increase, followed decrease MC	Amorim & Vasconcelos (1999)
Zebra mussels (Dreissena polymorpha)	MC	Natural routes in lake	0.2	91-820			Babcock-Jackson et al, 2002

(i) freshwater shrimps (Palemon modestus; Macrobrachium nipponensis) (ii) red swamp crayfish (Procambarus clarkii)	MC-LR and RR	Natural routes in lake	(i) P.modestus - stomach: 4.53 hepatopancreas: 4.29 gonads: 1.17 eggs: 2.34 muscle: 0.13 gills: 0.51 (ii) M. nipponensis - stomach: 2.92 hepatopancreas: 0.53 gonads: 0.48 eggs: 0.27 muscle: 0.04 gills: 0.05 (iii) P. clarkia - stomach: 9.97 hepatopancreas: 0.08 gonads: 0.93 muscle: 0.05 gills: 0.27	LC-MS	(i) Proportion of MC-LR of total MC varied with type of tissue and the species; ratio of MC-LR:MC: P. modestus decreased gonad (94 %) > stomach (61 %) > eggs (56 %) > HP (30 %) > muscle (5 %) > gills (0 %) M. nipponensis: stomach (71 %) > muscle (48 %) > HP (39 %) > eggs (39 %)> gonads (33 %) > gills (0 %) P. clarkia: muscle (100 %) > intestine (68 %) > stomach (58 %) > gonad (57 %) > gills(52 %) (ii) considerable part toxin burden in eggs (29 % in P. modestus), i.e.MC transferred to offspring	Chen & Xie, 2005
Freshwater snail (Bellamya aeruginosa)	MC-LR and RR	Natural routes in lake	(i) digestive track: 0.8- 240 4.54 –MC LR:RR=0.44 (ii) HP: 1.06-7.42 - LR:RR = 0.63 (iii)) gonad: 0-2.62 - LR:RR=0.96 (iv) foot: 0.01	HPLC	MC hepatopancreas > digestive track: selective bioaccumulation?	Chen & Xie, 2005

Zebra mussels (Dreissena polymorpha)	MC-LR	Feeding on toxic Microcystis cells	(i) 11 – feeding solely Microcystis (ii) 3.9 feeding on mixture Microcystis and green algae	3.1	LC-MS; MMPB	(i) maximum share covalently bound MC 38 % total (ii) only 0.5 % offered MC found in mussels (iii) rapid depuration, after 3 wk nearly complete	Dionisio Pires et al, 2004
Mysid shrimp (Mysis relicta)	NODLN	Feeding on copepods pre-exposed to cyanobacteria	(i) 0.74 (ELISA) (ii) 0.52 (PPase)		ELISA and PPase	(i) time dependent accumulation in shrimps, not in fish (results ELISA) (ii) copepods as vectors to higher trophic levels	Ensgtröm-Öst et al, 2002
Zebra mussels (Dreissena polymorpha)	MC	Natural routes in lake foodweb	1-30	7-3912	HPLC		Ibelings et al, 2005
Black tiger prawns (Penaeus monodon)	MC, NODLN	(i) natural routes in ponds (ii) oral uptake NODLN via food in experiments (iii) injection MC-LR	(i) ponds - HP (TEH) 0.006-0.08 (ii) experiment (NODLN) - brain and heart: 0.36 HP: 0.25 (0.83 peak level) gut: 0.1 gills: 0.014 muscle: 0.01 (iii) experiment (MC) - HP: 0.130 (peak)	(i) ponds: TEH: 1.2 (ii) exp: 10 μg kg bw^{-1}	ELISA; HPLC	Rapid depuration from prawns	Kankaanpää et al, 2005b
Baltic Sea zooplankton: Acartia tonsa, Eurytemora affinis, Strombium sulcatum	NODLN	Exposure to dissolved NODLN (^{3}H-dihydronodularin)	A. tonsa: 0.37 μg g^{-1} C E.affinis: 0.60 S. aulcatum: 1.55	5 μg L^{-1}		(i) minimum BCF 12 – 18 for copepods; (ii) max BCF ciliate 22 (iii) possible vectorial transport with significant sublethal effects	Karjalainen et al, 2003

(i) pike larvae (Esox lucius) (ii) mysid shrimps (Neomysis integer)	NODLN	Fed with zooplankton pre-exposed to (i) Nodularia extract (ii) purified NODLN (20 μg L^{-1})	(i) pike larvae (12h): 0.47 (ii) Neomysis: (12h): 0.31			Only 0.12 (pike) and 0.03 % (shrimps) of ingested toxin was detected in animals	Karjalainen et al, 2005
Gastropods (Lymnea stagnalis, Helisoma trivolis, Physa gyrina)	MC	Grazing on (settled) lake phytoplankton	11 – 121	1.2 – 6.1 μg L^{-1} ~1220 μg g^{-1} DW	HPLC	MC in seston not expressed on DW basis	Kotak et al, 1996a
Baltic clam (Macoma balthica)	NODLN	Exposure to dissolved toxin and Nodularia cells in tanks	0.16 - 16.6 (24 h) – 30.3 (96 h) (TEH); < 5 % of this NODLN	10-50 μg L^{-1} (cells) 4-20 μg L^{-1} (dissolved)	ELISA, HPLC, MALDI-TOF-MS	No NODLN-GSH conjugates detected	Lehtonen et al, 2003
Signal crayfish (Pacifastacus leniusculus)	MC	Exposure to toxic and non toxic Planktothrix agardhii	MC present but not quantified	3610		(i) crayfish ingested 430 μg MC – no effects (ii) possible vectorial transport to fish, birds, mink	Liras et al, 1998
Crab (unspecified)	MC	Natural routes in lake	Muscle: 0.02 – 1.0	0.12 – 0.78 μg L^{-1}	ELISA		Magalhães et al, 2003
Freswhater mussel (Alathyria condola)	PST	feeding on toxic Anabaena	5.7	1580			Negri & Jones, 1995
Pearl oyster (Pinctada maxima)	SAX	(i) natural routes in sea; (ii) exposure juvenile oyster to Trichodesmium	0.73 in viscera diseased oyster	Not detectable	HPLC; mouse assay; sodium channel and saxiphilin binding assay	no mortality juvenile oysters in experiment	Negri et al, 2004

(i) freshwater snail (Sinotaia histrica) (ii) freshwater clam (Corbicula sandai)	MC-LR and RR	(i) natural routes in lake (ii) feeding on toxic Microcystis cells in exp.	(i) S. histrica (lake) intestine 2.7 – 19.5 (MC-LR + RR) HP 0 – 3.2 (MC-RR only) (ii) C. sinai: nd (iii) S. histrica (exp.) HP max. 436	51.8 – 284 (lake) 20.1 μg L^{-1} (exp.)	HPLC	(i) lag phase in depuration from snail tissue; biological half life 8.4 d (ii) MC in lake snail still present next spring	Ozawa et al, 2003
Freshwater mussel (Anodonta cygnea)	PST	Feeding on toxic Aphanizomenon - accumulation 14d, depuration 14d	0.26	1.9-2.6	HPLC	(i) Anodonta exposed to $1.4e^9$ cells L^{-1} d^{-1}, removed 65 % these; clearance rate negatively related to PST content (ii) slow-fast-slow depuration; s-shaped kinetics (8.2 % d^{-1})	Pereira et al, 2004
Freshwater clam (Anodonta grandis)		(i) exposure dissolved MC (ii) natural routes in lake	Viscera 0.59 (i) gills 0.31 (ii) muscle 0.36	MC 51-55 μg L^{-1} dissolved up to 8.3 μg L^{-1} in lake		(i) toxin burden evenly distributed over three body parts (ii) rapid depuration first 6d (~70% gone), stable for 15d afterwards (iii) suggestion of bioconcentration	Prepas et al, 1997

Freshwater mussel (Anodonta cygnea)	CYN	Feeding on Cylindrospermopsis culture	(i) haemolymph 61.5 (= 408 μg L^{-1}) (ii) viscera: 5.9 (iii) mantle: 0.13 (iv) foot + gonads: 0.75 (v) whole body extract: 2.9 (all maximum conc. after 10-16 d accumulation)	14-90 μg L^{-1}	HPLC	(i) no adverse effects on mussels despite bioaccumulation CYN in haemolymph to conc. higher than in water (ii) bi-phasic depuration, increase in CYN content from day 22-28	Saker et al, 2004
Blue mussel (Mytilus edulis)	NODLN	Natural routes in the sea	Max 2.15	Max 2400			Sipiä et al, 2001
(i) balthic clam (Macoma balthica) (ii) blue mussel (Mytilus edulis)	NODLN and MC	Natural routes in the sea	(i) mussels 1.5 (max of TEH) (ii) clams 0.1-0.13	150-8700	ELISA and MALDI-TOF-MS	(i) NODLN-GSH conjugates confirmed with MS (ii) mussels MC 30 fold increase summer (iii) time lagged accumulation deep sites	Sipiä et al, 2002
Blue mussels (Mytilus edulis)	NODLN	(i) grazing on Nodularia (ii) coprophagy (exposure to feces of mussels grazing on Nodularia)	(i) Mussels pre-exposure to Nodularia: 0.05-0.1 (ii) Mussels grazing on Nodularia - digestive track 245 body 80 gills 2 (iiii) feces when feeding on Nodularia 95 (iv) Body after coprophagy 0.065 (v) Feces coprophagy 1 (vi) PF from (ii) 714	Nodularia culture 16 μg L^{-1}	LC-MS	(i) high NODLN in PF indicative of selective feeding; (ii) cells Nodularia in PF may survive but growth inhibited	Svensen et al, 2005

Mytilus galloprovincialis	MC-LR	Grazing on Microcystis cells	10.5 (of which 96 % in digestive gland + stomach)	28 µg 10^8 cells^{-1}		Depuration bi-phasic and fairly rapid (13d)	Vasconcelos 1995
Crayfish (Procamabarus clarkia)	MC	Feeding on toxic (and non toxic) Microcystis cells	2.9 (max after 11 d), of which 53 % in intestine, 38 % HP and 9 % rest body	2300	ELISA	Juvenile crayfish enhanced mortality on non-toxic Microcystis: compounds other than MC more relevant to invertebrates (?)	Vasconcelos et al, 2001
Dungeness crab larvae (Cancer magister)	MC-LR	Natural routes in foodweb	(i) 0.006 (after MeOH extraction) (ii) 84.4 (after Lemieux)		(i) PPase after MeOH extract. (ii) Lemieux (GC-MS)	10.000 fold greater MC concentration using MMPB after Lemieux extraction	Williams et al, 1997a
Blue mussel (Mytilus edulis)	MC-LR	(i) feeding on toxic Microcystis cells (ii) natural routes in foodweb	(i) 3369 decreasing to 113 after 4 d depuration (Lemieux) (ii) 2 - dropping to 0.14 over 53 d depuration (MeOH)		(i) PPase after MeOH extract. (ii) Lemieux (GC-MS)	Less than 0.1 % of MC extractable with MeOH	Williams et al, 1997b
Freshwater snail (Bellamya aeruginosa)	MC-LR, RR		(i) HP 1.06-7.42 (ii) digestive track 0.8-4.54 (iii) gonad 0-2.62 (iv) foot 0-0.06 (MC-LR only)		HPLC, LC-MS	(i) bioaccumulation in HP (ii) ratio LP:RR increased from digestive track to HP to gonad (iii) suggestion LR more resistant to degradation?	Xie et al, 2005
Freshwater mussel (Unio douglasiae)	MC-LR	Exposure to Microcystis cells	HP: (i) at 15 oC – 130 (ii) at 25 ° C - 250	27 - 50 µg L^{-1}		(i) accumulation temperature dependent (ii) depuration relatively fast; but slower at 15 than 25 °C; halted in winter (iii) no adverse effects on mussels	Yokoyama & Park (2003)

Pulmonate snails (Lymnea stagnalis, Helisoma trivolis, Physa gyrina)	MC	Natural exposure in lakes of varying trophic status	Up to 144	0 – 1526	HPLC	Possible uptake routes via food and directly from water	Zurawell et al, 1999
ZOOPLANKTON							
Echinogammarus ischnus	MC	Natural routes in lake	2	91-820		Exposure of this detrivore via pseudofaeces Dreis-sena?	Babcock-Jackson et al, 2002
Community of different spp, i.a. Daphnia galeata	MC	Natural routes in lake	0-1352		HPLC		Ibelings et al, 2005
(i) calenoid copepods (Eurytemora affinis, Acartia tonsa) (ii) ciliate (Strombidium sulcatum)	NODLN	Uptake of dissolved NODLN from water during 15 min - 6 d	(i) E.affinis 1.14 (ii) A. tonsa 0.66 (iii) S. sulcatum 4.8 (maxima)	5 µg L^{-1}		(i) BCF copepods 12-18 (ii) ciliate 22 (iii) vectorial transport to shrimps and planktiv. fish	Karjalainen et al, 2005
Community of different spp., a/o Daphnia pulex	MC	Grazing on lake phytoplankton	Up to 67	1.2 – 6.1 µg L^{-1}	HPLC		Kotak et al, 1996
(i) Thamnocephalus, platyurus (ii) Eudiaptomus gracili, (iii) Daphnia hyaline (iv) Cyclops abyssorum	MC	(i) grazing on mix-tures of Cryptomo-nas and Planktothrix (ii) Planktothrix at 0.05 or 0.1 mg C L^{-1} - MC(+) or MC(-)		4.7 ng MC-LR per µg C^{-1}	HPLC		Kurmayer & Jüttner, 1999

Calanoid copepod (Eurytemora affinis)	NODLN	(i) feeding on Nodularia and non-toxic flagellates (ii) natural seston	(i) fed with Nodularia: 0.032 ng copepod^{-1} (background) rising to 0.007 (ELISA) (ii) 0.0095 to 0.101 (PPase) (iii) feacal pellets - 0.0067 ng pellet^{-1} (ELISA) (iv) idem 0.0050 (PPase)		ELISA and PPase	(i) lower conc. when fed with natural seston (dominated by non-tox Aphanizomenon) (ii) ELISA and PPase different but not significant (iii) no acute effects on copepods despite accumulation (iv) vectorial transport to foodweb (v) toxin content feacal pellets low: no vector to coprophagous animals	Lehtiniemi et al, 2002
Daphnia magna	PST	(i) grazing on Aphanizomenon (1.2e^{6} cells mL^{-1}) (ii) lyophilized material (1 mg mL^{-1})	(i) exposure Apha: 0.065-0.378 pmol PST animal^{-1} (ii) exposure lyophilized material: 0.007	(i) cells: 643-1170 pmol mL^{-1} (ii) lyoph.: 2745 pmol mL^{-1}	HPLC-FLD	Bioaccumulation factor > 1 after 12h feeding	Nogueira et al, 2004
Daphnia magna	CYN	Grazing on Cylindrospermopsis strains (+/- CYN)	(i) 0.025 ng animal^{-1} (24 h) (ii) 0.020 ng animal^{-1} (48 h)	(i) intracellular toxin 219.6 – 236.4 ng mL^{-1} (ii) extracellular toxin 12.4 – 47.4 ng mL^{-1} (increase in time) (iii) total CYN 234.2 – 278.4 ng L^{-1}	HPLC-MS	Bioaccumulation factors smaller than unity: 0.71 (24 h) and 0.46 (48 h); levels not high enough to indicate bioaccumulation	Nogueira et al, 2004(b)

Daphnia magna	MC	(i) grazing on Microcystis alone or mixed with Scenedesmus (3.2 mg C L^{-1}) (ii) exposure to lake water enriched with Microcystis from enclosures (8.9 mg C L^{-1})	Daphnia 0.2-24.5 μg g^{-1} DW (highest when exposed solely to Microcystis)	(i) Microcystis: 2000 μg g DW^{-1} (ii) MC varied between 5 – 156 μg L^{-1} in enclosures	ELISA	(i) calculation shows MC really accumulated in body, not just gut content (ii) calculation shows transport to roach results only in sublethal effects	Thostrup & Christoffersen, 1999

Table A.2. Effects of various cyanobacterial toxins on aquatic biota. Where possible dose and effect are shown.

Exposed organism	Toxin	Exposure route	Concentration (dose) and effect	Remark	Reference
BIRDS					
FISH					
Zebra fish (Danio rerio)	MC–LR	Purified MC, dissolved in water	(i) 0.5, 5 µg L^{-1}: increased motility (ii) 15, 50 µg L-1: decreased motility (iii) 50 µg $L^{-1:}$ reduced spawning	(i) effects on behaviour dose dependent (reversal at higher conc.) (ii) LOEC < 0.5 µg L^{-1}	Baganz et al, 1998
Brown trout (Salmo trutta)	MC	i) purified MC ii) aqueous extracts of Microcystis	5, 50, 500 µg L^{-1} MC: cardiovascular effects	Purified MC no effect on (some) cardiac responses, in contrast to cell extracts – always effects; cardiac output only affected by MC at higher doses	Best et al, 2001
Zebra fish (Danio rerio)	MC–LR; LPS	In vivo exposure fish embryos to dissolved toxins; LPS different bacterial origins (including cyanobacterial)	LPS (up to 29.8 10^6 EU mg^{-1} DW) + MC (up to 12.6 ng mg DW^{-1}): reduced mGST + sGST	LPS from axenic bacteria or lake blooms reduced GST activity, interfered detoxication	Best et al, 2002
Rainbow trout (Oncorhynchus mykiss)	MC	Aqueous suspension whole or broken Microcystis cells	100 µg L^{-1} MC–LR: osmoregulatory imbalance, increased liver mass	Interaction MC with LPS	Best et al, 2003

Exposed organism	Toxin	Exposure route	Concentration (dose) and effect	Remark	Reference
Brown trout (Salmo trutta)	MC–LR	(i) purified MC (ii) lysed Microcystis cells	i) lysed tox Microcystis, containing 41 – 68 μg L^{-1} MC: ionic imbalance, reduced growth ii) purified toxin: effects on growth (less than toxic lysate) iii) lysed non–tox Microcystis: effects on growth (less than toxic lysate)	Ammonia similar effects to MC	Bury et al, 1995
Carp (Cyprinus carpio)	MC	Fish exposed to toxins via natural routes in the lake, including Microcystis scums	Up to 4000 μg g^{-1} DW: hepatic lesions	Lesions in > 50 % fish examined	Carbis et al, 1997
Carp (Cyprinus carpio)	MC–LR	Gavage freeze dried Microcystis cells (single sublethal bolus)	400 μg kg^{-1} bw: pathological changes in hepatopancreas, kidney, gastrointestinal tract (i.a. apoptosis)	In carp – cyprinid – compared to salmonids liver pathology develops faster and at lower MC concentrations	Fischer & Dietrich, 2000
Rainbow trout (Oncorhynchus mykiss)	MC	Gavage freeze dried Microcystis cells	5700 μg kg^{-1} bw: liver necrosis, hepatocyte apoptosis	Unbound MC results in fast effects, including PP inhibition and liver necrosis; covalently binding to MC results slower effects, i.a. apoptosis	Fischer et al, 2000
(i) Hypophthalmichthys molitrix (ii) Carassius gibelio	MC–LR; MC–RR	Feeding on cultured Microcystis cells	unspecified dose: reduced growth		Jang et al, 2004

Exposed organism	Toxin	Exposure route	Concentration (dose) and effect	Remark	Reference
Tilapia (Oreochromis sp)	MC–LR	Feeding on lyophilized Microcystis bloom (crushed and non–crushed cells)	60 µg $fish^{-1}$ day^{-1} MC–LR: oxidative stress in liver, kidney and gills	(i) crushed cells more effect non–crushed on LPO and levels antioxidant enzymes (ii) time dependent development oxidative stress: not yet present after 14d; 21 dlipid peroxidation + enhanced antioxidant enzyme activities	Jos et al, 2005
Sea trout (Salmo trutta)	NODLN	Gavage single dose Nodularia	210–620 µg kg^{-1} bw: rapid (1–2 d) but reversible (4–8 d) liver damage	(i) rapid detoxication, cross reaction conjugates in ELISA (ii) no effects on swimming	Kankaanpää et al, 2002
Rainbow trout (Oncorhynchus mykiss)	MC–LR	IP injection of purified MC	(i) 1000 µg kg^{-1}: 100 % mortality (ii) 400 µg kg^{-1}: no mortality; increased ratio liver to body mass, liver necrosis; kidney lesions	Fish less sensitive than mice: LD_{50} = 400 – 1000 µg kg^{-1}	Kotak et al, 1996b
Carp (Cyprinus carpio)	MC–LR	In vitro exposure hepatocytes to 10 µg L^{-1} MC–LR	Increase ROS, depletion GSH, increase SOD, CAT and GS–Px activity, GST unchanged: oxidative shock by exposure MC	Antioxoidant enxymes did not prevent oxidative shock: apoptosis and necrosis hepatocytes	Li et al, 2003
Carp (Cyprinus carpio)	MC	Feeding on Microcystis scum in tanks	50 µg kg^{-1} bw for 28 days: reduced growth, increased liver enzyme activity, damaged hepatocytes,	Long term subchronic effects	Li et al, 2004

Exposed organism	Toxin	Exposure route	Concentration (dose) and effect	Remark	Reference
Loach (Misguruns mizolepis)	MC–LR	Dissolved MC–LR	MC–LR at 1, 3, 10, 100, 1000 µg L^{-1}: (i) mortalilty, delayed hatching, liver (necrosis) and cardiotoxicity (ii) (late stage) embryos and larvae more sensitive than juvenile fish: LC_{50} larvae = 164; juveniles = 593 µg L^{-1}	(i) embryonic development loach affected by MC in contrast to zebrafish (ii) results dose dependent, i.e. increasing effects with increasing conc. MC. No mortality at the lower doses.	Liu et al, 2002
Goldfish (Carassius auratus)	MC–LR	IP injection	125 µg kg^{-1} bw: no change ionic homeostasis, liver lesions, changes liver enzyme activity	Liver damage reversible	Malbrouck et al, 2003
Goldfish (Carassius auratus)	MC–LR	IP injection of fed and fasted juvenile goldfish	(i) 125 µg kg^{-1} bw: inhibition PPase, complete after 6 h in fasted, less inhibition in fed fish (ii) recovery after 96 h (iii) GSH levels and GST unaffected	Feeding status has effect on toxicity MC, possibly by acting on bile formation and secretion	Malbrouck et al, 2004a
(i) silver carp (Hypophthalmichthys molitrix) (ii) carp (Cyprinus carpio)	MC–LR	(i) ip injection purified MC (ii) cyanobacterial biomass per os / anus	(i) 400 µg kg^{-1} bw (purified MC): no – minor effects (ii) 3 – 1,200 µg kg^{-1} bw (biomass): changes blood indices and immunological changes	(i) effects crude biomass much stronger purified toxin (ii) oxidative stress	Palikova et al, 1998

Exposed organism	Toxin	Exposure route	Concentration (dose) and effect	Remark	Reference
Seven fish (cyprinids) and three amphibian species	MC–LR, RR and YR, SAX, ANA(a)	(i) in vivo emersion (ii) exposure to dissolved purified toxin (iii) exposure crude cyanobacterial cell extracts	(i) MC 0 – 50 μg L^{-1}: no acute effects on embryonic development (ii) MC–RR > 0.5 μg L^{-1}; YR > 5 μg L^{-1}; LR > 50 μg L^{-1}: timing hatching affected (ii) morphological effects only at highest conc. MC–LR 10 mg L^{-1} (iii) SAX > 10 μg L^{-1} delayed hatching malformations at 500 μg L^{-1} (iv) ANA(a) 400 μg L^{-1} hearth rate affected, no chronic effects (v) far more pronounced effects with crude extracts: malformations and mortality	Crude cell extracts of several cyanobacteria gave severe effects (more so than purified toxin): cannot be attributed MC alone	Oberemm et al, 1999

Exposed organism	Toxin	Exposure route	Concentration (dose) and effect	Remark	Reference
Rainbow trout (Oncorhynchus mykiss)	MC–LR	(i) IP injection or oral dosage purified MC (ii) IP injection or oral dosage freeze dried Microcystis cells	(i) 550 $\mu g\ kg^{-1}$ bw MC (ip): severe liver damage, death (ii) 550 $\mu g\ kg^{-1}$ bw Microcystis (ip): severe liver damage, death (iii) 1200 $\mu g\ kg^{-1}$ bw MC (oral): no effects (iv)1700 $\mu g\ kg^{-1}$ bw Microcystis (oral): no effects (v) 6600 $\mu g\ kg^{-1}$ bw Microcystis (oral): severe liver damage, death (vi) 550 $\mu g\ kg^{-1}$ bw Microcystis (8 times oral dosage): modest – severe liver damage	Treatments included repeated (oral) exposure to MC, i.e. close to natural exposure but during 'limited' period of time (8 times at 12 h intervals)	Tencalla et al, 1994
Zebra fish (Danio rerio) embryos	MC–LR	Purified dissolved MC	(i) > 0.1 $\mu g\ L^{-1}$: increased sGST and GPx activity (ii) > 2.0 $\mu g\ L^{-1}$: effects on growth and survival	Dose dependent relationship MC~GST	Wiegand et al, 1999
Carp (Cyprinus carpio)	MC–LR	Purified dissolved MC	inhibition ATP–ase activity of $Na^{+}K^{+}$ pump in gills, disruption ion homeostasis	Harmful effects of dissolved MC (without ingestion of cells)	Zambrano & Canelo, 1996
MACRO–INVERTEBRATES					
Brine shrimp (Artemia salina)	MC–LR, MCHtyR and NODLN	Purified dissolved toxin	0.5 $\mu g\ L^{-1}$: elevation GST activity		Beattie et al, 2003

Exposed organism	Toxin	Exposure route	Concentration (dose) and effect	Remark	Reference
(i) pike larvae (Esox lucius) (ii) mysid shrimps (Neomysis integer)	NODLN	(i) purified NODLN (20 µg L^{-1}) (ii) crude extract Nodularia	(i) crude extract: pike larvae: decreased ingestion and faeces production rates shrimps: no effects on molting, faeces production, C:N or growth (ii) purified NODLN: no effects	Purified NODLN no effects on pike larvae; crude cell extracts stronger effects	Karjalainen et al, 2005
Baltic clam (Macoma balthica)	NODLN	(i) exposure to dissolved toxin (ii) exposure to toxic and non toxic Nodularia cells in tanks	4–20 µg NODLN per day: (i) conc. dependent neurotoxic effects (increase / decrease AChE activity when exposed to low and high NODLN respectively) (ii) some treatments low siphon activity	Abundant unidentified compound with NODLN like spectral characteristics (found in both toxic and non–toxic Nodularia treatments)	Lehtonen et al, 2003
Brine shrimp (Artemia salina)	CYN and MC	(i) purified dissolved CYN and MC (ii) extracts Cylindrospermposis, Microcystis	(i) CYN LC_{50} decreased from 4.48 to 0.71 µg mL^{-1} between 24 and 72 h (ii) likewise MC LC_{50} from 4.58 to 0.85 µg mL^{-1}	(i) dose and time dependent mortality (ii) LC_{50} cell extracts typically > than purified CYN (reduced bioavailabilty?)	Metcalf et al, 2002
Estuarine crab (Chasmagnathus granulatus)	MC	Cell extracts Microcystis	Injected daily for 4–7 d with 17.6 ng MC: (i) increased enzyme activity (GST, CAT) in hepatopancreas (ii) no change LPO – no oxidative damage (?) (iii) yet histological damage		Pinho et al, 2003

Exposed organism	Toxin	Exposure route	Concentration (dose) and effect	Remark	Reference
Estuarine crab (Chasmagnathus granulatus)	MC	Crabs injected twice with cell extracts Microcystis	(i) decreased Na^+, K^+–ATPase activity anterior gills (ii) increased GST posterior gills (iii) increased TOSC	Increased TOSC in response MC as protective mechanism against LPO (level unchanged by exposure to MC)	Vinagre et al, 2003
ZOOPLANKTON					
Moina macrocopa	Unidentified metabolites (possibly cyanopeptolins A–D)	Freeze dried Microcystis	Inhibition of proteases		Agrawal et al, 2005
Thamnocephalus platyurus	[D–Asp^3, (E)–Dhb^7]MC–RR, MC–LR, MC–YR, MC–RR, NODLN	Purified toxins	LC_{50} NODLN < [D–Asp^3, (E)–Dhb^7]MC–RR < other MC	LC_{50} insufficient to study response of organisms to exposure; LC_{10} + LC_{90} required to get slope	Blom et al, 2001

Exposed organism	Toxin	Exposure route	Concentration (dose) and effect	Remark	Reference
Daphnia pulicaria, D. Pulex, D. hyaline, Diaptomus birgei	MC–LR, NODLN, ANA(a)	Exposure to: (i) purified toxins (ii) cell extracts (iii) toxic or non–toxic cyanobacterial strains	(i) species specific responses to toxin exposure; not all daphnids equally sensitive in terms feeding inhibition (ii) LC_{50} for MC after 24 h (varied from > 50 μg L^{-1} in D. pulicaria to < 1.0 in D.birgei); for NODLN > 20 and < 0.6 respectively (iii) LC_{50} decreased with longer exposure	(i) feeding inhibition protects against toxic effects: less sensitive Daphnia stronger inhibiton (ii) zooplankters insensitive dissolved MC, more so dissolved ANA(a)	DeMott et al, 1991
Daphnia spp	MC	Feeding on mixtures Microcystis and Scenedesmus; 0, 50 or 80 % Microcystis in total food conc. of 0.5 mg C L^{-1}	(i) rapid feeding inhibition (but recovery after continued exposure to same mixture) (ii) reduced growth and reproduction (iii) reduction in growth/ingestion: direct toxic effects and feeding inhibition	Clear differences between Daphnia spp	DeMott, 1999
Temperate and tropical cladocerans (Ceriodaphnia cornuta, Daphnia pulex; D. pulicaria, D similes, Moina micrura, Moinodaphnia macleayi)	MC	(i) grazing on toxic Microcystis strains mixed with Ankistrodesmus, total conc. 1.0 mg C L^{-1}; (ii) acute and chronic exposure	MC contents 2810–4080 μg g^{-1} DW: (i) decreased survival in presence tox Microcystis (ii) toxic Microcystis inhibited feeding rate, even when just 5 % in mixture with greens (ii) non–toxic cyanobacterium as sole food:poor growth	(i) species from low productivity sites – adapted to starvation – showed lowest sensitivity to toxic Microcystis (ii) small and large bodied fast growing spp prone to starvation and most sensitive to MC; small bodied slow growing spp most resistant and least sensitive	Ferrão–Filho et al, 2000

Exposed organism	Toxin	Exposure route	Concentration (dose) and effect	Remark	Reference
Tropical cladocerans (Moina micrura, Ceriodaphnia cornuta)	MC	Grazing on: (i) large Microcystis colonies (ii) single cells and (small) colonies Microcystis	MC 2.0 – 16.0 μg L^{-1}: (i) toxic Microcystis inhibited growth reduced reproduction cladocerans; partly through feeding inhibition (ii) effects unicellular lab cultures stronger than colonial cultures or natural seston, especially with large colonies (although very toxic 3.9 mg MC g DW^{-1})	Colony forming cyanobacteria: little effect despite high toxicity; possible explanation why field studies fail to demonstrate effects of toxic blooms	Ferrão–Filho & Azevedo 2003
Daphnia magna	MC	Grazing on mixtures of Microcystis and Scenedesmus; max of 140,000 Microcystis cells mL^{-1} (1.0 mg C L^{-1})	(i) reduced growth, reproduction and survival, increasing effects with increased proportion Microcystis cells (0; 50 or 100 %)0 (ii) pre–exposure (acclimation) reduced harmful effects: development of tolerance		Gustafsson & Hansson, 2004
Daphnia pulex	MC–LR	(i) acute exposure to Microcystis cells 0–2.43 mg C L^{-1} = 0–360,000 cells mL^{-1} (ii) chronic exposure 30,000 cells mL^{-1}	MC–LR 7.6 10^{-5} ng $cell^{-1}$: (i) variation in acute tolerance (EC_{50}) to toxic Microcystis (ii) increase temperature: decrease EC_{50} (iii) chronic exposure, reduced survival and reproduction, clonal differences reversed compared to acute exposure	Suggestion made that more resistant clones show stronger feeding inhibition	Hietala et al, 1997

Exposed organism	Toxin	Exposure route	Concentration (dose) and effect	Remark	Reference
Calanoid copepods (Acartia bifilosa, Eurytemora affinis)	NODLN	Cultures of Nodularia added to enclosures	NODLN ~ 11µg mg DW^{-1} (3–4 µg mL^{-1}): Baltic copepds feed, ingest, reproduce and survive in presence toxic Nodularia, no negative effects	Perhaps hatching success more sensitive to toxins (not measured)	Koski et al, 2002
Eudiaptomus gracilis, Thamnocephalus platyurus, Daphnia hyaline, Cyclops abyssorum	MC	Grazing on (artificially shortened) filaments Planktothrix	(i) reduced survival Thamnocephalus (ii) survival naturally co–existing zooplankton unaffected; (iii) Eudiaptomus high sensitivity but also strict food avoidance (iv) Daphnia and Cyclops greater physiological resistance to MC, less avoidance – ingested filaments	(i) Daphnia feeding rates increased (not so for copepods) when prior toexposure MC were extracted from filaments (MC acts as feeding deterrent) (ii) unidentified lipophilic toxin present (iii) high avoidance linked to high sensitivity	Kurmayer & Jüttner (1999)
Daphnia pulex	MC–LR	Grazing on toxic Microcystis cells (0 – 320,000 cells mL^{-1}) + low or high density Scenedesmus (20,000 or 80,000 cells mL^{-1})	MC–LR 8.9 10^{-5} ng $cell^{-1}$: (i) decreased population density (ii) delayed maturity (iii) increased number ephippia ind^{-1}	Effects toxic Microcystis comparable to food of low quality or lack of food	Laurén Määttä et al, 1997

Exposed organism	Toxin	Exposure route	Concentration (dose) and effect	Remark	Reference
Daphnia magna	MC	Grazing on MC(+) wt and MC(–) mutant + Scenedesmus; Microcystis 0 –100 % of 5 mg C L^{-1}	(i) severe disturbance Daphnia population development (increased mortality, decreased reproduction) (ii) MC(+) killed Daphnia faster MC(–) (iii) also MC(–) had negative effects on survival and growth, even in presence Scenedesmus	(i) nutritional insufficiency Microcystis not responsible effects on Daphnia (ii) feeding inhibition and/or compounds other MC (cyanopeptolines?) may play role (iii) MC alone cannot explain harmful effects on Daphnia	Lürling, 2003
Daphnia magna	MC–LR	(i) grazing on MC(+) and MC(–) strains of Microcystis in max conc. of 5 mg C L^{-1} in mixtures with Scenedesmus + in some treatments in addition: (ii) exposure to purified, dissolved MC, max 3.5 μg L^{-1}	(i) no effects dissolved MC (ii) exposure cellbound MC: reduced feeding and growth (iii) reductions also in treatment 50 % Microcystis of MC(–) strain + 50 % Scenedesmus	Inhibition of Daphnia feeding and growth in presence of Scenedesmus and MC(–) strain Microcystis: unknown toxic compounds	Lürling & van der Grinten, 2003
Daphnia magna	PST	Grazing on Aphanizomenon	$1.2e^{6}$ cells mL^{-1} containing 643–1170 pmol mL^{-1} PST: (i) reduced fitness, growth and survival (ii) reduced activity cGST		Nogueira et al, 2004

Exposed organism	Toxin	Exposure route	Concentration (dose) and effect	Remark	Reference
Daphnia magna	CYN, unidentified hepatotoxin	Grazing on 2 different strains Cylindrospermopsis (+/– CYN); 1.8 – 3.6 10^6 cells mL^{-1};	CYN 4.78 ng mg cells DW^{-1}: (i) reduced growth and increased mortality, also in comparison to starvation treatment, (ii) above also true for CYN(-) strain (iii) sGST + mGST activity increased	Unknown toxic compounds present in strain that does not produce CYN	Nogueira et al, 2004(b)
Daphnia pulex; D. longispina	MC, unidentified toxins	Grazing on mixtures Scenedesmus and Microcystis cells (10,000 or 40,000 cells mL^{-1} ~ 0.076 – 0.304 mg C L^{-1})	(i) increased allocation resources to reproduction (ii) lower dose resulted smaller clutch size D. Pulex; D. longispina no effect (iiii) higher dose virtual inhibition reproduction D. Pulex; D. longispina reduced size neonates	(i) severe and dose dependent effects Microcystis on reproduction in Daphnia (ii) toxicity and food quality play a role	Reinikainen et al, 1999
Estuarine calanoid copepods (Eurytemora affinis; Acartia bifilosa)	MC–LR, ANA(a), NODLN	Purified dissolved toxins, single and in combination	(i) 1 µg mL^{-1} MC or ANA no effect on egg hatching (ii) 0, 0.25, 0.5 and 1 µg mL^{-1}l: reduced survival for MC > 0.1 µg mL^{-1} (iii) ANAa and NODLN only weak effects		Reinikainen et al, 2002
Daphnia galeata	MC	Daphnia feeding on toxic (wt) and non toxic (mutant does not produce MC) Microcystis strains	(i) wt toxic to Daphnia: decreased swimming + death, mutant not toxic (ii) both wt and mutant inhibit ingestion rate	Dose response relationship not between MC content of the food and effects in Daphnia but between ingestion rate and effects	Rohrlack et al, 1999

Exposed organism	Toxin	Exposure route	Concentration (dose) and effect	Remark	Reference
Daphnia pulicaria	Microviridin J	(i) Grazing on Microcystis strains containing 1.05 – 1.57 μg mm^{-3} microviridin (ii) Purified, dissolved microviridin (0–12 mg L^{-1})	Lethal molting disruption	Microviridin is a protease inhibitor	Rohrlack et al, 2004
Daphnia spp.	MC	Daphnia feeding on toxic MC(+) wt and non–toxic (MC(-) mutant Microcystis strains	(i) inhibition of feeding rate, equal inhibition for tox and non–tox Microcystis (ii) reduced survival time (LT_{50}) Daphnia when feeding on wt (iii) Daphnia feeding on mutant signs starvation	(i) MC major source of acute Daphnia poisoning (ii) clear relationship – dose–response – between LT_{50} and MC ingestion rate	Rohrlack et al, 2001
Daphnia galeata	MC	Daphnia feeding on MC(+) and MC(–) strain	Toxic strain 0.87 mg L^{-1} MC: (i) both tox and non–tox Microcystis negatively affect the cohesion of midgut epithelium (within 9 h) (ii) tox strain: uptake MC in blood, increase 0.25 to 1 ng L^{-1} (6–9 h) (iii) tox strain: decreased beat rates (5–9 h), constant contraction midgut, finally complete loss beat rates, death (32–41 h)	(i) midgut disrupting factor is not MC, disruption stimulates uptake bioactive compounds from cyanobacteria in blood Daphnia (ii) results not easily translated to field	Rohrlack et al, 2005
Daphnia magna	MC	(i) Daphnia exposed to suspensions of toxic Microcystis cells (ii) Daphnia exposed to lake water (filtered or not)	MC=2000 μg g DW^{-1}: reduced survival, fecundity and growth	Negative correlation between toxin content and growth (strong) and between toxin concentration and fecundity (weak)	Thostrup & Christoffersen, 1999

Chapter 33: Cyanobacteria blooms: effects on aquatic ecosystems

Karl E Havens

Department of Fisheries and Aquatic Sciences, University of Florida, 7922 NW 71st Street, Gainesville, FL 32653 E–mail: khavens@ufl.edu

Introduction

Lakes, rivers and estuaries that experience frequent and/or prolonged blooms of cyanobacteria display an array of ecosystem properties that may have impacts on water quality, biological communities and ecosystem services. Some impacts of blooms may be direct, including possible effects of toxins on fish, invertebrates, and other aquatic fauna, or indirect, including: a reduction of submerged plants when plankton biomass becomes very high; and changes in fish community structure if summer cold water refuges are lost due to hypolimnetic anoxia. This paper is a concise overview of cyanobacteria blooms, focusing on their relationship to trophic state, their temporal dynamics, and their potential impacts on ecosystem structure and function.

Cyanobacterial blooms and trophic state

When phytoplankton biomass increases during eutrophication, there are coincident changes in taxonomic structure. Most notably the relative biomass of cyanobacteria increases with eutrophication. This relationship has been documented in boreal, temperate, subtropical and tropical ecosystems, and a typical pattern of change in phytoplankton taxonomic structure along the trophic gradient can be found in Auer and Arndt (2004). In general, the potential for cyanobacteria dominance rises rapidly as total phosphorus (TP) increases from 30 to 100 μg L^{-1} (Downing et al. 2001); however, the response pattern in any given system also depends on other factors, such as mean depth, mixing regime, flushing rate and water temperature. Occurrence of blooms correlates positively with TN and TP in lakes, rivers and coastal waters (Paerl 1988), and these correlative relationships have been used in TMDL (Total Maximum Daily Load) development. For example, Havens and Walker (2002) identified a positive relationship between TP concentration and the risk of chlorophyll *a* exceeding 40 μg L^{-1} (a concentration associated with blooms in a Florida lake) and used this to identify the lake water TP goal to establish TMDL guidelines. Although the positive relationship between relative cyanobacteria biomass, blooms and nutrient concentrations holds true for most cases of eutrophication, a point can be reached along the trophic continuum where eukaryotic algae (flagellated chlorophytes) replace the cyanobacteria. This phenomenon is common in Denmark, where some hyper–eutrophic lakes have extremely high levels of TP and TN (TP > 1 mg L^{-1} and TN > 5 mg L^{-1}) and nutrients do not limit algal growth. At such high levels of nutrient availability, small r–selected green algae can out compete certain cyanobacteria (Jensen et al. 1994). This situation is not common in the USA.

Temporal dynamics of cyanobacteria blooms

Cyanobacteria blooms display a range of temporal dynamics. Some lakes, rivers and estuaries have seasonal blooms that start in summer and last into autumn, some have persistent blooms that encompass all seasons, and some have blooms that occur as extreme peaks and crashes lasting just days or weeks. The temporal aspect of a cyanobacterial bloom in a particular ecosystem depends on the extent to which different environmental factors influence bloom dynamics. In deep temperate eutrophic lakes with stable summer stratification, phytoplankton typically progresses through stages of diatom dominance in spring, followed by a relatively clear water phase and then cyanobacteria dominance in mid to late summer. Cyanobacteria dominance often occurs when water temperature rises above 20°C when there is depletion of dissolved inorganic N and free CO_2 from the water. This pattern also occurs in subtropical waters, including coastal systems. For example, a regular seasonal pattern is observed in Pensacola Bay, Florida, where there are summer blooms of cyanobacterial pico–plankton (Murrell and Lores 2004).

In extremely shallow lakes (mean depth < 2 m), dominance of cyanobacteria may persist for years if the ratio of photic depth to mixed depth never falls to levels that prevent net growth of low–light adapted taxa such as *Oscillatoria agardhii* (Berger 1989), but remains low enough to exclude other plankton. At the opposite end of the spectrum is one of the most extreme examples of temporal variation – eutrophic Hartbeespoort Dam (Zohary et al. 1995). In this system, high irradiance in the surface mixed layer and low wind velocities result in short–lived hyper-scums of *Microcystis aeruginosa*. The bloom is often followed by a population crash during periods of high outflow volume, and washed out to downstream systems, at which times other algae can become dominant. As noted below, there is uncertainty as to which factors are coincident with, vs. the actual cause of cyanobacteria blooms and their seasonal and inter–system variation.

Predicting cyanobacteria dominance and bloom occurrence

There has been a long–standing discussion about the relative importance of nutrient concentrations, nutrient ratios, and other factors such as light and water column stability in determining whether or not cyanobacteria dominate the plankton. This discussion relates back to a paper by Smith (1983) that suggested low TN:TP ratios are responsible for cyanobacteria dominance under eutrophic conditions. This hypothesis is consistent with resource ratio theory and was supported by observations from some temperate lakes with varying ratios of TN:TP. Since that time, certain observational and experimental studies have shown that cyanobacteria become increasingly dominant at low TN:TP (Smith and Bennett 1999), yet some scientists discount this as coincidental rather than causal (Reynolds 1999). Others question the value of nutrient ratios as a predictive tool for resource management. Downing et al. (2001), for example, note that concentration of TP is a better predictor of cyanobacteria dominance than TN:TP ratios. Studies have suggested other potential causal factors (summarized by Dokulil and Teubner 2000), including high pH and scarcity of free CO_2, which theoretically should favor cyanobacteria that have a low K_s for CO_2 uptake and can use bicarbonate as a C source. Others have suggested that greater resistance to zooplankton grazing may favor cyanobacteria, and that certain cyanobacteria produce allelopathic chemicals that inhibit growth of other algae. Accessory pigments that allow net growth to occur at low irradiance (Scheffer et al. 1997) and buoyancy that allows certain taxa to bloom at the water surface (Reynolds et al. 1987) also are considered important to cyanobacteria dominance. In eutrophic dimictic lakes, summer bloom development also has been linked with emergence of cyanobacteria from sediment akinete populations (Tsujimura and Okubo 2003).

All of these explanations may be correct under specific environmental conditions because cyanobacteria are a diverse group, as are the aquatic ecosystems in which they occur. There is greater certainty about controlling factors when one considers certain ecotypes of cyanobacteria in particular types of lakes, rivers or estuaries. In regard to ecotypes, a key difference is between taxa that form water blooms and have relatively low light requirements vs. taxa that form surface blooms and have high light requirements. In regard to ecosystems, key differences are shallow vs. deep and mixed vs. thermally stratified. In shallow mixed lakes with high TP, there is extreme and highly predictable dominance by *Oscillatoria* when a low ratio of euphotic to mixed depth (z_{eu}/z_{mix}) allows these cyanobacteria to out compete other algae (Scheffer et al. 1997). In contrast, predominance of high light requiring N_2 fixers such as *Anabaena circinalis* and *Aphanizomenon flos aquae* predictably is linked with stable water columns, depletion of dissolved inorganic N and high temperature. Both species are more common in relatively deep lakes, but can bloom in shallow lakes during calm summer periods. For example, Havens et al. (1998) observed a shift from *Oscillatoria* dominance under mixed / low irradiance conditions, to *Anabaena*, *Aphanizomenon* and *Microcystis* dominance during periods of water column stability and increased underwater irradiance. Phlips et al. (1997) documented that changes in biomass of N_2 fixing cyanobacteria and density of heterocysts were strongly coupled with depletion of dissolved inorganic N, N–limitation in bioassays, relatively high irradiance, and high N fixation rates. Thus, we can generally predict that if an ecosystem is enriched with P, it is likely to have cyanobacteria dominance. If it is a shallow mixed system, the z_{eu}/z_{mix} ratio is critical to determining whether *Oscillatoria* (low ratio) or other cyanobacteria will be dominant. An interesting exception to this pattern is *Cylindrospermopsis raciborskii*, which can develop blooms in shallow lakes under relatively high or low irradiance conditions, losing its terminal heterocysts in that later situation and functioning much like *Oscillatoria* (Phlips, personal communication).

Despite this knowledge, predicting the onset of a bloom at the daily or weekly timescale is very challenging because of the importance of stochastic variables such as wind and rain, which affect water column stability and underwater irradiance. For example, in Lake Mendota, Wisconsin, Soranno (1997) documented short–lived surface blooms throughout the summer and fall, and linked them with periods when there was a combination of low wind velocity, absence of rainfall and higher than average solar radiation. Surface blooms collapsed when wind velocities increased or cloudy weather or rainfall occurred. Yet even with this retrospective information it was not possible to accurately forecast the onset of blooms. Research must continue to focus on defining the underlying conditions that allow blooms to occur, to determine if controllable variables including P and N loading rates, N:P loading ratios, flushing rate, etc. can be manipulated to reduce the overall risk of blooms, even when atmospheric conditions are favorable for their occurrence. An over–arching premise must also be to determine whether occurrence of a particular bloom is natural or related to human activities.

Predicting toxin levels produced by cyanobacteria is even less certain than predicting cyanobacterial bloom occurrence. Toxin production is linked with dominance by particular taxa of cyanobacteria, including species of *Microcystis*, *Anabaena*, *Aphanizomenon*, *Oscillatoria*, *Nostoc*, and *Aphanocapsa*. However, factors controlling the amount of toxin produced during a bloom are not well understood. Studies have suggested links between toxin concentrations and ratios of particulate to dissolved nutrients (Oh et al. 2001), concentrations of soluble P (Jacoby et al. 2000), TP (Rapala et al. 1997), TN and irradiance (Rolland et al. 2005). It is not possible to say whether these relationships are causal or coincidental. This is an area where additional research is required.

It is important to recognize that conditions associated with blooms and toxins may lead to reduced competition or predation on the taxa forming the blooms. This positive feedback stimulates further development of the bloom until environmental conditions become unfavorable (e.g., multiple cloudy days, intense rainfall, dramatically increased flushing rate, or input of turbid or stained water) and the bloom collapses. Feedback loops add further complexity to prediction of bloom dynamics and ecological effects.

Biological effects of frequent or persistent cyanobacteria blooms

A simple conceptual model (Fig. 1) summarizes ecological effects of cyanobacteria blooms and their potential adverse impacts. When cyanobacterial blooms occur, irradiance is reduced in the water column, reducing the growth of producers that cannot maintain a position near the surface of the water, including epiphyton, benthic algae and rooted vascular plants. Thus, lakes with very dense blooms, especially if they are frequent or long–lasting, may not support large populations of other producers. In shallow eutrophic lakes, research has shown that the transition from plant to phytoplankton dominance can occur rapidly (Scheffer et al. 1993). Increased nutrient loading results in rather small increases in phytoplankton biomass when plants and periphyton are present, but when a critical turbidity is reached where net plant growth is negative, plants are replaced by phytoplankton. When this occurs, there is a rapid increase in the level of turbidity (phytoplankton biomass) with little or no further increase in nutrient loading. Furthermore, when that turbid state develops, the nutrient load required to bring plants back into dominance is lower than that at which they formerly occurred. This is a critical concept for management of cyanobacterial blooms and water quality in shallow eutrophic lakes. As illustrated by Scheffer et al. (1997), these alternative states also occur in shallow lakes that switch between *Oscillatoria* dominance and dominance by other algae – i.e., the phenomenon is not restricted to plant-algae switches. There remains some uncertainty about whether increased nutrient concentrations alone can bring about these changes, or whether they must be proximally driven by some additional forcing function, such as an intense wind storm uprooting plants or increasing water turbidity.

During intense blooms photosynthetic activity depletes free CO_2 from lake water and pH is driven up. Some have argued that this favors dominance of cyanobacteria, which for the most part are superior competitors when CO_2 is scarce. Low CO_2 also may stimulate formation of surface scums and extreme dominance by cyanobacteria taxa that can move to the air–water interface where CO_2 is most available, shading other algae in the process (Paerl and Ustach 1982). There is evidence that high pH during intense cyanobacteria blooms may be toxic to certain species of fish (Kann and Smith 1999), although this presumably might occur with blooms of any kind of phytoplankton (bacterial or algal) or in dense beds of plants. Oxygen depletion that occurs in the water during bloom senescence also can have biological impacts, the most visible being fish kills. There also are observations of adverse impacts of high levels of ammonia during bloom senescence. For example, during collapse of a dense *Anabaena circinalis* bloom in Lake Okeechobee, Florida, low oxygen levels and ammonia were considered the cause of mortality for snails and other macro–invertebrates (Jones 1987).

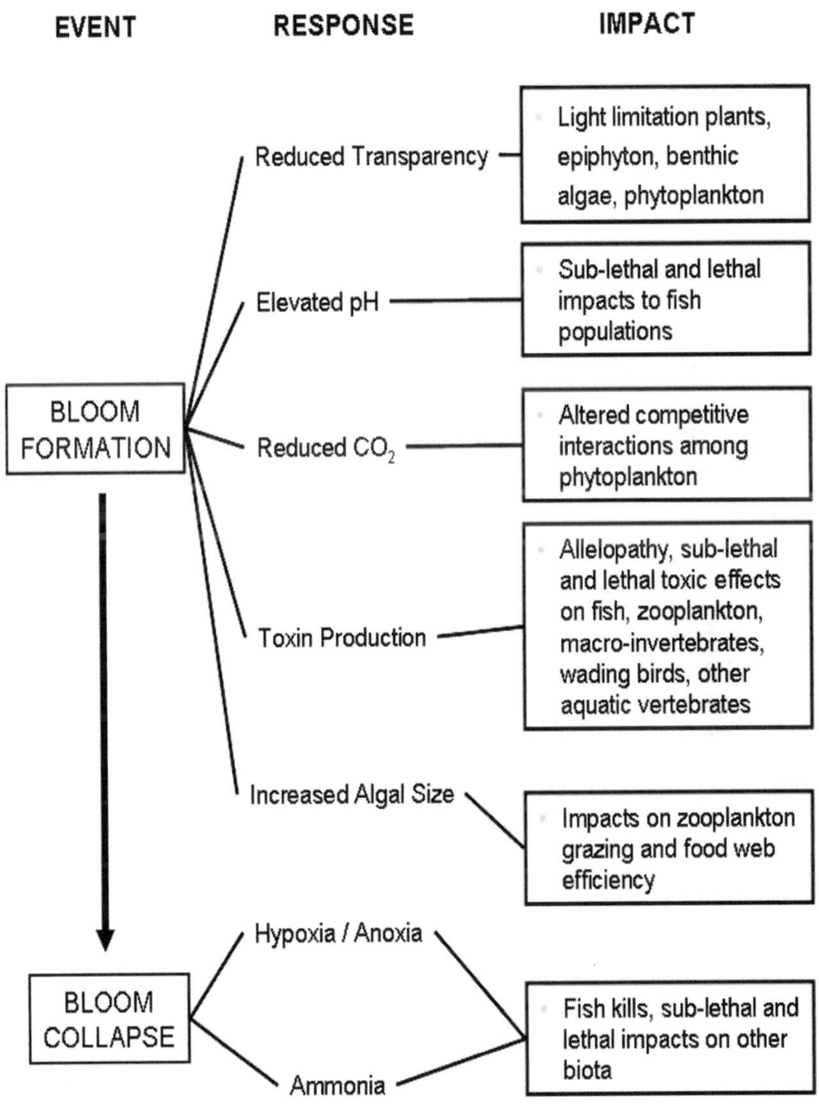

Fig. 1. A summary of ecological responses and impacts associated with blooms of cyanobacteria in lakes, rivers, and estuaries.

Toxin production by certain cyanobacteria (e.g., *Anabaena circinalis*, *Aphanizomenon flos–aquae*, *Cylindrospermopsis raciborskii*, *Microcystis aeruginosa*) may lead to a wide array of biological impacts, including: allelopathic effects on other phytoplankton (Suikkanen et al. 2004); suppression of zooplankton grazing, leading to reduced growth and reproductive rates and changes in dominance (Gilbert 1990, Ferrao–Filho et al. 2000, Ghadouani et al. 2003); hepatotoxic effects on fish (Andersen et al. 1993); and accumulation of toxins in tissues of invertebrates (Liras et al. 1998, Lehtiniemi et al. 2002) and fish (Magalhaes 2001). Toxins affected survival, growth and fecundity of snails in a laboratory exposure at typically observed concentrations (Gerard and Poullain 2005). Accumulation of toxins in tissues of freshwater clams has been suggested as a route of toxicity for muskrat (*Ondatra zibethicus*) and their predators (Prepas et al. 1997). There also are reports of impacts to waterfowl, such as the mass mortality of ducks that coincided with a *Microcystis* bloom in a Japanese lake (Matsunaga et al. 1999). These papers represent just a small percentage of the literature on potential effects of cyanobacterial toxins. Most studies are focused on single species and particular modes of exposure (e.g., feeding). At this time it is not possible to take this information and deduce with a high level of certainty what the actual impacts of toxins are on aquatic communities. Community–level responses depend on toxin concentration, frequency and duration of exposure, the combined outcome of direct and indirect effects, and possible synergistic effects with other natural and anthropogenic stressors. This represents a critical area for future research.

In addition to potential toxic effects, cyanobacteria blooms may affect grazing zooplankton by mechanical interference with the filtration apparatus (Gliwicz and Lampert 1990). It has been suggested that the high C:P ratios that occur during blooms may lead to growth limitation of zooplankton taxa like *Daphnia*, which have a high P requirement (Hessen et al. 2005). It is of interest that large *Daphnia*, which are generally considered the most effective grazers of algae and the taxa responsible for such things as the spring clear water phases in eutrophic lakes, are most sensitive to chemical stressors, including cyanobacterial toxins (Fulton 1988) and most sensitive to mechanical interference. As a result of this differential sensitivity, smaller zooplankton (e.g., *Bosmina* and rotifers) become increasingly dominant as lakes progress from mesotrophic to eutrophic. At the same time, the average size of phytoplankton increases. This convergence of zooplankton and phytoplankton size leads to an energetic bottleneck in the grazing food chain that restricts C and energy flow to higher trophic levels (Havens and East 1997). Microbial pathways become relatively more important and overall food web efficiency is reduced in eutrophic lakes with cyanobacterial blooms (Gliwicz 1969, Hillbricht–Ilkowska

1977). To a certain extent, the loss of *Daphnia* may be due to increased predation, because the biomass of planktivorous and omnivorous fish increases with eutrophication (Jeppesen et al. 2000).

Given these biological changes, it is important to consider what effects, if any, frequent or persistent cyanobacterial blooms have on biomass and taxonomic structure of fish, in particular the commercially and recreationally important species. The answer may depend on Latitude. In temperate and boreal regions, where piscivores (salmonids) require a cold water refuge during summer, eutrophication may eliminate those fish if the hypolimnion becomes anoxic (Colby et al. 1972). Temperate eutrophic lakes with dense cyanobacteria blooms in summer also may experience winterkill of fish under the ice cover in winter (Lee et al. 1991). In contrast, high densities of piscivores, such as largemouth bass (*Micropterus salmoides*), persist in highly eutrophic Florida, USA lakes, because fish in the subtropics have no cold water requirements and the winterkill situation cannot occur (Bachmann et al. 1996). A key factor determining success of bass and certain other sport fish is habitat structure – i.e., presence of a diverse assemblage of aquatic plants (Havens et al. 2005). As noted above, this habitat may be lost under highly eutrophic conditions if the lake is shallow and switches to a turbid state. Thus, eutrophication and dense algal blooms could lead to losses of economically important fish taxa in both temperate regions and the subtropics, although the underlying mechanisms differ. Additional research is needed to determine what additional effects, if any, the above mentioned declines in food web efficiency have on productivity of certain fish taxa. There is considerable evidence that fertilization can improve productivity of sport fish in ultra–oligotrophic lakes by increasing efficiency of plankton food webs, but the link between food web structure and fish productivity at the opposite end of the trophic spectrum is not as well documented, especially relative to other effects like hypolimnetic anoxia and loss of plant habitat.

Effects on sediments, nutrient cycling and internal loading

Physical, chemical and biological processes occurring in the water column of lakes, rivers and estuaries have large influences on the sediments and their associated biota and nutrient cycles (Palmer et al. 2000). Systems with frequent and/or prolonged blooms of cyanobacteria display benthic responses. Because the particle size, nutrient stoichiometry and other properties (e.g., presence of toxins) of settling seston is altered when cyanobacteria become dominant, compared with diatoms or other algae, this affects the benthic chemistry and the biota that use that settling organic material as substratum or food. The enhanced organic export to sediments that occurs in eutrophic lakes leads to sediment anoxia, and this alters the taxonomic structure of benthic invertebrates and reduces the extent to which Fe binds to PO_4 at the sediment water interface. This may lead to increased diffusive internal P loading. Cyanobacterial blooms themselves may directly enhance internal P loading to surface waters if vertically migrating algae pick up P near the sediment surface and then move up into the epilimnion. Estimates of P loading by this process range from 2.0 to 3.6 mg P m^{-2} d^{-1} (Barbiero and Kann 1994, Head et al. 1999). Where such loading occurs at a high rate, there may be long–term consequences for lake rehabilitation. Head et al. (1999) note that "following reduction of external P loading, utilization of such internal P sources may delay expected reductions of bloom–forming cyanobacterial communities, and consequently delay improvements in other aspects of water quality." Research is needed to determine the extent of that delay in recovery time in lakes with a long history of high external loads. Modeling of response time in most TMDL related studies does not take into consideration this potentially important stabilizing feedback loop.

Research needs and data gaps

Although the understanding of toxic cyanobacteria blooms has greatly increased in the last decade, there remain some major areas of uncertainty where additional research is required if scientists are to provide sound guidance to water resource managers. The following are some key questions that should be addressed with future research.

1. What physical, chemical and biological factors determine the level of toxins associated with a cyanobacteria bloom?
2. What are the realized ecological impacts of toxic cyanobacterial blooms, when one takes into consideration typically occurring levels of toxins, potential for both direct and indirect effects, and the possibility of synergistic effects of natural stressors?
3. What impacts, if any, do changes in food web function have on fish in nutrient–enriched lakes, rivers and estuaries that become dominated by cyanobacteria?

4. To what extent does dominance by vertically migrating cyanobacteria lengthen the response time of lakes, rivers or estuaries to external nutrient load reduction? Can this biological effect be incorporated into predictive models used to set a TMDL?

Conclusion and Summary

Cyanobacteria become increasingly dominant as concentrations of TP and TN increase during eutrophication of lakes, rivers and estuaries. Temporal dynamics of cyanobacteria blooms are variable – in some systems persistent blooms occur in summer to fall, whereas in other systems blooms are more sporadic. Cyanobacteria blooms have a wide range of possible biological impacts including potential toxic effects on other algae, invertebrates and fish, impacts to plants and benthic algae due to shading, and impacts to food web function as large inedible algae produce a bottleneck to C and energy flow in the plankton food web. In lakes with dense blooms of cyanobacteria, accumulation of organic material in lake sediments and increased bacterial activity also may lead to anoxic conditions that alter the structure of benthic macro–invertebrates. Diffusive internal P loading may increase, and hypolimnetic anoxia may lead to a loss of piscivorous fish that require a summer cold water refuge in temperate lakes. Ecosystem changes associated with frequent blooms may result in delayed response of lakes, rivers and estuaries to external nutrient load reduction. Despite numerous case studies and a vast literature on species–specific responses, community level effects of cyanobacterial blooms are not well understood – in particular the realized impacts of toxins and changes in food web structure/function. These areas require additional research given the prevalence of toxic blooms in the nation's lakes, rivers and coastal waters – systems that provide a wide range of valued ecosystem services.

Acknowledgements

Comments provided by Mike Coveney, Ed Phlips and John Burns were helpful to improving an earlier version of this manuscript.

References

Anderson RJ HA, Luu DZX, Chen CFB, Holmes ML, Kent M, Le Blanc FJR, Taylor and DE Williams (1993) Chemical and biological evidence links microcystins to salmon nepten liver disease. Toxicon 31: 1315–1323

Auer B, H Arndt (2004) Comparison of pelagic food webs in lakes along a trophic gradient and wth seasonal aspects: influence of resources and predation. Journal of Plankton Research 26: 697–709

Bachmann RW, BL, Jones DD, Fox M, Hoyer LA, Bull and DE Canfield Jr (1996) Relations between trophic state indicators and fish in Florida (USA) lakes. Canadian Journal of Fisheries and Aquatic Sciences 53: 842–855

Barbiero RP, Kann J (1994) The importance of benthic recruitment to the population development of Aphanizomenon flos–aquae and internal loading in a shallow lake. Journal of Plankton Research 16: 1561–1588

Berger C (1989) In situ primary productivity, biomass and light regime in the Wolderwijd, the most stable Oscillatoria agardhii lake in the Netherlands. Hydrobiologia 185: 233–244

Colby PJ, GR, Spangler D, Hurley and AM McCombie (1972) Effects of eutrophication on salmonid communities in oligotrophic lakes. Journal of the Fisheries Research Board of Canada 29: 975–983

Dokulil MT, Teubner K (2000) Cyanobacterial dominance in lakes. Hydrobiologia 438: 1–12

Downing JA, Watson SB, McCauley E (2001) Predicting cyanobacteria dominance in lakes. Canadian Journal of Fisheries and Aquatic Sciences 58: 1905–1908

Ernst B, Hitzfeld B, Dietrich D (2001) Presence of Planktothrix sp and cyanobacterial toxins in Lake Ammersee, Germany and their impact on whitefish (Coregonus lavaretus). Environmental Toxicology 16: 483–488

Ferrao–Filho A, Azevedo S, DeMott WR (2000) Effects of toxic and non–toxic cyanobacteria on the life history of tropical an temperate cladocerans. Freshwater Biology 45: 1–19

Fulton RS (1988) Resistance to blue–green algal toxins by Bosmina longirostris. Journal of Plankton Research 10: 771–778

Gerard C, Poullain V (2005) Variation in the response of the invasive species Potamopyrgus antipodarum (Smith) to natural (cyanobacterial toxin) and anthropogenic (herbicide atrizine) stressors. Environmental Pollution 138: 28–33

Ghadouani A, Pinel–Alloul B, Prepas EE (2003) Effects of experimentally induced cyanobacterial blooms on crustacean zooplankton communities. Freshwater Biology 48: 363–381

Gilbert JJ (1990) Differential effects of Anabaena affinis on cladocerans and rotifers: mechanisms and implications. Ecology 71: 1727–1740

Gliwicz ZM (1969) Studies on the feeding of pelagic zooplankton in lakes with varying trophy. Ekologia Polska 17: 663–708

Gliwicz ZM, Lampert W (1990) Food thresholds in Daphnia species in the absence and presence of blue–green filaments. Ecology 71: 691–702

Havens KE, East, TL (1997) Carbon dynamics in the grazing food chain of a subtropical lake. Journal of Plankton Research 19: 1687–1711

Havens KE, Fox D, Gornak S, Hanlon C (2005) Aquatic vegetation and largemouth bass population responses to water level variations in Lake Okeechobee, Florida (USA). Hydrobiologia 539: 225–237

Havens KE, Phlips EJ, Cichra MF, Li B–L (1998) Light availability as a possible regulator of cyanobacteria species composition in a shallow subtropical lake. Freshwater Biology 39:547–556

Havens KE, Walker WW (2002) Development of a total phosphorus concentration goal in the TMDL process for Lake Okeechobee, Florida (USA). Lake and Reservoir Management 18: 227–238

Head RM, Jones RI, Bailey–Watts AE (1999) Vertical movements by planktonic cyanobacteria and the translocation of phosphorus: implications for lake restoration. Aquatic Conservation Marine and Freshwater Ecosystems 9: 111–120

Hessen DO, Van Donk E, Gulati R (2005) Seasonal seston stoichiometry: effects on zooplankton in cyanobacteria–dominated lakes. Journal of Plankton Research 27: 449–460

Hillbricht–Ilkowska A (1977) Trophic relations and energy flow in pelagic plankton. Polish Ecological Studies 3: 3–98

Jacoby JM, Collier DC, Welch EB, Hardy J, Crayton M (2000) Environmental factors associated with a toxic bloom of Microcystis aeruginosa. Canadian Journal of Fisheries and Aquatic Sciences 57: 231–240

Jensen JP, Jeppesen E, Olrik K, Kristensen P (1994) Impact of nutrients and physical factors on the shift from cyanobacterial to chlorophyte dominance in shallow Danish lakes. Canadian Journal of Fisheries and Aquatic Sciences 51: 1692–1699

Jeppesen E, Lauridsen TL, Mitchell SF, Christoffersen K, Burns CW (2000) Trophic structure in the pelagial of 25 shallow New Zealand lakes: changes along nutrient and fish gradients. Journal of Plankton Research 22: 951–968

Jones BI (1987) Lake Okeechobee eutrophication research and management. Aquatics 9: 21–26

Kann J, Smith VH (1999) Estimating the probability of exceeding elevated pH values critical to fish populations in a hypereutrophic lake. Canadian Journal of Fisheries and Aquatic Sciences 56: 2262–2270

Lee GF, Jones RA (1991) Effects of eutrophication on fisheries. Reviews in Aquatic Sciences 5: 287–305

Lehtiniemi M, Engstrom–Ost J, Karjalainen M, Kozlowsky–Suzuki B, Viitasalo M (2002) Fate of cyanobacterial toxins in the pelagic food web: transfer to copepods or to fecal pellets? Marine Ecology Progress Series 241: 13–21

Liras V, Lindberg N, Nystom P, Annadotter H, Lawton LA, Graf B (1998) Can ingested cyanobacteria be harmful to the signal crayfish (Pacifastacus leniusculus)? Freshwater Biology 39: 233–242

Magalhaes VF, Soares RM, Azevedo SMFO (2001) Microcystin contamination in fish from Jacarepagua Lagoon: ecological implication and human health risk. Toxicon 39: 1077–1085

Matsunga H, Harada KI, Senma M, Ito Y, Yasuda N, Ushida S, Kimura Y (1999) Possible cause of unnatural mass death of wild birds in a pond in Nishinomiya, Japan: sudden appearance of toxic cyanobacteria. Natural Toxins 7: 81–84

Murrell MC, Lores EM (2004) Phytoplankton and zooplankton seasonal dynamics in a subtropical estuary: importance of cyanobacteria. Journal of Plankton Research 26: 371–382

Oh HM, Lee SJ, Kim JH, Kim HS, Yoon BD (2001) Seasonal variation and indirect monitoring of microcystin concentrations in Daechung Reservoir, Korea. Applied and Environmental Microbiology 67: 1484–1489

Paerl HW (1988) Nuisance phytoplankton blooms in coastal, estuarine and inland waters. Limnology and Oceanography 33: 823–847

Paerl HW, Ustach JF (1982) Blue–green algal scums: an explanation for their occurrence during freshwater blooms. Limnology and Oceanography 27: 212–217

Palmer MA, Covich AP, Lake S, Biro P, Brooks JJ, Cole J, Dahm C, Gibert J, Goedkoop W, Marten K, Verhoeven J, Van de Bund WJ (2000) Linkages between aquatic sediment biota and life above sediments as potential drivers of biodiversity and ecological processes. BioScience 50: 1062–1075

Phlips EJ, Cichra M, Havens KE, Hanlon C, Badylak S, Rueter B, Randall M, Hansen P (1997) Relationships between phytoplankton dynamics and the availability of light and nutrients in a shallow subtropical lake. Journal of Plankton Research 19: 319–342

Prepas EE, Kotak BG, Campbell LM, Evans JC, Hrudey SE, Holmes CFB (1997) Accumulation and elimination of cyanobacterial hepatotoxins by the freshwater clam Anodonta grandis simpsoniana. Canadian Journal of Fisheries and Aquatic Sciences 54: 41–46

Rapala J, Sivonen K, Lyra C, Niemela SI (1997) Variation of microcystins, cyanobacterial hepatotoxins, in Anabaena spp as a function of growth stimuli. Applied and Environmental Microbiology 63: 2206–2212

Reynolds CS (1984) The Ecology of Freshwater Phytoplankton. Cambridge University Press, UK, 384 pp

Reynolds CS (1999) Non–determinism to probability, or N:P in the community ecology of phytoplankton. Archiv fur Hydrobiologie 146: 23–35

Reynolds CS, Oliver RL, Walsby AE (1987) Cyanobacterial dominance: the role of buoyancy regulation in dynamic lake environments. New Zealand Journal of Marine and Freshwater Research 21: 379–390

Rolland A, Bird DF, Giani A (2005) Seasonal changes in composition of the cyanobacterial community and the occurrence of hepatotoxic blooms in the eastern townships, Quebec, Canada. Journal of Plankton Research 27: 683–694

Scheffer M, Hosper SH, Meijer ML, Moss B, Jeppesen E (1993) Alternative equilibria in shallow lakes. Trends in Ecology and Evolution 8: 275–279

Scheffer M, Rinaldi S, Gragnani A, Mur LR, van Nes EH (1997) On the dominance of filamentous cyanobacteria in shallow turbid lakes. Ecology 78: 272–282

Smith VH (1983) Low nitrogen to phosphorus ratios favor dominance by blue–green algae in lake phytoplankton. Science 221: 669–671

Smith VH, Bennet SJ (1999) Nitrogen:phosphorus supply ratios and phytoplankton community structure in lakes. Archiv fur Hydrobiologie 146: 37–53

Soranno PA (1997) Factors affecting the timing of surface scums and epilimnetic blooms of blue–green algae in a eutrophic lake. Canadian Journal of Fisheries and Aquatic Sciences 54: 1965–1975

Suikkanen S, Fistarol GOl, Graneli E (2004) Allelopathic effects of the Baltic cyanobacteria Nodularia spumigena, Aphanizomenon flos–aquae and Anabaena lemmermannii on algal monocultures. Journal of Experimental Marine Biology and Ecology 308: 85–101

Tsujimura S, Okubo T (2003) Development of *Anabaena* blooms in a small reservoir with dense sediment akinete population, with special reference to temperature and irradiance. Journal of Plankton Research 25: 1059–1067

Zhang Y, Prepas E (1996) Regulation of the dominance of planktonic diatoms and cyanobacteria in four eutrophic hardwater lakes by nutrients, water column stability and temperature. Canadian Journal of Fisheries and Aquatic Sciences 53: 621–633

Zohary T, Pais–Madeira AM, Robarts RD, Hambright D (1995) Cyanobacteria–phytoplankton dynamics of a hypereutrophic African lake. Water Science Technology 32: 103–104

Chapter 34: Ecosystem Effects Workgroup Poster Abstracts

Local adaptation of *Daphnia pulicaria* to toxic cyanobacteria

Sarnelle O[1] and Wilson AE[2]

[1]Department of Fisheries and Wildlife, Michigan State University, East Lansing, Michigan 48824-1222; [2]School of Biology, Georgia Institute of Technology, Atlanta, Georgia 30332-0230

Introduction

A well-established tenet in limnology holds that the taxonomic composition of summer phytoplankton assemblages shifts with phosphorus enrichment toward greater dominance by cyanobacteria. One important consequence of this shift toward cyanobacteria with eutrophication is that summer phytoplankton assemblages in eutrophic lakes are relatively resistant to zooplankton grazing. However, one recent study showed that zooplankton from a lake in Europe adapted to tolerate bloom-forming cyanobacteria in their diet after several decades of cultural eutrophication. This project aimed to determine if adaptation by grazers to toxic prey occurs across lakes that vary in nutrient concentration.

Hypothesis

Daphnia pulicaria clones isolated from high-nutrient, cyanobacteria-abundant lakes are more tolerant of toxic *Microcystis aeruginosa* than clones isolated from low-nutrient lakes.

Methods

Twenty-two *D. pulicaria* clones were isolated from six lakes in southern Michigan. The six lakes were grouped into two categories based on the midpoint of ranges of summer total phosphorus (TP) concentration, low TP (9 - 13 μg L^{-1}) and high TP (31 - 235 μg L^{-1}). Two juvenile growth experiments were conducted and data from the two experiments were pooled for all analyses. Neonates (<24 hours old) of each clone were transferred individually into 100 ml glass beakers filled with 80 ml of glass-fiber filtered lake water, and a random subset of neonates was transferred individually to a dried and tared weighing tin for initial mass estimates (W_i). Neonates were fed either *Ankistrodesmus falcatus*, a nutritious green alga, or a single-celled toxic strain of *Microcystis aeruginosa* (microcystin quota: 36 μg mg^{-1} C) at growth-saturating concentrations (1.5 mg C L^{-1}) and transferred to new beakers with fresh medium and food daily. On day 3 (W_f) of the experiments, each animal was transferred individually to a tared weighing tin, dried, and weighed. Instantaneous somatic growth rate (g, d^{-1}) was calculated for each beaker as: $\{[\ln(W_f) - \ln(W_i)]/3\}$. We calculated a relative index of growth inhibition by *Microcystis* for each clone as: $(g_a - g_m)/g_a$, where g_a is growth rate on *Ankistrodesmus* and g_m is growth rate on *Microcystis*. Growth responses were averaged across clones for each lake, and differences between low-TP and high-TP lakes assessed via two-tailed *t*-tests, with lakes as replicates.

Results

Daphnia clones generally grew well on a diet of *Ankistrodesmus*, and there was no significant difference in growth on this diet between lake categories ($p > 0.40$), despite substantial overall variation in growth rate. As expected, all clones grew poorly on the *Microcystis* diet. More importantly, *D. pulicaria* from high-TP lakes grew significantly better, on average, than *D. pulicaria* from low-TP lakes on the *Microcystis* diet ($p < 0.02$). On average, *D. pulicaria* from low-TP lakes lost weight when fed *Microcystis* (growth < 0, *t*-test, $p < 0.04$, $n = 3$), while *D. pulicaria* from high-TP lakes did not (growth <0, *t*-test, $p > 0.30$, $n = 3$).

Conclusions

We quantified within-species variation in the tolerance of the large lake-dwelling daphnid, *D. pulicaria*, to toxic cyanobacteria in the diet. Juvenile growth rates on diets consisting of 100% *Ankistrodesmus* or 100% toxic

Microcystis were compared for *D. pulicaria* clones isolated from lakes expected to have low and high levels of bloom-forming cyanobacteria during summer. Growth rates of clones isolated from high-nutrient lakes were higher, and showed less relative inhibition, on the cyanobacterial diet, compared to clones isolated from low-nutrient lakes. Our results suggest that *D. pulicaria* populations exposed to high cyanobacterial levels over long periods of time can adapt to being more tolerant of toxic cyanobacteria in the diet.

Cytotoxicity of microcystin-LR to primary cultures of channel catfish hepatocytes and to the channel catfish ovary cell line

Schneider JE Jr, Beck BH, Terhune JS, Grizzle JM

Department of Fisheries and Allied Aquacultures, 203 Swingle Hall, Auburn University, Auburn, Alabama 36849

Introduction

We observed responses of channel catfish (*Ictalurus punctatus*) hepatocytes in primary culture to microcystin-LR (MC-LR) in order to develop biomarkers of effect for cyanobacterial hepatoxins. Livers in live animals and freshly isolated liver cells (hepatocytes) are highly sensitive to microcystins because the toxin is concentrated into hepatocytes by the bile acid transporter system. Work with rat hepatocytes has shown that microcystin MC-LR is taken into the cell cytoplasm where it inhibits protein phosphatases leading to hyperphosphorylated cytoskeleton proteins leading to collapse of the cell structure, followed later by cell death. Although fish have not appeared to be as sensitive to the microcystin class of hepatotoxins as are rats, the following data indicate that channel catfish hepatocytes share some MC-LR toxic mechanisms with rat hepatocytes.

Hypothesis

We propose that (1) primary cultures of channel catfish hepatocytes respond to the presence of MC-LR with the formation of abnormal morphology before loss of viability; and (2) there is a significantly greater sensitivity of hepatocytes to MC-LR in comparison with the channel catfish ovary (CCO) cell line, which would not be expected to possess the bile acid transporter.

Methods

Hepatocyte culture Hepatocytes from channel catfish (75-120 g) maintained in aquaria at 20°C were prepared essentially as described by Seddon and Prosser (1999) except that cells were washed by centrifugation in phosphate buffered saline and resuspended in serum-free EMEM at a concentration of 1 million cells/mL after a selection for viable cells by centrifugation though 30% Percoll. The hepatocytes were incubated in Falcon Primaria 96-well plates, 100,000 cells per well in serum-free EMEM medium buffered with 25 mM HEPES in the dark at 20-22°C. The medium was not changed throughout an experiment.

Channel catfish ovary (CCO) cell culture. CCO cells obtained from the American Type Culture Collection were seeded into 96-well plates in EMEM supplemented with 10% fetal calf serum at a confluency of 20% and exposed to doses of MC-LR for 96 hours at 30°C.

XTT cytotoxicity assay. XTT, a tetrazolium salt that is converted to its formazan derivative by metabolic reduction due to cellular dehydrogenases, was used to measure the viability of hepatocytes. The assay was performed according to the manufacturer's instructions except for the extended incubation times at 20-22°C. Formazan was measured at 450 nm wavelength on a Synergy HT microplate reader. Hepatocytes cultured in 96-well plates were observed and photographed with the aid of a bright-field, inverted microscope.

Results

Viability measured as XTT reduction in primary cell cultures of channel catfish hepatocytes following exposure to microcystin for 5 days resulted in approximately 30% less XTT reduction for each 1 μg microcystin/mL compared with control cells without MC-LR. In addition, a dose response was evident for the concentrations of microcystin used in this experiment. After a 3-day exposure to 10 μg/mL microcystin, channel catfish hepatocytes in cell culture had rounded. This morphological change was observed before a change in XTT reduction was detectable. CCO cells required approximately 50 μg/mL MC-LR to attain a similar amount of XTT reduction as did 1 μg/mL in the hepatocyte culture.

Conclusion

Exposure of channel catfish hepatocytes in primary culture to microcystin-LR elicits a similar pattern of toxicity to that reported for rat hepatocytes, suggesting that there are common toxic mechanisms between the two systems upon which biomarkers of effect may be developed.

Mortality of bald eagles and american coots in southeastern reservoirs linked to novel epiphytic cyanobacterial colonies on invasive aquatic plants

Wilde SB,[1] Williams SK,[1,2] Murphy T,[3] Hope CP,[3] Wiley F,[4] Smith R,[4] Birrenkott A,[4] Bowerman W,[4] Lewitus AJ[1]

[1]Belle W. Baruch Institute for Marine Biology and Coastal Research, University of South Carolina, Georgetown, SC, USA; [2]University of College of Charleston, Masters of Environmental Studies, Charleston, SC, USA; [3]South Carolina Department of Natural Resources, Marine Resources Research Institute, Charleston, SC, USA; [4]Clemson University Department of Forestry and Natural Resources, Clemson University, Clemson, SC

Introduction

Invasive species compromise habitat and degrade the quality of the environment for many species and can even accelerate the decline of endangered species in these sites. Our field surveys and feeding studies implicate invasive aquatic plants and an associated epiphytic cyanobacteria species in an emerging avian disease to herbivorous waterfowl and their avian predators. The disease, Avian Vacuolar Myelinopathy (AVM), was first noted in 1994 and has been the cause of death for at least 100 American bald eagles (*Haliaeetus leucocephalus*) and 1000's of American coots (*Fulica americana*) and other waterfowl. The disease causes neurological dysfunction in the birds prior to death but no known neurotoxins or disease agents have been detected at the sites or within the birds.

Hypotheses

We propose that the agent for the disease is a neurotoxin produced by a novel cyanobacterial epiphyte of the order Stigonematales growing on invasive aquatic plants. Our working hypothesis is that the neurotoxin travels through the food chain from hydrilla and other aquatic plants to waterfowl who are consumed by predators including the American Bald Eagle and Great Horned Owls.

Methods

This research incorporated field site monitoring, laboratory experiments, and field trials. An extensive survey of documented AVM sites was conducted from 2001–2005 to monitor the field occurrence of the disease, the abundance of the invasive plant species and the density of the target cyanobacterial species. Laboratory and field trials were conducted to test hydrilla with an abundance of the targeted cyanobacteria would cause AVM lesions in the experimental mallards and triploid grass carp.

Results

In reservoirs where eagles and waterfowl deaths have been most prevalent, the novel Stigonematales species was dominant; and diversity of other groups (primarily diatoms and green algae) was lower. In reservoirs where bird deaths from AVM have not been diagnosed, epiphytic assemblages were diverse and abundant, but the suspect Stigonematales species was either rare or not present. Mallard and grass carp fed hydrilla dominated by the novel Stigonematalean species developed AVM lesions in the laboratory tanks and in the field trial. Those receiving the control hydrilla which contained abundant epiphytic algae, but none of the target Stigonematales species did not develop AVM lesions.

Conclusion

The agent responsible for AVM is associated with the hydrilla and other aquatic plants in the sites where the disease occurs. The most probable theory is that the epiphytic algae growing on these plants are producing a neurotoxin that causes brain lesions and death in the birds who consume it. The invasive potential of these exotic plants make it likely that the disease will expand to new sites and the impact of this disease on waterfowl and eagles will continue to increase.

Investigation of a novel epiphytic cyanobacterium associated with reservoirs affected by avian vacuolar myelinopathy

Williams SK,[1,2] Wilde SB,[1] Murphy TM,[3] Hope CP,[3] Birrenkott A,[4] Lewitus AJ[1,3]

[1]Belle W. Baruch Institute for Marine Biology and Coastal Research, University of South Carolina, Georgetown, SC, USA; [2]University of College of Charleston, Masters of Environmental Studies, Charleston, SC, USA; [3]South Carolina Department of Natural Resources, Marine Resources Research Institute, Charleston, SC, USA; [4]Clemson University Department of Forestry and Natural Resources, Clemson University, Clemson, SC

Introduction

The recovery and conservation of bald eagles (*Haliaeetus leucocephalus*) has been contested due to a newly identified fatal bird disease: Avian Vacuolar Myelinopathy (AVM). Since the discovery of the disease in 1994, AVM has caused mortality in at least 100 bald eagles, thousands of American coots, and other various species of bird throughout the southeastern US. AVM has been found in AR, TX, NC, SC, and GA. The cause of the disease has yet to be identified.

A strong association has been observed between the occurrence of AVM, *Hydrilla verticillata* (hydrilla), and a novel potentially toxic, epiphytic cyanobacterium on hydrilla. The correlation has led to the hypothesis that this epiphyte is the source of the neurotoxin causing AVM. During 2001–2004, the Stigonematales species was present on the surface of hydrilla at every site where AVM had been diagnosed, but was absent or scarcely found in areas where AVM was not observed.

Hypotheses

It is hypothesized that the proposed toxin of the cyanobacteria is bioaccumulated through the food chain from waterfowl (e.g. coots) ingesting the Stigonematales species growing on the hydrilla. The goals of the study included; establishing a culture of the targeted Stigonematales species, expanding morphological descriptions of the species, determining gene sequence data from material collected in the field, and developing a Real Time–PCR assay specific to the cyanobacterium.

Methods

A monoculture of the cyanobacterium was established on BG–11 medium at 27°C. The 16S rRNA sequence identity was determined from environmental isolates using DGGE and then "ground–truthed" with culture isolates. The 16S rRNA sequence data were aligned with additional cyanobacteria sequences to determine designations for probe development, and to use in phylogenetic analysis. Real–Time PCR assays were developed specific to the Stigonematales species.

Results

The Stigonematales species has been cultured in order to aide in development of species identity, genetic research, feeding trials, and toxin analysis. 16S rRNA sequence data were aligned with additional cyanobacteria sequences to advance understanding of the species' phylogeny, and to lay groundwork for its formal description. Phylogeny data confirmed that the species is in section V, order Stigonematales. Phylogeny also inferred that the species is novel and most genetically similar to a *Stigonema* sp. Based on sequence variability, a Real Time–PCR assay has been developed for rapid, specific detection of the Stigonematales species from environmental samples.

Conclusion

The dominant epiphyte found on collected *Hydrilla* is an undescribed species of cyanobacterium in the order Stigonematales. The genetic probe and taxonomic information produced by this study will help test the hypothetical link between these cyanobacteria and AVM, and therefore help guide decisions on managing hydrilla and other invasive macrophytes in AVM–affected waters.

Chapter 35: Risk Assessment Workgroup Report

Co–chairs: Joyce Donohue and Jennifer Orme–Zavaleta

Work Group Members[1]: Michael Burch, Daniel Dietrich, Belinda Hawkins, Tony Lloyd, Wayne Munns, Jeffery Steevens, Dennis Steffensen, Dave Stone, and Peter Tango

Disclaimer: The opinions expressed in this paper are those of the authors and not necessarily those of the Organizations they represent.

Introduction

Risk assessment is a four–stage process used in evaluating the impact of contaminants on the well being of individuals, populations and/or the physical environment. As defined by the National Academy of Sciences (1983), the four components are as follows: hazard identification, dose–response assessment, exposure assessment and risk characterization.

The goal of a risk assessment is to utilize existing information coupled with site specific data to quantitatively characterize the potential risk of a stressor to an identified receptor(s). Quantitative, risk–based estimates of dose–response relationships integrated with exposure scenarios and information on environmental conditions often become the basis for regulatory measures or management policies to protect the population or physical environment from harm. The precision of the guideline value is impacted by the quantity and quality of scientific data available because uncertainty factors are applied in its derivation to compensate for deficiencies in the database. The more comprehensive the database, the lower the uncertainty in the risk assessment and the more precise the value generated.

Risk assessments are one tool used by risk managers when choosing between various options for protecting human health and the environment. They play a significant role in risk management decisions. However, the physical and societal environment is complex. It includes a multitude of

[1] See workgroup member affiliations in Invited Participants section.

receptors, each of which may be impacted by any risk management decision. Management decisions almost always involve considerations of a variety of risk factors, competing priorities, societal value systems, and resource limitations. In addition, the decision process may need to consider balancing risks.

The Risk Assessment Work Group was given the overall charge to identify the research needs for both cyanobacteria and their toxins. In order to provide context and focus to their deliberations, the work group addressed the following six charge questions:

- What data are available to derive health–based guideline values (TDI's, RfD's) for cyanobacterial harmful algal blooms (CHABs)?
- What research is needed to reduce uncertainty in health based guidelines?
- What research is needed to minimize the cost and maximize the benefits of various regulatory approaches?
- What are the exposure pathways for the receptors of concern?
- What are the ecosystem–services we want to protect?
- How can regulators best devise a framework for making risk management determinations that incorporates consideration of the characteristics of CHABs, the risk to human health and ecosystem sustainability, and the costs and benefits of CHABs detection and management?

The report that follows will address each of the stated charge questions in sequence culminating with a management framework that integrates concerns for human health protection with those for environmental ecosystems.

Regulatory Context

Cyanobacteria produce toxins that have adverse effects on the health of humans, domestic animals and wild life. These effects range from mild cases of dermatitis to death. Overgrowth of cyanobacteria in surface waters can produce unsightly conditions along the shoreline and in open waters making them unsuitable for recreation (e.g., swimming, fishing, boating). Affected surface waters that are the source for drinking water lead to concern that the toxins may gain access to public drinking water supplies. These are the situations that give rise to the need to consider possible regulatory controls for the cyanobacteria and their toxins under the US Safe Drinking Water Act (SDWA) and Clean Water Act (CWA) statutes. Offensive taste and odors associated with cyanobacteria can also make water unsuitable for drinking.

Contaminant Candidate List (CCL) and National Primary Drinking Water Regulations (NPDWRs)

The SDWA, as amended in 1996, required the US Environmental Protection Agency (U.S. EPA) to establish a list of contaminants to aid the Agency in regulatory priority setting for the drinking water program and to reconstitute that list every five years. EPA published the first Contaminant Candidate List (CCL) on March 2, 1998 (63 FR 10273, U.S. EPA, 1998). The second CCL was published as final February 24, 2005 (70 FR:9071, US EPA 2005). Cyanobacteria and their toxins were included on the first CCL and carried over to CCL2.

The SDWA requires EPA to make regulatory determinations for no fewer than five contaminants from the CCL list within three years of its publication. The criteria established by the SDWA for a positive regulatory determination are as follows:

1. The contaminant may have an adverse effect on the health of persons.
2. The contaminant is known to occur or there is a substantial likelihood that the contaminant will occur in public water systems with a frequency and at levels of public health concern.
3. In the sole judgment of the Administrator, regulation of such contaminant presents a meaningful opportunity for health risk reduction for persons served by public water systems.

Positive findings for all three criteria must be met in order to make a determination on whether to regulate. A decision not to regulate is considered a final Agency action and is subject to judicial review.

The inclusion of cyanobacteria and their toxins on the first and second CCL is one factor that fuels the need for research. As indicated by the decision criteria, regulatory determination for contaminants requires the EPA to evaluate the health impact of the contaminants and quantify the dose–response relationship through a formal health risk assessment process. Monitoring data from public water systems must also be available along with effective treatment technologies.

There are major data deficiencies and barriers that prevent the US EPA from making regulatory determinations for the cyanobacterial toxins at this time. At present there are insufficient health effects data for many of the cyanobacterial toxins (although from a European standpoint, it could be argued that there are sufficient data to warrant a precautionary approach in the absence of comprehensive data). Analytical methods with the sensitivity to detect many of the contaminants at concentrations of possible health concern and suitable for national monitoring of public water systems in the US have yet to be developed. Accordingly, these data gaps and others have been highlighted as research needs for the SDWA.

The data on occurrence in drinking water are gathered through Unregulated Contaminant Monitoring Rules (UCMR). The SDWA grants the US EPA the authority to require large (serving >10,000) National Pubic Drinking Water Systems (NPDWS) and a representative sample of small systems to monitor for no more than 30 unregulated contaminants over a one year period. Samples are collected quarterly for surface water systems and semiannually for ground water systems. The monitoring results are reported to the EPA in the National Contaminant Database. Methods development and inclusion of a contaminant in the UCMR are closely coordinated with the CCL. EPA can issue a new list of contaminants for UCMR monitoring every 5 years. Methods development problems have thus far prevented inclusion of cyanobacterial toxins in the UCMR.

In cases where EPA determines under the CCL program that a regulation is necessary, the regulation should be proposed within 24 months of the regulatory determination and finalized within eighteen months of the proposal. As required by the SDWA, a decision to regulate commits the EPA to publication of a Maximum Contaminant Level Goal (MCLG), Maximum Contaminant Level (MCL), and promulgation of a National Primary Drinking Water Regulation (NPDWR) for that contaminant. EPA can also determine that there is no need for a regulation when a contaminant fails to meet one of the statutory criteria.

In addition to health effects studies and analytical method development, data needs that underlie the development of the NPDWR include suitable treatment technologies for large and small systems and the economic data required for cost–benefit assessments. While there are technologies available for treatment, data gaps exist in both the treatment technologies and cost–benefit areas as they apply to cyanobacteria and their toxins.

Clean Water Act Requirements for Ambient Waters

The objective of the US Clean Water Act (CWA) is to restore, maintain and protect the chemical, physical and biological integrity of the nation's waters. The nation's waters include navigable rivers, streams, lakes, natural ponds, wetlands, and marine waters. Under this statute, the US EPA sets water quality criteria and technology–based effluent guidelines to protect water quality. States set specific water quality–based standards. The standards provide a means for achieving the goals of the CWA.

There are 3 components of a state's water quality standards: uses, criteria, and an anti–degradation policy. States determine use designations for the protection and propagation of fish, shellfish and wildlife, recreation, drinking water, agricultural and industrial uses, as well as other uses such as navigation, special habitats such as coral reef protection, oceanographic research, aquifer protection, marinas, and hydroelectric power. Uses are determined through a use attainability analysis that involves a water–body survey, waste load allocation, and economic analysis.

Water quality criteria establish a limit on a pollutant or on a condition of a water body. The criteria are intended to protect the designated use of that water and will trigger a management action if exceeded. There are two types of water quality criteria: numeric and narrative. The numeric criteria are developed for specific chemicals or microbial agents. The narrative criteria are set for contaminants that are more difficult to quantify. For example, "surface water shall be free from floating, non–petroleum oils of vegetable or animal origin."

The types of criteria include:

- Aquatic life criteria for the protection of aquatic plants and animals
- Human health criteria protective for water and fish consumption
- Biological threshold or guideline levels describing the desired biological integrity of waters
- Sediment criteria to assess material that may pose a threat to human or ecological health.

An anti–degradation policy is designed to protect existing uses, describes water quality characteristics, and includes implementation measures to protect designated uses.

Existing Regulatory Guidelines

Presently there are no US regulations or guidelines that apply to cyanobacterial toxins under the SDWA or CWA. Several US States have implemented standards or guidelines that apply to recreational water uses. The World Health Organization has issued a guideline that applies to microcystin LR and guidelines or standards have been established by a number of countries around the globe. Relevant standards and guidelines are discussed below.

U.S. EPA Secondary Standards

The US EPA has established secondary Maximum Contaminant Levels (SMCLs) for Drinking Water Contaminants that apply to factors such as color, taste and odor which may be considered relevant to cyanobacteria. SMCLs are not regulatory; however, some may be adopted as regulations by individual states. Existing SMCLs for color and odor may have some utility as mechanisms to stimulate action by states in situations where cyanobacteria affect the color or odor of drinking water.

The SMCL for color is 15 color units (CUs). A CU is defined as a color that is objectionable to a significant number of users. For comparison, a CU of 5 represents color that can be detected in a bathtub and a CU of 30 can be detected by all users and is considered objectionable. The SMCL of 15 CU has been set to prevent the majority of consumer complaints regarding color.

The SMCL for odor is 3 threshold odor numbers (TON). A TON of water is the dilution factor required before the odor is minimally perceptible. A TON of 1 indicates odor–free water, while a TON of 3 indicates that a volume of the test water would have to be diluted to 3–times its volume before the odor became minimally perceptible. Some sources cause odors that may be considered by consumers to be less tolerable than others of equal intensity, and some affect taste as well as odor. Water that is relatively odor–free helps to maintain consumer confidence. The decay of algae in water can cause a disagreeable musty odor in the water. Oxidation and activated carbon are two treatment methods for controlling odors in drinking water.

State Guidelines

In the absence of U.S. EPA guidance values regarding cyanotoxins, most states have looked to the World Health Organization (WHO) and the latest research in Australia for suggested drinking and recreational water use guidelines. Water and algal testing, health alerts, and subsequent beach and lake closures involving cyanobacteria bloom waters have increased in recent years with widely publicized dog deaths in waters of New York, Nebraska, Wisconsin and Minnesota. States such as Maryland and Virginia have used WHO guidelines for cyanobacteria and microcystins in support of beach closures. Nebraska and Iowa have implemented 15 ppb microcystin guideline values for issuing recreational use health alerts on lakes with blooms. The Vermont State Health Department has set a standard of 6 ppb microcystin for reopening a beach after a toxic bloom event. Cyanobacteria derived food supplements are big business but no national guidance exists for acceptable contaminant levels such as microcystins in these food supplements. The Oregon Health Department has adopted a 1ppm maximum acceptable concentration.

World Health Organization (WHO)

There are insufficient data to determine health–based guidelines or standards for even a representative selection of the toxins. The best studied is microcystin LR, although uncertainties exist, particularly with regard to its tumour promoting capability. WHO proposed a provisional guideline value in 1998 for microcystin LR, based on the data generated by the United Kingdom (UK) National Research Programme.

The WHO will develop additional guidelines for other toxins when there are adequate data, but the production of guideline values for an increasing list of toxins is seen as potentially counter–productive. The WHO (WHO, 2003) paragraph in Volume 1 of the revised Guidelines reads as follows:

> Cyanobacteria occur widely in lakes, reservoirs, ponds and slow flowing rivers. Many species are known to produce toxins, a number of which are of concern for health. There are many cyanotoxins, which vary in structure and may be found within cells or released into water. There is wide variation in the toxicity of recognised toxins (including amongst different varieties of a single toxin, e.g., Microcystins) and it is likely that further toxins remain unrecognized.
>
> The health hazard is primarily associated with overgrowth, (bloom) events. Such blooms may develop rapidly and they may be of short duration. In most circumstances, but not all, they are seasonal.
>
> Analysis of these substances is also difficult although rapid methods are becoming available for a small number, e.g. microcystins, in addition analytical standards are frequently not available. The preferred approach is therefore, monitoring of source water for evidence of blooms, or bloom forming potential, and increased vigilance where such events occur.
>
> A variety of actions are available to decrease the probability of bloom occurrence and some effective treatments are available for removal of cyanobacteria or cyanotoxins. For these reasons, monitoring of cyanotoxins is not the preferred focus of routine monitoring and is primarily used in response to bloom events. Whilst guideline values are derived where sufficient data exist, they are intended to inform the interpretation of data from the above

monitoring and not to indicate that there is a requirement for routine monitoring by chemical analysis.

Australia and New Zealand

Cyanobacterial blooms are common problems in Australia and New Zealand. Accordingly, the Australian and New Zealand Governments have been leaders in establishing risk management policies for CHABS and guideline values for cyanobacterial toxins in recreational waters and drinking water (See Burch, this volume). Australia has a drinking water standard for total microcystins (1.3 µg L^{-1}) based on the toxicity of microcystin LR. New Zealand has a guideline for the presence of cyanobacteria in drinking water (less than 1 cyanobacterium per 10 ml of sample) and provisional values for several anatoxins (anatoxin = 6 µg L^{-1}, anatoxin–a = 1 µg L^{-1}, homoanatoxin = 2 µg L^{-1}) mycrocystin LR (1 µg L^{-1}), cylindrospermopsin (1 µg L^{-1}), nodularin (1 µg L^{-1}), and saxitoxin–equivalents (3 µg L^{-1}).

The Australian guidelines for recreational waters are based on total microcystins or cell counts. Beach closure is recommended if either of the two following conditions are met:

- **Condition 1**: total microcystins at a concentration of either 10 µg L^{-1} total microcystins or >50,000 cells mL^{-1} toxic M. aeruginosa or a biovolume equivalent of >4 mm^3 L^{-1} for the combined total of all cyanobacteria where a known toxin producer is dominant in the total biovolume.
- **Condition 2**: either the total biovolume of all cyanobacterial material exceeds 10 mm^3 L^{-1} or scums are consistently present.

United Kingdom

The water industry in England and Wales was privatized in 1989 and the Government's technical regulator for the industry is the Drinking Water Inspectorate (DWI). The Water Supply (Water Quality) Regulations (2000), which the DWI enforces, do not include algal toxins as a specific parameter. However, the Regulations require that no substance may be present in drinking waters at concentrations that would cause a risk to health. In this respect water utilities would be required to monitor for algal toxins, if a risk situation existed. In the UK, water utilities currently base that risk assessment on the potential for algal loadings to compromise treatment processes and contaminate supplies.

The current UK view is that setting a standard based on the few toxins for which there were adequate data could be construed as potentially misleading because the absence of a particular toxin does not indicate the absence of a problem. In addition the potential for changes in the presence and absence of toxins means that sampling to give an appropriate level of reassurance could be problematical. Prevention of bloom formation is the best way forward, although this may present some difficulties. Control of eutrophication is an important issue for the Environment Agency in the UK and at the European level. It will be an important consideration in the Implementation of the European Union's Water Framework Directive.

Other Countries

A number of countries have adopted the WHO drinking water guideline for microcystins (See Busch, this volume). Brazil also has guideline values for saxitoxin equivalents and cylindrospermopsin. Germany and the Netherlands have guidelines for recreational waters based on microcystin concentrations. France's guidelines for recreational waters follow the cell count approach recommended by the WHO (Level 1: <20,000 cells mL^{-1}, Level 2: 20,000 to 100,000 cells mL^{-1}, Level 3: Presence of scum). The risk to human health increases with the level.

Charge 1

What data are available to derive health–based guideline values (TDIs; RfDs) for Cyanobacterial Harmful Algal Blooms (CHABs)?

As discussed previously, dose–response assessment involves describing the quantitative relationship between the amount of exposure to a substance and the extent of toxic injury or disease. Data are derived from animal studies or, less frequently, from studies in exposed human populations. The risks of a substance cannot be described with any degree of confidence unless dose–response relations are quantified, even if the substance is known to be toxic.

Health–based guidelines are based on quantitative values that describe an estimate of the exposure to the human population (including susceptible subgroups) that is likely to be without an appreciable risk of adverse health effects over a lifetime. These values are generally derived from a statistical lower confidence limit on the benchmark dose (BMDL), a no–observed–adverse effect–level (NOAEL), a lowest–observed–adverse–effect level

(LOAEL), or another suitable point of departure, with uncertainty/variability factors applied to reflect limitations of the data used.

The data available for derivation of health–based guidelines for cyanobacterial toxins are very limited. Due to the stringent data quality requirements set forth by the US Information Quality Act for the derivation of quantitative values, many available toxicity studies are deemed inappropriate for consideration due to one or more data quality failures. Additionally, the US EPA follows published guidelines for quantitative dose–response assessment and much of the available toxicity data are inherently insufficient for guideline value determination. Many of the toxicity studies that have been conducted on cyanobacterial toxins utilized cell extract preparations with unquantified total toxin levels rather than employing known quantities of purified toxin. As most cell extracts contain more than one toxin and, at equivalent doses, have been shown to be more potent than purified toxin (most likely due to additive or synergistic effects), studies that employ cell extracts are deemed inappropriate for single–chemical quantitative dose–response assessment. The single–chemical toxicity data currently available for potential guideline values for oral exposure to anatoxin–a, cylindrospermopsin and microcystin LR are described in Table 1.

As discussed above, there are inherent limitations in establishing health–based guidelines for individual toxins. There is a wide variation in the toxicity of known toxins, multiple toxins are produced during a bloom event, and it is likely that previously unrecognized toxins will continue to be identified. It is important to recognize that the development of health–based guidelines for individual toxins is simply a first step in the overall risk assessment of CHABs. Further exploration into the potential use of approaches such as a Toxicity Equivalency Factor (TEF) or quantitative structure–activity relationship (QSAR) is warranted.

Table 1. Summary Results of Major Studies for Oral Exposure of Experimental Animals to Anatoxin–a, Cylindrospermopsin and Microcystin–LR

Species	Sex	Dose (µg kg^{-1}–day)	Exposure Duration	NOAEL (µg kg^{-1}–day)	LOAEL (µg kg^{-1}–day)	Responses	Comments	Reference
Anatoxin–a								
Short–term Exposure								
Mouse	M/F	1200, 2500, 6200, 12300	5 days	ND	ND	Mortality at doses of 6200 & 123000	Range–finding study for 28–day study (see below);No control group	Fawell and James, 1994; Fawell et al., 1999a
Mouse	M/F	0, 100, 500, 2500	28 days	100	500	Mortality; no other significant treatment–related effects	Treatment–related mortality can not be ruled out; true NOAEL may be 2500	Fawell and James, 1994; Fawell et al., 1999a
Subchronic Exposure								
Rat	F	0, 51, 510	7 weeks	510	ND		No changes in any monitored parameters were reported	Astrachan and Archer, 1981; Astrachan et al., 1980

Species	Sex	Dose (μg kg^{-1}–day)	Exposure Duration	NOAEL (μg kg^{-1}–day)	LOAEL (μg kg^{-1}–day)	Responses	Comments	Reference
Developmental Toxicity								
Mouse		0, 2500	GD 6–15	2500	ND			Fawell and James, 1994; Fawell et al., 1999a
Cylindrospermopsin								
Short–term Exposure								
Mouse	NR	NR	14 days	50	150	Lipid infiltration in liver	Report of study provides limited detail	Shaw et al., 2000, 2001
Subchronic Exposure								
Mouse	M	0, 30, 60,120, 240	11 weeks	30	60	Increased relative kidney weight		Humpage and Falconer, 2003
Microcystin–LR								
Short–term Exposure								
Rat	M	0, 50, 150	28 days	ND	50	Slight to moderate degenerative and necrotic hepato-cytes with hemor-rhages		Heinze, 1999

Species	Sex	Dose (μg kg^{-1}–day)	Exposure Duration	NOAEL (μg kg^{-1}–day)	LOAEL (μg kg^{-1}–day)	Responses	Comments	Reference
Subchronic Exposure								
Mouse	M/F	0, 40, 200, 1000	13 weeks	40	200	Minimal/slight chronic inflammation with haemosiderin deposits and single hepatocyte degeneration		Fawell et al., 1999b
Chronic Exposure								
Mouse	F	0, 3	18 months	3	ND	No effects on survival, body weight, hematology, serum biochemistry, organs, or histopathology	Minor changes in ALP and cholesterol deemed insignificant	Ueno et al, 1999
Mouse	NR	80	80–100x over 28 weeks	ND	ND	Light injuries to hepatocytes in the vicinity of the central vein	Only liver examined; only 3 control animals	Ito et al., 1997

Species	Sex	Dose (μg kg^{-1}–day)	Exposure Duration	NOAEL (μg kg^{-1}–day)	LOAEL (μg kg^{-1}–day)	Responses	Comments	Reference
Monkey	NR	20–80	47 weeks	ND	ND	No clinical signs or effects on hematology, serum biochemistry, histopathology	Report of study provides limited detail	Thiel , 1994

ND = Not determined
NR = Not reported

Charge 2

What research is needed to reduce uncertainty in health–based guidelines?

Hazard and Dose Response Data Needs

Hazard assessment is the characterization of the adverse effects on human health caused by oral, inhalation or dermal exposure. Effects can range from short–term reversible dermatitis to death from respiratory paralysis or cancer. Hazard identification is descriptive; dose–response assessment is quantitative. The hazard identification includes a description of all of the adverse health effects caused by a toxic substance, independent of the doses causing the effects. On the other hand, the dose–response assessment identifies whether or not effects are manifest at specific doses and the impact of an increase in the dose on the appearance and/or severity of the effects. It is rare for any single study to provide a complete picture of potential effects for any contaminant and the relationship of those effects to dose. Generally, a suite of studies is necessary to fully elucidate the potential for hazard and its relation to dose. At present there are numerous deficiencies in the database that impede a high confidence hazard and dose–response assessment for the cyanobacterial toxins (see Health Effects Work Group Report and Ecosystem Effects Work Group Report this volume). Accordingly, the Risk Assessment Work Group has focused on how filling critical data gaps in the hazard and dose–response database for the cyanobacterial toxins would reduce the uncertainty in the risk assessment (Table 2). This approach to research prioritization will help to improve the precision of the risk assessment, the efficiency of the research plan and the risk management costs.

After examining the available data on hazard, dose–response, and exposure pathways, the Work Group developed a matrix (Table 2) to illustrate how the execution of specific types of studies will contribute to reductions in uncertainty in the risk assessment. An "X" in a given cell designates the importance of the study to reducing uncertainty. A question mark in a cell suggests uncertainty in the need for the study at this time. Notes provide additional information on the type of study suggested and its contribution to the database needs.

Table 2. Cyanobacterial Toxins: Research Needs Categorized Based on Reducing Uncertainty in the Risk Assessment

Related Uncertainty Factor (UF)	**Intra– and Interspecies Factors**		**Duration Uncertainty**		**Data Deficiencies Uncertainty**	**Notes**
Toxin	***Kinetics***	***Dynamics***	***Acute Toxicity***	***Subchronic/ Chronic Toxicity***	***Developmental/ Reproductive/ Other Toxicity***	
Reduction in UF	*3→1*	*3→1*	*NA*	*10→3→1*	*10→3→1*	*Reduction in some UFs can be achieved in increments*
Microcystins	X	X		X	X	–Absorption, Distribution, Metabolism, and Excretion (ADME) studies needed. Only if data are adequate to model tissue dose for the target organ(s) will it be possible to reduce the toxicokinetic UF for inter– and/or intraspecies adjustments. –Data are needed regarding the kinetic and dynamic differences among individual microcystins (e.g. LA,RR,LI, RI, YR) –A cancer bioassay is needed –There are some developmental toxicity data. There are no reproductive toxicity studies

Related Uncertainty Factor (UF)	Intra– and Interspecies Factors		Duration Uncertainty		Data Deficiencies Uncertainty	Notes
Toxin	***Kinetics***	***Dynamics***	***Acute Toxicity***	***Subchronic/ Chronic Toxicity***	***Developmental/ Reproductive/ Other Toxicity***	
Reduction in UF	*3→1*	*3→1*	*NA*	*10→3→1*	*10→3→1*	*Reduction in some UFs can be achieved in increments*
Anatoxin A	X	X	X	X	X	– ADME Studies needed. Only if data are adequate to model tissue dose for the target organ(s) will it be possible to reduce the toxicokinetic UF for inter– and intraspecies adjustments. –A subchronic study is needed. For longer–term and lifetime risk values –The acute toxicity data are marginally adequate for a short term value. Identify whether the dog is a good model for acute toxicity.
Anatoxin A(s)	X					–Risk assessments, at least for short term effects can utilize data on organophoaphates and QSAR. ADME data would be helpful in utilizing the QSAR application to risk assessment

Related Uncertainty Factor (UF)	Intra– and Interspecies Factors		Duration Uncertainty		Data Deficiencies Uncertainty	Notes
Toxin	*Kinetics*	*Dynamics*	*Acute Toxicity*	*Subchronic/ Chronic Toxicity*	*Developmental/ Reproductive/ Other Toxicity*	
Reduction in UF	*3→1*	*3→1*	*NA*	*10→3→1*	*10→3→1*	*Reduction in some UFs can be achieved in increments*
Cylindro–spermopsin	X			X	X	–ADME Studies needed. Only if the data are adequate to model tissue dose for the target organ(s) will it be possible to reduce the toxicokinetic UF for inter– and intraspecies adjustments –Based on mutagenicity, a chronic bioassay is needed –Developmental data could help reduce the short term data uncertainty. –A reproductive toxicity study will allow for an additional decrease in database uncertainty
Saxitoxin	?	?	X	X	X	In general, data are limited making it difficult to conduct a risk assessment. While there are human intoxication data from marine exposures, a complete battery of studies is suggested.

Related Uncertainty Factor (UF)	**Intra– and Interspecies Factors**		**Duration Uncertainty**		**Data Deficiencies Uncertainty**	**Notes**
Toxin	***Kinetics***	***Dynamics***	***Acute Toxicity***	***Subchronic/ Chronic Toxicity***	***Developmental/ Reproductive/ Other Toxicity***	
Reduction in UF	*3→1*	*3→1*	*NA*	*10→3→1*	*10→3→1*	*Reduction in some UFs can be achieved in increments*
BMAA						Is important to study fate during drinking water treatment before investing in additional toxicological research

An "X" in a cell indicates that filling the indicated data need would have a strong potential to reduce uncertainty in the risk assessment. An "?" in a cell indicates that the impact of filling the indicated data deficiency on uncertainty cannot be determined at this time.

In the case of microcystin–LR, the chronic and reproductive toxicity studies will have the most significant impact on reducing uncertainty because between the two, they have the potential to reduce the overall uncertainty by a factor of 10. To the extent that studies on the dynamics of the toxicity were incorporated in the chronic and reproductive toxicity studies, additional reductions in uncertainty might be obtained. In the case of the other microcystin congeners the most productive research relative to reductions in uncertainty will be that supporting quantitative measures of toxic equivalence to microcystin–LR including kinetic and dynamic parameters, because the total data base for the other microcystin congeners is very limited compared to that for microcystin–LR (Dietrich et al. this volume).

Subchronic and developmental toxicity studies are those likely to have the most immediate impact on reducing the uncertainty for anatoxin A. There are several moderately informative studies of the acute neurotoxicity of this compound but studies that evaluate a more comprehensive set of health endpoints following moderate duration exposures will make a significant addition to the database. Anatoxin A(s)'s toxic activity appears to be qualitatively and quantitatively comparable to organophosphate cholinesterase inhibitors. Accordingly, the development of a QSAR model based on analysis of the structure and functional groups of organophosphate pesticides, would be a useful approach to predicting hazard and dose–response properties for this toxin.

Cylindrospermopsin tested positive for mutagenicity in several studies. Thus, completion of a long term cancer bioassay combined with analysis for other long term toxic effects is a definitive data need for this compound. Such a study has the potential to reduce a chronic duration uncertainty factor from a 10 to a 1. A reproductive study with integrated evaluation of developmental endpoints could produce an additional three–or ten–fold reduction in uncertainty.

The Work Group felt that the saxitoxins and beta–methylamino–L–alanine (BMAA) were presently of low priority for research on cyanotoxins. Regulatory and action limits for PSP toxins are well established in the international community (Anderson et al. 2001). The supporting work has been based primarily on shellfish poisoning concerns from estuarine and marine dinoflagellates producing chemicals of the saxitoxin family. However, freshwater cyanobacteria have been recognized to produce saxitoxin as well (e.g., Cylindrospermopsin, Aphanizomenon, Lyngbya). Because the database on fresh water saxitoxins is very limited, to single out one particular study type that would have the greatest impact of reducing uncertainty in the risk assessment is difficult. However, the use of state–of–the–art analytical methodology allows quantitation of saxitoxins and neosaxitoxins in the freshwater environment and thus the comparison with levels of concern for the marine environment. In the case of BMAA, its identification as a cyanobacterial toxin is quite recent (Cox et al. 2005). Thus, much more must be learned about its environmental fate and transport before singling out any particular type of study that would have the greatest impact on uncertainty reduction in the risk assessment process. However, most recent information does suggest BMAA may be contained in copious quantities in cyanobacteria food supplements, i.e. *Spirulina sp.* And *Aphanizaomenon flos–aquae* based products (Dietrich et al. this volume), thus suggesting that the prioritization of BMAA with regard to research efforts may have to be revisited if these findings are confirmed by other work groups.

One cross cutting problem in conducting toxicological research for all of the cyanotoxins in Table 2, is the difficulty and expense of obtaining sufficient pure toxin for use in short or long term animal studies. Both chronic and reproductive toxicity studies require as an absolute minimum 20 animals of each sex per dose group and sufficient toxin to dose the animals for up to two years. In addition, although chronic and reproductive studies with single toxins may improve the database on the single toxins species, they do not resolve the problems of potential additive or synergistic toxicity. Indeed, as pointed out in Dietrich et al. (this volume) exposure to multiple toxins in bloom events appears more likely the norm rather than the exception. Consequently, and in support of WHOs' stance on additional guideline values, frequent monitoring and vigilance with regard to blooms and presence of toxins may be a better approach for most risk scenarios (e.g. recreational or drinking water). However, because guideline values present authorities with possibilities of legal enforcement, lack thereof and substitution with monitoring and vigilance may not suffice for human health protection. This may be exemplified by cases where cyanobacterial toxin exposure of humans occurs via contaminated food and food supplements.

Contrary to the direct exposure of humans to cyanobacterial toxins via contaminated water, the risk situation involving exposure via food and food supplements is much more complex. Worst–case exposures can be interpolated from assumed daily or weekly consumption of specific food sources (e.g. fish, crayfish, shellfish, vegetables, salads, etc.) for the general populace as well as for populations at high risk (e.g. indigenous tribes predominantly existing on a specific food source) (Dietrich and Hoeger 2005). However, the potential human toxin exposure via food that provides the basis for risk calculations is also largely determined by the degree of toxin contamination of a given food source as well as by the bioavailability of the toxin from the food type. Furthermore, bioaccumulation of cyanotoxins in the food chain, as is the case with BMAA, may provide for an additional element of risk (Cox et al. 2005). The occurrence of multiple toxins within the same food chain and the potential for additive or synergistic effects complicates hazard identification. The lack of appropriate guidance by authorities (e.g. WHO or federal or state laws) will prohibit local authorities from implementation and enforcement of measures intended to reduce human health risks.

Analytical Methods Research Needs

The challenges posed by cyanotoxins in water are in many respects different from those posed by other chemical toxins. Whether the toxin is present in the source water or generated during treatment, occurrence of a concentration posing an acute risk is unlikely, unless a contamination event has occurred. Furthermore, once seasonal effects and the influence of treatment processes have been characterized, variations in the concentrations of many chemical toxins are reasonably predictable.

The cyanotoxins are possibly unique among chemical toxins in that they can cause serious illness or death rapidly at concentrations that occur naturally in the environment. Although their presence can be anticipated through surveys of algal populations, cyanotoxin concentrations in water are unpredictable and may change quickly.

Two distinct analytical requirements can be distinguished: (i) methods to characterize the concentrations of specific cyanotoxins or their congeners and (ii) methods to detect the toxins at levels to support assessment of a risk to health. These requirements coincide if there is only one cyanotoxin present. However, different cyanotoxins, or congeners of the same cyanotoxin type may be present and the risk posed by the different toxins or their congeners may be different. Furthermore, where mixtures of toxins are involved, an assessment of the overall risk to health may be of more immediate interest (toxic equivalency concept) than quantification of individual compounds.

Requirement (i) applies in studies of removal or inactivation of cyanotoxins in water treatment processes, in surveys of concentrations in environmental waters, or in checking compliance against guidelines or standards for specific cyanotoxins. Quantitative analysis for cyanotoxins has been an active branch of analytical chemistry since the mid 1980s. In Australia and the UK, compendiums of standard methods have now been published (Anon, 1998; Brenon and Burch, 2001) and an output from the European Union's Framework Research Programmes includes a monograph on monitoring and analysis (Meriluoto and Codd, 2005). Nevertheless, the extent of validation of methods of analysis varies.

Although the performance of methods for microcystins, nodularins, cylindrospermopsin and saxitoxins have been demonstrated in inter–laboratory studies, there is a need for better characterization of the performance of methods for anatoxins and BMAA. Confidence in analytical methods would be further improved by the application of standard protocols to assess the performance characteristics of the methods.

Requirement (ii) is more likely to be of interest when exposure to cyanotoxins through recreational use of water or through consumption of fish and shellfish is being considered. If water treatment processes are absent or have been compromised in some way, there may be a concern for health risks.

The HPLC and MS based methods that have been developed for individual cyanotoxins and their isomers are characterized by low daily throughput. The rate determining steps are the sample transport time from remote locations and the time needed to prepare extracts of samples for analysis. The timescale between commissioning the taking of a sample and receiving the results of analysis is typically days to weeks. This may be unacceptable if health risks are involved and especially so if the result could determine whether restriction of access to water or sale of food is necessary. The problems of poor speed of response are compounded if there is change in the toxicity characteristics of algal blooms, for example, the species(s) of algae predominating in the bloom and consequently the type(s) of toxin(s) present change over relatively short periods of time.

Where a rapid speed of response is essential, analysis will need to be carried out onsite, or in an adjacent location where facilities may fall far short of what is expected in a laboratory environment. This creates a demand for simple to use kits for specific cyanotoxins, or the entity that confers toxicity (e.g. the alanine, aspartate, alanine, aspartate (ADAD) amino acid components of microcystins and nodularins) (Fischer et al. 2001; Zeck et al. 2001). Other possibilities include *in vitro* systems such as the acetylcholine esterase or protein phosphatase inhibition assays. A promising format for rapid screening tests would appear to be broad spectrum Enzyme–Linked Immunosorbent Assay (ELISA) techniques with universal cross–reactivity to the numerous toxin congeners. ELISA test kits are already available for Microcystins and for toxins causing Amnesic Shellfish Poisoning.

There is a need for research to support development of a wider range of rapid test systems to provide the data necessary for managing exposure to cyanotoxins. Managers will need to be confident about the consistency and comparability of data generated by different operators in different locations. This implies the need for independent assessment of the performance of test kits using recognised test protocols and for the results of these assessments to be placed in the public domain.

In order to evaluate the performance of test kits it will also be necessary to develop stable standards suitable for distribution in performance studies.

Research Prioritization to Reduce Uncertainty in Health–Based Guidelines

The Work Group recognized that the prioritization of research needs is as important as their identification. Accordingly the group further characterized the hazard, dose–response, analytical method, and treatment technology needs identified above according to whether they should be targeted for immediate study or classified as longer term research needs. The Work Group suggestions are summarized below.

The Work Group suggestions were selected with the objective of obtaining the maximum research output with the smallest monitary investment by answering those questions, on exposure and toxicity that, at the moment appear to be the most pressing. Each study suggested will provide some answers and undoubtedly also raise new questions. Accordingly, the suggestions must be revised and reordered as additional data become available.

Near-term Research Priorities

- Microcystins
  - Kinetic and Dynamic equivalences between congeners
  - Certified analytical methods for monitoring
  - Monitoring of finished drinking water
- Anatoxins
  - Subchronic study for Anatoxin a
  - QSAR for Anatoxin A(s) based on organophosphate data
  - Impact of treatment technologies on removal
- Cylindrospermopsin
  - Occurrence data for ambient and drinking water
  - Developmental effects

- General
  - Kinetic studies
  - Suitability of extract studies for Clean Water Act guidelines

Long-term Research Priorities

- Microcystins
  - Preparation of enough pure material to conduct a long term study
  - Chronic cancer bioassay
- Anatoxins
  - Evaluation of dogs as an appropriate model for human toxicity
  - Long term effects of A(s) variant
- Cylindrospermopsin
  - Preparation of enough pure material to conduct a long tern study
  - Chronic cancer bioassay
  - Bioconcentration studies
- General
  - Toxin interactions in mixtures
  - Factorial design studies

Charge 3

What research is needed to minimize the costs and maximize the benefits of the various regulatory approaches?

The costs and benefits of preventing cyanobacterial blooms is very dependant on the nature of the water body and its uses. It is therefore difficult to generalize (See Steffenson this volume). However, the efficacy of watershed management techniques have been demonstrated (See Piehler this volume).

The cost of engineering works to control or prevent blooms, while site specific, is generally the easiest component to assess. Assessing the impact of blooms on aesthetic and environmental aspects and the value of controlling or preventing those impacts is the most difficult area. The impacts may affect a wide range of activities including tourism, fishing and agriculture. Further research is required to quantify these costs and benefits. There are also less tangible aspects such as the value one places on maintaining natural ecosystems. The willingness of the community to pay for the preventative measures may be the best guide of the value of preventing cyanobacterial blooms.

When comparing the relative costs and benefits of alternative control measures, the outcome will be influenced by how broadly the assessment is made. Increasingly there will be an expectation that the cost benefit analysis includes the broader social and environmental aspects and consideration of the sustainability of the options. Issues such as energy use and green house gas production may become more important in the future and may make some of the engineering options less attractive.

Charge 4

What are the exposure pathways for the receptors of concern?

In addressing this question, the workgroup felt it was important to first articulate what constitutes a bloom as a way of providing context for various exposure pathways (see also Fig. 2a and 2b).

What is a bloom?

There have been continuing efforts to develop a definition for what designates an algal bloom. A bloom as an ecological phenomenon has characteristics of magnitude (biomass and abundance), duration, frequency, spatial extent, and composition. Blooms collectively represent part of a trophodynamic process with regional, seasonal and species–specific issues (Smayda 1997). In a traditional sense of the plankton science, 'bloom' has reflected the historical focus of marine phytoplankton ecologists on the annual, high biomass, diatom dominated spring (upwelling) abundances or biomass (Smayda 1997). 'Harmful Algal Bloom' can refer to "blooms of toxic and non–toxic algae that discolor the water, as well as to blooms which are not sufficiently dense to change water color but which are dangerous because of the algal toxins they contain or the physical damage they cause to other biota." (Anderson et al. 2001). This definition reflects the diversity of phytoplankton now recognized for harmful effects and focuses on population phenomena being observed. We can extend the concept to include cyanobacteria as Falconer (1998) noted that when the body of water is visibly colored by cyanobacteria, then is it considered a bloom and cyanobacteria probably number more than 10,000 cells/ml.

While the discussion and debate continues on an all–encompassing definition for bloom, we can functionally apply suggested guidance values available or being developed for the species of interest. With specificity toward species, habitats, regions, population and trophodynamics involved, and no one definition yet suitable to all bloom conditions, we increasingly find the use of abundance (cell density) and toxin thresholds reflected in natural resource management programs. Cell counts and toxin concentrations for cyanobacteria linked with no effect, sub–chronic, chronic and lethal thresholds are of interest in managing waterways for protecting human health. Potential impacts are increasingly being defined with respect to counts that trigger toxin testing in shellfish, restricting recreational activity or limiting agricultural uses such as cattle watering. Cyanotoxin thresholds are increasingly desired or available for guidance with drinking water, fish or shellfish harvest and their consumption.

Threshold definitions are most frequently developed for the protection for human health. Definitions of thresholds protecting ecosystem integrity and services, however, further challenge our research needs. Notable consequences of blooms have included wildlife, fish, shellfish and human health effects both sublethal and lethal. Indirect effects of blooms are many such as reductions in water clarity that impact light to submerged aquatic vegetation, effects on the dissolved oxygen dynamics that can lead to fish kills in shallow water zones, organic matter sinking and leading to hypoxic or anoxic conditions developing in deep water, biogeochemical changes in nutrient pathways, and synergistic or allelopathic effects of toxins. Gastrich and Wazniak (2002) provide an example and potential model of categorizing bloom effects on natural resources without human health implications for the golden–brown algae *Aureoccous anophagefferens* (Table 3). Species, toxins and effects pathways within the ecosystem continue to be evaluated. Linkage with risk assessment research is likely to provide additional guidance for threshold developments in ecosystem management.

Table 3. Brown Tide Bloom Index

Category	**Cell Count cells/ml**	**Impact**
1	<35,000	No observed impact
2	≥ 35,000 to < 200,000	Reduction in growth of juvenile hard clams, (Mercenaria mercenaria). Reduced feeding rates in adult hard clams; Growth reduction in mussels (Mytilus edulis) and bay scallops (Argopecten irradians).

Category	Cell Count cells/ml	Impact
3	≥ 200,000	Water becomes discolored yellow–brown; Feeding rates of mussels severely reduced; Recruitment failures of bay scallops; No significant growth of juvenile hard clams; Negative impacts to eelgrass due to algal shading; Copepod production reduced and negative impacts to protozoa.

Ingestion Pathway

Cyanobacterial–supplements

Food supplements made from cyanobacteria (blue–green alga supplements; BGAS) can concentrate toxins and result in human exposure (See Dietrich et al. this volume). The levels of algal toxins in food supplements are unregulated at the Federal level in the United States because they fall outside the purview of the US Food and Drug Administration. However, Oregon has set limits on microcystins in food supplements.

Regulatory approaches to BGAS products based on toxicity have not yet been developed and limit the management options for insuring safety. BGAS are generally produced from three cyanobacteria species: Spirulina maxima, Spirulina platensis or Aphanizomenon flos–aquae. Analysis of BGAS for the presence of toxins is not wide spread, but low levels of anatoxins, microcystins, and/or saxitoxins have been found in some BGAS samples (See Dietrich et al. this volume). There is also the possibility that BGAS supplements may contain the neurotoxic amino acid BMAA (See Dietrich et al. this volume). Since supplements can contain one or more of the toxins produced by the species used, issues of potential additivity and synergy must be considered in the risk assessment for BGAS products.

Drinking Water

At present there are no monitoring data from public water systems in the United States for individual cyanobacterial toxins. The lack of data is due, in part, to the absence of standardized analytical methods for individual toxins that can be utilized in a national monitoring program. Problems with cyanobacterial toxins in drinking water, including some human deaths have been reported in the United States, Australia, South America, China, and other countries, but are infrequent (Hitzfeld et al. 2000). In one incident, several dozen individuals died as a result of dialysis with contaminated water (Jochimsen et al. 1998). Although blooms in source water cannot always be detected visually, they may be detected through inspection of filters at water treatment facilities. Such detections indicate that the source water may be contaminated with algal toxins. Water treatment processes can be initiated to eliminate the toxins from finished water. However, the efficacy of treatment processes is dependent upon many factors (see Causes, Prevention, and Mitigation Work Group Report this volume). Successful treatment may be dependent upon the identification of toxin type and data on the efficacy of treatment techniques for the toxins identified. Research is needed to better describe the efficacy of treatment techniques by toxin type.

Fish and Shellfish Consumption

Consumption of CHABs through contaminated shellfish and fish can lead to impacts on the liver and the nervous system. Microcystins affect the liver and can promote tumor growth. Cylindrospermopsin also produces liver toxins. Anatoxins produced by *Anabaena* and *Oscillatoria spp* are acutely neurotoxic through interaction with cholinergic mechanisms. Saxitoxins, the cause of Paralytic shellfish poisoning (PSP) are also neurotoxic. Freshwater CHABs such as *Lyngbya wollei* and *Aphanizomenon flos–aquae* produce neurotoxins similar to saxitoxins. For further information on poisonings related to contaminated fish and shellfish consumption see Carmichael et al. (1997), Carmichael (2001); and Van Dolah et al. (2001).

Dermal Contact

Dermal contact with cyanobacteria and their toxins can occur through a variety of water–related recreational activities, most notably swimming at CHAB impacted beaches (salt or fresh water). There have been case reports of skin rashes and dermal or ocular irritation from recreational exposures (Queensland Health 2001; WHO 2003), but controlled toxicity studies of dermal and ocular responses are largely lacking.

Showering and bathing

The use of treated water for showering or bathing minimizes concern for contact with the cyanobacteria because most treatment processes would remove or reduce cyanobacteria in the filtration process, although dense blooms may overwhelm filtration units allowing cells or cell fragments to pass through. However, the toxins could still be present in treated water allowing for exposure through dermal uptake and inhalation of aerosol during showering. To the extent that cells were carried through the treatment process, heating of the water for bathing and showering would lyse the cells, releasing the toxins.

The use of untreated water for showering or bathing increases the risk for toxin exposure since higher levels of cells and toxin are likely to be present. In one case, after the use of cyanobacteria–contaminated water for a sauna in Finland, 48 people developed gastrointestinal, dermal and neurological symptoms that could have been related to toxin exposure (Hoppu et al. 2002 as cited in Dietrich et al. this volume)

Direct contact with ambient water

Water–sports (e.g. swimming, boating, fishing, etc.) in fresh, estuarine, and ocean water are popular recreational activities. When water bodies are impacted by CHABs, water–sports can be an important exposure route. For swimming and boating, the peak season for these activities tends to parallel that for the cyanobacterial blooms, increasing the risk of exposure. Enjoyment of recreational water sports tends to be a series of episodes that vary in frequency; causing concern for both higher level acute and lower level repeated exposures.

Most case reports of dermal irritation (contact dermatitis, eye irritation) due to cyanobacteria are related to swimming exposures. It has been suggested that the toxins responsible for skin and eye irritation are lipopolysaccharides, endotoxins, the blue–green pigment of the cyanotoxins (phycosyanin) and dermal toxins produced by Lyngbya and Planktothrix species (Queensland Health 2001). There are differences in sensitivity to these toxins; some individuals respond to very low concentrations while others are much more tolerant to exposures from swimming in CHAB impacted waters. Sensitive individuals can experience symptoms ranging from mild contact dermatitis to blistering and peeling of the skin (Queensland Health 2001). Prolonged contact through wet bathing suits increases the risk for dermal effects.

Charge 5

What are the ecosystem–services we want to protect?

Ecosystem services are processes by which the environment produces resources. Such services and their related resources may be affected by cyanobacteria abundance and biomass as well as toxins. Significant ecosystem services may therefore be protected through guidance values regarding cyanobacterial abundance and toxin levels in the environment. The following is a discussion of some of the ecosystem services potentially affected by cyanobacterial blooms.

Nutrient cycling

High biomass cyanoblooms can drive short and long term fluctuations in dissolved oxygen resources. Dissolved oxygen availability plays a critical role in nutrient cycling in the water column and the sediments where aerobic conditions favor biogeochemistry that will sequester phosphorus; anaerobic conditions promote liberation and greater availability of phosphorus. Phosphorus availability is frequently the critical limiting nutrient affecting bloom development, magnitude and persistence. Other nutrients, however, such as nitrogen can also play a concomitant critical role with bloom dynamics often determining whether cyanobacteria with heterocysts for fixing nitrogen or those without heterocysts predominate in a bloom. Bloom conditions can further lead to increases in pH affecting conditions that vary the nutrient cycling pathways, particularly with respect to phosphorus dynamics. High pH promotes dissociation of bound phosphorus, again altering source–sink dynamics of a system and making the phosphorous available and to perpetuating the longevity of blooms. Limiting cyanobacteria blooms can be one factor promoting environmental conditions more suitable to effective nutrient processing in the ecosystem.

Hydrologic cycle effects– Contamination of water sources.

While groundwater is frequently the source of public water supplies, surface water sources are usually those that serve the largest populations and are slated for additional development in some regions affected by blooms. Cyanobacteria can impart unfavorable taste and odors to tap water but additional risks are present from a diversity of cyanotoxins. Preventing blooms in surface waters also has beneficial implications for livestock, pets and aquatic dependent wildlife including plants. Research is needed to assess the ability of cyanotoxins to accumulate in ground water.

Energy conversion

Production of safe food

The accumulation of cyanobacteria biomass promotes the risk that toxins could be concentrated and bioavailable. Controlling blooms protects the service of uncontaminated surface water used for agricultural irrigation, watering livestock, and/or growing fish in aquaculture. The accumulation of cyanotoxins in the food web could impact subsistence and recreational harvest of fish and shellfish but is poorly characterized at this time. For example, microcystins can accumulate readily in the liver and significantly less in the muscle. Saxitoxin in shellfish is known to persist but there appears to be little evidence so far for issues of cyanobacterially–derived saxitoxin being problematic in freshwater environments. Additionally, there are reports of fish tasting musty when harvested from cyanobloom waters, reducing their desirability as a food source.

Trophic transfer of energy through the ecosystem

Cyanobacteria are not frequently considered favorable primary producers toward passing energy efficiently through the food web. Microzooplankton for instance may track *Microcystis* populations; however, under the same environmental conditions, larval and juvenile fish growth rates feeding on microzooplankton can be reduced over fish feeding on mesozooplankton due to energy density per food item consumed. Such effects on energetic pathways affect growth and survival of organisms throughout the food web, year class strength of populations and therefore community dynamics in the ecosystem. Such effects may ultimately have implications in the availability of harvestable fish.

Maintenance of ecological diversity and integrity

Extensive bloom conditions effectively block light needed to support survival of submerged aquatic vegetation. Some toxins or chemicals associated with the blooms may also act to inhibit growth of submerged aquatic vegetation. Thus ecosystem integrity is impacted by species specific toxins and species nonspecific shading factors. Indirect effects of cyanoblooms on habitat complexity (light limitation to submerged aquatic vegetation or dissolved oxygen conditions stressful or lethal to aquatic life) can affect spatial and temporal distribution of refuges that affect predatory–prey relationships. Aquatic community composition or the protection of threatened and endangered fauna can be impacted. Many disease fighting drugs available and under development have been mined from the available diversity. Conditions that promote lower diversity on a local to global scale would be expected to further limit the possibilities of culturally valuable mining of natural resources for their disease treatment and other properties.

Disease vectoring

Disease prevalence has been correlated with quantities of clean water available for personal and domestic hygiene (Chorus and Bartram 1999). Controlling blooms and their toxicity can therefore provide ecosystem services that aid in regulating disease and mortality. For example, human skin irritations are common through cyanobloom water contact. Skin irritations, related allergic reactions, skin, eye and ear infections compromise natural defense mechanisms of animals and humans. Bloom affected waters have promoted conditions for increased prevalence of such health impairments (Chorus and Bartram 1999).

Disease effects may also impact the condition of natural resources via indirect pathways. Biomass of cyanobacteria can accumulate along the windward shorelines of a waterbody. Decomposition of this organic–rich biomass can produce indirect effects of hypoxic (low oxygen) and anoxic (no oxygen or anaerobic) environments. Such environments typically occur in mid–late summer with temperatures favorable to germination of the *Clostridium botulinum* bacteria associated with botulinum toxins. The toxin can be inadvertently ingested by waterfowl leading to a potential botulism outbreak. Maggots feeding on a dead carcass in such an environment can accumulate the toxin and are ingested by other waterfowl and shorebirds promoting sickness and death in those populations. Hypoxic and anoxic environments lead to habitat impairments increasing stress on fish and shellfish compromising their immune defenses, and allowing access of disease vectors into the organism and population.

Transmission of cyanotoxins through the food web is a concern to natural resource and human health management agencies. There is a long history of livestock and pet deaths associated with consumption of bloom waters containing cyanotoxins (Chorus and Bartram 1999). Necropsies of Great Blue Herons (*Ardea herodias*) from a waterbird kill in a Chesapeake Bay–related event showed they exhibited a condition known as steatitis, excessive fat production (Driscoll et al. 2002). A leading hypothesis is that microcystin toxicosis may be a precursor to the development of this condition. It was determined in the analyses of liver tissue that microcystin levels were sufficient to account for the observed toxicosis (W. Carmichael pers. comm.). Understanding the transmission of such toxins through the food web and their potential to impact the expression of other disease conditions is poorly understood.

Health and wellness through leisure services provided by the ecosystem

Cyanobacteria bloom impacts can reduce the effectiveness of leisure services provided by the ecosystem that contributes to human wellbeing and quality of life. "Healthy" refers not only to physical well–being but also to the status of a number of related processes (Heintzman 1999). It involves a holistic integration of the physical, emotional, spiritual, intellectual, and social dimensions of people's lives (Bensley 1991; Crompton 1998; Ellison 1983; Ellison and Smith 1991). As an integrative component of holistic wellness, spiritual wellness needs to be an important consideration in leisure services that can enhance the quality of life for persons who have disabilities or who are devalued (Heintzman 1999).

Although coastal counties (excluding Alaska) account for only 11% of the land area in the United States, they are home to 53% of the population (Hunter 2001). Populations in proximity to coastal water resources as well as inland water bodies increase the demand for outdoor experiences dependent upon water quality. Unfortunately, many waterbodies are increasingly eutrophic and can be suitable for cyanobacteria bloom conditions. Chorus and Bartram (1999) cite a 1990's survey that showed large percentages of lakes already classified as eutrophic (Asia Pacific region (54%), Europe (53%), Africa (28%), North America (48%) and South America (41%). Bloom conditions for example have increasingly led to beach closures (Chorus and Bartram 1999) affecting recreational opportunities we frequently associate with leisure activities. In 2001, more than 82 million U.S. residents fished, hunted and watched wildlife (USDI et al. 2002). These activities bring recreationalists into contact with waterways that are or can be directly and indirectly affected by bloom waters. Guidelines that may be translated into water quality standards would aid the protection of such leisure services valuable to individual and social well–being.

Charge 6

How can regulators best devise a framework for making risk management determinations that incorporates consideration of the characteristics of CHABs, the risk for human health and ecosystem sustainability, and the costs and benefits of CHABs detection and management.

Cyanobacterial harmful algal blooms lead to a broad spectrum of public health, environmental protection and economic concerns. Stressors associated with these blooms can pollute drinking water supplies, degrade ecological services, and decrease agricultural productivity. Effective management of CHABs and the problems they create will require a comprehensive decision–support framework that addresses all facets of bloom occurrence, ecological and human health risks, and the control options for prevention and mitigation of those risks. This framework can be used to inform development of guidelines and standards for human exposure to cyanotoxins, to understand and control environmental impacts, and to support evaluation of the relative benefits and costs of alternative risk management options.

To maximize its utility, the decision–support framework must be able to accommodate the range of considerations relevant to bloom formation and occurrence, the causal pathways and mechanisms leading from blooms to ecological and human health effects, management actions to prevent blooms and minimize their impacts, and the costs associated with bloom occurrence and management. It should reflect the current state of knowledge regarding cyanobacteria ecology, the hazards of the cyanobacteria present, and the technologies available to address those hazards. Ideally, the framework also should be flexible with respect to incorporating new knowledge and technologies as these are developed.

Risk assessment has been adopted internationally as an important decision–support tool informing policy and the management of stressors affecting human health and ecological vitality. Because CHABs can pose risks simultaneously to a wide variety of assessment endpoints (valued components of the combined ecological–human–socioeconomic system potentially impacted by CHABs), and those risks likely are interconnected, an integrated approach to risk assessment (Suter et al. 2003; See Orme–Zavaleta and Munns this volume) is an attractive alternative to separate human health and ecological risk assessments. Furthermore, as multiple toxin exposures during CHABs are highly likely, an integrated risk assessment could provide additional information. When deployed with other technologies, such as multi–criteria decision analysis and benefit–cost analysis, and used in conjunction with approaches proven to be effective for managing CHAB risks, integrated risk assessment provides a logical cornerstone for an effective CHAB decision–support framework.

The workgroup recommends an overall decision–support framework with six basic elements (Fig. 1). The first two of these focus on integrated conceptual models that relate CHAB formation and occurrence to environmental and human health risks generically and comprehensively (Element 1), and on a site and situation–specific basis (Element 2). Reflected in the conceptual models are options for CHAB prevention and mitigation, and the socioeconomic costs of CHAB impacts. Element 3 utilizes these models to plan and perform risk assessments. It is important to understand the likelihood of adverse effects of CHABs on assessment endpoints relevant the specific problem at hand, be it development of national guidelines for cyanotoxins in drinking water, or prevention of blooms in livestock tanks. The concepts and approaches of multi–criteria decision analysis are used in Element 4 to help evaluate the attractiveness of alternatives for managing the risks characterized for the specific problem. Element 5 uses the collective information from the previous elements to construct management plans to control site and situation–specific risks. These plans identify control options, methods to monitor the effectiveness of controls, and the costs and benefits of options to assist in real–time decision–making. Finally, Element 6 evaluates the effectiveness of the overall framework for CHAB detection and management. Each of these elements is outlined below, together with the research and development activities needed to implement that element and the overall framework.

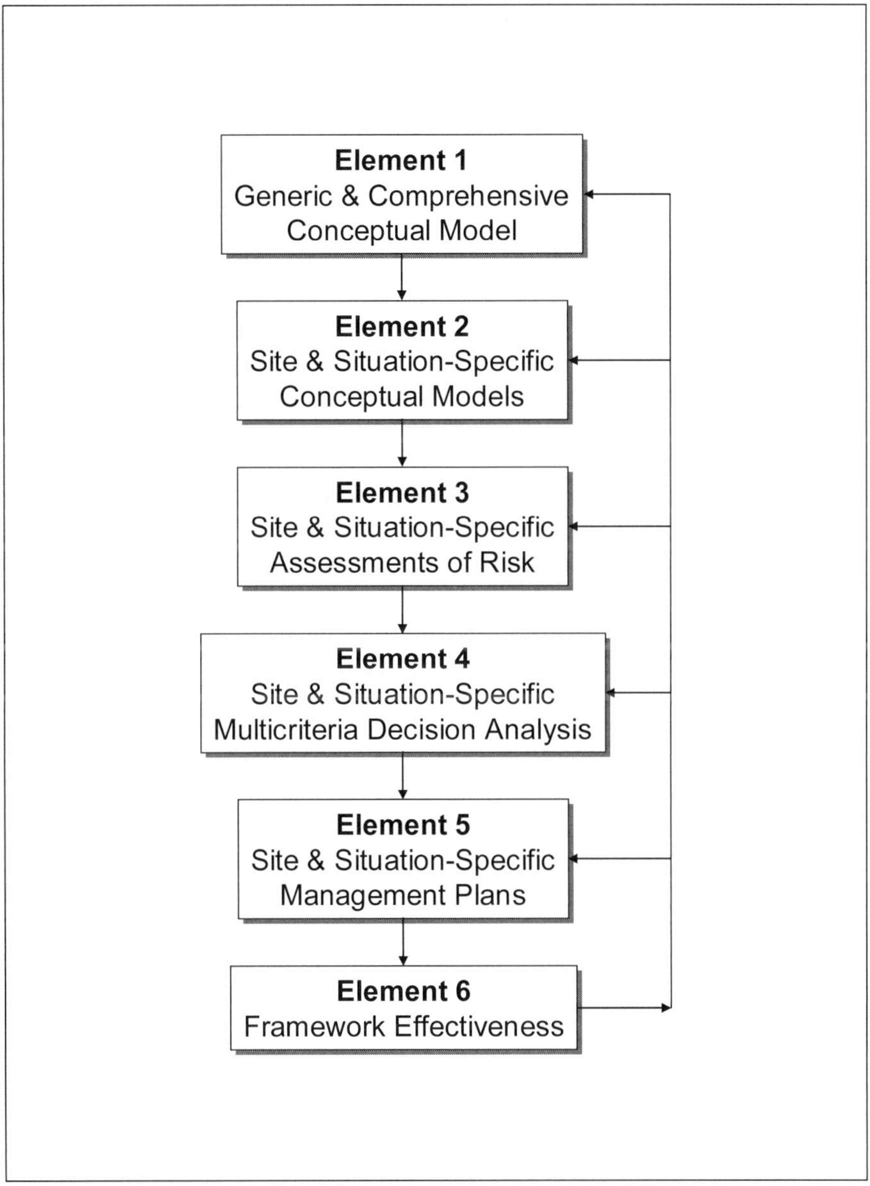

Fig. 1. Multidecision Framework

Generic and Comprehensive Conceptual Model

As evidenced during the symposium, the environmental factors influencing formation of cyanobacterial HABs are complex and incompletely understood. Similarly, the risks posed by CHABs to humans and valued ecological receptors and services are diverse, resulting from interconnected exposure pathways, environmental processes, and mechanisms of effect. Effective identification of hazards and assessment of risk requires a system–wide conceptualization of the environmental factors, processes, and social behaviors that influence the occurrence and possible outcomes of cyanobacteria blooms. This conceptual model should be comprehensive with respect to the state of knowledge, reflecting current technical understanding as a series of working hypotheses that describe formation of blooms, pathways of exposure to human and ecological receptors for stressors associated with blooms (toxins, biomass, etc.), biological and ecological effects resulting from those exposures, the factors that amplify or moderate these effects (e.g., presence of other stressors, conditions that affect receptor susceptibility), and relationships among system elements that directly or indirectly influence risks to important assessment endpoints. The generic conceptual model also should identify the costs incurred by CHABs and the various management actions that can be taken to prevent or mitigate the effects of blooms.

An initial construct for the generic conceptual model is illustrated in Fig. 2a and 2b. This model attempts to capture current understanding of CHAB occurrence, and the exposure media and pathways through which human and ecological receptors come into contact with stressors associated with CHABs (e.g., ingestion of toxins in drinking water). It also reflects key interactions among system components and the factors that modify the nature and intensity of effects, and ultimately risk. To facilitate its development and use, the generic conceptual model is organized into sub–models, each describing an important component of the overall CHAB problem or an expected pathway leading to risk. Thus, an Occurrence Sub–model encompasses the important environmental factors and processes, including human activity in the landscape, as they affect CHAB development and persistence. A Toxin Effects Sub–model describes exposure pathways relevant to human and ecological receptors, and begins to lay out the nature of effects that could be experienced as a result of exposure to cyanotoxins. A Cost Sub–model identifies in a cursory way the many effects that CHAB occurrence, prevention and mitigation have on social and economic systems. These can range from lost revenues and opportunities for recreation and tourism, to the emotional costs associated with loss of pets and even livelihoods.

The generic conceptual model communicated in Fig. 2a and 2b is incomplete with respect to important effects sub–models and specific descriptions of causal pathways and mechanisms associated with exposure and effect. For this reason, an Algal Biomass Effects Sub–model is included solely as a placeholder to indicate the need to describe fully the multitude of issues associated with the CHAB problem. Further, salient details that relate, for example, to costs associated with prevention or mitigation of cyanobacteria blooms, an element of the Cost Sub–model, and to potential control points in the Occurrence Sub–model are omitted due to ignorance of those relationships, as well as to preserve the communication value of Fig. 2a and 2b. An important development activity with respect to the implementing the decision–support framework will be to complete this model to the extent current understanding permits. The deliberations of the other workgroups in this symposium can contribute to the model's completion.

Although informal guidance is available for development of conceptual models (e.g., U.S. EPA 1998; Harwell and Gentile 2000), their construction is as much an art as it is a science. To be fully supportive of CHAB risk management needs, the conceptual model is best developed in a group exercise that involves diverse disciplines, vocations and stakeholders. Members of this group should include scientists and public health specialists, regulatory analysts and managers, water distribution and treatment specialists, environmental economists, and representatives of key stakeholder groups. This group would focus on the realism, accuracy and completeness of the generic conceptual model as a system–wide representation of the CHAB problem. Its deliberations would be critical to identification of assessment endpoints against which risks are to be assessed, considering the myriad regulatory, economic, ecological and social factors associated with CHABs. To help ensure its credibility, the model should be independently reviewed by similar experts. Further, the model should be revisited periodically and refined with new technical understanding and the lessons learned from its application in management of CHAB issues.

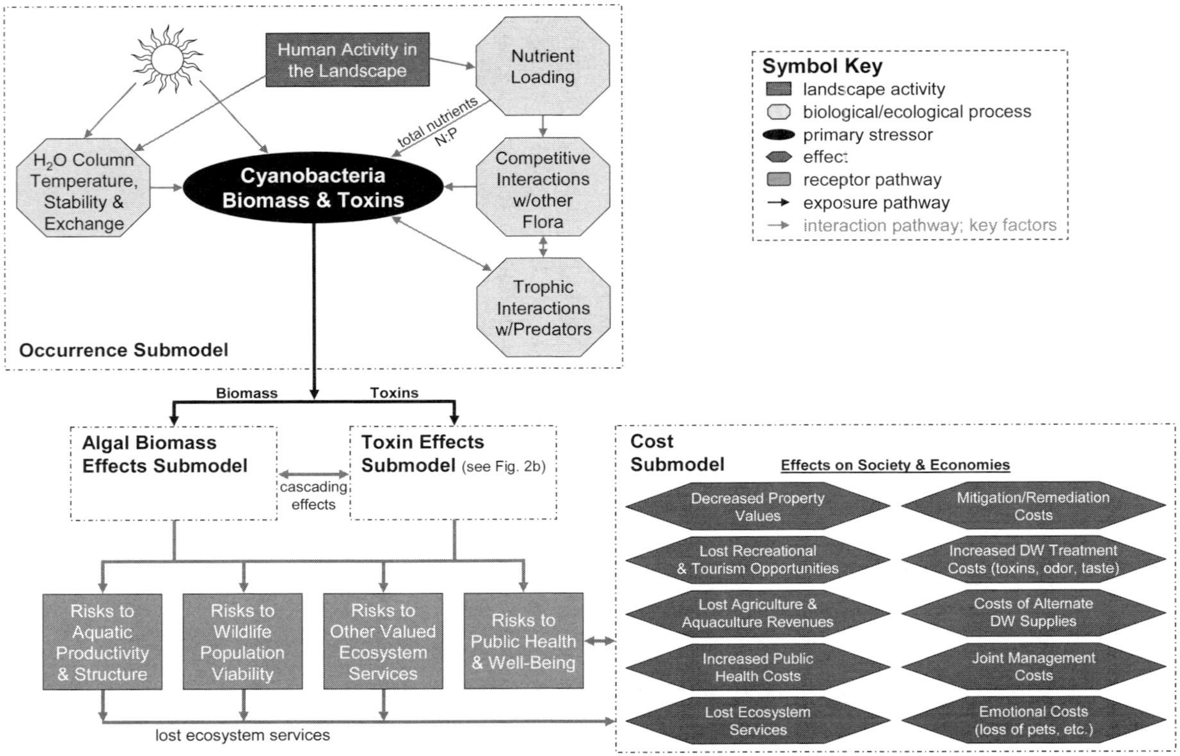

Fig. 2a. Conceptual model of cyanobacteria integrated risk: overall model;

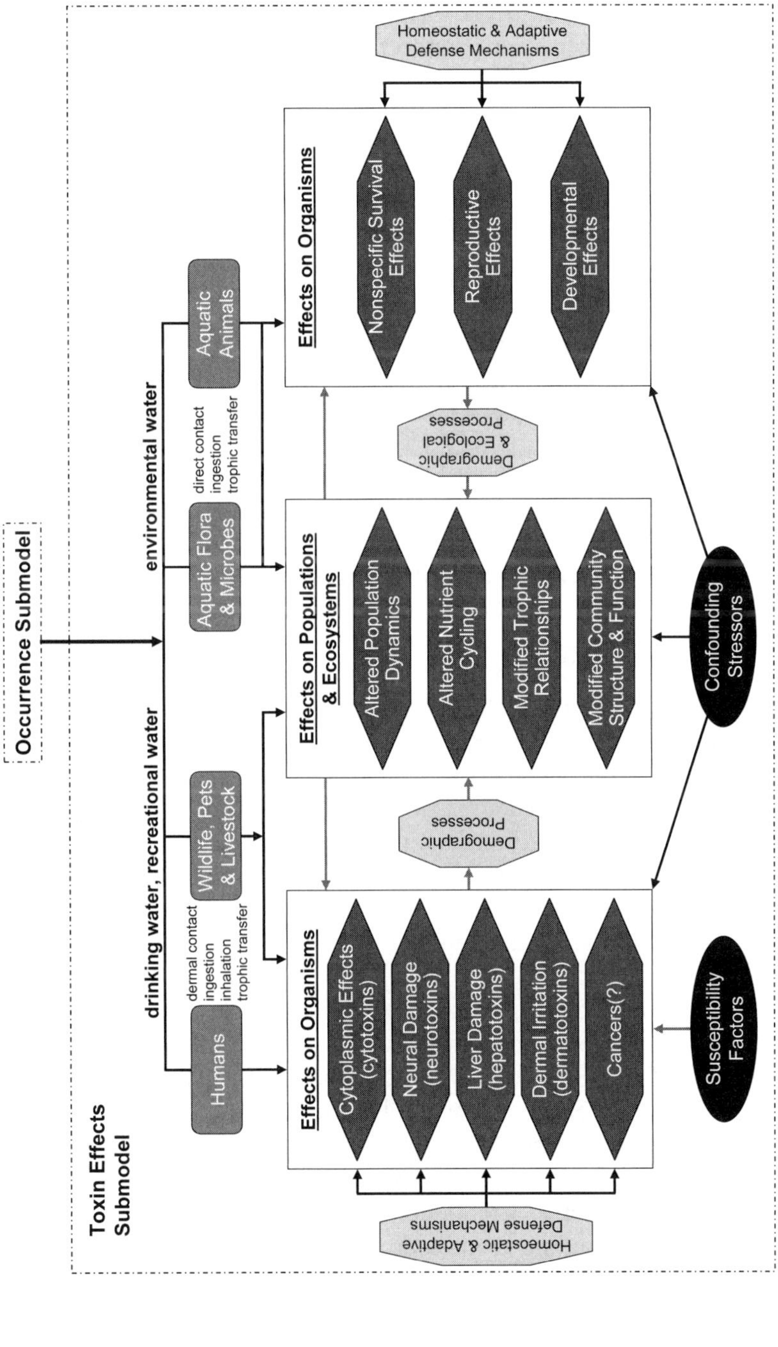

Fig. 2b. Expansion of the Toxins Effects Submodel shown in 2a.

An issue relevant to development of the generic conceptual model is the level of detail and complexity that it should reflect. Although a comprehensive model might seem overly complex and cumbersome, it does provide great value. First, it will support understanding of the full range of risks posed by CHABs, and therefore can help to inform regulatory and risk management actions through recognition of the connectivity of various system components. This will help to minimize unintended consequences associated with those actions. Second, it will facilitate recognition of critical prevention and mitigation control points and the evaluation of the effectiveness those controls through enhanced understanding of the relationships and factors influencing risks. Third, as a reflection of the state of knowledge, it can be used to identify critical research and information that are needed to manage CHAB risks effectively and efficiently. Fourth, it can serve as a useful tool for communicating the CHAB problem and its management to various stakeholder groups and the general public. And finally, it will facilitate development of the site and situation–specific conceptual models of Element 2 of the framework.

Site and Situation–Specific Conceptual Models

The issues associated with CHABs range from broad problems that are national in scope to localized ones, with characteristics that depend on context, scale, location and specific circumstances. Although comprehensive, the conceptual model of Element 1 is generic in its description of assessment endpoints, environmental circumstances, relevant stressors, and so on. Management action to address almost all CHAB problems will require additional specificity in defining the problem to be effective. To meet this need, Element 2 involves refinement of the generic, conceptual model on a site and situation–specific basis.

The conceptual models developed in this element would reflect the specific circumstances and factors relevant to particular CHAB prevention and management problems. Using a generalized case of drinking water distribution in South Australia to illustrate, the model would account for local factors affecting the occurrence of species of cyanobacteria within the source reservoir. Because the combination of species and the environmental factors that contribute to their bloom dynamics are somewhat unique to this situation, the conceptual model would focus specifically on those species and factors. The model would need to describe the relevant exposure pathways leading from source water to tap, identifying various control options that might be employed to detect and minimize drinking water contamination. It would account for possible effects linked to individual cyanotoxins (and their combinations) prevalent in South Australian species, reflecting the specific modifying factors (e.g., confounding stressors) operative at the site. Importantly, the model would identify those assessment endpoints important to local municipalities and stakeholders. Such a model might also describe inadvertent exposure of wildlife and domestic livestock to the source water and its consequences, if those issues are pertinent. Management options reflected in the model for detecting and controlling CHABs and their toxins would be those feasible for the particular water distribution system, and their associated costs would be grounded in the local economies. The conceptual model constructed for this South Australian drinking water illustration likely would differ in key aspects from one developed for, CHAB risks relevant to recreational uses of water bodies in Australia or another country. The degree of site specificity will also vary across models in Element 2. The conceptual model for a localized drinking water distribution system will contain much more detail about local conditions and factors than would one supporting establishment of national guidelines for cyanotoxins in drinking water.

The site and situation–specific conceptual models of Element 2 are refinements of the comprehensive model of Element 1. We recommend that they be developed by stripping away irrelevant or unimportant causal pathways and assessment endpoints from the comprehensive model, and adding detail relevant to the particular circumstances of the CHAB problem being addressed. Obviously, this requires in–depth understanding of circumstances and processes important to each problem, suggesting the need for additional information–gathering and research to fill key knowledge gaps. Once developed, the resulting site and situation–specific conceptual models will be important tools that help to focus the analytical activities of Elements 3 and 4, and creation of management plans (Element 5) that are responsive to the CHAB problem at hand.

Site and Situation–Specific Assessments of Risk

Element 3 of the decision–support framework utilizes the site and situation–specific conceptual model(s) to frame quantitative assessments of CHAB risks. The results of these risk assessments can be used to: 1) inform the development of protective guidelines and standards for cyanotoxin exposure in drinking water and recreational waters (reviewed by Burch, this volume); 2) understand the nature and magnitude of adverse ecological effects potentially resulting from CHABs; 3) diagnose potential causes of public health and ecological problems; and 4) facilitate evaluation of management alternatives and control options. For reasons outlined elsewhere (See Orme–Zavaleta and Munns this volume), we recommend that the risk assessments supporting CHABs management be integrated.

Risk assessment as a technology is fairly well developed; its concepts and uses for supporting policy and management need not be detailed here. Yet, several research and developmental needs remain with respect to its application to specific CHABs management problems. Many of these are identified earlier in this chapter and elsewhere in this volume as they apply to key knowledge gaps in CHAB occurrence, exposure and effects. Application of an integrated approach for risk assessment also will require development and testing of system–wide modeling techniques that reflect the connectivity of system components and therefore risks (See Orme–Zavaleta and Munns this volume). This need is not unique to the CHABs problem, rather being one of integrated risk assessment in general. As applied to CHABs , methods for integrated risk assessment will need to account for the complex interactions that occur within ecological–human–socioeconomic systems that are potentially affected by the stressors associated with blooms. Also required is the ability to accommodate a multitude of assessment endpoints salient to public health, ecological sustainability and services, domestic production and human well being. Until the science of integrated risk assessment is fully developed, this element of the decision–support framework may need to rely on independent assessments of health and ecological risks.

It is likely that some CHAB management determinations can be made effectively in a context narrower than that afforded by integrated risk assessment. For instances where regulatory or other requirements focus singularly on some component or aspect of the overall problem (e.g., mitigating risk of livestock mortality caused by cyanobacteria in an isolated pond), reasonable decisions can be made relative to such requirements without the need to evaluate the overall problem comprehensively. Thus, the risk assessments of Element 3 can be performed with this singular aspect as their objective. While permitting implementation of the decision–support framework in advance of full development of integrated risk assessment methods, the decisions that result may fail to acknowledge the implicit tradeoffs involved.

The degree of conservatism in assumptions taken in the risk assessments of Element 3 will depend upon the management decisions supported. Application of the framework to national–scale issues, such as establishing protective guidelines for drinking water, likely will require use of uncertainty or safety factors to ensure protection of especially sensitive or susceptible receptors. For some localized issues, such as prevention of blooms in water bodies used primarily for recreation, lower levels of conservatism may be advantageous as the benefits of recreational use are weighed against the costs of preventative measures. As with that for performing integrated risk assessments, the research needed to ensure appropriate conservatism in risk assessment is not unique to the CHAB problem, but this problem provides a distinct context within which to conduct that research.

Site and Situation–Specific Multi–criteria Decision Analysis

As described during this symposium and reflected in the comprehensive conceptual model of Element 1, the ecological–human–socioeconomic systems potentially affected by CHABs are complex. The information used to evaluate these systems is diverse, as often are the stakeholders affected by decisions made to manage CHAB risks. Because of this, it might be argued that policy and decision making can only be accomplished by partitioning the problem into more tractable subsets. To do so, however, may reduce the effectiveness of management determinations through failure to recognize the tradeoffs inherent to those decisions and the unintended consequences that may result. In Element 4, we recommend applying the concepts and methods of multi–criteria decision analysis (Belton and Steward, 2002) to facilitate informed decision making in the complex context of CHABs problems.

Multi–criteria decision analysis is designed to support selection among alternatives in situations involving potentially conflicting objectives or decision criteria. Approaching such problems in a systematic fashion, multi–criteria decision analysis involves the key steps of: 1) structuring decision making goals in terms of defined hierarchies of criteria; 2) evaluating decision alternatives in terms of the extent to which they satisfy each of the identified criteria; and 3) aggregating across criteria to measure the extent to which each alternative satisfies the overall goals represented by the criteria. The result of its application is an ordered ranking of decision alternatives that communicates the best option while considering a number of factors (e.g., risk, benefits, costs, option effectiveness, stakeholder values, etc.).

Multi–criteria decision analysis is an evolving technology, receiving increasing attention for managing complex environmental problems. Examples of multicriteria–decision analysis used by Federal Agencies are available (Kiker et al. 2005). Multi–criteria decision analysis would utilize the assessments of risk from the previous element, together with an understanding of the alternatives for management action, estimates of associated costs, and the values expressed by stakeholders, to provide a decision–support framework for addressing the site and situation–specific CHAB issue. There are numerous specific methods available to accomplish the three key steps of multi–criteria decision analysis (Belton and Steward, 2002 and See Linkov and Steevens this volume). Finding no examples of their application to CHABs, a substantial research need for the framework will be to explore and refine approaches for multi–criteria decision analysis for use with the types of problems and issues associated with CHABs. This approach will facilitate decision making in the face of the seemingly overwhelming complexity of some CHAB problems, thus supporting development of management plans to address those problems.

Site and Situation–Specific Management Plans

In general, managers can exert a greater influence on physical rather than chemical factors controlling CHABs. Many of the actions are directly related to watershed management alternatives (Fig. 3). For example, water resource managers may be able to alter water flow and thus decrease residence time in a reservoir, or affect vertical mixing in the water column by controlling water intake. Similarly, the ability to release water from different depths behind dams could affect algal blooms by changing the temperature profile of water downstream. Consistent forceful mixing prevents algae from maintaining optimal water depth, slowing their growth. Additionally, shear disrupts the filaments which hold together heterocysts, the nitrogen–fixing cells formed by some cyanobacteria. Although mixing the water in a reservoir may not be a practical option for many managers, other options may be available such as controlled downstream releases to reduce cell and toxin concentrations or dredging to lower nutrient and trace metal concentrations. Our literature review indicates that even though managers may be able to influence multiple factors associated with algal blooms, little work has been done to study the actual impact of water management options on CHABs and subsequent toxin production.

Armed with an understanding of the risks posed by CHABs in specific situations, and of effective alternatives for managing those risks, Element 5 of the decision–support framework consists of management plans for controlling risks on a prospective and real–time basis. These plans would focus on CHAB prevention, detection, response and mitigation as appropriate to the situational context, all as reflected in the relevant conceptual model. Various versions of such plans currently are being used, particularly in Australia. The recommendations for management plans in Element 5 rely heavily on the best–practice experiences gained through their use.

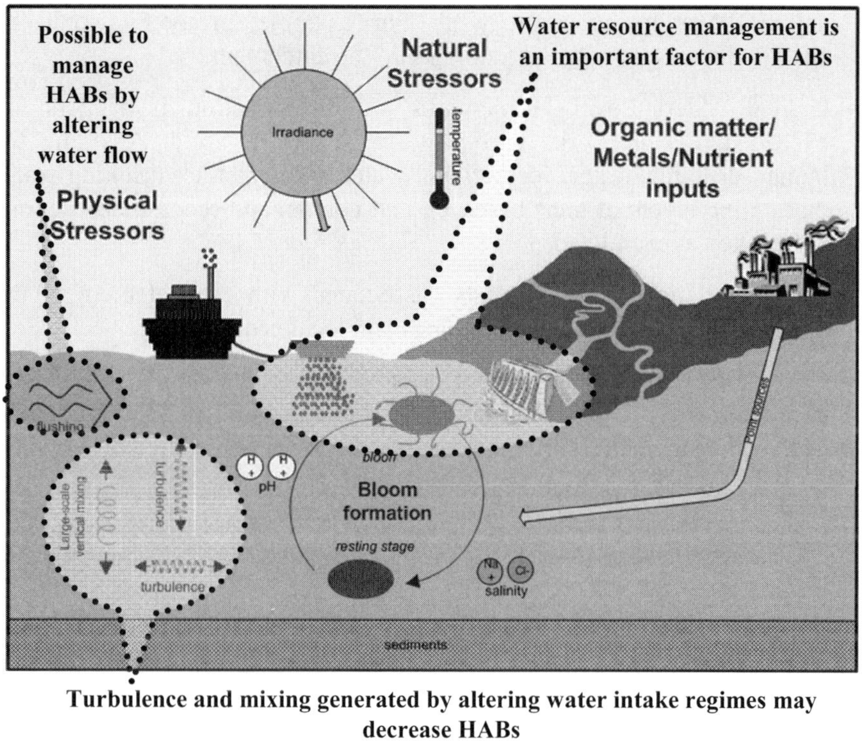

Fig. 3. HABs and water resource management actions.

Each management plan would identify the particular actions to be taken to manage aspects of CHAB risk. An outline of a management plan for prevention of CHABs in a hydrological system providing water to a municipal water supply is provided in Fig. 3 as an illustration. Identified in this outline are the:

- "Critical Control Points" for management action within the system (e.g., catchment, rivers, etc. Fig. 3). These identify where, and perhaps when, management action is to be taken to control some factor or environmental process contributing to risk.
- Environmental factors or processes to be managed and approaches for their management ("Management Control"). Targets and controls for nitrogen and phosphorous loading into the catchment would be specified in the relevant critical control point example reflected in Fig. 4.
- Protocols to provide the data both to monitor the effectiveness of controls and to identify the data elements that act as triggers for further

action ("Monitoring Strategy"). For example, protocols would be specified to assess nutrient loading into the catchment.

- Graded "Alert Level Frameworks" to direct action if triggers are exceeded. Frameworks currently are in use in South Australia for human drinking water, recreational water and livestock drinking water that define levels of alert based on cell counts, and recommend actions to be taken at each level.
- Characterizations of the "Costs" associated with each element of the plan, based on available technologies and local economies.
- Characterizations (to the extent possible) of the "Benefits" gained by each element of the plan, including benefits not directly related to CHAB management (e.g., improved aesthetics as an ecological service).

Aspects of management plan development would be informed by the multi–criteria decision analyses of Element 5. Current understanding of the environmental and social context of the situation, plus transfer of best–practice knowledge gained from local experience, are critical to effective plans. In addition to the research identified elsewhere in this volume, site–specific understanding of local hydrologic cycles, sources of nutrient input, cyanobacteria dynamics, social behaviors, and other factors may be needed on a site and situation–specific basis to tailor the management plans to specific problems. The site and situation–specific conceptual models of Element 2 can be used to guide investments in this research, focusing on their largest uncertainties.

Critical Control Point	Management Control	Monitoring Strategy	Alert Level Framework	Costs	Benefits
1. Catchment	nitrogen & phosphorus	nutrient loading	Alert Level 1 • triggers • actions Alert Level 2 • triggers • actions etc.	A$	A$, improved aesthetics
2. Rivers	flow	stratification, cell counts	Alert Level 1 • triggers • actions Alert Level 2 • triggers • actions etc.	A$	A$, improved productivity
3. Reservoir	nitrogen & phosphorous, stratification	nutrient concentrations, stratification, cell counts	Alert Level 1 • triggers • actions Alert Level 2 • triggers • actions etc.	A$	A$, reduced livestock risks

Treat after extraction is a further control point:

Critical control point	Management control	Monitoring strategy	Alert level framework	Cost	Benefits
Treatment plant	cyanobacterial cell and metabolites	Removal of cells and metabolites		$	$ reduced human health risk

Fig. 4. Example of Critical Control Point

The details and uses of management plans likely will vary across CHAB problems. While the example outlined above has obvious application for municipal and private authorities providing drinking water, a management plan appropriate for national implementation of a guideline would take a very different form, one likely grounded in the established regulatory and compliance structure of that country (or other authority). A research need for the decision–support framework therefore relates to optimizing the nature of the management plans for various CHAB issues.

Framework Effectiveness

Effective decision–making requires an explicit structure for jointly considering the environmental, ecological, technological, economic, and socio–political factors relevant to evaluating alternatives and making a decision. Integrating this heterogeneous information with respect to human aspirations and technical applications demands a systematic and understandable framework to organize the people, processes, and tools for making a structured and defensible decision.

For further information on MCDA see Appendix A of this report (Linkov and Steevens this volume).

Summary and Conclusions

The Risk Assessment Work Group focused on six charge questions related to CHABS, cyanobacteria and their toxins. The charge questions covered the following topics:

- Research needed to reduce uncertainty in establishing health based guidelines
- Research that minimize the cost and maximize the benefits of various regulatory approaches
- Exposure pathways for receptors of concern
- Data available to support the derivation of health–based guideline values for harmful cyanobacterial algal blooms
- Ecological services that guidelines or regulations should protect?
- A framework for making risk management determinations that incorporates consideration of the characteristics of CHABs, the risk for human health, ecosystem viability, and the costs and benefits of CHABs detection and management?

The Work Group concluded that there is a considerable amount of human case-study data and information from animal studies to demonstrate that cyanobacterial toxins pose a hazard to humans, domestic animals, wildlife, and the ecosystem. However, the data on dose–response are limited and confounded by a lack of sufficient pure toxin to conduct most of the toxicological studies that will be needed in order to answer remaining questions on risk, and to provide the data for quantitative dose–response analysis. The Work Group recommended that research on purification or synthesis of pure toxin must be accomplished before the large scale studies to establish dose–response relationships will be possible. As the necessary–pure toxins become available, the Work Group recommended that studies be prioritized by the impact that they will have on reducing the uncertainty in the risk assessment in order to minimize the research costs and maximize the risk assessment benefits. Use of quantitative structure activity relationships (QSAR) and toxicity equivalency factor studies are also recommended as approaches for filling dose–response data gaps.

The Work Group recognized that CHABs rarely introduce single toxins into the water supply. Under CHAB conditions, affected water is likely to contain a variety of toxins in varying concentrations that may change over the duration of the bloom. Accordingly, research on cyanotoxin interactions is needed, along with the development of risk assessment approaches for CHAB mixtures.

The development of simple, accurate analytical methods that can be utilized by most analytical laboratories or used in the field was recognized as a major data need for establishing exposure potential and monitoring bloom conditions. Most currently available methods are time–consuming and/or costly.

Human exposure to cyanobacterial toxins can occur through ingestion of contaminated drinking water, plus dermal contact and/or inhalation of aerosols while bathing and showering in tap water. Treatment can reduce the concentrations of both the toxins and the bacteria in the treated water but there is still much to be learned about the effectiveness of most treatment technologies on cyanobacteria and toxin removal.

Human exposure to cyanobacteria and their toxins also occurs through incidental ingestion, dermal contact, and inhalation of aerosols during recreational use of surface waters, ingestion of contaminated fish and other foods of aquatic origin, and/or BGAS supplements. Establishing intakes and duration parameters for these exposure scenarios will facilitate the application of risk assessment approaches to these situations.

References

Anderson DM, Anderson P, Bricelj VM, Cullen JJ, Rensel JE (2001) Monitoring and Management Strategies for harmful algal blooms in coastal waters. APEC#201–MR–01.1, Asia

Anon (1998) The Determination of Microcystin Algal Toxins in Raw and Treated Waters by High Performance Liquid Chromatography. Environment Agency Bristol

Astrachan NB, Archer BG (1981) Simplified monitoring of anatoxin–a by reverse–phase high performance liquid chromatography and the sub–acute effects of anatoxin–a in rats. In: The Water Environment: Algal Toxins and Health WW Carmichael edn. Plenum Press, New York NY pp 437–446

Astrachan NB, Archer BG, Hilbelink DR (1980) Evaluation of the subacute toxicity and teratogenicity of anatoxin–a. Toxicon 18(5–6):684–688

Belton V, Steward T (2002) Multicriteria decision analysis: An integrated approach. Kluwer, Boston MA USA

Bensley RJ (1991) Defining spiritual health: A review of the literature. Journal of Health Education 22(5) pp 287–290

Brenon C, Burch MD (2001) Evaluation of Analytical Methods for Detection and Quantification of Cyanotoxins in Relation to Australian Drinking Water Guidelines. Legislative Services Ausinfo

Burch M (this volume) Effective doses, guidelines and regulations. Advances in Experimental Medicine and Biology

Carmichael WW, Evans WR, Yin QQ, Bell P, Moczydlowski E (1997) Evidence for paralytic shellfish poisons in freshwater cyanobacterium, Lyngbya wollei (Farrow ex Gomont) comb. Nov Appl Environ Microbiol 63:3104

Carmichael W (2001) Health effects of toxin–producing cyanobacteria: The CyanoHABs. Human & Ecol Risk Assessment 7:1393–1407

Chorus and Bartram (eds) (1999) Toxic cyanobacteria in water: a guide to their public health consequences, monitoring and management. E& FN Spon for the World Health Organization pp 416

Cox PA, Banack SA, Murch SJ, Rasmussen U, Tien G, Bidigare RR, Metcalf JS, Morrison LF, Codd GA, Bergman B (2005) Diverse taxa of cyanobacteria produce b–N–methylamino–L–alanine, a neurotoxic amino acid, PNAS 102(14): 5074–5078

Crompton JL (1998) Financing and acquiring park, recreation and open space resources. Champaign, IL: Human Kinetics

Dietrich, D.R., Fischer, A., Michel, C., and Hoeger, S.J., (this volume). Toxin mixture in cyanobacterial blooms – a critical comparison of reality with current procedures employed in human health risk assessment

Dietrich DR, Hoeger S (2005) Guidance values for microcystins in water and cyanobacterial products (blue–green algal supplements): a reasonable or misguided approach? Toxicol Appl Pharmacol 203(3):273–289

Driscoll CP, McGowan PC, Miller EA, Carmichael WW (2002) Case Report: great blue heron (Ardea herodias) morbidity and mortality investigation in Maryland's Chesapeake Bay. Proceedings of the Southeast Fish and Wildlife Conference Baltimore, Maryland Oct. 24, 2002 (Poster)

Ellison CW (1983) Spiritual well–being: Conceptualization and measurement. Journal of Psychology and Theology 11, 330–340

Ellison CW, Smith J (1991) Toward an integrative measure of health and wellbeing. Journal of Psychology and Theology 19(1) pp 35–48

Falconer IR (1998) Algal toxins and human health. In Hubec J (ed) Handbook of Environmental Chemistry vol 5 Part C Quality and Treatment of Drinking Water. pp 53–82

Fawell JF, James HA (1994) Toxins from Blue–green Algae: Toxicological Assessment of Anatoxin–a and a Method for its Determination in Reservoir Water. Foundation for Water Research, Marlow, UK. FWR Report No. FR0434/DoE 3728

Fawell JK, Mitchell RE, Hill RE, Everett DJ (1999a) The toxicity of cyanobacterial toxins in the mouse: II Anatoxin–a. Hum Exp Toxicol 18:168–173

Fawell JK, Mitchell RE, Everett DJ, Hill RE (1999b) The toxicity of cyanobacterial toxins in the mouse. 1. Microcystin–LR. Human Exp Toxicol 18(3):162 167

Fischer WJ, Garthwaite I, Miles CO, Ross KM, Aggen JB, Chamberlin RA, Towers NR, Dietrich DR (2001) A Congener–Independent Immunoassay for Microcystins and Nodularins. Environ Sci Technol 35(24): 4753–4757

Gastrich MD, Wazniak CE (2002) A brown tide bloom index based on the potential harmful effects of brown tide alga, Aureococcosu anophagefferens. Aq. Ecosys. Health and Manage 5(4):435–441

Harwell MA, Gentile JH (2000) Environmental decision–making for multiple stressors: framework, tools, case studies, and prospects. In: Ferenc SA, Foran JA edn, Multiple stressors in ecological risk and impact assessment: Approaches to risk estimation. SETAC Press, Pensacola FL USA

Heinze R (1999) Toxicity of the cyanobacterial toxin microcystin–LR to rats after 28 days intake with the drinking water. Environ Toxicol 14(1):57–60

Heintzman P (1999) Spiritual wellness: theoretical links with leisure. Journal of Leisurability 26(2)

Hitzfeld B, Höger SJ, Dietrich DR (2000) Cyanobaterial Toxins: Removal during Drinking Water Treatment, and Human Risk Assessment. Environmental Health Perspectives vol 108 Supplement 1 March pp 113–122

Hoppu K, Salmela J, Lahti K (2002) High risk for symptoms from use of water contaminated with cyanobacteriae in sauna. Clinical Toxicology 40 309–310

Humpage AR, Falconer IR (2003) Oral toxicity of the cyanobacterial toxin cylindrospermopsin in male Swiss albino mice: Determination of no observed adverse effect level for deriving a drinking water guideline value. Environ Toxicol 18:94–103

Hunter LM (2001) The environmental implications of population dynamics. RAND publishing. http://www.rand.org/publications/MR/MR1191/

Ito E, Kondo F, Harada KI (1997) Hepatic necrosis in aged mice by oral administration of microcystin–LR. Toxicon 35(2):231–239

Jochimsen EM, Carmichael WW, An J, Cardo DM, Cookson ST, Holmes CEM, Antunes de MBC, Filho DAM, Lyra TM, Barreto VST, Azevedo SMFO, Jarvis WR (1998) Liver failure and death after exposure to microcystins at a haemodialysis center in Brazil. New Engl J Med 338(13):873–878

Kiker GA, Bridges TS, Varghese A, Seager TP, Linkov I (2005) Application of multicriteria decision analysis in environmental decision making. Integrated Environmental Assessment and Management 1:95–108

Linkov and Steevens (this volume) Appendix A to Risk assessment of Cyanobacterial Harmful Algal Blooms.

Meriluoto J, Codd GA (2005) Cyanobacterial Monitoring and Cyanotoxin Analysis. Abo Akademi University Press

Orme–Zavaleta J, Munns WR Jr (this volume) Integrating human and ecological risk assessment for the cyanobacterial harmful algal bloom problem. Advances in Experimental Medicine and Biology

Piehler, MF (this volume) Watershed management strategies to prevent and control cyanobacterial harmful algal blooms. Advances in Experimental Medicine and Biology

Queensland Health (2001) Environmental health assessment guidelines: Cyanobacteria in recreational and drinking waters. Environmental Health Unit Brisbane, Australia

Shaw GR, Seawright AA, Moore MR, Lam PKS (2000) Cylindrospermopsin, a cyanobacterial alkaloid: Evaluation of its toxicologic activity. Ther Drug Monit 22(1):89–92

Shaw GR, Seawright AA, Moore MR (2001) Toxicology and human health implications of the cyanobacterial toxin cylindrospermopsin. In: Mycotoxins and Phycotoxins in Perspective at the Turn of the Millennium, WJ Dekoe, RA Samson, HP van Egmond et al. edn IUPAC & AOAC International, Brazil, pp 435–443

Smayda T (1997) What is a bloom? A commentary. Limnol Oceanogr 42(5, part2) 1132–1136

Steffenson D (this volume) Economic costs of cyanobacterial blooms. Advances in Experimental Medicine and Biology

Suter GW II, Vermeire T, Munns WR Jr, Sekizawa J (2003) Framework for the integration of health and ecological risk assessment. Human and Ecological Risk Assessment 9:281–301

Thiel P (1994) The South African contribution to studies on the toxic cyanobacteria and their toxins. In: Toxic Cyanobacteria: Current Status of Research and Management. Proceedings of an International Workshop. Adelaide, Australia, March 22–26

Ueno, Y., Y. Makita, S. Nagata et al. (1999) No chronic oral toxicity of a low–dose of microcystin-LR, a cyanobacterial hepatoxin, in female Balb/C mice. Environ. Toxicol. 14(1):45–55

U.S. DI (Department of the Interior, Fish and Wildlife Service) and U.S. Department of Commerce, U.S. Census Bureau (2002) 2001 National survey of fishing, hunting and wildlife–associated recreation. United States Fish and Wildlife Service, pp 170

U.S. EPA (U.S. Environmental Protection Agency) (1998) Guidelines for ecological risk assessment. EPA/630/R–95/002F. Washington, DC, USA

U.S. EPA (U.S. Environmental Protection Agency) (1998) Announcement of the drinking water contaminant candidate list: NoticeFinal Notice. Federal Register 63(40):10274–10287 March 2 1998

U.S. EPA (U.S. Environmental Protection Agency) (2005) Drinking water contaminant candidate list 2: Final Notice. Federal Register 70(36):9071–9077 February 24 2005

Van Dolah FM, Roelke D, Greene R (2001) Health and ecological impacts of harmful algal blooms: Risk assessment needs. Human and Ecol Risk Assessment 7:1329–1345

Water Supply Regulations (2000) Statutory Instrument 2000 No. 3184. The Stationery Office Limited. United Kingdom

World Health Organization (2003) Guidelines for safe recreational waters, Volume 1– Coastal and fresh waters, Chapter 8: Algae and cyanobacteria in fresh water. WHO Publishing, Geneva, pp 136–158 [Available at: http://www.who.int/water_sanitation_health/bathing/srwe1/en/]

Zeck A, Weller MG, Bursill D, Niessner R (2001) Generic microcystin immunoassay based on monoclonal antibodies against Adda. Analyst 126, 2002–2007

Chapter 35 Appendix A: Multi-Criteria Decision Analysis

Linkov I, Steevens J

A detailed analysis of the theoretical foundations of different MCDA methods and their comparative strengths and weaknesses is presented in Belton and Stewart (2002). MCDA methods utilize a decision matrix to provide a systematic analytical approach for integrating risk levels, uncertainty, and valuation, which enables evaluation and ranking of many alternatives. MCDA overcomes the limitations of less structured methods such as comparative risk assessment (CRA), which suffers from the unclear way in which it combines performance on criteria (see Bridges et al. 2005 for more information on CRA). Within MCDA, almost all methodologies share similar steps of organization and decision matrix construction, but each methodology synthesizes information differently (Yoe 2002). Different methods require diverse types of value information and follow various optimization algorithms. Some techniques rank options, some identify a single optimal alternative, some provide an incomeplete ranking, and others differentiate between acceptable and unacceptable alternatives.

Elementary MCDA methods can be used to reduce complex problems to a singular basis for selection of a preferred alternative. However, these methods do not necessarily weight the relative importance of criteria and combine the criteria to produce an aggregate score for each alternative. While elementary approaches are simple and can, in most cases, be executed without the help of computer software, these methods are best suited for single-decision maker problems with few alternatives and criteria, a condition that is rarely characteristic of environmental projects.

Table A1 summarizes a number of more sophisticated MCDA methods. Multi-attribute utility theory (MAUT), multi-attribute value theory (MAVT), and the analytical hierarchy process (AHP) are more complex methods that use optimization algorithms, whereas outranking eschews optimization in favor of a dominance approach. The optimization approaches employ numerical scores to communicate the merit of each option on a single scale. Scores are developed from the performance of alternatives with respect to individual criteria and then aggregated into an overall score. Individual scores may be simply summed or averaged, or a weighting mechanism can be used to favor some criteria more heavily than others. The goal of MAUT is to find a simple expression for the net benefits of a decision. Through the use of utility or value functions, the MAUT method transforms diverse criteria into one common scale of utility or value. MAUT relies on the assumptions that the decision-maker is rational (preferring more utility to less utility, for example), that the decision-maker has perfect knowledge, and that the decision-maker is consistent in his judgments. The goal of decision-makers in this process is to maximize utility or value. Because poor scores on criteria can be compensated for by high scores on other criteria, MAUT is part of a group of MCDA techniques known as "compensatory" methods.

Similar to MAUT, AHP (Saaty 1994) aggregates various facets of the decision problem using a single optimization function known as the objective function. The goal of AHP is to select the alternative that results in the greatest value of the objective function. Like MAUT, AHP is a compensatory optimization approach. However, AHP uses a quantitative comparison method that is based on pair-wise comparisons of decision criteria, rather than utility and weighting functions. All individual criteria must be paired against all others and the results compiled in matrix form. For example, in examining the choices in the selection of a non-lethal weapon, the AHP method would require the decision-maker to answer questions such as, "With respect to the selection of a weapon alternative, which is more important, the efficiency or the reduction of undesired effects (e.g., health impacts)?" The user uses a numerical scale to compare the choices and the AHP method moves systematically through all pair-wise comparisons of criteria and alternatives. The AHP technique thus relies on the supposition that humans are more capable of making relative judgments than absolute judgments. Consequently, the rationality assumption in AHP is more relaxed than in MAUT.

Unlike MAUT and AHP, outranking is based on the principle that one alternative may have a degree of dominance over another (Kangas et al. 2001). Dominance occurs when one option performs better than another on at least one criterion and no worse than the other on all criteria (ODPM 2004). However, outranking techniques do not presuppose that a single best alternative can be identified. Outranking models compare the performance of two (or more) alternatives at a time, initially in terms of each criterion, to identify the extent to which a preference for one over the other can be asserted. Outranking techniques then aggregate the preference information across all relevant criteria and seek to establish the strength of evidence favoring selection of one alternative over another. For example, an outranking technique may entail favoring the alternative that performs the best on the greatest number of criteria. Thus, outranking techniques allow inferior performance on some criteria to be compensated for by superior performance on others. They do not necessarily, however, take into account the magnitude of relative underperformance in a criterion versus the magnitude of over-performance in another criterion. Therefore, outranking models are known as "partially compensatory." Outranking techniques are most appropriate when criteria metrics are not easily aggregated, measurement scales vary over wide ranges, and units are incommensurate or incomparable (Seager 2004).

Table A1. Comparison of Critical Elements, Strengths and Weaknesses of Several Advanced MCDA Methods: MAUT, AHP, and Outranking (after [19]).

Method	Important elements	Strengths	Weaknesses
Multi-attribute utility theory	• Expression of overall performance of an alternative in a single, non-monetary number representing the utility of that alternative • Criteria weights often obtained by directly surveying stakeholders	• Easier to compare alternatives whose overall scores are expressed as single numbers • Choice of an alternative can be transparent if highest scoring alternative is chosen • Theoretically sound — based on utilitarian philosophy • Many people prefer to express net utility in non-monetary terms	• Maximization of utility may not be important to decision makers • Criteria weights obtained through less rigorous stakeholder surveys may not accurately reflect stakeholders' true preferences • Rigorous stakeholder preference elicitations are expensive
Analytical hierarchy process	• Criteria weights and scores are based on pairwise comparisons of criteria and alternatives, respectively	• Surveying pairwise comparisons is easy to implement	• The weights obtained from pairwise comparison are strongly criticized for not reflecting people's true preferences • Mathematical procedures can yield illogical results. For example, rankings developed through AHP are sometimes not transitive

Method	Important elements	Strengths	Weaknesses
Outranking	• One option outranks another if : 1. "it outperforms the other on enough criteria of sufficient importance (as reflected by the sum of criteria weights)" and 2. it "is not outperformed by the other in the sense of recording a significantly inferior performance on any one criterion" • Allows options to be classified as "incomparable"	• Does not require the reduction of all criteria to a single unit • Explicit consideration of possibility that very poor performance on a single criterion may eliminate an alternative from consideration, even if that criterion's performance is compensated for by very good performance on other criteria	• Does not always take into account whether over-performance on one criterion can make up for under-performance on another • The algorithms used in outranking are often relatively complex and not well understood by decision makers

Example Ahp Application Framework

As an illustrative example of the analytical hierarchy process, consider the selection of a harmful algal bloom management strategy. Three options are available to the hypothetical managers:

- Algaecides
- Flushing
- Detoxification

The first step is to decide upon the objectives or criteria by which the alternative management techniques will be measured. As an example, we select the following criteria: (1) the strategy's human health impacts, (2) its environmental impacts, and (3) its social impacts.

The second step is to weight the importances of these criteria for the decision maker. Although in this simple scenario it would be possible to assign weights directly, in many practical applications it may be difficult because of the multitude of criteria and subcriteria that the decision maker may face. Therefore, in AHP, the decision-maker does not give importance weightings directly; rather, the category weightings are derived from a series of relative judgments. In this scenario, the decision-maker has input three relative judgments, in the form of weightings ratios. He has, for example, weighted human health impacts as four times more important than social impacts (see Table A2). From these relative weightings, AHP derives normalized weightings for the three criteria (see Table A3).

Table A2. Relative importance weightings, in the ratio form of row element / column element.

Main criteria table	Human Health Impacts	Environmental Impacts	Social Impacts
Human Health Impacts		4.0	4.0
Environmental Impacts			1.0
Social Impacts			

Table A3. Importance weightings for main criteria categories.

Main criteria weightings	
Human Health Impacts	0.667
Environmental Impacts	0.167
Social Impacts	0.167

Additionally, even in this simple case, because the main criteria categories are too broad to be used directly in evaluating management alternatives, sub-criteria within each of these categories should be developed. Within the Human Health Impacts category, for instance, one might consider drinking water quality, dermal effects, and inhalation effects. Similarly, sub-criteria may be developed for the other two criteria categories – such as the strategy's effects on fish, its birds, and mammals, or its cost and public acceptability (see Table A4). Sub-criteria are compared and weighted in a pairwise manner similar to that for the main criteria (see Table A5, Table A6, and Table A7).

Table A4. Sub-criteria for each main criteria category.

Goal: Identify best management techniques for harmful algal blooms

Main criteria category	**Sub-criteria**
Human Health Impacts	• Drinking water quality • Dermal effects • Inhalation effects
Environmental Impacts	• Effects on fish • Effects on birds • Effects on mammals
Social Impacts	• Cost • Public acceptability

Table A5. Importance weightings for Human Health Impacts sub-criteria.

Human Health Impacts sub-table	Drinking water quality	Dermal effects	Inhalation effects
Drinking water quality		7.0	5.0
Dermal effects			1.0
Inhalation effects			

Table A6. Importance weightings for Environmental Impacts sub-criteria.

Environmental Impacts sub-table	Effects on fish	Effects on birds	Effects on mammals
Effects on fish		1.0	7.0
Effects on birds			8.0
Effects on mammals			

Table A7. Importance weightings for Social Impacts sub-criteria.

Social Impacts sub-table	Cost	Public acceptability
Cost		6.0
Public acceptability		

Once relative weightings have been given for each of the sub-criteria, normalized weightings may be calculated for use in scoring different harmful algal bloom management alternatives (see breakdown in Table A8).

Table A8. Importance weightings for both main criteria categories and embedded sub-criteria.

Goal: Select harmful algal bloom management response	**Weighting**	**Sub-weighting**
Human Health Impacts	0.667	
• Drinking water quality		0.747
• Dermal effects		0.119
• Inhalation effects		0.134
Environmental Impacts	0.167	
• Effects on fish		0.458
• Effects on birds		0.479
• Effects on mammals		0.063
Social Impacts	0.167	
• Cost		0.857
• Public acceptability		0.143

The third step is to measure relative performance of each management option on each criteria. Again, the decision-maker inputs a relative ranking – only now it is a preference ranking between alternatives rather than an importance ranking among criteria. If a quantitative answer is not given, a qualitative statement may be transformed into a numerical value through a standardized system (i.e. the numbers 1, 3, 5, 7, and 9 correspond to the judgments "equally important," "moderately more," "strongly more," "very strongly more," and "extremely more," respectively). Once the decision-maker gives inputs for each alternative under each sub-criteria, he may use the previously obtained weightings to calculate scores for each main criteria, followed by an overall score for each alternative (see Table A9). The highest scoring alternative is, according to the rankings and preferences given by the decision-maker throughout the analytic hierarchy process, the best strategy for the situation.

Table A9. Score breakdown for example decision.

Goal: Select harmful algal bloom management response	**Algaecides**	**Flushing**	**Detoxification**
Human Health Impacts	**0.061**	**0.332**	**0.607**
• Drinking water quality	0.061	0.353	0.586
• Dermal effects	0.060	0.249	0.691
• Inhalation effects	0.062	0.285	0.653
Environmental Impacts	**0.779**	**0.112**	**0.109**
• Effects on fish	0.783	0.174	0.043
• Effects on birds	0.778	0.042	0.180
• Effects on mammals	0.761	0.191	0.048
Social Impacts	**0.100**	**0.320**	**0.581**
• Cost	0.089	0.323	0.588
• Public acceptability	0.163	0.297	0.540
OVERALL SCORE	**0.187**	**0.293**	**0.520**

Many software packages exist to assist the decision-maker with implementation of the above process.

Framework Effectiveness

Effective decision-making requires an explicit structure for jointly considering the environmental, ecological, technological, economic, and socio-political factors relevant to evaluating alternatives and making a decision. Integrating this heterogeneous information with respect to human aspirations and technical applications demands a systematic and understandable framework to organize the people, processes, and tools for making a structured and defensible decision. Based on our review of MCDA, we have synthesized our understanding into a systematic decision framework (Fig. A1). This framework is intended to provide a generalized road map to the decision-making process.

Having the right combination of people is the first essential element in the decision process. The activity and involvement levels of two basic groups of people (decision-makers and scientists & engineers) are symbolized in Fig A1 by dark lines for direct involvement and dashed lines for less direct involvement. While the actual membership and the function of these groups may overlap or vary, the roles of each are essential in maximizing the utility of human input into the decision process. Each group has its own way of viewing the world, its own method of envisioning solutions, and its own societal responsibility. Policy- and decision-makers spend most of their effort defining the problem context and the overall constraints on the decision. In addition, they may have responsibility for the selection of the final decision and its implementation. Scientists and engineers have the most focused role in that they provide the measurements or estimations of the desired criteria that determine the success of various alternatives. While they may take a secondary role as decision-makers, their primary role is to provide the technical input as necessary in the decision process.

The framework places process in the center (Fig. A1). While it is reasonable to expect that the decision-making process may vary in specific details among regulatory programs and project types, emphasis should be given to designing an adaptable structure so that participants can modify aspects of the project to suit local concerns, while still producing a structure that provides the required outputs. The process depicted follows two basic themes: 1) generating alternatives, success criteria, and value judgments and 2) ranking the alternatives by applying the value weights. The first part of the process generates and defines choices, performance levels, and preferences. The latter section methodically prunes non-feasible alternatives by first applying screening mechanisms (for example, overall cost, technical feasibility, possible undesired consequences, or general societal acceptance) followed by a more detailed ranking of the remaining options by decision analytical techniques (AHP, MAUT, outranking) that utilize the various criteria levels generated by tools such as modeling, monitoring, or stakeholder surveys.

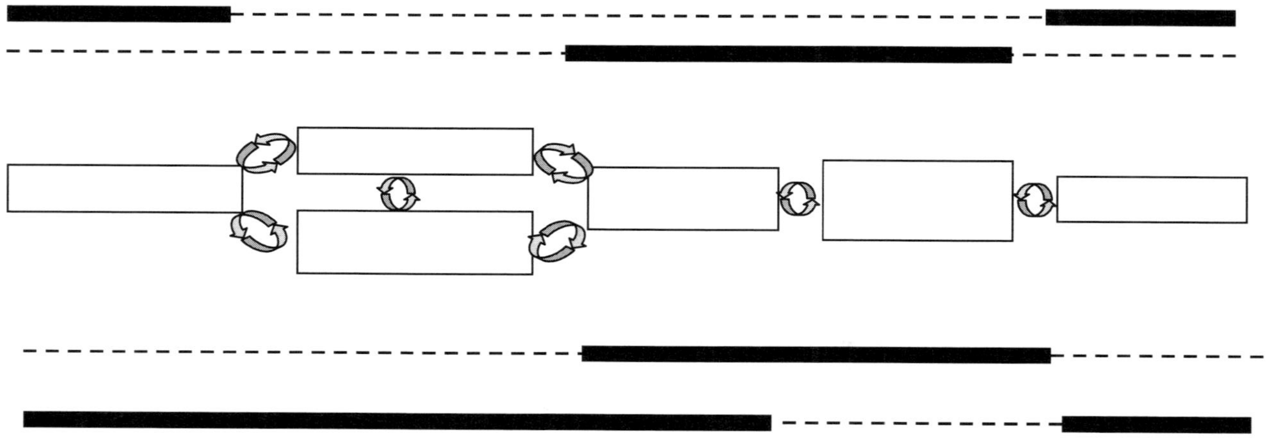

Fig. A1. General MCDA framework. Solid lines symbolize direct group involvement; dashed lines symbolize less direct involvement.

As shown in Fig. A1, the tools used within group decision-making and scientific research are essential elements of the overall decision process. As with people, the applicability of the tools is symbolized by solid lines (direct or high utility) and dotted lines (indirect or lower utility). Decision analysis tools help to generate and map value judgments into organized structures that can be linked with the other technical tools from risk analysis, modeling and monitoring, and cost estimations. Decision analysis software can also provide useful graphical techniques and visualization methods to express the gathered information in understandable formats. When changes occur in the requirements or decision process, decision analysis tools can respond efficiently to reprocess and iterate with the new inputs. The framework depicted in Fig. A1 provides a focused role for the detailed scientific and engineering efforts invested in experimentation, monitoring, and modeling that provide the rigorous and defendable details for evaluating criteria performance under various alternatives. This integration of decision and scientific and engineering tools allows each to have a unique and valuable role in the decision process without attempting to apply either type of tool beyond its intended scope.

As with most other decision processes, it is assumed that the framework in Fig. A1 is iterative at each phase and can be cycled through many times in the course of complex decision-making. A first-pass effort may efficiently point out challenges that may occur or modeling studies that should be initiated. As these challenges become more apparent, one iterates again through the framework to explore and adapt the process to address the more subtle aspects of the decision, with each iteration giving an indication of additional details that would benefit the overall decision.

Conclusions

The end result of the application of multi-criteria decision analysis is a comprehensive, structured process for selecting the optimal alternative in any given situation, drawing from stakeholder preferences and value judgments as well as scientific modeling and risk analysis. This structured process would be of great benefit to decision-making for homeland security, where there is currently no structured approach for making justifiable and transparent decisions with explicit trade-offs between social and technical factors. The MCDA framework links technological performance information with decision criteria and weightings elicited from decision-makers, allowing visualization and quantification of the trade-offs involved in the decision-making process. As demonstrated above, it is of great utility in applications such as management techniques for HABs.

Chapter 35 Appendix A References

Ame MV, Diaz MD, Wunderlin DA (2003) Occurrence of toxic cyanobacterial blooms in San Roque Reservoir (Cordoba, Argentina): A field and chemometric study. Environmental Toxicology 18(3):192-201

Bridges T, Kiker G, Cura J, Apul D, Linkov I (2005) Towards Using Comparative Risk Assessment To Manage Contaminated Sediments. In: Levner E, Linkov I, Proth JM (eds) Strategic Management of Marine Ecosystems. Springer, Amsterdam

Chorus I, Bartram J (1999) Toxic cyanobacteria in water: A guide to their public health consequences. World Health Organization, Geneva http://www.who.int/docstore/water_sanitation_health/toxicyanobact/begin.htm. Accessed 24 October 2005

Cole RA, Scodari PF, Franklin MA (2005) Impacts and management costs caused by freshwater aquatic nuisance species at civil works projects operated by the United States Army Corps of Engineers. Alexandria VA: USACE Institute for Water Resources

Kangas J, Kangas A, Leskinen P, Pykalainen J (2001) MCDM methods in strategic planning of forestry on state-owned lands in Finland: applications and experiences. Journal of Multi-Criteria Decision Analysis 10:257-271

Landsberg JH (2002) The effects of harmful algal blooms on aquatic organisms. Reviews in Fisheries Science 10(2):113-390

Nelson TA, Nelson AV, Tjoelker M (2003) Seasonal and spatial patterns of "Green tides" (Ulvoid algal blooms) and related water quality parameters in the coastal waters of Washington State USA. Botanica Marina 46(3):263-275

Office of the Deputy Prime Minister (ODPM) DLTR Multi-Criteria Decision Analysis Manual. Downloaded June 16 2004 http://www.odpm.gov.uk/stellent/groups/odpm_about/documents/page/odpm_about_608524-02.hcsp

Paerl HW (1996) A comparison of cyanobacterial bloom dynamics in freshwater, estuarine and marine environments. Phycologia 35(Supplement):25-35

Pertola S, Kuosa H, Olsonen R (2005) Is the invasion of Prorocentrum minimum (Dinophyceae) related to the nitrogen enrichment of the Baltic Sea? Harmful Algae 4(3):481-492

Rhudy KB, Sharma VK, Lehman RL, et al. (1999) Seasonal variability of the Texas 'brown tide' (Aureoumbra lagunensis) in relation to environmental parameters. Estuarine Coastal and Shelf Science 48(5):565-574

Saaty TL (1994) Fundamentals of Decision Making and Priority Theory with the Analytic Hierarch Process. The Analytic Hierarch Process Series: vol VI Pittsburgh PA: RWS

Seager TP (2004) Understanding industrial ecology and the multiple dimensions of sustainability. In Strategic Environmental Management by O'Brien and Gere Engineers. New York: John Wiley & Sons
Woods Hole Oceanographic Institution (2005) Red tide and harmful algal blooms. http://www.whoi.edu/redtide/. Accessed 24 October 2005
Yoe C (2002) Trade-Off Analysis Planning and Procedures Guidebook. Prepared for Institute for Water Resources, US Army Corps of Engineers April 2002

Chapter 36: Effective doses, guidelines & regulations

Michael D Burch

Australian Water Quality Centre, PMB 3, Salisbury, Adelaide, SA 5108, Australia.

CRC for Water Quality and Treatment, PMB 3, Salisbury, SA, 5108, Australia.

Introduction

Cyanobacteria are an important ecological component of all freshwater, estuarine and marine ecosystems worldwide. They contribute significantly to ecosystem productivity – sometimes excessively. When this occurs they also form water 'blooms' which are now recognised as a serious water quality problem with our use of water both for drinking water supply, recreational amenity and for agricultural use. Blooms are a symptom of eutrophication and are evidence of the deterioration of our water resources as a result of effluent discharge, poor land and catchment management, and often also of poor water allocation practices in rivers. This is now becoming better understood and acknowledged both in the United States and worldwide (Burns 2005, Chorus and Bartram, 1999).

The conditions which favour the growth of cyanobacteria and lead to blooms are nutrient enrichment (largely phosphorus but also nitrogen), warm temperatures, and calm stable water conditions such as those occurring in thermally stratified lakes and slow-flowing rivers. The latter may be the result of hydrology altered by abstraction practices. These conditions are often caused by human actions and activities, but can also be associated with natural climatic cycles such as droughts.

There is an international consensus that there has been an increase in frequency and severity of harmful algal blooms in both the marine and freshwater environments. In the case of the fresh water environment, the occurrence of toxic cyanobacterial blooms (CyanoHABs) presents problems for treatment, management and regulation of the quality of drinking water supplies. In regard to cyanotoxins a number of countries however have developed regulations or guidelines for cyanotoxins and cyanobacteria in drinking water, and in some cases in water used for recreation and agriculture.

There are currently no federal regulations or guidelines in the US for protecting human health and ecosystem viability from cyanobacterial harmful algal blooms (CyanoHABs) that occur in fresh, estuary, and marine water environments. This paper will explore the regulations and guidelines that have been developed around the world, identify others that may be needed, identify research needed to support their development, and identify factors that would be needed in a model to predict the need for revised and/or additional regulations or guidelines concerning CHABs.

Regulations and Guidelines

A number of countries have developed regulations or guidelines for cyanotoxins and cyanobacteria in drinking water, and in some cases in water used for recreational activity and agriculture. The approaches taken and the degree of adoption of guidelines are summarised in a comprehensive recent international compilation by Chorus (2005a). The main focus internationally has been upon microcystin toxins, produced by *Microcystis aeruginosa* and *Planktothrix agardhii*. This is because microcystins are widely regarded as the most significant potential source of human injury from cyanobacteria on a world-wide scale. Many international guidelines have taken their lead from the World Health Organization's (WHO) provisional guideline of 1 $\mu g\ L^{-1}$ for microcystin-LR in drinking-water released in 1998 (WHO 2004). The WHO guideline value is stated as being 'provisional', - "because it covers only microcystin-LR, for reasons that the toxicology is limited and new data for toxicity of cyanobacterial toxins are being generated". A comprehensive list of guidelines from various countries for toxins in drinking water is given in Table 1, and for recreational water in Table 2.

Table 1. Status of guidelines or standards for cyanobacterial toxins in drinking water for various countries. Information is also included on countries that are considering guidelines. (Information derived from websites and Chorus, 2005; Codd et al., 2005).

Country	Guideline Value/Standard	Comments/Explanations
Argentina	Under review	(Codd et al., 2005)
Australia	1.3 μg L^{-1} Total Microcystins, expressed as toxicity equivalents of microcystin-LR	From the Australian Drinking Water Guidelines see text for explanation on the process of guideline derivation. (NHMRC/NRMMC, 2004). http://www.nhmrc.gov.au/publications/synopses/eh19syn.htm
Brazil	1.0 μg L^{-1} for microcystins 3.0 μg L^{-1} for saxitoxins (equivalents) 15 μg L^{-1} for cylindrospermopsin	Guideline values for microcystins, saxitoxins and cylindrospermopsin, along with biomass monitoring programs. Guideline value for microcystins adopted as mandatory. Guideline values for equivalents of saxitoxins and for cylindrospermopsin included as recommendations. Use of algicides prohibited and toxicity testing/toxin analysis is required when cell counts exceed 10,000 cells mL^{-1} or 1mm^3 L^{-1} biovolume. (Codd et al., 2005)

Country	Guideline Value/Standard	Comments/Explanations
Canada	1.5 µg L^{-1} cyanobacterial toxins as microcystin-LR MAC	Canada uses guidelines as the standard of water quality. The guidelines are expressed with the unit of Maximum Acceptable Concentration (MAC). These are derived from tolerable daily intake (TDI), which in turn is derived from a calculated no-observed adverse effect level (NOAEL) from data from human or animal studies. To derive a MAC from a TDI, adjustments are made for average body weight and drinking water consumption, as well as other considerations. In terms of health the guidelines ensure that the MACs are far below exposure levels at which adverse effects have been observed. For the case of cyanobacterial toxins the guideline is considered protective of human health against exposure to other microcystins (total microcystins) that may also be present. (Health Canada, 2002) http://www.hc-sc.gc.ca/ewh-semt/pubs/water-eau/doc_sup-appui/sum_guide-res_recom/index_e.html
Czech Republic	1 µg L^{-1} microcystin-LR	Value as national legislation, follows WHO provisional guideline value. (Codd et al., 2005)
China	1 µg L^{-1} microcystin-LR	WHO provisional guideline for microcystin-LR. (Codd et al., 2005)
France	1 µg L^{-1} microcystin-LR	Drinking water decree. (Codd et al., 2005)
Italy	0.84 µg L^{-1} total microcystins	WHO provisional guideline for microcystin-LR used as a reference by local authorities. (Codd et al., 2005)
Japan	1 µg L^{-1} microcystin-LR	WHO provisional guideline for microcystin-LR. (Codd et al., 2005)
Korea	1 µg L^{-1} microcystin-LR	WHO provisional guideline for microcystin-LR. Algal alarming system – based on cell numbers and chlorophyll a. (Codd et al., 2005)

Country	Guideline Value/Standard	Comments/Explanations
New Zealand	For cyanobacteria: <1 potentially toxic cyanobacterium present in 10 mL of sample. PMAV for cyanobacterial toxins: Anatoxin: 6.0 μg L^{-1} Anatoxin-a (S): 1.0 μg L^{-1} Cylindrospermopsin: 1.0 μg L^{-1} Homoanatoxin-a: 2.0 μg L^{-1} Microcystin-LR Toxicity Eq: 1.0 μg L^{-1} Nodularin: 1.0 μg L^{-1} Saxitoxins (as STX-eq) :3.0 μg L^{-1}	Maximum acceptable values (MAVs) are applied for micro-organisms or organic contaminants of health significance. Provisional MAV (PMAV) have been recommended for cyanobacterial toxins. MAVs are based on the WHO 'Guidelines for Drinking Water Quality'. They are the concentration of a determinand, which is not considered to cause any significant risk to the consumer over a lifetime of consumption of water. The method of derivation varies according to NZ conditions and the way in that the determinand presents a risk. However they are derived with the use of a TDI. The MAVs are standards in NZ. The Standards provide compliance criteria and compliance is routinely monitored. (Kouzminov, 2005). http://www.moh.govt.nz/moh.nsf/c7ad5e032528c34c4c2566690076db9b/70727db605b9f56a4c25696400802887?OpenDocument
Norway	1 μg L^{-1} microcystin-LR	Provisional WHO guideline for drinking water adopted. (Codd et al. 2005)
Poland	1 μg L^{-1} microcystin-LR	National legislation for guideline value in drinking water.
South Africa	0-0.8 μg L^{-1} for microcystin-LR	Guideline levels for microcystins in potable water as a "Target Water Quality Range". http://www.dwaf.gov.za/IWQS/wq_guide/
Spain	1 μg L^{-1} microcystins	National legislation, maximum permissible amount in drinking water. (Codd et al., 2005)
Thailand	No guideline currently	Awareness of the need for guidelines. (Codd et al. 2005)

Country	Guideline Value/Standard	Comments/Explanations
United States of America	No guideline currently.	Maximum Contaminant Levels (MCLs) are the highest level of a contaminant that is allowed in drinking water. They are enforceable standards. Cyanobacteria and their toxins are listed as microbiological contaminants on the contaminant candidate list (CCL). This means that they are currently recognised as unregulated contaminants, but are known to occur in public water systems and may require regulation under the Safe Drinking Water Act. Contaminants on the CCL are a priority for the US Environmental Protection Agency with the aim to set MCLs. http://www.epa.gov/safewater/mcl.html
Uruguay	Under review	(Codd et al. 2005)
World Health Organization	1 μg L^{-1} for microcystin-LR GV	World Health Organization (2004) http://www.who.int/water_sanitation_health/dwq/gdwq3/en/

Table 2. Status of guidelines or standards for cyanobacterial toxins in water used for recreation or bathing water for various countries.

Country	Guideline /Standard	Comments
Australia	Two Level Guideline Level 1: 10 μg L^{-1} total microcystins or >50,000 cells mL^{-1} toxic *M. aeruginosa* or biovolume equivalent of >4 mm^3 L^{-1} for the combined total of all cyanobacteria where a known toxin producer is dominant in the total biovolume. Level 2: either the total biovolume of all cyanobacterial material exceeds 10 mm^3 L^{-1} or scums are consistently present	The definitions of the levels are as follows: Level 1: Probability of adverse health effects due to known toxins Level 2: Probability of adverse health effects due to high levels of cyanobacterial material where known toxins are not present Closure is recommended at either level A 3-stage Alert Level Framework is used for situation assessment up to the guideline level. (NHMRC, 2006)
Germany	Three Level Guideline Level 1: <10 μg L^{-1} microcystins Level 2: >10 - <100 μg L^{-1} microcystins Level 3: >100 μg L^{-1} microcystins	The definitions and recommended actions for each of the levels are as follows: Level 1: Monitor cyanobacteria in a routine surveillance program (14-d) Level 2: Publish warnings, discourage bathing, and consider temporary closure. Level 3: Publish warnings, discourage bathing, temporary closure recommended. (Chorus, 2005a)

Country	Guideline /Standard	Comments
Netherlands	20 μg microcystin-LR L^{-1}	Note that this value is not a mandatory standard. Provinces use the guideline in a management framework as follows: MC-LR > 10 μg L^{-1}: issue warning MC-LR > 20 μg L^{-1}: issue warning and continue monitoring; if levels are persistently high close the bathing facility Presence of scums: at least a warning and continued monitoring (Ibelings, 2005)
World Health Organization	Three Level Guideline Level 1: 20,000 cyanobacterial cells mL^{-1} Level 2: 100,000 cyanobacterial cells mL^{-1} Level 3: Presence of scums	The definitions of the levels are as follows: Level 1: Relatively low probability of adverse health effects Level 2: Moderate probability of adverse health effects Level 3: High probability of adverse health effects The recommendation at Level 3 is for 'Immediate action to control scum contact' (Chorus & Bartram, 1999)
France	Three Level Guideline Level 1: 20,000 cyanobacterial cells mL^{-1} Level 2: >20,000 -<100,000 cyanobacterial cells mL^{-1} Level 3: Presence of scums	The actions the levels are as follows: Level 1: Monitoring intensified to fortnightly Level 2: Microcystins analysed. If > 25 μg MC-LR eq L^{-1}, swimming is prohibited Level 3: All activities are prohibited

The countries that have adopted the WHO provisional guideline for microcystin-LR for drinking water directly include the Czech Republic, France, Japan, Korea, New Zealand, Norway, Poland, Brazil and Spain.

Chorus (2005b) provides a summary of the application of the guideline in a range of European countries: "Polish regulations require a limit of 1 μg L^{-1} to be met for microcystin-LR in drinking-water. Czech legislation requires monitoring of tap water for microcystin-LR with a limit of 1 μg L^{-1}, and an update of the ordinance is expected in 2005 which will include alternatives to microcystin analysis such as quantification of cyanobacterial biomass in raw water or bioassays in conjunction with cell counts, requiring toxin analyses only if thresholds for cyanobacterial biomass are exceeded. The French Drinking-water Decree includes a maximum limit of 1 μg L^{-1} microcystin-LR, with analyses being required in the event of cyanobacterial proliferation in the raw water. The Spanish decree establishing the water quality criteria for human consumption includes a limit for "microcystin" (variants not specified) of 1 μg L^{-1}, to be reviewed at 5-year intervals, with sampling regimes specified in relation to size of population served.

In two countries, the provisional WHO Guideline value for microcystin-LR is important for implementation of regulations which do not explicitly address microcystins: In Germany, and very similarly in Finland, the national Drinking-water Ordinances stipulate that drinking-water should contain no substances in concentrations that may be harmful to human health, and the provisional WHO value for microcystin-LR provides an important definition of such concentrations. A prerequisite for this approach was that drinking-water suppliers using surface water run long-established phytoplankton monitoring programmes as basis for adapting treatment to raw water quality, usually have effective treatment in place, and are aware of the cyanotoxin hazard. In Italy also, no limit value has been implemented, but the national drinking-water decree considers algae as an accessory parameter to be monitored in case local authorities suspect a risk to human health, with the provisional WHO Guideline of 1μg L^{-1} micocystin-LR used as basis for this assessment.

In Hungary: the decree on drinking-water quality and the ordinance on monitoring include cyanobacteria among biological parameters to be monitored by microscopy, though no limit is given for cyanotoxins, only for the number of cyanobacterial cells. In Finland, starting in the late 1980's the waterworks have also been advised to monitor cyanobacteria microscopically, and if cyanobacterial cells occur in raw or treated water, to analyse toxins."

In addition, other countries (e.g. Australia and Canada) have decided to develop slight variants of the guideline based upon their local requirements. For example, Australia has developed a guideline for Total Microcystins of 1.3 μg L^{-1}, expressed as toxicity equivalents of microcystin-LR (NHMRC/NRMMC, 2004). This derivation has essentially used the same animal studies, the TDI and derivation convention as WHO which treats microcystins as a threshold (non-cancer) toxicant. The Australian Guidelines however is for total microcystins and the rationale for this is that blooms of *Microcystis aeruginosa,* which is the most common toxin-producing cyanobacterium in Australia, can contain a wide range of variants of microcystin in varying amounts. Experience indicates that the number of variants in an individual sample can range from a few to up to more than 20 in some cases. It is the cumulative toxicity of the microcystins in total that represents the potential hazard to human health from ingestion via drinking water. Therefore the unit recommended for the quantitative expression of this cumulative toxicity in the guideline is total microcystins expressed as toxicity equivalents of microcystin-LR. There are some issues for compliance monitoring in relation to this guideline, particularly in relation to the availability of analytical standards for microcystins and for the selection of appropriate analytical methods, and these are discussed in Nicholson and Burch (2001).

The WHO recognized that the recommendation of a guideline for a single microcystin congener (microcystin-LR) where more than 80 variants are known could make it problematic to express and interpret quantitative results from analysis or assays for these other toxins in relation to the guideline value. A discussion about the calculation and expression and limitations of the interpretation of microcystin concentrations other than microcystin-LR, in terms of "concentration equivalents (CE)" and "toxicity equivalents (TE)" is given by Falconer, et al. (1999).

In 1999, Canada set a *maximum accepted concentration* (MAC) for microcystin-LR in drinking-water of 1.5 μg L^{-1}, and research by Health Canada has been addressing the need to comprehensively include other microcystin variants in surveys and in monitoring (Health Canada, 2002). Brazilian Federal legislation is perhaps the most comprehensive and includes a mandatory standard of 1 μg L^{-1} for microcystins (variants not specified), and recommendations are given for saxitoxins (3 μg L^{-1}) and for cylindrospermopsin (15 μg L^{-1}) (Azevedo 2005).

Guidelines for cyanotoxins and/or cyanobacterial cell numbers for recreational waters are in place in a number of countries (Table 2). The World Health Organization considered that for recreational waters a single guideline value for cyanobacteria or cyanotoxins is not appropriate (Chorus and Bartram 1999; WHO 2003). Rather, a series of guideline values associated with incremental severity and probability of health effects were defined at three levels based upon cyanobacterial cell densities and the presence of scums at the upper level (Table 2).

Other national guidelines for recreational water are also related to risk of adverse outcomes from ingestion of known toxins (Netherlands: 20 μg microcystin L^{-1}). On the other hand, the new Australian guideline has 2 levels defined by probability of adverse outcomes based upon microcystin ingestion (Level 1: 10 μg microcystin L^{-1}) and also from the probability of adverse health effects due to high levels of cyanobacterial material where known toxins are not present (Level 2: Total biovolume of all cyanobacterial material exceeds 10 mm^3 L^{-1} or scums) (Table 2).

Australia has also developed Livestock Drinking Water Guidelines for Cyanobacteria. These form part of the Australian and New Zealand Guidelines for Fresh and Marine Water Quality (ANZECC/ARMCANZ 2000). The guidelines for livestock are referred to as trigger values, which have the following definition and application: "Below the trigger value there should be little risk of adverse effects on animal health. Above the trigger value, investigations are recommended (e.g. of other factors such as age, condition, other dietary sources) to further evaluate the situation" (ANZECC/ARMCANZ 2000). The trigger values were developed using data on chronic and toxic effect levels on animals, taking into consideration animal weights, percentage intake from water, and safety factors for data not specific to the species. The summarised advice in the livestock drinking water guidelines is as follows: "Algal blooms should be treated as possibly toxic and the water source should be withdrawn from stock until the algae are identified and the level of toxin determined. An increasing risk to livestock health is likely when cell counts of *Microcystis* exceed 11,500 cells mL^{-1} and/or concentrations of microcystins exceed 2.3 μg L^{-1} expressed as microcystin-LR toxicity equivalents. There are insufficient data available to derive trigger values for other species of cyanobacteria" (ANZECC/ARMCANZ 2000). The guidelines provide individual derivations of trigger values for cattle, sheep, pigs, chickens and horses which take into account interspecies sensitivity.

Although there are no national guidelines for cyanotoxins in fish or shellfish in Australia, a 'health alert' level has been derived for toxins in fish, prawns and mussels in the state of Victoria, Australia (Van Buynder et al. 2001). The health alert level for microcystins and nodularin toxins in seafood is as follows: fish (250 μg kg^{-1}), prawns (1,100 μg kg^{-1}) and mussels (1,500 μg kg^{-1}). They were derived by determining a tolerable daily intake level for adults and modified to be protective for short-term exposure.

Role of Risk Assessment in Guideline Development

All of the international guidelines developed so far for microcystin in drinking-water have been based upon the World Health Organization's (WHO) provisional guideline of 1 μg L^{-1} for microcystin-LR released in 1998.

The derivation of this guideline is based upon data that there is reported human injury related to consumption of drinking water containing cyanobacteria, or from limited work with experimental animals. It was also recognised that at present the human evidence for microcystin tumor promotion is inadequate and animal evidence is limited. As a result the guideline is based upon the model of deriving a Tolerable Daily Intake (TDI) (i.e., Reference Dose; RfD) from an animal study No Observed Adverse Effects Level (NOAEL), with the application of appropriate safety or uncertainty factors. The resultant WHO guideline by definition is the concentration of a toxin that does not result in any significant risk to health of the consumer over a lifetime of consumption (WHO, 2004). Briefly the details of the calculation of the guideline are given in Chorus and Bartram (1999):

> "A 13-week mouse oral (by gavage) study with pure microcystin-LR is considered the most suitable for the derivation of a guideline value for microcystin-LR. In that study, a NOAEL of 40 μg/kg bw/day was determined, based on liver histopathology and serum enzyme level changes (Fawell et al., 1994). By applying a total uncertainty factor of 1000 (10 for intra-species variability, 10 for inter-species variability and 10 for lack of data on chronic toxicity and carcinogenicity), a provisional TDI of 0.04 μg/kg bw/day is determined for microcystin-LR....which....was used in deriving a provisional guideline value".

As indicated, where other countries have developed guidelines for microcystins that differ numerically from the WHO provisional guideline, this is due to the application of different scaling factors in the calculation. For example the same ingestion study in mice was used to calculate the Australian guideline of 1.3 μg L^{-1} total microcystin (Table 1) the WHO provisional guideline of 1 μg L^{-1} microcystin-LR. The guidelines differ due to the incorporation of a different average body weight for an adult (70 kg versus 60 kg), and to a difference with regard to the proportion of the daily intake of microcystin being attributed to the consumption of drinking water. The proportion for the Australian situation is regarded to be 0.9, which is higher than 0.8 selected by WHO. This is due to what was regarded as lower potential for exposure from other environmental sources, such as contaminated bathing water or via dietary supplements potentially containing microcystins. Similarly the Canadian guideline of 1.5 μg L^{-1}

(Table 1) results from incorporation of different scaling factors in the calculation for body-weight (70 kg versus 60 kg), and assumption of daily water consumption of 1.5 L as opposed to 2 L used by WHO.

In relation to recreational water guidelines for cyanotoxins and cyanobacteria, a number of countries have developed guidelines based either upon microcystin concentrations or equivalent cell densities of cyanobacteria. These guidelines essentially use a similar derivation process or are a translation from the drinking water guidelines for microcystin LR, while accounting for different exposures in recreational or bathing situations.

The World Health Organization was first to review the basis for guidelines for recreational water environments (WHO 2003). They recommended that in developing guidelines for cyanobacteria in freshwater the approach should consider:

1. The occurrence of cyanobacteria in general (in addition to known toxins) as part of the hazard: as it is not clear that all known toxic components have been identified and irritation symptoms reported may be due to these unknown substances.
2. The particular hazard due to the well known microcystin toxins.
3. The hazard associated with the characteristic or tendency for the occurrence of a heterogeneous distribution of many cyanobacterial populations in freshwater environments, which can result in the potential for scum formation.

The potential health effects were considered to be in two classes:

- chiefly irritative symptoms caused by unknown cyanobacterial substances
- the potentially more severe hazard of exposure to high concentrations of known cyanotoxins, particularly microcystins.

The result of this review was that they indicated that a single guideline value is not appropriate. Rather, a series of guideline values associated with incremental severity and probability of health effects were defined at three Levels (see Table 2). While these cell densities for the three Levels are not specifically indicated as being for toxic *Microcystis aeruginosa*, the Level 2 densities are cross-correlated to microcystin concentrations at ‘the level of 20 μg microcystin Litre^{-1} which is equivalent to 20 times the WHO provisional guideline value concentration for microcystin-LR in drinking-water (WHO 1998) and would result in consumption of an amount close to the tolerable daily intake (TDI) for a 60 kg adult consuming 100 mL of water while swimming (rather than 2 litres of drinking-water)’.

A similar approach has been followed by the Netherlands, which is described by Ibelings (2005). This involved using the tolerable daily intake for microcystins (MC-LR < 0.04 μg per kg bodyweight), and "assuming that a swimmer ingests 100 mL of water and thereby calculating an exposure limit of 20 μg MC-LR L^{-1} of bathing water (Table 2). While the derivation assumes that bathing occurs 365 days per year – it is recognised that a more likely exposure would be less than 35 days (Ibelings 2005).

The recent Australian approach for guideline derivation (NHMRC, 2006) is somewhat different from the WHO and recommends a two-level guideline based upon:

1. Level 1: The probability of adverse health effects from ingestion of known toxins, in this case based upon the toxicity of microcystins.
2. Level 2: The probability of increased likelihood of non-specific adverse health outcomes, principally respiratory, irritation and allergy symptoms, from exposure to very high cell densities of cyanobacterial material irrespective of the presence of toxicity or known toxins.

The Level 1 guideline uses animal toxicity data for microcystin toxins and conventional toxicological calculations to derive a guideline for short-term (14-day) exposure to microcystins via ingestion for both children and adults based upon typical bodyweights (Table 2). The derivation of the guideline selects the LOAEL from the 44-day pig study (Kuiper-Goodman et al, 1999) as the most suitable data to derive a shorter-term exposure LOAEL (i.e., 14 days) that is representative of a period of repeated daily exposure for an uninterrupted period of up to 2 weeks. This is regarded as a likely, but albeit rather intense, continuous exposure for swimming and aquatic recreation in a summer season (e.g., a holiday period exposure).

The second guideline level (Level 2) is recommended for circumstances where high cell densities or scums of 'non-toxic' cyanobacteria are present, i.e. where the population has been tested and shown not to contain known toxins (microcystin, nodularin, cylindrospermopsin or saxitoxins). In this case where the microcystin-related biovolume guideline is exceeded, and there are no microcystins or other toxins present, it is felt appropriate to issue warning where either the total biovolume of all cyanobacterial material exceeds 10 mm^3 L^{-1} or scums are consistently present, i.e. are seen at some time each day at the recreational bathing site. This guideline level is recommended based upon the work of Stewart (2004), where it was shown that there was an increase in likelihood of symptom reporting in bathers above this approximate biovolume. The potential symptoms reported above this level of cyanobacteria were primarily mild respiratory symptoms Stewart (2004).

Need for Additional Guidelines

As discussed, it is clear that considerable and growing international attention has been given to development of guidelines for cyanotoxins in drinking water and to a lesser extent for cyanobacterial numbers and toxins in recreational waters. Internationally there has been a "pronounced demand" for the WHO to develop a guideline for cylindrospermopsin (Chorus, 2005a), to provide authoritative guidance for countries to develop a risk-based approach to health hazards from cyanotoxins. The WHO guidelines provide a scientifically-substantiated target with a transparent explanation of the health-based considerations that lead to derivation of the value.

In relation the need for development or adoption of regulations for the US, an important part of the process is to carry out further surveys as part of hazard and exposure assessments for the various toxin types that have been found to occur. The information base for these assessments is currently being assembled from a range of studies in the US. In the context of Hazard Identification, a recent review of cyanotoxin occurrence in the US by Burns (2005) indicated that cyanobacteria are common in surface waters throughout the USA and that toxins have been detected in water bodies in the states of Florida, Indiana, Nebraska, Missouri, New York and Wisconsin, and in the Great Lakes. In addition a survey over 1996-1998, twenty-four public water systems in the USA and Canada were surveyed for microcystins (AwwaRF 2001). The results indicated that 80% of the 677 samples tested were positive for microcystin. A range of other investigations and surveys (e.g. SJRWMD 2001) have detected microcystin, cylindrospermopsin and anatoxin-a in finished drinking water, whereas saxitoxins have been found in freshwater cyanobacteria (Burns 2005). The surveys of toxin occurrence from New York State and the lower Great Lakes System reported by Boyer (see this volume) also support this incidence trend, where microcystin were by far the most common toxin (38% of samples tested), followed by anatoxin (8%), with a relatively low incidence of cylindrospermopsin (2%).

Certainly with regard to the need for guidelines, the incidence of microcystins, cylindrospermopsin and anatoxin in public water supply would suggest these are priority compounds for consideration for regulation in the US.

In relation to the use of water for other than drinking, some countries have developed guidelines for livestock-drinking protection (e.g., Australia: "trigger levels": for microcystins only) and local regulations for seafood harvesting (Victoria, Australia) (Burch and House, 2005). It is considered however, that there is currently insufficient information to derive sound guidelines for the broader range of toxins that could be present for the use of water contaminated by cyanobacteria or toxins for irrigated agricultural production, fisheries and aquaculture. The potential for algal toxin, odour or tainting residues in certain produce or commodities is an issue for agricultural activities which use contaminated water for irrigation or processing. Aquaculture and fish harvesting are particularly susceptible to residues, and this industry needs to be aware of algal blooms and their potential importance. The knowledge gaps in this area pose a business risk for these industries. The probability of this contamination may be quite low, and any possible public health risk may be small to negligible, however the damage to reputation is potentially more significant. This is because of the negative emotive image of produce contaminated with toxins and the absence of any guidance on acceptable background levels of residues or MCLs for these compounds in food (with the exception of PSPs in shellfish from a marine situation).

The other important area that requires further research is the issue of toxins and ecosystem protection. This is a developing research area, where more information on impact of toxins is required (See Ibelings this volume).

Research Needs

Additional research is required to support guideline development, including both short-term and whole-of–life animal studies with each of the known cyanotoxins. In view of the animal studies that indicate that microcystins may act as tumor promoters, and also some evidence of genotoxicity and carcinogenicity for cylindrospermopsin, it may be appropriate to carry out whole-of-life animal studies with both toxicity and carcinogenicity as end-points.

As part of the expert review by WHO in 1997, prior to the development of the guideline for microcystin-LR, it was revealed that there was insufficient information available for calculation of a TDI for most of the cyanotoxins including nodularin, cylindrospermopsin, anatoxin-a, homoanatoxin-a, anatoxin-a(S), and saxitoxins (Kuiper-Goodman et al. 1999). There has however been a recent sub-chronic, repeat animal oral-toxicity assessment study for cylindrospermopsin which can be used to determine a NOAEL, which in turn has been used to calculate a TDI and propose a drinking water guideline for cylindrospermopsin (Humpage and Falconer 2003).

In relation to microcystins, it is known that there are a large number of congeners, and the toxico-dynamics and kinetics of these variants are not well understood. Further research is needed to consider the approach to take in formulating health advisories or regulations for toxin mixtures, i.e. multiple microcystins, or mixtures of toxin types.

Applicability of International Guidelines to the United States

The cyanotoxins must be subjected to the same complete risk assessment procedure as any other environmental contaminant prior to the development of regulations in the US. This risk assessment includes the four steps of

1. Hazard identification;
2. Dose-response assessment;
3. Exposure assessment; and
4. Risk characterization,

with risk characterization being the transitional step to risk management (ref - NCEA website). The suitability of existing regulations for use in the US needs to be considered in the context of the above assessment process.

Current efforts and investigations in the US are strongly focussed on collecting information and developing knowledge in the area of hazard identification. In addition proposals are well advanced for research on dose-response assessment. However, very little has been done in a coordinated way with regard to exposure assessment and risk characterization.

In the above context then, none of the current international guidelines for toxins in drinking water would appear to be immediately suitable for adoption as regulations in the US. The major limitations that need to be overcome include: the capacity to deal with multiple toxin congeners, the absence of robust analytical methods for compliance monitoring, and the absence of certified toxin standards to support analyses.

The limitation of the WHO guideline for Microcystin-LR in relation to its use as a regulation is that it does not deal with congeners other than microcystin-LR. This is of course why WHO have recommended it as 'provisional'. Consideration needs to be given to the issue of mixtures and the incidence of the most common microcystin congeners in the US for regulation development.

The second issue for both the US and indeed international context is the requirement for appropriate and robust compliance monitoring and analytical protocols to accompany to any regulations. Although there is considerable analytical capability for cyanotoxins within the research community, it cannot be said that there are that rapid and economical screening or quantitative analytical methods that are readily available to the water industry. Organisations such as AwwaRF and the EPA are supporting research to develop validated methods. In this context there are also no validated analytical standards available commercially for the major toxins, i.e. microcystins, cylindrospermopsin and Anatoxin-a, although this may change in the near future.

The current WHO provisional guideline for microcystin, or the other national guideline variants that are based upon it, (e.g., Canadian, Australian) may be appropriate to adopt as a health advisory. The advantage of having a health advisory as a target is that it will assist in the development of a database on occurrence of cyanotoxins and contribute complementary information on exposure assessment, where local or state-based monitoring programs become established.

The bathing and recreational water guidelines developed in other countries could also be translated for the protection of public health in recreational situation in the US, by whatever state mechanisms are appropriate. The WHO guidelines that are widely used in Europe have provided a useful target for local authorities to manage hazards from cyanobacterial blooms in recreational water situations. The recent Australian guidelines for cyanobacteria in bathing waters (NHMRC 2006) are the most recent review of recreational water hazards from toxic cyanobacteria. These guidelines are evidence–based and deal with exposure to both toxic and non-toxic cyanobacteria, and also provide recommendations on sampling strategies and development of staged monitoring programs.

Predictive Models

The occurrence of CHABs and their toxins are stressors that are multifacetted in impact. They impact upon humans, ecosystems and biota, and natural resources in interdependent ways. Regulations and guidelines need to consider the impact upon human health via various exposure pathways (drinking water, food, recreational exposure). Similarly the integrated economic, environmental and social impact of CHABs should somehow be reflected in the formulation of ecosystem protection guidelines for CHABs. In this context the integrated human health and ecological risk assessment models proposed by Orme-Zavaleta and Munns (2007) offer potential to characterise the human-environment risk assessment for a stressors such as CHABs and toxins. The models offer ways to consider linkages of stressors and impacts in complex systems, and may provide a useful framework to apply to CHABs.

Conclusion and Summary

A number of countries have developed regulations or guidelines for cyanotoxins and cyanobacteria in drinking water, and in some cases in water used for recreational activity and agriculture. The main focus internationally has been upon microcystin toxins, produced predominantly by *Microcystis aeruginosa*. This is because microcystins are widely regarded as the most significant potential source of human injury from cyanobacteria on a world-wide scale. Many international guidelines have taken their lead from the World Health Organization's (WHO) provisional guideline of 1 μg L^{-1} for microcystin-LR in drinking-water released in 1998 (WHO 2004).

The WHO guideline value is stated as being 'provisional', because it covers only microcystin-LR, for reasons that the toxicology is limited and new data for toxicity of cyanobacterial toxins are being generated. The derivation of this guideline is based upon data that there is reported human injury related to consumption of drinking water containing cyanobacteria, or from limited work with experimental animals. It was also recognised that at present the human evidence for microcystin tumor promotion is inadequate and animal evidence is limited. As a result the guideline is based upon the model of deriving a Tolerable Daily intake (TDI) from an animal study No Observed Adverse Effects Level (NOAEL), with the application of appropriate safety or uncertainty factors. The resultant WHO guideline by definition is the concentration of a toxin that does not result in any significant risk to health of the consumer over a lifetime of consumption.

Following the release of this WHO provisional guideline many countries have either adopted it directly (e.g., Czech Republic, France, Japan, Korea, New Zealand, Norway, Poland, Brazil and Spain), or have adopted the same animal studies, TDI and derivation convention to arrive at slight variants based upon local requirements (e.g., Australia, Canada). Brazil currently has the most comprehensive federal legislation which includes a mandatory standard of 1 μg L^{-1} for microcystins, and also recommendations for saxitoxins (3 μg L^{-1}) and for cylindrospermopsin (15 μg L^{-1}).

Although guidelines for cyanotoxins and cyanobacterial cell numbers for recreational waters are in place in a number of countries, it is considered that there is currently insufficient information to derive sound guidelines for the use of water contaminated by cyanobacteria or toxins for agricultural production, fisheries and ecosystem protection.

In relation to the need for specific regulations for toxins for the US, the surveys that have been carried out to date would indicate that the priority compounds for regulation, based upon their incidence and distribution, are microcystins, cylindrospermopsin and Anatoxin-a.

Additional research is required to support guideline development, including whole-of-life animal studies with each of the known cyanotoxins. In view of the animal studies that indicate that microcystins may act as tumor promoters, and also some evidence of genotoxicity and carcinogenicity for cylindrospermopsin, it may be appropriate to carry out whole-of-life animal studies with both toxicity and carcinogenicity as end-points. In relation to microcystins, it is known that there a large number of congeners, and the toxico-dynamics and kinetics of these variants are not well understood. Further research is needed to consider the approach to take in formulating health advisories or regulations for toxin mixtures, i.e. multiple microcystins, or mixtures of toxin types.

An important requirement for regulation is the availability of robust monitoring and analytical protocols for toxins. Currently rapid and economical screening or quantitative analytical methods are not available to the water industry or natural resource managers, and this is a priority before the release of guidelines and regulations.

There is insufficient information available in a range of the categories usually required to satisfy comprehensive risk assessment process for the major toxins to currently adopt any of the international guidelines as regulations in the US. The major limitations that need to be overcome include: the capacity to deal with multiple toxin congeners, the absence of robust analytical methods for compliance monitoring, and the absence of certified toxin standards to support analyses.

However, the current WHO provisional guideline for microcystin-LR, or the other national guideline variants that are based upon it, (e.g., Canadian, Australian) may be appropriate to adopt as a health advisory in the short-term, while regulations are developed. The bathing and recreational water guidelines developed in other countries could also be translated for use as recreational water guidelines situation in the US.

Acknowledgements

I would like to thank Jenny House for assistance with collating the information for Table 1 for toxin guidelines. Andrew Humpage provided valuable information and discussion on the current state of toxicological studies with cyanotoxins. The CRC for Water Quality and Treatment, Australia has provided financial support over a number of years, through a range of projects, which has allowed for the input to the development of various Australian guidelines for cyanobacteria and their toxins.

References

ANZECC/ARMCANZ (2000) Australian and New Zealand Guidelines for Fresh and Marine Water Quality, Volume 3, Primary Industries - Rationale and Background Information (Chapter 9). Australian and New Zealand Environment and Conservation Council (ANZECC) and Agriculture and Resource Management Council of Australia and New Zealand (ARMCANZ.). Available at: http://www.mincos.gov.au/publications/

AwwaRF (2001) Assessment of Blue-Green Algal Toxins in Raw and Finished Drinking Water. Final report #256, AwwaRF, Denver.

Azevedo SMFO (2005) Brazil: Management and regulatory approaches for cyanobacteria and cyanotoxins. pp 27-30 in Chorus, I. (Ed.), 2005. Current Approaches to Cyanotoxin Risk Assessment, Risk Management and Regulations in Different Countries. 117 pp. Federal Environment Agency (Unweltbundesamt), Berlin.

Boyer GL this volume

Burns JW (2005) United States of America: Cyanobacteria and the status of regulatory approaches. pp 111-117, in Chorus, I. (Ed.), 2005. Current Approaches to Cyanotoxin Risk Assessment, Risk Management and Regulations in Different Countries. 117 pp. Federal Environment Agency (Unweltbundesamt), Berlin.

Burch M, House J (2005) Australasia and Oceania: Cyanobacteria, Cyanotoxins and their management Pp 47–70. CYANONET. A global network for cyanobacterial bloom and toxin risk management. Initial Situation Assessment and Recommendations IHP-VI Tech document in hydrology No 76 UNESCO, Paris.

Chorus I (2005a) Current Approaches to Cyanotoxin Risk Assessment, Risk Management and Regulations in Different Countries. 117 pp.Federal Environment Agency (Unweltbundesamt), Berlin. Available at: http://www.umweltdaten.del/publikationen/fpdf-1/2910.pdf

Chorus I (2005b) Germany: Approaches to assessing and managing the cyanotoxins risk. Pp 59-67. in Chorus, I. (Ed.), 2005. Current Approaches to Cyanotoxin Risk Assessment, Risk Management and Regulations in Different Countries. 117 pp. Federal Environment Agency (Unweltbundesamt), Berlin.

Chorus I, Bartram J (1999) Toxic Cyanobacteria in Water: a Guide to Public Health Significance, Monitoring and Management. Published on behalf of WHO by E & FN Spon /Chapman & Hall, London, 416 pp.

Codd GA, Azevedo SMFO, Bagchi SN, Burch MD, Carmichael WW, Harding WR, Kaya K, Utkilen HC, (2005) CYANONET, a Global Network for Cyanobacterial Bloom and Toxin Risk Management: Initial Situation Assessment and Recommendations. UNESCO IHP-VI.

Falconer I, Bartram J, Chorus I, Kuiper-Goodman T, Utkilen H, Burch M, Codd GA (1999) Safe Levels and Safe Practices. pp 155-178 in Toxic Cyanobacteria in Water. A guide to their public health consequences, monitoring and management. I. Chorus and J. Bartram. (Eds.), E&FN Spon publishers, London.

Fawell JK, James CP, James HA (1994) Toxins from blue-green algae: toxicological assessment of microcystin-LR and a method for its determination in water. WRc, 1-46.

Health Canada (2002) Guidelines for Canadian Drinking Water Quality: Supporting Documentation — Cyanobacterial Toxins — Microcystin-LR. Water Quality and Health Bureau, Healthy Environments and Consumer Safety Branch, Health Canada, Ottawa, Ontario. Available at: http://www.hc-sc.gc.ca/

Humpage AR, Falconer IR (2002) Oral toxicity of cylindrospermopsin: No observed adverse effect level determination in Swiss albino mice. Research Report 13. CRC for Water Quality and Treatment, Adelaide.

Humpage AR, Falconer IR (2003) Oral toxicity of the cyanobacterial toxin cylindrospermopsin in male Swiss albino mice: Determination of No Observed Adverse Effect Level for deriving a drinking water Guideline Value. *Environmental Toxicology* 18: 94-103.

Ibelings BW (2005) Netherlands: Risks of toxic cyanobacterial blooms in recreational waters: guidelines. pp 85-91 in Chorus, I. (Ed.), 2005. Current Approaches to Cyanotoxin Risk Assessment, Risk Management and Regulations in Different Countries. 117 pp. Federal Environment Agency (Unweltbundesamt), Berlin.

Ibelings BW, Havens KH this volume

Kouzminov A (2005) New Zealand: Risk assessment, management and regulatory approach for cyanobacteria and cyanotoxins in drinking-water pp 95-100 in Chorus, I. (Ed.), 2005. Current Approaches to Cyanotoxin Risk Assessment, Risk Management and Regulations in Different Countries. 117 pp. Federal Environment Agency (Unweltbundesamt), Berlin.

Kuiper-Goodman T, Falconer I, Fitzgerald J (1999) Human health aspects. In: Chorus I and Bartram J [Eds] *Toxic Cyanobacteria in Water: A Guide to their public health consequences, monitoring and management.* 115-153. London: E&FN Spon.

NHMRC (2006) Guidelines for Managing Risks in Recreational Water. National Health and Medical Research Council, Canberra.
Available at: http://www.nhmrc.gov.au/publications/

NHMRC/NRMMC (2004) Australian Drinking Water Guidelines. National Health and Medical Research Council/Natural Resource Management Ministerial Council, Canberra. Available at: http://www.nhmrc.gov.au/publications/

Nicholson BC, Burch MD (2001) Evaluation of Analytical Methods for Detection and Quantification of Cyanotoxins in Relation to Australian Drinking Water Guidelines. Occasional Paper. National Health and Medical Research Council. Canberra, Australia. Available at: http://www.nhmrc.gov.au/publications/pdf/eh22.pdf

Orme-Zavaleta J, Munns W this volume

SJRWMD (2001) Assessment *of Cyanotoxins in Florida's Lakes, Reservoirs, and Rivers*. Report to the Florida Harmful Algal Bloom Task Force, Florida Fish and Wildlife Conservation Commission, St. Petersburg, Florida.

Stewart I (2004) *Recreational Exposure to Freshwater Cyanobacteria: Epidemiology, Dermal Toxicity and Biological Activity of Cyanobacterial Lipopolysaccharides*. PhD Thesis, School of Population Health, The University of Queensland.

Van Buynder PG, Oughtred T, Kirby B, Phillips S, Eaglesham G, Thomas K, Burch M (2001) Nodularin uptake by seafood during a cyanobacterial bloom. *Environmental Toxicology*, 16: 468-471.

World Health Organization (2003) Guidelines for safe recreational waters, Volume 1 – Coastal and fresh waters, Chapter 8: Algae and cyanobacteria in fresh water. pp. 136-158. WHO Publishing, Geneva. Available at: http://www.who.int/water_sanitation_health/bathing/srwe1/en/

World Health Organization (2004) Guidelines for Drinking-water Quality. Volume 1. Recommendations, 3rd Edition, WHO Publishing, Geneva. Available at: http://www.who.int/water_sanitation_health/dwq/gdwq3/en/

World Health Organization (1998) Guidelines for Drinking-water Quality. Second Edition, Addendum to volume 2, Health criteria and other supporting information. World Health Organization, Geneva.

Chapter 37: Economic cost of cyanobacterial blooms

Dennis A Steffensen

Cooperative Research Centre for Water Quality & Treatment and Australian Water Quality Centre, Private Mail Bag Salisbury Australia 5108.

Abstract

Cyanobacterial blooms impact upon the water quality, environmental and ecological status of water bodies and affect most of the uses we make of water. The extent of the impact depends upon the type, size and frequency of the blooms, the size of the water body affected, the uses made of the water and the treatment options available to respond to the blooms. The impacts therefore vary considerably from place to place. Overall costs should also account for the planning and remedial actions taken to prevent future blooms.

Problem

Safe and aesthetically acceptable water is a critical need in a modern society. Good water quality is also a key prerequisite for sustainable environments. Cyanobacterial blooms impact the environmental health of water resources and effect how the water can be used. In particular cyanobacteria may damage human and animal health and impair the recreational value of the water bodies.

Benefits of Bloom reduction

The benefits of reducing cyanobacterial blooms are the reduction in damages and adverse effects. The benefits can be valued by the costs that may be avoided or by using willingness to pay estimates.

Improved Human Health

It is the fundamental requirement of a water utility to ensuring that the water provided for drinking is safe. Cyanotoxins clearly have the potential to impact on the health of consumers and are therefore not acceptable in drinking water supplies or in water used to irrigate crops and water stock. Cyanobacteria can also produced unacceptable taste and odors. As prevention of health impacts is so fundamental to water supplies it is difficult to put a figure value of the benefits of preventing cyanobacterial blooms. The best estimate is therefore the cost of controlling blooms as detailed in section III.

Improved Recreational Opportunities

Water–based recreation is an increasingly important consideration and in many areas forms the basis of the tourist industry. Cyanobacterial blooms can render a water body unsuitable for swimming, fishing, water skiing. This will impact on the busineses that cater for those activities with knock on effects for other businesses. The blooms can therefore impact negatively on the attraction of the general area as a tourist destination. Isolated blooms may result in short–term losses. Recurring blooms impact on the reputation of the area resulting in long–term decline in tourism.

Walker and Greer (1992) conducted detailed studies of the impact of cyanobacterial blooms on recreational activities in New South Wales in 1991/92. The methodology considered the following costs:

- those associated with recreation and tourism such as accommodation, transport and tourism;
- those associated with commercial recreation facilities such as caravan and tourist parks;
- those associated with the amenity value including aesthetics; and
- long–term costs related to permanent loss of trade.

Case study 1 Darling River 1991

In 1991 a bloom of neuro–toxic *Anabaena* covered 1000 km of the Darling River in central Australia. The affected region is a sparsely populated agricultural area in central Australia, however, the area is popular with tourists who enjoy pursuits such as fishing, swimming, camping, sight–seeing and hunting. All of these activities are impacted by cyanobacterial blooms. From surveys of two representative towns it was estimated that losses to the tourist industry were around $1.5 million.

Case study 2 Nepean/Hawkesbury River

The Hawkesbury Nepean River is located near Sydney in the state of New South Wales. The area supports a number of aquatic recreational facilities with activities including swimming, fishing, water skiing, canoeing, camping and picnicking. The proximity of the area to Australia's largest city ensures that the region receives high numbers of visitors. A series of blooms of cyanobacteria occurred over the 1991/1992 summer between Windsor and Wiseman Ferry. The assessment of their impact included estimates of the costs to consumers of traveling to other sites. From a survey of tourist facilities it was estimated that the revenue was $6.7 million lower than the previous year when no blooms had been present. It is noteworthy that the blooms were not classified as toxic and the reduction in revenue was the result of negative publicity about the blooms.

Case study 3 Various storages in New South Wales

In 1991, nine water storages in New South Wales that are used for recreation were seriously affected by algal blooms. The economic loss was estimated at $1.2 million.

A number of other cyanobacterial blooms had significant impacts on tourism but those costs have not been established. Examples include:

- Lower River Murray – Periodic *Anabaena* blooms ;
- Lake Alexandrina– *Nodularia* blooms, 1989 – 1992
- East Gippsland Lakes – *Nodularia* bloom in 1987/88;
- Peel Harvey Inlet – Periodic *Nodularia* blooms;
- Lance Creek – *Anabaena* bloom 1990;
- Candowire Reservoir Philip Island – *Anabaena* bloom 1991;
- Paskeville Reservior Yorke Peninsula – *Phormidium* bloom 2000.

All of the above incidences occurred during peak tourism periods and caused considerable disruption. The publicity has political as well as economic consequences.

Improved agriculture and fishing

There is a considerable body of evidence for the death of livestock as a result of drinking cyanobacterial contaminated water. There have been thirteen documented cases of stock deaths related to cyanobacterial blooms in Australia (Steffensen *et al* 1999). During the *Anabaena* bloom in the Darling River, 1600 livestock deaths were reported (Dept Water resources 1992). Using contaminated water for crop irrigation is also a concern as the impact of toxins on the crops and also on livestock who consume the crops is uncertain.

Toxins can accumulate in fish and shellfish with potential health risks to consumers. Humpage et al (1993) discussed the possible accumulation of saxitoxins from *Anabaena* in freshwater mussels. Falconer et al (1992) reported toxicity in mussels in the Peel Harvey Inlet at the time of a *Nodularia* bloom. However, there is a paucity of information on this issue and assessment of the economic impact is not possible.

Improved Environment

Cyanobacterial blooms are an indicator of environmental degradation and are often associated with reduced bio–diversity and greater instability. While it is difficult to place a value on the maintenance of a natural ecosystems there may be impacts fisheries, agriculture and tourism.

Control Costs

Immediate costs of cyanobacterial blooms

Monitoring and testing

Critical factors to be considered when assessing the risks related to a cyanobacterial bloom include the type and amount of toxin present and the likely progression of the bloom. Investigation of these factors requires intensive monitoring and testing. The costs involved in assessing the nature and extent of the bloom will depend on its size and the facilities available for testing the toxins. Toxicity tests may cost over $1,000 per sample. Furthermore, cyanobacterial blooms can be highly patchy and unpredictable. Intensive monitoring may be required to predict the course of a bloom.

Blooms in small isolated storages may be relatively easily monitored and cost a few thousand dollars per annum. Blooms in large water bodies that connect with other water bodies may cost hundreds of thousands to adequately assess. An example of the latter is Lake Alexandrina in South Australia which is subject to toxic *Nodularia* blooms, some lasting several months. The Lake has an area of 75,000 hectares (180,000 acres) and has a number of water extraction points for domestic and agricultural use. It is also heavily used for recreation including swimming, boating and fishing. In 1989, Lake Alexandrina experienced the first major toxic cyanobacterial bloom in South Australia. The size of the lake, the number of extraction points, the variety of water uses and inexperience in dealing with blooms of this size caused difficulties in assessing and controlling the bloom. Assessing the risks posed by the blooms involved intensive sampling over a wide area using boats and shore based personnel. Aerial photography was also used to map the extent of the bloom and proved to be a very valuable tool.

Another example of a large toxic bloom occurred in the Darling River in Australia in 1991. This bloom of toxic *Anabaena* covered a 1000 km length of the river and affected water off–takes for a number of towns and numerous agricultural extraction points. The remoteness of the river increased the monitoring costs. It has been estimated that the cost of monitoring for cyanobacteria and for contingency planning to deal with blooms in Australia is $8.7 million per year (Atech, 2000).

Risk Assessment

Health risks from cyanotoxins are the major concern for water utilities and health authorities. Water quality guidelines have been proposed for some toxins and others are under consideration. These provide a basis for assessing immediate risks to health but are of little value in assessing the risk of an incident occurring. Also, the focus on toxins ignores other significant issues such as taste and odours, filter clogging and oxygen depletion which are also caused by cyanobacterial blooms. Furthermore these problems are often associated with other groups of phytoplankton. One can look at a hierarchy of risk assessment as follows.

- What is the risk of excessive phytoplankton growth?
- What is the risk that they will be cyanobacteria?
- What is the risk that they will be toxic?

This hierarchy moves from general environmental considerations to more specific factors that influence the occurrence of toxic species and the degree of toxicity. This broad risk assessment approach fits into the Water Quality Management Framework that has been adopted in the 2004 Australian Drinking Water Guidelines. A similar approach has been adopted by the World Health Organization. The main elements of the framework are:

- assessment of the likely severity and frequency of the impacts from all possible risks;
- selection of critical control points;
- development of management plans to mitigate those risks and
- a monitoring program to assess the success of those mitigation strategies.

This provides multiple barriers for the protection of the water supply. Cost benefit analysis can be used to prioritise the management options.

Control Measures

The actions taken to manage a toxic bloom can involve measures taken within the water body, after extraction of the water or the provision of alternative supplies.

In–water measures

The in–water measures may include artificial mixing to disperse the bloom, the use of booms to protect water off–takes, application of algicides, and release of water up stream to flush out blooms. The cost of these measures is very site specific and difficult to generalize.

Algicides are commonly used in some areas but banned in other areas. The most readily used algicides are copper based. In South Australia treatment of a bloom in a water supply reservoir with algicides may cost $20,000 to $50,000 for reservoirs ranging from 1,300 to 26,000 ML. SA Water spends in excess of $1 million a year using algicides to treat blooms and dispose of the copper contaminated water treatment sludge. As copper has impacts on a wide range of aquatic organisms it is not recommended for use in natural water bodies.

Stratification of water bodies creates favourable conditions for the growth of cyanobacteria. Artificial mixing is a common measure used to disrupt stratification and can be useful in dispersing cyanobacterial blooms. The most common approach is bubbling compressed air into the bottom of the water bodies but mechanical mixers can also be used. In small water bodies it may be possible to install temporary systems as an immediate response. For larger water bodies or for permanent systems the lead time precludes artificial mixing as a reactionary procedure for rapid response. It is also necessary to understand the factors that influence the stratification of the particular water body which requires research into the local conditions. The costs for destratification vary according the situation but large systems can cost several hundred thousand dollars in capital with running and maintenance costs about 10% of the capital costs.

In regulated rivers there may be scope for releasing water from up–stream storages to flush out blooms. Sydney Water has released water from Lake Burragorang to flush blooms from the Nepean/Hawkesbury River. Releases of up to 70,000ML of water have been made but this is regarded as an extreme action. Releases have also been made from the Menindee Lakes to the Murrumbidgee and Darling Rivers. In 1999, 480ML of water was released from Kangaroo Creek Dam to flush a *Microcystis* bloom out of Lake Torrens in Adelaide. This approach has not been used in subsequent blooms in Lake Torrens as it was not regarded as an appropriate use of water. In recent years frequent droughts in Australia have required careful use of water resources. These conditions, combined with the competition for water from agricultural users, have limited the use of water for flushing toxic blooms.

Water Treatment after extraction

Untreated water supplies are the most vulnerable to the impacts of cyanobacterial blooms. Supplies that are disinfected with an oxidant may give some protection against some toxins. Conventional water treatment such as flocculation, filtration and disinfection will remove the cyanobacterial cells and the toxins they contain provided that the cells are removed intact and a chlorine residual is maintained after treatment. Use of oxidants prior to flocculation and filtration is not recommended as it lyses the cells and releases the toxins. However, conventional treatment will have little effect on dissolved toxins. Additional treatment such as activated carbon or powerful oxidants such as ozone are needed to remove or destroy dissolved toxins. Most water treatment plants in parts of Australia that are subject to cyanobacterial blooms now have the capacity to dose with powdered activated carbon (PAC) and in some cases have granular activated carbon (GAC) filters. The use of these additional treatments increases the capital and operating costs compared with conventional treatment.

Provision of alternative supplies

If adequate treatment is not available alternative supplies may need to be provided. This may involve isolation of the contaminated source and the re–routing of other sources into the distribution system. Where that option is not available water may need to be tankered in or bottled water provided for drinking. In South Australia this has happened in two areas. During the *Nodularia* blooms in Lake Alexandra mentioned above, water was tankered in and made available at distribution points or used to fill existing water tanks. In 2000 a bloom of toxic *Phormidium* a reservoir servicing a small town in South Australia was forced off–line during a busy holiday period. Bottled water was provided to consumers and temporary treatment facilities installed at some commercial businesses.

Long Term Costs

Prevention of future problems can involve dealing with the factors that cause blooms or improving the treatment of the water once the bloom occurs.

Environmental Flows

Increasing flows in regulated rivers to flush out existing blooms has been successful in some circumstances. Consideration has also been given to using increased flow to prevent blooms from forming. The feasibility of using this approach largely depends on the volumes of water required and competition for water from other users. Maier *et al (*2001) reviewed the options for managing cyanobacteria in the River Murray in South Australia. Blooms in that area of the river are associated with periods of thermal stratification. It was estimated that additional flows of 10,000 ML per day would be required to halve the risk of thermal stratification. If the flow was required for 1 month, the cost of the water would be $15 million. Mitrovic *et al.* (2003) reported on a similar study in the Darling River. That study indicated that flows of 0.05 ms^{-1} would preclude *Anabaena* blooms in the weir pools. In the reach of river studied, this related to discharge rates between 100 and 450 ML $day^{-}1$. Flow rates often fall below those stated in summer resulting in frequent blooms. However, it is not clear whether additional flows would be available to prevent blooms in summer. While environmentalists are a calling or greater environmental flows for rivers in Australia the competition from other users is rising making increasing flow to control cyanobacterial blooms more difficult. The introduction of water trading has further increased the competition for water.

Covering storages

For small storages cyanobacteria can be eliminated by removing light. Following a major cyanobacterial incident in small open storages, SA Water covered three storages at a total cost of $7.1 Million. The capacities ranged from 64 ML to 150 ML.

Environmental improvement

In Australia the increase in cyanobacterial blooms has intensified the attention on eutrophication and on nutrient reduction. Significant expenditure has been allocated to improve waste water treatment and to find alternative routes that will divert discharges away from rivers and lakes. Nutrient reduction, especially phosphorus is the main driver for treatment works upgrades. Improved land management including better management of riparian strips is also relevant. It is not always clear to what extent the environmental improvement programs are due to cyanobacterial blooms or the result of more general concerns about eutrophication. Atech (2000) estimated that in Australia the cost of environmental protection schemes attributable to cyanobacteria was $121 million per year. This included urban sewage and stormwater ($43 mill), agriculture and industrial waste water ($33 mill), and rehabilitation of land and water resources ($45 mill).

Overall costs

Atech (2000) put the overall costs in Australia related to cyanobacteria in $ millions per year as:

Cost Category	Cost ($)
Joint Management costs	9
Urban extractive users	35
Rural extractive users	30
Non–extractive users	76–136
Total	180–240

Confidence levels for cost estimates

Direct costs to the agencies responsible for managing blooms such as monitoring, treatment and the provision of alternative water sources can be determined relatively easily. Costs to tourism and agriculture are more difficult to assess due to the range of people and activities that may be affected. Estimating the value of aesthetic appeal and environmental values is especially difficult. What value do we place on clean water as opposed to unsightly smelly scums?

The cost of engineering works for prevention or management of blooms such as additional treatment, installation of mixers and roofing of storages can be estimated with reasonable confidence. It is more difficult to assess the costs and benefits of environmental measures such as improved waste water treatment. The extent to which these programs are initiated by concerns about cyanobacteria and the whether they will resolve the problem is often unclear.

Atech (2000) argued that the willingness of the community to pay for prevention of blooms is an indicator of the economic impact. This study also suggested that the cost of the environmental improvement programs is a conservative estimate. The estimate was based on the consideration that the programs should partially reverse costs currently incurred and would not completely resolve the problem. They proposed a multiplier of 1.5 to 2 which places the annual economic impact at $180 million to $240 million.

References

Atech (2000) Cost of algal blooms. Report to Land and Water Resources Research and Development Corporation, Canberra, ACT 2601 ISBN 0 642 76014 4

Australian Drinking Water Guidelines (2004) National Health and Medical Research Council and the National Resoures Management Ministerial Council

Department of Water Resources (1992) Blue–green Algae Final report of the NSW Blue–Green Algae Task Group. ISBN 0 7395 7886 0

Falconer IR, Choice A, Hosja W (1992) Toxicity of the edible mussel (Mytilus edulis) growing naturally in an estuary during a bloom of Nodularia spumigena. Journal of Environmental Toxicology and Water Quality 7:119–123

Maier HR, Burch MD, Bormans M (2001) Flow management strategies to control blooms of the cyanobacterium, Anabaena circinalis, in the River Murray at Morgan, South Australia. Regulated Rivers. Research & Management 17:637–650

Mitrovic SM, Oliver RL, Rees C, Bowling LC, Buckney RT (2003) Critical flow velocities for the growth and dominance of Anabaena circinalis in some turbid freshwater rivers. Freshwater Biology 48:164–174

Steffensen DA, Nicholson BC, Burch MD, Drikas M, Baker PD (1999) Ecology and management of blue–green algae (cyanobacteria) In Australia. Environmental Toxicology and Water Quality 14 (1) 183–195

Walker C, Greer L (1992) The Economic Costs Associated with Lost Recreation Benefits due to Blue–Green Algae in New South Wales: Three case studies. in Hassall and Associates, Blue Green Algae: Final report of the New South Wales Blue–green Algal Task force, Department of Water Resources, Parramatta

Chapter 38: Integrating human and ecological risk assessment: application to the cyanobacterial harmful algal bloom problem

Jennifer Orme-Zavaleta, Wayne R Munns Jr.

USEPA National Health and Environmental Effects Research Laboratory, Office of Research and Development

Abstract

Environmental and public health policy continues to evolve in response to new and complex social, economic and environmental drivers. Globalization and centralization of commerce, evolving patterns of land use (e.g., urbanization, deforestation), and technological advances in such areas as manufacturing and development of genetically modified foods have created new and complex classes of stressors and risks (e.g., climate change, emergent and opportunist disease, sprawl, genomic change). In recognition of these changes, environmental risk assessment and its use are changing from stressor-endpoint specific assessments used in command and control types of decisions to an integrated approach for application in community-based decisions. As a result, the process of risk assessment and supporting risk analyses are evolving to characterize the human-environment relationship. Integrating risk paradigms combine the process of risk estimation for humans, biota, and natural resources into one assessment to improve the information used in environmental decisions (Suter et al. 2003b). A benefit to this approach includes a broader, system-wide evaluation that considers the interacting effects of stressors on humans and the environment, as well the interactions between these entities. To improve our understanding of the linkages within complex systems, risk assessors will need to rely on a suite of techniques for conducting rigorous analyses characterizing the exposure and effects relationships between stressors and biological receptors. Many of the analytical techniques routinely employed are narrowly focused and unable to address the complexities of an integrated assessment. In this paper, we describe an approach to integrated risk assessment, and discuss qualitative community modeling and Probabilistic Relational Modeling techniques that address these limitations and evaluate their potential for use in an integrated risk assessment of cyanobacteria.

Introduction

Cyanobacterial blooms occur in both fresh water and marine environments, producing a variety of toxins, and posing risks to humans and animals through recreational and drinking water use as well as consumption of contaminated fish and shellfish (Codd et al. 2005). As a result of their complex ecology, involving multiple endpoints we propose an integrative approach in assessing the risks posed by cyanobacterial blooms.

The environmental risk assessment paradigm is shifting from independent analyses of human health or ecological effects to a more integrative, or unified, approach. The idea of integrating risk assessment has been the topic of extensive discussion over the past decade (e.g., Harvey et al. 1995; WHO 2000). Integration ideally combines the process of risk estimation for humans, biota, and natural resources into one assessment to improve the information used in environmental decisions, resulting in more effective protection of both humans and the environment (Suter et al. 2003b). A benefit to this approach is a broader, system-level evaluation that considers the interactions of the effects of stressors on humans and the environment, as well the interactions between these entities. In addition, stressors other than chemicals need to be considered. The basis for such an integrated approach would be the perspective that ecosystems serve as part of the foundation defining human well-being and vice versa.

Risk assessments are important tools for informing public health and environmental protection decisions. They constitute the scientific reasoning for estimating the likelihood of an adverse human or ecological effect resulting from exposure to a stressor. Although the human health and ecological risk assessment paradigms were developed independently, they are related (Suter et al. 2003a). In both paradigms, risk characterization is a key step providing a description of the evidence concerning the hazard, potential exposures, and the uncertainties, variability, and assumptions used in the assessment. Thus, the integration of risk assessment approaches is encapsulated in the analytical processes it entails.

The shift in risk assessment to an integrated approach is consistent with changes in the scientific approach to complex problems. In many instances, a multidisciplinary approach is a necessity to evaluate cause and effects relationships fully. Wilson (Wilson 1998) noted that science is no longer a specialized activity, but involves the synthesis of causal explanations. Thus, scientific research is shifting towards understanding linkages within highly complex systems (Vitousek et al. 1997; Wilson, 1998; NAS, 2000; Forget and Lebel, 2001).

To improve our understanding of the linkages of complex systems as part of an integrated risk assessment, risk assessors must rely on a suite of techniques for conducting rigorous analyses characterizing exposure and effects relationships among stressors and biological receptors. Current analytical techniques have been criticized as inadequate and irrelevant; they can be misinterpreted due to a lack of understanding of the problem and the inability to deal with uncertainty (NRC 1996; Peterman and Anderson 1999). Further, many of the commonly used techniques are narrow in focus and unable to evaluate complex systems adequately. In this paper, we describe integrated risk assessment and review community-level modeling techniques that account for current limitations. Lastly, we evaluate their potential for integrated risk assessment of the cyanobacterial harmful algal bloom (CHAB) problem.

Integrated Risk Assessment Paradigms

Over the past decade, several frameworks for integrating risk have been proposed that are based on existing approaches for human health and ecological risk assessments. Some approaches view integration in the context of chemical exposures, combining acute and chronic risks to organisms and considering exposures from different sources, pathways and routes (Gurjar and Mohan 2003; Bridges and Bridges 2004). Harvey and coworkers (Harvey et al. 1995) developed a 'holistic' approach that consisted of concurrent *and* integrated health and ecological assessments. Their process followed the steps originally outlined by the NAS (NAS 1983) conducting human health and ecological assessments in parallel. A series of risk choices is produced for the risk manager by integrating the results of two parallel assessments during the risk characterization step. Using mercury as a case study, they developed a risk characterization consisting of a series of risk estimates for humans exposed through inhalation or ingestion that address neurological and reproductive effects, and for wildlife exposed through the aquatic food chain addressing reproductive success and decreased species distribution. The authors suggested that the series of risk estimates would provide options for risk managers to choose from in making a decision (Harvey et al. 1995). Although cast as a holistic process, the Harvey et al. (Harvey et al. 1995) approach is not really integrative, but rather a comparison of different risk values generated for different exposure scenarios and toxicity endpoints; protective of different species. Thus, this approach may be too generic and unresponsive to a particular problem or management decision.

A special forum of the World Health Organization's International Programme on Chemical Safety (IPCS) developed another approach (Munns et al. 2003). They outlined an integrated process combining elements of both human health and ecological processes (WHO 2001; Suter et al. 2003b). This paradigm (Figure 1) is more closely aligned with the concepts of the *Guidelines for Ecological Risk Assessment* (USEPA 1998). Here, hazard identification becomes an element of problem formulation, and dose response assessment occurs as part of the effects characterization. Most importantly, this approach considers the interactions among stressors and receptors such as wildlife or humans, and the abiotic environment.

One distinct difference of the IPCS integrated approach from the Harvey et al. (Harvey et al. 1995), NAS (NAS 1983) and ecological risk paradigms (USEPA 1998) is the involvement of stakeholders and risk managers in the process. The human health and ecological risk paradigms were designed to be independent from risk management so that their outcomes reflect scientific analyses that are not influenced by socio-political bias. In the IPCS approach, stakeholder and risk management involvement throughout the process is viewed as essential to ensure buy-in and responsiveness of the assessment to the specific problem, considering both human and ecological risks where applicable (Suter et al. 2003b). While this, in and of itself, does not ensure integration, it increases the potential depending on how the problem is defined at the onset of the risk assessment.

The IPCS approach combines the process of risk estimation for humans, biota, and natural resources into one assessment for the purpose of improving the information used in environmental decisions, resulting in more effective protection of resources valued by society (Miranda et al. 2002; Suter et al. 2003b). Integration is achieved through all phases of the risk assessment process (Suter et al. 2003b). Under problem formulation, integration entails the development of stressor-driven assessment questions common to both health and environmental problems that focus on potential susceptible human and ecological endpoints. Exposure and effects characterizations are integrated through an evaluation of all the possible sources of exposure and an understanding of common modes of toxic action in humans and other organisms. Similar to the holistic approach (Harvey et al. 1995), the IPCS risk characterization includes multiple estimates of risk from which a best estimate of human and ecological risk is selected using a common and consistent approach (Suter et al. 2003b). The authors go on to indicate that evidence for health and ecological risks would be integrated when appropriate but do not describe how this would be achieved.

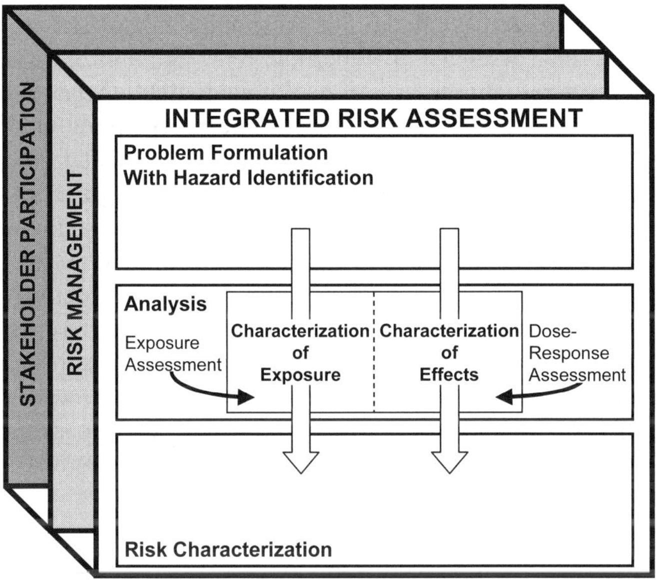

Fig. 1 Integrated Risk Assessment Paradigm. Adapted from WHO 2001.

The IPCS integrated approach was applied to several complex environmental problems (Table 1). The case studies developed using the integrated approach identify aspects of where integration can or should occur with respect to exposure and effects characterization, but they do not actually conduct integrated assessments. Rather, they illustrate how such assessments might be conducted. The risk characterization section in each of the case studies largely reflects parallel risk comparisons. Two studies (Ross and Birnbaum 2003; Vermeire et al. 2003) propose a common quantitative approach, a Toxic Equivalency Factor (TEF) approach as a means of integrating risks. It is not clear, however, that having a common quantitative approach to estimate risks for different species is actually integrative, but rather reflects the commonalities in the toxic endpoints and mechanisms of toxicity for the exposures and species of interest. Thus, the IPCS approach goes beyond Harvey et al.'s (Harvey et al. 1995) holistic approach in describing levels of integration throughout the risk assessment process. However, the information included in the risk characterization step largely presents parallel risk estimates for human and ecological endpoints under different exposure and effect scenarios. The responsiveness of the assessment to a particular problem is likely to be greater under the IPCS approach given the interaction with risk managers and stakeholders throughout the process.

Other approaches to integrative assessments have been proposed that focus on human and environmental linkages including socioeconomic and political factors (e.g., Bruins and Heberling 2005; Stahl et al. in press), or have focused more broadly on human health-ecological integrity reflecting dimensions of both the natural and social systems (Miranda et al. 2002). Epstein (Epstein 1994) developed an integrated assessment framework of climate change and ecosystem vulnerability. His generalized framework depicted overlapping and interacting climate and social systems with ecosystems whose intersection directly or indirectly produced various outcomes ranging from changes in health, crop yields, and demography to economic productivity. Epstein noted that integration was dependent on the use of specific biological, social or geochemical indicators depicting the functions of complex systems. Referring to the complex relationship between disease emergence and changes in climate and ecosystems, Epstein (Epstein 1994) proposed a number of principles for modeling and monitoring complex ecosystems. He emphasized the need to account not only for direct impacts to the different systems but also those indirect effects resulting from the interactions among factors within the three overlapping systems. He noted that those diseases transmitted directly from person to person reflect changes in population density with little interaction among the three systems, while vector-borne diseases reflect environmental changes involving all three systems in his integrated model. Integration in Epstein's approach also occurs through scientific and political collaborations. He did not present an overall assessment of risk but suggested guidelines for identifying system vulnerabilities affecting overall stability and resilience; key elements in his view for mitigating disease emergence.

Table 1 Summary of IPCS integrated risk assessment case studies.

Environmental Problem	Assessment Endpoints	Areas of Integration	Proposed Risk Characterization	Reference
"Dioxin-like" Persistent Organic Pollutants	Humans and upper trophic level wildlife	· Route of exposure · Mode of action · Toxicity	Apply Toxic Equivalency Approach (TEF) to both humans and wildlife	Ross and Birnbaum 2003.
Tributyl- and triphenyltins	Humans and piscivorous wildlife	· Route of exposure · Mode of action · Toxicity	Species and exposure-specific human and ecological risk estimates	Sekizawa et al. 2003.
UV-Radiation	Amphibians, coral, humans, and oceanic primary productivity	· Exposure pathways · Mechanistic pathways	Parallel characterization of risk across assessment endpoints.	Hansen et al. 2003.
Organophospherous pesticides	Humans and wildlife	· Exposure pathways · Toxicity	Species-specific TEFs	Vermeire et al. 2003.
Nonylphenol	Humans (occupational and environmental exposure), wildlife, and aquatic organisms	· Exposure pathways · Mechansim of action · Toxicity	Species and exposure-specific human and excological risk estimates	Bontje et al. 2004

Vanleewen et al. (Vanleewen et al. 1999) presented a conceptual 'butterfly' model that focuses on human health in an ecosystem context. Human health is determined from the intersection of biophysical socioeconomic environments. The boundaries of the butterfly could be at the community, watershed, or population level and include the interactions between humans and the nonhuman environment. Their model is not an approach for assessing risk *per se* but can be viewed as a mechanism for determining risk factors influencing human health. As the authors noted, this model focuses only on human health and does not determine health for other species in the ecosystem, limiting its utility for comprehensive assessment of risk.

Integrative Analytical Approaches to Risk Assessment

The integrated paradigms described above provide frameworks for considering human and environmental interactions but fall short of demonstrating specific analytical techniques for conducting an integrated risk analysis. The examples include a mix of conceptual, integrated approaches that are either descriptive or consist of parallel risk assessments. Considering the models presented by Epstein (Epstein 1994) and VanLeeuwen et al. (VanLeeuwen et al. 1999), it is clear that an evaluation of interactions among human populations, their environment, and other important ecological factors are needed in conducting an integrated analysis. This type of evaluation is similar to that encompassed by an ecoepidemiological approach. Similar to human epidemiology, ecoepidemiology has been used to study the ecological effects that are prevalent in certain areas among population groups, communities and ecosystems and their potential causes (Bro-Rasmussen and Løkke 1984; Martens 1998). This approach focuses on a description of the effects, identification of causes, and understanding their linkages. Humans are considered as part of the environment in these analyses.

An ecoepidemiological approach is similar to community and systems-level ecological risk assessment with respect to understanding relationships between biotic and abiotic factors. Levins (Levins 1973) noted that addressing more complex systems required breaking down disciplinary boundaries to create an integrated process that addresses management goals in which community structure and other mechanistic factors could be examined as a whole. A system in this context is defined as a habitat, geographic area, human community or network of communities (Levins 1998). As complexity increases, the ability to gather quantitative information is complicated by the impracticality of the number of parameters to measure and the loss of realism (Levins 1966; Puccia and Levins 1991).

Qualitative models can simplify complex systems without sacrificing realism (MacArthur and Levins 1965; Levins 1966) and enable an integrated analysis of a system. Qualitative modeling in the form of signed digraphs, 'loop analysis,' and matrix analysis facilitates the understanding of a system where there is incomplete information. Because qualitative models involve only the signs of the interactions among variables (positive, negative, or no change), variables representing poorly quantified aspects of the system can be included in the analysis (Puccia and Levins 1991). Such variables represent not only different species, but also resources, climate, or socioeconomic variables that influence community structure and function. When constructing models, qualitative modeling methods can help determine which variables should be included, what should be measured, and how system dynamics might be affected under different perturbation (stresses that result in a permanent change in a growth parameter) scenarios (Levins 1998).

Loop analysis and the corresponding community matrix is a useful analytical tool for exploring and understanding the effects of natural and anthropogenic stress on a system. Dambacher et al. (Dambacher et al. 1999) used this modeling procedure to characterize a predator-prey system involving snowshoe hare and arctic fox. This technique also proved useful in predicting the impact of species introductions into a community (Li et al. 1999; Castillo et al. 2000) and explaining complex transitions in community composition over time (Bodini 1998; Ortiz and Wolff 2002). Loiselle et al. (2000; 2002)) used loop analysis to examine different economically-based management scenarios in a wetland ecosystem to identify management options and guide monitoring programs.

In the context of integrated risk, Levins (Levins 1998) extended qualitative modeling to the problem of vector-borne disease. In his system, he identified the invasiveness of vectors and disease reservoirs as core variables that would be important in an epidemic, adding vector habitat requirements, vector and host behaviour, host health status, and economic variables as other factors to be considered. With an increasing 'web of causation,' Levins (Levins 1998) argued that internal processes critical to community function could be examined. On further analysis of this problem, Orme Zavaleta and Rossignol (Zavaleta and Rossignol 2004) developed a procedure to predict disease risk that combines recent developments in qualitative community modeling with biomathematical theory of vector-borne disease transmission. This procedure predicts the change in risk of vector-borne disease following perturbations such as increases in vector abundance, animal control measures, habitat alteration, or global warming.

Like Levin's postulated epidemic-disease community, this procedure allows the consideration of a complex community structure linking ecological factors to human disease. This procedure results in a rigorous prediction of an ecological community response to a perturbation with minimal to no quantitative parameterization. It generates focused hypotheses to guide data collection and control management strategies as interventions.

Bayesian analyses in the form of Bayesian networks are another tool that can be useful in an integrated risk analysis. A Bayesian approach is based on probability theory and is a useful decision-making or inferential technique when there is incomplete information or it is not possible to gather enough information to reduce uncertainties (Reckhow 2003). A Bayesian network is used to model a system containing uncertainty. It offers both qualitative and quantitative information in the form of conditional probabilities and can be applied to multivariate problems involving complex relationships among variables (Reckhow 2003). A Bayesian network consists of a directed acyclic graph and a probability distribution. The network characterizes variable relationships through interrelated nodes and arcs. The nodes represent variables and the arcs represent conditional dependencies between the nodes. Bayesian networks are used to identify those key variables influencing relationships within a system, and thus are an integrative analytical tool.

The use of Bayesian networks is increasing in scientific analyses of complex problems. Crome et al. (Crome et al. 1996) applied a Bayesian approach to evaluate the impact of logging on bird and mammal species in rain forests. The investigators had too few data to detect potential impacts using traditional statistical analysis. However, results of a Bayesian analysis suggested a correlation between canopy cover and impacted bird species that was not previously apparent. Further, of the 76 species of birds in question, only four species were identified as having a high probability of being adversely impacted by logging.

Bayesian networks have also been used to guide such diverse analyses as land management decisions (Marcot et al. 2001), fish stock assessment (Varis et al. 1993; Hammond and Ellis 2002), and potential risk factors associated with heart disease (Buntine 1991). Each of these cases started with a hypothesized model that could be updated as additional information became available, and involved large uncertainties, the pooling of information from different datasets, and expert judgment in the analysis.

When a specific model is not known, a data discovery technique, Probablistic Relational Modeling (PRM), can conduct a heuristic search of independent data sets to generate data-derived models (Jorgensen 2003). This technique involves machine learning guided by expert judgment to develop a probabilistic model. The PRM extends Bayesian networks to the relational level, modeling uncertainty related to variables, their properties, and relationships among them (Getoor et al. 2001). The probabilistic relationship between variables is such that a change in any one variable affects all the others. Thus, PRMs are well suited for application to complex systems.

There are a few examples of where PRM has been used to evaluate complex problems. Getoor et al. (Getoor et al. 2001) described a PRM analysis to determine possible probabilistic relationships between patients from a tuberculosis clinic, certain risk factors, and specific strains of tuberculosis. In a second example, Jorgensen et al. (Jorgensen et al. 2003) used a PRM approach to explore the long-term changes in the clarity of Crater Lake using information summarized in multiple databases. The PRM analysis enabled the investigators to construct multiple, complex hypotheses concerning the entire lake ecosystem given data obtained from the long-term studies of the lake.

Probablistic Relational Modeling was also used by Orme Zavaleta et al. (Zavaleta in review) to identify probabilistic relationships associated with the transmission of West Nile virus in Maryland. Similar to the Crater Lake study (Jorgensen et al. 2003), the RBM approach was used to explore relationships among multiple, independent databases. Multiple hypotheses were generated suggesting spatial and temporal relationships between key vector, host and habitat variables related to disease transmission.

Thus, the PRM technique appears to be an effective means of conducting an integrated risk analysis through the qualitative and quantitative evaluation of complex community interactions. The hypotheses generated by the PRM analysis can be used to guide further quantitative testing of specific relationships between probabilistically linked variables.

Integrated Risk of CHABs

The concepts of integrated risk assessment can be applied to the problem of CHABs to help define the specific information, tools and research needed for effective decision-making and action. Although it would be presumptuous to attempt a fully developed assessment in this paper, communication of an initial conceptual model can facilitate the discussion and additional analyses required to advance such an assessment. The conceptual model in Figure 2 reflects existing knowledge about the factors contributing to blooms, the health, ecological and socioeconomic effects of blooms, and the linkages among important components of this multifaceted system. Different pathways of exposure and effect are shown for humans, wildlife, and aquatic plants and animals, as the processes influencing these groups of receptors vary. However, the pathways intersect to illustrate system linkages, or when biological processes are common to multiple receptor groups. Along the bottom row of the model are loose expressions of candidate assessment endpoints for an integrated risk assessment, reflecting some of the values whose protection may underlie the need for management action to control or mitigate CHABs.

Additional refinement of the conceptual model is required to advance an integrated assessment of the risks of cyanobacterial blooms. Are the important environmental processes and factors controlling blooms captured? Is the array of assessment endpoints important to the CHABs problem articulated fully? Are the key system components and their linkages described adequately? With agreement on the adequacy of the conceptual model, planning discussions can address the availability of tools and information required to evaluate critical risk hypotheses represented in the model, potentially leading to identification of additional data and research needed to complete the assessment. Though challenging, performance of a thoroughly-planned integrated risk assessment would support comprehensive decisions for managing the risks of CHABs.

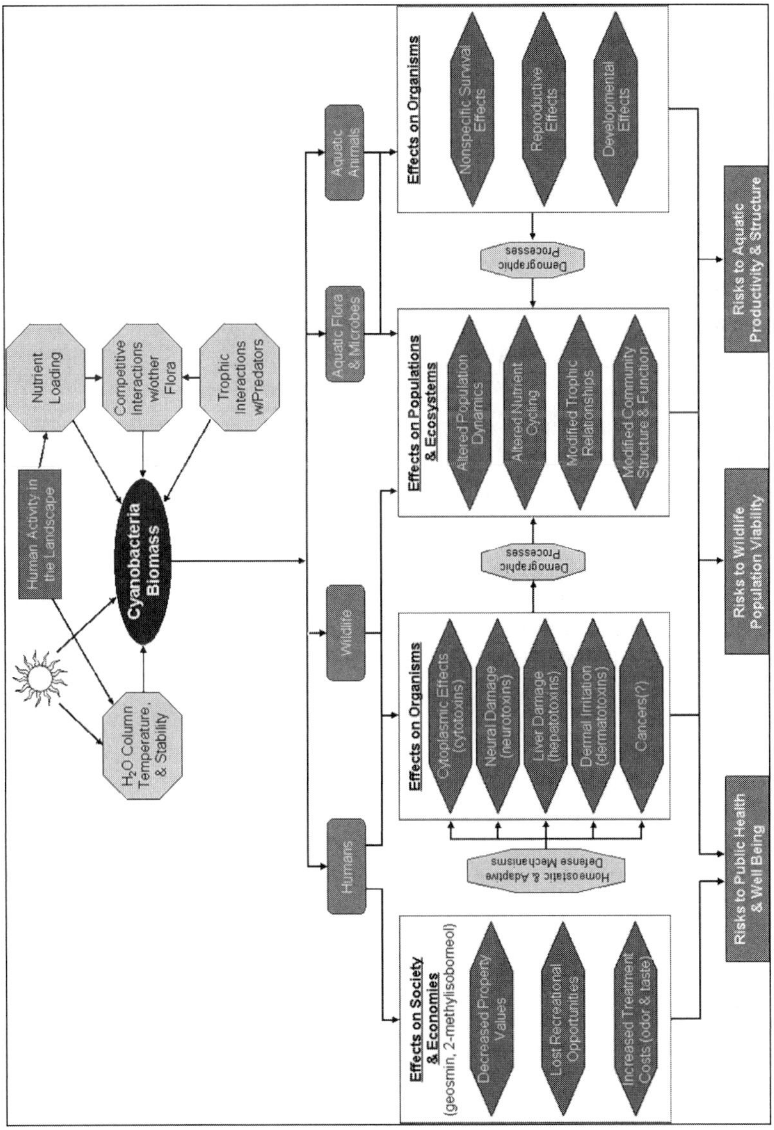

Fig. 2. Conceptual model for cyanobacterial harmful algal blooms.

Discussion

To conduct an integrated risk assessment of CHABs, a suite of tools is needed that integrates human and environmental health in the problem formulation (for hypothesis generation) and analysis phases of the assessment, not simply during the risk characterization phase. Such tools should consider the interacting system as a whole, from the environmental processes that influence CHAB formation to the changes caused by those blooms in the combined human and ecological system. Although this adds complexity in the analysis, models and other decision support methods are available that can simplify and reduce complexity.

The 'integrative' models reviewed in this paper may not be robust enough to integrate multiple stressors or multiple endpoints in their use of either parallel assessments or deductive reasoning to remove stressor interactions from consideration. The analytical techniques employed in these models to characterize risk are applied to either human health or ecological assessments. Qualitative modeling and Probablistic Relational Models provide an integrated risk analysis framework that identifies relationships important in the system and thus, guide the application of quantitative models or provide sufficient information for management decisions. Both techniques rely on community structure for generating hypotheses and testing predictions. Experimental comparison of various community theories suggests that loop analysis may be the theoretical approach best suited for predicting the behaviour of complex community structures following a perturbation (Hulot et al. 2000). Used in conjunction with mechanistic models, the integrated analytical techniques provide a balanced, iterative approach for not only assessing risk, but evaluating possible consequences of different CHABs management scenarios.

References

Bodini A (1998) Representing ecosystem structure through signed digraphs; model reconstruction, qualitative predictions and management: The case of freshwater ecosystem. Oikos 83:93-106

Bontje D, Hermens J, Vermeire T, Damstra T (2004) Integrated Risk Assessment: Nonylphenol Case Study. International Programme on Chemical Safety (IPCS). WHO/IPCS/IRA/12/04 Geneva, Switzerland

Bridges JW, Bridges O (2004) Integrated risks assessment and endocrine disrupters. Toxicol 205:11-15

Bro-Rasmussen F, Løkke H (1984) Ecoepidemiology – A casuistic discipline describing ecological disturbances and damages in relation to their specific causes: Exemplified by chlorinated phenols and chlorophenoxy acids. Regulatory Tox and Pharm 4:391-399

Bruins RJF, Heberling MT (2005) Economics and Ecological Risk Assessment. CRC Press, New York

Buntine W (1991) Theory refinement on Bayesian networks. IN: Proceedings of the Seventh Conference of Uncertainty in Artificial Intelligence. Morgan Kaufmann (eds). San Fransisco pp 52-60

Castillo GC, Li HW, Rossignol PA (2000) Absence of overall feedback in a benthic estuarine community: A system potentially buffered from impacts of biological invasions. Estuaries 23:275-291

Codd GA, Lindsay J, Yound FM, Morrison LF, Metcalf JS (2005) Chapter 1. Harmful Cyanobacteria. From mass mortalities to management measures. In: Huisman J, Matthijs HCP, Visser PM (eds) Harmful Cyanobacteria. Springer, Netherlands, pp 1-23

Crome FHJ, Thomas MR, Moore LA (1996) A novel Bayesian approach to assess the impacts of rain forest logging. Ecological Applications 6:1104-1123

Dambacher JM, Li HW, Wolff JO, Rossignol PA (1999) Parsimonious interpretation of the impact of vegetation, food, and predation on the snowshoe hare. Oikos 84: 530-532

Epstein PR (1994) Framework for an integrated assessment of climate change and ecosystem vulnerability. In: Wilson ME, Levins R, Spielman A (eds) Disease in Evolution: Global Changes and Emergence of Infectious Diseases. New York Academy of Science, New York, pp 423-435

Forget G, Lebel J (2001) An ecosystem approach to human health. International J of Occup and Env Health Supplement to 7:3-36

Gurjar BR, Mohan M (2003) Integrated risk analysis for acute and chronic exposure to toxic chemicals. J Haz Mat A102:25-40

Hammond TR, Ellis JR (2002) A meta-assessment for elasmobranches based on dietary data and Bayesian networks. Ecological Indicators 1:197-211

Hansen L, Hedtke SF, Munns WR Jr (2003) Integrated human and ecological risk assessment: A case study of ultraviolet radiation effects on amphibians, coral, humans, and oceanic primary productivity. Human and Ecol Risk Assessment 9: 359-377

Harvey T, Mahaffey KR, Velazquez S, Dourson M (1995) Holistic risk assessment: An emerging process for environmental decisions. Reg Tox and Pharm 22:110-117

Hulot FD, Lacroix G, Lescher-Moutoue F, Loreau M (2000) Functional diversity governs ecosystem response to nutrient enrichment. Nature 405:340-344

Jorgensen J, D'Ambrosio B, Rossignol PA (2003) Data-driven construction of community models of Crater Lake. NSF Biocomplexity Workshop -The vertical organization of energy, carbon, and nutrient cycles in an ultraoligotrophic ecosystem: A workshop on Crater Lake, Oregon February 16 -18

Lackey R (1997) If ecological risk assessment is the answer, what is the question? Human and Ecol Risk Assessment 3:921-928

Levins R (1966) The strategy of model building in population biology. Am Nat 54:421-431

Levins R (1973) Fundamental and applied research in agriculture. Science 181:523-524

Levins R (1998) Chapter 11 Qualitative mathematics for understanding, prediction, and intervention in complex ecosystems. IN: Approaches to Assessing Ecosystem Health. Rapport D, Costanza R, Epstein P, Gaudet C, Levins R (eds). Blackwell MA, pp 178-204

Li HW, Rossignol PA, Castillo G (1999) Chapter 30: Risk analysis of species introductions: Insights from qualitative modeling. In *Nonindigenous Freshwater Organisms*. Vectors, Biology, and Impacts. Claudi R, Leach JH (eds). Lewsi Publ, Boca Raton pp 431-447

Loiselle S, Carpenato GM, Hull V, Waller T, Rossi C (2000) Feedback analysis in reserve management: studying local myths using qualitative models. Ecol Modelling 129:25-37

Loiselle S, Hull V, Permengeat E, Falcucci M, Rossi C (2002) Qualitative models to predict impacts of human interventions in a wetland ecosystem. Web Ecology 3:56-69

MacArthur RH, Levins R (1965) Competition, habitat selection, and character displacement in a patchy environment. Proc National Academy Science 51:1207-1210

Marcot BG, Holthausen RS, Raphael MG, Rowland MM, Wisdom MJ (2001) Using Bayesian belief networks to evaluate fish and wildlife population viability under land management alternatives from an environmental impact statement. For Ecol and Management 153:29-42

Martens P (1998) Health and Climate Change. Earthscan Publs Ltd, London

Munns WR Jr, Kroes R, Veith G, Suter II GW, Damstra T, Waters M (2003) Approaches for integrated risk assessment. Human and Ecol Risk Assessment 9: 267-272

National Research Council (NRC), National Academy of Sciences (1983_ Risk Assessment in the Federal Government. National Academy Press, Washington DC

National Research Council (NRC), National Academy of Sciences (1996) Understanding Risk: Informing Decisions in a Democratic Society. National Academy Press, Washington DC pp 249

National Research Council (NRC) (2000) Grand Challenges in Environmental Sciences. National Academy Press, Washington DC

Orme Zavaleta J, Rossignol PA (2004) Community-Level Analysis of Risk of Vector-Borne Disease. Trans Royal Soc Trop Med & Hyg 98(10): 610-618

Orme Zavaleta J, Jorgensen J, D'Ambrosio B, Altendorf E, Rossignol PA (in press) Discovering Spatio-Temporal Models of the Spread of West Nile Virus. Risk Analysis

Ortiz M, Wolff M (2002) Application of loop analysis to benthic systems in northern Chile for the elaboration of sustainable management strategies. Marine Ecol Progress Series 242:15-27

Peterman RM, Anderson J (1999) Decision analysis: A method for taking uncertainties into account in risk-based decision making. Human and Ecological Risk Assessment 5:231-244

Puccia CJ, Levins R (1991) Chapter 6. Qualitative modeling in ecology: Loop analysis, signed digraphs, and time averaging. IN: Qualitative Simulation Modeling and Analysis. Fishwick PA, Luker PA (eds). Springer-Verlag Publ, New York pp 119-143

Reckhow KH (2003) Bayesian approaches in ecological analysis and modeling. IN:The Role of Models in Ecosystem Science. Canham CD, Cole JJ, Lauenroth WK (eds). Princeton Univ Press, In press

Ross P, Birnbaum L (2003) Integrated human and ecological risk assessment: A case study of persistent organic pollutant (POPs) risk to humans and wildlife. Human and Ecol Risk Assessment 9: 303-324

Sekizawa J, Suter GW, Birnbaum L (2003) Integrated human and ecological risk assessment: A case study of tributyltin and triphenyltin compounds. Human and Ecol Risk Assessment 9: 325-342

Stahl RG Jr, Kapustka L, Bruins RJF, Munns, WR Jr (eds) (in press) Valuation of Ecological Resources: Integration of Ecological Risk Assessment and Socioeconomics to Support Environmental Decisions. SETAC Press, Pensacola, FL

Suter GW II, Norton SB, Barnthouse LW (2003a) The evolution of frameworks for ecological risk assessment from the Red Book ancestor. Human and Ecol Risk Assessment 9: 1349-1360

Suter GW II, Vermeire T, Munns W Jr, Sekizawa J (2003b) Framework for the integration of health and ecological risk assessment. IPCS Workshop summary. Human and Ecol Risk Assessment 9: 281-301

US Environmental Protection Agency (USEPA) (1995) Guidance for Risk Characterization. Office of Science Policy. February

Varis O, Kuikka S, Kettunen J (1993) Belief networks in fish stock assessment – The Baltic salmon case. ICES Statutory Meeting Ref M Statistics Committee Ref Anacat Committee pp 1-18

Vermeire T, MacPhail R, Waters M (2003) Integrated human and ecological risk assessment: A case study of organophospherous pesticides in the environment. Human and Ecol Risk Assessment 9: 343-357

Vitousek PM, Mooney HA, Lubchenco J, Mellilo LM (1997) Human Domination of Earth's Ecosystems. Science 277:494-499

Wilson EO (1998) Integrated science and the coming century of the environment. Science 279:2048-2049

World Health Organization (2001) Integrated Risk Assessment. International Programme on Chemical Safety (IPCS). WHO/IPCS/IRA/01/12 Geneva, Switzerland.

Chapter 39: Toxin mixture in cyanobacterial blooms – a critical comparison of reality with current procedures employed in human health risk assessment

Daniel R Dietrich, Fischer A, Michel C, Hoeger SJ

Abstract

Cyanobacteria are the oldest life forms on earth known to produce a broad spectrum of secondary metabolites. The functions/advantages of most of these secondary metabolites (peptides and alkaloids) are unknown, however, some of them have adverse effects in humans and wildlife, especially when ingested, inhaled or upon dermal exposure. Surprisingly, some of these cyanobacteria are ingested voluntarily. Indeed, for centuries mankind has used cyanobacteria as a protein source, primarily *Spirulina* species. However, recently also *Aphanizomenon flos–aquae* are used for the production of so called blue green algae supplements (BGAS), supposedly efficacious for treatment of various diseases and afflictions. Unfortunately, traces of neurotoxins and protein phosphatases (inhibiting compounds) have been detected in BGAS, making these health supplements a good example for human exposure to a mixture of cyanobacterial toxins in a complex matrix. The discussion of this and other possible exposure scenarios, e.g. drinking water, contact during recreational activity, or consumption of contaminated food, can provide insight into the question of whether or not our current risk assessment schemes for cyanobacterial blooms and the toxins contained therein suffice for protection of human health.

Cyanobacterial metabolites: Health hazards for humans?

Cyanobacteria exist worldwide ubiquitously, including in extreme environments (Hitzfeld et al. 2000a; Wynn–Williams 2000). Some of these cyanobacterial species produce complex compounds at great energy expense. The ecological or physiological function or advantage of these compounds for these cyanobacteria species is yet unknown. Toxic cyanobacterial blooms, whether on the surface, benthic growths or as subsurface layers of fresh, brackish or marine water bodies have been responsible for numerous acute and subacute intoxication incidents of humans and animals. Most of these toxic metabolites are very small alkaloids or peptides, with a molecular weight of <2kD, a very complex structure and, in case of the peptides, composed of uncommon amino acids. Amongst the alkaloids synthesized by cyanobacteria are e.g. saxitoxins or anatoxins, known to be toxic for humans and other organisms at very low concentrations (Hitzfeld et al. 2000b). The cyanobacterial peptides are not metabolized in the mammalian digestive tract due to the presence of D–amino acids and a cyclic structure. Consequently, either local toxicity, e.g. diarrhea, or systemic toxicity, e.g. hepatotoxicity, neurotoxicity, renal toxicity i.e. morbidity or even mortality are the direct consequences of acute exposure. However, the uptake of these toxins and their organ distribution appears to be mediated by specific transporters, of which most, if not all, remain unidentified in humans and other mammalian species (Fischer et al. 2005). An exception is the uncommon, non–protein amino acid β–N–methylamino–L–alanine (BMAA), a neurotoxin, which is probably produced by most likely all known groups of cyanobacteria (Cox et al. 2005).

In order to provide a reasonable hazard and risk assessment for these toxins in conjunction with acute or chronic exposures, it is prerequisite to characterize their uptake and distribution (kinetics) and consequently the toxic dose that induces *in–situ* effects (dynamics). Moreover, it is essential to understand the kinetics and dynamics and thus the hazard and risk of single toxins before the hazard and risk of toxin mixtures can be evaluated and meaningful risk management strategies can be developed.

In the environmental setting, not every cyanobacterial family or even strain within a family synthesizes highly toxic alkaloids and peptides, meaning that the simple occurrence of cyanobacteria in a water body does not mandate the presence of cyanobacterial toxins. In order to ensure that a given water body is free of or has only a limited contamination with toxins, specific toxin analyses should be carried out. However, to do this on a routine basis can be financially problematic, thus calling for a prioritization of water bodies and definition of sampling routines that would ensure the highest degree of safety at reasonable cost. First and foremost, all water bodies in use as drinking water resources must be monitored routinely, the frequency of analyses should be determined based on the cyanobacterial history of the water body and the observation of the water body by qualified personal. However, even oligotrophic lakes can contain blooms of toxic cyanobacteria that are not readily and routinely observable. Especially *Planktothrix rubescens*, the species having the highest microcystin concentration/cell of all known cyanobacteria, occurs specifically in the metalimnion of oligotrophic and deep lakes, sometimes even during winter. Second, water bodies used for aquaculture (fish, shrimp and shellfish) must be monitored and the products harvested must be controlled for contamination with known cyanobacterial toxins, since for some of them (e.g. BMAA) bioaccumulation and trophic transfer has been demonstrated. Especially lakes in use for harvesting and the production of cyanobacterial food supplements must be monitored and a product and consumption derived limit of acceptable contamination defined. Finally, water bodies used for recreational purposes must be monitored and the bathing areas must be closed, if the concentration of cyanobacterial toxins is considered to endanger human health, especially children and toddlers.

For risk assessment and risk management purposes one would be inclined to try to differentiate and classify species that occur specifically only in one given environment, and thus, to try to deduce the possible type(s) of toxins that could potentially evolve from a bloom of this species in the given environment. This approach, however, has so far proven to be unsatisfactory due to the following observations:

- While some cyanobacterial species appear to prefer specific environments (e.g. *Nodularia spumigena*, *Planktothrix rubescens*), others are found nearly in all environments (*Lyngbia* sp.)
- Cyanobacterial blooms may be predominated by one given cyanobacterial species, however this does not exclude the quantitatively important presence of another toxin producing species
- Blooms of different cyanobacterial species (e.g. *Anabaena circinalis* and *Microcystis aeruginosa*) have been observed to occur in the very same water body at different time–points but also sometimes overlapping when occurring in rapid succession to one another
- Many cyanobacterial species have been observed to be capable of producing several different toxin types as well as different toxin congeners (e.g. microcystin congeners)
- Bloom size, expressed as cells density (cells/ml), are not a reliable indicator for toxin production especially as large blooms have been reported that did not produce extensive toxin amounts, while conversely blooms containing low to moderate cell densities have been demonstrated to produce copious amounts of toxin.

The following two examples may serve as an illustration of the above points:

1. In 1996 an intensive phytoplankton bloom was observed in the Montargil reservoir, Portugal. From May to June *Aphanizomenon flos–aquae* had been the predominant species and bloom extracts exhibited clear neurotoxic symptoms in a mouse bioassay. Pereira et al. (2000) analytically identified five saxitoxin variants following the isolation of an *A. flos–aquae* strain predominating the bloom. From July to August *Microcystis aeruginosa* predominated the bloom with extracts showing high hepatotoxicity in the mouse bioassay. The observed hepatotoxicity was attributed to the presence of high concentrations of microcystins. Although the two cyanobacteria species occurred in the bloom with alternating predominance, *A. flos–aquae* was still present in considerable densities during *M. aeruginosa* predominance. However, despite the presence of large numbers of *A. flos*–aquae no neurotoxicological activity (AOAC mouse bioassay) was observable suggesting that the *A. flos–aquae* strains co–occurring within the *M. aeruginosa* bloom did not produce saxitoxin at all or, at least not in relevant quantities. It seems that in the Montargil reservoir several saxitoxin producing and non–producing strains coexist within the same bloom but with varying densities. Alternatively, the *A. flos–aquae* in the Montargil reservoir was only one strain of *A. flos–aquae* with rapidly changing capabilities of producing saxitoxins.
2. In a drinking water reservoir in Australia where *M. aeruginosa* and *A. circinalis* were the predominant cyanobacteria, Hoeger et al. (2004) found that the highest concentrations of microcystins produced by *M. aeruginosa* did not coincide with the highest cell counts. This could, as suggested above, indicate the simultaneous presence of non–toxin–producing and toxin–producing strains of *M. aeruginosa.* In contrast, saxitoxin(s) concentrations found appeared to be highly correlated with the cell densities of *A. circinalis,* suggesting that only one continuously toxin–producing strain of *A. circinalis* was present in this specific water body.

The two examples above strongly illustrate the importance of detailed understanding of the limnological and toxicological characteristics of a specific water body, i.e. the spatial and temporal occurrence of toxin(s) and cyanobacterial species present in a given bloom. Consequently, generalized statements regarding the species and toxins occurring in a specific water body, or water bodies in general, is difficult if not impossible. Reliable risk characterization of possibly toxic blooms demands assessment of the cyanobacterial species and toxins by specialists or trained personnel, and a thorough understanding of the limnological history of the respective water body concerned.

Taking the above points regarding the characterization of bloom events into consideration, it becomes stringently clear that each and every bloom event in a given water body represents a unique risk scenario for humans and animals. In order to reduce complexity and facilitate understanding of the main issues at hand, the following discussion primarily focuses on human health risks. The discussion of the possible exposure scenarios in conjunction with the toxin specific health risk extrapolations will provide insight into the question as to whether or not our current risk assessment schemes for cyanobacterial blooms suffice for protection of human health as well as pin–point areas where more profound understanding of the underlying toxicity mechanisms is needed.

General risk scenarios (cyanobacterial blooms)

Massive toxic cyanobacterial blooms of concern for humans predominantly occur in fresh water bodies used for the preparation of drinking water and recreational activities. Consequently, human cyanobacterial toxin exposure can occur either via contaminated drinking water or during recreational activities in cyanobacteria blooming water. The exposure scenarios that follow are therefore artificially divided into the two predominant human exposure routes, namely oral and inhalation exposure. Although dermal exposure and subsequent effects have been described in the literature (Pilotto et al. 1997; Soong et al. 1992; Stewart et al. 2001; Torokne et al. 2001), primarily as anecdotal reporting of incidences without a experimental follow–up and/or proof of principle analysis, the cyanobacteria and toxin(s) causally responsible for the reported dermal effects have not been established with the exception of *Lyngbya majuscule*. The contact with this species causes dermatitis, and accidental oral consumption led to an burning sensation in the mouth (Osborne et al. 2001).

Oral exposure

The oral exposure scenario assumes that the person is exposed voluntarily or accidentally. In both situations, it is the goal of the risk assessment to elucidate the highest toxin concentration in the water that has no health consequences for the exposed person ingesting a given volume of water in an acute (one time ingestion) or chronic (continuous ingestion) situation.

Drinking water

For drinking water, one of the primary concerns is whether water treatment can eliminate or reduce the concentration of the cyanobacterial toxins present during a bloom event. Although this generally appears to be the case for most technically moderately modern water treatment systems (Hitzfeld et al. 2000b; Hoeger et al. 2005) and bloom situations, exceptions to the rule do occur (Hoeger 2005).

In the years 1985 –1988 recurring cases of severe gastro–enteritis were observed in Bahia State, Brazil, apparently closely associated with the consumption of treated and untreated water from the Itaparica Dam reservoir (Teixera et al. 1993). Although cases of gastro–enteritis and associated diarrhea were observed throughout the years at low incidence, increased numbers of cases with concurrent hospitalization and subsequent mortalities occurred regularly between January to May of each year. In 1988 the most severe outbreak of gastro–enteritis and associated diarrhea with fatal outcome occurred between February and May. Of approximately 2392 reported cases of gastro–enteritis, 368 were immediately hospitalized. The average monthly mortality among the hospitalized patients was 24% with a peak of 45.1% in May. No abnormalities in the frequency of diarrhea incidences arose in adjacent regions with a different drinking water source. Clinical/microbial investigations of the water from the Itaparica dam reservoir treated water did not identify the presence of fecal coliforms, salmonella, shigella, adenovirus or rotavirus. Boiling and filtration of the treated water had no impact on the occurrence of severe gastro–enteritis. Accordingly, it appeared that the agent in the treated water was heat and filtration resistant, possibly pointing to cyanobacterial toxins e.g. microcystins. Limnological analyses of the dam water in April 1988 revealed the presence of *Anabaena* and *Microcystis* species at densities exceeding the WHO maximum acceptable levels for untreated drinking water by a factor 4 – 33. Seventy percent of the hospitalized gastroenteritis cases were children under the age of 5 in the period between March and May, 1988. The high incidence of illness among children suggests that risk calculations based on data from human adults or animal studies may not be sufficiently protective for children.

In 1999 and 2000 a survey of cyanobacterial toxins in surface waters and drinking water supplies was carried out across Florida, USA, revealing the following concentrations of cyanotoxins in raw and treated/drinking water (Burns 2004):

- concentrations of microcystins ranged from below the detection limit to 12.5 μg/l in raw and treated/drinking water samples
- concentrations of anatoxin–a ranged from below the detection limit to 8.46 μg/l in three treated drinking water samples
- concentrations of cylindrospermopsin ranged from 8.07 to 97.12 μg/l in nine treated drinking water samples. In some cases, trichomes of *Cylindrospermopsis raciborskii* in the raw water survived the treatment process and were present in the tap water.

An acute maximum dose that is not toxic to infants, small children or adults can be extrapolated from the lowest acute toxic doses of cyanotoxins observed in mouse studies using body weight assumptions (Dietrich and Hoeger 2005; Fromme et al. 2000). For example, a single intake of 12.5, 50 or 150 μg micorcystin–LR (MC–LR) equiv. are assumed not to produce adverse effects in toddlers (< 5 kg bw), small children (>20 kg bw) and adults (60 kg bw), respectively. However, such calculations are fraught with difficulties as it is unclear how much of the compounds are actually biologically available in humans, i.e. whether or not direct extrapolations from mouse studies to humans are reliable, in view of the fact that multiple toxins may occur in a bloom simultaneously, because there is lack of knowledge regarding combinatorial toxicity, and there are vast species differences in toxin specific transporter expression in the gastro–intestinal tract (see below).

For the calculation of safe levels for chronic exposure, estimates of drinking water intakes and average body weights must be used. The average daily consumption of water varies between 1500 and 2000 ml of treated (drinking) water per day for adults with an average weight between 50 – 80 kg bw. For infants and small children average daily water consumption is assumed on average to be 750 ml and 1000 ml per day, respectively (Dietrich and Hoeger 2005). In all calculations the assumptions are made that treated water is consumed, that consumption will occur nearly for a life–time, and that only one of the variety of cyanobacterial toxins will occur at any one given time. However, toxic cyanobacteria are unlikely to be continuously present, different cyanotoxins toxins can be present at the same time in varying concentrations, and water treatment procedures do not always guarantee cyanotoxin removal.

Recreational Activities

All activities at, in, and on the surface of the water body that bring humans into direct contact with water are considered to be recreational activities.

These include bathing, swimming incl. contests e.g. triathlons, water skiing, boating, etc. While pursuing these activities it is assumed that some toxin–contaminated water will be accidentally and involuntarily ingested. The amount of water ingested is – in a worst–case situation – approximated as ranging from 50 – 120 ml per episode. As this is usually an intermittent rather than a continuous occurrence, with the exception of people continuously swimming and bathing in a water body chronically contaminated with toxic cyanobacteria blooms, all risk calculations are driven toward an acute exposure scenario. The following episode is an example of this type of exposure:

Turner et al. (1990) reported cyanobacterial poisoning of two 16 – year old army recruits. Both became ill after a canoe exercise on a freshwater reservoir in Staffordshire, England, experiencing a massive *M. aeruginosa* bloom containing high concentrations of MC–LR. Both swallowed some of the water; both recruits developed pneumonia and other symptoms including blistering around the mouth, sore throat, dry cough, pleuritic pain, vomiting and abdominal pains requiring medical attention. One of the patients presented with temporary difficulties in walking, suggesting possibly neurotoxic effects. Subsequently another sixteen soldiers received medical attendance also after canoeing. Eight exhibited some symptoms similar to the first two cases. The disease pattern was considered to stem from microcystin intoxication, even though two weeks later a critical contamination of the reservoir with *Escherichia coli* was confirmed. The coincident presence of *M. aeruginosa* and *E. coli* could suggest a simultaneous exposure of the recruits to MC–LR and lipopolysaccharides (LPS) because the patterns of symptoms, with possibly the exception of the transient neurological disorder, are characteristic for a MC and LPS exposure. The neurological symptoms could be due either to MC (Fischer et al. 2005) or other neurotoxic cyanotoxins. In addition, augmentation and exacerbation (additive or synergistic) of effects and symptoms due to simultaneous exposure to different toxins (MC and LPS) cannot be excluded (Best et al. 2002).

Specific health risks from contaminated food and cyanobacterial food supplements

Contrary to the direct exposure of humans to cyanobacterial toxins via contaminated water, the risk situation involving exposure via food and food supplements is much more complex. Worst–case exposures can be interpolated from assumed daily or weekly consumption of specific food sources e.g. fish, crayfish, shellfish, vegetables, salads, etc. for the general populace as well as for populations at high risk, e.g. indigenous tribes predominantly existing on a specific food source (Dietrich and Hoeger 2005;

Ernst et al. 2005). However, the potential human toxin exposure via food that provides the basis for risk calculations is also largely determined by the degree of toxin contamination of a given food source as well as by the bioavailability of the toxin from the food type. Furthermore, although most likely an exception to the rule, bioaccumulation of cyanotoxins in the food chain, as is the case with beta–N–methylamino–L–alanine (BMAA), may provide ~~for~~ an additional element of risk (Cox et al. 2005).

Exposure considerations similar to those for drinking water also apply to the voluntary consumption of cyanobacteria (blue–green alga) food supplements on the basis of cyanobacteria, also know as blue–green alga supplements (BGAS). The main difference is that the daily consumption of pills, powders and drinks is extremely difficult to estimate. Furthermore, these products are often handled as if they were pharmaceuticals. Therefore, the consumption per person does not correspond to bodyweight as is the case for real food sources, e.g. fish, crayfish etc. A detailed discussion of this issue can be found in Dietrich & Hoeger (2005), a publication that also estimates the effects of varying levels of microcystin contamination of food and food supplements and the corresponding limit amount that can be safely consumed by infants, children and adults (Table 1).

BGAS are usually produced from *Spirulina maxima* or *platensis* and *Aphanizomenon flos–aquae*. Although *Spirulina maxima* is considered as being non–toxic (Salazar et al. 1996; Salazar et al. 1998), Draisci et al. (2001) identified Epoxyanatoxin–a and Dihydrohomoanatoxin–a at concentrations ranging from non detectable to 19 $\mu g\ g^{-1}$ dw in *Spirulina*–based BGAS.

In alkaline crater lakes in Kenia, *Arthrospira fusiformis* (syn. *Spirulina fusiformis* = *platensis*) was found to produce small amounts of both, microcystins and anatoxin–a (Ballot et al. 2004; Ballot et al. 2005). Positive ELISA– results were evidence for ~~a~~ possible contamination of Spirulina supplements with microcystins (Gilroy et al. 2000). In addition, Spirulina based BGAS is suspected to be responsible for the liver injury of a 52–year old Japanese (Iwasa et al. 2002). Further corroboration of these data is necessary before proper evaluation of the potential hazard to human health in the context of a risk assessment can be carried out.

Contrary to *Spirulina*, where direct production of toxin is yet uncorroborated, *A. flos–aquae* is known to be capable of producing the neurotoxic alkaloids anatoxin–a (Rapala et al. 1993) and saxitoxins (Adelman et al. 1982; Ferreira et al. 2001; Mahmood and Carmichael 1986; Pereira et al. 2000), as well as the neurotoxic, nonprotein amino acid BMAA (Cox et al. 2005). Maatouk et al. (2002) assumed that a mono–specific *Aph. flos–aquae* bloom was responsible for the microcystin content of a *Aph. flos–aquae* sample from Saint–Caprais reservoir in France. The cytotoxic and most likely carcinogenic alkaloid cylindrospermopsin has yet not been detected in *Aph. flos–aquae*, however its presence in *Aph. ovalisporum* is confirmed (Banker et al. 1997; Shaw et al. 1999).

In contrast to Spirulina, *Aph. flos–aquae* is, to our knowledge, generally harvested from natural lakes. One of its biggest sources is Lake Klamath, Oregon, where *Aph. flos–aquae* co–exists and coincides with *Microcystis sp.* blooms (Carmichael et al. 2000), this coexistence can also be observed in other lakes (Ekman–Ekebom et al. 1992; Teubner et al. 1999). Consequently, consumers of Spirulina– and *Aph. flos–aquae* – BGAS are potentially exposed to toxins produced by all three of the cyanobacteria species.

Although neurotoxic alkaloids (anatoxins and saxitoxins) have not been identified to date in Klammath Lake, Sawyer et al. (1968) reported on an aqueous extract of an Klamath Lake *Aph. flos–aquae* bloom that was nearly instantaneously lethal to mice after i.p. injection. In addition, *Aph. flos–aquae* was reported to be responsible for massive fish kills (Barica 1978) and being generally highly toxic to members of the freshwater fauna (Gentile and Maloney 1969). Out of 88 bloom samples from Finland in which *Aph. flos–aquae* was one of the predominating species, 11 were determined to be neurotoxic and 25 to be hepatotoxic (Sivonen et al. 1990). Acute toxicity testing of 23 populations of *Aph. flos–aquae* from 12 localities of inland waters in South Norway, resulted in protracted toxic response in the test animals with 60% of the samples tested (Underdal et al. 1999).

Several investigations report high concentrations of microcystins in BGAS products (Hoeger and Dietrich 2004; Lawrence et al. 2001; Saker et al. 2005; Yu et al. 2002), often dramatically exceeding the provisional guidance value of 1 µg/g dw set by the Oregon Health Division and the Oregon Department of Agriculture (Gilroy et al. 2000). Based on some major weaknesses in the assumptions underlying the risk assessment leading to this provisional guidance value, Dietrich and Hoeger (2005) concluded that BGAS could pose a serious health risk to consumers, especially children. One of the main weaknesses in the risk calculation process is the assumption that all congeners of a toxin, in this case microcystins have the same toxicokinetics and dynamics, i.e. that all congener concentrations in a given example can be added up to give a $MC\text{–}LR_{equiv.}$ concentration. Another weakness is that all of the risk calculations are focused on risk of a single compound exposure situation but not, as was shown above, on the potential simultaneous exposure to multiple toxins of vastly different structure, kinetics and dynamics, ignoring possible additive or even synergistic effects, as will be discussed later in the text. The toxicity of BGAS supplements is particularly worrisome because the benefits of the consumption of BGAS are still unclear and could not be confirmed scientifically (Vitale et al. 2004).

Inhalation exposure through drinking/hygienic water

Inhalation exposure to cyanobacterial toxins stems primarily from three exposure situations, namely habitation near chronically contaminated surface waters, use of contaminated water for hygienic purposes (showering and sauna), and finally during recreational activities as already indicated above. The problem is that good exposure models are lacking. Only rough assumptions can be made for the amount of toxin and toxic cyanobacterial cells available in an aerosolic form and for the average inhalation rate (liters minute^{-1}) of the persons potentially exposed to the cyanobacteria and their toxins. Needless to say that the average inhalation rate of a person involved in a high–energy recreational activity e.g. swimming/triathlon competition, water skiing or those involved in a moderate activity e.g. swimming leisurely, boating, etc or those living at the coast–line of a chronically bloom contaminated surface water will be entirely different. The following examples illustrate situations in which adverse effects from inhalation exposure was very likely although proper documentation is missing:

Hoppu et al. (2002) reported cases where cyanobacterial contaminated water was used for washing and for producing hot steam in a sauna in Finland. By pouring cyanbacteria containing water on heated stones, cyanobacterial cells may be lysed, toxins released and aerosolized. As a consequence, people could experience both dermal and inhalation exposure to cyanotoxins via contaminated water and aerosols. Subsequent to sauna and bathing, 48 persons developed gastrointestinal, dermal and neurological symptoms most likely attributable to cyanobacterial toxin exposure. Unfortunately no information was available on the cyanobacteria species or the cyanotoxins present in the water used for bathing and the sauna.

A similar situation could have occurred in Florida, where copious amounts of microcystins, anatoxin–a and cylindrospermopsin as well as trichomes of *Cylindrospermopsis raciborskii* were reported in raw and treated water (Burns 2004). Inhalation exposure could have taken place especially in showers, as shower water usually exceeds 70°C during the heating process, cyanotoxins are released from cells due to heat lysis, and aerosolized with the steam in the shower cabin. Up to 10^4 cells ml^{-1} could be counted in the tap water in a town in Queensland/Australia (Hoeger et al. 2004). Although no exposure related effects were reported for residents in either case, the above examples demonstrate that high cell numbers of potentially toxic cyanobacteria can pass water treatment and provide a basis for inhalation, dermal and oral toxin exposure.

Furthermore, the above examples demonstrate that exposure to multiple toxins via inhalation can most likely occur. Heating of cyanobacteria contaminated water and aerosolization of cyanobacterial components may provide a form of exposure unlikely to occur under other circumstances (bathing, etc.). Consequently it is at present unclear whether the risk assessment calculations carried out with known cyanobacterial toxins offer any safety for the inhalation exposure scenario.

Human health risk assessment single vs. multiple compounds: toxicokinetic and –dynamic considerations

Investigations of cyanobacteria intoxication events have so far focused on MCs due to their frequent occurrence, high toxic potential, and clear association, if not causal relationship, with severe poisonings of humans (Kuiper–Goodman et al. 1999) and animals (Landsberg 2002). As most of these incidences were related primarily to ingestion of contaminated water, the WHO (Chorus and Bartram, 1999) developed a provisional guidance value of 1 µg $MC\text{–}LR_{equiv}$ L^{-1} for drinking water, based on data from a mouse study using purified MC–LR. A similar study with pigs and MC containing extracts resulted a nearly the same no/lowest observed adverse effect value (Dietrich and Hoeger 2005).

Toxicological studies using MC containing extracts inherently appear to have a greater relevance for assessment of potential toxicological hazards to human health than those using one purified congener. To date numerous studies on the toxicity of purified MCs, MC–LR in particular, and of MC containing extracts have been carried out using different test organisms and different routes of administration. Despite that microcystins have generally been characterized as hepatotoxins, there is also evidence for neurotoxicity, as suggested in the Caruaru incident, where more than 60 dialysis patients died following i.v. treatment with MC contaminated dialysis water (Azevedo et al. 2002; Fischer et al. 2005; Jochimsen et al. 1998; Pouria et al. 1998), and for nephrotoxicity, as indicated by several animal experiments (Bhattacharya et al. 1997; Fischer and Dietrich 2000; Khan et al. 1996; Milutinovic et al. 2002; Milutinovic et al. 2003; Moreno et al. 2005; Stotts et al. 1997). Studies on organ distribution support these findings and suggest that, beside liver and kidney, the brain and the lungs (Picanco et al. 2004) can also be affected to a varying extent (Dietrich and Hoeger 2005;

Meriluoto et al. 1990; Pace et al. 1990; Robinson et al. 1989; Robinson et al. 1991; Stotts et al. 1997).

Since microcystins are charged and spatially large molecules they cannot cross cell membranes via simple diffusion, thus cellular uptake is most likely mediated by energy dependent transporters. Indeed, Runnegar (1995) assumed sinusoidal, bile–salt, organic–anion transporters, including members of the superfamily of organic anion transporting polypeptides (human: OATP; rodent: oatp), are the unlikely transporting route for mediating the uptake of microcystins into hepatocytes, because rifamycin inhibited the MC uptake in hepatocytes (Runnegar et al. 1995), whereas rifampicin, a member of the same antibiotic family as rifamycin, did not inhibit the transport capacity of the investigated oatp (Jacquemin et al. 1994; Jacquemin et al. 1991). However, more detailed investigation by Fischer et al. (2005) demonstrated the uptake of [^{3}H]–dihydro–microcystin–LR in *Xenopus laevis* oocytes via human OATP1B1 and OATP1B3, both located at the basolateral (sinusoidal) membrane of hepatocytes, as well as OATP1A2, located in liver and brain, possibly responsible for the observed hepatotoxicity and neurotoxicity, respectively. Transport could be inhibited by co–incubation with the known OATP/oatp substrates taurocholate and bromosulfophthalein.

OATPs/oatps could be detected in nearly all tissues of humans and rodents (Hagenbuch and Meier 2003; Meier and Stieger 2002; van Montfoort et al. 2003a) and therefore might be responsible for the observed microcystin mediated effects in the various organs of the respective species. Whether only OATPs/oatps mediate the transport of microcystins is presently unknown. However, this appears rather unlikely as not every organ is observed to express OATPs, at least according to current knowledge. Generally OATPs are responsible for importing substrates into cells, although bidirectional transport was also observed in some cases. Multidrug Resistance Proteins (MDRs) and Multidrug Resistance associated Proteins (MRPs), both members of the ATP Binding Cassette superfamily (ABC transporters), are other potential mediators for the export of microcystins for the following reasons:

- MCs fit into the molecular weight spectrum of MDRs and MRPs substrates (table 1)
- The uptake of MCs was shown to be inhibited after protein phosphatase inhibition, suggesting that at least some of the responsible carriers are controlled via phosphorylation (Runnegar et al. 1995)

- Most of the MRPs are known to transport glutathione conjugates and MCs were demonstrated to be conjugated with glutathione (Pflugmacher et al. 1998)
- Glutathione conjugated microcystins were observed in the bile of microcystin exposed animals (Sahin et al. 1994)
- MRD and MRP are expressed at the apical side (hepatocyte/bile membrane) of hepatocytes as well as the basolateral side of bile cells (Hagenbuch et al. 2002; Hagenbuch and Meier 2003, 2004; van Montfoort et al. 2003a).

The assumption that these ABC transporters and OATPs (Ballatori et al. 2005; Faber et al. 2003; Hagenbuch and Meier 2003; Ho and Kim 2005; Kunta and Sinko 2004; Meier and Stieger 2002; Miyazaki et al. 2004; Nies et al. 2004; Shitara et al. 2005; van Montfoort et al. 2003b) are the responsible carriers for MCs would potentially explain some of the organ distribution characteristics of MC–LR after oral intoxication. However, none of these investigations was carried out using other MC congeners. Indeed, *in vivo* evidence from Milutinovic et al. (2002; 2003) strongly suggests that transport of microcystins via ABC transporters and OATPs is largely congener dependent, and thus, results in a congener dependent organ distribution and distribution–mediated intensities of adverse effects. Further characterization of the transport of microcystins via OATPs, as well as other potential transport systems, is of prime importance for improving better understanding of the species– and congener–specific kinetics and thus for improved risk extrapolation and human risk assessment.

Based on the current state of the science, the risk assessment of microcystins is hampered by the following inadequacies as summarized below:

- 80 different microcystins congeners are known to date and most likely more will be discovered in the near future. However, the risk assessment calculations are based on the toxicokinetic– and dynamic properties of one single MC congeners, namely MC–LR
- Multiple MC congeners can occur simultaneously in a cyanobacterial bloom, emphasizing the uncertainty in estimating true toxic potency in a bloom situation
- MC–congeners display distinct differences in organ distribution and location of toxic effects (Gupta et al. 2003; Milutinovic et al. 2002; Milutinovic et al. 2003) strongly suggesting that these differences are governed primarily by the affinities of the respective MC congeners for

the transporting proteins present in the different organs, i.e. also on their expression levels ((Fischer et al. 2005), see above)

- The comparison of different species (rodents versus humans) strongly suggests that microcystin transporting proteins are neither identical nor comparable amongst species; neither in their level of expression and distribution amongst organs of a given species nor in their respective microcystin transporting capabilities. This strongly indicates that risk extrapolation from rodents to the humans, despite the inclusion of a safety factor of 10 for inter–species extrapolation (Dietrich and Hoeger 2005), may underestimate the potential risk posed by the different toxins to humans.
- Additive, synergistic or antagonistic effects of the various MC congers have so far not been investigated.
- Current risk assessment schemes, although incorporating allocation factors for contribution to total toxin exposure from sources other than the main source (drinking water), fail to consider route and matrix differences in bioavailability that may exist.

In order to understand the toxicities, bioavailabilities and toxicokinetics of single toxins and extracts, additional studies are needed and are essential for an improved risk assessment.

Obviously, the above uncertainties in the risk assessment of microcystins also emphasize the problems of establishing highly reliable guidance values for drinking water, inhalation exposure, or voluntary exposure via consumption of BGAS. Moreover, similar risk calculations for other toxic compounds e.g. cylindrospermopsin (CYN), anatoxins (ANA) saxitoxins (STX) and β–methylamino–L–alanine (BMAA), as well as the ubiquitously present lipopolysaccharides (LPS) are still under development but also fraught with the problems described for microcystins. Indeed, with the exception of BMAA (see below), a good understanding of the uptake and distribution kinetics of the other cyanobacterial toxins is nearly completely lacking to date. Thus, all calculations in the risk assessment process would have to be based on physico–chemical characteristics of the toxins. Anatoxin–a, homoanatoxin–a, anatoxin–a(S), saxitoxins and cylindrospermopsin are very small molecules, with molecular weights of 165, 179, 252, 299 and 415, respectively, and thus are assumed to be able to cross the cell membranes via diffusion or facilitated diffusion. However, to date there is no proof for these assumptions, emphasizing again the dearth of information available for risk assessment purposes.

BMAA (β–N–methylamino–L–alanine), acts as an agonist of animal glutamate receptors (Brenner et al. 2003) and is chemically related to excitant amino acids. Its uptake into brain is most likely mediated by the large neutral amino acid carrier of the blood–brain barrier (Smith and Shine 1992). BMAA, a cyanobacterial, neurotoxic, nonprotein amino acid has been associated with the significantly increased incidences of amyotrophic lateral sclerosis/parkinsonism–dementia complex (ALS/PDC) among the Chamorro people in Guam and other islands of the Pacific. BMAA, originating from cyanobacterial cycad root symbionts of the genus Nostoc, was demonstrated to bioaccumulate at various trophic levels including–in some of the traditional foods of these inhabitants of these islands (Cox et al. 2003). However, recently Cox et al. (2005) revealed that BMAA production by cyanobacteria appeared most likely to be a general phenomenon, as BMAA was detected it in 95% of 21 cyanobacterial genera. Furthermore, most recent GC–MS/MS analyses of Aph. flos–aquae and Spirulina spec. both sold as BGAS demonstrated large quantities of BMAA in these products (Dietrich unpublished data). If, BMAA is indeed involved in the aetiology of ALS/PDC, the findings of BMAA in BGAS may potentially explain the detection of BMAA in brain tissues of Canadians with Alzheimer's disease (Cox 2005). Although it would have to be established whether or not these patients had consumed BGAS at any given time. The presence of BMAA, although not highly toxic in an acute scenario, profusely demonstrates that indeed, cyanobacterial bloom toxicity is NOT a single toxin scenario but most likely a combinatorial toxicity of many toxins present. Consequently, risk assessments must take these issues into consideration, even though to do so may prove extremely difficult.

Extract toxicity and synergisms

As a consequence of the limited knowledge on cyanotoxins other than MCs and BMAA, e.g. cylindrospermopsin, anatoxin A, etc., the additional factors influencing potential risk i.e. antagonistic, additive or synergistic toxicity of simultaneously occurring cyanobacterial toxins has largely been ignored. Although antagonistic effects have so far not been reported they have also not been the subject of toxicological studies. Additive or synergistic toxicity has been suggested by a number of observations. Indeed, cyanobacterial extracts have shown much greater toxicity then what would have been expected from the amounts of toxins (e.g. MCs) contained in the respective samples.

The following examples may serve as an illustration the potential presence of additive or synergistic toxicity of extracts: Majsterek et al. (2004) reported on the increased toxicity of a MC–LR containing extract compared to a microcystin–LR standard employing a cytochrome c oxidase assay using mammalian mitochondria from *Bos taurus*. The extract inhibited the activity of mitochondrial oxidase to a much higher extent than the same concentration of purified MC–LR standard. These authors assumed that the extract contained additional congeners of MC not detected with their means of chemical analysis.

Another potential explanation for the increased extract toxicity was provided by Best et al. (2002), who investigated the effect of lipopolysaccharides (LPS) from axenic cyanobacteria and from natural blooms on the activity of microsomal (m) and soluble (s) glutathione *S*–transferase (GST) of zebra fish embryos (*Danio rerio*) *in vivo*. Both activities were significantly reduced by cyanobacterial LPS, as well as by co–exposure to cyanobacterial LPS and MC–LR. Hence, they concluded that LPS may potentially cause inhibition of the GST catalyzed conjugation of MCs to glutathione, which represents the first step in detoxification of MCs and therefore may prolong residence time of MCs in the zebrafish resulting in a much higher toxicity and what was expected from pure MC–LR.

Fitzgeorge et al. (1994) compared the toxicities of anatoxin–a and MC–LR with the toxicity of both toxins administered simultaneously. An intranasal LD_{50} of 2000 μg kg^{-1} bw for anatoxin–a and a non–lethal dose of 31.3 μg kg^{-1} bw for MC–LR were determined in mice. Upon administration of 31.3 μg kg^{-1} bw MC–LR 30 minutes prior to anatoxin–a, the LD_{50} for anatoxin–a was lowered to 500 μg kg^{-1} bw. This potential synergistic effect of anatoxin–a and MC–LR was further investigated by Rogers et al. (2005) using oral toxin administration. Mice were gavaged with either 0, 500 or 1000 μg microcystin–LR kg^{-1} bw, followed by 0, 500, 1000 or 2500 μg anatoxin–a kg^{-1} bw 50 minutes later. Despite the high concentrations used, no deaths, no clinical signs of intoxication and no differences in weight between pre– and post–treatment were reported. It was concluded that the failure in demonstrating the synergistic effect was most likely due to the different routes toxin administration. Indeed, intranasal application of toxins resulted in similar LD_{50}s as did i.p. application, while the oral LD_{50} of MC–LR in mice was approximately 12–fold higher than the i.p. or intranasal administration (Fitzgeorge et al. 1994). These findings, although yet uncorroborated in a more wide assessment of toxin interactions, suggest that the route of exposure as well as the toxin composition may be critical for the onset of adverse effects. In view of the vast differences of toxin transporter presence and level of expression in the different organs, as is the case for MCs and BMAA, it is not surprising to see route of administration dependent differences in LD_{50}s as well as differences in the potential for additive or synergistic effects. Consequently, extrapolation from various routes of exposure as a means for an overall risk assessment may be quite problematic.

Conclusions

The present analysis of the data on cyanobacterial toxins and their use within the context of risk assessment demonstrates that the dearth of information precludes development of simple safety assumptions. Although the WHO drinking water guidance value for MCs may provide a little more certainty due to the physiologically limited quantity of drinking water per person per day, the simultaneous presence of other toxins in the drinking water questions the reliability of this guidance value as it is expressively specified for only one (MC–LR) of the nearly 80 presently known toxin congeners and does not take other congeners or other toxins types into consideration. The provisional guidance value for BGAS as proposed by the Oregon Department of Health of 1 μg $MC\text{–}LR_{equiv.}$ g^{-1} dw BGAS, when considering the potential for additive or synergistic effects stemming from the presence of other cyanobacterial toxins in bloom situations in Klamath Lake as well as the nearly unlimited voluntary daily uptake of these compounds by the public and especially children is even more problematic. The latter situation appears even more of a concern because recent analyses confirmed the presence of BMAA and microcystins in these BGAS products, thus clearly emphasizing the high potential for the onset of hepatic, renal and neurological disorders in these children, whether or not this may be of the subacute/subchronic (MCs), or the delayed (MCs and BMAA) type. While sale of BGAS should be severely controlled or even restricted, authorities and academia should be supported in developing additional data pertaining to the inhalation and the food contamination exposure scenarios.

Table 1. Calculated possible daily ingestion to avoid acute health problems according to the calculations of Fromme et al. (2000). For details, see Dietrich et al (2005)

Ingestion route	MC concentrations	Infants 5 kg = 12.5µg	Children 20 kg = 50 µg	Adults 60 kg = 150 µg
Food	100 µg kg^{-1}	125 g	500 g	1,500 g
	10000 µg kg^{-1}	1.25 g	5 g	15 g
Cyanobacterial bloom in lake/river	100 µg l^{-1}	125 ml	500 ml	1,500 ml
	1000 µg l^{-1}	12.5 ml	50 ml	150 ml
Drinking water	1.0 µg l^{-1}	12,500 ml	50,000 ml	150,000 ml
	100 µg l^{-1}	125 ml	500 ml	1,500 ml
BGAS	1.0 µg g^{-1}	12.5 g	50 g	150 g
	10 µg g^{-1}	1.25 g	5 g	15 g

Table 2. Known and possible transporters of microcystins and their known distribution in human and rodent organ system. oatp: organic anion transporting polypeptide; mdr: multidrug resistance proteins; mrp: multidrug resistance associated Proteins, human transporters are generally written in capitals, rodent transporters lower case.

Organs	Transporter	Human	Rat/ Mouse	References
Gastro Intestinal	OATPs/Oatps	2A1, (2B1), 3A1, 4A1	1a5	(Cheng et al. 2005; Hagenbuch and Dawson 2004; Ho and Kim 2005)
	MDR/Mdr	1	1a, 1b	(Faber et al. 2003; Ho and Kim 2005)
	MRP/Mrp	1–3, 5, 7–9	1–3	(Faber et al. 2003; Glavinas et al. 2004; Ho and Kim 2005)
Liver	OATPs/Oatps	1A2*, 1B1*, 1B3*, 2A1, (2B1), 3A1, 4A1	(1a1), (1a4), 1b2*, 2b1	(Cheng et al. 2005; Hagenbuch and Dawson 2004)
	MDR/Mdr	1, 2/3	1b, 1a, 2	(Faber et al. 2003; Glavinas et al. 2004; Ho and Kim 2005; van Montfoort et al. 2003a)
	MRP/Mrp	1–3, 5–9	1–3, 6	(Faber et al. 2003; Glavinas et al. 2004; Ho and Kim 2005; van Montfoort et al. 2003a)

Organs	Transporter	Human	Rat/ Mouse	References
Kidney	OATPs/Oatps	1A2*, 2A1, 3A1, 4A1, 4C1	(1a1), 1a3, 1a6, 3a1, 4c1	(Cheng et al. 2005; Hagenbuch and Dawson 2004)
	MDR/Mdr	1	1a, 1b	(Glavinas et al. 2004; Ho and Kim 2005; van Montfoort et al. 2003a)
	MRP/Mrp	1–9	2	(Glavinas et al. 2004; Ho and Kim 2005; Robertson and Rankin 2005; van Montfoort et al. 2003a)
Blood–Brain	OATPs/Oatps	1A2*, 1C1, 2A1, 3A1, 4A1	(1a4), 1c1	(Cheng et al. 2005; Hagenbuch and Dawson 2004)
	MDR/Mdr	1	1a	(Glavinas et al. 2004; Hagenbuch et al. 2002; Ho and Kim 2005)
	MRP/Mrp	1, 5, 7–9	2	(Glavinas et al. 2004; Hagenbuch and Dawson 2004; Hagenbuch et al. 2002; Ho and Kim 2005)

* dihydro – MC–LR transport demonstrated by Fischer et al (Fischer et al. 2005)

() no dihydro – MC–LR transport

References

Adelman WJ Jr, Fohlmeister JF, Sasner JJ Jr, Ikawa M (1982) Sodium channels blocked by aphantoxin obtained from the blue–green alga, Aphanizomenon flos–aquae. Toxicon 20 513–6

Azevedo SM, Carmichael WW, Jochimsen EM, Rinehart KL, Lau S, Shaw GR, Eaglesham GK (2002) Human intoxication by microcystins during renal dialysis treatment in Caruaru–Brazil. Toxicology 181–182 441–6

Ballatori N, Hammond CL, Cunningham JB, Krance SM, Marchan R (2005) Molecular mechanisms of reduced glutathione transport: role of the MRP/CFTR/ABCC and OATP/SLC21A families of membrane proteins. Toxicology and Applied Pharmacology 204 238– 255

Ballot A, Krienitz L, Kotut K, Wiegand C, Metcalf JS, Codd GA, Pflugmacher S (2004) Cyanobacteria and cyanobacterial toxins in three alkaline rift valley lakes of Kenya – Lakes Bogoria, Nakuru and Elmenteita. Journal of Plankton Research 26 925–935

Ballot A, Krienitz L, Kotut K, Wiegand C, Pflugmacher S (2005) Cyanobacteria and cyanobacterial toxins in the alkaline crater lakes Sonachi and Simbi, Kenya. Harmful Algae 4 139–150

Banker R, Carmeli S, Hadas O, Teltsch B, Porat R, Sukenik A (1997) Identification of cylindrospermopsin in Aphanizomenon ovalisporum (Cyanophyceae) isolated from Lake Kinneret, Israel. Journal of Phycology 33 613–616

Barica J (1978) Collapses of Aphanizomenon flos–aquae blooms resulting in massive fish kills in eutrophic lakes: effect of weather. Verh Internationale Vereinigung für Limonologie 20 208–213

Best J, Pflugmacher S, Wiegand C, Eddy F, Metcalf J, Codd G (2002) Effects of enteric bacterial and cyanobacterial lipopolysaccharides, and of microcystin–LR, on glutathione S–transferase activities in zebra fish (Danio rerio). Aquatic Toxicology 60 223

Bhattacharya R, Sugendran K, Dangi RS, Rao PV (1997) Toxicity evaluation of freshwater cyanobacterium Microcystis aeruginosa PCC 7806: II. Nephrotoxicity in rats. Biomedical and Environmental Sciences 10 93–101

Brenner ED, Stevenson DW, McCombie RW, Katari MS, Rudd SA, Mayer KF, Palenchar PM, Runko SJ, Twigg RW, Dai G, Martienssen RA, Benfey PN, Coruzzi GM (2003) Expressed sequence tag analysis in Cycas, the most primitive living seed plant. Genome Biol 4 R78

Burns JW (2004) Cyanotoxins in Floridas (USA) surface waters: considerations for water supply planning. In ICTC 6th, Bergen/Norway, pp 4

Carmichael WW, Drapeau C, Anderson DM (2000) Harvesting of Aphanizomenon flos–aquae Ralfs ex Born & Flah var flos–aquae (Cyanobacteria) from Klamath Lake for human dietary use. Journal of Applied Phycology 12 585–595

Cheng X, Maher J, Chen C, Klaassen CD (2005) Tissue distribution and ontogeny of mouse organic anion transporting polypeptides (Oatps). Drug Metab Dispos 33 1062–73

Chorus I, Bartram J (1999) Toxic cyanobacteria in water. A guide to their public health consequences, monitoring and management. World Health Organization, E & FN Spon London, pp 416

Cox P, Banack S, Murch S (2003) Biomagnification of cyanobacterial neurotoxins and neurodegenerative disease among the Chamorro people of Guam. Proceedings of the National Academy of Sciences of the United States of America 100 13380–13383

Cox PA, Banack SA, Murch SJ, Rasmussen U, Tien G, Bidigare RR, Metcalf JS, Morrison LF, Codd GA, Bergman B (2005) Diverse taxa of cyanobacteria produce beta–N–methylamino–L–alanine, a neurotoxic amino acid. Proc Natl Acad Sci U S A 102 5074–8

Dietrich DR, Hoeger SJ (2005) Guidance values for microcystins in water and cyanobacterial supplement products (blue–green algal supplements): a reasonable or misguided approach? Toxicology and Applied Pharmacology 203 273–289

Draisci R, Ferretti E, Palleschi L, Marchiafava C (2001) Identification of anatoxins in blue–green algae food supplements using liquid chromatography–tandem mass spectrometry. Food Additives and Contaminants 18 525–31

Ekman–Ekebom M, Kauppi M, Sivonen K, Niemi M, Lepisto L (1992) Toxic Cyanobacteria in Some Finnish Lakes. Environmental Toxicology & Water Quality 7 201–213

Ernst B, Dietz L, Hoeger SJ, Dietrich DR (2005) Recovery of MC–LR in fish liver tissue. Environmental Toxicology 20 449–58

Faber KN, Müller M, Jansen PLM (2003) Drug transport proteins in the liver. Advanced Drug Delivery Reviews 55 107–124

Ferreira FMB, Soler JMF, Fidalgo ML, Fernandez–Vila P (2001) PSP toxins from Aphanizomenon flos–aquae (cyanobacteria) collected in the Crestuma–Lever reservoir (Douro river, northern Portugal). Toxicon 39 757–761

Fischer WJ, Altheimer S, Cattori V, Meier PJ, Dietrich DR, Hagenbuch B (2005) Organic anion transporting polypeptides expressed in liver and brain mediate uptake of microcystin. Toxicology and Applied Pharmacology 203 257–263

Fischer WJ, Dietrich DR (2000) Pathological and biochemical characterization of microcystin–induced hepatopancreas and kidney damage in carp (Cyprinus carpio). Toxicology and Applied Pharmacology 164 73–81

Fitzgeorge RB, Clark SA, Keevil CW (1994) Routes of intoxication. In Detection methods for cyanobacterial toxins, Codd GA, Jefferies TM, Keevil CW, Potter E (eds) vol 149 Royal Society of Chemistry, Cambridge, UK

Fromme H, Koehler A, Krause R, Fuehrling D (2000) Occurrence of cyanobacterial toxins– microcystins and anatoxin–a–in Berlin water bodies with implications to human health and regulations. Environmental Toxicology 15 120–130

Gentile JH, Maloney TE (1969) Toxicity and environmental requirements of a strain of Aphanizomenon flos–aquae (L.) Ralfs. Can J Microbiol 15 165–73

Gilroy DJ, Kauffman KW, Hall RA, Huang X, Chu FS (2000) Assessing potential health risks from microcystin toxins in blue–green algae dietary supplements. Environmental Health Perspectives 108 435–9

Glavinas H, Krajcsi P, Cserepes J, Sarkadi B (2004) The Role of ABC Transporters in Drug Resistance, Metabolism and Toxicity. Current Drug Delivery 1 27–42

Gupta N, Pant SC, Vijayaraghavan R, Rao PV (2003) Comparative toxicity evaluation of cyanobacterial cyclic peptide toxin microcystin variants (LR, RR, YR) in mice. Toxicology 188 285–96

Hagenbuch B, Dawson P (2004) The sodium bile salt cotransport family SLC10. Pflugers Arch 447 566–70

Hagenbuch B, Gao B, Meier PJ (2002) Transport of xenobiotics across the blood–brain barrier. News Physiol Sci 17 231–4

Hagenbuch B, Meier PJ (2003) The superfamily of organic anion transporting polypeptides. Biochimica et Biophysica Acta 1609 1–18

Hagenbuch B, Meier PJ (2004) Organic anion transporting polypeptides of the OATP/ SLC21 family: phylogenetic classification as OATP/ SLCO superfamily, new nomenclature and molecular/functional properties. Pflugers Arch 447 653–65

Hitzfeld B, Lampert C, Späth N, Mountfort D, Kaspar H, Dietrich D (2000a) Toxin production in cyanobacterial mats from ponds on the McMurdo Ice Shelf, Antarctica. Toxicon 38 1731–1748

Hitzfeld BC, Hoeger SJ, Dietrich DR (2000b) Cyanobacterial toxins: removal during drinking water treatment, and human risk assessment. Environmental Health Perspectives 108 Suppl 1 113–122

Ho RH, Kim RB (2005) Transporters and drug therapy: Implications for drug disposition and disease. Clin Pharmacol Ther 78 260–77

Hoeger SJ, Dietrich DR (2004) Possible health risks arising from consumption of blue–green algae food supplements. Sixth International Conference on Toxic Cyanobacteria, Bergen, Norway

Hoeger SJ, Hitzfeld, BC, Dietrich DR (2005) Occurrence and elimination of cyanobacterial toxins in drinking water treatment plants. Toxicology and Applied Pharmacology 203 231–242

Hoeger SJ, Shaw G, Hitzfeld BC, Dietrich DR (2004) Occurrence and elimination of cyanobacterial toxins in two Australian drinking water treatment plants. Toxicon 43 639–49

Hoppu K, Salmela J, Lahti K (2002) High risk for symptoms from use of water contaminated with cyanobacteriae in sauna. Clinical Toxicology 40 309–310

Iwasa M, Yamamoto M, Tanaka Y, Kaito M, Adachi Y (2002) Spirulina–associated hepatotoxicity. Am J Gastroenterol 97 3212–3

Jacquemin D, Hagenbuch B, Stieger B, Wolfkoff AW, Meier PJ (1994) Expression cloning of a rat liver Na+–independent organic anion transporter. Proceedings of the National Academy of Sciences of the USA 91 133–137

Jacquemin E, Hagenbuch B, Stieger B, Wolkoff AW, Meier PJ (1991) Expression of the hepatocellular chloride–dependent sulfobromophthalein uptake system in Xenopus laevis oocytes. Journal of clinical investigations 88 2146–2146

Jochimsen EM, Carmichael WW, An JS, Cardo DM, Cookson ST, Holmes CE, Antunes MB, de Melo Filho DA, Lyra TM, Barreto VS, Azevedo SM, Jarvis WR (1998) Liver failure and death after exposure to microcystins at a hemodialysis center in Brazil [published erratum appears in N Engl J Med 1998 Jul 9;339(2):139]. New England Journal of Medicine 338 873–8

Khan SA, Wickstrom M, Haschek W, Schaeffer S, Ghosh S, Beasley V (1996) Microcystin–LR and kinetics of cytoskeletal reorganization in hepatocytes, kidney cells, and fibroblasts. Natural Toxins 4 206–214

Kuiper–Goodman T, Falconer IR, Fitzgerald DJ (1999) Human Health Aspects. In Toxic Cyanobacteria in Water: A Guide to their Public Health Consequences, Monitoring and Management, Chorus I, Bartram J (eds) E & FN Spon, London, pp 114–153

Kunta JR, Sinko PJ (2004) Intestinal Drug Transporters: In Vivo Function and Clinical Importance. Current Drug Metabolism 5 109–124

Landsberg JH (2002) The Effect of Harmful Algal Blooms on Aquatic Organisms. Reviews in Fisheries Sciences 10 113–390

Lawrence JF, Niedzwiadek B, Menard C, Lau BP, Lewis D, Kuiper–Goodman T, Carbone S, Holmes C (2001) Comparison of liquid chromatography/mass spectrometry, ELISA, and phosphatase assay for the determination of microcystins in blue–green algae products. Journal of the AOAC International 84 1035–1044

Maatouk I, Bouaicha N, Fontan D, Levi Y (2002) Seasonal variation of microcystin concentrations in the Saint–Caprais reservoir (France) and their removal in a small full–scale treatment plant. Water Research 36 2891–2897

Mahmood NA, Carmichael WW (1986) Paralytic shellfish poisons produced by the freshwater cyanobacterium Aphanizomenon flos–aquae NH–5. Toxicon 24 175–186

Majsterek I, Sicinska P, Tarczynska M, Zalewski M, Walter Z (2004) Toxicity of microcystin from cyanobacteria growing in a source of drinking water. Comparative Biochemistry and Physiology C 139 175–179

Meier PJ, Stieger B (2002) Bile salt transporters. Annual Review of Physiology 64 635–61

Meriluoto JA, Nygard SE, Dahlem AM, Eriksson JE (1990) Synthesis, organotropism and hepatocellular uptake of two tritium–labeled epimers of dihydromicrocystin–LR, a cyanobacterial peptide toxin analog. Toxicon 28 1439–46

Milutinovic A, Sedmak B, Horvat–Znidarsic I, Suput D (2002) Renal injuries induced by chronic intoxication with microcystins. Cellular and Molecular Biology Letters 7 139–41

Milutinovic A, Zivin M, Zorc–Pleskovic R, Sedmak B, Suput D (2003) Nephrotoxic effects of chronic administration of microcystins –LR and –YR. Toxicon 42 281–8

Miyazaki H, Sekine T, Endou H (2004) The multispecific organic anion transporter family: properties and pharmacological significance. Trends in Pharmacological Sciences 25

Moreno I, Pichardo S, Jos A, Gomez–Amores L, Mate A, Vazquez CM, Camean AM (2005) Antioxidant enzyme activity and lipid peroxidation in liver and kidney of rats exposed to microcystin–LR administered intraperitoneally. Toxicon 45 395–402

Nies AT, Jedlitschky G, König J, Herold–Mende C, Steiner HH, Schmitt H–P, Keppler D (2004) Expression and Immunolocalization of the Multidrug Resistance Proteins, MRP1–MRP6 (ABCC1–ABCC6), in human brain. Neuroscience 129 349–360

Osborne NJ, Webb PM, Shaw GR (2001) The toxins of Lyngbya majuscula and their human and ecological health effects. Environ Int 27 381–92

Pace JG, Robinson NA, Miura GA, Lynch TG, Templeton CB (1990) Pharmacokinetics, metabolism and distribution of microcystin ([3H]Mcyst–LR) in the rat. The Toxicologist 10 219

Pereira P, Onodera H, Andrinolo D, Franca S, Araujo F, Lagos N, Oshima Y (2000) Paralytic shellfish toxins in the freshwater cyanobacterium Aphanizomenon flos–aquae, isolated from Montargil reservoir, Portugal. Toxicon 38 1689–1702

Pflugmacher S, Wiegand C, Oberemm A, Beattie KA, Krause E, Codd GA, Steinberg CE (1998) Identification of an enzymatically formed glutathione conjugate of the cyanobacterial hepatotoxin microcystin–LR: the first step of detoxication. Biochim Biophys Acta 1425 527–33

Picanco MR, Soares RM, Cagido VR, Azevedo SM, Rocco PR, Zin WA (2004) Toxicity of a cyanobacterial extract containing microcystins to mouse lungs. Braz J Med Biol Res 37 1225–9

Pilotto L, Douglas R, Burch M, Cameron S, Beers M, Rouch G, Robinson P, Kirk M, Cowie C, Hardiman S, Moore C, Attewell R (1997) Health effects of exposure to cyanobacteria (blue–green algae) during recreational water–related activities. Australian and New Zealand Journal of Public Health 21 562–566

Pouria S, de Andrade A, Barbosa J, Cavalcanti R, Barreto V, Ward C, Preiser W, Poon G, Neild G, Codd G (1998) Fatal microcystin intoxication in haemodialysis unit in Caruaru, Brazil. The Lancet 352 21–26

Rapala J, Sivonen K, Luukkainen R, Niemela SI (1993) Anatoxin–a concentration in Anabaena and Aphanizomenon under different environmental conditions and comparison of growth by toxic and non–toxic Anabaena–strains: A laboratory study. Journal of Applied Phycology 5 581–591

Robertson EE, Rankin GO (2005) Human renal organic anion transporters: Characteristics and contributions to drug and drug metabolite excretion. Pharmacol Ther

Robinson NA, Miura GA, Matson CF, Dinterman RE, Pace JG (1989) Characterization of chemically tritiated microcystin–LR and its distribution in mice. Toxicon 27 1035–1042

Robinson NA, Pace JG, Matson CF, Miura GA, Lawrence WB (1991) Tissue distribution, excretion and hepatic biotransformation of microcystin–LR in mice. Journal of Pharmacology and Experimental Therapeutics 256 176–182

Rogers EH, Hunter ES 3rd, Moser VC, Phillips PM, Herkovits J, Munoz L, Hall LL, Chernoff N (2005) Potential developmental toxicity of anatoxin–a, a cyanobacterial toxin. J Appl Toxicol

Runnegar M, Berndt N, Kaplowitz N (1995) Microcystin uptake and inhibition of protein phosphatases: effects of chemoprotectants and self–inhibition in rela-

tion to known hepatic transporters. Toxicology and Applied Pharmacology 134 264–272
Saker M L, Jungblut A–D, Neilan BA, Rawn DFK., Vasconcelos VM (2005) Detection of microcystin synthetase genes in health food supplements containing the freshwater cyanobacterium Aphanizomenon flos–aquae. Toxicon 46 555–562
Salazar M, Chamorro GA, Salazar S, Steele CE (1996) Effect of Spirulina maxima consumption on reproduction and peri– and postnatal development in rats. Food Chem Toxicol 34 353–9
Salazar M, Martinez E, Madrigal E, Ruiz LE, Chamorro GA (1998) Subchronic toxicity study in mice fed Spirulina maxima. Journal of Ethnopharmacology 62 235–41
Sawyer PJ, Gentile JH, Sasner JJ Jr (1968) Demonstration of a toxin from Aphanizomenon flos–aquae (L.) Ralfs. Canadian Journal of Microbiology 14 1199–204
Shaw GR, Sukenik A, Livne A, Chiswell RK, Smith MJ, Seawright AA, Norris RL, Eaglesham GK, Moore MR (1999) Blooms of the cylindrospermopsin containing cyanobacterium, Aphanizomenon ovalisporum (Forti), in newly constructed lakes, Queensland, Australia. Environmental Toxicology 14 167–177
Shitara Y, Sato H, Sugiyama Y (2005) Evaluation of drug–drug interaction in the hepatobiliary and renal transport of drugs. Annu Rev Pharmacol Toxicol 45 689–723
Sivonen K, Niemelä SI, Niemi RM, Lepistö L, Luoma TH, Räsänen LA (1990) Toxic cyanobacteria (blue–green) algae in Finnish fresh and coastal waters. Hydrobiologia 190 267–275
Smith GM, Shine HD (1992) Immunofluorescent labeling of tight junctions in the rat brain and spinal cord. Int J Dev Neurosci 10 387–92
Soong FS, Maynard E, Kirke K, Luke C (1992) Illness associated with blue–green algae. Medical Journal of Australia 156 67
Stewart I, Webb PM, Schluter PJ, Shaw GR (2001) A prospective epidemiological study of recreational exposure to cyanobacteria in fresh and brackish waters in some Queensland and NSW lakes. In Fifth International Conference on Toxic Cyanobacteria, Noosa, Australia
Stotts RR, Twardock AR, Haschek WM, Choi BW, Rinehart KL, Beasley VR (1997) Distribution of tritiated dihydromicrocystin in swine. Toxicon 35 937–953
Teixera M, Costa M, Carvalho V, Pereira M, Hage E (1993) Gastroenteritis epidemic in the area of the Itaparica Dam, Bahia, Brazil. Bulletin of the Pan–American Health Organization 27 244–253
Teubner K, Feyerabend R, Henning M, Nicklisch A, Woitke P, Kohl JG (1999) Alternative blooming of Aphanizomenon flos–aquae or Planktothrix agardhii induced by the timing of the critical nitrogen: Phosphorus ratio in hypertrophic riverine lakes. Ergebnisse der Limnologie 54 325–344

Torokne A, Palovics A, Bankine M (2001) Allergenic (sensitization, skin and eye irritation) effects of freshwater cyanobacteria—experimental evidence. Environmental Toxicology 16 512–516

Underdal B, Nordstoga K, Skulberg OM (1999) Protracted toxic effects caused by saline extracts of Aphanizomenon flos–aquae (Cyanophyceae/Cyanobacteria). Aquatic Toxicology 46 269–278

van Montfoort JE, Hagenbuch B, Groothuis GM, Koepsell H, Meier PJ, Meijer DK (2003a) Drug uptake systems in liver and kidney. Current Drug Metabolism 4 185–211

van Montfoort JE, Hagenbuch B, Groothuis GM, Koepsell H, Meier PJ, Meijer DK (2003b) Drug uptake systems in liver and kidney. Current Drug Metabolism 4 185–211

Vitale S, Miller NR, Mejico LJ, Perry JD, Medura M, Freitag SK, Girkin C (2004) A randomized, placebo–controlled, crossover clinical trial of super blue–green algae in patients with essential blepharospasm or Meige syndrome. Am J Ophthalmol 138 18–32

Wynn–Williams DD (2000) Cyanobacteria in the Deserts – Life at the Limit? In The Ecology of Cyanobacteria, Whitton BA, Potts M (eds) Kluwer Academic Publishers, Dordrecht/London/Boston pp 341–366

Yu FY, Liu BH, Chou HN, Chu FS (2002) Development of a sensitive ELISA for the determination of microcystins in algae. Journal of Agricultural and Food Chemistry 50 4176–82

Index

Printed In The United States Of America